Introduction to
Hydrology

Fourth Edition

Warren Viessman, Jr.
University of Florida

Gary L. Lewis
Consulting Engineer

HarperCollins*College*Publishers

To Bette and Gayle for their patience, understanding, and help

Sponsoring Editor: T. Michael Slaughter
Project Editor: Lisa A. De Mol
Design Administrator: Jess Schaal
Cover Design: Jeanne Calabrese
Cover Illustration/Photo: Water Beads/Daniel Furon/PHOTONICA
Production Administrator: Randee Wire
Compositor: Interactive Composition Corporation
Cover Printer: Phoenix Color Corporation
Printer and Binder: R. R. Donnelley and Sons

Introduction to Hydrology, fourth edition

Library of Congress Cataloging-in-Publication Data

Viessman, Warren.
   Introduction to hydrology.—4th ed. / Warren Viessman, Jr., Gary
L. Lewis
    p. cm
    ISBN 0-673-99165-2
    1. Hydrology. I. Lewis, Gary L. II. Title.
GB661.2.V44 1995
551.48—dc20   94-42677
                                   CIP

    96   97   98 9 8 7 6 5 4 3 2

Contents

CHAPTER 3
Interception and Depression Storage   40

CHAPTER 4
Infiltration   52

CHAPTER 5
Evaporation and Transportation   82

CHAPTER 6
Streamflow   111

PART TWO
HYDROLOGIC MEASUREMENTS AND MONITORING   121

PART THREE
SURFACE WATER HYDROLOGY   151

PART FOUR
GROUNDWATER HYDROLOGY   425

CHAPTER 17
Groundwater, Soils, and Geology   427

CHAPTER 18
Mechanics of Flow   435

CHAPTER 19
Wells and Collection Devices   460

Preface

Water management is taking on new dimensions. New federal thrusts, the growing list of global issues, and strong public sentiment regarding environmental protection have been the principal driving forces.

In the early years of the 20th century, water resources development and management were focused almost exclusively on water supply and flood control. Today, these issues are still important, but protecting the environment, ensuring safe drinking water, and providing aesthetic and recreatioinal experiences compete equally for attention and funds. Furthermore, an environmentally conscious public is pressing for greater reliance on improved management practices, with fewer structural components, to solve this nation's water problems. The notion of continually striving to provide more water has been replaced by one of husbanding this precious natural resource.

There is a growing constituency for allocating water for the benefit of fish and wildlife, for protection of marshes and estuary areas, and for other natural system uses. But estimating the quantities of water needed for environmental protection and for maintaining and/or restoring natural systems is difficult, and there are still many unknowns. Scientific data are sparse, and our understanding of the complex interactions inherent in ecosystems of all scales is rudimentary. Indeed, this is a critical issue, since the quantities of water involved in environmental protection can be substantial and competition for these waters from traditional water users is keen. The nations of the world are facing major decisions regarding natural systems—decisions that are laden with significant economic and social impacts. Thus there is an urgency associated with developing a better understanding of ecologic systems and of their hydrologic components.

Water policies of the future must therefore take on broader dimensions. More emphasis must be placed on regional planning and management, and regional institutions to accommodate this must be devised. Water management must be practiced at, and between, all levels of government. Land use and water use planning must be more tightly coordinated as well.

Water scientists and engineers of tomorrow must be equipped to address a diversity of issues such as: the design and operation of data retrieval and storage systems; forecasting; developing alternative water use futures; estimating water requirements for natural systems; exploring the impacts of climate change; developing more efficient systems for applying water in all water-using sectors; and analyzing and designing water management systems incorporating technical, economic, environmental, social, legal, and political elements. A knowledge of hydrologic principles is a requisite for dealing with such issues.

This fourth edition has been designed to meet the contemporary needs of water scientists and engineers. It is organized to accommodate students and practitioners who are concerned with the development, management, and protection of water resources. The format of the book follows that of its predecessor, providing material for both an introductory and a more advanced course.

Parts One through Four provide the basics for a beginning level course, while Parts Five and Six may be used for a more advanced course on hydrologic modeling. This fourth edition has been updated throughout, and many solved examples have been added. In addition, new computer approaches have been introduced and problem-solving techniques include the use of spreadsheets as appropriate. New features of each chapter include an introductory statement of contents and, at the conclusion of the chapter, a summary of key points.

Many sources have been drawn upon to provide subject matter for this book, and the authors hope that suitable acknowledgment has been given to them. Colleagues and students are recognized for their helpful comments and reviews, particularly the following reviewers.

Gert Aron, *The Pennsylvania State University*
John W. Bird, *University of Nevada-Reno*
Istvan Bogardi, *University of Nebraska*
Ronald A. Chadderton, *Villanova University*
Richard N. Downer, *University of Vermont*
Bruce E. Larock, *University of California–Davis*
Frank D. Masch, *University of Texas–San Antonio*
Philip L. Thompson, *Federal Highway Administration*

A special note of thanks is due to Dr. John W. Knapp, President of the Virginia Military Institute, coauthor of previous editions of this book, for his past contributions and valuable guidance.

Warren Viessman, Jr.
Gary L. Lewis

PART ONE

THE HYDROLOGIC CYCLE

Chapter 1

Introduction

■ Prologue

The purpose of this chapter is to:

- Define *hydrology*.
- Give a brief history of the evolution of this important earth science.
- State the fundamental equation of hydrology.
- Demonstrate how hydrologic principles can be applied to supplement decision support systems for water and environmental management.

1.1 HYDROLOGY DEFINED

Hydrology is an earth science. It encompasses the occurrence, distribution, movement, and properties of the waters of the earth. A knowledge of hydrology is fundamental to decisionmaking processes where water is a component of the system of concern. Water and environmental issues are inextricably linked, and it is important to clearly understand how water is affected by and how water affects ecosystem manipulations.

1.2 A BRIEF HISTORY

Ancient philosophers focused their attention on the nature of processes involved in the production of surface water flows and other phenomena related to the origin and occurrence of water in various stages of the perpetual cycle of water being conveyed from the sea to the atmosphere to the land and back again to the sea. Unfortunately, early speculation was often faulty.[1-7]* For example, Homer believed in the existence of large subterranean reservoirs that supplied rivers, seas, springs, and deep wells. It is interesting to note, however, that Homer understood the dependence of flow in the

*Superior numbers indicate references at the end of the chapter.

Greek aqueducts on both conveyance cross section and velocity. This knowledge was lost to the Romans, and the proper relation between area, velocity, and rate of flow remained unknown until Leonardo da Vinci rediscovered it during the Italian Renaissance.

During the first century B.C. Marcus Vitruvius, in Volume 8 of his treatise *De Architectura Libri Decem* (the engineer's chief handbook during the Middle Ages), set forth a theory generally considered to be the predecessor of modern notions of the hydrologic cycle. He hypothesized that rain and snow falling in mountainous areas infiltrated the earth's surface and later appeared in the lowlands as streams and springs.

In spite of the inaccurate theories proposed in ancient times, it is only fair to state that practical application of various hydrologic principles was often carried out with considerable success. For example, about 4000 B.C. a dam was constructed across the Nile to permit reclamation of previously barren lands for agricultural production. Several thousand years later a canal to convey fresh water from Cairo to Suez was built. Mesopotamian towns were protected against floods by high earthen walls. The Greek and Roman aqueducts and early Chinese irrigation and flood control works were also significant projects.

Near the end of the fifteenth century the trend toward a more scientific approach to hydrology based on the observation of hydrologic phenomena became evident. Leonardo da Vinci and Bernard Palissy independently reached an accurate understanding of the water cycle. They apparently based their theories more on observation than on purely philosophical reasoning. Nevertheless, until the seventeenth century it seems evident that little if any effort was directed toward obtaining quantitative measurements of hydrologic variables.

The advent of what might be called the "modern" science of hydrology is usually considered to begin with the studies of such pioneers as Perrault, Mariotte, and Halley in the seventeenth century.[1,4] Perrault obtained measurements of rainfall in the Seine River drainage basin over a period of 3 years. Using these and measurements of runoff, and knowing the drainage area size, he showed that rainfall was adequate in quantity to account for river flows. He also made measurements of evaporation and capillarity. Mariotte gauged the velocity of flow of the River Seine. Recorded velocities were translated into terms of discharge by introducing measurements of the river cross section. The English astronomer Halley measured the rate of evaporation of the Mediterranean Sea and concluded that the amount of water evaporated was sufficient to account for the outflow of rivers tributary to the sea. Measurements such as these, although crude, permitted reliable conclusions to be drawn regarding the hydrologic phenomena being studied.

The eighteenth century brought forth numerous advances in hydraulic theory and instrumentation. The Bernoulli piezometer, the Pitot tube, Bernoulli's theorem, and the Chézy formula are some examples.[8]

During the nineteenth century, experimental hydrology flourished. Significant advances were made in groundwater hydrology and in the measurement of surface water. Such significant contributions as Hagen–Poiseuille's capillary flow equation, Darcy's law of flow in porous media, and the Dupuit-Thiem well formula were evolved.[9-11] The beginning of systematic stream gauging can also be traced to this period. Although the basis for modern hydrology was well established in the nineteenth century, much of the effort was empirical in nature. The fundamentals of physical hydrology had not yet been well established or widely recognized. In the early years of the twentieth century the inadequacies of many earlier empirical formulations became well known. As a result, interested governmental agencies began to develop their own programs of hydrologic research. From about 1930 to 1950, rational analyses began to replace empiricism.[3] Sherman's unit hydrograph, Horton's infiltration theory, and Theis's nonequilibrium approach to well hydraulics are outstanding examples of the great progress made.[12-14]

Since 1950 a theoretical approach to hydrologic problems has largely replaced less sophisticated methods of the past. Advances in scientific knowledge permit a better understanding of the physical basis of hydrologic relations, and the advent and continued development of high-speed digital computers have made possible, in both a practical and an economic sense, extensive mathematical manipulations that would have been overwhelming in the past.

For a more comprehensive historical treatment, the reader is referred to the works of Meinzer, Jones, Biswas, and their co-workers.[1, 2, 4, 5, 15]

1.3 THE HYDROLOGIC CYCLE

The hydrologic cycle is a continuous process by which water is transported from the oceans to the atmosphere to the land and back to the sea. Many subcycles exist. The evaporation of inland water and its subsequent precipitation over land before returning to the ocean is one example. The driving force for the global water transport system is provided by the sun, which furnishes the energy required for evaporation. Note that the water quality also changes during passage through the cycle; for example, sea water is converted to fresh water through evaporation.

The complete water cycle is global in nature. World water problems require studies on regional, national, international, continental, and global scales.[16] Practical significance of the fact that the total supply of fresh water available to the earth is limited and very small compared with the salt water content of the oceans has received little attention. Thus waters flowing in one country cannot be available at the same time for use in other regions of the world. Raymond L. Nace of the U.S. Geological Survey has aptly stated that "water resources are a global problem with local roots."[16] Modern hydrologists are obligated to cope with problems requiring definition in varying scales of order of magnitude difference. In addition, developing techniques to control weather must receive careful attention, since climatological changes in one area can profoundly affect the hydrology and therefore the water resources of other regions.

1.4 THE HYDROLOGIC BUDGET

Because the total quantity of water available to the earth is finite and indestructible, the global hydrologic system may be looked upon as closed. Open hydrologic subsystems are abundant, however, and these are usually the type analyzed. For any system, a water budget can be developed to account for the hydrologic components.

Figures 1.1, 1.2, and 1.3 show a hydrologic budget for the coterminous United States, a conceptualized hydrologic cycle, and the distribution of a precipitation input, respectively. These figures illustrate the components of the water cycle with which a hydrologist is concerned. In a practical sense, some hydrologic region is dealt with and a budget for that region is established. Such regions may be topographically defined (watersheds and river basins are examples), politically specified (e.g., county or city limits), or chosen on some other grounds. Watersheds or drainage basins are the easiest to deal with since they sharply define surface water boundaries. These topographically determined areas are drained by a river/stream or system of connecting rivers/streams such that all outflow is discharged through a single outlet. Unfortunately, it is often necessary to deal with regions that are not well suited to tracking hydrologic components. For these areas, the hydrologist will find hydrologic budgeting somewhat of a challenge.

The primary input in a hydrologic budget is precipitation. Figures 1.1–1.3 illustrate this. Some of the precipitation (e.g., rain, snow, hail) may be intercepted by trees, grass, other vegetation, and structural objects and will eventually return to the atmosphere by evaporation. Once precipitation reaches the ground, some of it may fill depressions (become depression storage), part may penetrate the ground (infiltrate) to replenish soil moisture and groundwater reservoirs, and some may become surface runoff—that is, flow over the earth's surface to a defined channel such as a stream. Figure 1.3 shows the disposition of infiltration, depression storage, and surface runoff.

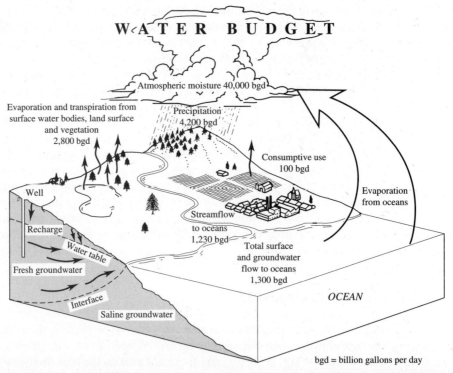

Figure 1.1   Hydrologic budget of coterminous United States. (U.S. Geological Survey.)

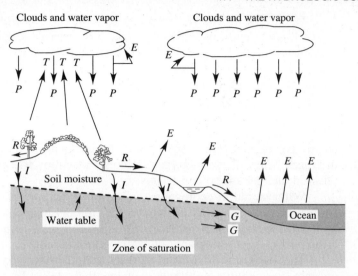

Figure 1.2   The hydrologic cycle: T, transpiration; E, evaporation; P, precipitation; R, surface runoff; G, groundwater flow; and I, infiltration.

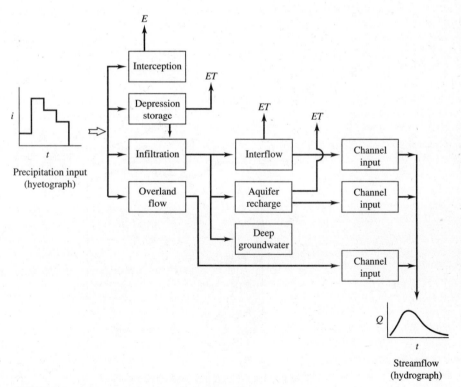

Figure 1.3   Distribution of precipitation input.

Water entering the ground may take several paths. Some may be directly evaporated if adequate transfer from the soil to the surface is maintained. This can easily occur where a high groundwater table (free water surface) is within the limits of capillary transport to the ground surface. Vegetation using soil moisture or groundwater directly can also transmit infiltrated water to the atmosphere by a process known as *transpiration*. Infiltrated water may likewise replenish soil moisture deficiencies and enter storage provided in groundwater reservoirs, which in turn maintain dry weather streamflow. Important bodies of groundwater are usually flowing so that infiltrated water reaching the saturated zone may be transported for considerable distances before it is discharged. Groundwater movement is subject, of course, to physical and geological constraints.

Water stored in depressions will eventually evaporate or infiltrate the ground surface. Surface runoff ultimately reaches minor channels (gullies, rivulets, and the like), flows to major streams and rivers, and finally reaches an ocean. Along the course of a stream, evaporation and infiltration can also occur.

The foregoing discussion suggests that the hydrologic cycle, while simple in concept, is actually exceedingly complex. Paths taken by particles of water precipitated in any area are numerous and varied before the sea is reached. The time scale may be on the order of seconds, minutes, days, or years.

A general hydrologic equation can be developed based on the processes illustrated in Figs. 1.2 and 1.3. Consider Fig. 1.4. In it, the hydrologic variables P, E, T, R, G, and I are as defined in Fig. 1.2. Subscripts s and g are introduced to denote vectors originating above and below the earth's surface, respectively. For example, R_g

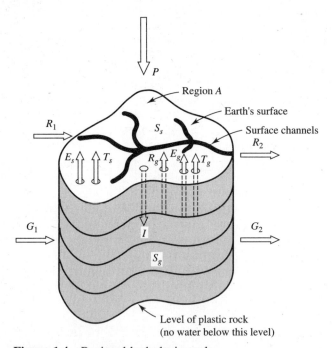

Figure 1.4 Regional hydrologic cycle.

signifies groundwater flow that is effluent to a surface stream, and E_s represents evaporation from surface water bodies or other surface storage areas. Letter S stands for storage. The region under consideration specified as A has a lower boundary below which water will not be found. The upper boundary is the earth's surface. Vertical bounds are arbitrarily set as projections of the periphery of the region. Remembering that the water budget is a balance between inflows, outflows, and changes in storage, Fig. 1.4 can be translated into the following mathematical statements, where all values are given in units of volume per unit time:

1. Hydrologic budget above the surface

$$P + R_1 - R_2 + R_g - E_s - T_s - I = \Delta S_s \qquad (1.1)$$

2. Hydrologic budget below the surface

$$I + G_1 - G_2 - R_g - E_g - T_g = \Delta S_g \qquad (1.2)$$

3. Hydrologic budget for the region (sum of Eqs. 1.2 and 1.3)

$$P - (R_2 - R_1) - (E_s + E_g) - (T_s + T_g) - (G_2 - G_1) = \Delta(S_s + S_g)$$
$$(1.3)$$

If the subscripts are dropped from Eq. 1.3 so that letters without subscripts refer to total precipitation and net values of surface flow, underground flow, evaporation, transpiration, and storage, the hydrologic budget for a region can be written simply as

$$P - R - G - E - T = \Delta S \qquad (1.4)$$

This is the basic equation of hydrology. For a simplified hydrologic system where terms G, E, and T do not apply, Eq. 1.4 reduces to

$$P - R = \Delta S \qquad (1.5)$$

Equation 1.4 is applicable to exercises of any degree of complexity and is therefore basic to the solution of all hydrologic problems.

The difficulty in solving practical problems lies mainly in the inability to measure or estimate properly the various hydrologic equation terms. For local studies, reliable estimates often are made, but on a global scale quantification is usually crude. Precipitation is measured by rain or snow gauges located throughout an area. Surface flows can be measured using various devices such as weirs, flumes, velocity meters, and depth gauges located in the rivers and streams of the area. Under good conditions these measurements are 95 percent or more accurate, but large floods cannot be measured directly by current methods and data on such events are sorely needed. Soil moisture can be measured using neutron probes and gravimetric methods; infiltration can be determined locally by infiltrometers or estimated through the use of precipitation–runoff data. Areal estimates of soil moisture and infiltration are generally very crude, however. The extent and rate of movement of groundwater are usually exceedingly difficult to determine, and adequate data on quantities of groundwater are not always available. Knowledge of the geology of a region is essential for groundwater estimates if they are to be more than just rough guides. The determination of the quantities of water evaporated and transpired is also extremely difficult under the present state of development of the science. Most estimates of evapotranspiration are obtained by using evaporation pans, energy budgets, mass transfer methods, or empirical relations. A predicament inherent in the analysis of large drainage basins is the fact that rates of evaporation, transpiration, and groundwater movement are often assumed to be highly heterogeneous.

The hydrologic equation is a useful tool; the reader should understand that it can be employed in various ways to estimate the magnitude and time distribution of hydrologic variables. An introductory example is given here, and others will be found throughout the book.

EXAMPLE 1.1 _____

In a given year, a 10,000-mi^2 watershed received 20 in. of precipitation. The average rate of flow measured in the river draining the area was found to be 700 cfs (cubic feet per second). Make a rough estimate of the combined amounts of water evaporated and transpired from the region during the year of record.

Solution.   Beginning with the basic hydrologic equation

$$P - R - G - E - T = \Delta S \tag{1.4}$$

and since evaporation and transpiration can be combined,

$$ET = P - R - G - \Delta S \tag{1.6}$$

The term ET is the unknown to be evaluated and P and R are specified. The equation thus has five variables and three unknowns and cannot be solved without additional information.

In order to get a solution, two assumptions are made. First, since the drainage area is quite large (measured in hundreds of square miles), a presumption that the groundwater divide (boundary) follows the surface divide is probably reasonable. In this case the G component may be considered zero. The vector R_g exists but is included in R. The foregoing assumption is usually not valid for small areas and must therefore be used carefully. It is also presupposed that $\Delta S = 0$, thus implying that the groundwater reservoir volume has not changed during the year. For such short periods this assumption can be very inaccurate, even for well-watered regions with balanced withdrawals and good recharge potentials. In arid areas where groundwater is being mined (ΔS consistently negative), it would be an unreasonable supposition in many cases. Nevertheless, the assumption is made here for illustrative purposes and qualified by saying that past records of water levels in the area have revealed an approximate constancy in groundwater storage. Hydrology is not an exact science, and reasonable well-founded assumptions are required if practical problems are to be solved.

Using the simplifications just outlined, the working relation reduces to

$$ET = P - R$$

which can be solved directly. First, change R into inches per year so that the units are compatible:

$$R, \frac{ft^3}{sec} \times \frac{1}{area\ (in\ ft^2)} \times \frac{sec}{yr} \times \frac{in.}{ft} = R,\ in.$$

$$R = \frac{700 \times 86,400 \times 365 \times 12}{10^4 \times (5280)^2} = 0.95\ in.$$

Therefore, $ET = 20 - 0.95 = 19.05$ in./yr.

The amount of evapotranspiration for the year in question is estimated to be 19.05 in. This is admittedly a crude approximation but could serve as a useful guide for water resources planning. ■■

1.5 HYDROLOGIC MODELS

Hydrologic systems are generally analyzed by using mathematical models. These models may be empirical, statistical, or founded on known physical laws. They may be used for such simple purposes as determining the rate of flow that a roadway grate must be designed to handle, or they may guide decisions about the best way to develop a river basin for a multiplicity of objectives. The choice of the model should be tailored to the purpose for which it is to be used. In general, the simplest model capable of producing information adequate to deal with the issue should be chosen.

Unfortunately, most water resources systems of practical concern have physical, social, political, environmental, and legal dimensions, and their interactions cannot be exactly described in mathematical terms. Furthermore, the historical data necessary for meaningful hydrologic analyses are often lacking or unreliable. And when one considers that hydrologic systems are generally probabilistic in nature, it is easy to understand that the modeler's task is not a simple one. In fact, it is often the case that the best that can be hoped for from a model is an enhanced understanding of the system being analyzed. But this in itself can be of great value, leading, for example, to the implementation of data collection programs that can ultimately support reliable modeling efforts.

For the most part, mathematical models are designed to describe the way a system's elements respond to some type of stimulus (input). For example, a model of a groundwater system might be developed to demonstrate the effects on groundwater storage of various schemes for pumping. Equations 1.1 and 1.2 are mathematical models of the hydrologic budget, and Figure 1.3 can be considered a pictorial model of the rainfall-runoff process. In later chapters, a variety of hydrologic models will be presented and discussed. These models provide the basis for informed water management decisions.

1.6 HYDROLOGIC DATA

Hydrologic data are needed to describe precipitation; streamflows; evaporation; soil moisture; snow fields; sedimentation; transpiration; infiltration; water quality; air, soil, and water temperatures; and other variables or components of hydrologic systems. Sources of data are numerous, with the U.S. Geological Survey being the primary one for streamflow and groundwater facts. The National Weather Service (NOAA or National Oceanic and Atmospheric Administration) is the major collector of meterologic data. Many other federal, state, and local agencies and other organizations also compile hydrologic data. For a complete listing of these organizations see Refs. 3 and 17.

1.7 COMMON UNITS OF MEASUREMENT

Stream and river flows are usually recorded as cubic meters per second (m^3/sec), cubic feet per second (cfs), or second-feet (sec-ft); groundwater flows and water supply flows are commonly measured in gallons per minute, hour, or day (gpm, gph, gpd), or millions of gallons per day (mgd); flows used in agriculture or related to water storage are often expressed as acre-feet (acre-ft), acre-feet per unit time, inches (in.) or centimeters (cm) depth per unit time, or acre-inches per hour (acre-in./hr).

Volumes are often given as gallons, cubic feet, cubic meters, acre-feet, second-foot-days, and inches or centimeters. An acre-foot is equivalent to a volume of water 1 ft deep over 1 acre of land (43,560 ft^3). A second-foot-day (cfs-day, sfd) is the accumulated volume produced by a flow of 1 cfs in a 24-hr period. A second-foot-hour (cfs-hr) is the accumulated volume produced by a flow of 1 cfs in 1 hr. Inches or centimeters of depth relate to a volume equivalent to that many inches or centimeters of water over the area of concern. In hydrologic mass balances, it is sometimes useful to note that 1 cfs-day = 2 acre-feet with sufficient accuracy for most calculations.

Rainfall depths are usually recorded in inches or centimeters whereas rainfall rates are given in inches or centimeters per hour. Evaporation, transpiration, and infiltration rates are usually given as inches or centimeters depth per unit time. Some useful constants and tabulated values of several of the physical properties of water are given in Appendix A at the end of the book.

1.8 APPLICATION OF HYDROLOGY TO ENVIRONMENTAL PROBLEMS

It is true that humans cannot exist without water; it is also true that water, mismanaged, or during times of deficiency (droughts), or times of surplus (floods), can be life threatening. Furthermore, there is no aspect of environmental concern that does not relate in some way to water. Land, air, and water are all interrelated as are water and all life forms. Accordingly, the spectrum of issues requiring an understanding of hydrologic processes is almost unlimited.

As water becomes more scarce and as competition for its use expands, the need for improved water management will grow. And to provide water for the world's expanding population, new industrial developments, food production, recreational demands, and for the preservation and protection of natural systems and other purposes, it will become increasingly important for us to achieve a thorough understanding of the underlying hydrologic processes with which we must contend. This is the challenge to hydrologists, water resources engineers, planners, policymakers, lawyers, economists, and others who must strive to see that future allocations of water are sufficient to meet the needs of human and natural systems.

■ Summary

Hydrology is the science of water. It embraces the occurrence, distribution, movement, and properties of the waters of the earth. In a mathematical sense, an accounting may be made of the inputs, outputs, and water storages of a region so that a history of water movement for the region can be estimated.

After reading this chapter you should be able to understand the hydrologic budget and make a simple accounting of water transport in a region. You should also have gained an understanding of how hydrologic analyses can be used to facilitate design and management processes for water resources systems.

PROBLEMS

1.1. One-half inch of runoff results from a storm on a drainage area of 50 mi². Convert this amount to acre-feet and cubic meters.

1.2. Assume you are dealing with a vertical walled reservoir having a surface area of 500,000 m² and that an inflow of 1.0 m³/sec occurs. How many hours will it take to raise the reservoir level by 30 cm?

1.3. Consider that the storage existing in a river reach at a reference time is 15 acre-ft and at the same time the inflow to the reach is 500 cfs and the outflow from the reach is 650 cfs. One hour later, the inflow is 550 cfs and the outflow is 680 cfs. Find the change in storage during the hour in acre-feet and in cubic meters.

1.4. During a 24-hr time period, the inflow to a 500-acre vertical walled reservoir was 100 cfs. During the same interval, evaporation was 1 in. Was there a rise or fall in surface water elevation? How much was it? Give the answer in inches and centimeters.

1.5. The annual evaporation from a lake is 50 in. If the lake's surface area is 3000 acres, what would be the daily evaporation rate in acre-feet and in centimeters?

1.6. A flow of 10 cfs enters a 1-mi² vertical walled reservoir. Find the time required to raise the reservoir level by 6 in.

1.7. A drainage basin has an area of 4571 mi². If the average annual runoff is 5102 cfs and the average rainfall is 42.5 in., estimate the evaportranspiration losses for the area in 1 year. How reliable do you think this estimate is?

1.8. The storage in a reach of a river is 16.0 acre-ft at a given time. Determine the storage (acre-feet) 1 hr later if the average rates of inflow and outflow during the hour are 700 and 650 cfs, respectively.

1.9. Rain falls at an average intensity of 0.4 in./hr over a 600-acre area for 3 days. (a) Determine the average rate of rainfall in cubic feet per second; (b) determine the 3-day volume of rainfall in acre-feet; and (c) determine the 3-day volume of rainfall in inches of equivalent depth over the 600-acre area.

1.10. The evaporation rate from the surface of a 3650-acre lake is 100 acre-ft/day. Determine the depth change (feet) in the lake during a 365-day year if the inflow to the lake is 25.2 cfs. Is the change in lake depth an increase or a decrease?

1.11. One and one-half inches of runoff are equivalent to how many acre-feet if the drainage area is 25-mi²? (Note: 1 acre = 43,560 ft².)

1.12. One-half inch of rain per day is equivalent to an average rate of how many cubic feet per second if the area is 500 acres? How many meters per second?

REFERENCES

1. P. B. Jones, G. D. Walker, R. W. Harden, and L. L. McDaniels, "The Development of the Science of Hydrology," Circ. No. 60-03, Texas Water Commission, Apr. 1963.
2. W. D. Mead, *Notes on Hydrology*. Chicago: D. W. Mead, 1904.
3. Ven Te Chow (ed.), *Handbook of Applied Hydrology*. New York: McGraw-Hill, 1964.
4. O. E. Meinzer, *Hydrology,* Vol. 9 of *Physics of the Earth*. New York: McGraw-Hill, 1942. Reprinted by Dover, New York, 1949.
5. P. D. Krynine, "On the Antiquity of Sedimentation and Hydrology," *Bull. Geol. Soc. Am.* **70,** 1721–1726(1960).
6. Raphael G. Kazmann, *Modern Hydrology*. New York: Harper & Row, 1965.
7. H. Pazwosh and G. Mavrigian, "A Historical Jewelpiece—Discovery of the Millennium Hydrologic Works of Karaji," *Water Resources Bull.* **16**(6), 1094–1096(Dec. 1980).
8. Hunter Rouse and Simon Ince, *History of Hydraulics,* Iowa Institute of Hydraulic Research, State University of Iowa, 1957.
9. G. H. L. Hagen, "Ueber die Bewegung des Wassers in engen cylindrischen Rohren," *Poggendorfs Ann. Phys. Chem.* **16,** 423–442(1839).
10. Henri Darcy, *Les fontaines publiques de la ville de Dijon*. Paris: V. Dalmont, 1856.
11. J. Dupuit, *Études théoriques et practiques sur le mouvement des eaux dans les canaux découverts et à travers les terrains perméables,* 2nd ed. Paris: Dunod, 1863.
12. L. K. Sherman, "Stream Flow from Rainfall by the Unit-Graph Method," *Eng. News-Rec.* **108**(1932).
13. R. E. Horton, "The Role of Infiltration in the Hydrologic Cycle," *Trans. Am. Geophys. Union* **14,** 446–460(1933).
14. C. V. Theis, "The Relation Between the Lowering of the Piezometric Surface and the Rate and Duration of a Well Using Ground Water Recharge," *Trans. Am. Geophys. Union* **16,** 519–524(1935).
15. Asit K. Biswas, "Hydrologic Engineering Prior to 600 B.C.," *Proc. ASCE J. Hyd. Div.,* Proc. Paper 5431, Vol. 93, No. HY5 (Sept. 1967).
16. Raymond L. Nace, "Water Resources: A Global Problem with Local Roots," *Environ. Sci. Technol.* **1**(7) (July 1967).
17. D. K. Todd (ed.), *The Water Encyclopedia*. New York: Water Information Center, 1970.

Chapter 2

Precipitation

■ **Prologue**

The purpose of this chapter is to:

- Define *precipitation,* discuss its forms, and describe its spatial and temporal attributes.
- Illustrate techniques for estimating areal precipitation amounts for specific storm events and for maximum precipitation-generating conditions.

Precipitation replenishes surface water bodies, renews soil moisture for plants, and recharges aquifers. Its principal forms are rain and snow. The relative importance of these forms is determined by the climate of the area under consideration. In many parts of the western United States, the extent of the snowpack is a determining factor relative to the amount of water that will be available for the summer growing season. In more humid localities, the timing and distribution of rainfall are of principal concern.

Precipitated water follows the paths shown in Figs. 1.2 and 1.3. Some of it may be intercepted, evaporated, infiltrated, and become surface flow. The actual disposition depends on the amount of rainfall, soil moisture conditions, topography, vegetal cover soil type, and other factors.

Hydrologic modeling and water resources assessments depend upon a knowledge of the form and amount of precipitation occurring in a region of concern over a time period of interest.

2.1 WATER VAPOR

The fraction of water vapor in the atmosphere is very small compared to quantities of other gases present, but it is exceedingly important to our way of life. Precipitation is derived from this atmospheric water. The moisture content of the air is also a significant factor in local evaporation processes. Thus it is necessary for a hydrologist to be acquainted with ways for evaluating the atmospheric water vapor content and to understand the thermodynamic effects of atmospheric moisture.[1]

Under most conditions of practical interest (modest ranges of pressure and temperature, provided that the condensation point is excluded), water vapor essentially obeys the gas laws. Atmospheric moisture is derived from evaporation and transpiration, the principal source being evaporation from the oceans. Precipitation over the United States comes largely from oceanic evaporation, the water vapor being transporated over the continent by the primary atmospheric circulation system.

Measures of water vapor or atmospheric humidity are related basically to conditions of evaporation and condensation occurring over a level surface of pure water. Consider a closed system containing approximately equal volumes of water and air maintained at the same temperature. If the initial condition of the air is dry, evaporation takes place and the quantity of water vapor in the air increases. A measurement of pressure in the airspace will reveal that as evaporation proceeds, pressure in the airspace increases because of an increase in partial pressure of the water vapor (vapor pressure). Evaporation continues until vapor pressure of the overlying air equals the surface vapor pressure [a measure of the excess of water molecules leaving (evaporating from) the water surface over those returning]. At this point, evaporation ceases, and if the temperatures of the air space and water are equal, the airspace is said to be *saturated*. If the container had been open instead of closed, the equilibrium would not have been reached, and all the water would eventually have evaporated. Some commonly used measures of atmospheric moisture or humidity are vapor pressure, absolute humidity, specific humidity, mixing ratio, relative humidity, and dew point temperature.

Amount of Precipitable Water

Estimates of the amount of precipitation that might occur over a given region with favorable conditions are often useful. These may be obtained by calculating the amount of water contained in a column of atmosphere extending up from the earth's surface. This quantity is known as the *precipitable water W*, although it cannot all be removed from the atmosphere by natural processes. Precipitable water is usually expressed in centimeters or inches.

An equation for computing the amount of precipitable water in the atmosphere can be derived as follows. Consider a column of air having a square base 1 cm on a side. The total water mass contained in this column between elevation zero and some height z would be

$$W = \int_0^z \rho_w dz \qquad (2.1)$$

where ρ_w = the absolute humidity and W is the depth of precipitable water in centimeters. The integral can be evaluated graphically or by dividing the atmosphere into layers of approximately uniform specific humidities, solving for these individually, and then summing. Figure 2.1 illustrates the average amount of precipitable water for the continental United States up to an elevation of 8 km.[2]

Geographic and Temporal Variations

The quantity of atmospheric water vapor varies with location and time. These variations may be attributed mainly to temperature and source of supply considerations. The greatest concentrations can be found near the ocean surface in the tropics, the

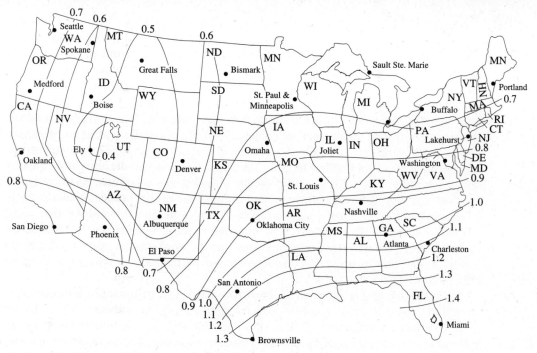

Figure 2.1 Mean precipitable water, in inches, to an elevation of 8 km. (U.S. Weather Bureau.)[2]

concentrations generally decreasing with latitude, altitude, and distance inland from coastal areas.

About half the atmospheric moisture can be found within the first mile above the earth's surface. This is because the vertical transport of vapor is mainly through convective action, which is slight at higher altitudes. It is also of interest that there is not necessarily any relation between the amount of atmospheric water vapor over a region and the resulting precipitation. The amount of water vapor contained over dry areas of the Southwest, for example, at times exceeds that over considerably more humid northern regions, even though the latter areas experience precipitation while the former do not.

2.2 PRECIPITATION

Precipitation is the primary input vector of the hydrologic cycle. Its forms are rain, snow, and hail and variations of these such as drizzle and sleet. Precipitation is derived from atmospheric water, its form and quantity thus being influenced by the action of other climatic factors such as wind, temperature, and atmospheric pressure. Atmospheric moisture is a necessary but not sufficient condition for precipitation. Continental air masses are usually very dry so that most precipitation is derived from moist maritime air that originates over the oceans. In North America about 50 percent of the evaporated water is taken up by continental air and moves back again to the sea.

Formation of Precipitation

Two processes are considered to be capable of supporting the growth of droplets of sufficient mass (droplets from about 500 to 4000 μm in diameter) to overcome air resistance and consequently fall to the earth as precipitation. These are known as the *ice crystal process* and the *coalescence process.*

The coalescence process is one by which the small cloud droplets increase their size due to contact with other droplets through collision. Water droplets may be considered as falling bodies that are subjected to both gravitational and air resistance effects. Fall velocities at equilibrium (terminal velocities) are proportional to the square of the radius of the droplet; thus the larger droplets will descend more quickly than the smaller ones. As a result, smaller droplets are often overtaken by larger droplets, and the resulting collisions tend to unite the drops, producing increasingly larger particles. Very large drops (order of 7 mm in diameter) break up into small droplets that repeat the coalescence process and produce somewhat of a chain effect. In this manner, sufficiently large raindrops may be produced to generate significant precipitation. This process is considered to be particularly important in tropical regions or in warm clouds.

An important type of growth is known to occur if ice crystals and water droplets are found to exist together at subfreezing temperatures down to about $-40°$ C. Under these conditions, certain particles of clay minerals and organic and ordinary ocean salts serve as freezing nuclei so that ice crystals are formed. The vapor pressure under these conditions is higher over the water droplets than over the ice crystals, and thus condensation occurs on the surface of the crystals. The ice crystals grow in size, and uneven particle size distributions develop, which further favor growth through contact with other particles. This is considered to be a very important precipitation-producing mechanism.

The artificial inducement of precipitation has been studied extensively, and these studies are continuing. It has been demonstrated that condensation nuclei supplied to clouds can induce precipitation. The ability of humans to ensure the production of precipitation or to control its geographic location or timing has not yet been attained, however.

Many legal as well as technological problems are associated with the prospects of "rain-making" processes. Of interest here is the impact on hydrologic estimates that uncontrolled or only partially controlled artificial precipitation might have. Many naturally occurring hydrologic variables are considered as statistical variates that are either randomly distributed or distributed with a random component. If the distribution or time series of the variable can be modeled, an inference as to the frequency of occurrence of significant hydrologic events of a given magnitude (such as precipitation) can be made. If, however, artificial controls are used and if the effects of these cannot be reliably predicted, frequency analyses may prove to be totally unreliable tools.

Precipitation Types

Dynamic or adiabatic cooling is the primary cause of condensation and is responsible for most rainfall. Thus it can be seen that vertical transport of air masses is a requirement for precipitation. Precipitation may be classified according to the conditions that generate vertical air motion. In this respect, the three major categories of precipitation type are *convective, orographic,* and *cyclonic.*

Convective Precipitation   Convective precipitation is typical of the tropics and is brought about by heating of the air at the interface with the ground. This heated air expands with a resultant reduction in weight. During this period, increasing quantities of water vapor are taken up; the warm moisture-laden air becomes unstable; and pronounced vertical currents are developed. Dynamic cooling takes place, causing condensation and precipitation. Convective precipitation may be in the form of light showers or storms of extremely high intensity (thunderstorms are a typical example).

Orographic Precipitation   Orographic precipitation results from the mechanical lifting of moist horizontal air currents over natural barriers such as mountain ranges. This type of precipitation is very common on the West Coast of the United States where moisture laden air from the Pacific Ocean is intercepted by coastal hills and mountains. Factors that are important in this process include land elevation, local slope, orientation of land slope, and distance from the moisture source.

In dealing with orographic precipitation, it is common to divide the region under study into zones for which influences aside from elevation are believed to be reasonably constant. For each of these zones, a relation between rainfall and elevation is developed for use in producing isohyetal maps (see Section 2.5).

Cyclonic Precipitation   Cyclonic precipitation is associated with the movement of air masses from high-pressure regions to low-pressure regions. These pressure differences are created by the unequal heating of the earth's surface.

Cyclonic precipitation may be classified as frontal or nonfrontal. Any barometric low can produce nonfrontal precipitation as air is lifted through horizontal convergence of the inflow into a low-pressure area. Frontal precipitation results from the lifting of warm air over cold air at the contact zone between air masses having different characteristics. If the air masses are moving so that warm air replaces colder air, the front is known as a *warm front;* if, on the other hand, cold air displaces warm air, the front is said to be *cold.* If the front is not in motion, it is said to be a *stationary front.* Figure 2.2 illustrates a vertical section through a frontal surface.

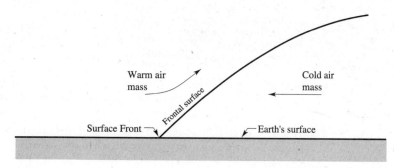

Figure 2.2   Vertical cross-section through a frontal surface.

Thunderstorms

Many areas of the United States are subjected to severe convective storms, which are generally identified as thunderstorms because of their electrical nature. These storms, although usually very local in nature, are often productive of very intense rainfalls that are highly significant when local and urban drainage works are considered.

Thunderstorm cells develop from vertical air movements associated with intense surface heating or orographic effects. There are three primary stages in the life history of a thunderstorm. These are the *cumulus stage,* the *mature stage,* and the *dissipating stage.* Figure 2.3 illustrates each of these stages.

All thunderstorms begin as cumulus clouds, although few such clouds ever reach the stage of development needed to produce such a storm. The cumulus stage is characterized by strong updrafts that often reach altitudes of over 25,000 ft. Vertical wind speeds at upper levels are often as great as 35 mph. As indicated in Fig. 2.3a, there is considerable horizontal inflow of air (entrainment) during the cumulus stage. This is an important element in the development of the storm, as additional moisture is provided. Air temperatures inside the cell are greater than those outside, as indicated by the convexity of the isotherms viewed from above. The number and size of the water droplets increase as the stage progresses. The duration of the cumulus stage is approximately 10–15 min.

The strong updrafts and entrainment support increased condensation and the development of water droplets and ice crystals. Finally, when the particles increase in size and number so that surface precipitation occurs, the storm is said to be in the mature stage. In this stage strong downdrafts are created as falling rain and ice crystals cool the air below. Updraft velocities at the higher altitudes reach up to 70 mph in the early periods of the mature stage. Downdraft speeds of over 20 mph are

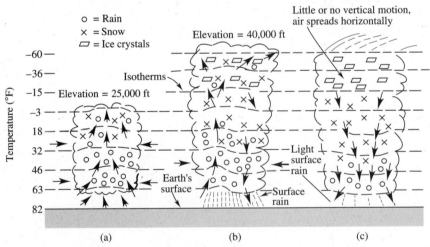

Figure 2.3 Cumulus, mature, and dissipating stages of a thunderstorm cell. (Department of the Army.)

usual above about 5000 ft in elevation. At lower levels, frictional resistance tends to decrease the downdraft velocity. Gusty surface winds move outward from the region of rainfall. Heavy precipitation is often derived during this preiod, which is usually on the order of 15–30 min.

In the final or dissipating stage, the downdraft becomes predominant until all the air within the cell is descending and being dynamically heated. Since the updraft ceases, the mechanism for condensation ends and precipitation tails off and ends.

Precipitation Data

Considerable data on precipitation are available in publications of the National Weather Service.[4,5] Other sources include various state and federal agencies engaged in water resources work. For critical regional studies it is recommended that all possible data be compiled; often the establishment of a gauging network will be necessary (see also Chapters 7–9).

Precipitation Variability

Precipitation varies geographically, temporally, and seasonally. Figure 2.4 indicates the mean annual precipitation for the continental United States, while Fig. 2.5 gives an example of seasonal differences. It should be understood that both regional and temporal variations in precipitation are very important in water resources planning and hydrologic studies. For example, it may be very important to know that the cycle of minimum precipitation coincides with the peak growing season in a particular area, or that the period of heaviest rainfall should be avoided in scheduling certain construction activities.

Precipitation amounts sometimes vary considerably within short distances. Records have shown differences of 20 percent or more in the catch of rain gauges less that 20 ft apart. Precipitation is usually measured with a rain gauge placed in the open so that no obstacle projects within the inverted conical surface having the top of the gauge as its apex and a slope of 45°. The catch of a gauge is influenced by the wind, which usually causes low readings. Various devices such as Nipher and Alter shields have been designed to minimize this error in measurement. Precipitation gauges may be of the recording or nonrecording type. The former are required if the time distribution of precipitation is to be known. Information about the features of gauges is readily available.[3]

Because precipitation varies spatially, it is usually necessary to use the data from several gauges to estimate the average precipitation for an area and to evaluate its reliability (see Chapter 27). This is especially important in forested areas where the variation tends to be large.

Time variations in rainfall intensity are extremely important in the rainfall-runoff process, particularly in urban areas (see Fig. 2.6a). The areal distribution is also significant and highly correlated with the time history of outflow (see Fig. 2.6b). These considerations are discussed in greater detail in following chapters.

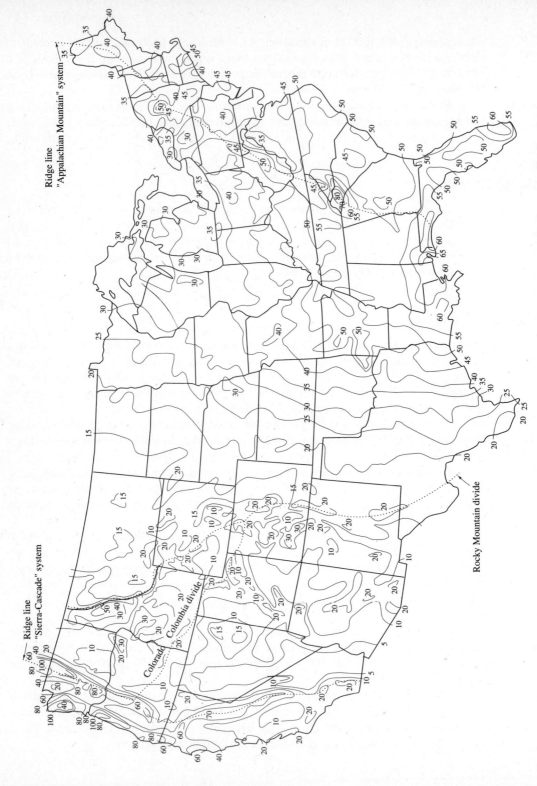

Figure 2.4 Mean annual precipitation in inches, 1899–1939. (U.S. Department of Agriculture, Soil Conservation Service.)

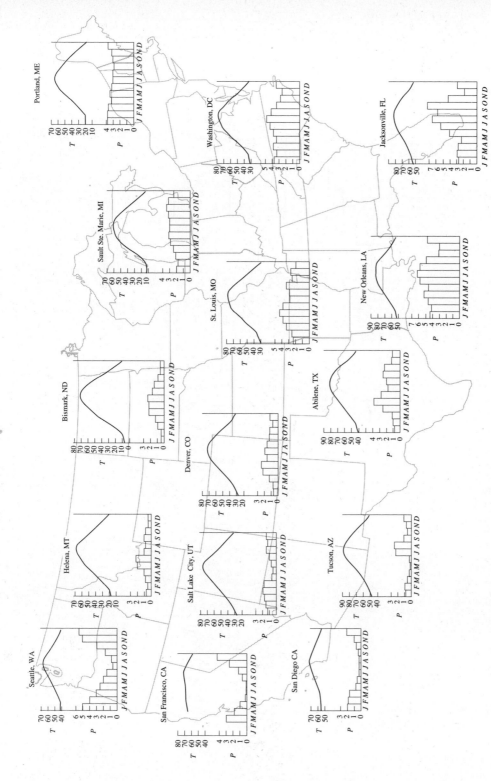

Figure 2.5 Precipitation and temperature distributions: T, mean monthly temperature (°F); P, mean monthly precipitation (in.). (U.S. Department of Agriculture, Soil Conservation Service.)

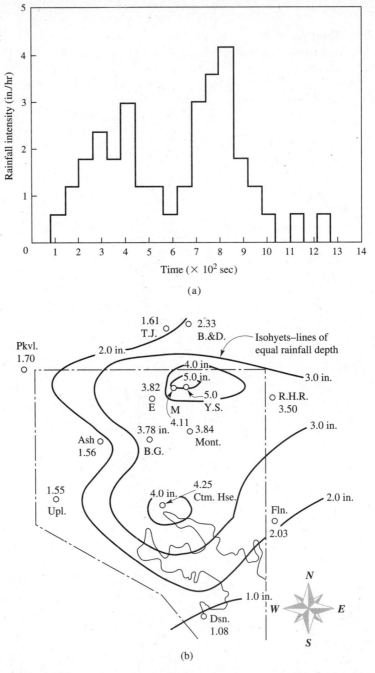

Figure 2.6 (a) Rainfall distribution in a convective storm June 1960, Baltimore, Maryland. (b) Isohyetal pattern, storm of September 10, 1957, Baltimore, Maryland. ○, recording rain gauge.

2.3 DISTRIBUTION OF THE PRECIPITATION INPUT

Total precipitation is distributed in numerous ways. That intercepted by vegetation and trees may be equivalent to the total precipitation input for relatively small storms. Once interception storage is filled, raindrops begin falling from leaves and grass, where water stored on these surfaces eventually becomes depleted through evaporation. Precipitation that reaches the ground may take several paths. Some water will fill depressions and eventually evaporate; some will infiltrate the soil. Part of the infiltrated water may strike relatively impervious strata near the soil surface and flow approximately parallel to it as interflow until an outlet is reached. Other portions may replenish soil moisture in the upper soil zone, and some infiltrated water may reach the groundwater reservoir that sustains dry weather streamflow. The component of the precipitation input that exceeds the local infiltration rate will develop a film of water on the surface (surface detention) until overland flow commences. Detention depths varying from $\frac{1}{8}$ to $1\frac{1}{2}$ in. for various conditions of slope and surface type have been reported.[3] Overland flow ultimately reaches defined channels and becomes streamflow.

Figure 2.7 illustrates in a general way the disposition of a uniform storm input to a natural drainage basin. Although such an input is not to be expected in nature, the indicated relations are representative of actual conditions. Modifications resulting from nonuniform storms will be discussed as they arise.

In Fig. 2.7a note that the storm input is distributed uniformly over time t_a at a rate equal to i (dimensionally equal to LT^{-1}). This input is dissected into components i_1 through i_4, the sum of which is equal to i at any time t. Figure 2.7b illustrates the manner in which infiltrated water is further subdivided into interflow, groundwater, and soil moisture. Figure 2.7c shows the transition from overland flow supply into streamflow. The mechanics of these processes will be treated in detail in later sections. The nature of the curves presented depicts the general runoff process. It should be realized, however, that actual graphs of infiltration and/or other factors versus time might appear quite different in form and relative magnitude when compared with these illustrations because of the effects of nonuniform storm patterns, antecedent conditions, and other factors.

The rate and areal distribution of runoff from a drainage basin are determined by a combination of physiographic and climatic factors. Important climatic factors include the form of precipitation (rain, snow, hail), the type of precipitation (convective, orographic, cyclonic), the quantity and time distribution of the precipitation, the character of the regional vegetative cover, prevailing evapotranspiration characteristics, and the status of the soil moisture reservoir. Physiographic factors of significance include geometric properties of the drainage basin, land-use characteristics, soil type, geologic structure, and characteristics of drainage channels (geometry, slope, roughness, and storage capacity).

Large drainage basins often react differently from smaller ones when subjected to a precipitation input. This can be explained in part by such factors as geologic age, relative impact of land-use practices, size differential, variations in storage characteristics, and other causes. Chow defines a small watershed as a drainage basin whose

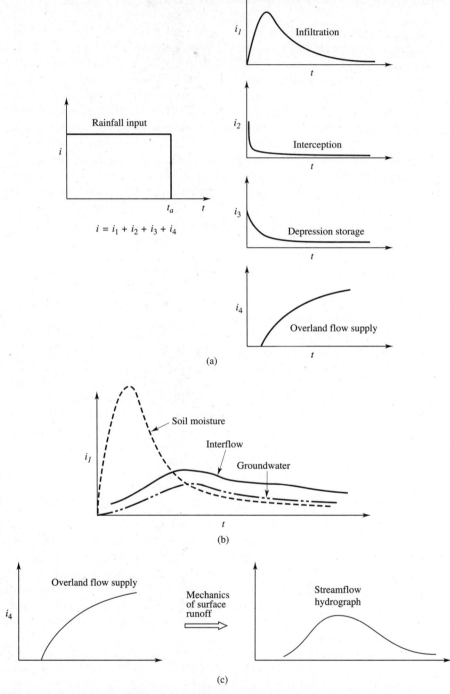

Figure 2.7 The runoff process: (a) disposition of precipitation, (b) components of infiltration, and (c) disposition of overland flow supply.

characteristics do not filter out (1) fluctuations characteristic of high-intensity, short-duration storms; or (2) the effects of land management practices.[6] On this basis, small basins may vary from less than an acre up to 100 mi². A large basin is one in which channel storage effectively filters out the high frequencies of imposed precipitation and effects of land-use practices.

2.4 POINT PRECIPITATION

Precipitation events are recorded by gauges at specific locations. The resulting data permit determination of the frequency and character of precipitation events in the vicinity of the site. Point precipitation data are used collectively to estimate areal variability of rain and snow and are also used individually for developing design storm characteristics for small urban or other watersheds. Design storms are discussed in detail in Chapter 16.

Point rainfall data are used to derive intensity–duration–frequency curves such as those shown in Fig. 2.8. Such curves are used in the rational method for urban storm drainage design (Chapter 25); their construction is discussed in Chapter 27. In applying the rational method, a rainfall intensity is used which represents the average intensity of a storm of given frequency for a selected duration. The frequency chosen should reflect the economics of flood damage reduction. Frequencies of up to 100 years are commonly used where residential areas are to be protected. For higher-value districts and critical facilities, up to 500 years or higher return periods are often selected. Local conditions and practice normally dictate the selection of these design criteria. (Executive Order 11988, Floodplain Management, 1977).

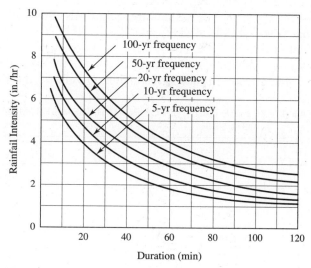

Figure 2.8 Typical intensity-duration-frequency curves for Baltimore, Maryland, and vicinity.

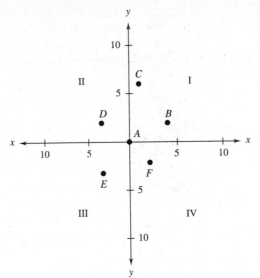

Figure 2.9 Four quadrants surrounding precipitation station *A*.

It is occasionally necessary to estimate point rainfall at a given location from recorded values at surrounding sites. This can be done to complete missing records or to determine a representative precipitation to be used at the point of interest. The National Weather Service has developed a procedure for this which has been verified on both theoretical and empirical bases.[7]

Consider that rainfall is to be calculated for point *A* in Fig. 2.9. Establish a set of axes running through *A* and determine the absolute coordinates of the nearest surrounding points *B*, *C*, *D*, *E*, and *F*. These are recorded in columns 3 and 4 of Table 2.1. The estimated precipitation at *A* is determined as a weighted average of the other five points. The weights are reciprocals of the sums of the squares of ΔX and ΔY; that is, $D^2 = \Delta X^2 + \Delta Y^2$, and $W = 1/D^2$. The estimated rainfall at the point of interest is given by $\Sigma (P \times W)/\Sigma W$. In the special case where rainfall is known in only two adjacent quadrants (e.g., I and II), the estimate is given as $\Sigma (P \times W)$. This has the effect of reducing estimates to zero as the points move from an area of

TABLE 2.1 DETERMINATION OF POINT RAINFALL FROM DATA AT NEARBY GAUGES

(1) Point	(2) Rainfall (in.)	(3) ΔX	(4) ΔY	(5) (D^2)	(6) $W \times 10^3$	(7) $P \times W \times 10^3$
A	—	—	—	—	—	—
B	1.60	4	2	20	50	80.0
C	1.80	1	6	37	27.0	48.6
D	1.50	3	2	13	76.9	115.4
E	2.00	3	3	18	55.6	111.2
F	1.70	2	2	8	125.0	212.5
Sums	—	—	—	—	334.5	567.7

Note: Estimated precipitation (P) at $A = 567.7/334.5$; $P = 1.70$ in.

precipitation to one with no records. This is considered to be the most logical procedure for handling this unusual case.[7] The estimated result will always be less than the greatest and greater than the smallest surrounding precipitation. For special effects such as mountain influences, an adjustment procedure can be applied.

2.5 AREAL PRECIPITATION

For most hydrologic analyses, it is important to know the areal distribution of precipitation. Usually, average depths for representative portions of the watershed are determined and used for this purpose. The most direct approach is to use the arithmetic average of gauged quantities. This procedure is satisfactory if gauges are uniformly distributed and the topography is flat. Other commonly used methods are the isohyetal method and the Thiessen method. The reliability of rainfall measured at one gauge in representing the average depth over a surrounding area is a function of (1) the distance from the gauge to the center of the representative area, (2) the size of the area, (3) topography, (4) the nature of the rainfall of concern (e.g., storm event versus mean monthly), and (5) local storm pattern characteristics.[8] For more information on errors of estimation, the reader should consult Refs. 7 and 8. Chapter 27 also contains a discussion of areal variability of precipitation.

Figures 2.10 and 2.11 illustrate how the measured rainfall at a single gauge relates to the average rainfall over a watershed with change in (1) the relative position of the gauge in the watershed and (2) the time period over which the average is calculated. In the first case it is clear that the more central the gauge location, the more closely its observations will match the average for a representative area, providing that the region is not too large. Figure 2.11 shows, not surprisingly, that areal averages

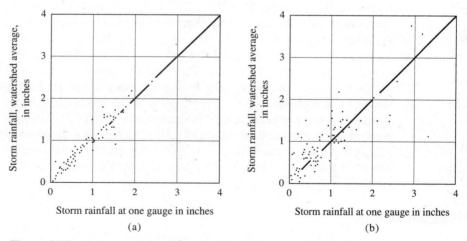

Figure 2.10 Errors resulting from use of a single gauge to estimate watershed average (gauge location effect, Soil Conservation Service). (a) Watershed area is 0.75 mi^2 and gauge is near the center. (b) Watershed area is 0.75 mi^2 and gauge is 4 mi outside the watershed boundary.

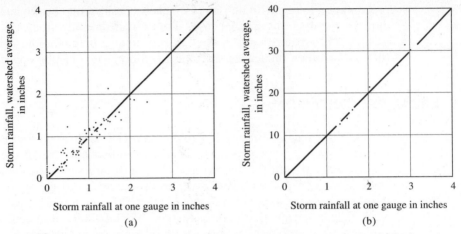

Figure 2.11 Errors resulting from use of a single gauge to estimate watershed average (time period effect, Soil Conservation Service). (a) Watershed area is 5.45 mi^2 and the gauge is on the boundary. (b) Watershed area is 5.45 mi^2 and the gauge is on the boundary.

over long time periods, in this case one year, may be expected to conform more closely to a single gauge average than those for an individual storm event. This suggests that the design of gauging networks should be tempered with both space and time considerations.

Isohyetal Method

The two principal methods for determining areal averages of rainfall are the isohyetal method and the Thiessen method. The isohyetal method is based on interpolation between gauges. It closely resembles the calculation of contours in surveying and mapping. The first step in developing an isohyetal map is to plot the rain gauge locations on a suitable map and to record the rainfall amounts (Fig. 2.12). Next, an interpolation between gauges is performed and rainfall amounts at selected increments are plotted. Identical depths from each interpolation are then connected to form isohyets (lines of equal rainfall depth). The areal average is the weighted average of depths between isohyets, that is, the mean value between the isohyets. The isohyetal method is the most accurate approach for determining average precipitation over an area, but its proper use requires a skilled analyst and careful attention to topographic and other factors that impact on areal variability. Figure 2.13 illustrates the representation of a major storm event in North Carolina by an isohyetal map.

Thiessen Method

Another method of calculating areal rainfall averages is the Thiessen method. In this procedure the area is subdivided into polygonal subareas using rain gauges as centers. The subareas are used as weights in estimating the watershed average depth. Thiessen diagrams are constructed as shown in Fig. 2.14. This procedure is not suitable for

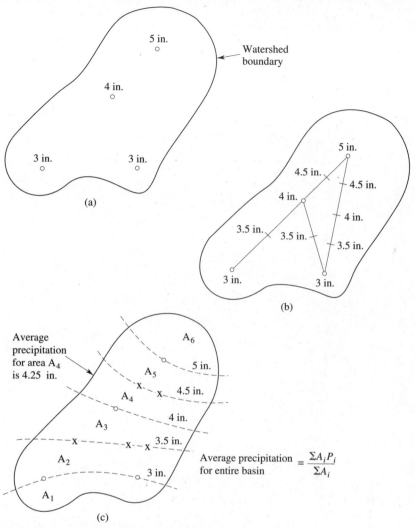

Figure 2.12   Construction of an isohyetal map: (a) locate rain gauges and plot values; (b) interpolate between gauges; and (c) plot isohyets.

mountainous areas because of orographic influences. The Thiessen network is fixed for a given gauge configuration, and polygons must be reconstructed if any gauges are relocated.

Accuracy

Irrespective of the method used for estimating areal precipitation, the location of the gauge used in deriving the estimate relative to the point of application of the estimate must be taken into consideration. In mountainous localities, vertical distances may be more important than horizontal ones. For gentle landscapes, horizontal spacings are

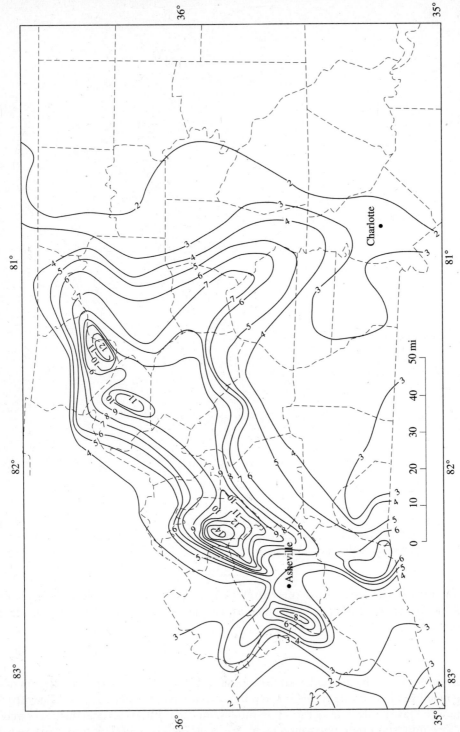

Figure 2.13　Map of Asheville-Statesville area showing the precipitation that caused the flood of November 1977. Total precipitation is given in inches for the period from Friday, November 4 at 7 a.m. to Monday, November 7 at 7 a.m. (Map prepared by the National Weather Service.)

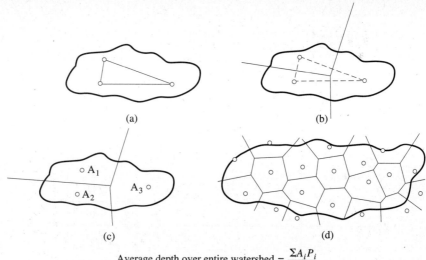

(a)　　　　　　　　　　　　(b)

(c)　　　　　　　　　　　　(d)

$$\text{Average depth over entire watershed} = \frac{\Sigma A_i P_i}{\Sigma A_i}$$

Figure 2.14 Construction of a Thiessen diagram: (a) connect rain gauge locations; (b) draw perpendicular bisectors; and (c) calculate Thiessen weights (A_1, A_2, A_3). (d) A completed network.

the most important. When a precipitation gauging network is to be developed, both spacing and arrangement of gauges must be considered.

EXAMPLE 2.1 _____

Given the drainage area of Fig. 2.15 and the rainfall data displayed in column 3 of Table 2.2, calculate the average rainfall over the area using (a) the arithmetic mean, and (b) the Thiessen polygon weighting system.

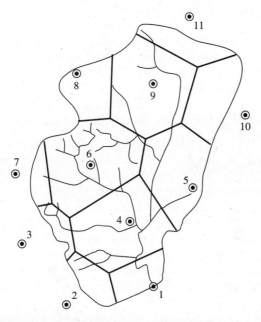

Figure 2.15 Thiessen diagram for Example 2.1.

TABLE 2.2 DATA AND THIESSEN POLYGON
CALCULATION FOR EXAMPLE 2.1.

(1)	(2)	(3)	(4)
Gauge No.	% Area	Precip.-in.	(2) × (3)
1	5	1.56	0.08
2	4	2.95	0.12
3	3	3.44	0.10
4	15	2.91	0.44
5	11	4.17	0.46
6	19	4.21	0.80
7	4	2.7	0.11
8	7	2.45	0.17
9	21	3.88	0.81
10	6	3.98	0.24
11	5	2.51	0.13
Total	100		3.45

Solution.

a. Identify those gauges falling within the area boundary. They include gauges 1, 4 through 6, 8, and 9. Averaging the values for these six gauges yields an estimated mean areal rainfall of 3.20 inches.

b. Following the Thiessen method as described in Section 2.5, construct polygons using triangles to connect gauge points. These polygons are shown on Fig. 2.15. Calculate the percent of the total area associated with each gauge and record as in column 2 of Table 2.2. The Thiessen weighted average is obtained by multiplying the values in column 2 by the values in column 3. The Thiessen average is computed as 3.45 inches of rainfall. The use of a spreadsheet (Table 2.2) facilitates computations and aids in organizing data. ■■

2.6 PROBABLE MAXIMUM PRECIPITATION

The probable maximum precipitation (PMP) is the critical depth-duration-area rainfall relation for a given area and season which would result from a storm containing the most critical meteorological conditions considered probable.[9] Such storm events are used in flood flow estimates by the U.S. Corps of Engineers and other water resources agencies. The critical meteorological conditions are based on analyses of air–mass properties (effective precipitable water, depth of inflow layer, wind, temperature, and other factors), synoptic situations during recorded storms in the region, topography, season, and location of the area. The rainfall derived is termed *probable maximum precipitation* since it is subject to limitations of meteorological theory and data and is based on the most effective combination of factors controlling rainfall

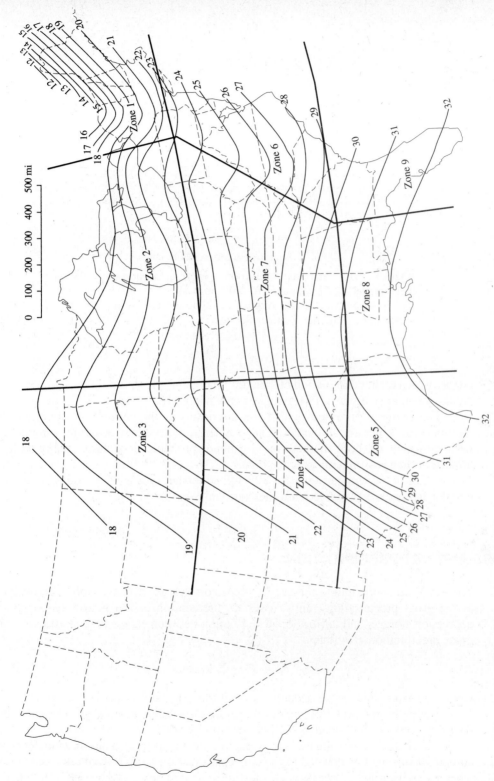

Figure 2.16 Probable maximum precipitation (in inches) for 200 mi² in 24 hr. (U.S. Department of Commerce, National Weather Service.)

35

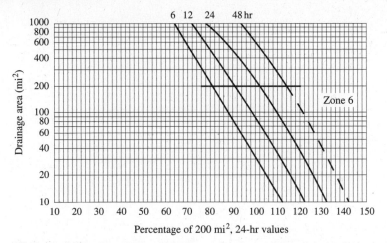

Figure 2.17 Seasonal variation, depth-area-duration relations; percentage to be applied to 200 mi^2-24 hr probable maximum precipitation values for August in Zone 6. (U.S. Department of Commerce, National Weather Service.)

intensity.[9] An earlier designation of "maximum possible precipitation" is synonymous. Additional information on PMP is given in Chapter 16.

The seasonal variation of PMP is important in the design and operation of multipurpose structures and in flooding considerations that may occur in combination with snowmelt. In both of these cases, annual probable maximums might be less important than seasonal maximums. Figures 2.16 and 2.17 display 24-hr PMP for the eastern half of the United States for 200-mi^2 watersheds during the month of August (similar figures are available from the National Weather Service).

2.7 GROSS AND NET PRECIPITATION

The net (excess) precipitation that contributes directly to surface runoff is equivalent to the gross precipitation minus losses to interception, storm period evaporation, depression storage, and infiltration. The relation between excess precipitation P_e and gross precipitation P is thus

$$P_e = P - \Sigma \text{ losses} \tag{2.2}$$

where the losses include all deductions from the gross storm input.

The paths that water precipitated over an area may take can be represented by flow diagrams of the type given in Fig. 1.3 and by equations of the form of Eq. 2.2. Models such as these are the basis for most hydrologic investigations, and much of the content of this book is devoted to the conceptualization of individual components of the various hydrologic processes and to synthesizing these components into holistic representations of hydrologic events.

■ Summary

Precipitation is the source of fresh water replenishment for the planet Earth. Too much or too little can mean the difference between prosperity and disaster. In between these extremes are the normal precipitation events that are experienced with a frequency and intensity related mainly to geographic position and topographic features.

After reading this chapter you should understand that both the timing and amount of precipitation occurring over an area are important and that there is considerable geographic variability in precipitation. You should be able to estimate areal precipitation amounts from gauge data and conceptualize simple hydrologic process models. It should be recognized that average values of precipitation for a region shed some light on the quantity of water that might be made available for various uses, while a knowledge of the time-distribution and time-disposition of precipitation are requisites for developing management plans for periods of excess and shortage.

PROBLEMS

2.1. Rain gauge X was out of operation for a month during which there was a storm. The rainfall amounts at three adjacent stations A, B, and C were 4.2, 3.7, and 4.9 in., respectively. The average annual precipitation amounts for the gauges are $X = 36.5$, $A = 42.1$, $B = 37.1$, and $C = 39.8$. The delta x and delta y values respectively for each station are $X - 0, 0$; $A - 3, 7$; $B - 4, 6$; and $C - 5, 9$. Using a weighted average, estimate the amount of rainfall for gauge X.

2.2. Compute the rainfall for gauge X in Problem 2.1 if the storm readings at A, B, and C were 3.7, 4.1, and 4.8 in., respectively.

2.3. Compute the mean annual precipitation for the watershed in the following figure using the arithmetic mean, the Thiessen polygon method, and the isohyetal method. The

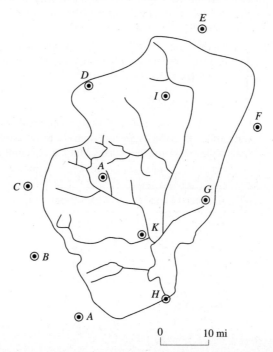

gauge readings for gauges A–K, respectively, are: 29.79, 34.97, 25.6, 24.27, 24.6, 42.61, 42.35, 15.51, 39.99, 43.04, and 28.41.

2.4. Compute the mean annual precipitation for the watershed in the figure for Problem 2.3 using the arithmetic mean and the Thiessen polygon method. The gauge readings for gauges *A–K,* respectively, are: 28.1, 33.7, 25.6, 23.9, 24.6, 40.7, 41.3, 37.2, 38.7, 41.1, and 29.3.

2.5. The chart from a rain gauge shown in the sketch represents a record that you must interpret. Find the average rainfall intensity (rate) between 6 A.M. and noon on August 10. Find also the total precipitation on August 10 and August 11.

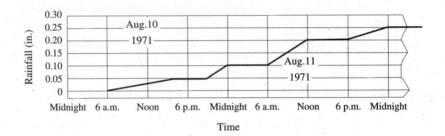

2.6. Refer to the chart of Problem 2.5. Calculate the rainfall intensity for the period between 6 A.M. and noon on August 11. Would you consider this to be a period of intense rainfall?

2.7. Use the map of Fig. 2.6 and from it construct a set of Thiessen polygons. Using these, estimate the mean rainfall for the region.

2.8. A mean draft of 100 mgd is produced from a drainage area of 200 mi². At the flow line the reservoir is estimated to cover 4000 acres. The annual rainfall is 37 in., the mean annual runoff is 10 in., and the mean annual lake evaporation is 30 in. Find the net gain or loss in storage. Compute the volume of water evaporated. How significant is this amount?

2.9. A mean draft of 380,000 m³/day is produced from a drainage area of 330 km². At the flow line, the reservoir is estimated to cover about 1600 hectares. The annual rainfall is 96.5 cm, the mean annual runoff is 22.8 cm, and the mean annual lake evaporation is 77.1 cm. Find the net gain or loss in storage and compute the volume of water evaporated. Calculate volumes in m³.

2.10. Drainage areas within each of the isohyetal lines for a storm are tabulated for a watershed. Use the isohyetal method to determine the average precipitation depth within the basin for the storm. Make a conceptual sketch.

Isohyetal interval (in.)	Area (acres)
0–2	2700
2–4	1900
4–6	1000
6–8	0

2.11. Rework Problem 2.10 if the values in the second column of the table are 2,500, 2,100 1,200, and 300, respectively.

2.12. Discuss how you would go about collecting data for analysis of the water budget of a region. What agencies would you contact? What other sources of information would you seek out?

2.13. For an area of your choice, plot the mean monthly precipitation versus time. Explain how this fits the pattern of seasonal water uses for the area. Will the form of precipitation be an important consideration?

REFERENCES

1. Tennessee Valley Authority, "Heat and Mass Transfer Between a Water Surface and the Atmosphere," Lab. Rep. No. 14, TVA Engineering Lab, Norris, TN, Apr. 1972.
2. A. L. Shands, "Mean Precipitable Water in the United States," U.S. Weather Bureau, Tech. Paper No. 10, 1949.
3. R. K. Linsley, Jr., M. A. Kohler, and J. L. H. Paulhus, *Applied Hydrology.* New York, McGraw-Hill, 1949.
4. D. W. Miller, J. J. Geraghty, and R. S. Collins, *Water Atlas of the United States.* Port Washington, NY: Water Information Center, 1963.
5. U.S. Weather Bureau, Tech. Papers 1–33. Washington, DC: U.S. Government Printing Office.
6. Ven Te Chow (ed.), *Handbook of Applied Hydrology.* New York, McGraw-Hill, 1964.
7. Staff, Hydrologic Research Laboratory, "National Weather Service River Forecast System Forecast Procedures," NOAA Tech. Mem. NWS HYDRO 14, National Weather Service, Silver Spring, MD, Dec. 1972.
8. V. Mockus, Sec. 4, in *SCS National Engineering Handbook on Hydrology,* Washington, DC: Soil Conservation Service, Aug. 1972.
9. J. T. Riedel, J. F. Appleby, and R. W. Schloemer, "Seasonal Variation of the Probable Maximum Precipitation East of the 105th Meridian for Areas from 10 to 1000 Square Miles and Durations of 6, 12, 24, and 48 Hours," Hydrometeorological Rept. No. 33, U.S. Weather Bureau, Washington, D.C., 1967.

Chapter 3

Interception and Depression Storage

■ Prologue

The purpose of this chapter is to:

- Define *interception* and *depression storage*.
- Define the roles these abstracting mechanisms play in affecting the amount of precipitated water ultimately available for other distribution.
- Provide some approaches to estimating the quantities of water intercepted and stored in depressions during precipitation events.

Figure 1.3 indicates the paths that precipitated water may take as it reaches the earth. The first encounters are with intercepting surfaces such as trees, plants, grass, and structures. Water in excess of interception capacity then begins to fill surface depressions. A film of water also builds up over the ground surface. This is known as surface detention. Once this film is of sufficient depth, surface flow toward defined channels commences, providing that the rate at which water seeps into the ground is less than the rate of surface supply. This chapter deals with the first two mechanisms by which the gross precipitation input becomes transformed into net precipitation.

3.1 INTERCEPTION

Part of the storm precipitation that occurs is intercepted by vegetation and other forms of cover on the drainage area. Interception can be defined as that segment of the gross precipitation input which wets and adheres to aboveground objects until it is returned to the atmosphere through evaporation. Precipitation striking vegetation may be retained on leaves or blades of grass, flow down the stems of plants and become stemflow, or fall off the leaves to become part of the throughfall. The modifying effect that a forest canopy can have on rainfall intensity at the ground (the throughfall) can be put to practical use in watershed management schemes.

The amount of water intercepted is a function of (1) the storm character, (2) the species, age, and density of prevailing plants and trees, and (3) the season of the year. Usually about 10–20 percent of the precipitation that falls during the growing season is intercepted and returned to the hydrologic cycle by evaporation. Water losses by interception are especially pronounced under dense closed forest stands—as much as 25 percent of the total annual precipitation. Schomaker has reported that the average annual interception loss by Douglas fir stands in western Oregon and Washington is about 24 percent.[1] A 10-year-old loblolly pine plantation in the South showed losses on a yearly basis of approximately 14 percent, while Ponderosa pine forests in California were found to intercept about 12 percent of the annual precipitation. Mean interception losses of approximately 13 percent of gross summer rainfall were reported for hardwood stands in the White Mountains of New Hampshire. Additional information given in Table 3.1 includes some data on interception measurements obtained in Maine from a mature spruce-fir stand, a moderately well-stocked white and gray birch stand, and an improved pasture.[1]

Lull indicates that oak or aspen leaves may retain as much as 100 drops of water.[2] For a well-developed tree, interception storage on the order of 0.06 in. of precipitation could therefore be expected on the basis of an average retention of about

TABLE 3.1 WEEKLY AVERAGE PRECIPITATION CATCH OF STANDARD U.S. WEATHER BUREAU-TYPE RAIN GAUGES LOCATED IN A SPRUCE-FIR STAND, A HARDWOOD STAND, AND A PASTURE DURING THE WINTER OF 1965–1966

Measuring date[a]	Weekly average precipitation catch (in. of equivalent rain)			Percent interception by forest cover	
	Spruce-fir	Birch	Pasture	Spruce-fir	Birch
11/9/65	0.24	0.33	0.39	38	15
11/16/65	1.01	1.25	1.45	30	14
11/23/65	1.01	1.23	1.36	26	10
12/10/65[b]	1.41	1.65	1.79	21	8
12/17/65	0.55	0.81	0.87	37	7
12/30/65	0.66	0.95	1.08	39	12
1/4/66	0.20	0.25	0.26	23	4
1/12/66	0.36	0.55	0.61	41	10
1/18/66	Trace	Trace	Trace	—	—
1/25/66	0.25	0.58	0.59	58	2
2/1/66	1.38	1.91	1.96	30	3
2/8/66	0.05	0.07	0.06	17	16
2/11/66	0.29[c]	0.02	Trace	—	—
2/15/66	0.76	0.81	0.98	22	17
2/21/66	0.17	0.22	0.22	23	0
3/2/66	0.86	1.23	1.45	41	16
3/7/66	0.76	0.84	0.97	22	13
3/15/66	0	0	0	—	—
3/29/66	0.73	1.13	1.27	43	11
Total	10.69	13.83	15.31	30.2	9.5

[a] The period between measuring dates is 7 days, except when precipitation occurred on the seventh day. In this event, measurement was postponed until precipitation ceased.

[b] Measurements were delayed until a method was devised to melt frozen precipitation on the site.

[c] This measurement in the spruce stand was the result of foliage drip during a thaw from previously intercepted snow.

Source: After C. E. Schomaker, "The Effect of Forest and Pasture on the Disposition of Precipitation," *Maine Farm Res.* (July 1966).

20 drops per leaf. For light showers (where gross precipitation $P < 0.01$ in.) 100 percent interception might occur, whereas for showers where $P > 0.04$ in., losses in the range of 10–40 percent are realistic.[3]

Figure 3.1 illustrates the general time distribution pattern of interception loss intensity. Most interception loss develops during the initial storm period and the rate of interception rapidly approaches zero thereafter.[1-6] Potential storm interception losses can be estimated by using[2,3,6]

$$L_i = S + KEt \tag{3.1}$$

where L_i = the volume of water intercepted (in.)
   S = the interception storage that will be retained on the foliage against the forces of wind and gravity (usually varies between 0.01 and 0.05 in.)
   K = the ratio of surface area of intercepting leaves to horizontal projection of this area
   E = the amount of water evaporated per hour during the precipitation period (in.)
   t = time (hr)

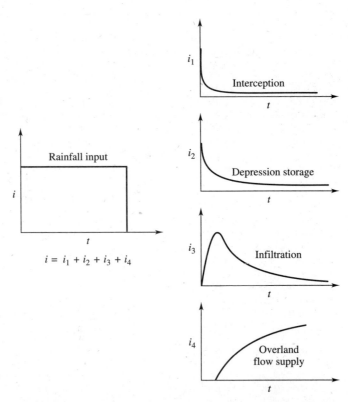

Figure 3.1 Disposition of rainfall input in terms of interception, depression storage, infiltration, and overland flow.

Equation 3.1 is based on the assumption that rainfall is sufficient to fully satisfy the storage term S. The following equation was designed to account for the rainfall amount[7-9]

$$L_i = S(1 - e^{-P/S}) + KEt \qquad (3.2)$$

where P = rainfall and e is the base of natural logarithms. Note in Eqs. 3.1 and 3.2 that the storm time duration t is given in hours, while L_i, S, and E are commonly measured in in. or mm.

It is important to recognize that forms of vegetation other than trees can also intercept large quantities of water. Grasses, crops, and shrubs often have leaf-area to ground-area ratios that are similar to those for forests. Table 3.2 summarizes some observations that have been made on crops during growing seasons and on a variety of grasses. Intercepted amounts are about the same as those for forests, but since some of these types of vegetation exist only until harvest, their annual impact on interception is generally less than that of forested areas.

Precipitation type, rainfall intensity and duration, wind, and atmospheric conditions affecting evaporation are factors that serve to determine interception losses. Snow interception, while highly visible, usually is not a major loss since much of the intercepted snowfall is eventually transmitted to the ground by wind action and melt. Interception during rainfall events is commonly greater than for snowfall events. In both cases, wind velocity is an important factor.

The importance of interception in hydrologic modeling is tied to the purpose of the model. Estimates of loss to gross precipitation through interception can be significant in annual or long-term models, but for heavy rainfalls during individual storm events, accounting for interception may be unnecessary. It is important for the modeler to assess carefully both the time frame of the model and the volume of precipitation with which one must deal.

TABLE 3.2   OBSERVED PERCENTAGES OF INTERCEPTION BY VARIOUS CROPS AND GRASSES[a]

Vegetation type	Intercepted (%)	Comments
Crops		
Alfalfa	36	
Corn	16	
Soybeans	15	
Oats	7	
Grasses[b]		
Little bluestem	50–60	
Big bluestem	57	
Tall panic grass	57	Water applied at rate of $\frac{1}{2}$ in. in 30 min
Bindweed	17	
Buffalo grass	31	
Blue grass	17	Prior to harvest
Mixed species	26	
Natural grasses	14–19	

[a] Values rounded to nearest percent. Data for table were obtained from Refs. 2, 4, and 5.
[b] Grass heights vary up to about 36 in.

Equations 3.1 and 3.2 can be used to estimate total interception losses, but for detailed analyses of individual storms, it is necessary to deal with the areal variability of such losses. General equations for estimating such losses are not available, however. Most research has been related to particular species or experimental plots strongly associated with a given locality. In addition, the loss function varies with the storm's character. If adequate experimental data are available, the nature of the variance of interception versus time might be inferred. Otherwise, common practice is to deduct the estimated volume entirely from the initial period of the storm (*initial abstraction*).

EXAMPLE 3.1

Using the following equations developed by Horton[6] for interception by ash and oak trees, estimate the interception loss beneath these trees for a storm having a total precipitation of 1.5 in.

Solution

1. For ash trees,

$$L_i = 0.015 + 0.23P$$
$$= 0.015 + 0.23(1.5) = 0.36 \text{ in.}$$

2. For oak trees,

$$L_i = 0.03 + 0.22P$$
$$= 0.03 + 0.22(1.5) = 0.36 \text{ in.} \quad \blacksquare\blacksquare$$

3.2 THROUGHFALL

A number of relationships for estimating throughfall for a variety of forest types have been developed.[9-12] Determining factors for throughfall quantities include canopy coverage, total leaf area, number and type of layers of vegetation, wind velocity, and rainfall intensity. The areal variability of these factors results in little or no throughfall in some locations and considerable throughfall in others. In general, prediction equations for throughfall must include measures of canopy surface area and cover as prime variables. An example of a throughfall relationship for an eastern United States hardwood forest follows.[12]

For the growing season

$$T_h = 0.901P - 0.031n \qquad (3.3)$$

For the dormant season

$$T_h = 0.914P - 0.015n \qquad (3.4)$$

where T_h = throughfall (in.)
   P = total precipitation (in.)
   n = number of storms

3.3 DEPRESSION STORAGE

Precipitation that reaches the ground may infiltrate, flow over the surface, or become trapped in numerous small depressions from which the only escape is evaporation or infiltration. The nature of depressions, as well as their size, is largely a function of the original land form and local land-use practices. Because of extreme variability in the nature of depressions and the paucity of sufficient measurements, no generalized relation with enough specified parameters for all cases is feasible. A rational model can, however, be suggested.

Figure 3.1 illustrates the disposition of a precipitation input. A study of it shows that the rate at which depression storage is filled rapidly declines after the initiation of a precipitation event. Ultimately, the amount of precipitation going into depression storage will approach zero, given that there is a large enough volume of precipitation to exceed other losses to surface storage such as infiltration and evaporation. Ultimately, all the water stored in depressions will either evaporate or seep into the ground. Finally, it should be understood that the geometry of a land surface is usually complex and thus depressions vary widely in size, degree of interconnection, and contributing drainage area. In general, depressions may be looked upon as miniature reservoirs and as such they are subject to similar analytical techniques.

According to Linsley et al.[13] the volume of water stored by surface depressions at any given time can be approximated using

$$V = S_d(1 - e^{-kP_e}) \tag{3.5}$$

where V = the volume actually in storage at some time of interest

S_d = the maximum storage capacity of the depressions

P_e = the rainfall excess (gross rainfall minus evaporation, interception, and infiltration)

k = a constant equivalent to $1/S_d$

The value of the constant can be determined by considering that if $P_e \approx 0$, essentially all the water will fill depressions and dV/dP_e will equal one. This requires that $k = 1/S_d$. Estimates of S_d may be secured by making sample field measurements of the area under study. Combining such data with estimates of P_e permits a determination of V. The manner in which V varies with time must still be estimated if depression storage losses are to be abstracted from the gross rainfall input.

One assumption regarding dV/dt is that all depressions must be full before overland flow supply begins. Actually, this would not agree with reality unless the locations of depressions were graded with the largest ones occurring downstream. If the depression storage were abstracted in this manner, the total volume would be deducted from the initial storm period such as shown by the shaded area in Fig. 3.2. Such postulates have been used with satisfactory results under special circumstances.[14]

Depression storage intensity can also be estimated using Eq. 3.5. If the overland flow supply rate σ plus depression storage intensity equal $i - f$, where i is the rainfall intensity reaching the ground and f is the infiltration rate, then the ratio of overland flow supply to overland flow plus depression storage supply can be proved equal to

$$\frac{\sigma}{i - f} = 1 - e^{-kP_e} \tag{3.6}$$

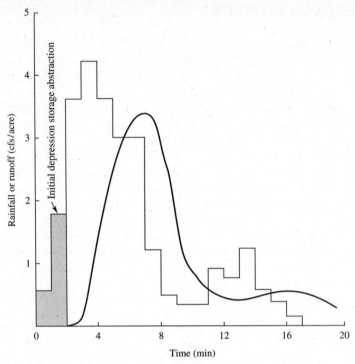

Figure 3.2 Simple depression storage abstraction scheme.

This expression can be derived by adjudging

$$\frac{\sigma}{i - f} = \frac{i - f - v}{i - f} \qquad (3.7)$$

and noting that v is equal to the derivative of Eq. 3.5 with respect to time. Then

$$v = \frac{d}{dt} S_d(1 - e^{-kP_e}) \qquad (3.8)$$

$$v = (S_d k e^{-kP_e})\frac{dP_e}{dt} \qquad (3.9)$$

It was shown that $k = 1/S_d$ so that

$$v = e^{-kP_e}\frac{dP_e}{dt} \qquad (3.10)$$

The excess precipitation P_e equals the gross rainfall minus infiltrated water, and since the derivative with respect to time can be replaced by the equivalent intensity $(i - f)$, the intensity of depression storage becomes

$$v = e^{-kP_e}(i - f) \qquad (3.11)$$

Inserting in Eq. 3.7, we obtain

$$\frac{\sigma}{i - f} = \frac{(i - f) - (i - f)e^{-kP_e}}{i - f} \qquad (3.12)$$

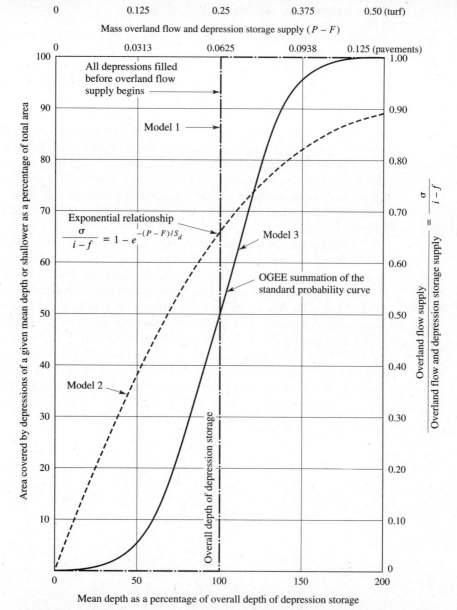

Figure 3.3 Depth distribution curve of depression storage. Enter graph from top, read down to selected curve, and project right or left as desired. (After Tholin and Kiefer.[15])

and

$$\frac{\sigma}{i - f} = \frac{(i - f)(1 - e^{-kP_e})}{i - f} \tag{3.13}$$

$$\frac{\sigma}{i - f} = 1 - e^{-kP_e} \tag{3.14}$$

Figure 3.3 illustrates a plot of this function versus the mass overland flow and depression storage supply $(P - F)$, where F is the accumulated mass infiltration[15] and

P is the gross precipitation. In the plot mean depths of 0.25 in. for turf and 0.0625 in. for pavements were assumed. Maximum depths were 0.50 and 0.125 in., respectively.

The figure also depicts the effect on estimated overland flow supply rate, which is derived from the choice of the depression storage model. Three models are shown in the figure: the first one assumes that all depressions are full before overland flow begins. For a turf area having depressions with a mean depth of 0.25 in., the figure shows that for $P - F$ values less than 0.25 in., there is no overland flow supply, while for $P - F$ values greater than 0.25 in., the overland flow supply is equal to $i - f$.

For the exponential model (Model 2), σ always will be greater than zero. Tholin and Kiefer have recommended that a relation between those previously mentioned is likely more representative of fully developed urban areas.[15] A cumulative normal probability curve was selected for this representation and is also described in Fig. 3.3 (Model 3).

Depression storage deductions are usually made from the first part of the storm as illustrated in Fig. 3.2. The amount to be deducted is a function of topography, ground cover, and extent and type of land development. During major storms, this loss is often considered to be negligible. Some guidelines for estimating depression storage losses have been developed based on studies of experimental and other watersheds. Values for depression storage losses from intense storms reported by Hicks are 0.20 in. for sand, 0.15 in. for loam, and 0.10 in. for clay.[16] Tholin and Kiefer have used values of 0.25 in. in pervious urban areas and 0.0625 in. for pavements.[15] Studies of four small impervious drainage areas by Viessman yielded the information shown in Fig. 3.4, where mean depression storage loss is highly correlated with slope. This is

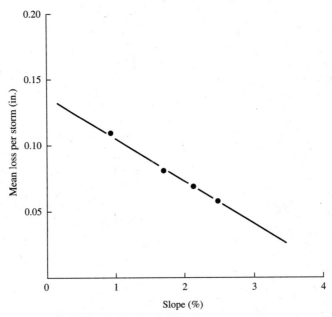

Figure 3.4 Depression storage loss versus slope for four impervious drainage areas. (From Viessman.[14])

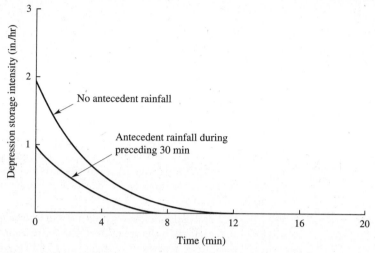

Figure 3.5   Depression storage intensity versus time for an impervious area. (After Turner.[17])

easily understood, since a given depression will hold its maximum volume if horizontally oriented. Using very limited data from a small, paved-street section, Turner devised the curves shown in Fig. 3.5.[17] Other sources of data related to surface storage are available in the literature.[2, 18, 19]

■ Summary

Accounting for the disposition of precipitation is an important part of the hydrologic modeling process. Two abstractions from the precipitation input, interception, and depression storage were covered in this chapter.

Interception losses during the course of a year may be substantial, but during intense storms, they may be sufficiently small to neglect. Precipitation type, rainfall intensity and duration, wind, and atmospheric conditions affecting evaporation are factors that serve to determine interception losses for a given forest stand or ground cover configuration. Interception during rainfall events is commonly greater than for snowfall events.

Depression storage deductions occur early in a storm sequence and they are a function of topography, ground cover, and extent and type of land development. During major storms, this loss is often considered to be negligible.

PROBLEMS

3.1.   Using Fig. 3.2, estimate the volume of depression storage for a 3-acre paved drainage area. State the volume in cubic feet and cubic meters. Convert it to equivalent depth over the area in in. and cm.

3.2. Estimate the percentage of the total volume of rainfall that is indicated as depression storage in Fig. 3.2.

3.3. Using the average annual precipitation for your state, estimate the annual amount of interception loss.

3.4. Refer to Fig. 2.4 and estimate the annual interception losses in Illinois, Florida, California, and New Mexico. How good do you think these estimates are? In which estimates do you have the most confidence? Why? In which of these states would the water budget be most affected by interception?

3.5. Using Fig. 3.4, estimate the percentage of rainfall that would be lost to depression storage for a 10-acre parking lot having a mean slope of 1 percent. Repeat for a slope of 3 percent. Using the total rainfall volume determined in Problem 3.2, estimate the equivalent depth over the area of the depression storage loss for both slopes. State depths in mm and in.

3.6. Refer to Fig. 3.3 and estimate the ratio of overland flow supply to overland flow and depression storage supply if the area is turf, the OGEE summation curve is the model, and the mean depth of depression storage is (a) 75 percent and (b) 125 percent.

3.7. Explain how a relation such as that given in Fig. 3.3 could be used in a simulation model of the rainfall-runoff process.

3.8. Using Eqs. 3.3 and 3.4, estimate the throughfall in in. for 28 in. of rainfall during the growing season (21 events), and 17 in. of rainfall during the dormant season (13 events).

3.9. Using Horton's equations given in Example 3.1, estimate the interception losses by ash and oak trees for a storm having a total precipitation of 1.33 in.

REFERENCES

1. C. E. Schomaker, "The Effect of Forest and Pasture on the Disposition of Precipitation," *Maine Farm Res.* (July 1966).
2. Ven Te Chow (ed.), *Handbook of Applied Hydrology.* New York: McGraw-Hill, 1964.
3. Joseph Kittredge, *Forest Influences.* New York: McGraw-Hill, 1948.
4. O. R. Clark, "Interception of Rainfall by Herbaceous Vegetation," *Science* **86**(2243), 591–592(1937).
5. J. S. Beard, "Results of the Mountain Home Rainfall Interception and Infiltration Project on Black Wattle, 1953–1954," *J. S. Afr. Forestry Assoc.* **27,** 72–85(1956).
6. R. E. Horton, "Rainfall Interception," *Monthly Weather Rev.* **47,** 603–623(1919).
7. R. A. Meriam, "A Note on the Interception Loss Equation," *J. Geophys. Res.* **65,** 3850–3851(1960).
8. D. M. Gray (ed.), *Handbook on the Principles of Hydrology.* National Research Council, Canada, Port Washington: Water Information Center, Inc., 1973.
9. K. N. Brooks, P. F. Folliott, H. M. Gregersen, and J. L. Thames, *Hydrology and the Management of Watersheds.* Ames, IA: Iowa State University Press/Ames, 1991.
10. G. J. Blake, "The Interception Process." In *Prediction in Catchment Hydrology,* National Symposium on Hydrology, eds. T. G. Chapman and F. X. Dunin, Melbourne Aust. Acad. Sci., 59–81, 1975.
11. F. A. Roth, II, and M. Chang, "Throughfall in Planted Stands of Fourth Southern Pines Species in East Texas," *Water Resources Bulletin* **17,** 880–885(1981).

12. J. D. Helvey and J. H. Patric, "Canopy and Litter Interception by Hardwoods of Eastern United States," *Water Resour. Res.* **1,** 193–206(1965).

13. R. K. Linsley, Jr., M. A. Kohler, and J. L. H. Paulhus, *Applied Hydrology*. New York: McGraw-Hill, 1949.

14. Warren Viessman, Jr., "A Linear Model for Synthesizing Hydrographs for Small Drainage Areas," paper presented at the Forty-eighth Annual Meeting of the American Geophysical Union, Washington, D.C., Apr. 1967.

15. A. L. Tholin and C. J. Kiefer, "The Hydrology of Urban Runoff," *Trans. ASCE* **125,** 1308–1379(1960).

16. W. I. Hicks, "A Method of Computing Urban Runoff," *Trans. ASCE* **109,** 1217–1253(1944).

17. L. B. Turner, "Abstraction of Depression Storage from Storms on Small Impervious Areas," Master's thesis, University of Maine, Orono, ME, Aug. 1967.

18. R. E. Horton, *Surface Runoff Phenomena: Part I, Analysis of the Hydrograph*. Horton Hydrol. Lab. Pub. 101. Ann Arbor, MI: Edwards Bros., 1935.

19. L. F. Huggins and E. J. Monke, "The Mathematical Simulation of the Hydrology of Small Watersheds," Tech. Rept. 1, Purdue University Water Resources Center, Lafayette, IN, Aug. 1966.

Chapter 4

Infiltration

■ **Prologue**

The purpose of this chapter is to:

- Define *infiltration.*
- Indicate the role infiltration plays in affecting runoff quantities and in replenishing soil moisture and groundwater storages.
- Present models for estimating infiltration and provide examples of how they can be used.

Infiltration is that process by which precipitation moves downward through the surface of the earth and replenishes soil moisture, recharges aquifers, and ultimately supports streamflows during dry periods. Along with interception, depression storage, and storm period evaporation, it determines the availability, if any, of the precipitation input for generating overland flows (Fig. 1.3). Furthermore, infiltration rates influence the timing of overland flow inputs to channelized systems. Accordingly, infiltration is an important component of any hydrologic model.

The rate f at which infiltration occurs is influenced by such factors as the type and extent of vegetal cover, the condition of the surface crust, temperature, rainfall intensity, physical properties of the soil, and water quality.

The rate at which water is transmitted through the surface layer is highly dependent on the condition of the surface. For example, inwash of fine materials may seal the surface so that infiltration rates are low even when the underlying soils are highly permeable. After water crosses the surface interface, its rate of downward movement is controlled by the transmission characteristics of the underlying soil profile. The volume of storage available below ground is also a factor affecting infiltration rates.

Considerable research on infiltration has taken place, but considering the infinite combinations of soil and other factors existing in nature, no perfectly quantified general relation exists.

4.1 MEASURING INFILTRATION

Commonly used methods for determining infiltration capacity are hydrograph analyses and infiltrometer studies. Infiltrometers are usually classified as rainfall simulators or flooding devices. In the former, artifical rainfall is simulated over a small test plot and the infiltration calculated from observations of rainfall and runoff, with consideration given to depression storage and surface detention.[1] Flooding infiltrometers are usually rings or tubes inserted in the ground. Water is applied and maintained at a constant level and observations made of the rate of replenishment required.

Estimates of infiltration based on hydrograph analyses have the advantage over infiltrometers of relating more directly to prevailing conditions of precipitation and field. However, they are no better than the precision with which rainfall and runoff are measured. Of particular importance in such studies is the areal variability of rainfall. Several methods have been developed and are in use. Reference 1 gives a good description of these methods.

4.2 CALCULATION OF INFILTRATION

Infiltration calculations vary in sophistication from the application of reported average rates for specific soil types and vegetal covers to the use of differential equations governing the flow of water in unsaturated porous media. For small urban areas that respond rapidly to storm input, more precise methods are sometimes warranted. On large watersheds subject to peak flow production from prolonged storms, average or representative values may be adequate.

The infiltration process is complicated at best. Even under ideal conditions (uniform soil properties and known fluid properties), conditions rarely encountered in practice, the process is difficult to characterize. Accordingly, there has been considerable study of the infiltration process. Most of these efforts have related to the development of (1) empirical equations based on field observations and (2) the solution of equations based on the mechanics of saturated flow in porous media.[1,2]

Later in this chapter, several commonly used infiltration models are discussed. As a preface to that discussion, a brief description of the infiltration process follows. It reviews the principal factors affecting infiltration and points out some of the problems encountered by hydrologic modelers.

We begin our discussion with an ideal case, one in which the soil is homogeneous throughout the profile and all the pores are directly interconnected by capillary passages. Furthermore, it is assumed that the rainfall is uniformly distributed over the area of concern. Under these conditions, the infiltration process may be characterized as one dimensional and the major influencing factors are therefore soil type and moisture content.[3]

The soil type characterizes the size and number of the passages through which the water must flow while the moisture content sets the capillary potential and relative conductivity of the soil. Capillary potential is the hydraulic head due to capillary forces. Capillary suction is the same as capillary potential but with opposite sign.

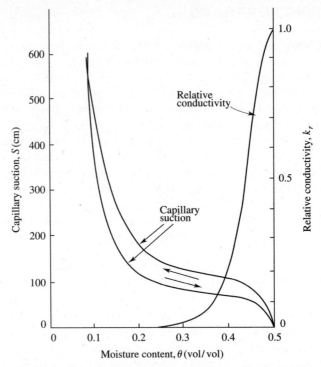

Figure 4.1   Typical capillary suction-relative conductivity-moisture content relation. (After Mein and Larson.[9])

Capillary conductivity is the volume rate of flow of water through the soil under a gradient of unity (dependent on soil moisture content). Relative conductivity is the capillary conductivity for a specified moisture content divided by the saturated conductivity. Figure 4.1 illustrates the relations among these variables. Note that at low moisture contents, capillary suction is high while relative conductivity is low. At high moisture contents the reverse is true.

With this background, an infiltration event can be examined. Consider that rainfall is occurring on an initially dry soil. As shown in Fig. 4.1, the relative conductivity is low at the outset due to the low soil moisture conditions. Thus, for the water to move downward through the soil, a higher moisture level is needed. As moisture builds up, a wetting front forms with the moisture content behind the front being high (essentially saturated) and that ahead of the front being low. At the wetting front, the capillary suction is high due to the low moisture content ahead of the front.

At the beginning of a rainfall event, the potential gradient that drives soil moisture movement is high because the wetting front is virtually at the soil surface. Initially, the infiltration capacity is higher than the rainfall rate and thus the infiltration rate cannot exceed the rainfall rate. As time advances and more water enters the soil, the wetting zone dimension increases and the potential gradient is reduced. Infiltration capacity decreases until it equals the rainfall rate. This occurs at the time the soil at the land surface becomes saturated. Figures 4.2 and 4.3 illustrate

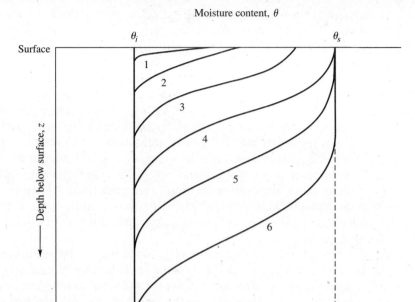

Figure 4.2 Typical moisture profile development with a constant rainfall rate.

these conditions. Figure 4.2 shows how a moisture profile might develop when a rainstorm of constant intensity occurs. In the diagram the soil moisture at the surface is shown to range from its initial value at the top left to its saturated value at the top right. Thus in moving downward on the left-hand side of the diagram, one can trace the downward progression of the wetting front for varying levels of soil moisture

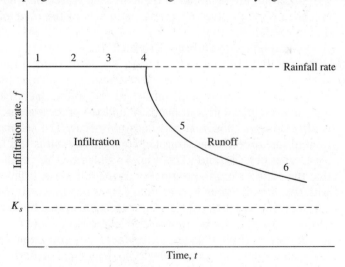

Figure 4.3 Infiltration rate versus time for a given rainfall intensity. (After Mein and Larson.[9])

content at the land surface. Figure 4.3 indicates that until saturation is reached at the surface, the infiltration rate is constant and equal to the rainfall application rate at the surface. At Point 4, a point that corresponds to the time at which saturation occurs at the surface, the infiltration rate begins to proceed at its capacity rate, the maximum rate at which the soil can transmit water across its surface. As time goes on, the infiltration capacity continues to decline until it becomes equal to the saturated conductivity of the soil, the capillary conductivity when the soil is saturated. This ultimate infiltration rate is shown by the dashed line to the right of K_s in Fig. 4.3.

Of particular interest is the determination of Point 4 on the curve of Fig. 4.3. This is the point at which runoff would begin for the conditions specified above. It is also the point at which the actual infiltration rate f becomes equal to the infiltration capacity rate f_p rather than the rainfall intensity rate i. The time of occurrence of this point depends, for a given soil type, on the initial moisture content and the rainfall rate. The shape of the infiltration curve after this point in time is also influenced by these factors.

Another factor that must be reckoned with in the infiltration process is that of hysteresis. In Fig. 4.1 it can be seen that the plot of capillary suction versus soil moisture is a loop. The curve is not the same for wetting and drying of the soil. The curves shown on the figure are the boundary wetting and boundary drying curves, curves applicable under conditions of continuous wetting or drying. Between these curves, an infinite number of possible paths exist that depend on the wetting and drying history of the soil. A number of approaches to the hysteresis problem have been reported in the literature.[3]

The illustration of the infiltration process presented was based on an ideal soil. Unfortunately, such conditions are not replicated in natural systems. Natural soils are highly variable in composition within regions and soil cover conditions are also far-ranging. Because of this, no simple infiltration model can accurately portray all the conditions encountered in the field. The search has thus been for models that can be called upon to give acceptable estimates of the rates at which infiltration occurs during rainfall events.

Mein and Larson have described three general cases of infiltration associated with rainfall.[3] The first case is one in which the rainfall rate is less than the saturated conductivity of the soil. Under this condition, shown as (4) in Fig. 4.4, runoff never occurs since all the rainfall infiltrates the soil surface. Nevertheless, this condition must be recognized in continuous simulation processes since the level of soil moisture is affected even though runoff does not occur. The second case is one in which the rainfall rate exceeds the saturated conductivity but is less than the infiltration capacity. Curves (1), (2), and (3) of Fig. 4.4 illustrate this condition. It should be observed that the period from the beginning of rainfall to the time of surface saturation varies with the rainfall intensity. The final case is one in which the rainfall intensity exceeds the infiltration capacity. This condition is illustrated by the infiltration capacity curve of Fig. 4.5 and those portions of infiltration curves (1), (2), and (3) of Fig. 4.4 that are in their declining stages. Only under this condition can runoff occur. All three cases have relevance to hydrologic modeling, particularly when it is continuous over time.

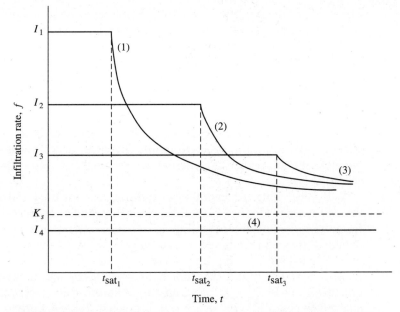

Figure 4.4 Infiltration curves for several rainfall intensities. (After Mein and Larson.[9])

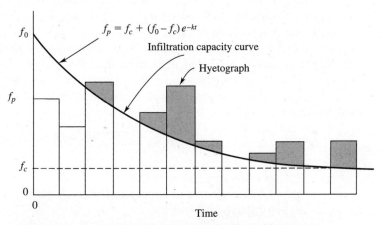

Figure 4.5 Horton's infiltration curve and hyetograph.

4.3 HORTON'S INFILTRATION MODEL

The infiltration process was thoroughly studied by Horton in the early 1930s.[4] An outgrowth of his work, shown graphically in Fig. 4.1, was the following relation for determining infiltration capacity:

$$f_p = f_c + (f_0 - f_c)e^{-kt} \tag{4.1}$$

where f_p = the infiltration capacity (depth/time) at some time t
   k = a constant representing the rate of decrease in f capacity
   f_c = a final or equilibrium capacity
   f_0 = the initial infiltration capacity

It indicates that if the rainfall supply exceeds the infiltration capacity, infiltration tends to decrease in an exponential manner. Although simple in form, difficulties in determining useful values for f_0 and k restrict the use of this equation. The area under the curve for any time interval represents the depth of water infiltrated during that interval. The infiltration rate is usually given in inches per hour and the time t in minutes, although other time increments are used and the coefficient k is determined accordingly.

By observing the variation of infiltration with time and developing plots of f versus t as shown in Fig. 4.5, we can estimate f_0 and k. Two sets of f and t are selected from the curve and entered in Eq. 4.1. Two equations having two unknowns are thus obtained; they can be solved by successive approximations for f_0 and k.

Typical infiltration rates at the end of 1 hr (f_1) are shown in Table 4.1. A typical relation between f_1 and the infiltration rate throughout a rainfall period is shown graphically in Fig. 4.6a; Fig. 4.6b shows an infiltration capacity curve for normal antecedent conditions on turf. The data given in Table 4.1 are for a turf area and must be multiplied by a suitable cover factor for other types of cover complexes. A range of cover factors is listed in Table 4.2.

Total volumes of infiltration and other abstractions from a given recorded rainfall are obtainable from a discharge hydrograph (plot of the streamflow rate versus time) if one is available. Separation of the base flow (dry weather flow) from the discharge hydrograph results in a direct runoff hydrograph (DRH), which accounts for the direct surface runoff, that is, rainfall less abstractions. Direct surface runoff or precipitation excess in inches uniformly distributed over a watershed can readily be calculated by picking values of DRH discharge at equal time increments through the hydrograph and applying the formula[5]

$$P_e = \frac{(0.03719)(\Sigma\ q_i)}{An_d} \tag{4.2}$$

where P_e = precipitation excess (in.)
   q_i = DRH ordinates at equal time intervals (cfs)
   A = drainage area (mi^2)
   n_d = number of time intervals in a 24-hr period

For most cases the difference between the original rainfall and the direct runoff can be considered as infiltrated water. Exceptions may occur in areas of excessive subsurface drainage or tracts of intensive interception potential. The calculated value of infiltration can then be assumed as distributed according to an equation of the form of Eq. 4.1 or it may be uniformly spread over the storm period. Choice of the method employed depends on the accuracy requirements and size of the watershed.

To circumvent some of the problems associated with the use of Horton's infiltration model, some adjustments can be made.[6] Consider Fig. 4.5. Note that where the infiltration capacity curve is above the hyetograph, the actual rate of infiltration

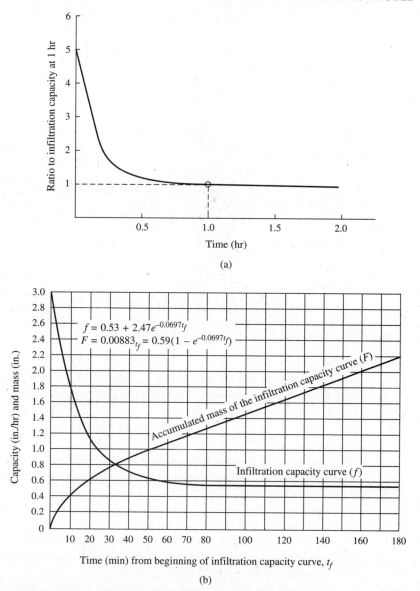

Figure 4.6   (a) Typical infiltration curve. (b) Infiltration capacity and mass curves for normal antecedent conditions of turf areas. [After A. L. Tholin and Clint J. Kiefer, "The Hydrology of Urban Runoff," *Proc. ASCE J. Sanitary Eng. Div.* **84**(SA2), 56 (Mar. 1959).]

TABLE 4.1   TYPICAL f_1 VALUES

Soil group	f_1 (in./hr)	f_1 (mm/h)
High (sandy soils)	0.50–1.00	12.50–25.00
Intermediate (loams, clay, silt)	0.10–0.50	2.50–12.50
Low (clays, clay loam)	0.01–0.10	0.25–2.50

Source: After *ASCE Manual of Engineering Practice*, No. 28.

TABLE 4.2   COVER FACTORS

	Cover	Cover factor
Permanent forest and grass	Good (1 in. humus)	3.0–7.5
	Medium ($\frac{1}{4}$–1 in. humus)	2.0–3.0
	Poor ($<\frac{1}{4}$ in. humus)	1.2–1.4
Close-growing crops	Good	2.5–3.0
	Medium	1.6–2.0
	Poor	1.1–1.3
Row crops	Good	1.3–1.5
	Medium	1.1–1.3
	Poor	1.0–1.1

Source: After *ASCE Manual of Engineering Practice*, No. 28.

is equal to that of the rainfall intensity, adjusted for interception, evaporation, and other losses. Consequently, the actual infiltration is given by

$$f(t) = \min \left[f_p(t), i(t) \right] \tag{4.3}$$

where $f(t)$ is the actual infiltration into the soil and $i(t)$ is the rainfall intensity. Thus the infiltration rate at any time is equal to the lesser of the infiltration capacity, $f_p(t)$ or the rainfall intensity.

Commonly, the typical values of f_0 and f_c are greater than the prevailing rainfall intensities during a storm. Thus, when Eq. 4.1 is solved for f_p as a function of time alone, it shows a decrease in infiltration capacity even when rainfall intensities are much less than f_p. Accordingly, a reduction in infiltration capacity is made regardless of the amount of water that enters the soil.

To adjust for this deficiency, the integrated form of Horton's equation may be used,

$$F(t_p) = \int_0^{t_p} f_p \, dt = f_c t_p + \frac{f_0 - f_c}{k} (1 - e^{-kt_p}) \tag{4.4}$$

where F is the cumulative infiltration at time t_p, as shown in Fig. 4.7. In the figure, it is assumed that the actual infiltration has been equal to f_p. As previously noted, this is not usually the case, and the true cumulative infiltration must be determined. This

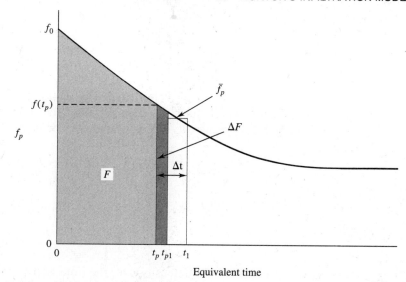

Figure 4.7 Cumulative infiltration.

can be done using

$$F(t) = \int_0^t f(t)\, dt \tag{4.5}$$

where $f(t)$ is determined using Eq. 4.3.

Equations 4.4 and 4.5 may be used jointly to calculate the time t_p, that is, the equivalent time for the actual infiltrated volume to equal the volume under the infiltration capacity curve (Fig. 4.7). The actual accumulated infiltration given by Eq. 4.5 is equated to the area under the Horton curve, Eq. 4.4, and the resulting expression is solved for t_p. This equation,

$$F = f_c t_p + \frac{f_0 - f_c}{k}(1 - e^{-kt_p}) \tag{4.6}$$

cannot be solved explicitly for t_p, but an iterative solution can be obtained. It should be understood that the time t_p is less than or equal to the actual elapsed time t. Thus the available infiltration capacity as shown in Fig. 4.7 is equal to or exceeds that given by Eq. 4.1. By making the adjustments described, f_p becomes a function of the actual amount of water infiltrated and not just a variable with time as is assumed in the original Horton equation.

In selecting a model for use in infiltration calculations, it is important to know its limitations. In some cases a model can be adjusted to accommodate shortcomings; in other cases, if its assumptions are not realistic for the nature of the use proposed, the model should be discarded in favor of another that better fits the situation.

Part One of this book deals with the principal components of the hydrologic cycle. In later chapters, the emphasis is on putting these components together in various hydrologic modeling processes. When these models are designed for continuous simulation, the approach is to calculate the appropriate components of the hydrologic equation, Eq. 1.4, continuously over time. A discussion of how infiltration could be incorporated into a simulation model follows. It exemplifies the use of Horton's equation in a storm water management model (SWMM).[6]

First, an initial value of t_p is determined. Then, considering that the value of f_p depends on the actual amount of infiltration that has occurred up to that time, a value of the average infiltration capacity, $\bar{f}_p$, available over the next time step is calculated using

$$\bar{f}_p = \frac{1}{\Delta t} \int_{t_p}^{t_1 = t_p + \Delta t} f_p \, dt = \frac{F(t_1) - F(t_p)}{\Delta t} \qquad (4.7)$$

Equation 4.3 is then used to find the average rate of infiltration, $\bar{f}$.

$$\bar{f} = \begin{cases} \bar{f}_p & \text{if } \bar{i} \geq \bar{f}_p \\ \bar{i} & \text{if } \bar{i} < \bar{f}_p \end{cases} \qquad (4.8)$$

where $\bar{i}$ is the average rainfall intensity over the time step.

Following this, infiltration is incremented using the expression

$$F(t + \Delta t) = F(t) + \Delta F = F(t) + \bar{f} \, \Delta t \qquad (4.9)$$

where $\Delta F = \bar{f} \, \Delta t$ is the added cumulative infiltration (Fig. 4.7).

The next step is to find a new value of t_p. This is done using Eq. 4.6. If $\Delta F = \bar{f}_p \, \Delta t$, $t_{p1} = t_p + \Delta t$. But if the new t_{p1} is less than $t_p + \Delta t$ (see Fig. 4.7), Eq. 4.6 must be solved by iteration for the new value of t_p. This can be accomplished using the Newton–Raphson procedure.[6]

When the value of $t_p \geq 16/k$, the Horton curve is approximately horizontal and $f_p = f_c$. Once this point has been reached, there is no further need for iteration since f_p is constant and equal to f_c and no longer dependent on F.

EXAMPLE 4.1 ——

Given an initial infiltration capacity f_0 of 3.0 in./hr and a time constant k of 0.29 hr^{-1}, derive an infiltration capacity vs. time curve if the ultimate infiltration capacity is 0.55 in./hr. For the first ten hours, estimate the total volume of water infiltrated in inches over the watershed.

Solution. Using Horton's equation (4.1), values of infiltration can be computed for various times. The equation is as follows:

$$f = f_c + (f_0 + f_c)e^{-kt}$$

Substituting the appropriate values into the equation yields

$$f = 0.55 + (3.0 - 0.55)e^{-0.29t}$$

Then for the times shown in spreadsheet Table 4.3, values of f are computed and entered into the table. Using the spreadsheet graphics package, the curve of Fig. 4.8 is derived.

TABLE 4.3   CALCULATIONS FOR EXAMPLE 4.1

Time (hr)	Infiltration (in./hr)	Time (hr)	Infiltration (in./hr)
0.00	3.00	5.00	1.12
0.10	2.93	6.00	0.98
0.25	2.83	7.00	0.87
0.50	2.67	8.00	0.79
1.00	2.38	9.00	0.73
2.00	1.92	10.00	0.68
3.00	1.58	15.00	0.58
4.00	1.32	20.00	0.56

To find the volume of water infiltrated during the first 10 hours, Eq. 4.1 can be integrated over the range of 0–10

$$V = \int [0.55 + (3.0 - 0.55)e^{-0.29t}]\, dt$$
$$V = [0.55t + (2.45/-0.29)e^{-0.29t}]_0^{10}$$
$$V = 12.47 \text{ in.}$$

The volume in inches over the watershed is thus 12.47 in. ∎∎

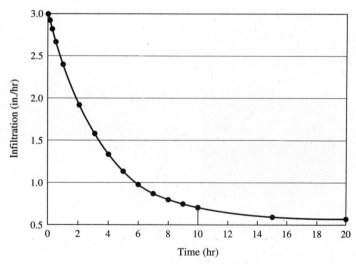

Figure 4.8   Graph for Example 4.1.

4.4 GREEN–AMPT MODEL

The Green–Ampt infiltration model, originally proposed in 1911, has had a resurgence of interest.[3,6-11] This approach is based on Darcy's law (see Chapter 18). In its original form, it was intended for use where infiltration resulted from an excess of water at the ground surface at all times. In 1973, Mein and Larson presented a methodology for applying the Green–Ampt model to a steady rainfall input.[9] They also developed a procedure for determining the value of the capillary suction parameter used in the model. In 1978, Chu demonstrated the applicability of the model for use under conditions of unsteady rainfall.[10] As a result of these and other efforts, the Green–Ampt model is now employed as an option in such widely used continuous simulation models as SWMM.[6]

The original formulation by Green and Ampt assumed that the soil surface was covered by ponded water of negligible depth and that the water infiltrated a deep homogenous soil with a uniform initial water content (see Fig. 4.9). Water is assumed to enter the soil so as to define sharply a wetting front separating the wetted and unwetted regions as shown in the figure. If the conductivity in the wetted zone is defined as K_s, application of Darcy's law yields the equation

$$f_p = \frac{K_s(L + S)}{L} \tag{4.10}$$

where L is the distance from the ground surface to the wetting front and S is the capillary suction at the wetting front. Referring to Fig. 4.9, it can be seen that the cumulative infiltration F is equivalent to the product of the depth to the wetting front L and the initial moisture deficit, $\theta_s - \theta_i = IMD$. Making these substitutions in

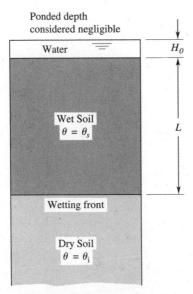

Figure 4.9 Definition sketch for Green-Ampt model.

Eq. 4.10 and rearranging, we obtain

$$f_p = K_s\left(1 + \frac{S \times IMD}{F}\right) \tag{4.11}$$

Considering that $f_p = dF/dt$, we can state

$$\frac{dF}{dt} = K_s\left(1 + \frac{S \times IMD}{F}\right) \tag{4.12}$$

Integrating and substituting the conditions that $F = 0$ at $t = 0$, we obtain

$$F - S \times IMD \times \log_e\left(\frac{F + IMD \times S}{IMD \times S}\right) = K_s t \tag{4.13}$$

This form of the Green–Ampt equation is more convenient for use in watershed modeling processes than Eq. 4.10 because it relates the cumulative infiltration to the time at which infiltration began. The derivation of this equation assumes a ponded surface so that the actual rate of infiltration is equal to the infiltration capacity at all times. Using Eq. 4.13, we can determine the cumulative infiltration at any time, a feature desirable for continuous systems modeling. All the parameters in the equation are physical properties of the soil–water system and are measurable. The determination of suitable values for the capillary suction S is often difficult, however, particularly for relations such as that shown for a clay–type soil in Fig. 4.10. It can be observed from the figure that for this curve there is a wide variation of capillary suction with soil moisture content.[3]

The Mein–Larson formulation using the Green–Ampt model incorporates two stages.[3,6] The first stage deals with prediction of the volume of water that infiltrates before the surface becomes saturated. The second stage is one in which infiltration capacity is calculated using the Green–Ampt equation. In the widely used storm water management model, the modified Green–Ampt model of infiltration is one of the options that can be employed to estimate infiltration.[6] Computations are made using

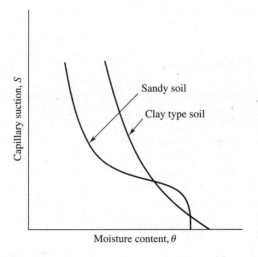

Figure 4.10 Capillary suction versus moisture content curves.

the following equations: for $F < F_s(f = i)$,

$$F_s = \frac{S \times IMD}{i/K_s - 1} \quad \text{for } i > K_s \tag{4.14}$$

and there is no calculation of F_s for $i \leq K_s$; for $F \geq F_s(f = f_p)$:

$$f_p = K_s\left(1 + \frac{S \times IMD}{F}\right) \tag{4.11}$$

where
    f = actual infiltration rate (ft/sec)
    f_p = infiltration capacity (ft/sec)
    i = rainfall intensity (ft/sec)
    F = cumulative infiltration volume in the event (ft)
    F_s = cumulative infiltration volume required to cause surface saturation (ft)
    S = average capillary suction at the wetting front (ft of water)
    IMD = initial moisture deficit for the event (ft/ft)
    K_s = saturated hydraulic conductivity of soil (ft/sec)

Equation 4.10 shows that the volume of rainfall needed to saturate the surface is a function of the rainfall intensity. In the modeling process, for each time step for which $i > K_s$, the value of F_s is computed and compared with the volume of rainfall infiltrated to that time. If F equals or exceeds F_s, the surface saturates and calculations for infiltration then proceed using Eq. 4.14. Note that by substituting f for i in Eq. 4.14 and rearranging, the equation takes the same form as Eq. 4.11.

For rainfall intensities less than or equal to K_s, all the rainfall infiltrates and its amount is used only to update the initial moisture deficit, IMD.[6] The cumulative infiltration volume F_s is not altered.

After saturation is achieved at the surface, Eq. 4.11 shows that the infiltration capacity is a function of the infiltrated volume, and thus of the infiltration rates during previous time steps. To avoid making numerical errors over long time steps, the integrated form of the Green–Ampt equation (Eq. 4.13) is used. This equation takes the following form as it is used in SWMM:

$$K_s(t_2 - t_1) = F_2 - C \ln (F_2 + C) - F_1 + C \ln (F_1 + C) \tag{4.15}$$

where
    $C = IMD \times S$ (ft of water)
    t = times (sec)
    1, 2 = subscripts indicating the starting and ending of the time steps.

Equation 4.15 must be solved iteratively for F_2, the cumulative infiltration at the end of the time step. A Newton–Raphson routine is used.[6]

In the SWMM model, infiltration during time step $t_2 - t_1$ is equal to $(t_2 - t_1)i$ if the surface is not saturated and is equal to $F_2 - F_1$ if saturation has previously occurred and there is a sufficient water supply at the surface. If saturation occurs during an interval, the infiltrated volumes over each stage of the process within the time steps are computed and summed. When the rainfall ends or becomes less than the infiltration capacity, any ponded water is allowed to infiltrate and is added to the cumulative infiltration volume.

4.5 HUGGINS–MONKE MODEL

Several investigators have circumvented the time dependency problem by introducing soil moisture as the dependent variable.[2,10–13] The following equation proposed by Huggins and Monke is an example:[2]

$$f = f_c + A\left(\frac{S - F}{T_p}\right)^P \qquad (4.16)$$

where   A and P = coefficients
   S = the storage potential of a soil overlying the impeding layer (T_p minus antecedent moisture)
   F = the total volume of water that infiltrates
   T_p = the total porosity of soil lying over the impeding stratum

The coefficients are determined using data from sprinkling infiltrometer studies. The variable F must be calculated for each time increment in the iteration process. At the beginning of a storm $F = 0$ and f is therefore known. In essence the continuity equation is solved for a block of soil with an inflow rate f (or smaller if the rainfall is less) and an outflow determined according to Eq. 4.17. Expression $(dS/dt)\Delta t$ then gives the change in storage of the soil. When added to the storage at the beginning of the time increment, the total storage is obtained. Equation 4.16 is a modification of one originated by Holtan and Overton[12,13] and appears to have merit over the form of Eq. 4.1 if the rate of infiltration supply is less than infiltration capacity.

In order to use this relation when the water supply rate only intermittently exceeds the infiltration capacity, the rate at which water drains from the "control zone," which determines the soil moisture content $(S - F)$, must be found. It is evaluated as follows:[10] (1) where the moisture content of the control zone is less than the field capacity (amount of water held in the soil after excess gravitational water has drained), the drainage rate is considered zero; (2) the drainage rate is assumed equal to the infiltration rate when the soil is saturated and the infiltration rate becomes constant; and (3) if the water content is between the field capacity and saturation, the drainage rate is computed as

$$\text{drainage rate} = f_c\left(1 - \frac{P_u}{G}\right)^3 \qquad (4.17)$$

where P_u = the unsaturated pore volume
   G = maximum gravitational water, that is, the total porosity minus the field capacity

Data from sprinkling infiltrometer studies of various watersheds of interest are used to estimate the coefficients in Eq. 4.16.[2]

4.6 HOLTAN MODEL

Another equation for infiltration capacity has been developed by Holtan:[14,15]

$$f = aS_a^{1.4} + f_c \tag{4.18}$$

where f = the infiltration capacity (in./hr)
   a = the infiltration capcity [(in./hr)/in.$^{1.4}$] of the available storage (index of surface-connected porosity)
   S_a = available storage in the surface layer (A-horizon in agricultural soils, that is, about first 6 in.) in inches of water equivalent
   f_c = the constant rate of infiltration after long wetting (in./hr)

This equation has been modified somewhat for use in the USDAHL-70 watershed model.[16]

$$f = GI \times aS_a^{1.4} + f_c \tag{4.19}$$

where a is a vegetation parameter and GI is a growth index (see Chapter 5). Information about a is given in Table 4.4.

In Eq. 4.19, it is assumed that the portion of the available storage connected to the surface is a function of the density of plant roots. This is given by the vegetation parameter a, which has been determined at plant maturity as the percentage of the ground surface area occupied by plant stems or root crowns. In this manner, the fraction of porosity in the agricultural A-horizon that is surface connected by mature plant roots to form conduits for air or water is represented.

TABLE 4.4   TENTATIVE ESTIMATES OF THE VEGETATION PARAMETER a IN INFILTRATION EQUATION $f = GI \times aS_a^{1.4} + f_c$

	Basal area rating[a]	
Land use or cover	Poor condition	Good condition
Fallow[b]	0.10	0.30
Row crops	0.10	0.20
Small grains	0.20	0.30
Hay (legumes)	0.20	0.40
Hay (sod)	0.40	0.60
Pasture (bunchgrass)	0.20	0.40
Temporary pasture (sod)	0.40	0.60
Permanent pasture (sod)	0.80	1.00
Woods and forests	0.80	1.00

[a] Adjustments needed for weeds and grazing.
[b] For fallow land only, poor condition means after row crop, and good condition means after sod.
Source: U.S. Department of Agriculture, Agricultural Research Service, 1975.

4.7 RECOVERY OF INFILTRATION CAPACITY

The infiltration capacity curve of Fig. 4.5 illustrates that the ability of a soil to infiltrate water decays over time, providing that rainfall is continuous and that it exceeds infiltration capacity. In nature, once rainfall ceases, there is a recovery of infiltration capacity with time. The extent of recovery at any point in time depends on the dryness of the period. This condition of recovery must be recognized and incorporated in continuous simulation modeling processes. For the SWMM model, Huber and co-workers have developed an approach that follows the notation of Fig. 4.11.[6] The SWMM model regenerates infiltration capacity whenever there are dry time steps, that is, during periods when there is no precipitation or surface water ponding. The equation used in the model is

$$f_p = f_0 - (f_0 - f_c)e^{-k_d(t-t_w)} \tag{4.20}$$

where k_d = a decay coefficient for the recovery curve (sec^{-1})
   t_w = a hypothetical projected time at which $f_p = f_c$ on the recovery curve (sec)

In the SWMM model, k_d is assumed to be a constant fraction or multiple of k,

$$k_d = Rk \tag{4.21}$$

where R is a constant ratio, considered to be much less than 1.0. This implies a longer drying curve than a wetting curve.[6]

Following the sequence shown in Fig. 4.11, new values of t_p are generated. For example, along the recovery curve,

$$f_1 = f_p(t_{w1}) = f_0 - (f_0 - f_c)e^{-k_dT_{w1}} \tag{4.22}$$

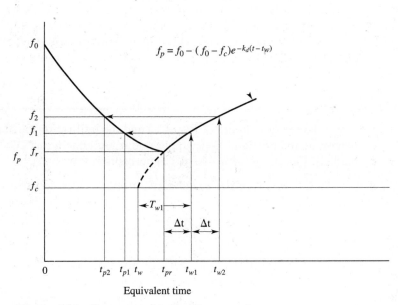

Figure 4.11   Recovery of infiltration capacity.

where $T_{w1} = t_{w1} - t_w$

$$T_{w2} = t_{w2} - t_w$$

Then, solving Eq. 4.22 for the initial time difference, T_{wr},

$$T_{wr} = t_{pr} - t_w = \frac{1}{k_d} \ln \left(\frac{f_0 - f_c}{f_0 - f_r} \right) \tag{4.23}$$

$$T_{w1} = T_{wr} + \Delta t \tag{4.24}$$

where t_{pr} = the value of t_p at the beginning of recovery (sec)
f_r = the corresponding value of f_p (ft/sec)

The value of f_1 (see Fig. 4.11) is found using Eq. 4.22. Then t_{p1} is obtained by the application of Eq. 4.1:

$$t_{p1} = \frac{1}{k} \ln \left(\frac{f_0 - f_c}{f_1 - f_c} \right) \tag{4.25}$$

4.8 TEMPORAL AND SPATIAL VARIABILITY OF INFILTRATION CAPACITY

The infiltration capacity generally varies both in space and time within a given drainage basin.[17-19] Spatial variations occur because of differences in soil types and vegetation. The usual procedure used to accommodate this type of variation is to subdivide the total region into components having approximately uniform soil and vegetal cover properties.

The infiltration capacity at a given location in a watershed varies with time as shown in Fig. 4.5. The initial infiltration capacity is a function of antecedent conditions and can be estimated from a knowledge of the area's soil moisture or from an antecedent precipitation index. If precipitation occurs at a rate less than the f capacity rate, the change in f capacity with time will not be that given by the f capacity curve; during periods of no precipitation, the infiltration capacity will recover. Example 4.2 illustrates these concepts.

EXAMPLE 4.2 _____

Given the rainfall pattern of Fig. 4.12 and the infiltration capacity curve of Fig. 4.13, determine the overland flow supply rate σ. Assume a turf cover and that the OGEE curve of Fig. 3.3 governs. Neglect interception losses.

Solution. In order to solve the problem, it is necessary to determine P, F, i, and f.

1. Construct a curve of mass infiltration F versus f capacity. This is done by calculating the areas under the curve in Fig. 4.13a at given times and plotting them versus f capacity as shown in Fig. 4.13b. Calculations to determine cumulative infiltration are shown in Table 4.5. Note that the F values are plotted versus f capacity at the end of the corresponding time

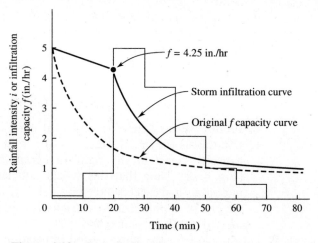

Figure 4.12   Storm infiltration capacity curve constructed from original f capacity curve.

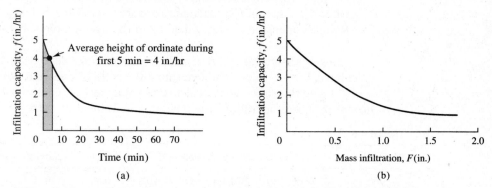

Figure 4.13   (a) Infiltration capacity curve and (b) mass infiltration versus f curve for Example 4.1.

TABLE 4.5

Time increment (min)	Average height of ordinate (in./hr)	Cumulative infiltration, F (in.)
0–5	4.00	4 $\times$ 5/60 = 0.33
5–15	2.50	2.5 $\times$ 10/60 + 0.33 = 0.75
15–30	1.50	1.5 $\times$ 15/60 + 0.75 = 1.13
30–60	1.15	1.15 $\times$ 30/60 + 1.13 = 1.7

interval. For example, the first value, 0.33. is plotted versus $f = 3.4$, which occurs at the end of the 5-min interval.

2. Determine the storm period infiltration. The storm pattern and original f capacity curve are plotted as shown in Fig. 4.12.

   a. In the first 20 min $f > i$; therefore, all the rainfall is infiltrated.

$$F = 0.1 \times \tfrac{1}{6} + 0.8 \times \tfrac{1}{6} = 0.15 \text{ in.}$$

   b. From F versus f curve (Fig. 4.13b), for $F = 0.15$, $f = 4.25$ in./hr.

   c. Use this as the initial value of f at $t = 20$ min and shift the original f capacity curve to the right to obtain the storm infiltration curve (Fig. 4.12). Note that this would not have been done if all rainfall intensities had exceeded the original f capacity curve ordinates. Since at the end of 20 min, some f capacity remained unfilled, the curve shift is carried out to accommodate this.

3. Having plots for the storm period infiltration and the rainfall versus time, values of P, F, i, and f can be determined. Calculations for P and F are listed in Table 4.6. Note that the curve of F versus f (Fig. 4.13b) relates to the original f capacity curve and is used to aid in constructing the storm f curve, while the values of F calculated above are related to actual storm conditions. Rainfall intensities (i) are taken from the hyetograph of Fig. 4.12.

Having determined F, P, i, and f, it is now possible to enter Fig. 3.3 using calculated $P - F$ values and determine the ratio of overland flow supply σ to $i - f$. Using this ratio and the calculated values of $i - f$ permits the determination of σ. These operations are tabulated in Table 4.7. ■■

TABLE 4.6

Time (min)	i (in./hr)	Cumulative precipitation, P (in.)	f (in./hr)	Cumulative infiltration, F (in.)
10	0.10	$0.10 \times 10/60 \quad = 0.02$	0.10	$0.10 \times 10/60 \quad = 0.02$
20	0.80	$0.80 \times 10/60 + 0.02 = 0.15$	0.80	$0.8 \times 10/60 + 0.02 = 0.15$
30	5.00	$5 \quad \times 10/60 + 0.15 = 0.98$	2.90	$2.9 \times 10/60 + 0.15 = 0.63$
40	3.70	$3.7 \times 10/60 + 0.98 = 1.60$	1.80	$1.8 \times 10/60 + 0.63 = 0.93$
50	2.00	$2 \quad \times 10/60 + 1.60 = 1.93$	1.40	$1.4 \times 10/60 + 0.93 = 1.17$
60	1.10	$1.1 \times 10/60 + 1.93 = 2.12$	1.10	$1.1 \times 10/60 + 1.17 = 1.35$
70	0.50	$0.5 \times 10/60 + 2.12 = 2.20$	0.50	$0.5 \times 10/60 + 1.35 = 1.43$

TABLE 4.7   DETERMINATION OF THE OVERLAND FLOW SUPPLY RATE

Time (min)	P (in.)	F (in.)	$P - F$ (in.)	i (in./hr)	f (in./hr)	$i - f$ (in./hr)	$\dfrac{\sigma^a}{i - f}$	σ (in./hr)
10	0.02	0.02	0	0.1	0.1	—	0	0
20	0.15	0.15	0	0.8	0.8	—	0	0
30	0.98	0.63	0.35	5.0	2.9	2.1	0.91	1.9
40	1.60	0.93	0.67	3.7	1.8	1.9	1.0	1.9
50	1.93	1.17	0.76	2.0	1.4	0.6	1.0	0.6
60	2.12	1.35	0.77	1.1	1.1	—	1.0	0
70	2.20	1.43	0.77	0.5	0.5	—	1.0	0

^{a} From Fig. 3.3.

4.9 SCS RUNOFF CURVE NUMBER PROCEDURE

The Soil Conservation Service (SCS) has developed a widely used *curve number procedure* for estimating runoff.[20] The effects of land use and treatment, and thus infiltration, are embodied in it. The procedure was empirically developed from studies of small agricultural watersheds. While the SCS procedure is not designed to estimate infiltration directly, a look at Fig. 4.14 shows that the method does embody an infiltration estimate.

The SCS procedure consists of selecting a storm and computing the direct runoff by the use of curves founded on field studies of the amount of measured runoff from numerous soil cover combinations. A runoff curve number (CN) is extracted from Table 4.8. Selection of the runoff curve number is dependent on antecedent conditions and the types of cover. Soils are classified A, B, C, or D according to the following criteria:

A. (low runoff potential) Soils having high infiltration rates even if thoroughly wetted and consisting chiefly of deep well to excessively drained sands or gravels. They have a high rate of water transmission.

B. Soils having moderate infiltration rates if thoroughly wetted and consisting chiefly of moderately deep to deep, moderately well to well-drained soils with moderately fine to moderately coarse textures. They have a moderate rate of water transmission.

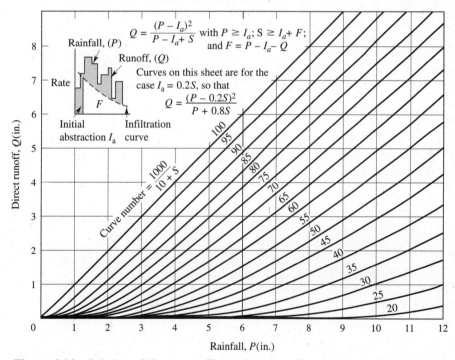

Figure 4.14 Solution of direct runoff equation. $S = S' + I_a$, where S is watershed storage in inches, I_a is initial abstraction, and S' is potential maximum retention exclusive of I_a.

TABLE 4.8 RUNOFF CURVE NUMBERS FOR HYDROLOGIC SOIL-COVER COMPLEXES[a]

Land use or cover	Treatment or practice	Hydrologic condition	Hydrologic soil group			
			A	B	C	D
Fallow	Straight row	—	77	86	91	94
Row crops	Straight row	Poor	72	81	88	91
	Straight row	Good	67	78	85	89
	Contoured	Poor	70	79	84	88
	Contoured	Good	65	75	82	86
	Contoured and terraced	Poor	66	74	80	82
	Contoured and terraced	Good	62	71	78	81
Small grain	Straight row	Poor	65	76	84	88
		Good	63	75	83	87
	Contoured	Poor	63	74	82	85
		Good	61	73	81	84
	Contoured and terraced	Poor	61	72	79	82
		Good	59	70	78	81
Close-seeded legumes[b] or rotation meadow	Straight row	Poor	66	77	85	89
	Straight row	Good	58	72	81	85
	Contoured	Poor	64	75	83	85
	Contoured	Good	55	69	78	83
	Contoured and terraced	Poor	63	73	80	83
	Contoured and terraced	Good	51	67	76	80
Pasture or range		Poor	68	79	86	89
		Fair	49	69	79	84
		Good	39	61	74	80
	Contoured	Poor	47	67	81	88
	Contoured	Fair	25	59	75	83
	Contoured	Good	6	35	70	79
Meadow		Good	30	58	71	78
Woods		Poor	45	66	77	83
		Fair	36	60	73	79
		Good	25	55	70	77
Farmsteads		—	59	74	82	86
Roads (dirt)[c]		—	72	82	87	89
(hard surface)[c]		—	74	84	90	92

[a] Antecedent moisture condition II and $I_a = 0.2S$.

[b] Close drilled or broadcast.

[c] Including right-of-way.

Source: After "Hydrology," Suppl. A to Sec. 4, *Engineering Handbook,* U.S. Department of Agriculture, Soil Conservation Service, 1968.

C. Soils having slow infiltration rates if thoroughly wetted and consisting chiefly of soils with a layer that impedes the downward movement of water, or soils with moderately fine to fine texture. They have a slow rate of water transmission.

D. (High runoff potential) Soils having very slow infiltration rates if thoroughly wetted and consisting chiefly of clay soils with a high swelling potential, soils with a permanent high water table, soils with a claypan or clay layer at or near the surface, and shallow soils over nearly impervious material. They have a very slow rate of water transmission.

A composite curve number (CN) for a watershed having more than one land use, treatment, or soil type can be found by weighting each curve number according to its area. If, for example, 80 percent of a watershed has a CN of 75 and the remaining 20 percent is impervious $(CN = 100)$, then the weighted $CN = 0.80 \times 75 + 0.20 \times 100 = 80$.

The curve numbers in Table 4.8 are applicable to average antecedent moisture conditions. Other antecedent moisture conditions (AMC) are as follows:

AMC I. A condition of watershed soils where the soils are dry but not to the wilting point, and when satisfactory plowing or cultivation takes place. (This condition is not considered applicable to the design flood computation methods presented in this text.)

AMC II. The average case for annual floods, that is, an average of the conditions that have preceded the occurrence of the maximum annual flood on numerous watersheds.

AMC III. If heavy rainfall or light rainfall and low temperatures have occurred during the 5 days previous to the given storm and the soil is nearly saturated.

The corresponding curve numbers for Condition I and Condition III can be obtained from Table 4.9 if the CN for AMC II is known.

The SCS has developed two synthetic 24-hr rainfall distributions from Weather Service rainfall frequency data. The Type I distribution is representative of the maritime climate, including Hawaii, Alaska, and the coastal side of the Sierra Nevada and Cascade mountains in California, Oregon, and Washington. The Type II distribution represents the remainder of the United States where high runoff rates are generated from summer thunderstorms. The procedure used in developing the SCS rainfall distributions is given in Ref. 21.

Once a rainfall amount has been determined, the direct runoff resulting from this precipitation can be estimated using an appropriate curve number and Figs. 4.14 and 4.15. These figures are applicable for areas up to 2000 acres. In Fig. 4.14, S is a retention index reflecting the potential storage of the watershed in inches.[20]

TABLE 4.9 CURVE NUMBERS (*CN*) FOR WET (AMC III) AND DRY (AMC I) ANTECEDENT MOISTURE CONDITIONS CORRESPONDING TO AN AVERAGE ANTECEDENT MOISTURE CONDITION

	Corresponding *CN*s	
CN for AMC II	AMC I	AMC III
100	100	100
95	87	98
90	78	96
85	70	94
80	63	91
75	57	88
70	51	85
65	45	82
60	40	78
55	35	74
50	31	70
45	26	65
40	22	60
35	18	55
30	15	50
25	12	43
20	9	37
15	6	30
10	4	22
5	2	13

AMC I: Lowest runoff potential. Soils in the watershed are dry enough for satisfactory plowing or cultivation.

AMC II: The average condition. AMC III: Highest runoff potential. Soils in the watershed are practically saturated from antecedent rains.

Source: After "Hydrology," Suppl. A to Sec. 4, *Engineering Handbook,* U.S. Department of Agriculture, Soil Conservation Service, 1968.

4.10 ϕ INDEX

Infiltration indexes generally assume that infiltration occurs at some constant or average rate throughout a storm. Consequently, initial rates are underestimated and final rates are overstated if an entire storm sequence with little antecedent moisture is considered. The best application is to large storms on wet soils or storms where infiltration rates may be assumed to be relatively uniform.[1]

The most common index is termed the ϕ *index* for which the total volume of the storm period loss is estimated and distributed uniformly across the storm pattern. Then the volume of precipitation above the index line is equivalent to the runoff (Fig. 4.16). A variation is the *W* index, which excludes surface storage and retention. Initial abstractions are often deducted from the early storm period to exclude initial depression storage and wetting.

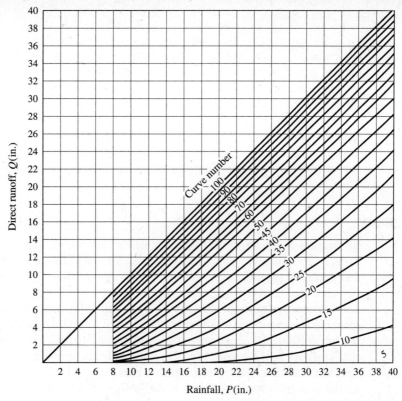

Figure 4.15 Solution of direct runoff equation. (After Ref. 20.)

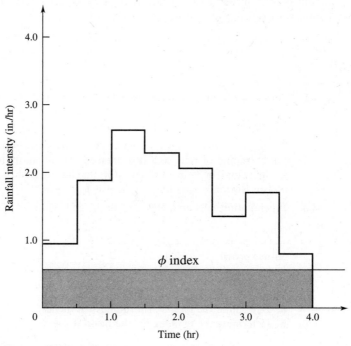

Figure 4.16 Representation of a φ index.

To determine the ϕ index for a given storm, the amount of observed runoff is determined from the hydrograph and the difference between this quantity and the total gauged precipitation is then calculated. The volume of loss (including the effects of interception, depression storage, and infiltration) is distributed uniformly across the storm pattern as shown in Fig. 4.16.

Use of the ϕ index for determining the amount of direct runoff from a given storm pattern is essentially the reverse of this procedure. Unfortunately, the ϕ index determined from a single storm is not generally applicable to other storms, and unless it is correlated with basin parameters other than runoff, it is of little value.

■ Summary

Infiltration is an important element in the hydrologic process. Soils have varying capacity to infiltrate water. Influencing factors include soil type, degree of saturation, and nature of ground cover. Activities that change the soil surface or alter its properties also have a modifying effect.

When the rainfall intensity is less than infiltration capacity, all of the water reaching the ground can infiltrate. But if rainfall intensity exceeds infiltration capacity, infiltration will occur only at the infiltration capacity rate, and water in excess of that capacity will be stored in depressions, become surface runoff, or evaporate. In general, the initial infiltration capacity of a dry soil is high. As rainfall continues, and as the soil becomes saturated, it diminishes to a relatively constant rate (ultimate capacity).

Infiltration rates have been determined for a variety of soils and ground cover conditions. A number of equations have been developed to serve as models for the infiltration process. They are exemplified by Eqs. 4.1, 4.11, and 4.16.

PROBLEMS

4.1. Gross rain intensities during each hour of a 5-hr storm over a 1000-acre basin were 5, 4, 1, 3, and 2 in./hr, respectively. The direct surface runoff from the basin was 375 acre-ft. Determine the basin ϕ index.

4.2. The infiltration rate for excess rain on a small area was observed to be 4.5 in./hr at the beginning of rain, and it decreased exponentially toward an equilibrium of 0.5 in./hr. A total of 30 in. of water infiltrated during a 10-hr interval. Determine the value of k in Horton's equation $f = f_c + (f_0 - f_c)e^{-kt}$.

4.3. Precipitation falls on a 500-acre drainage basin according to the following schedule:

30-min period	1	2	3	4
Intensity (in./hr)	4.0	2.0	6.0	5.0

a. Determine the total storm rainfall (in inches).
b. Determine the ϕ index for the basin if the net storm rain is 3.0 in.

4.4. Rework Problem 4.1 assuming that the storm occurred over a basin of 2.5 km², the rainfall intensities were 12, 10, 3, 8, and 5 cm/hr, and the direct surface runoff was 463,000 m³.

4.5. Rework Problem 4.2 assuming that the initial infiltration capacity was 10 cm/hr, the ultimate capacity was 1.2 cm/hr, and a total of 33 cm of water infiltrated during the 10-hr interval.

4.6. The direct surface runoff volume from a 4.40-mi² drainage basin is determined by planimeter from the area under the hydrograph to be 10,080 cfs-hr. The hydrograph was produced by a 1.71-in./hr rain storm with a duration of 5 hr. Determine (a) the net rain and (b) the ϕ index.

4.7. The following table lists the storm rainfall data and infiltration capacity data for a 24-hr storm beginning at midnight on April 14 of the current year.
   a. Plot the rainfall hyetograph and the f capacity curve on rectangular coordinate paper.
   b. Determine the total storm precipitation in inches.
   c. By counting squares or by planimeter, determine the net storm rain by the f capacity method.

RAINFALL DATA FOR A HYPOTHETICAL STORM ON APRIL 15 OF THE CURRENT YEAR
(Beginning at Midnight on April 14)

Hour	Rainfall intensity (in./hr)	Infiltration capacity at beginning of hour (in./hr)	Hourly deduction for depression storage (in./hr)
1	0.41	0.200	0.20
2	0.49	0.160	0.14
3	0.32	0.125	0.04
4	0.31	0.100	0.02
5	0.22	0.085	0.00
6	0.08	0.070	0.00
7	0.07	0.065	
8	0.09	0.057	
9	0.08	0.052	
10	0.06	0.047	
11	0.11	0.044	
12	0.12	0.040	
13	0.15	0.037	
14	0.23	0.036	
15	0.28	0.035	
16	0.26	0.034	
17	0.21	0.033	
18	0.09	0.033	
19	0.07	0.033	
20	0.06	0.032	
21	0.03	0.032	
22	0.02	0.032	
23	0.01	0.031	
24	0.01	0.031	

4.8. Tabulated below are total rainfall intensities during each hour of a frontal storm over a drainage basin.
  a. Plot the rainfall hyetograph (intensity versus time).
  b. Determine the total storm precipitation amount in inches.
  c. If the net storm rain is 2.00 in., determine the exact ϕ index (in./hr) for the drainage basin. (Note that by definition the area under the hyetograph above the ϕ index line must be 2.00 in.)
  d. Determine the area of the drainage basin (acres) if the net rain is 2.00 in. and the measured volume of direct surface runoff is 2015 cfs-hr.
  e. Using the ϕ index calculated in part c, determine the volume of direct surface runoff (acre-ft) that would result from the following:

Hour	1	2	3	4
Intensity	0.40	0.05	0.30	0.20 in./hr

Hour	Rainfall intensity (in./hr)	Hour	Rainfall intensity (in./hr)
1	0.41	13	0.15
2	0.49	14	0.23
3	0.22	15	0.28
4	0.31	16	0.26
5	0.22	17	0.21
6	0.08	18	0.09
7	0.07	19	0.07
8	0.09	20	0.06
9	0.08	21	0.03
10	0.06	22	0.02
11	0.11	23	0.01
12	0.12	24	0.01

4.9. The SCS curve number method of estimating rainfall excess (net rain) is based on the assumption that the curve number for a watershed depends on several factors. Name or describe that factors that are considered when a curve number is determined.

4.10. A 7-mi² drainage basin has a composite curve number CN of 50. According to the SCS, exactly how much rain must fall before the direct runoff commences?

4.11. Determine a composite SCS runoff curve number for a 600-acre basin that is totally within soil group C. The land use is 40 percent contoured row crops in poor hydrologic condition and 60 percent native pasture in fair hydrologic condition.

4.12. Which SCS-classified soil would have the highest infiltration rate, A, B, C, or D?

4.13. Rework Problem 4.10 assuming that the drainage basin is 12 km² and for a CN of 55 and 80. Give your answers in cm.

4.14. Rework Problem 4.11 assuming that the soil group is B and the land use is 50 percent straight row crops under poor conditions and 50 percent meadow under good conditions.

4.15. For the composite curve number found in Problem 4.14, estimate the amount of runoff if the direct rainfall is 20 cm.

REFERENCES

1. Ven Te Chow (ed.), *Handbook of Applied Hydrology*. New York: McGraw-Hill, 1964.
2. L. F. Huggins and E. J. Monke, "The Mathematical Simulation of the Hydrology of Small Watersheds," Tech Rept. 1, Purdue University Water Resources Center, Lafayette, IN, Aug. 1966.
3. R. G. Mein and C. L. Larson, "Modeling the Infiltration Component of the Rainfall-Runoff Process," Bull. 43, Water Resources Research Center, University of Minnesota, Minneapolis, MN, Sept. 1971.
4. R. E. Horton, *Surface Runoff Phenomena: Part I, Analysis of the Hydrograph.* Horton Hydrol. Lab. Pub. 101. Ann Arbor, MI: Edwards Bros., 1935.
5. W. D. Mitchell, "Unit Hydrographs in Illinois," Illinois Waterways Division, 1968.
6. W. C. Huber, J. P. Heaney, S. J. Nix, R. E. Dickinson, and D. J. Polmann, "Storm Water Management User's Manual, Version III," Department of Environmental Engineering Sciences, University of Florida, Gainesville, Nov. 1981.
7. H. T. Haan, H. P. Johnson, and D. L. Brakensiek (eds.), *Hydrologic Modeling of Small Watersheds*, ASAE Monograph No. 5. St. Joseph, MI: American Society of Agricultural Engineers, 1982.
8. W. H. Green and G. A. Ampt, "Studies on Soil Physics, 1. The Flow of Air and Water Through Soils," *J. Agric. Sci.* **4,** 11–24(1911).
9. R. G. Mein and C. L. Larson, "Modeling Infiltration During a Steady Rain," *Water Resources Res.* **9**(2), 384–394(Apr. 1973).
10. S. T. Chu, "Infiltration During an Unsteady Rain," *Water Resources Res.* **14**(3), 461–466(June 1978).
11. B. J. Knapp, "Infiltration and Storage of Soil Water," in *Hillslope Hydrology,* M. J. Kirby (ed.). New York: John Wiley, 1978.
12. H. N. Holtan, "A Concept for Infiltration Estimates in Watershed Engineering," U.S. Department of Agriculture, Agricultural Research Service, 1961, pp. 41–51.
13. D. C. Overton, "Mathematical Refinement of an Infiltration Equation for Watershed Engineering," ARS 41-99. U.S. Department of Agriculture, Washington, D.C., 1964.
14. H. N. Holtan, "A Concept for Infiltration Estimates in Watershed Engineering," ARS 41–51. U.S. Department of Agriculture, Agricultural Research Service, Washington, D.C., 1961.
15. H. N. Holtan, "A Model for Computing Watershed Retention from Soil Parameters," *J. Soil Water Conserv.* **20**(3), 91–94(1965).
16. H. N. Holtan, G. J. Stiltner, W. H. Henson, and N. C. Lopez, "USDAHL-74 Revised Model of Watershed Hydrology," ARS Tech. Bull. No. 1518. U.S. Department of Agriculture, Washington, D.C., 1975.
17. N. H. Crawford and R. K. Linsley, Jr., "Digital Simulation in Hydrology: Stanford Watershed Model IV," Tech. Rept. 39, Department of Civil Engineering, Stanford University, July 1966.
18. Staff, Hydrologic Research Laboratory, "National Weather Service—River Forecast System Forecast Procedures," NWS HYDRO 14, U.S. Department of Commerce, Washington, D.C., Dec. 1972.
19. A. M. Lumb et al., "GTWS: Georgia Tech Watershed Simulation Model," School of Civil Engineering, ERC-0175, Georgia Institute of Technology, Atlanta, GA, 1975.
20. "Hydrology," Suppl. A to Sec. 4, *Engineering Handbook*. U. S. Department of Agriculture, Soil Conservation Service, 1968.
21. Soil Conservation Service, "A Method for Estimating Volume and Rate of Runoff in Small Watersheds," Technical Paper No. 149, USDA-SCS, Washington, D.C., 1973.

Chapter 5

Evaporation and Transpiration

▪ Prologue

The purpose of this chapter is to:

- Define *evaporation* and *transpiration*.
- Describe methods for estimating these major hydrologic abstractions.
- Illustrate the application of *ET* models.

Evaporation is the process by which water is transferred from the land and water masses of the earth to the atmosphere. Transpiration is the evaporation counterpart for plants. It is the process by which soil moisture taken up by vegetation is eventually evaporated as it exits at plant pores. Evaporation and transpiration combined (evapotranspiration) generally constitute the largest component of losses in rainfall-runoff sequences. Accordingly, good estimates of evapotranspiration are a requisite for hydrologic modeling.

On average, about 40,000 billion gallons per day (bgd) of water moves across the conterminous U.S. in the form of water vapor.[1] Of this amount, approximately 10 percent is precipitated. The remainder continues to move in atmospheric suspension. Of the precipitated amount (about 4,200 bgd), about two thirds (2,750 bgd) is evaporated from wet surfaces or transpired from vegetation (see also Fig. 1.1). Evaporation is particularly significant over large bodies of water such as lakes, reservoirs, and the ocean. And estimates of evaporation are critical elements in the design and operation of reservoirs. In the cool humid northeastern United States, annual evaporation amounts range from about 20 to 30 inches, while in the warm dry southwest, annual figures are on the order of 80 or more inches per year (Fig. 5.1). Transpiration is an important component of the water budget of heavily vegetated areas and is of particular concern to the producers of agricultural products.

Evaporation and transpiration rates depend upon temperature, vapor pressure, wind velocity, and the nature of the evaporating surface. Figure 5.1 gives mean annual evaporation for shallow lakes and reservoirs, Table 5.1 gives the adjusted mean monthly Class A pan evaporation for 40 selected stations in the United States, and Fig. 5.2 shows the mean monthly percent of annual evaporation for the stations given in Table 5.1.[2,3]

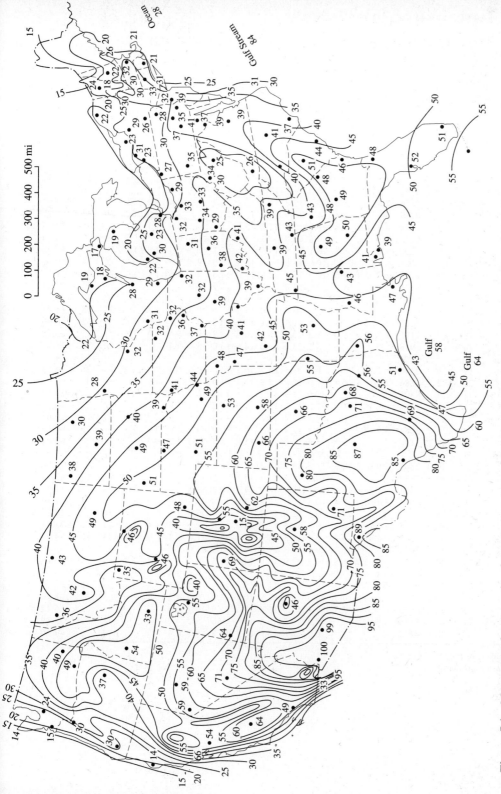

Figure 5.1 Mean annual evaporation from shallow lakes and reservoirs, in inches. NOTE: Evaporation from large deep lakes and reservoirs, particularly in arid regions, will be substantially less in spring and summer, greater in fall and winter, and less for the year than the values shown here. Evaporation from the surfaces of soil and vegetation immediately after rains or irrigation will begin at greater rates and diminish rapidly with the supply of available moisture. Significant local differences in topography and climate in mountainous regions cause large local differences in evaporation not adequately shown here, particularly in the western states. (U.S. Department of Agriculture, Soil Conservation Service.)

TABLE 5.1   ADJUSTED MEAN MONTHLY CLASS A PAN EVAPORATION FOR SELECTED STATIONS 1956–1970

Station Name	Map ID*	State Index No.**	Station Index No.**	Percent of Annual												May thru Oct	Nov thru Apr	Annual Inches
				Jan	Feb	Mar	Apr	May	Jun	Jul	Aug	Sep	Oct	Nov	Dec			
Fairhope 2NE, Ala.	1	1	2813	3.7	4.8	7.8	9.8	12.5	12.5	12.3	11.1	9.3	7.6	4.8	3.8	65	35	50.97
Bartlett Dam, Ariz.	2	2	0632	3.5	4.0	6.1	8.7	12.0	13.8	13.7	11.6	10.1	7.9	4.9	3.9	69	31	121.3
Bacus Ranch, Calif.	3	4	418	3.0	3.5	6.6	8.7	11.5	14.0	14.5	14.7	10.0	7.1	3.6	2.7	72	28	120.56
Sacramento, Calif. (Met)	4	4	7630	1.8	3.1	5.4	8.4	11.9	15.4	16.2	14.5	11.0	7.2	3.3	1.8	76	24	69.70
Wagon Wheel Gap, Colo.	5	5	8742				14.0	16.0	14.1	12.0	10.7	7.1				74	26	50.95
Hartford, Conn. (Met)	6	6	3456	2.6	3.1	5.8	10.1	13.3	14.3	15.1	13.7	9.0	6.4	4.0	2.5	72	28	42.52
Tamiami Trail, Fla.	7	8	8780	5.3	5.9	8.4	10.4	10.9	10.2	10.6	10.1	8.8	8.2	6.0	5.2	59	41	56.48
Experiment, Ga.	8	9	3271	4.1	4.5	7.3	10.0	12.3	12.6	12.4	11.4	9.3	6.7	5.1	4.2	65	35	64.65
Moscow, U of I, Idaho	9	10	6152				6.8	12.0	14.1	19.3	17.7	11.6	6.0			81	19	45.25
Pocatello, Idaho	10	10	7211	1.6	2.3	5.8	8.1	11.9	14.5	19.1	15.1	10.5	6.5	2.9	1.7	78	22	60.98
Ames, Iowa	11	13	205				10.0	14.6	15.8	15.5	13.3	9.3	7.6	3.4		76	24	50.10
Toronto Dam, Kans.	12	14	8191	2.3	3.4	6.6	10.3	12.6	12.5	15.0	14.3	9.5	7.6	4.1	1.7	72	28	61.19
Tribune, Kans.	13	14	8235				9.0	11.8	13.9	15.7	13.9	9.9				73	27	92.98
Madisonville, Ky.	14	15	5067				11.1	13.1	13.9	14.6	13.2	9.6	7.8			72	28	55.26
Urbana, Ill.	15	11	8750				8.6	13.3	15.0	15.2	13.6	10.3	7.3	3.8		75	25	49.46
Woodworth State Forest, La.	16	16	9865	3.4	4.4	7.3	9.4	12.1	13.1	13.0	12.5	9.2	7.7	4.5	3.4	68	32	48.86
Caribou, Maine (Met)	17	17	1175	1.8	2.4	5.0	8.3	15.4	16.0	16.4	13.9	9.0	6.5	3.2	2.1	77	23	22.25
Rochester, Mass.	18	19	6938				8.1	13.0	15.0	14.6	13.0	8.7	5.4			70	30	35.71
East Lansing Hort. Farm, Mich.	19	20	2395				9.4	13.7	15.3	16.2	14.0	9.6	6.4	2.3		75	25	44.53
Scott, Miss.	20	22	7886	3.0	3.4	6.8	9.6	12.9	13.8	13.4	11.9	9.2	7.0	4.3	3.1	68	32	60.99
Weldon Springs Farm, Mo.	21	23	8805				9.5	11.9	13.7	14.5	13.5	10.5	7.5	4.0		72	28	48.08
Bozeman Agric. Col., Mont.	22	24	1044				7.8	12.6	13.9	19.0	16.6	10.3	5.9			78	22	47.06
Medicine Creek Dam, Nebr.	23	25	5388				9.9	12.4	14.2	15.5	14.4	10.5	7.5			74	26	70.60
Boulder City, Nev.	24	26	1071	3.1	3.7	6.4	8.9	12.4	14.3	14.8	12.9	9.9	6.9	3.8	2.8	71	29	109.73
Topaz Lake, Nev.	25	26	8186				8.4	11.8	13.6	15.6	14.5	10.9	7.2	3.3		74	26	82.07
Elephant Butte Dam, N. Mex.	26	29	2848	2.9	4.3	7.5	11.1	13.7	14.8	12.5	10.6	8.5	6.8	4.2	2.8	67	33	116.86
El Vado Dam, N. Mex.	27	29	2837			9.9	10.4	15.1	14.4	14.5	11.5	9.3	6.1			71	29	57.91
Aurora Research Farm, N.Y.	28	30	331				12.5	15.4	16.7	14.3	10.1	6.8				76	24	41.08
Chapel Hill, N.C.	29	31	1677	3.1	4.7	7.8	10.5	12.3	12.6	13.2	11.8	9.3	6.9	4.7	3.2	66	34	52.89
Wooster Exp. Sta., Ohio	30	33	9312				9.1	12.6	15.1	15.5	13.7	9.9	7.1			74	26	46.12
Canton Dam, Okla.	31	34	1445	2.6	4.0	6.8	9.9	11.5	12.5	14.2	13.6	9.3	7.5	4.6	3.4	69	31	77.51
Detroit Power House, Oreg.	32	35	2292	.4	2.2	4.4	6.4	11.8	15.7	21.8	17.9	11.0	5.2	2.4	1.1	83	17	39.74
Redfield, S. Dak.	33	39	7052				9.6	13.3	14.5	16.9	15.9	11.0	7.2			79	21	51.83
Neptune, Tenn.	34	40	6454	2.4	3.7	6.8	10.5	12.0	13.8	14.0	12.5	9.3	7.1	4.2	3.5	69	31	46.47
Grapevine, Tex.	35	41	3691	3.1	4.0	7.2	8.7	10.3	12.4	14.5	13.9	9.8	7.4	4.9	3.9	68	32	84.81
Welasco, Tex.	36	41	9588	4.1	4.8	7.3	9.3	10.7	11.3	13.2	12.8	9.4	7.3	5.4	4.2	65	35	85.70
Ysletta, Tex.	37	41	9966	3.6	4.9	7.7	13.3	13.9	12.9	10.1	8.8	6.6	4.3	3.1		65	35	108.76
Utah Lake, Utah	38	42	8973			5.7	9.1	13.3	15.4	17.7	15.3	10.7	6.6			79	21	56.12
Templeau Dam, Wis.	39	47	8589				14.3	15.8	16.5	13.6	9.6	8.2				78	22	39.29
Heart Mountain, Wyo.	40	48	4411				6.9	13.5	13.9	16.3	14.8	9.5	6.4			74	26	49.36

* Plot identification number for Fig. 5.2.

** NOAA-EDIS *Climatological Data*

Source: Reference 2

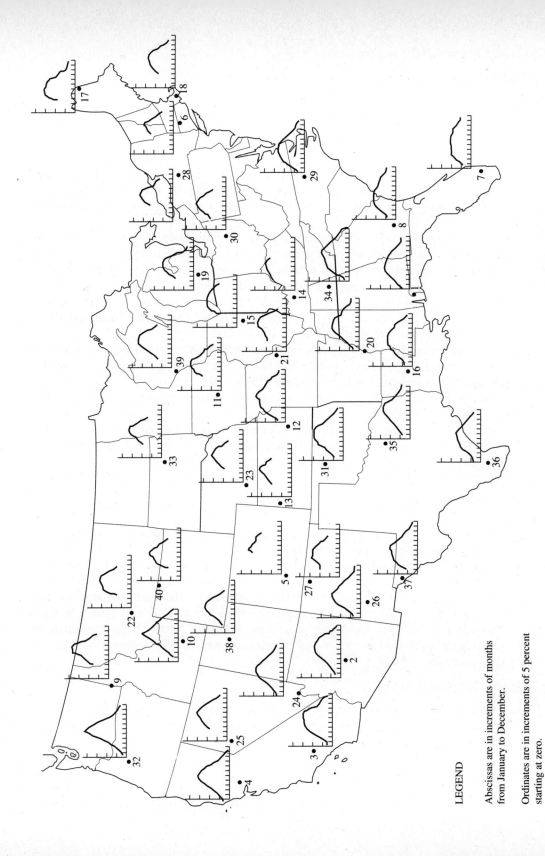

LEGEND

Abscissas are in increments of months from January to December.

Ordinates are in increments of 5 percent starting at zero.

Figure 5.2 Mean monthly percent of annual precipitation for the 40 stations shown in Table 5.1.[2]

5.1 EVAPORATION

Because there is a continuous exchange of water molecules between an evaporating surface and its overlying atmosphere, it is common in hydrologic practice to define evaporation as the *net rate of vapor transfer*. It is a function of solar radiation, differences in vapor pressure between a water surface and the overlying air, temperature, wind, atmospheric pressure, and the quality of evaporating water. Conversion of snow or ice into water vapor is in reality sublimation rather than evaporation, since water molecules do not pass through a liquid phase. Otherwise, the effects of these two processes are the same.

Evaporation from a particular surface is directly related to the opportunity for evaporation (availability of water) provided by that surface. For open bodies of water, evaporation opportunity is 100 percent, while for soils it varies from a high of 100 percent when the soil is highly saturated—for example, during storm periods—to essentially zero at stages of very low moisture content. Other types of surface provide diverse degrees of evaporation opportunity and, except in rare cases, these will almost always vary widely with time.

Direct measurements of evaporation are not easily obtained for large bodies of water because of the extensive surfaces involved. In fact, of all variables included in the general hydrologic equation, surface runoff is the only one that readily permits direct evaluation, since it is confined within well-defined geometric boundaries that permit determination of both rate and cross-sectional area of flow. The choice of method used to determine evaporation depends on the required accuracy of results and the type of instrumentation available. Accuracy is related to the varying degree of reliability with which the method's parameters can be determined.

5.2 ESTIMATING EVAPORATION

The methods applicable to estimating evaporation are the water budget, the energy budget, mass transfer techniques, and the use of pans. Usually, instrumentation for energy budget and mass transfer methods is quite expensive and the cost to maintain observations is substantial. For these reasons, the water budget method and use of evaporation pans are more common. The pan method is the least expensive and will frequently provide good estimates of annual evaporation. Any approach selected is dependent, however, on the degree of accuracy required. As our ability to evaluate the terms in the water budget and energy budget improves, so will the resulting estimates of evaporation.

Water Budget Calculations

The water budget method for determining evaporation is a very simple procedure, but it seldom produces reliable results. In this method, reservoir (lake, pool) evaporation E_s can be computed by rearranging Eq. 1.1:

$$E_s = P + R_1 - R_2 + R_g - T_s - I - \Delta S_s \qquad (5.1)$$

It is useful to deal with the net transfer of seepage through the ground, $R_g - I = O_s$, and consider that the transpiration term T_s equals zero. With these few modifications,

Eq. 5.1 becomes

$$E_s = P + R_1 - R_2 + O_s - \Delta S_s \tag{5.2}$$

All the terms are in volume units for a time period of interest, and Δt should be at least a week. In general, however, the method would more likely be used to estimate monthly or annual evaporation from a particular reservoir. Note that all errors in measuring inflow, precipitation, net seepage, and change in storage are reflected in the final estimate of evaporation. Precipitation, runoff, and changes in storage can often be determined within reasonable limits of accuracy, but evaluation of the net seepage O_s is frequently subject to appreciable errors; if the magnitude of O_s is on the order of E_s, very large errors are possible. Seepage estimates usually come from measurements of groundwater levels and/or soil permeability. The water budget is usable on a continuous basis if a stage-seepage relation for the lake can be established.

In cases where the water budget for a lake is defined by only two unknowns, net seepage and evaporation, these losses can be separated and evaluated in a relatively simple manner by assuming that evaporation is proportional to the product $u(e_0 - e_a)$. This is the mass transfer product described later in the section on mass transfer techniques. The variables are wind velocity u, saturation vapor pressure e_0 (related to water surface temperature), and vapor pressure of the air e_a. When the product $u(e_0 - e_a)$ is zero, evaporation may be neglected.

Periods of no surface inflow or outflow are desirable for net seepage determination, since during such intervals the only losses are evaporation and seepage. Under these conditions, whenever the mass transfer product is equal to zero, the change in water elevation is considered equivalent to the net seepage loss. Normally, a daily plot of change in elevation versus $u(e_0 - e_a)$ is obtained and a best-fitting line constructed. The intercept of this line on the change in stage axis is the net seepage rate. Values of net seepage estimated in this manner can be used on a long-term basis in the water budget equation if the net seepage does not change appreciably over extended periods. Unfortunately, this condition is rarely representative, since net seepage is a function of reservoir stage and season of the year in many cases. Unless these effects can be calculated, net seepage values determined from limited data have little utility.

Good estimates of evaporation using the water budget equation have been obtained, as exemplified by research conducted on Lake Hefner in Oklahoma.[4] Under optimal conditions, the order of accuracy of the method is about 10 percent.

Example 1.1 illustrated the use of the water budget for estimating basin evapotranspiration. For such estimates, reliable values can be expected if the period of time chosen is 1 year or longer. Short-period values may also be obtained if observations are adequate. Mean annual evapotranspiration is successfully judged by using long-time averages of precipitation and surface flows and credible information on the fluctuation of storage. Adequate short-period estimates are also possible if variables in the budget equation can be satisfactorily quantified on a short-term basis.

Energy Budget Method

The energy budget method illustrates an application of the continuity equation written in terms of energy. It has been employed to compute the evaporation from oceans and lakes, that is, for Lake Hefner in Oklahoma and at Elephant Butte Reservoir in New Mexico.[4,5] The equation accounts for incoming and outgoing energy balanced by the amount of energy stored in the system. The accuracy of estimates of evaporation using the energy budget is highly dependent on the reliability and preciseness of measurement data. Under good conditions, average errors of perhaps 10 percent for summer periods and 20 percent for winter months can be expected.

The energy budget equation for a lake may be written as

$$Q_0 = Q_s - Q_r + Q_a - Q_{ar} + Q_v - Q_{bs} - Q_e - Q_h - Q_w \tag{5.3}$$

where Q_0 = increase in stored energy by the water
$\quad\quad Q_s$ = solar radiation incident at the water surface
$\quad\quad Q_r$ = reflected solar radiation
$\quad\quad Q_a$ = incoming long-wave radiation from the atmosphere
$\quad\quad Q_v$ = net energy advected (net energy content of incoming and outgoing water) into the water body
$\quad\quad Q_{ar}$ = reflected long-wave radiation
$\quad\quad Q_{bs}$ = long-wave radiation emitted by the water
$\quad\quad Q_e$ = energy used in evaporation
$\quad\quad Q_h$ = energy conducted from water mass as sensible heat
$\quad\quad Q_w$ = energy advected by evaporated water

All the terms are in calories per square centimeter per day (cal/cm²-day). Heating brought about by chemical changes and biological processes is neglected as is the energy transfer that occurs at the water-ground interface. The transformation of kinetic energy into thermal energy is also excluded. These factors are usually very small, in a quantitative sense, when compared with other terms in the budget if large reservoirs are considered. As a result, their omission has little effect on the reliabiliy of results.

During winter months when ice cover is partial or complete, the energy budget only occasionally yields adequate results because it is difficult to measure reflected solar radiation, ice-surface temperature, and the areal extent of the ice cover. Daily evaporation estimates based on the energy budget are not feasible in most cases because reliable determination of changes in stored energy for such short periods is impractical. Periods of a week or longer are more likely to provide satisfactory measurements.

In using the energy budget approach, it has been demonstrated that the required accuracy of measurement is not the same for all variables.[5] For example, errors in measurement of incoming long-wave radiation as small as 2 percent can introduce errors of 3 to 15 percent in estimates of monthly evaporation, while errors on the order of 10 percent in measurements of reflected solar energy may cause errors of only 1 to 5 percent in calculated monthly evaporation.

To permit the determination of evaporation by Eq. 5.3, it is common to use the following relation:

$$B = \frac{Q_h}{Q_e} \tag{5.4}$$

where B is known as Bowen's ratio,[6] and

$$Q_w = \frac{c_p Q_e (T_e - T_b)}{L} \tag{5.5}$$

where c_p = the specific heat of water (cal/g-°C)
  T_e = the temperature of evaporated water (°C)
  T_b = the temperature of an arbitrary datum usually taken as 0°C
  L = the latent heat of vaporization (cal/g)

Introducing these expressions in Eq. 5.3 and solving for Q_e, we obtain

$$Q_e = \frac{Q_s - Q_r + Q_a - Q_{ar} - Q_{bs} - Q_0 + Q_v}{1 + B + c_p(T_e - T_b)/L} \tag{5.6}$$

To determine the depth of water evaporated per unit time, the following expression may be used:

$$E = \frac{Q_e}{\rho L} \tag{5.7}$$

where E = evaporation (cm³/cm²-day)
  ρ = the mass density of evaporated water (g/cm³)

The energy budget equation thus becomes

$$E = \frac{Q_s - Q_r + Q_a - Q_{ar} - Q_{bs} - Q_0 + Q_v}{\rho[L(1 + B) + c_p(T_e - T_b)]} \tag{5.8}$$

The Bowen ratio can be computed using

$$B = 0.61 \frac{p}{1000} \frac{(T_0 - T_a)}{(e_0 - e_a)} \tag{5.9}$$

where p = the atmospheric pressure (mb)
  T_0 = the water surface temperature (°C)
  T_a = the air temperature (°C)
  e_0 = the saturation vapor pressure at the water surface temperature (mb)
  e_a = the vapor pressure of the air (mb)

This expression circumvents the problem of evaluating the sensible heat term, which does not lend itself to direct measurement.

Mass Transfer Techniques

Mass transfer equations are based primarily on the concept of the turbulent transfer of water vapor (by eddy motion) from an evaporating surface to the atmosphere. Many equations, both theoretical and empirical, have been developed. Most are similar in form to a relation between evaporation and vapor pressure first recognized by Dalton,[7]

$$E = \kappa(e_0 - e_a) \tag{5.10}$$

where E = direct evaporation
  κ = a coefficient dependent on the wind velocity, atmospheric pressure, and other factors
  e_0, e_a = the saturation vapor pressure at water surface temperature and the vapor pressure of air, respectively

Theoretical mass transfer equations are based on the concepts of discontinuous and continuous mixing at the air-liquid interface.

Empirical approaches often require exacting and costly instrumentation and observations, so their general utility is limited. The complexity of the equations varies from simple expressions such as Eq. 5.10 to complex relations like Sutton's equation[8] for a circular lake of radius r.

$$E = \frac{0.623}{p} G' \rho u^{(2-n)/(2+n)} r^{(4+n)/(2+n)} (e_0 - e_a) \qquad (5.11)$$

where
E = evaporation (cm/day)
ρ = mass density of the air (g/cm³)
u = average wind velocity (cm/sec)
r = radius of circular lake (cm)
p = atmospheric pressure (mb)
e_0, e_a = as previously defined
n = an empirical constant
G' = a complex function

A commonly used empirical equation has been developed by Meyer.[9] This equation takes the form

$$E = C(e_0 - e_a)\left(1 + \frac{W}{10}\right) \qquad (5.12)$$

where
E = the daily evaporation in inches depth
e_0, e_a = as previously defined but in units of in. Hg
W = the wind velocity in mph measured about 25 ft above the water surface
C = pan empirical coefficient

For daily data on an ordinary lake, C is about 0.36. For wet soil surfaces, small puddles, and shallow pans, the value of C is approximately 0.50.

Another mass transfer equation used to estimate the rate of evaporation is one developed by Dunne.[10, 11] It takes the following form

$$E = (0.013 + 0.00016u_2)e_a[(100 - R_h)/100] \qquad (5.13)$$

where E is the evaporation rate in cm/day, u_2 is the wind velocity measured at two m above the surface in km/day, e_a is defined as before but in millibars, and R_h is the relative humidity given in percent.

EXAMPLE 5.1

Using the Meyer and Dunne equations, find the daily evaporation rate for a lake given that the mean value for air temperature was 87°F and for water temperature was 63°F, the average wind speed was 10 mph, and the relative humidity was 20 percent. Refer to Appendix Table A.2 for vapor pressure values.

Solution.   Interpolating from Table A.2, we find that

$$e_0 = 0.58 \text{ in. Hg}$$
$$e_a = 1.29 \times 0.20 = 0.26 \text{ in. Hg} = 8.75 \text{ mb}$$

Using Eq. 5.12, we obtain

$$E = 0.36(0.58 - 0.26)\left(1 + \frac{10}{10}\right)$$
$$= 0.36 \times 0.32 \times 2$$
$$= 0.23 \text{ in./day}$$

Using Eq. 5.13, after converting wind speed to metric units, we obtain

$$E = (0.013 + 0.00016 \times 386) \times 8.75 \times ((100 - 20)/100)$$
$$= 0.075 \times 8.75 \times 0.8$$
$$= 0.524 \text{ cm/day, or } 0.21 \text{ in./day}$$

The two estimates are comparable. ■■

Investigations of the utility of mass transfer equations conducted at Lake Hefner indicated that a simple equation using wind speed and vapor pressure differences yielded as good results as any that were tested.[4] The equation was

$$E = Nu(e_0 - e_a) \tag{5.14}$$

where E = the evaporation (cm/day)
    N = a coefficient
    u = the wind velocity at 2 m above the water surface (m/sec)
e_0, e_a = as previously defined (mb)

The value of N can be determined by comparative studies between mass transfer and energy budget methods. This is the preferred approach. If such an evaluation is not available, it can be approximated using

$$N = \frac{0.0291}{A^{0.05}} \tag{5.15}$$

where A is the surface area of the water surface in square meters. For values of A less than about 4×10^6 m^2, variations in wind exposure may become important and Eq. 5.15 should be used with caution. When N is based on comparative studies using the energy budget, average errors in evaporation estimates of about 15 percent can be expected, while errors of roughly 30 percent would likely obtain if Eq. 5.15 were employed.

Use of Evaporation Pans

The most widely used method of finding reservoir evaporation is by means of evapo-ration pans.[12-14] The standard National Weather Bureau Class A pan, built of un-painted galvanized iron, is currently the most popular. It is 4 ft in diameter, 10 in. deep, and mounted 12 in. above the ground on a wooden frame. Relations developed between pan and actual evaporation from large bodies of water such as lakes indicate multiplying the former by a factor of 0.70–0.75 (pan coefficient) gives the equivalent lake evaporation. Ratios of annual reservoir evaporation to pan evaporation are consistent from year to year and region to region, while monthly ratios often show considerable variation.

Estimates of reservoir evaporation based on short-period pan observations (less than 1 year) may be seriously in error. The use of a pan coefficient to estimate evaporation from an ungauged location should reflect the geographic variability in heat transfer through the sides of the Class A pan. For lakes subjected to significant amounts of advected energy, local pan–lake relations should be established.

Available data indicate that the annual ratio of lake evaporation to Class A pan evaporation is essentially 0.70, provided that net advection is balanced by the change in energy stored, conduction through the pan is negligible, and the pan is located so that its exposure conditions are representative of the body of water being considered. Considerable care must be taken, however, in installing and using evaporation pans. Pans may be sunken, floating, or set above the ground surface. Sunken pans tend to have fewer boundary heat transfer problems but are more difficult to gauge and they are more prone to collect trash. While floating pan observations more closely approximate lake evaporation than shore installations, such pans are not without problems. Commonly there are appreciable boundary effects and splashing often negates the validity of observations. Pans located above ground exhibit heat exchange problems related to side walls and the bottom, but these difficulties may be largely overcome by the use of insulation. Pans installed above ground are easily erected, gauged, and maintained. Such installations are the most common, characterized by the Class A pan.

An equation for daily Class A pan evaporation, E_a, assuming that air and water temperatures are equal, is

$$E_a = (e_0 - e_a)^{0.88}(0.42 + 0.0029u_p) \tag{5.16}$$

where the daily pan evaporation is given in millimeters per day, and u_p is the wind movement 150 mm above the rim of the pan in kilometers per day. The vapor pressure difference term $(e_0 - e_a)$ can be determined using the following equation.[14]

$$e_0 - e_a = 33.86[(0.00738T_a + 0.8072)^8 - (0.00738T_d + 0.8072)^8] \tag{5.17}$$

where vapor pressures are measured in millibars and the dew point temperature T_d and air temperature T_a are measured in degrees Celsius. Note that this equation is valid as long as T_d equals or exceeds $-27°C$.

Penman, through a simultaneous solution of an aerodynamic equation and an energy balance equation, derived the following equation for daily evaporation E:[15]

$$E = \frac{1}{\Delta + \gamma}(Q_n\Delta + \gamma E_a) \tag{5.18}$$

where Δ is the slope of the saturation vapor pressure versus temperature curve at the air temperature T_a; E_a is the pan evaporation given by Eq. 5.15; Q_n is the net radiant energy expressed in the same units as those of E; and γ is the $[0.61 p/1000]$ term in Bowen's ratio (Eq. 5.9) where p is the atmospheric pressure in millibars.

Equation 5.18 can be used to estimate lake evaporation after introducing an appropriate pan coefficient. For practical purposes, Class A pan evaporation is estimated to be approximately 0.70 of pan evaporation provided that (1) any net advection into the lake is balanced by the change in energy storage, (2) the net transfer of sensible heat through the pan is negligible, and (3) the pan exposure is representative.[15,16] If these conditions are met, annual lake evaporation can be estimated from the following equation:

$$E_L = 0.70\left(\frac{Q_n\Delta + E_a\gamma}{\Delta + \gamma}\right) \qquad (5.19)$$

where E_L is the average daily lake evaporation in units of length, and the other terms are as previously defined. By making allowances for advected energy and heat transfer through the pan, Kohler and co-workers have developed the following expression for calculating annual lake evaporation in inches per day:

$$E_L = 0.70[E_p + 0.00051P\alpha_p(0.37 + 0.0041u_p)(T_0 - T_a)^{0.88}] \qquad (5.20)$$

where α_p is the proportion of advected energy (Class A pan) used for evaporation. Equation 5.20 assumes that any energy advected into the lake is balanced by a change in energy storage and that the pan exposure is representative.[12] A graphical solution of Eq. 5.20 is given in Fig. 5.3; values of α_p can be obtained using Fig. 5.4. An

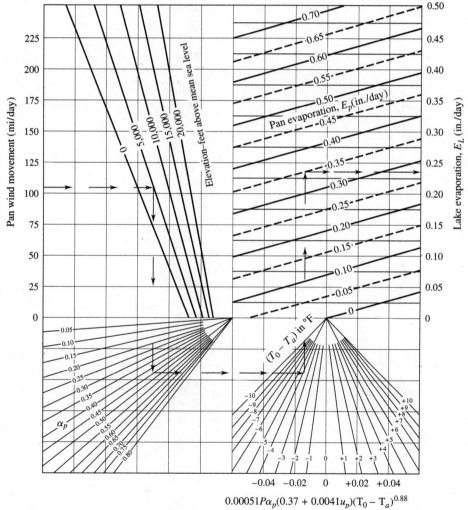

Figure 5.3 Graphical solution of Eq. 5.20. (U.S. Weather Bureau Res. Paper No. 38.)

$$\alpha_p = \frac{[(e_0^* - e_a)^{0.88} - (e_0 - e_a)^{0.88}](0.37 + 0.0041u_p)}{[(e_0^* - e_a)^{0.88} - (e_0 - e_a)^{0.88}](0.37 + 0.0041u_p) + (7.6 \times 10^{-11})[(K_0^*)^4 - K_0^4] + 0.000367P[(T_0^* - T_a)^{0.88} - (T_0 - T_a)^{0.88}](0.37 + 0.0041u_p)}$$

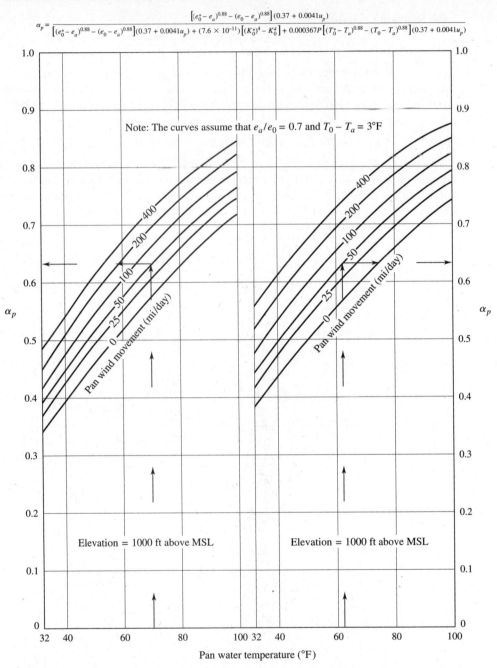

Figure 5.4 Proportion of advected energy (into a Class A pan) used for evaporation. (U.S. Weather Bureau Res. Paper No. 38.)

approximation for estimating α_p for use in computer simulations is the following:

$$\alpha_p = 0.13 + 0.0065T_0 - 6.0 \times 10^{-8}T_0^3 + 0.016u_p^{0.36} \tag{5.21}$$

where T_0, the outerface temperature of the pan, is in degrees Fahrenheit and the wind velocity is in miles per day.[14] Equation 5.20 is considered to be reliable where water temperature data are available along with appropriate Class A pan observations.[12]

5.3 EVAPORATION CONTROL

Evaporation losses can be greatly significant at any location. Consequently, the concept of evaporation reduction is receiving widespread attention. Evaporation losses from soils can be controlled by employing various types of mulch or by chemical alteration. They may be reduced from open waters by (1) storing water in covered reservoirs, (2) making increased use of underground storage, (3) controlling aquatic growths, (4) building storage reservoirs with minimal surface area, (5) the use of chemicals, and (6) conveying in closed conduits rather than open channels. Some of these approaches may be impractical (covering large reservoirs) or uneconomical (large-scale vegetation control). All have potential advantages, however, under the proper circumstances.

The first four approaches need no explanation of the mechanism expected to control evaporation. The fifth method, chemical means, requires further comment. Research has shown that certain types of organic compounds such as hexadecanol and octadecanol form monomolecular films that are effective as evaporation inhibitors.[5,17] Studies by the Bureau of Reclamation indicate that evaporation may be suppressed by as much as 64 percent with hexadecanol films in 4-ft diameter pans under controlled conditions. Actual reductions on large bodies of water would be significantly less than this, however, due to problems in maintaining films against wind and wave action. Evaporation reduction in the range of 22–35 percent has been observed for some studies on small lakes of roughly 100 acres in size with reductions of 9 to 14 percent reported for Lake Hefner in Oklahoma (2500 acres).[18] Wind was a major problem at Lake Hefner, however.

In Australia, extended tests on medium-sized lakes (less than 2500 acres) have indicated savings of 30 to 50 percent, although adverse winds were generally not encountered.[18] Although considerable research and development work remain, the use of monomolecular films to control evaporation appears promising. Franzini has indicated that the cost per acre-foot of water saved by evaporation suppression is, in fact, competitive with various alternate means of increasing local water supplies, and these costs will likely decrease with advances in research.[17]

5.4 TRANSPIRATION

Root systems of plants absorb water in varying quantities. Most of this water is transmitted through the plant and escapes through pores in the leaf system. This is known as *stomatal transpiration*. Plants also lose water by other mechanisms, but usually this is negligible compared with that lost through the microscopic leaf apertures. Transpiration is basically a process by which water is evaporated from the airspaces in plant leaves. Therefore, it is controlled essentially by the same factors that dominate evaporation, namely, solar radiation, temperature, wind velocity, and vapor pressure gradients. In addition, transpiration is affected to some degree by the character of the plant and plant density.

Soil moisture content, when reduced to the wilting point (stage at which plants wilt and do not recover in a humid atmosphere) also affects transpiration. The effects of decreased soil moisture above the wilting point are not clearly established and are somewhat controversial. Nevertheless, it appears that as long as soil moisture lies between the limits of the wilting point and field capacity (the amount of water retained in a soil against gravity after percolation ceases) transpiration is not materially affected. Saturated soils can sometimes adversely affect plant life.

Diffusion of water vapor from plant leaves to the atmosphere is proportional to the vapor pressure gradient at the leaf-atmosphere interface. Upon absorbing solar radiation, leaves tend to become warmer than the surrounding air (often by as much as 5–10°F). The amount of water vapor held by the air at the leaf-air interface thus increases; more rapid water losses are favored; and transpiration follows a diurnal cycle, which is approximately that of light intensity. It has also been demonstrated that transpiration and the rate of plant growth are related. Below a temperature of about 40°F the amount of water transpired is considered negligible.

Different species and types of plants often display considerably different demands on soil moisture even if the same environmental conditions prevail. For example, an oak tree may transpire as much as 170 qts of water a day, whereas a corn plant will transpire only about 2 qts. The area covered by the two root systems is, of course, significantly different. Various species also indicate different patterns in seasonal demands for water. Agricultural products obviously have their periods of greatest transpiration at the peak of the growing season.

Precise values for quantities of water transpired are difficult to acquire, since many variables are active in the process and these range widely from one region to another. Available estimates should be used with caution, and the conditions under which they were obtained should be determined before applying the data. Adequate relations between climatic factors and transpiration become prerequisites if data derived in one climatic region are to have general utility.

Transpiration may be measured in the laboratory by using tanks wherein evaporation is eliminated and water losses are found by weighing. Coefficients must be derived before such data can be applied to field conditions, and even then the observations usually provide little more than an index to field water use. Large-scale field measurements of transpiration are virtually impossible under prevailing field conditions so it is common to find measures of consumptive use (combined evaporation plus transpiration) more widely adopted and of greater value to the practicing hydrologist. Most field observations are made by using lysimeters (grass or crop-covered containers for which a water budget is maintained). Table 5.2 gives some values of consumptive use for several crops in the Montrose area of Colorado.[20] These values are presented only to indicate their order of magnitude in this area during the growing or irrigation season. More complete information on consumptive use by various crops can be found elsewhere.[19–24]

TABLE 5.2 CONSUMPTIVE USE FOR CROPS IN THE MONTROSE, COLORADO, AREA DURING THE IRRIGATION OR GROWING SEASON

Crop	Consumptive use (in.)
Alfalfa	26.5
Corn	19.7
Small grain	14.9
Grass hay	23.3
Natural vegetation	37.3

Source: H. F. Blaney, "Water and Our Crops," in *Water, The Yearbook of Agriculture.* Washington, D.C.: U.S. Department of Agriculture, 1955.

For many small local projects it is not possible to carry out detailed field studies to determine the consumptive use of crops. In such cases it is common to use either the Blaney–Criddle or Penman method for estimating seasonal consumptive use.[23,24] The Blaney–Criddle method is briefly described here.

The seasonal consumptive use for a particular crop can be computed using the relation

$$U = k_s B \tag{5.22}$$

where U = the consumptive use of water during the growing season (in.)
  k_s = a seasonal consumptive use coefficient applicable to a particular crop, empirically derived (Table 5.3)
  B = the summation of monthly consumptive use factors for a given season

The term B can be expressed as

$$B = \sum \left(\frac{tp}{100} \right) \tag{5.23}$$

where t = the mean monthly temperature (°F)
  p = the monthly daytime hours given as percentage of the year (Table 5.4)

If monthly values for the consumptive use coefficient k are available, monthly consumptive use can be found by using

$$u = \frac{ktp}{100} \tag{5.24}$$

where u is the monthly consumptive use (in.) and the other terms are as previously defined. Selected values of p and k are available in the literature.[21,23] An example illustrates the use of this equation.

TABLE 5.3   SEASONAL CONSUMPTIVE USE CROP COEFFICIENTS (k_s) FOR IRRIGATED CROPS, FOR USE IN EQUATION 5.22

Crop	Length of normal growing season or period[a]	Consumptive use coefficient k_s[b]	Maximum monthly k[c]
Alfalfa	Between frosts	0.80–0.90	0.95–1.25
Bananas	Full year	0.80–1.00	—
Beans	3 months	0.60–0.70	0.75–0.85
Cocoa	Full year	0.70–0.80	—
Coffee	Full year	0.70–0.80	—
Corn (maize)	4 months	0.75–0.85	0.80–1.20
Cotton	7 months	0.60–0.70	0.75–1.10
Dates	Full year	0.65–0.80	—
Flax	7–8 months	0.70–0.80	—
Grains, small	3 months	0.75–0.85	0.85–1.00
Grain, sorghums	4–5 months	0.70–0.80	0.85–1.10
Oilseeds	3–5 months	0.65–0.75	—
Orchard crops:			
Avocado	Full year	0.50–0.55	—
Grapefruit	Full year	0.55–0.65	—
Orange and lemon	Full year	0.45–0.55	0.65–0.75[d]
Walnuts	Between frosts	0.60–0.70	—
Deciduous	Between frosts	0.60–0.70	0.70–0.95
Pasture crops:			
Grass	Between frosts	0.75–0.85	0.85–1.15
Ladino white clover	Between frosts	0.80–0.85	—
Potatoes	3–5 months	0.65–0.75	0.85–1.00
Rice	3–5 months	1.00–1.10	1.10–1.30
Soybeans	140 days	0.65–0.70	—
Sugar beets	6 months	0.65–0.75	0.85–1.00
Sugarcane	Full year	0.80–0.90	—
Tobacco	4 months	0.70–0.80	—
Tomatoes	4 months	0.65–0.70	—
Truck crops, small	2–4 months	0.60–0.70	—
Vineyard	5–7 months	0.50–0.60	—

[a]Length of season depends largely on variety and time of year when the crop is grown. Annual crops grown during the winter period may take much longer than if grown in the summertime.

[b]The lower values of k_s for use in the Blaney-Criddle formula, $U = k_s B$, are for more humid areas and the higher values are for more arid climates.

[c]Dependent on mean monthly temperature and crop growth stage.

[d]Given by Criddle as "citrus orchard."

Source: From *Irrigation Water Requirements*, Technical Release no. 21, Soil Conservation Service, USDA, September 1970.

EXAMPLE 5.2 ———

Determine the monthly consumptive use of an alfalfa crop grown in southern California for the month of July if the average monthly temperature is 72°F, average daytime hours in percentage of the year is 9.88, and the mean monthly consumptive use coefficient for alfalfa is 0.85.

TABLE 5.4 DAYTIME HOURS COEFFICIENT (p) FOR USE IN EQUATION 5.23

Latitude North South	Jan Jul	Feb Aug	Mar Sep	Apr Oct	May Nov	Jun Dec	Jul Jan	Aug Feb	Sep Mar	Oct Apr	Nov May	Dec Jun
60°	0.15	0.20	0.26	0.32	0.38	0.41	0.40	0.34	0.28	0.22	0.17	0.13
50°	0.19	0.23	0.27	0.31	0.34	0.36	0.35	0.32	0.28	0.24	0.20	0.18
40°	0.22	0.24	0.27	0.30	0.32	0.34	0.33	0.31	0.28	0.25	0.22	0.21
30°	0.24	0.25	0.27	0.29	0.31	0.32	0.31	0.30	0.28	0.26	0.24	0.23
20°	0.25	0.26	0.27	0.28	0.29	0.30	0.30	0.29	0.28	0.26	0.25	0.25
10°	0.26	0.27	0.27	0.28	0.28	0.29	0.29	0.28	0.28	0.27	0.26	0.26
0°	0.27	0.27	0.27	0.27	0.27	0.27	0.27	0.27	0.27	0.27	0.27	0.27

Note: Values for (p) are determined by dividing the mean daily daytime hours for a specified month by the total daytime hours in a year and then multiplying the ratio by 100.

Solution. Using Eq. 5.24 we find that

$$u = \frac{ktp}{100}$$

$$= 0.85 \times 72 \times \frac{9.88}{100}$$

$$= 6.05 \text{ in. of water} \quad ■■$$

EXAMPLE 5.3 _____

Determine the seasonal consumptive use of a tomato crop grown in New Jersey if the mean monthly temperatures for May, June, July, and August are 61.6, 70.3, 75.1, and 73.4°F, respectively, and the percent daylight hours for the given months are 10.02, 10.08, 10.22, and 9.54 as percent of the year, respectively.

Solution

1. From Table 5.3, the growing season for tomatoes is four months and the range of the consumptive use coefficient is 0.65 to 0.70. Since New Jersey is a humid area, choose the lower value of $k_s = 0.65$.
2. The term B is calculated using Eq. 5.23 as:

   $B = (61.6 \times 10.02/100) + (70.3 \times 10.08/100) + (75.1 \times 10.22/100)$
   $+ (73.4 \times 9.54/100) = 27.9$

3. Seasonal consumptive use is determined using Eq. 5.22:

   $U = k_s B$
   $U = 0.65 \times 27.9$
   $= 18.1$ in. of water for the four-month growing season.

Note that the total amount of water to be applied to an irrigated area must include consumptive use plus conveyance and other losses. Thus the amount of water allocated at the source may have to be considerably more than the consumptive use expectation.   ■■

5.5 TRANSPIRATION CONTROL

Water conservation through transpiration reduction is being seriously studied, and certain preventative practices are presently in use. Methods of control include the use of chemicals to inhibit water consumption (analogous to the use of films to control surface evaporation except that chemicals are applied in the root zone), harvesting of plants, improved irrigation practices, and actual removal or destruction of certain vegetative types.[25]

In arid regions of the Southwest, certain plants known as *phreatophytes* (plants capable of tapping the water table or capillary fringe) transpire enormous quantities of water each year without providing any particular apparent benefit (although this statement is open to question). Many of these plants, such as the salt cedar, grow in stream channels and tend to create flood control problems by restricting channels in addition to using valuable underground water supplies. In New Mexico there have been as many as 43,000 acres of salt cedar along the Pecos River alone. Control of these phreatophytes could result in estimated savings of over 200,000 acre-ft of water in a critically water-short region of the United States.[26] Conservation through transpiration control may be important, but the ecologic consequences of such control practices should be given careful consideration.

5.6 EVAPOTRANSPIRATION

In most cases of practical interest to the hydrologist, only total evaporation from an area—combined evaporation plus transpiration (consumptive use)—is of real interest. Various methods for determining evapotranspiration have been proposed, but there is no one system generally acceptable under all conditions. Basically, there are three major approaches:

1. Theoretical, based on physics of the process
2. Analytical, based on energy or water budgets
3. Empirical

The water budget method was illustrated in Example 1.1. Its adequacy is dependent on the accuracy with which the several terms in the budget equation can be evaluated. The energy budget can also be used to calculate field evapotranspiration in a manner similar to that described previously for lakes. For this application, however, the soil's thermal properties must be known, and temperature and vapor pressure gradients measured at two levels above the ground are needed in Bowen's ratio. For field plots the amount of energy advected usually can be neglected.

Mass transfer equations of the form previously discussed can also be used to estimate evapotranspiration. The Thornthwaite–Holzman equation is a good example of a mass transfer equation that has often been employed for this purpose. However, Linsley and co-workers indicate that there is some question as to the adequate verification of this model to estimate evapotranspiration losses.[27] The equation is expressed as

$$E = \frac{833\kappa^2(e_1 - e_2)(V_2 - V_1)}{(T + 459.4)\log_e(z_2/z_1)^2} \tag{5.25}$$

where
E = evaporation (in./hr)
κ = von Kármán's constant (0.4)
e_1, e_2 = vapor pressures (in. Hg)
V_1, V_2 = wind speeds (mph)
T = the mean temperature (°F) of the layer between the lower level z_1 and the upper level z_2

It is assumed in Eq. 5.25 that the atmosphere is adiabatic and the wind speed and moisture are distributed logarithmically in a vertical direction. In view of the small differences between wind and vapor pressure to be expected at two levels so closely spaced, and since these gradients are directly related to the sought-after evaporation, highly exacting instrumentation is required to get reliable results.

Potential Evapotranspiration

Thornthwaite defined potential evapotranspiration as "the water loss which will occur if at no time there is a deficiency of water in the soil for the use of vegetation." In a practical sense, however, most investigators have assumed that potential evapotranspiration is equal to lake evaporation as determined from National Weather Service Class A pan records. This is not theoretically correct because the albedo (amount of incoming radiation reflected back to the atmosphere) of vegetated areas and soils ranges as high as 45 percent.[28] As a result, potential evapotranspiration should be somewhat less than free water surface evaporation. Errors in estimating free water evapotranspiration from pan records are such, however, as to make an adjustment for potential evapotranspiration of questionable value.

An equation for estimating potential evapotranspiration developed by the Agricultural Research Service (ARS) illustrates efforts to include vegetal characteristics and soil moisture in such a calculation. The evapotranspiration potential for any given day is determined as follows:[29]

$$ET = GI \times k \times E_p \times \left(\frac{S - SA}{AWC}\right)^x \tag{5.26}$$

where
ET = evapotranspiration potential (in./day)
GI = growth index of crop in percentage of maturity
k = ratio of GI to pan evaporation, usually 1.0–1.2 for short grasses, 1.2–1.6 for crops up to shoulder height, and 1.6–2.0 for forest
E_p = pan evaporation (in./day)
S = total porosity
SA = available porosity (unfilled by water)
AWC = porosity drainable only by evapotranspiration
x = AWC/G (G = moisture freely drained by gravity)

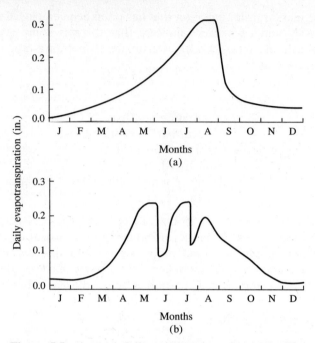

Figure 5.5 Average daily consumption of water: (a) for year 1953 by corn, followed by winter wheat under irrigation; (b) for year 1955, with irrigated first-year meadow of alfalfa, red clover, and timothy. Both measurements taken on lysimeter Y 102 C at the Soil and Water Conservation Research Station, Coshocton, Ohio. (After Holtan et al.[29])

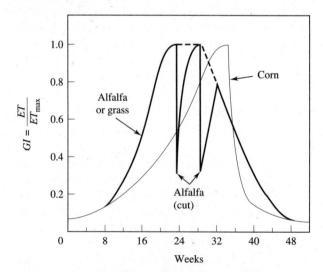

Figure 5.6 Growth index $GI = ET/ET_{max}$ from lysimeter records, irrigated corn, and hay for 1955, from Coshocton, Ohio. (After Holtan et al.[29])

TABLE 5.5 HYDROLOGIC CAPACITIES OF SOIL TEXTURE CLASSES

Texture class	S^a (%)	G^b (%)	AWC^c (%)	x AWC/G
Coarse sand	24.4	17.7	6.7	0.38
Coarse sandy loam	24.5	15.8	8.7	0.55
Sand	32.3	19.0	13.3	0.70
Loamy sand	37.0	26.9	10.1	0.38
Loamy fine sand	32.6	27.2	5.4	0.20
Sandy loam	30.9	18.6	12.3	0.66
Fine sandy loam	36.6	23.5	13.1	0.56
Very fine sandy loam	32.7	21.0	11.7	0.56
Loam	30.0	14.4	15.6	1.08
Silt loam	31.3	11.4	19.9	1.74
Sandy clay loam	25.3	13.4	11.9	0.89
Clay loam	25.7	13.0	12.7	0.98
Silty clay loam	23.3	8.4	14.9	1.77
Sandy clay	19.4	11.6	7.8	0.67
Silty clay	21.4	9.1	12.3	1.34
Clay	18.8	7.3	11.5	1.58

$^a S$ = total porosity − 15 bar moisture %.
$^b G$ = total porosity − 0.3 bar moisture %.
$^c AWC = S − G$.

Source: Adapted from C. B. England, "Land Capability: A Hydrologic Response Unit in Agricultural Watersheds," U.S. Department of Agriculture, ARS 41–172, Sept. 1970. After H. N. Holtan et al.[29]

The *GI* curves have been developed by expressing experimental data on daily evapotranspiration for several crops (Fig. 5.5) as a percentage of the annual maximal daily rate (Fig. 5.6). Equation 5.26 is used by the Agricultural Research Service in its USDAHL-74 model of watershed hydrology in combination with *GI* curves to calculate daily evapotranspiration. Representative values for *S, G,* and *AWC* are given in Table 5.5.

5.7 ESTIMATING EVAPOTRANSPIRATION

Transpiration is an important component in the hydrologic budget of vegetated areas, but it is a difficult quantity to measure because of its dependence on phytological variables. It is a function of the number and types of plants, soil moisture and soil type, season, temperature, and average annual precipitation. As noted previously, evaporation and transpiration are commonly estimated in their combined evapotranspiration form.

If the precipitation and net runoff for an area are known, and estimates of groundwater flow and storage can be made, rough estimates of *ET* can be had using the basic hydrologic equation, Eq. 1.1. A more sophisticated approach developed by Penman follows.[15] It is representative of the methods most often used.

The Penman Method

Both the energy budget and mass transport methods for estimating evapotranspiration (ET) have limitations due to the difficulties encountered in estimating parameters and in making other required assumptions. To circumvent some of these problems, Penman developed a method to combine the mass transport and energy budget theories. This widely used method is one of the more reliable approaches to estimating ET rates using climatic data.[13,15,23,30]

The Penman equation is of the form of Eq. 5.18; it is theoretically based and shows that ET is directly related to the quantity of radiative energy gained by the exposed surface. In its simplified form, the Penman equation is[15]

$$ET = \frac{\Delta H + 0.27E}{\Delta + 0.27} \tag{5.27}$$

where Δ = the slope of the saturated vapor pressure curve of air at absolute temperature (mm Hg/°F)

H = the daily heat budget at the surface (estimate of net radiation) (mm/day)

E = daily evaporation (mm)

ET = the evapotranspiration or consumptive use for a given period (mm/day)

The variables E and H are calculated using the following equations:

$$E = 0.35(e_a - e_d)(1 + 0.0098u_2) \tag{5.28}$$

where e_a = the saturation vapor pressure at mean air temperature (mm Hg)

e_d = the saturation vapor pressure at mean dew point (actual vapor pressure in the air) (mm Hg)

u_2 = the mean wind speed at 2 m above the ground (mi/day)

The equation used to determine the daily heat budget at the surface, H, is

$$H = R(1 - r)(0.18 + 0.55S) - B(0.56 - 0.092e_d^{0.5})(0.10 + 0.90S) \tag{5.29}$$

where R = the mean monthly extraterrestrial radiation (mm H_2O evaporated per day)

r = the estimated percentage of reflecting surface

B = a temperature-dependent coefficient

S = the estimated ratio of actual duration of bright sunshine to maximum possible duration of bright sunshine.

The empirical reflective coefficient r is a function of the time of year, the calmness of the water surface, wind velocity, and water quality. Typical ranges for r are 0.05 to 0.12.[31] Values of e_a and Δ can be obtained from Figs. 5.7 and 5.8, those for R and B can be obtained from Tables 5.6 and 5.7. The use of Penman's equation requires a knowledge of vapor pressures, sunshine duration, net radiation, wind speed, and mean temperature. Unfortunately, regular measurements of these parameters are often unavailable at sites of concern and they must be estimated. Another complication is making a reduction in the value of ET when the calculations are for vegetated surfaces. While results of experiments to quantify reduction factors have not completely resolved the problem, there is evidence that the annual reduction factor is close to

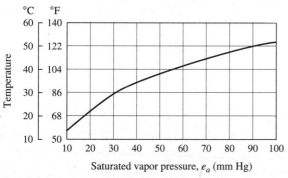

Figure 5.7 Relation between temperature and saturated vapor pressure.

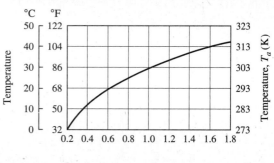

Figure 5.8 Temperature versus Δ relation for use with the Penman equation. (After Criddle.[23])

unity.[32-34] Thus, unless there is evidence to support another value, it appears that using a value of 1 for the reduction coefficient may give satisfactory results for surfaces having varied vegetal covers. Accordingly, any estimate of free water evaporation could be used to estimate *ET*, providing it is modified by an appropriate reduction coefficient.

EXAMPLE 5.4

Using the Penman Method, Eqs. 5.27 to 5.29, estimate *ET*, given the following data: temperature at water surface = 20 degrees C, temperature of air = 30 degrees C, relative humidity = 40 percent, wind velocity = 2 mph (48 mi/day), the month is June at latitude 30 degrees north, *r* is given as 0.07, and *S* is found to be 0.75.

Solution

1. Given the data for temperature, the values of e_a and e_d can be determined. Using Fig. 5.7 or Appendix Table A.2, the saturated vapor pressures are found to be 17.53 and 31.83 mm Hg respectively. Thus e_a = 31.83, and for a relative humidity of 40 percent, e_d = 31.83 × 0.4 = 12.73. Then, using Eq. 5.28,

$$E = 0.35(31.83 - 12.73)(1 + 0.0098 \times 48)$$
$$E = 9.83 \text{ mm/day}$$

2. The value of Δ is found using Fig. 5.8; for the given latitude and month, *R* is obtained from Table 5.6; and *B* is gotten from Table 5.7 for a temperature of 30°C. The values found are Δ = 1.0, *R* = 16.5, and *B* = 17.01. Then, using Eq. 5.29,

$$H = 16.5(1 - 0.07)(0.18 + 0.55 \times 0.75)$$
$$- 17.01(0.56 - 0.092 \times 12.73^{0.5})(0.10 + 0.90 \times 0.75)$$
$$H = 6.04 \text{ mm/day}$$

TABLE 5.6 TABULATED VALUES OF R, MEAN MONTHLY INTENSITY OF SOLAR RADIATION ON A HORIZONTAL SURFACE,[a] FOR USE IN THE PENMAN EQUATION

	Latitude (deg)	J	F	M	A	M	J	J	A	S	O	N	D
North	60	1.3	3.5	6.8	11.1	14.6	16.5	15.7	12.7	8.5	4.7	1.9	0.9
	50	3.6	5.9	9.1	12.7	15.4	16.7	16.1	13.9	10.5	7.1	4.3	3.0
	40	6.0	8.3	11.0	13.9	15.9	16.7	16.3	14.8	12.2	9.3	6.7	5.5
	30	8.5	10.5	12.7	14.8	16.0	16.5	16.2	15.3	13.5	11.3	9.1	7.9
	20	10.8	12.3	13.9	15.2	15.7	15.8	15.7	15.3	14.4	12.9	11.2	10.3
	10	12.8	13.9	14.8	15.2	15.0	14.8	14.8	15.0	14.9	14.1	13.1	12.4
	0	14.5	15.0	15.2	14.7	13.9	13.4	13.5	14.2	14.9	15.0	14.6	14.3
South	10	15.8	15.7	15.1	13.8	12.4	11.6	11.9	13.0	14.4	15.3	15.7	15.8
	20	16.8	16.0	14.6	12.5	10.7	9.6	10.0	11.5	13.5	15.3	16.4	16.9
	30	17.3	15.8	13.6	10.8	8.7	7.4	7.8	9.6	12.1	14.8	16.7	17.6
	40	17.3	15.2	12.2	8.8	6.4	5.1	5.6	7.5	10.5	13.8	16.5	17.8
	50	17.1	14.1	10.5	6.6	4.1	2.8	3.3	5.2	8.5	12.5	16.0	17.8
	60	16.6	12.7	8.4	4.3	1.9	0.8	1.2	2.9	6.2	10.7	15.2	17.5

[a]Measured in mm H_2O evaporated per day.
Source: After Criddle.[23]

TABLE 5.7 VALUES OF TEMPERATURE–DEPENDENT
COEFFICIENT B FOR USE IN THE PENMAN
EQUATION

T_a (K)	B (mm H_2O/day)	T_a (°F)	B (mm H_2O/day)
270	10.73	35	11.48
275	11.51	40	11.96
280	12.40	45	12.45
285	13.20	50	12.94
290	14.26	55	13.45
295	15.30	60	13.96
300	16.34	65	14.52
305	17.46	70	15.10
310	18.60	75	15.65
315	19.85	80	16.25
320	21.15	85	16.85
325	22.50	90	17.46
		95	18.10
		100	18.80

Source: After Criddle.[23] Note that $B = \sigma T_a^4$ where σ is the Boltzmann
constant, 2.01×10^{-9} mm/day.

3. Using Eq. 5.27,

$$ET = (1.0 \times 6.04 + 0.27 \times 9.83)/(1 + 0.27)$$
$$ET = 6.85 \text{ mm/day}$$

Thus the estimated evapotranspiration is 6.85 mm/day. ■■

Simulating Evapotranspiration

The volume of water evaporated or transpired from a watershed over time can be
substantial. Accordingly, continuous hydrologic modeling processes should incorporate an ET component. The models given in this chapter typify such an approach (see
also the flow chart of Fig. 1.3). References abound on this subject.[28,29,35,36,37]

■ Summary

Figure 1.1 and Table 5.1 show the overall importance of ET in the hydrologic budget.
In many regions of the United States, annual ET exceeds annual precipitation by a
significant amount. As a result, plans for water resources development and use must
incorporate estimates of ET losses. Where irrigated agriculture is practiced, these
estimates are especially important.

A number of approaches to estimating ET have been developed. They generally
fall into the following classes: theoretical, based on the physics of the process; analytical, based on energy or water budgets; and empirical, based on observations. Equations used in making ET calculations are usually of the type illustrated by Eqs. 5.1,
5.8, 5.10, 5.19, 5.22, and 5.26.

PROBLEMS

5.1. An 8000-mi^2 watershed received 20 in. of precipitation in a 1-year period. The annual streamflow was recorded as 5000 cfs. Roughly estimate the combined amounts of water evaporated and transpired. Qualify your answer.

5.2. Find the daily evaporation from a lake during which the following data were obtained: air temperature 90°F, water temperature 60°F, wind speed 20 mph, and relative humidity 30 percent.

5.3. Find the monthly consumptive use of an alfalfa crop when the mean temperature is 70°F, the average percentage of daytime hours for the year is 10, and the monthly consumptive use coefficient is 0.87.

5.4. During a given month a lake having a surface area of 350 acres has an inflow of 20 cfs, an outflow of 18 cfs, and a total seepage loss of 1 in. The total monthly precipitation is 1.5 in. and the evaporation loss is 4.0 in. Estimate the change in storage.

5.5. What are two methods that might be used to reduce evaporation from a small pond?

5.6. Compute the daily evaporation from a Class A pan if the amounts of water required to bring the level to the fixed point are as follows:

Day	1	2	3	4	5
Rainfall (in.)	0	0.65	0.12	0	0.01
Water added (in.)	0.29	0.55	0.07	0.28	0.10
Evaporation					

5.7. For Problem 5.6, the pan coefficient is 0.70. What is the lake evaporation (in inches) for the 5-day period for a lake with a 250-acre surface area?

5.8. The pan coefficient for a Class A evaporation pan located near a lake is 0.7. A total of 0.50 in. of rain fell during a given day. Determine the depth of evaporation from the lake during the same day if 0.3 in. of water had to be added to the pan at the end of the day in order to restore the water level to its original value at the beginning of the day.

5.9. A 2500 mi^2 drainage basin receives 25 in./yr rainfall. The discharge of the river at the basin outlet is measured at an average of 650 cfs. Assuming that the change in storage for the system is essentially zero, estimate the *ET* losses for the area in inches and cm for the year. State your assumptions.

5.10. Determine the daily evaporation from a lake for a day during which the following mean values were obtained: air temperature 78°F; water temperature 62°F; wind speed, 8 mph; and relative humidity, 45 percent.

5.11. Using the Meyer and Dunne equations, find the daily evaporation rate for a lake given that the mean value for air temperature was 80°F, for water temperature 60°F, the average wind speed was 10 mph, and the relative humidity was 25 percent. Refer to Appendix Table A.2 for vapor pressure values.

5.12. Determine the seasonal consumptive use of truck crops grown in Pennsylvania if the mean monthly temperatures for May, June, July, and August are 62, 71, 76, and 75°F respectively and the percent daylight hours for the given months are 10.02, 10.1, 10.3, and 9.6 as percent of the year respectively.

5.13. Using the Penman Method, Eqs. 5.27 to 5.29, estimate *ET,* given the following data: temperature at water surface = 20 degrees C, temperature of air = 32 degrees C, relative humidity = 45 percent, wind velocity = 3 mph, the month is June at latitude 30 degrees north, *r* is given as 0.08, and *S* is found to be 0.73.

REFERENCES

1. Water Resources Council, *The Nation's Water Resources: 1975–2000*, U.S. Govt. Print. Off., Washington, D.C., 1978.
2. U.S. Department of Commerce, National Oceanic and Atmospheric Administration, National Weather Service, NOAA Technical Report 33, *Evaporation Atlas of the Contiguous 48 United States,* Washington, D.C., June 1982.
3. U.S. Department of Commerce, National Oceanic and Atmospheric Administration, National Weather Service, NOAA Technical Report NWS 34, "Mean, Monthly Seasonal, and Annual Pan Evaporation for the United States," Washington, D.C., June 1982.
4. "Water-Loss Investigations," Vol. 1, Lake Hefner Studies, U.S. Geologic Survey Professional Paper No. 269 (1954). (Reprint of U.S. Geological Survey Circ. 229, 1952.)
5. N. N. Gunaji, "Evaporation Investigations at Elephant Butte Reservoir, New Mexico," *Int. Assoc. Sci. Hydrol. Pub.* **18,** 308–325(1968).
6. I. S. Bowen, "The Ratio of Heat Losses by Conduction and by Evaporation from Any Water Surface," *Phys. Rev.* **27,** 779–787(1926).
7. E. R. Anderson, L. J. Anderson, and J. J. Marciano, "A Review of Evaporation Theory and Development of Instrumentation," Lake Mead Water Loss Investigation; Interim Report, Navy Electronics Lab. Rept. No. 159 (Feb. 1950).
8. O. G. Sutton, "The Application to Micrometeorology of the Theory of Turbulent Flow over Rough Surfaces," *R. Meteor. Soc. Q. J.* **75**(No. 236), 335–350(Oct. 1949).
9. A. F. Meyer, "Evaporation from Lakes and Reservoirs," Minnesota Resources Commission, St. Paul, June 1944.
10. T. Dunne and L. B. Leopold, *Water in Environmental Planning,* San Francisco: Freeman and Co., 1978.
11. V. M. Ponce, *Engineering Hydrology: Principles and Practices.* Englewood Cliffs. New Jersey: Prentice Hall, 1989.
12. M. A. Kohler, T. J. Nordenson, and W. E. Fox. "Evaporation from Pans and Lakes," U.S. Department of Commerce, Weather Bureau, Res. Paper No. 38, Washington, D.C., 1955.
13. H. T. Haan, H. P. Johnson, and D. L. Brakensiek (eds.), *Hydrologic Modeling of Small Watersheds*, ASAE Monograph No. 5. St. Joseph, MI: American Society of Agricultural Engineers, 1982.
14. R. K. Linsley, M. A. Kohler, and J. L. H. Paulhus, *Hydrology for Engineers,* 3rd ed. New York: McGraw-Hill, 1982.
15. H. L. Penman, "Natural Evaporation from Open Water, Bare Soil, and Grass," *Proc. R. Soc. London Ser. A* **193**(1032), 120–145(Apr. 1948).
16. F. G. Millar, "Evaporation from Free Water Surfaces," Canada Department of Transport, Division of Meteorological Services, *Can. Meteor. Mem.* vol. 1, No. 2, 1937.
17. J. B. Franzini, "Evaporation Suppression Research," *Water and Sewage Works* (May 1961).
18. Victor K. La Mer, "The Case for Evaporation Suppression," *Chem. Eng.* (June 10, 1963).
19. D. R. Maidment (ed.), *Handbook of Hydrology.* New York: McGraw-Hill, 1993.
20. H. F. Blaney, "Water and Our Crops," in *Water, the Yearbook of Agriculture.* Washington, D.C.: U.S. Department of Agriculture, 1955.

21. H. F. Blaney, "Monthly Consumptive Use Requirements for Irrigated Crops," *Proc. ASCE, J. Irrigation Drainage Div.* **85**(IR1), 1–12(Mar. 1959).

22. Ven Te Chow (ed.), *Handbook of Applied Hydrology*. New York: McGraw-Hill, 1964.

23. W. D. Criddle, "Methods of Computing Consumptive Use of Water." *Proc. ASCE J. Irrigation Drainage Div.* **84**(IR1), 1–27(Jan. 1958).

24. H. L. Penman, "Estimating Evaporation," *Trans. Am. Geophys. Union* **37**(1), 43–50(1956).

25. N. J. Roberts, "Reduction of Transpiration," *J. Geophy. Res.* **66**(10), 3309–3312(Oct. 1961).

26. E. H. Hughes, "Research on Control of Phreatophytes," Proc. Ninth Annual Water Conference, New Mexico State University, Las Cruces, Mar. 1964.

27. R. K. Linsley, Jr., M. A. Kohler, and J. L. H. Paulhus, *Hydrology for Engineers*. New York: McGraw-Hill, 1958.

28. Staff, Hydrologic Research Laboratory, "National Weather Service River Forecast System Forecast Procedures," NWS HYDRO 14. U.S. Department of Commerce, Washington, D.C., Dec. 1972.

29. H. N. Holtan, G. J. Stiltner, W. H. Henson, and N. C. Lopez, "USDAHL-74 Revised Model of Watershed Hydrology," ARS Tech. Bull. No. 1518. U.S. Department of Agriculture, Washington, D.C., 1975.

30. N. J. Rosenberg, H. E. Hart, and K. W. Brown, "Evapotranspiration—Review of Research," MP 20, Agricultural Experiment Station, University of Nebraska, Lincoln, 1968.

31. R. H. McCuen, *Hydrologic Analysis and Design*. Englewood Cliffs, New Jersey: Prentice Hall, 1989.

32. W. O. Pruitt and F. J. Lourence, "Correlation of Climatological Data with Water Requirements of Crops." University of California Water Science Eng. Paper 9001. Davis, June 1968.

33. C. H. M. Van Bavel, "Potential Evaporation: The Combination Concept and Its Experimental Verification," *Water Resources Res.* **2**, 455–467(1966).

34. H. F. Blaney, "Discussion of Paper by H. L. Penman, 'Estimating Evaporation,'" *Trans. Am. Geophys. Union* **37,** 46–48(Feb. 1956).

35. N. H. Crawford and R. K. Linsley, Jr., "Digital Simulation in Hydrology: Stanford Watershed Model IV," Tech. Rept. 39, Department of Civil Engineering, Stanford University, July 1966.

36. W. C. Huber, J. P. Heaney, S. J. Nix, R. E. Dickinson, and D. J. Polmann, *Storm Water Management Model User's Manual, Version III,* EPA-600/2-84-109a (NTIS PB84–198423). Cincinnati, OH: Environmental Protection Agency, Nov. 1981.

37. L. A. Roesner, R. P. Shubinski, and J. A. Aldrich, *Storm Water Management Model User's Manual, Version III: Addendum I, Extran,* EPA-600/2-84-109b (NTIS PB84-198431). Cincinnati, OH: Environmental Protection Agency, Nov. 1981.

Chapter 6

Streamflow

■ Prologue

The purpose of this chapter is to:

- Introduce the concept of streamflow.
- Describe the characteristics of a hydrograph.
- Present approaches to measuring streamflow.

The amount of water flowing in surface water courses at any instant of time is small in terms of the earth's total water budget, but it is of considerable importance to those concerned with water resources development, supply, and management. A knowledge of the quantity and quality of streamflows is a requisite for: municipal, industrial, agricultural, and other water supply endeavors; flood control; reservoir design and operation; hydroelectric power generation; water-based recreation; navigation; fish and wildlife management; drainage; the management of natural systems such as wetlands; and water and wastewater treatment.

Streamflow is generated by precipitation during storm events and by groundwater entering surface channels. During dry periods, streamflows are sustained by groundwater discharges. Where groundwater reservoirs are below stream channels—often the case in arid regions—streams cease to flow during protracted precipitation-free periods. Relations between precipitation and streamflow are complex, being influenced by the factors discussed in the foregoing chapters. As a result, many approaches to relating these important hydrologic variables have been developed.[1-3] Several of them are discussed in detail in Part Three of the text.

Field measurements of streamflow are based on the use of flow-measuring devices such as weirs and flumes and on the measurement of channel cross-sections along with streamflow velocities (see Chapter 8).

6.1 DRAINAGE BASIN EFFECTS

The quality and quantity of streamflow generated in a drainage basin are affected by the basin's physical, vegetative, and climatic features.[4-9] Accordingly, it is important that the hydrologist have a good understanding of the soils, rocks, plants, topography, land-use patterns, and other basin characteristics that influence the sequence of events separating precipitation and runoff. It should be pointed out, however, that while natural basin features are very important elements in the runoff process, land-use features created by humans (e.g., housing developments, parking lots, agricultural patterns) may, in some cases, be the dominant ones. Land management practices can be beneficial, such as in retarding erosion, and they can also be detrimental when they function to accelerate natural hydrologic processes. In Chapter 10, the principal basin characteristics of concern to the hydrologist are discussed.

6.2 THE HYDROGRAPH

Streamflow, at a given location on a water course, is represented by a hydrograph. This continuous graph displays the properties of streamflow with respect to time, normally obtained by means of a continuous recorder that shows stage (depth) versus time (stage hydrograph), and is then transformed into a discharge hydrograph by application of a rating curve. In general, the term *hydrograph* as used herein means a discharge hydrograph.

As was shown in Fig. 1.3, the hydrograph produced in a stream is the result of various hydrologic processes that occur during and after any precipitation event. A more complete discussion of these processes is given in Chapter 11. A hydrograph has four component elements: (1) direct surface runoff, (2) interflow, (3) groundwater or base flow, and (4) channel precipitation. The rising portion of a hydrograph is known as the *concentration curve;* the region in the vicinity of the peak is called the *crest segment;* and the falling portion is the *recession.*[10] The shape of a hydrograph depends on precipitation pattern characteristics and basin properties. Figure 6.1 illustrates the definitions presented.

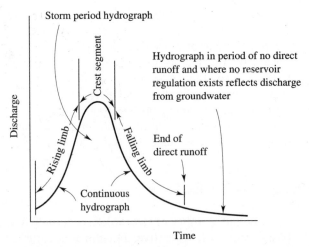

Figure 6.1   Hydrograph definition.

6.3 UNITS OF MEASUREMENT FOR STREAMFLOW

Two types of units are used in measuring water flowing in streams. They are units of discharge and units of volume. Discharge, or rate of flow, is the volume of water that passes a particular reference point in a unit of time. The basic units used in connection with stream gauging in the United States are the foot and meter for measurements of dimension and the second for measurements of time. Commonly used units of discharge measurement are cubic feet per second (cfs) and cubic meters per second (m³/sec). Other units of discharge in use are second-foot per square mile (sec-ft/mi²), for expressing the average rate of discharge from a drainage basin or defined area, and million gallons per day (mgd), commonly used in water supply calculations. Units of volume used are the cubic foot, cubic meter, liter, gallon, and acre-foot (a volume equivalent to 1 ft of water over an acre, 43,560 ft², of land). The latter unit is commonly used in irrigation practice in the western United States. Irrespective of whether English or metric units are used for dimensions, the standard unit of time for streamflow observations is the second.

6.4 MEASURING AND RECORDING STREAMFLOW

Streamflow rates may be determined using gauging devices such as flumes, weirs, and control sections, or they may be calculated from measurements of head (depth), velocity, and cross-sectional area.[1-3] Usually, specific devices such as flumes are

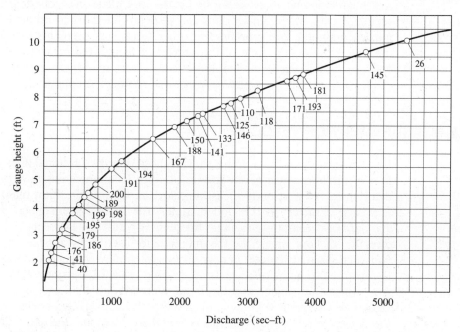

Figure 6.2   Station rating curve for Raquette River at Piercefield, New York. (U.S. Geological Survey.)

limited to small streams because of problems of scale. For large stream systems, discharge is normally estimated by measuring velocity and using the cross-sectional area to translate this measurement into discharge. Where flow-measuring devices are used, it is customary to observe the head and use a rating curve to translate this into discharge. When direct flow measurements cannot be made, discharge calculations are often facilitated by use of velocity–area relations, chemical tracers, electrical methods, and empirical equations such as the Manning formula.

In the United States, the primary responsibility for gauging major streams lies with the U.S. Geological Survey, with a systematic record of streamflow in terms of mean daily discharge being the norm. Usually, a stage recording is obtained at a gauging site and this record is converted into discharge by one or more of several methods. Rating curves, tables, and formulas are used for this purpose (see Chapter 11). Figure 6.2 shows a typical stage–discharge rating curve. The instruments and methods used to convert stage recordings to discharge must be carefully adapted to the natural or artificial conditions encountered at the gauging site so as to ensure the reliability of conversions.[1]

6.5 MEASUREMENTS OF DEPTH AND CROSS-SECTIONAL AREA

Unless a direct flow-measuring device can be installed in a stream, a rarity for streams of any scale, measurements of depth of flow and cross-sectional area will be needed to permit discharge to be calculated. Depth measurements may be taken using weighted sounding lines, calibrated rods, and ultrasonic sounding devices. Cross-sectional areas at stream sections can be determined using ordinary surveying techniques combined with soundings or other depth measurements that are taken below the water level at the time of the survey. Although the measurement of depth and cross-sectional area seems simple, accurate determinations require careful calibration of instruments and the ability to deal with submerged conditions.

6.6 MEASUREMENT OF VELOCITY

Velocity measurements, combined with those of cross-sectional area, permit calculation of discharge at a given stream or river location. Point flow velocities can be determined using velocity-measuring devices such as the Pitot tube, dynamometer, and current meter. In the United States, the Price current meter has long been a standard in streamflow gauging. This device operates by exposing cupped vanes to the direction of flow, much like the anemometer used in measuring wind velocity. The cup-vane assembly rotates in near proportion to flow velocity and the rate of rotation is converted to point velocity using a rating table or appropriate equation.

Various chemical and electrical methods are also employed in determining velocities. Commonly used chemical methods include salt velocity, salt dilution, and the detection of radioactive tracers. Of these methods, the salt velocity method is perhaps the most widely used. It is based on the principle that salt introduced into the stream will increase its electrical conductivity. Electrodes placed downstream of the

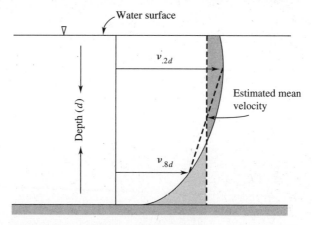

Figure 6.3   Vertical velocity profile

salt injection sense conductivity changes and these can be translated into velocity by knowing the spacing of electrodes and keeping track of time. Electrical methods include the use of hot wire anemometers, electromagnetic voltage generation, oxygen polarography, and supersonic waves.[4] Streamflow conditions vary widely and thus no specific approach to velocity determination is universally suitable. The best method for use at a given site must be determined on the basis of the characteristics of streamflow at that site.

Field observations have shown that the mean velocity in a vertical stream section is closely approximated by the average of the velocities occurring at depths of 20 percent and 80 percent of the total section depth respectively (see Fig. 6.3).[11, 12] Where depths are very shallow, on the order of 0.5 feet, single meter readings at about the 50 percent point have been shown to yield good results.[12] Velocity measurements and geometric definitions of stream channel cross-sections permit estimating channel flows at locations where velocity and depth measurements have been made.

6.7   RELATING POINT VELOCITY TO CROSS-SECTIONAL FLOW VELOCITY

While point velocity measurements are important, what is desired is a method to translate them into the average cross-sectional flow velocity. This average velocity, when multiplied by the cross-sectional area, yields the discharge at a given stream section. One procedure is to take point velocity measurements at numerous vertical and horizontal positions in a cross section, plot them, and then determine velocity contours. By calculating the areas between the contours and assigning the average of the flow velocities of the two confining contours to these areas, a determination of mean velocity can be made. Once this is accomplished, discharge is easily calculated.

Other approaches make use of the geometric properties of stream channel cross-sections. One such technique is the mean-section method. To use this approach, it is necessary to divide the stream channel cross-section at a gauging location into a

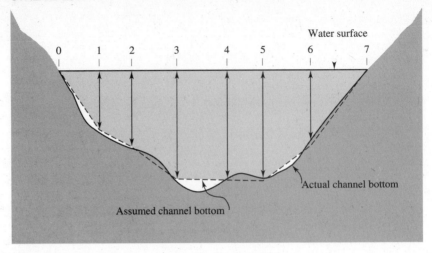

Figure 6.4   Channel cross-section for Example 6.1

series of geometric shapes (see Fig. 6.4). At each vertical location along the cross-section, the mean velocity is estimated from measurements. The average velocity of flow for the area between two verticals is considered to be equal to the average of the mean velocities for each of the bordering verticals. The discharge between two verticals is thus the average velocity for the section multiplied by the area of the section. The individual discharges are then summed to provide an estimated total flow for the channel at that location. Note that it is important to have enough measurements to characterize the cross-section. The procedure is illustrated in Example 6.1.

EXAMPLE 6.1

Calculate the discharge at the section given in Fig. 6.4. Data from field observations are shown in Tables 6.1 and 6.2.

TABLE 6.1   DATA FOR EXAMPLE 6.1

Vertical section #	Depth (ft.)	Avg. vel. (fps)
0	0	0
1	4	2.1
2	5	2.3
3	7.2	2.7
4	7.4	2.8
5	7.1	2.5
6	4.7	2.2
7	0	0

TABLE 6.2   DATA FOR EXAMPLE 6.1

Horizontal segment	Segment width–ft
0–1	4.2
1–2	3.3
2–3	4.8
3–4	5.2
4–5	3.7
5–6	5.1
6–7	5.9

TABLE 6.3   CALCULATIONS FOR
EXAMPLE 6.1

Area	Area (sq. ft.)	Vel. (fps)	Flow (cfs)
A1	8.40	1.05	8.82
A2	14.85	2.20	32.67
A3	29.28	2.50	73.20
A4	37.96	2.75	104.39
A5	26.83	2.65	71.09
A6	30.09	2.35	70.71
A7	13.87	1.10	15.25
	161.27		376.13

Total estimated discharge is 376.13 cfs.

Solution.   The first step is to calculate the individual section areas (A1, A2, etc.). The calculated areas (using triangular or trapezoidal formulas) are shown on spreadsheet Table 6.3. Next, the estimated mean velocities at the verticals are multiplied by the section areas to obtain the individual area discharges (see Table 6.3). These discharges are summed to yield an estimated 376.13 cfs of flow being delivered by the full channel width. ■■

6.8 THE SLOPE-AREA METHOD FOR DETERMINING DISCHARGE

In some cases it is difficult to make velocity or other measurements needed to determine discharge. This is often the case during large flood events. Under such circumstances, it is sometimes possible to estimate the flow by taking measurements of high water lines (after the flood event), cross-sectional areas, and channel slopes and then using these data in an equation such as Manning's to estimate the flow. The applicable Manning equation is

$$Q = (1.49/n)AR^{2/3}S^{1/2} \qquad (6.1)$$

where Q = discharge (cfs)
   n = Manning's roughness coefficient
   A = cross-sectional area (ft^2)
   R = the hydraulic radius
   S = the head loss per unit length of channel

For streamflows, Manning's n values may range between about 0.03 and 0.15. When reasonable determinations can be made of n, A, R, and S, Eq. 6.1 can be used to estimate the streamflow that occurred during the high-water period. For a more complete discussion of this and other streamflow determination methods, the references at the end of the chapter should be consulted. References 4 and 10 give an excellent overview of techniques and include a valuable list of references on streamflow.

■ Summary

Streamflow is the result of storm-period precipitation, snowmelt, and groundwater discharge.[13] It is a primary source of water for a host of instream and offstream uses. The graphical representation of streamflow is the hydrograph, a plot of flow versus time at a prescribed location along the water course of interest. As illustrated by Fig. 1.3, the end product of many hydrologic modeling processes is a hydrograph which is derived from a precipitation input, modified appropriately by various abstractions such as infiltration. Methods for measuring streamflow in the field were presented in this chapter. In later sections of the book, a variety of techniques for deriving hydrographs from precipitation and other hydrologic data are covered (see Part Three).

PROBLEMS

6.1. Consider that you have obtained a gauge height reading of 4 ft at a gauging site on the Raquette River (Fig. 6.2). What would you estimate the discharge to be in cfs and in m^3/sec? If the gauge height had been 9 ft, what would the discharge be? Which of the two estimates do you think would be the most reliable? Why?

6.2. Solve Problem 6.1 if the gauge height readings were 5 ft and 7 ft.

6.3. Consult a USGS Water Supply paper and plot the streamflow data for a drainage basin of interest. Discuss the factors that you believe influenced the shape of the hydrograph.

6.4. For the major surface water course in your locality, discuss the value of making streamflow forecasts.

6.5. Calculate the discharge at the section given in Fig. 6.4 if the depth measurements at the verticals were: 0, 3.8, 5.4, 7.7, 8.1, 7.0, 4.5, and 0 ft respectively. Give results in cfs and m^3/s.

6.6. Calculate the discharge at the section given in Fig. 6.4 if the velocities were: 0, 2.3, 2.6, 3.1, 2.9, 2.7, 2.5, and 0 fps respectively, and the depths of Problem 6.5 applied. Give results in cfs and m^3/s.

REFERENCES

1. N. C. Grover and A. W. Harrington, *Stream Flow*. New York: Wiley, 1943.
2. United States Department of the Interior, Bureau of Reclamation, *Water Measurement Manual*. Washington, D.C.: U.S. Government Printing Office, 1967.
3. I. E. Houk, "Calculation of Flow in Open Channels," State of Ohio, The Miami Conservancy District, Tech. Rept. Part IV, Dayton, OH, 1918.
4. Ven Te Chow (ed.), *Handbook of Applied Hydrology*. New York: McGraw-Hill, 1964.
5. American Society of Civil Engineers, "Hydrology Handbook," Manuals of Engineering Practice, No. 28, New York: ASCE, 1957.

6. O. E. Meinzer, *Hydrology*. New York: Dover, 1942.

7. A. N. Strahler, "Geology—Part II," *Handbook of Applied Hydrology*. New York: McGraw-Hill, 1964.

8. R. E. Horton, "Drainage Basin Characteristics," *Trans. Am. Geophys. Union* **13,** 350–361(1932).

9. W. B. Langbein et al., "Topographic Characteristics of Drainage Basins," U.S. Geological Survey, Water Supply Paper, 968-c, 1947.

10. R. K. Linsley, Jr., M. A. Kohler, and J. L. H. Paulhus, *Applied Hydrology*. New York: McGraw-Hill, 1949.

11. S. S. Butler, *Engineering Hydrology*. Englewood Cliffs, New Jersey: Prentice-Hall, Inc., 1957.

12. C. H. Pierce, "Investigation of Methods and Equipment Used in Stream Gauging," Water Supply Paper 868-A, U.S. Geological Survey, Washington, D.C.: Government Printing Office, 1941.

13. D. R. Maidment (ed.), *Handbook of Hydrology*. New York: McGraw-Hill, 1993.

PART TWO

HYDROLOGIC MEASUREMENTS AND MONITORING

Chapter 7

Hydrologic Data Sources

■ Prologue

The purpose of this chapter is to:

- Describe the principal sources of data for hydrologic investigations.

Data on hydrologic variables are fundamental to analyses, forecasting, and modeling. Such data may be found in numerous publications of state and federal agencies, research institutes, universities, and other oganizations. Several of the most significant sources of hydrologic data are described briefly in this chapter.[1-3]

7.1 GENERAL CLIMATOLOGICAL DATA

The most readily available sources of data on temperature, solar radiation, wind, and humidity are *Climatological Data* bulletins published by the Environmental Data Service of the National Oceanic and Atmospheric Administration (NOAA), and *Monthly Summary of Solar Radiation Data,* published by the National Climatic Data Center. The Environmental Data Service, in cooperation with the World Meteorological Organization (WMO), also publishes *Monthly Climatic Data for the World.* A 1968 publication of the Environmental Sciences Service Administration, entitled *Climatic Atlas of the United States,* summarizes wind, temperature, humidity, evaporation, precipitation, and solar radiation on a series of maps. In addition to these federal sources of data, state environmental, geologic, water resources, and agricultural agencies should be consulted. Most state universities also publish a variety of hydrologic data through their research centers and extension programs.

7.2 PRECIPITATION DATA

There are probably more records of precipitation than of most other hydrologic variables. The principal federal source of data on precipitation is NOAA. *Climatological Data,* published monthly and annually for each state or combination of states, the Pacific area, Puerto Rico, and the Virgin Islands by the Environmental Data Service, presents a table of monthly averages, departures from normal, and extremes of precipitation and temperature as well as tables of daily precipitation, temperature, snowfall, snow on ground, evaporation, wind, and soil temperature. *Hourly Precipitation Data* is issued monthly and annually for each state or combination of states and presents alphabetically by station the hourly and daily precipitation amounts for stations equipped with recording gauges. A station location map is also included. This publication is available from the Environmental Data Service. Another publication, *World Weather Records,* is issued by geographic regions for 10-year periods. Data are listed by country or area name, station name, latitude and longitude, and elevation. Monthly and annual mean values of station pressure, sea-level pressure, and temperature, and monthly and annual total precipitation are given in sequential order. Aside from NOAA, other federal and state agencies and universities publish precipitation data at varying intervals, often in a storm or site-specific context. In addition, many municipalities and water and wastewater utilities also collect and maintain precipitation and other related data. Computerized precipitation data are available from the National Climatic Data Center in Asheville, North Carolina.

7.3 STREAMFLOW DATA

The principal sources of streamflow data for the United States are the U.S. Geological Survey (USGS), U.S. Soil Conservation Service (SCS), U.S. Forest Service, and U.S. Agricultural Research Service (ARS). In addition, the U.S. Army Corps of Engineers (COE), the Tennessee Valley Authority (TVA), and the U.S. Bureau of Reclamation (USBR) make some streamflow measurements and tabulate streamflow data relative to their missions. State agencies, universities, and various research organizations also compile and publish a variety of streamflow data.

The USGS Water Supply Papers (WSP) are the benchmark for referencing streamflow data. Furthermore, computerized data are also available from the USGS. *Publications of the Geological Survey,* published every 5 years and supplemented annually, are an excellent source of information on that agency's reports. The SCS historically published data on streamflow from small watersheds and plots in its *Hydrologic Bulletin* series, but much of the data have been republished by ARS. Records from SCS "pilot watersheds" are published in cooperation with the USGS. U.S. Forest Service streamflow data are published at irregular intervals in various technical bulletins and professional papers.

7.4 EVAPORATION AND TRANSPIRATION DATA

Monthly and annual issues of *Climatological Data,* published by NOAA, include pan evaporation and related data. The ARS, agricultural colleges, and water utilities are other sources of information. In particular, data on evapotranspiration are often obtained by university researchers working through their Agricultural Experiment Stations.

■ Summary

Climatic and other data are keystones in hydrologic modeling processes. Numerous sources of data exist and may be accessed to support model development and verification, statistical analyses, and special studies.

PROBLEM

7.1 Develop a list of data sources in your state or locality by visiting the library or through other channels.

REFERENCES

1. Soil Conservation Service, U.S. Department of Agriculture, *SCS National Engineering Handbook,* "Hydrology", Sec. 4. Washington, D.C.: U.S. Government Printing Office, Aug. 1972.
2. J. F. Miller, "Annotated Bibliography of NOAA Publications of Hydrometeorological Interest," NOAA Tech. Mem. NWS HYDRO-22, National Oceanic and Atmospheric Administration, Washington, D.C., May 1975.
3. D. R. Maidment (ed.), *Handbook of Hydrology.* New York: McGraw-Hill, 1993.

Instrumentation

■ **Prologue**

The purpose of this chapter is to:

- Describe instruments used in measuring hydrologic variables.
- Indicate ways in which data are recorded and transmitted.
- Present limitations on measurements.

The data needed to support hydrologic analyses must be obtained in sufficient quantity, with adequate frequency, and in an appropriate form if they are to be of value. A variety of instruments are used to obtain and transmit the data. They are the subject of this chapter.

8.1 INTRODUCTION

Hydrologic instrumentation supports areal investigations, problem analyses, research, planning, and environmental policymaking and analysis. A host of measurements are needed to support efforts in water resources planning, management, design, and construction related to such subjects as aquifer systems analysis, solid waste management, flood hazard assessment, water supply availability, water quality management, groundwater recharge, protection of fish and wildlife, and navigation.

Historically, instruments were often used to obtain cumulative rather than continuous information about hydrologic variables such as rainfall and evaporation. Furthermore, there was often no attempt to correlate water quality constituent loadings, for example, with rates of water flow. Consequently, many historic data have limited utility, not so much because of lack of adequate instrumentation, but rather from using available instruments to measure the wrong thing or in too limiting a fashion. Today it is widely recognized that it is important not only to select appropriate instruments but to select them in the context of data networks that meet the needs of modern times. More will be said about this in Chapter 9. In Section 8.2, instruments for measuring hydrologic variables and ways in which they can be used jointly to create a complete representation of a functioning hydrologic system are discussed.

8.2 HYDROLOGIC INSTRUMENTS

Good sources of information about hydrologic instruments are the National Weather Service, U.S. Geological Survey, U.S. Bureau of Reclamation, U.S. Army Corps of Engineers, Soil Conservation Service, and instrument manufacturers. These agencies and industries have long been in the business of measuring hydrologic variables and they can provide detailed descriptions of state-of-the-art measuring devices. Some of the major types of measuring instruments are described here, but the coverage is far from exhaustive and the interested reader should consult the appropriate references.[1-3]

Precipitation

Gauges for measuring rainfall and snowfall may be recording or nonrecording. The most common nonrecording gauge is the U.S. Weather Service standard 8-in. gauge. The gauge may be read at any desirable interval but often this is daily. The gauge is calibrated so that a measuring stick, when inserted, shows the equivalent rainfall depth. Such gauges are useful when only periodic volumes are required, but they cannot be used to indicate the time distribution of rainfall.

Recording gauges continuously sense the rate of rainfall and its time of occurrence. These gauges are usually either of the weighing-recording type or the tipping bucket type. Weighing-type gauges usually run for a period of 1 week, at which time their charts must be changed. The figure associated with Problem 2.5 is typical of the recorded output. A mass curve of rainfall depth versus time is the product, and this curve can be translated into an intensity–time graph by calculating the ratios of accumulated rainfall to time for whatever time step is desired. Tipping bucket gauges, on the other hand, sense each consecutive rainfall accumulation when it reaches a prescribed amount, usually 0.01 in. or 1 mm of rain. A small calibrated bucket is located below the rainfall entry port of the gauge. When it fills to the 0.01-in. increment it tips over, bringing a second bucket into position. These two small buckets are placed on a swivel and the buckets tip back and forth as they fill. Each time a bucket spills it produces an indication on a strip chart or other recording form. In this way a record of rainfall depth versus time (intensity) is the outcome. For rain gauges to record snow accumulations, some modifications must be made. Usually these involve providing a melting agent so that the snow can be converted into measurable water.

Figure 8.1a is the diagram of a self-reporting rain gauging station. The tipping bucket mechanism generates a digital input signal whenever 1 mm of rainfall drains through the funnel assembly. The signal from the gauge is automatically transmitted to a receiving station where it records the station ID number and an accumulated amount of rainfall. The receiving station records the time at which the message was received and rainfall rates for desired periods can be calculated accordingly. Figure 8.1b shows a similar gauge equipped to measure snow. In this case, a glycometh solution is used to melt the snow. The melt water overflows through a temperature-compensating mechanism and is measured by the tipping bucket, which operates the station's transmitter. Gauges of the type shown in Fig. 8.1 can easily be incorporated into real-time monitoring systems that can be used in a variety of forecasting and operating modes.

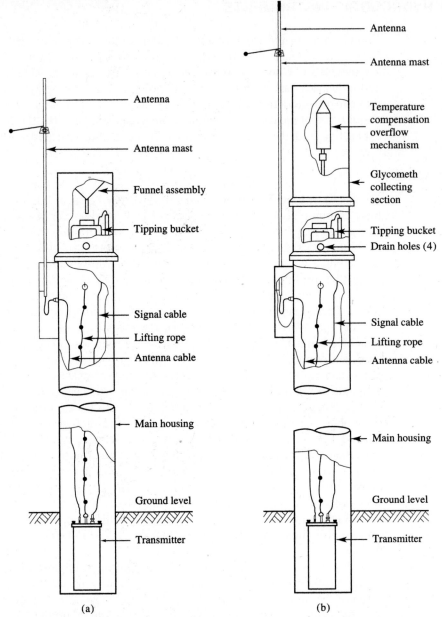

Figure 8.1 Self-reporting (a) rain and (b) snow stations. (Courtesy of Sierra-Misco, Inc., Environmental Products, Berkeley, CA.)

Evaporation and Transpiration

Evaporation pans have been widely used for estimating the amount of evaporation from free water surfaces. Devices such as that depicted in Fig. 8.2 are easy to use, but relating measurements taken from them to actual field conditions is difficult and the data they produce are often of questionable value for making areal estimates. A

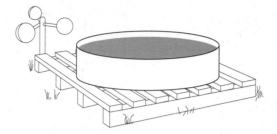

Figure 8.2 U.S. Weather Bureau Class A pan.

variety of pan types have been developed but the U.S. Weather Bureau Class A pan is the standard in the United States.[4] Pan evaporation observations have been used to estimate both free water (lake) evaporation and evapotranspiration from well-watered vegetation. Field experiments have shown a high degree of correlation of pan data with evapotranspiration from surrounding vegetation when there is full cover and good water supply.[4] As in the case of precipitation gauges, pan data can be recorded and transmitted continuously to a central receiving station.

Evapotranspiration measurements are often made using lysimeters. These devices are containers placed in the field and filled with soil, on which some type of vegetative growth is maintained. The object is to study soil–water–plant relations in a natural surrounding. The main feature of a weighing lysimeter is a block of undisturbed soil, usually weighing about 50 tons, encased in a steel shell that is 10 ft by 10 ft by 8 ft. The lysimeter is buried so that only a plastic border marks the top of the contained soil. The entire block of soil and the steel casing are placed on an underground scale sensitive enough to record even the movement of a rabbit over its surface. The soil is weighed at intervals, often every 30 min around the clock, to measure changes in soil water level. The scales are set to counterbalance most of the dead weight of the soil and measure only the active change in weight of water in the soil.[5] The scales can weigh accurately about 400 g (slightly under 1 1b), which is equivalent to 0.002 in. of water. The weight loss from the soil in the lysimeter represents water used by the vegetative cover plus any soil evaporation. Added water is also weighed and thus an accounting of water content can be kept. Crops or cover are planted on the area surrounding the lysimeter to provide uniformity of conditions surrounding the instrument. Continuous records at the set weighing intervals provide almost continuous monitoring of conditions. The data obtained can be transmitted to any desirable location for analysis and/or other use. Weighing lysimeters can produce accurate values of evapotranspiration over short periods of time. But they are expensive. Nonweighing types of lysimeters, which are less costly, have also been used, but unless the soil moisture content can be measured reliably by some independent method, the data obtained from them cannot be relied on except for long-term measurements such as between precipitation events.[5]

Wind, Temperature, and Humidity

Measurements of wind, temperature, and humidity are needed to support many types of hydrologic analyses. Wind is commonly measured using an anemometer, a device that has a wind-propelled element such as a cup (Fig. 8.2) or propeller whose speed is calibrated to reflect wind velocity. Wind direction is obtained using a vane, which orients itself with the direction of the wind.

Temperature measurements are made using standard thermometers of various types, while humidity is measured using a psychrometer. A psychrometer consists of two thermometers, one called a wet bulb, the other a dry bulb. Upon ventilation the thermometers measure differently, and this difference is called the wet-bulb depression. By using appropriate tables, dew point, vapor pressure, and relative humidity can be determined.[6]

Figure 8.3 depicts a complete weather station incorporating measurements of precipitation, wind, temperature, barometric pressure, and humidity. Such a station can automatically report weather data from remote sites on either an event and/or

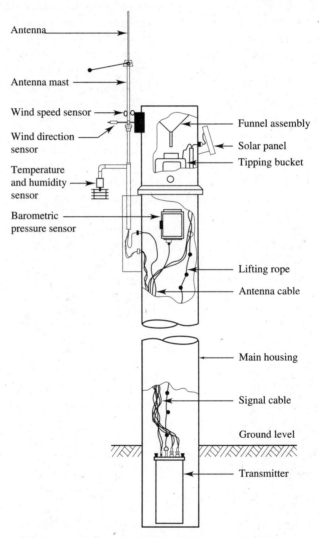

Figure 8.3   Self-reporting weather station. (Courtesy of Sierra-Misco, Inc., Environmental Products, Berkeley, CA.)

timed basis to a central site. A station such as this can be used for marine weather forecasting, quantitative determinations of oncoming storms, determination of wind effect on tidal areas, and establishing a data base for irrigation.

Open Channel Flow

Measurements of open channel (natural and created) flow are made using standard measuring devices such as flumes and weirs, and they are also made by calibrating special control sections along rivers and streams such that measurements of depth (stage) of flow can be related to discharge. Flow-measuring devices are designed so that sensing some parameter such as depth automatically translates the observation into units of flow (discharge). When a control section is used, observations of cross-sectional area for various depths must be obtained, and average flow velocities must be ascertained for various stages so that a section rating curve can be established. In the United States, the U.S. Geological Survey, the U.S. Bureau of Reclamation, the Soil Conservation Service, and the U.S. Army Corps of Engineers have done extensive flow measuring and have been active in developing instruments and procedures for ascertaining rates of flow.[2,6]

Weirs Weirs are common water-measuring devices. When they are properly installed and maintained they can be a very simple and accurate means for gauging discharge. The most often used weir types are the rectangular weir and the V-notch weir (Fig. 8.4). To be effective, weirs usually require a fall of about 0.5 ft or more in the channel in which they are placed. Basically, a weir is an overflow structure placed across an open channel. For a weir of specific size and shape with free-flow steady-state conditions and a proper weir-to-pool relation, only one depth of water can exist in the upstream pool for a particular discharge. Flow rate is determined by measuring the vertical distance from the crest of the overflow part of the weir to the water surface in the upstream pool. The weir's calibration curve then translates this recorded depth into rate of flow at the device.

Parshall Flumes A Parshall flume is a specially shaped open channel flow section that can be installed in a channel section. Figure 8.5 depicts one of these devices. The flume has several major advantages: (1) it can operate with a relatively small head loss; (2) it is fairly insensitive to the approach velocity; (3) it can be used even under submerged conditions; and (4) its flow velocity is usually sufficient to preclude sediment deposits in the structure.[2] The Parshall flume was developed by the late Ralph L. Parshall and it is a particular form of venturi flume. The constricted throat of the flume produces a differential head that can be related to discharge. Thus, as in the case of the weir, an observation of depth (head) is all that is required to determine the rate of flow at the control point. Weirs and flumes are generally best suited to gauging small streams and open channels, although large broad-crested weirs can be installed at dam sites as part of overflow structures. For major rivers, other measuring approaches such as developing field ratings at a specified control section must be relied on.

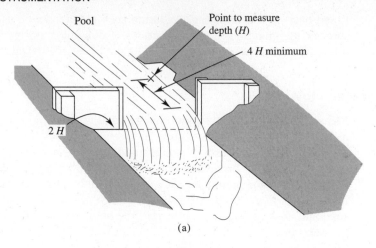

(a)

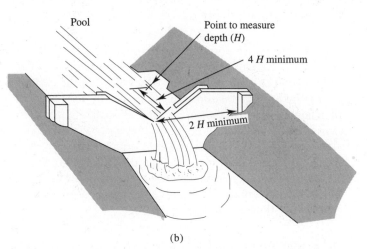

(b)

Figure 8.4 Field installation of weirs: (a) rectangular and (b) V-notch. USDA Cooperative Extension Service, Mountain States Area.

Control Sections   Where the installation of a weir, flume, or some other flow-measuring device is impractical, it is sometimes possible to develop a rating curve at some location along a stream by taking measurements of depth, cross-sectional area, and velocity and calculating the rate of flow for a particular stage at the location. By doing this for a range of depths of flow, a station rating curve can be developed. Instruments required to develop such a curve are depth-sensing devices, surveying instruments, and velocity meters. The velocity meter is similar to an anemometer. It is placed at various positions in the channel and a velocity is recorded. By doing this at a number of locations, a velocity profile for a given depth can be developed. From this an average flow velocity can be computed, and by using that determination and

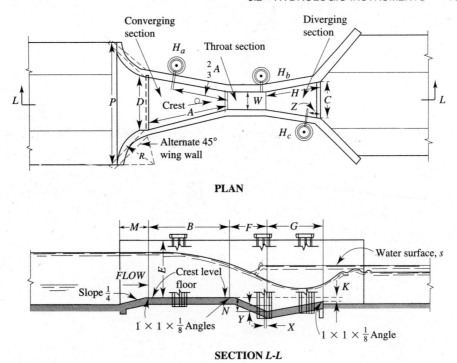

PLAN

SECTION *L-L*

Figure 8.5 Parshall flume. (U.S. Soil Conservation Service.)

the cross-sectional area, discharge can be calculated as the product of mean velocity and cross-sectional area. If observations can be made for a range of depths, a rating curve can be developed for the control section so that only measurements of depth will be needed to estimate rate of flow at some later time. Additional information on this procedure may be found in Refs. 2 and 6.

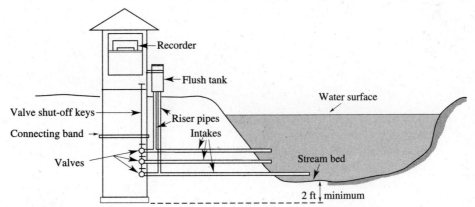

Figure 8.6 Recorder house and stilling well for a stream gauging station. (U.S. Bureau of Reclamation.)

Depth (Stage) Measurements   Most depth measurements are made using a float and cable arrangement in a stilling well or a bubbler gauge. In the first instance, a stilling well connected to the channel (Fig. 8.6) is used to house a float device that activates a recorder as it moves up and down. Figure 8.7 illustrates a self-reporting stilling well liquid-level station. Data from this station can be transmitted to any central location for analysis and/or other use. A bubbler-type installation makes use of dry air or nitrogen as a fluid for bubbling through an orifice into a channel bed. As the depth of flow changes, the change in head above the bubbler orifice causes a

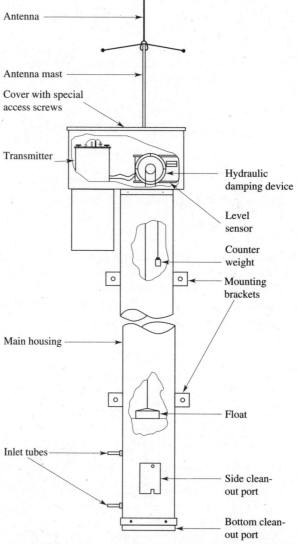

Figure 8.7   Self-reporting stilling well liquid-level station. (Courtesy of Sierra-Misco, Inc., Environmental Products, Berkeley, CA.)

corresponding pressure change. This results in a fluid-level change in the manometer connected to the gas supply and this in turn is used to reflect stage variation over time.

The foregoing descriptions are of a few of the instruments used in hydrologic work. Both the limitations associated with their use and their reliability must be understood if they are to be used correctly and their outputs are to be considered credible.

8.3 TELEMETRY SYSTEMS

Historically, many gauges were read periodically by an individual making the rounds of installations. This served well when the purpose of the data was to establish a base record of some variable such as rainfall. But today, under many circumstances, it has become necessary to make continuous recordings of rainfalls, streamflows, and evaporation rates and to have these data available for the real-time operation of water management systems and for forecasting hydrologic events. Some examples of activities requiring real-time hydrologic data are managing reservoirs, issuing flood warnings, allocating water for various uses such as irrigation, monitoring streamflows to ensure that treaties and pacts are honored, and monitoring the quality and quantity of water for regulatory and environmental purposes. Accordingly, gauging stations capable of electronically transmitting their data to a central location for immediate use have now become common. The advantages of such stations include providing information to users in a time frame that meets management needs, reducing the costs of collecting data, and providing a continuous and synchronous record of hydrologic events. Figure 8.8 shows a stream gauge reporting station using radio transmission. Figure 8.9 illustrates a satellite data collection and transmitting operation.[7-12]

8.4 REMOTE SENSING

Since the 1960s, remote sensing has become a common hydrologic tool. Examples of aircraft and satellite data collection and transmission abound.[13-16] Figure 8.10 illustrates the use of aircraft and satellites in a snow survey system. Other types of surveys such as those related to determining impervious areas, classifying land uses for assessing basin-wide runoff indexes, determining lake evaporation, and groundwater prospecting can be depicted in similar fashion. Table 8.1, which summarizes operational uses of satellite data in hydrology circa 1981, shows the great diversity of remote sensing and data transmission options that can be exercised.[16]

The principal value of remote sensing is its ability to provide regional coverage and at the same time provide point definition. Furthermore, satellite communications can be digitized and are thus compatible with the transfer of computerized information. Following the evolution of linkages between computer and communications technology, new software systems incorporating powerful data management systems have been developed. These systems facilitate the storage, compaction, and random access of large data banks of information. One data management option, geographic information systems (GIS), allows the overlaying of many sets of data (particularly satellite-derived data) for convenient analysis. Versatile color pictorial and graphic display systems are also becoming attractive as their costs have decreased.[14]

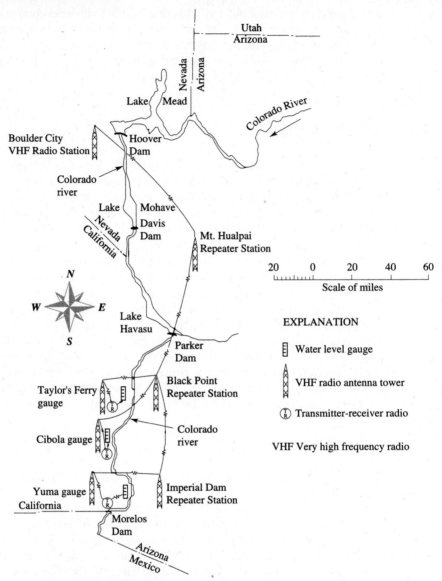

Figure 8.8 Stream gauge reporting system using radio transmission. Water stage information is requested from the gauging stations by VHF radio signal. In turn, this water stage information is obtained from the stream gauges and automatically encoded and transmitted to the Boulder City receiving station. All downstream releases from Hoover Dam are determined and integrated with this streamflow information in controlling the flow of the lower Colorado River. (U.S. Bureau of Reclamation.)

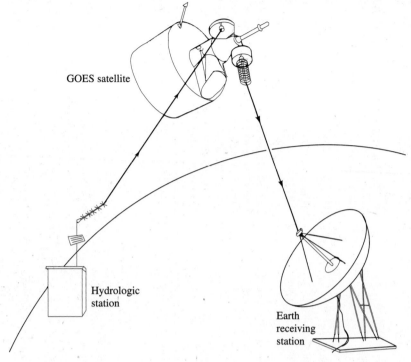

Figure 8.9   Hydrologic data collection by satellite. (U.S. Geological Survey.)

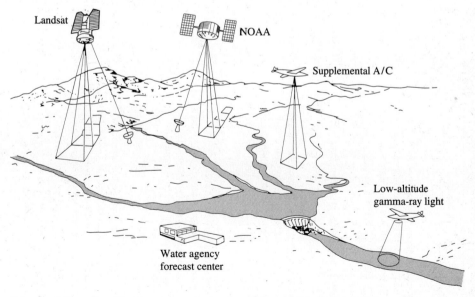

Figure 8.10.   Satellite snow survey system. (After Calabrese and Thome.[15])

TABLE 8.1   OPERATIONAL USES OF SATELLITE DATA IN HYDROLOGY

Application	Data sources	Remarks
Precipitation		
Estimate regional precipitation from cloud area and albedo, cloud temperature, and decrease in microwave brightness temperature of ground.	GOES, NOAA 5, TIROS N, and Nimbus 6 and 7. Passive microwave data are available only from Nimbus.	Ground radar observations generally are necessary for accurate estimates of local precipitation.
Snow and ice		
Snow line and snow area mapping.	Landsat images and digital data or meteorological satellite images and digital data.	Frequent cloud cover is a serious handicap for operational use in mountainous regions.
Estimate time and rate of snowmelt in plains areas from snowpack temperature and microwave brightness temperature.	Thermal infrared data from HCMM, GOES, NOAA 5, TIROS N, and Nimbus 6 and 7. Passive microwave data from Nimbus.	Principles of this application are established, but additional research is needed in mountainous terrain, where the application is important for streamflow prediction. In plains areas, further research is needed on detection of frozen ground for flood prediction.
Map and monitor sea ice area, concentration, and morphology. Assess river, lake, and estuary ice conditions. Glacier inventory and environmental monitoring.	Landsat, HCMM, and meteorological satellite data. Passive microwave data from Nimbus for sea ice.	Application is important for navigation and energy budget studies.
Surface water		
Inventory location and area of reservoirs, lakes, and ponds larger than 1–2 hectares. Determine seasonal and annual changes in area; estimate changes in water volume.	Landsat images and digital data.	Detection and measurement accuracy is limited by 80-m satellite resolution, the occurrence of vegetated water bodies, and confusion with wet soils or terrain shadows.
Delineate drainage patterns for fluvial morphology studies.	Landsat band 7 and color composite images; Landsat RBV images.	Drainage pattern and density generally are shown in more detail on Landsat images than on small-scale maps.

138

Application	Data	Remarks
Monitor shoreline and stream channel positions and migrations.	Landsat RBV and MSS data.	Monitoring capability is limited by the 30-m (RBV) or 80-m (MSS) resolution.
Regional overview of flood impact and damage.	Landsat and meteorological satellite images.	Satellite resolution is inadequate for hydraulic studies and detailed damage estimates, but satellite images may be used to plan other data acquisition.
Determine clear water depth by light absorption method.	Landsat digital data.	A few depth measurements and observations of bottom conditions are necessary.
Detect large underwater springs.	Landsat RBV and band 4 MSS images or digital data.	Only a few large springs can be detected unless outflows produce water turbidity plumes.

Water surface features

Application	Data	Remarks
Detect some oil spills. Inventory and monitor large areas of floating and emergent vegetation.	Landsat MSS data for vegetation and some oil slicks; HCMM and meteorological satellite data for detection of large oil spills on thermal infrared imagery.	Landsat spectral bands are poor for detection of oil slicks, but some oil slicks have been mapped with Landsat images.
Detect and study differences in sea state as shown by sun glint; sun elevation generally must be more than 55° above horizon.	Landsat and meteorological satellite data in the visible and near-infrared bands.	

Physical water quality

Application	Data	Remarks
Delineate water color and water turbidity patterns; interpret water circulation and current patterns; evaluate marine fish habitat; study fate of pollutants, biological productivity, and the sediment budgets of estuaries and coastal zones; inventory and monitor the trophic state of lakes and reservoirs; monitor reservoir filling.	Landsat MSS images and digital data. Nimbus 7 data from the Coastal Zone Color Scanner.	The synoptic view of satellites in obtaining optical measurements of water color and turbidity has not been fully realized and exploited.
Quantitative measurement of water turbidity, including turbidity caused by plankton and colloids.	Landsat and Nimbus 7 digital data. A few concurrent ground measurements generally are necessary.	If various methods of solar and atmospheric data correction are proven practical, all satellite data will have value as a record of water turbidity.
Monitor thermal current patterns in large lakes, estuaries, and coastal zones; evaluate marine fish habitat.	Thermal infrared imagery from HCMM, Nimbus 7, GOES, NOAA 5, and TIROS N.	Digital data and a few concurrent ground measurements are necessary for quantitative results.

TABLE 8.1 *(Continued)*

Application	Data sources	Remarks
Groundwater		
Detect, delineate, and interpret regional geologic structures.	Landsat images and mosaics for regional structure. Meteorological satellite images for continental structure.	The data are widely used for this purpose. Interpretations eventually should result in a better understanding of diagenetic and tectonic processes as well as the effect of these processes on groundwater occurrence and movement.
Shallow groundwater exploration by interpretation of landforms, drainage patterns and density (texture), and vegetation types and patterns.	Landsat images and mosaics.	Landsat images are used as a tool for groundwater exploration in most areas of the world.
Estimates of shallow groundwater salinity from salt crusts and location, type, and density of vegetation.	Landsat images and digital data. Some ground measurements and observations are desirable.	
Detect and inventory large groundwater seeps and shallow geothermal groundwater by anomalous snow melt patterns.	Landsat images and digital data.	Melt patterns show the integrated result of all heat sources, including earth heat flow and the presence of groundwater at a shallow depth.
Evapotranspiration		
Estimate evapotranspiration by atmospheric model, relative biomass measurements, or relative surface albedo and temperature measurements.	Atmospheric water vapor content from TIROS N. Thermal data from meteorological satellites. Albedo data from Landsat or meteorological satellites. Relative biomass measurements from Landsat data. Concurrent ground measurements are desirable to necessary.	Additional research is needed to determine best approach and to improve accuracy of estimates.

Estimate moisture content of surface and near-surface soils from thermal and microwave measurements.	Passive microwave data from Nimbus 7. Thermal images and digital data from GOES, NOAA 5, TIROS N, and Nimbus. A few concurrent ground measurements are desirable or necessary.	Additional research is needed to evaluate fully the effects of vegetation and to improve accuracy of estimates.

Land use/land cover

Inventory and map land cover and land use, including bare soils, cultivated cropland, center pivot irrigation systems, and wetlands. Monitor environmental effects of water development and management. Determine increase in water use for irrigation. Estimate percent impervious area. Inventory sediment source areas. Locate nonpoint source pollution.	Landsat images and digital data.	Classification accuracy can be improved by (1) combining Landsat and topographic data—correcting Landsat radiance values for surface slope and aspect, (2) merging Landsat scenes of the same area at different times during a year—using eight bands of Landsat data for machine classification instead of four bands, (3) stratification and separate classification of spectrally similar cover types, (4) correcting Landsat data for solar and atmospheric conditions, and (5) using a combination of machine processing and manual interpretation to obtain the advantages of each procedure. Considerable ground information is also needed.

Satellite data relay

Satellite relay of physical, chemical, and biological water data.	Landsat, GOES, and commercial communications satellites.	Landsat has experimental status and cannot be used for operational data relay.

Source: After Moore.[16]

With the advancement of satellite technology, the use of satellites as remote sensor platforms has spread. Currently available sensors can operate in a multitude of electromagnetic radiation wavelengths and the information content of their signals can include surface temperatures, radiation, atmospheric pollutants, and other types of meteorological data. As remote sensors are improved to permit the attainment of greater radiometric and geographic resolution, and as computer image-enhancing techniques become more sophisticated, it is certain that this powerful water management tool will see even greater and more diversified use.

■ Summary

Hydrologic data are important components of model design and testing and of a variety of statistical analyses. The quality of data obtained relate to attributes of measuring instruments and to the features of gauging sites. It is important to understand the pros and cons of various instruments and to know how they can best be used.

REFERENCES

1. "Irrigation Water Measurement," Mountain States Regional Publ. 1, revision of Extension Circ. 132, Irrigation Water Measurement, University of Wyoming, Laramie, June 1964.
2. U.S. Bureau of Reclamation, *Water Measurement Manual,* 2nd ed. Washington, D.C.: U.S. Government Printing Office, 1967.
3. Leupold & Stevens, Inc. *Stevens Water Resources Data Book,* 4th ed. Beaverton, OR: Leupold and Stevens, Jan. 1987.
4. W. Brutsaert, *Evaporation into the Atmosphere.* London: D. Reidel Publishing, 1982.
5. "Texas Lab Installs Weighing Lysimeters," *Irrigation J.* **37**(3), 8–12(May/June 1987).
6. R. K. Linsley, Jr., M. A. Kohler, and J. L. H. Paulhus, *Applied Hydrology.* New York: McGraw-Hill, 1982.
7. R. J. C. Burnash and T. M. Twedt, "Event-Reporting Instrumentation for Real-Time Flash Flood Warning," American Meteorological Society, Preprints, Conference on Flash Floods: Hydro-meteorological Aspects, May 1978.
8. D. E. Colton and R. J. C. Burnash, "A Flash-Flood Warning System," American Meteorological Society, Preprints, Conference on Flash Floods: Hydro-meteorological Aspects, May 1978.
9. R. J. C. Burnash, "Automated Precipitation Measurements," Aug. 1980.
10. R. J. C. Burnash and R. L. Ferral, "A Systems Approach to Real Time Runoff Analysis with a Deterministic Rainfall–Runoff Model," International Symposium on Rainfall–Runoff Modeling, University of Mississippi, May 18–21, 1981.
11. Hydrologic Services Division, National Weather Service, Western Region, "Automated Local Evaluation in Real Time: A Cooperative Flood Warning System for Your Community," Feb. 1981.
12. R. J. C. Burnash and R. L. Ferral, "Examples of Benefits and the Technology Involved in Optimizing Hydrosystem Operation Through Real-Time Forecasting," Conference on Real-Time Operation of Hydrosystems, Waterloo, Ontario, June 24–26, 1981.
13. M. Deutsch, D. R. Wiesnet, and A. Rango (eds.), *Satellite Hydrology.* Bethesda, MD: American Water Resources Association, 1981.

14. J. F. Bartholic, "Agricultural Meteorology: Systems Approach to Weather and Climate Needs for Agriculture, Forestry, and Natural Resources," in *Proceedings, Thirty-Second Meeting, Agricultural Research Institute.* Bethesda, MD: Agricultural Research Institute, 1983, pp. 75–85.

15. M. A. Calabrese and P. G. Thome, "NASA Water Resources/Hydrology Remote Sensing Program in the 1980's," in *Satellite Hydrology* (M. Deutsch, D. R. Wiesnet, and A. Rango, eds.). Bethesda, MD: American Water Resources Association, 1981, pp. 9–15.

16. G. K. Moore, "An Introduction to Satellite Hydrology," in *Satellite Hydrology* (M. Deutsch, D. R. Wiesnet, and A. Rango, eds.). Bethesda, MD: American Water Resources Association, 1981, pp. 37–41.

Chapter 9

Monitoring Networks

■ **Prologue**

The purpose of this chapter is to:

- Describe elements of systems for monitoring hydrologic variables.
- Indicate the importance of real-time and continuous recording of hydrologic events.

Information (data) is the requisite foundation for designing schemes for manipulating (managing) hydrologic systems, for evaluating the efficacy of actions taken to correct problem situations, and for identifying trouble spots deserving attention. But to be useful, the data must be of the right type, in the right form, and appropriately representative of critical space and time dimensions.

Modeling hydrologic systems requires an understanding of how these systems actually function; cleaning up a toxic waste discharge requires tracking the effects of remedial actions; enforcing environmental regulations requires knowledge of what has happened since the rules were implemented; and regulating reservoir releases to meet specified targets requires a continuous understanding of the state of the system being operated. The key to meeting such requirements lies in the products of carefully designed and managed monitoring networks. Developing such networks is no small task, however, as the number of variables that must be observed may be very large, the instruments to measure them costly to install and operate, and the data storage and management requirements extensive. Accordingly, a monitoring network's design must begin with a thorough understanding of its purpose so that the degree of resolution provided by its observations is adequate, but not excessive, for the task at hand. A good rule is to keep the network as simple as possible, within the constraints of what must be accomplished.

9.1 THE PURPOSE OF MONITORING

The purpose of monitoring is to gather information in a continuum such that the dynamics of the system can be ascertained. According to Dressing, objectives of monitoring for nonpoint source pollution control include development of baseline information, generating data for trend analysis, developing and/or verifying models, and investigating single incidents or events.[2] These objectives are also valid for hydrologic monitoring in general, but they should be supplemented by the following objectives: planning, real-time system operating, enforcing regulatory programs, and environmental policymaking. In the final analysis, the ultimate purpose of monitoring is to enhance decisionmaking, whether it be for development, management, regulation, or research aims.

9.2   SPECIAL CONSIDERATIONS

Before an acceptable monitoring plan can be devised, there must be a full understanding of the hydrologic system to be monitored and of the objectives to be met by monitoring. The costs of monitoring can be very high and thus it is essential that monitoring networks be efficient and cost effective.[1-5]

Time and Space Variability

In general, monitoring networks are designed to have both spatial and temporal dimensions. Although monitoring a specific point location may be all that is necessary under some circumstances, it is more common that what is happening in a regional setting is of importance. The temporal aspect is similar. While a snapshot at some point in time may suffice for some purposes, the time variance of conditions to be tracked is usually critical for effective analyses and/or decisionmaking. Both the short-term and long-term variabilities of many targets of monitoring must be ascertained. For example, water quality in a stream can change rapidly with time, while changes in lake levels, such as those experienced in the Great Lakes in the 1980s, are the result of long-term hydrologic variability.

Spatial variability must also be represented in a monitoring network: for example, infiltration rates may vary considerably within a region, rainfall intensities may be quite different within even short distances, and water quality in a river might be different in upstream and downstream locations. Topography, soils, vegetal covers, and many other factors affecting the performance of a hydrologic system are also distributed differently in space, and these differences must be recognized in the monitoring plan. The trick is to develop a monitoring system that can (1) provide the needed data, (2) recognize regional and temporal variabilities, and (3) keep installation, operation, and maintenance costs to a minimum. To do this requires a comprehensive knowledge of the system to be monitored, an understanding of what the data obtained by the system will be used for, and a knowledge of the level of detail in collecting the data that must be exercised in space and time.

Data Requirements

The amount and type of data to be generated by a monitoring system must be carefully considered in its design. Selecting appropriate instruments, determining sampling frequency, and setting data formats are elements that must be considered. Questions such as how much do we need to know and when do we need to know it must be answered. The form and extensiveness of data must be tightly related to monitoring

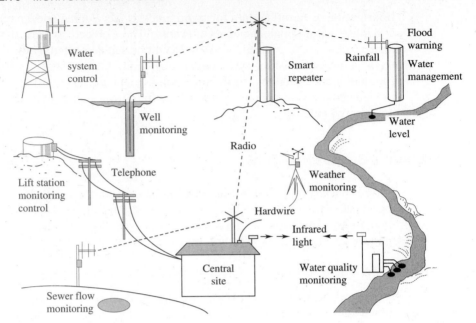

Figure 9.1   A telemetry monitoring system. (Courtesy of Sierra-Misco, Inc., Environmental Products, Berkeley, CA.

objectives. Furthermore, it might be necessary to monitor surrogates instead of the condition to be tracked.[2] For example, if lake eutrophication is the issue, phosphorus and chlorophyll concentrations might be surrogate measures. If this approach is taken, selection of appropriate surrogates is very important and the foregoing comments about data formats and so on are also applicable. Hydrologic, water quality, land use and treatment, topographic, soils, vegetative cover, meteorologic, and many other types of data may be required in combination or separately in a monitoring plan. It is easy to see that the amount of data required for a monitoring program can be enormous. Consequently, great care must be taken to see that the data collection effort is not in excess of the objectives of the monitoring program. Figure 9.1, depicting a telemetry monitoring system, gives an indication of the variety of data that might be collected in a monitoring program.

Quality Control and Quality Assurance

The costs of monitoring are usually substantial and thus it is essential that the data generated be of consistently high quality. Accordingly, most monitoring systems include quality control and quality assurance (QA/QC) elements. Quality control is a planned system of activities designed to produce a quality product (data in this case) that meets the needs of the user. Quality assurance is a planned system of activities designed to guarantee that the quality control program is being carried out properly. A quality management plan should be part of the overall monitoring program and should be prepared when the monitoring program is being developed to ensure that the data collected will be of a satisfactory nature for the monitoring program's objectives.[3]

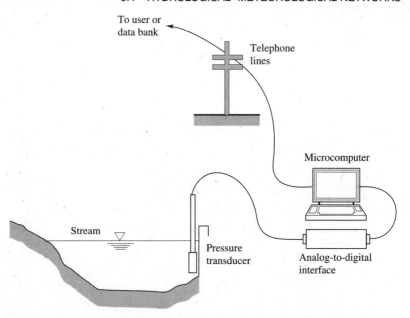

Figure 9.2   Microcomputer use in stream gauging.

9.3 USE OF COMPUTERS IN MONITORING

With the rapid technological development of computers, especially inexpensive microcomputers, the opportunities for automated collection of all types of hydrologic and water quality data have increased substantially. Microcomputers, used with analog-to-digital converters, pressure or liquid-level sensors, and the appropriate software can, for example, be used in hydrologic monitoring systems as flow metering/data acquisition systems (Fig. 9.2).[4] Furthermore, such systems are highly versatile and they are relatively inexpensive. Computer systems can be custom-designed for almost any data acquisition application and they are often less costly than other commercially available hardware systems designed for the same purpose. Computers can convert raw data into other more useful forms, store data for later use, and communicate with other computer terminals if necessary. As such, they are a powerful and important component of modern hydrologic monitoring systems. Figure 9.3 illustrates the use of computers in a real-time telemetry system.

9.4 HYDROLOGICAL–METEOROLOGICAL NETWORKS

Most modern hydrologic–meteorologic networks are designed to provide real-time information for purposes such as hydropower scheduling, releasing flows for irrigation, developing and testing hydrologic system models, regulating reservoir discharges, allocating water from multiple sources, streamflow forecasting, tracking pollutant transport, and enforcing environmental regulations. Hydrological–meteorological

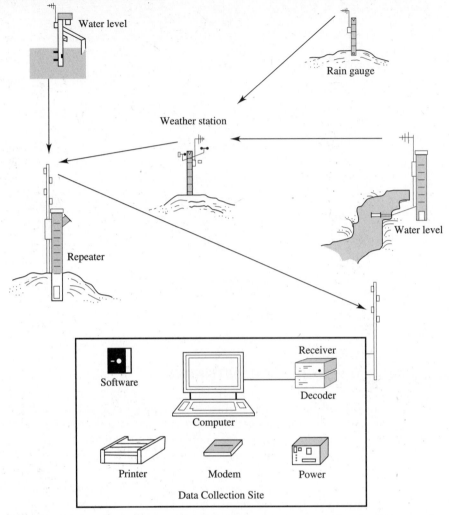

Figure 9.3   Computer use in a real-time telemetry system. (Courtesy of Sierra-Misco, Inc., Environmental Products, Berkeley, CA.)

networks may be designed to monitor physiographic, climatic, hydrologic, biologic, and chemical features, or combinations of these, in a region or river basin. They must have gauge densities and distributions that are sufficient to permit interpolation between gauge sites in a manner permitting valid conclusions to be drawn for the entire area covered by the network. Typically, measurements are made of such variables as precipitation, solar radiation, temperature, relative humidity, barometric pressure, snow depth, soil moisture, wind speed, streamflow, and water quality. In any event, special basin or regional climatic factors must be given due consideration. Each hydrological–meteorological network is different in its purpose and setting and thus its design must reflect both the spatial and temporally varying features at the locality to be monitored along with the objectives of the monitoring program.[5]

■ **Summary**

Monitoring of hydrologic systems is essential to better understanding of system inter-actions and to the design and testing of hydrologic models. It is also the means by which a determination can be made of the effectiveness of measures taken to alter watershed performance.

REFERENCES

1. S. J. Nix and P. E. Black (eds.), *Proceedings of the Symposium on Monitoring, Modeling, and Mediating Water Quality*. Bethesda, MD: American Water Resources Association, 1987.
2. S. A. Dressing, "Nonpoint Source Monitoring and Evaluation Guide," in Ref. 1, pp. 69–78.
3. J. Lawrence and A. S. Y. Chau, "Quality Assurance for Environmental Monitoring," in Ref. 1, pp. 165–176.
4. H. E. Post and T. J. Grizzard, "The Monitoring of Stream Hydrology and Quality Using Microcomputers," in Ref. 1, pp. 199–208.
5. P. J. Gabrielsen and A. J. Carmeli, "Operation of a Hydrologic–Meteorologic Monitoring Network in a Severe Winter Environment," in Ref. 1, pp. 113–122.

PART THREE

SURFACE WATER HYDROLOGY

Chapter 10

Runoff and the Catchment

■ Prologue

The purpose of this chapter is to:

- Expand on definitions of terms frequently used in describing the runoff process.
- Present concepts of surface runoff and drainage basin discharge.
- Introduce elements of drainage basin geomorphology.
- Describe quantitative measures of watershed characteristics.
- Familiarize the reader with elements of frequency analysis.

Surface water hydrology deals with the movement of water along the earth's surface as a result of precipitation and snow melt. Detailed analysis of surface water flow is highly important to such fields as municipal and industrial water supply, flood control, streamflow forecasting, reservoir design, navigation, irrigation, drainage, water quality control, water-based recreation, and fish and wildlife management.

The relation between precipitation and runoff is influenced by various storm and basin characteristics. Because of these complexities and the frequent paucity of adequate runoff data, many approximate formulas have been developed to relate rainfall and runoff. The earliest of these were usually crude empirical statements, whereas the trend now is to develop descriptive equations based on physical processes.

10.1 CATCHMENTS, WATERSHEDS AND DRAINAGE BASINS

Runoff occurs when precipitation or snowmelt moves across the land surface—some of which eventually reaches natural or artificial streams and lakes. The land area over which rain falls is called the *catchment* and the land area that contributes surface runoff to any point of interest is called a *watershed*. This can be a few acres in size or thousands of square miles. A large watershed can contain many smaller subwatersheds.

Streams and rivers convey both surface water and groundwater away from high water areas, preventing surface flooding and rising groundwater problems. The tract of land (both surface and subsurface) drained by a river and its tributaries is called a *drainage basin.* A watershed supplies surface runoff to a river or stream, whereas a drainage basin for a given stream is the tract of land drained of both surface runoff and groundwater discharge.

Rain falling on a watershed in quantities exceeding the soil or vegetation uptake becomes *surface runoff.* Water infiltrating the soil may eventually return to a stream and combine with surface runoff in forming the total *drainage* from the basin. The network of overland flow courses and defined drainage channels comprise the watershed. Surface runoff from tracts of land begins its journey as *overland flow,* often called *sheet flow,* before it reaches a defined swale or channel, usually before flowing more than a few hundred feet. The lines separating the land surface into watersheds are called *divides.* These normally follow ridges and mounds and can be delineated using contour maps, field surveys, or stereograph pairs of aerial photographs to identify gradient directions.

Contributing Area

In the majority of hydrologic analyses, the magnitude of total surface area contributing direct runoff to some point of interest is needed. Because of variations in topography, the true surface area cannot be easily measured. The horizontal projection of land area is easily obtained and normally adopted in hydrologic calculations. This results in an error in actual watershed area wherever the projected area is less than the actual. Some surface area in watersheds may not contribute to surface runoff, so the error in using the projected watershed area is somewhat offset.

Partial Area Hydrology

For light storms, or for some flat areas, portions of the catchment do not contribute to runoff. Precipitation falling on or flowing into depressed or blocked areas can exit only by seepage or evaporation, or by transpiration if vegetated. If sufficient rainfall occurs, such areas may overflow and contribute to runoff. Thus the total area contributing to runoff varies with the intensity and duration of the storm. Methods for incorporating this phenomenon in hydrologic studies are categorized under procedures for *partial area hydrology.*

In partial area hydrology, watershed areas are divided by one of several methods into contributing (active) and noncontributing (passive) subareas. For infrequent (severe) storms, larger percentages of the watershed surface may contribute to the peak flow and volume of runoff, which are the primary variables of interest to design engineers. For more frequent storms, significantly smaller portions of some watersheds may contribute. As a consequence, partial area hydrology is seldom incorporated in hydraulic structure design, and is of greater interest in water supply and water quality studies. As will be shown later (Chapter 12) unit hydrograph theory and runoff curve number methods are based on linearity of rainfall and runoff, and assume that the full watershed contributes to runoff in all storms and in proportional amounts at different times in the same storm. Application of these methods to watersheds that have significant noncontributing zones could, and do, introduce error if the zones are not first delineated and the distributed effects properly modeled.

Subdivision of contributing from noncontributing areas has traditionally been subjectively accomplished from site inspection, topographic and soils maps, and aerial photos. Soils having good drainage classifications, or dark tones or colors on aerial photos, can often be considered as passive areas. Other signs of noncontributing areas would include presence of wetlands, grassed areas, rooftops (unless connected to the drainage), terraces, erosion control structures, stock watering ponds, and flood control dams.

Boughton[1] developed a quantitative method of determining the proportion of a watershed that contributes surface runoff in different storms, and at different times during the same storm, by analyzing rainfall and runoff records. His logic is as follows:

1. Watersheds can be idealized as a group of "surface storage capacity" cells, each representing a fraction of the watershed area and each having some capacity to abstract rainfall into storage, infiltration, or evapotranspiration.
2. Runoff from each cell occurs when rain fills the surface storage capacity.
3. Runoff occurs from the cell with the smallest capacity before flowing from the cell with the next largest capacity (this is an assumption by Boughton that has not been fully verified).
4. Using these principles, storm data for the watershed are evaluated first to find those in which runoff occurs only from the area of smallest capacity. This is done using a graphical method outlined in the article that looks at slope changes in the rainfall-runoff graph. Both the capacity of the cell and its area as a percentage of the watershed are estimated.
5. After subtracting the contribution to runoff from the smallest capacity cell, the capacity and contributing area for the second smallest capacity cell are determined by the same procedure.
6. The process is repeated until all the runoff is accounted for, or until 100 percent of the watershed is contributing, whichever occurs first.

When Boughton applied the procedure to a test watershed,[1] it was found that runoff occurred from the entire watershed on only 3 of 30 events in the 15-year study period. In about two-thirds of the runoff events, discharge occurred only from the cell with the smallest surface storage capacity.

10.2 BASIN CHARACTERISTICS AFFECTING RUNOFF

The nature of streamflow in a region is a function of the hydrologic input to that region and the physical, vegetative, and climatic characteristics. As indicated by the hydrologic equation, all the water that occurs in an area as a result of precipitation does not appear as streamflow. Fractions of the gross precipitation are diverted into paths that do not terminate in the regional surface transport system. Precipitation striking the ground can go into storage on the surface or in the soil and into groundwater reservoirs beneath the surface. The character of the soil and rocks determines to a large extent the storage system into which precipitated water will enter. Opportunity for evaporation and transpiration will also be affected by the geologic and topographic nature of the area.

Stream Patterns

Wind, ice, and water act on land surfaces to create several types of drainage patterns seen in nature. The particular design that results is a function of several factors including slope, underlying soil and rock properties, and the histories of hydraulic action, freeze–thaw activity, and sediment transport.

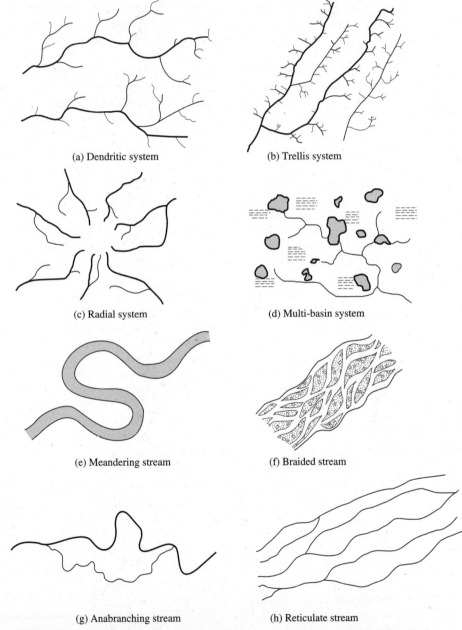

(a) Dendritic system

(b) Trellis system

(c) Radial system

(d) Multi-basin system

(e) Meandering stream

(f) Braided stream

(g) Anabranching stream

(h) Reticulate stream

Figure 10.1   Stream patterns (combined systems and individual stream shapes).

Typical stream patterns are shown in Fig. 10.1. The most common, *dendritic,* is characterized by numerous small tributaries joining at right angles into higher-order streams, eventually forming the major rivers in the region. The smaller tributaries often occur in sufficient quantities that little land surface area is left unintercepted by a defined channel of some form. The maximum overland distance between channels in these areas seldom exceeds a few hundred feet.

Trellis patterns are characterized by long main streams intercepted by numerous shorter right-angle tributaries. They are common in the Appalachians (Eastern United States) and are also seen in the Rocky Mountains along the foothills. *Multi-basin* patterns, also called *deranged* systems, occur in low gradient swampy areas with numerous surface depressions and normally have only a few tributaries. These occur in glaciated, windblown, and permafrost areas, and are common in plains and mountain valley regions of the United States. *Radial* patterns are typically found in foothill areas or mountain areas with more advanced soil development.

Individual streams favor one or more of the four patterns shown on the lower portion of Fig. 10.1. Streams are rarely straight except on steep slopes in homogeneous materials. Braided streams are characterized by numerous interconnected channels flowing around and over islands and bars, inundating most during high flows. Braided streams are generally transporting large amounts of sediment, but often less than the amount supplied. They have been called incipient forms of meandering streams due to the fact that many revert to meandering or other forms when sediment supplies or other factors change. Meandering in an otherwise straight channel occurs as a result of transverse currents. These currents are considered to be the result of forces acting on the stream particles, including bed and bank shear forces and coriolis effects.

In evaluating the effects of changes in streamflows, a relationship known as Lane's Law is often applied. It states that the product of bed slope and water discharge is proportional to the product of sediment size and transport rate. Changing any one of these four terms results in the likelihood of a shift in one or more of the others. Constructing a reservoir, for example, reduces the sediment transported into the reach just downstream. By Lane's Law, either the slope or discharge must decrease or sediment particle size must increase to offset the change in sediment transport. Most often, the slope decreases when the sediment-hungry flows deepen the bed in the reach. Degradation (downward cutting) of the bed of the Missouri River, for example, has occurred below some of its upstream reservoirs, isolating boat marinas and water intake structures in some locations.

Geomorphology of Drainage Basins

The principal geologic factors that affect surface waters are classified as lithologic and structural. Lithologic effects are associated with the composition, texture, and sequence of the rocks, whereas structural effects relate mainly to discontinuities such as faults and folds. A fault is a fracture that results in the relative displacement of rock that was previously continuous. Folds are geologic strata that are contorted or bent. Variations in the erodibility of the different strata can easily lead to the creation of distinctive forms of drainage systems.

Both large-scale and local effects on the storage and movement of surface waters exist because of geologic activity and structure. For example, drainage patterns are determined to a large extent by the nature of land forms. On the other hand, flowing surface waters also affect the surface geometry through the process of erosion. Thus significant land forms resulting from volcanic activity, folding, and faulting affect drainage, whereas drainage patterns, having been generated, can also modify the land forms by creating valleys, deltas, and other geomorphic features.

Streams are classified as being young, mature, or old on the basis of their ability to erode channel materials. Young streams are highly active and usually flow rapidly so that they are continually cutting their channels. The sediment load imposed on these streams by their tributaries is transported without deposition. Mature streams are those in which the channel slope has been reduced to the point where flow velocities are just able to transport incoming sediment and where the channel depth is no longer being modified by erosion. A stream is classified as old when the channels in its system have become aggraded. The flow velocities of old streams are low due to gentle slopes that prevail. Wide meander belts, broad flood plains, and delta formation are also characteristic of old streams. The lower reaches of the Mississippi, Rhine, and Nile are examples. Flows in young river basins are often "flashy," whereas sluggish flows are common to older streams.

The description of a drainage basin in quantitative terms was an important forward step in hydrology and can be traced back in large part to the efforts of Robert E. Horton.[2] Strahler, Langbein, and others have expanded Horton's original work.[3-4]

To quantify the geometry of a basin, the fundamental dimensions of length, time, and mass are used. Many drainage basin features that are important to the hydrologist can be quantified in terms of length, length squared, or length cubed. Examples are elevation, stream length, basin perimeter, drainage area, and volume. The concept of geometric similarity can be applied to drainage basins just as it is to many other systems.[3] Most readers will be aware of model-prototype studies of aircraft, dams, and turbomachinery. Such studies involve considerations of geometric as well as dynamic similarity. In the same manner that inferences as to the operation of a prototype can sometimes be drawn from a geometrically similar model, inferences can also be drawn about the operation of one drainage area on the basis of information obtained from a similar one. Perfect similarity will never be realized if natural drainage systems are compared, but striking similarities have been observed which can often be put to practical use.

Measures of Drainage Basin Characteristics

Important measures of drainage basin characteristics include overland flow lengths and stream lengths. The concept of stream order is often associated with the dimension of stream length.

If the stream system in a drainage basin is clearly defined on a topographic map, the smallest tributaries are classified as Order 1.[5] This is illustrated in Fig. 10.2. The point at which two first-order streams join is the beginning of a second-order segment. Third-order segments initiate where two second-order streams join, and so on. The main stream channel that carries the flow from the entire tributary area upstream of a point of interest will necessarily be the highest-order stream in that system.

The practical utility of the stream order system is based on the hypothesis that the size of the watershed, its channel dimensions, and streamflow are all proportional

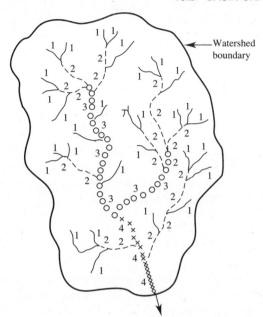

Figure 10.2   Sketch indicating definition of stream order.

to the stream order, provided that a large enough sample is investigated. The order number permits comparisons of drainage systems that are quite different in size because the number is a dimensionless quantity. Such comparisons should be made at locations in the two systems that have a similar geometry; that is, second-order streams, third-order streams, and so forth.

Stream lengths are determined by the measurement of their projections onto a horizontal plane. Topographic maps are useful for obtaining such measurements. If the mean length of a stream segment L_u of order u is defined as $\bar{L}_u$, then it is possible to determine $\bar{L}_u$ using[2]

$$\bar{L}_u = \frac{\sum_{i=1}^{N_u} L_{ui}}{N_u} \tag{10.1}$$

where N_u is the number of stream segments of stream order u.

Another measure related to stream length is the distance L_{ca} from a point of interest on the main stream to a point on the primary channel that is nearest the center of gravity of the drainage area (center of gravity of the plane area of the drainage basin). Studies of basin lag (time between the centers of mass of effective storm input and the resulting runoff) have made use of this dimension.

Of particular significance in the physiographic development of a drainage basin is the *overland flow length L_0*. This is the distance from the ridge line or drainage divide, measured along the path of surface flow which is not confined in any defined channel, to the intersection of this flow path with an established flow channel. If a drainage basin of the first order is the basic element of a larger drainage system, then a representative overland flow length can be determined for these first-order basins. One approach is to measure a number of possible flow paths from a map of the area and to average these. In some cases (for example, with the rational method, Chapter

15), the use of the longest overland flow length is prescribed, measured from the upstream end of the first-order stream to the most remote point of flow that will terminate at this point.

Areal Measurements

Just as linear measures relate to many factors of hydrologic interest, so do areal measures. For example, the quantity of discharge from any drainage basin is obviously a function of the areal extent of that basin.

Correlations have been observed between the average area, $\bar{A}_u$, of basins of order u, and the average length of stream segments, $\bar{L}_u$. These variables are often related by an exponential function. For example, studies of seven streams in the Maryland-Virginia area by Hack have produced the relationship[6]

$$L = 1.4A^{0.6} \tag{10.2}$$

where L = the stream length measured in miles to the drainage divide
         A = the drainage area (mi^2)

Hack's observations indicate that as the drainage basin increases in size, it becomes longer and narrower; thus precise geometric similarity is not preserved.

Drainage area has long been used as a parameter in precipitation–runoff equations or in simple equations indexing streamflow to area or other parameters. Many early empirical equations are of the form[3]

$$Q = cA^m \tag{10.3}$$

where Q = a measure of flow such as mean annual runoff
         A = the size of the contributing drainage area

Values of c and m are determined by regression analysis (see Chapter 26); Fig. 10.3 illustrates a relation of this form.

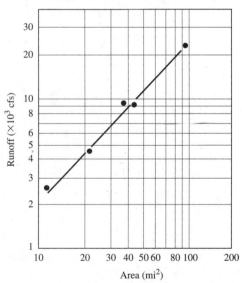

Figure 10.3   Runoff–drainage area correlation for five Maryland streams (1933 storm data).

Other areal measures include definitions of the basin shape and the density of the drainage network or *drainage density,* defined as the ratio of total channel segment lengths cumulated for all stream orders within a basin to the basin area. The *stream frequency* is defined as the summation of all segments in a drainage basin (total number of segments of all orders) divided by the drainage area.

Channel and Basin Gradients

The slopes of a drainage basin and its channels have a very strong effect on the surface runoff process of that region. Most stream channel profiles exhibit the characteristic of decreasing slope proceeding in a downstream direction. Figure 10.4 illustrates this particular trait. Also illustrated in the figure are the gross slope, which is the total elevation drop divided by the channel length, and the mean slope, which is determined such that the areas between the average slope line and the stream profile are equal; that is, $A_1 = A_2$ in the figure. The gross slope and the mean slope are not very useful as parameters to describe drainage character due to their generality; Fig. 10.4 should make this clear. Some mathematical functions that are used to more fully describe stream profiles are linear, exponential, logarithmic, and power forms. A single numerical value to represent the primary channel slope has been used by Taylor and Schwartz.[7] This factor, known as the equivalent main stream slope S_{st}, is the slope of a uniform channel that is equivalent in length to the longest water course and has the same travel time. This factor has been found to be related to unit hydrograph lag (time from the center of mass of rainfall excess to the peak rate of runoff) and maximum discharge.

In addition to the slope of the stream channel, the overall land slope of the basin is an important topographic factor. A quantitative relation between valley wall slopes and stream channel slopes has been derived by Strahler.[3] A commonly used method of determining the slopes of a basin has been presented by Horton.[8] The method involves superimposing a transparent grid over a topographic map of the drainage area in question. Each grid line is measured between its intersections with the drainage divide; the number of intersections of each grid line with a contour line is also needed. A determination of the land slope can then be made using

$$S = \frac{n \sec \theta}{l} h \qquad (10.4)$$

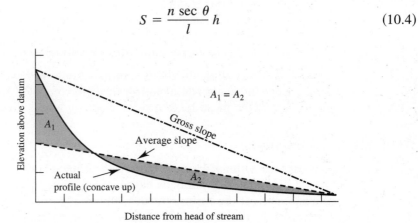

Figure 10.4 Typical stream profile.

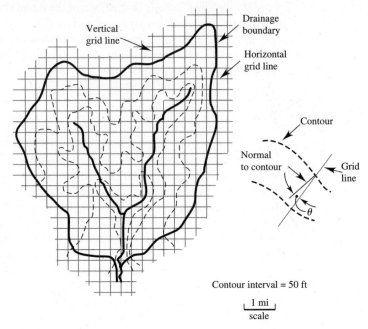

Figure 10.5   Determination of mean land slope: number of vertical intersections = 72; number of horizontal intersections = 120; total length of vertical grid segments = 103,900 ft; total length of horizontal grid segments = 101,200 ft.

$$S_v = \frac{72 \times 50}{103,900} = 0.035 \text{ ft/ft} \qquad S_H = \frac{120 \times 50}{101,200} = 0.059 \text{ ft/ft}$$

$$\text{Mean slope} = \frac{S_v + S_H}{2} = \frac{0.035 + 0.059}{2} = 0.047 \text{ ft/ft}$$

where n = the total number of contour intersections by the horizontal and vertical grid lines

l = the total length of grid line segments (horizontal and vertical)

h = the contour interval

θ = the angle measured between a normal to the contours and the grid line

Because θ is very difficult to measure it is often neglected, and separate values of average slope in the horizontal and vertical are computed and then averaged to obtain an estimate of the mean land slope. This procedure is illustrated in Fig. 10.5.

Area–Elevation Relation

How the area within a drainage basin is distributed between contours (Fig. 10.6) is of interest for comparing drainage basins and gaining insight into the storage and flow characteristics of the basin. For such studies, an area distribution curve such as that shown in Fig. 10.7 is used. The curve can be obtained by planimetering the areas

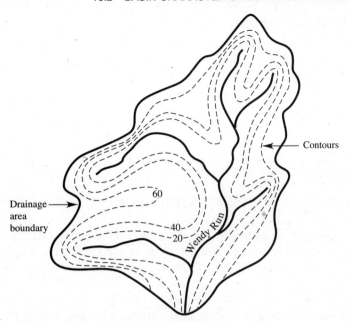

Figure 10.6 Topographic map of Wendy Run drainage area showing 20-, 40-, and 60-ft contour lines.

between adjacent contours or by using a grid as in Fig. 10.5 and forming the ratio of the number of squares between contours to the total number of squares contained within the drainage boundaries. The mean elevation is determined as the weighted average of elevations between adjacent contours. The median elevation can be determined from the area–elevation curves as the elevation at 50 percent.

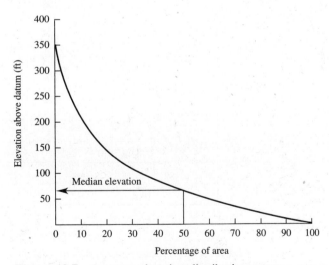

Figure 10.7 An area–elevation distribution curve.

Drainage Basin Dynamics

Geomorphology, like hydrology, was largely qualitative in nature in its formative years. With the passing of time and the greater need for reliable quantitative information, the science has progressed to the point where rational relations between variables are being developed. These relations are usually intended to quantify the interactions between the factors that modify the land form and the land form itself. In addition, equations relating the geomorphic properties to hydrologic, climatologic, or vegetative factors are being sought. Some of the functional relations of particular significance to the hydrologist will be discussed in the following chapters.

10.3 RUDIMENTARY PRECIPITATION–RUNOFF RELATIONSHIPS

A common approach in correlating precipitation and runoff has been to plot annual or monthly values, find the slope of the trend line, and estimate the percentage of rainfall appearing as runoff. Runoff quantities determined this way, however, are crude and subject to large errors. The degree of reliability is higher for drainage areas whose properties are least subject to seasonal or other types of variation, that is, an impervious area. Figure 10.8 illustrates the procedure. The resulting equation takes the form

$$Q = \left(\frac{1}{S}\right)(P - P_b) \tag{10.5}$$

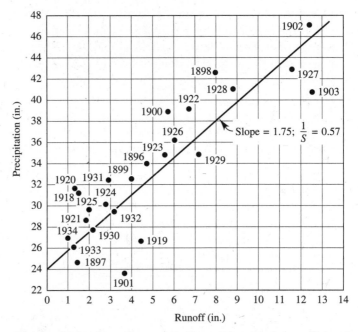

Figure 10.8 Annual precipitation and annual runoff in the Neosho River basin above Iola, Kansas. (U.S. Geological Survey Data.)

S = the slope of the line ($\Delta P/\Delta Q$)
P_b = a base precipitation value below which Q is zero

From Fig. 10.8 the relation for the example would be

$$Q = 0.57(P - 24)$$

where Q and P are the annual runoff and precipitation, respectively, in inches. Considerable scatter of several data points from the assumed relation indicates that this type of computation should be used with care. For rough approximations in preliminary planning studies, such methods are frequently helpful, however. Equations resembling Eq. 10.5 are improved if other parameters such as antecedent precipitation, soil moisture, season, and storm characteristics are included. Such relations can be described using multiple regression techniques or graphical methods. Linsley and co-workers present a very complete treatment of methods for developing correlations involving several variables.[9]

Soil moisture relations normally have a soil moisture index as the independent variable, since direct measurements of actual antecedent soil moisture are not generally practical. Indexes that have been inserted are groundwater flow at the beginning of the storm, antecedent precipitation, and basin evaporation.[10] Groundwater values should be weighted to reflect the effects of precipitation occurring within a few days of the storm because soil moisture changes from previous rains may affect results. Pan-evaporation measurements can be employed to estimate soil moisture amounts, since evaporation is related to soil moisture depletion.[11] Antecedent precipitation indexes (API) have probably received the widest use because precipitation is readily measured and relates directly to moisture deficiency of the basin.

A typical antecedent precipitation index is

$$P_a = aP_0 + bP_1 + cP_2 \tag{10.6}$$

where P_a = the antecedent precipitation index (in.)
P_0, P_1, P_2 = the amounts of annual rainfall during the present year and for 2 years preceding the year in question

This index links annual rainfall and runoff values.[12] Coefficients a, b, and c are found by trial and error or other fitting techniques to produce the best correlation between the runoff and the antecedent precipitation index. The sum of the coefficients must be 1.

Kohler and Linsley[13] have proposed the following API for use with individual storms:

$$P_a = b_1 P_1 + b_2 P_2 + \cdots + b_t P_t \tag{10.7}$$

where the t subscript on P refers to precipitation which occurred that many days prior to the given storm, and the constants b (less than unity) are assumed to be a function of t. Values for the coefficients can be determined by correlation techniques. In daily evaluation of the index, b_t is considered to be related to t by

$$b_t = K^t \tag{10.8}$$

where K is a recession constant normally reported in the range 0.85–0.98. The initial value of the API (P_{a0}) is coupled to the API t days later (P_{at}) by

$$P_{at} = P_{a0} K^t \tag{10.9}$$

To evaluate the index for a particular day based on that of the preceding one, Eq. 10.9 becomes

$$P_{a1} = KP_{a0} \tag{10.10}$$

because $t = 1$.

Various empirical relations for API have been proposed. Most are based on correlating two or three variables and at best yield only rough approximations. In many cases these were developed without considering physical principles or dimensional homogeneity. An added shortcoming is that many formulas fit only a specific watershed and have little general utility. Empirical equations demand great caution and an understanding of their origin.

EXAMPLE 10.1 ————————————————————————————————————

Precipitation depths P_i for a 14-day period are listed in Table 10.1. The API on April 1 is 0.00. Use $K = 0.9$ and determine the API for each successive day.

Solution.   Equation 10.9 reduces to

$$\text{API}_i = K(\text{API}_{i-1}) + P_i$$

which was applied in developing the successive values of API_i in Table 10.1.

TABLE 10.1

Date	Precipitation (P_i)	API_i
April 1	0.0	0.00
2	0.0	0.00
3	0.5	0.50
4	0.7	1.15
5	0.2	1.24
6	0.1	1.22
7	0.0	1.10
8	0.1	1.09
9	0.3	1.28
10	0.0	1.15
11	0.0	1.04
12	0.6	1.54
13	0.0	1.39
14	0.0	1.25

10.4 STREAMFLOW FREQUENCY ANALYSIS

Hydrologists estimate streamflows. Two approaches are employed. The first is a physical processes approach in which runoff is computed on the basis of observed or expected precipitation. The second is founded on statistical analyses of runoff records without resort to precipitation data. Such investigations usually include frequency

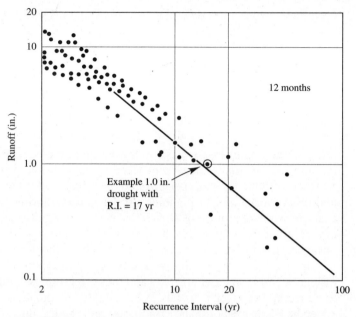

Figure 10.9   Low-flow frequency data consolidated for Five Rivers. (After Whipple.[15])

studies (Chapter 27) to indicate the likelihood of certain runoff events taking place. A knowledge of the frequency of runoff events is helpful in determining risks associated with proposed designs or anticipated operating schemes. Frequency analyses are usually directed toward studies of maximum (flood) and minimum (drought) flows.[14,15] Figure 10.9 illustrates a typical drought frequency analysis. Unfortunately, many existing runoff records are short-term; as a result they limit utility for reliable frequency analyses. Few adequate records are available earlier than about 1900. In some cases, sequential generating techniques (Chapter 22) can be used to develop synthetic records.

Time-series analyses are particularly pertinent to the problem of estimating trends, cycles, and fluctuations in hydrologic data. They also permit derivation of generating processes by which synthetic records of runoff can be developed.

Recurrence Interval and Frequency

The *recurrence interval* (R.I.) is defined as the average interval over a long period of years during which a corresponding magnitude of some hydrologic variable is at least met. This parameter is also called the *return period,* and sometimes, though less appropriately, the *frequency* of the event. For the example in Fig. 10.9, droughts less than 1 in. occurred in 8 of the 136 years of records. The 1.0 in. drought has an average recurrence interval of about 17 years. Stated another way, on the average, one year of every 17-year sequence is expected to experience a drought of at most 1.0 inch. Similarly, each year the probability of a 1.0-in. drought is 8/136 = 0.059, or about 6 percent. This is defined as the *exceedence probability* or *frequency,* and is the reciprocal of the return period. It should be obvious that the 1.0-in. drought could occur in any year, or in several consecutive years. This type of analysis cannot tell the investigator what will happen this year or next, and allows only an estimate of the average recurrence interval and the probability of occurrence in any given year. This subject is fully developed in Chapter 27.

10.5 STREAMFLOW FORECASTING

Surface water hydrology is basic to the design of many engineering works and important in water quality management schemes. In addition, the ability to provide reliable forecasts of flows for short periods into the future is of great value in operating storage and other works and in planning proper actions during times of flood.[9,16] A good example is the operation of a reservoir with an uncontrolled inflow but with a means of regulating the outflow. If information on the nature of the inflow is determinable in advance, then the reservoir can be operated by some decision rule to minimize downstream flood damage. Such operations can be computerized to continually improve estimates based on incoming data and thus offer direction on the nature of the releases to be made. For river forecasts to be reliable, adequate, dependable data on various watershed and meteorologic conditions are needed on a continuing basis. Modern monitoring stations capable of telemetering data to computer control centers provide an important support function for forecasting. The methods used to forecast flows are basically the same ones employed in design: precipitation–runoff equations, unit hydrographs, watershed models, and flow-routing techniques.

■ Summary

Runoff is probably the most complex yet most important hydrologic process to understand. It has attracted the attention and focus of engineers and scientists and comprises the greatest percentage by far of most hydrology textbooks and publications. The concepts introduced here will be more fully developed in the next six chapters, as well as in significant portions of Parts Five and Six.

PROBLEMS

10.1. For a drainage basin of your choice, plot the annual precipitation in inches versus the runoff in inches. Does the relation appear to be strong? Under what conditions and for what purposes might you use this?

10.2. Select a rain gauge record of interest. Use the annual values as data to calculate the coefficients of an antecedent precipitation index of the form of Eq. 10.8.

10.3. Determine the drainage density of the basin shown. Area = 6400 acres. Lengths are in miles.

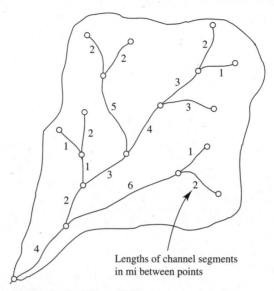

Figure for Problem 10.3

10.4. Using any dictionary, plus indexes or glossaries from one or two other hydrology texts, find and compare definitions of the following terms: *runoff, direct runoff, direct surface runoff, surface runoff, surface water, overland flow, streamflow, drainage, watershed, catchment, drainage basin, subbasin, drainage divide.*

REFERENCES

1. Boughton, W. C., "Systematic Procedure for Evaluating Partial Areas of Watershed Runoff," Proc. *ASCE J. Irrigation and Drainage Engineering* **116,** 1 (February 1990).
2. R. E. Horton, "Drainage Basin Characteristics," *Trans. Am. Geophys. Union* **13**(1932).
3. A. N. Strahler, "Geology—Part II," in *Handbook of Applied Hydrology.* New York: McGraw-Hill, 1964.
4. W. B. Langbein et al., "Topographic Characteristics of Drainage Basins," U.S. Geological Survey, Water Supply Paper, 968-c(1947).
5. R. E. Horton, "Erosional Development of Streams and Their Drainage Basins: Hydro-physical Approach to Qualitative Morphology," *Bull. Geol. Soc. Am.* **56**(1945).
6. J. T. Hack, "Studies of Longitudinal Stream Profiles of Small Watersheds," Tech. Rept. 18, Columbia University, Department of Geology, New York, 1959.
7. A. B. Taylor and H. E. Schwartz, "Unit Hydrograph Lag and Peak Flow Related to Basin Characteristics," *Trans. Am. Geophys. Union* **33**(1952).
8. R. E. Horton, "Discussion of Paper, Flood Flow Characteristics by C. S. Jarvis," *Trans. ASCE* **89**(1926).
9. R. K. Linsley, Jr., M. A. Kohler, and J. L. H. Paulhus, *Applied Hydrology,* 2nd Ed. New York: McGraw-Hill, 1975.
10. Ven Te Chow (ed.), *Handbook of Applied Hydrology.* New York: McGraw-Hill, 1964.
11. R. K. Linsley, Jr., and W. C. Ackerman, "Method of Predicting the Runoff from Rainfall," *Trans. ASCE* **107**(1942).
12. S. S. Butler, *Engineering Hydrology.* Englewood Cliffs, NJ: Prentice-Hall, 1957.

13. M. A. Kohler and R. K. Linsley, Jr., "Predicting the Runoff from Storm Rainfall," U.S. Weather Bureau, Res. Paper 34, 1951.

14. Leo R. Beard, "Statistical Methods in Hydrology," U.S. Army Engineer District, Sacramento, CA, 1962.

15. William W. Whipple, Jr., "Regional Drought Frequency Analysis," *Proc. ASCE J. Irrigation Drainage Div.* **92**(IR2), 11–31(June 1966).

16. Michael C. Quick, "River Flood Flows: Forecasts and Probabilities," *Proc. ASCE J. Hyd. Div.* **91**(HY3) (May 1965).

Hydrographs

■ **Prologue**

The purpose of this chapter is to:

- Characterize a hydrograph as a time plot of the discharge of surface runoff and groundwater from drainage basins.
- Introduce the components of hydrographs so that the reader can relate them to the quantitative assessments of runoff presented in subsequent chapters.
- Describe the time relationships most commonly used in hydrograph analysis.

Hydrograph analysis is the most widely used method of analyzing surface runoff. Its presentation in most textbooks is normally confined to one chapter. Because of numerous developments in hydrograph analyses, three chapters are dedicated to the subject in this text. Chapter 11 defines hydrographs and expands the concepts introduced in Chapter 6. The concepts referred to as unit hydrograph techniques are packaged together in Chapter 12. Individual streamflow hydrograph shapes vary as flow travels downstream, and the concepts for analyzing these changes, called hydrograph routing methods, are presented in Chapter 13.

11.1 STREAMFLOW HYDROGRAPHS

A *streamflow hydrograph* provides the rate of flow at all points in time during and after a storm or snowmelt event.[1] Hydrologists depend on measured or computed (synthesized) hydrographs to provide peak flow rates so that hydraulic structures can be designed to accommodate the flow safely. Because a hydrograph plots volumetric flow rates against time, integration of the area beneath a hydrograph between any two points in time gives the total volume of water passing the point of interest during the time interval. Thus, in addition to peak flows, hydrographs allow analysis of sizes of reservoirs, storage tanks, detention ponds, and other facilities that deal with volumes of runoff. A knowledge of the magnitude and time distribution of streamflow is essential to many of these aspects of water management and environmental planning.

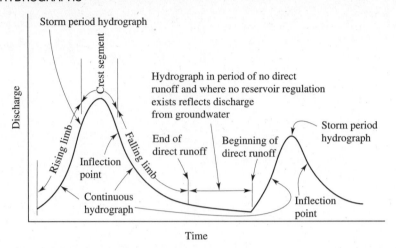

Figure 11.1　Hydrograph definition.

11.2 FACTORS AFFECTING HYDROGRAPH SHAPE

A hydrograph has four component elements: (1) direct surface runoff, (2) interflow, (3) groundwater or base flow, and (4) channel precipitation.[2] The rising portion of a hydrograph is known as the *concentration curve;* the region in the vicinity of the peak is called the *crest segment;* and the falling portion is the *recession.*[3] The shape of a hydrograph depends on precipitation pattern characteristics and basin properties. Figure 11.1 illustrates the definitions presented.

Precipitation-Streamflow Processes

During a given rainfall, water is continually being abstracted to saturate the upper levels of the soil surface; however, this saturation or infiltration is only one of many continuous abstractions.[4-6] Rainfall is also intercepted by trees, plants, and roof surfaces, and at the same time is evaporated. Once rain falls and fulfills initial requirements of infiltration, natural depressions collect falling rain to form small puddles, creating *depression storage.* In addition, numerous pools of water forming *detention storage* build up on permeable and impermeable surfaces within the watershed. This stored water gathers in small rivulets, which carry the water originating as *overland flow* into small channels, then into larger channels, and finally as *channel flow* to the watershed outlet. Figure 11.2a illustrates the distribution of a prolonged uniform rainfall. Although such an event is not the norm, the concept is useful for showing the manner in which detention and depression storage would be distributed.

In general, the channel of a watershed possesses a certain amount of *base flow* during most of the year. This flow comes from groundwater or spring contributions and may be considered as the normal day-to-day flow. Discharge from precipitation excess—that is, after abstractions are deducted from the original rainfall—constitutes the direct runoff hydrograph (DRH). Arrival of direct runoff at the outlet accounts for an initial rise in the DRH. As precipitation excess continues, enough time elapses for progressively distant areas to add to the outlet flow. Consequently, the

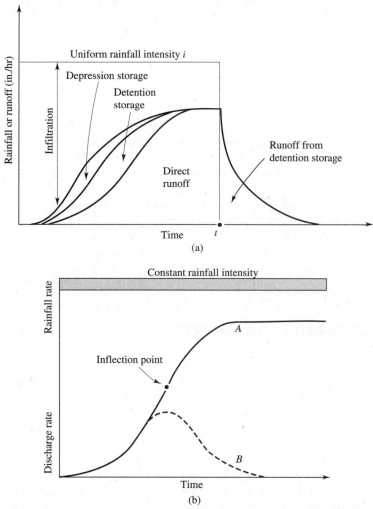

Figure 11.2   (a) Distribution of a uniform storm rainfall for condition of no interception loss. Note that all water stored in depressions is ultimately evaporated or infiltrated while some detention storage is also subjected to these losses. (b) Equilibrium discharge hydrograph.

duration of rainfall dictates the proportionate area of the watershed amplifying the peak, and the intensity of rainfall during this period of time determines the resulting greatest discharge.

Hydrograph Shapes

If the rainfall maintains a constant intensity for a long enough period of time, a state of equilibrium discharge is reached, as depicted by curve A in Fig. 11.2b. The inflection point on curve A often indicates the time at which the entire drainage area contributes to the flow. At this time maximum storage of the watershed is only partially complete. As rainfall continues, maximum storage capacity is attained and equilibrium [inflow (rainfall) equals outflow (runoff)] is reached. The condition of maximum storage and equilibrium is seldom if ever attained in nature. Extended rainfall may occur, but variations in intensity throughout its duration negate any possibility of a DRH of the theoretical shape for constant rainfall intensity.

A normal single-peak DRH generally possesses the shape shown by curve B in Fig. 11.2b rather than by the curve in Fig. 11.2a. The time to peak magnitude of this hydrograph depends on the intensity and duration of the rainfall, and the size, slope, shape, and storage capacity of the watershed. Once peak flow has been reached for a given isolated rainstorm, the DRH begins to descend, its source of supply coming largely from water accumulated within the watershed such as detention and channel storage.

Processes involved in forming the DRH can be better understood by visualizing the precipitation excess as partially disposed of immediately by surface runoff while a portion remains held within the watershed boundaries and is released later from storage. Thus the shape and timing of the DRH are integrated effects of the duration and intensity of rainfall and other hydrometeorological factors as well as the effect of the physiographic factors of the watershed upon the storage capacity.

11.3 HYDROGRAPH COMPONENTS

It is important to understand how the hydrograph can be subdivided into its component parts and to look at the effect on hydrograph shape of precipitation and watershed features. Figures 11.3 and 11.4 are used for this purpose.

A hydrograph is a continuous graph showing the rate of streamflow with respect to time, normally obtained by means of a continuous strip recorder that indicates stage versus time (stage hydrograph), which is then transformed to a discharge hydrograph by application of a rating curve. Hereafter, the term *hydrograph* is generally taken to indicate a discharge hydrograph.

Figure 11.3a illustrates the hydrograph of a permanent stream during a period between precipitation events, known as a *base flow hydrograph* because groundwater sustains the flow. Four general conditions cause modification of the base flow hydrograph shape. They are described by Horton[7] using the following sets of inequalities:

$$\begin{array}{llll} \text{Set 1} & i < f & \text{Set 3} & i > f \\ & F < S_D & & F < S_D \\ \text{Set 2} & i < f & \text{Set 4} & i > f \\ & F > S_D & & F > S_D. \end{array}$$

where Set 1 would produce a hydrograph similar to that shown in Fig. 11.3a except for a very small rise due to direct channel precipitation.

Interflow is that part of the subsurface flow that moves at shallow depths and reaches the surface channels in a relatively short period of time and therefore is commonly considered part of the direct surface runoff. No overland flow occurs in Fig. 11.3a since $i < f$, and since the soil's field capacity F is not reached, no interflow

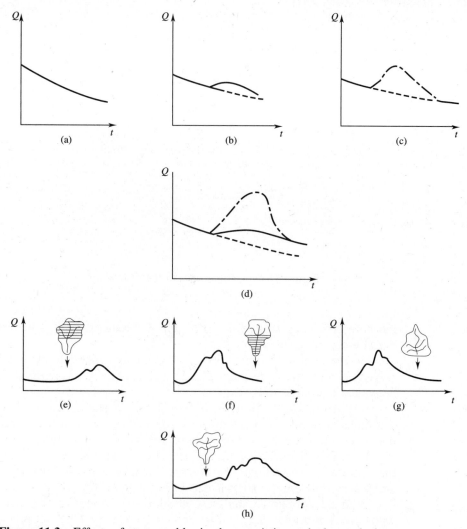

Figure 11.3 Effects of storm and basin characteristics on hydrograph shape.

or added groundwater components develops. The entire effect of the storm would be to slightly reduce the soil moisture deficiency S_D. The field capacity is the amount of water held in the soil after excess gravitational water has drained.

The conditions described by Set 2 still do not produce direct surface runoff, although the components of interflow and groundwater flow are added to channel precipitation. The initial hydrograph would be modified, since the field capacity of the soil is exceeded. Figure 11.3b illustrates this condition. Note that deviation of the hydrograph from the original base flow curve is likely to be very small under these conditions.

Figure 11.3c illustrates a case where surface runoff becomes a component of flow because $i > f$. In this example, interflow and groundwater flow are zero, as soil

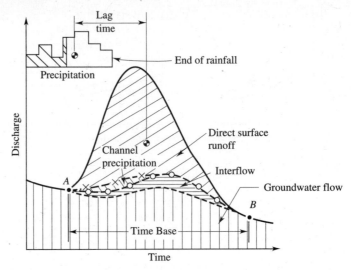

Figure 11.4   Components of the hydrograph.

moisture deficiency still exists, although at a reduced level. Channel precipitation likewise constitutes a component.

In the final set, Fig. 11.3d, all four components exist with rainfall intensity exceeding infiltration rate and the field capacity of the soil is reached. This case would be typical of a large storm event.

Figures 11.3e–h illustrate how hydrograph shape can be modified by areal variations in rainfall and rainfall intensity and by watershed configuration.[8] Minor fluctuations shown in these hydrographs are linked to variations in storm intensity. In Fig. 11.3e only the delaying effects pertinent to a storm over the upstream section of the area are indicated. Figure 11.3f shows the reverse of this condition. Figures 11.3g and h depict the comparative effects of basin geometry.

In most hydrograph analyses, interflow and channel precipitation are grouped with surface runoff rather than treated independently. Channel precipitation begins with inception of rainfall and ends with the storm. Its distribution with respect to time is highly correlated with the storm pattern. The relative volume contribution tends to increase somewhat as the storm proceeds, since stream levels rise and the water surface area tends to increase. The fraction of watershed area occupied by streams and lakes is generally small, usually on the order of 5 percent or less, so the percentage of runoff related to channel precipitation is usually minor during important storms. Distribution of interflow is commonly characterized by a slowly increasing rate up to the end of the storm period, followed by a gradual recession that terminates at the intersection of the surface flow hydrograph and base flow hydrograph. Figure 11.4 illustrates the approximate nature of the components of channel precipitation and interflow.

The base flow component is composed of the water that percolates downward until it reaches the groundwater reservoir and then flows to surface streams as ground-water discharge. The groundwater hydrograph may or may not show an increase during the actual storm period. Groundwater accretion resulting from a particular storm is normally released over an extended period, measured in days for small watersheds and often in months or years for large drainage areas.

The surface runoff component consists of water that flows overland until a stream channel is reached. During large storms it is the most significant hydrograph component. Figure 11.4 illustrates the surface runoff and groundwater components of a hydrograph. As pointed out in Fig. 11.3, the relative magnitude of each component for a given storm is determined by a combination of many factors. Hydrographs are analyzed to provide knowledge of the way precipitation and watershed characteristics interact to form them. The degree of hydrograph separation required depends on the objective of the study. For most practical work, surface runoff and groundwater components only are required. Research projects or more sophisticated analyses may dictate consideration of all components. When multiple storms occur within short periods, it is sometimes necessary to separate the overlapping parts of consecutive surface runoff hydrographs.

11.4 BASE FLOW SEPARATION

Several techniques are used to separate a hydrograph's surface and groundwater flows. Most are based on analyses of groundwater recession or depletion curves. If there is no added inflow to the groundwater reservoir, and if all groundwater discharge from the upstream area is intercepted at the stream-gauging point of interest, then groundwater discharge can be described by either [9,10]

$$q_t = q_0 K^t \quad \text{or} \quad q_t = q_0 e^{-Kt} \tag{11.1}$$

where q_0 = a specified initial discharge
       q_t = the discharge at any time t after flow q_0
       K = a recession constant
       e = base of natural logarithms

Time units frequently used are days for large watersheds and hours or minutes for small basins. A plot of either yields a straight line on semilogarithmic paper by plotting t on the linear scale.

For most watersheds, groundwater depletion characteristics are approximately stable, since they closely fit watershed geology. Nevertheless, the recession constant varies with seasonal effects such as evaporation and freezing cycles and other factors. Because $q_t \, dt$ is equivalent to $-dS$, where S is the quantity of water obtained from storage, integration of Eq. 11.1 produces

$$S = \frac{q_t - q_0}{\log_e K} \tag{11.2}$$

This equation determines the quantity of water released from groundwater storage between the times of occurrence of the two discharges of interest, or it can be used to calculate the volume of water still in storage at a time some chosen value of flow occurs. To get the latter, q_t is set equal to zero and q_0 becomes the reference discharge. Figure 11.5a is a plot of Eqs. 11.1 and 11.2 and provides additional definition.

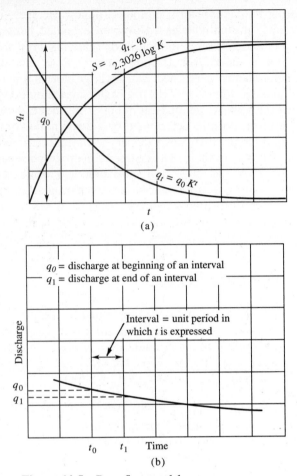

$$S = \frac{q_t - q_0}{2.3026 \log K}$$

q_t

q_0

$q_t = q_0 K^t$

t

(a)

Discharge

q_0 = discharge at beginning of an interval
q_1 = discharge at end of an interval

Interval = unit period in
which t is expressed

q_0
q_1

t_0 t_1 Time

(b)

Figure 11.5 Base flow model.

Groundwater depletion curves can be analyzed by various methods to evaluate the recession constant K. One of these will be described. Data from a stream-gauging station are a prerequisite and should reflect rainless periods with no upstream regulation, such as a reservoir, to affect flow at the gauging point. Otherwise an adjustment with its own errors is introduced.

From the streamflow data, plot a portion of the recession hydrograph (Fig. 11.5b) to find values of discharge at the beginning and end of selected time intervals. Flows at the beginning of each interval are analogous to q_0, whereas those at the end are analogous to q_1. Next, select several time intervals and plot corresponding q_0's versus q_1's shown in Fig. 11.6. The time period between consecutive values of q should be identical for each datum set. Figures taken from recession curves of times that still reflect surface runoff will usually fall below and to the right of a 45° line drawn on the plot. These values will also be associated with larger numbers for q. Points taken from true groundwater recession periods should approximately

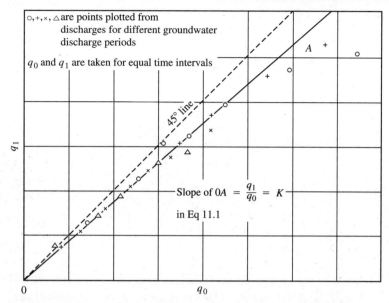

Figure 11.6   Graphical method for determining recession constant K. (U.S. Department of Agriculture, Soil Conservation Service.)

describe a straight line. Because $q_1 = q_0 = 0$ when $q_0 = 0$, a straight line can be fitted graphically to the data points. The slope of this line is $q_1/q_0 = K$. Using this value, the depletion curve plots as a straight line on semilogarithmic paper (t is the linear scale variable) or as a curve on arithmetic paper, Fig. 11.5a.

Separation Techniques

Several methods for base flow separation are used when the actual amount of base flow is unknown. During large storms, the maximum rate of discharge is only slightly affected by base flow, and inaccuracies in separation may not be important.

The simplest base flow separation technique is to draw a horizontal line from the point at which surface runoff begins, Point A in Fig. 11.7, to an intersection with the hydrograph recession where the base flow rate is the same as at the beginning of direct runoff as indicated by Point B. A second method projects the initial recession curve downward from A to C, which lies directly below the peak rate of flow. Then point D on the hydrograph, representing N days after the peak, is connected to point C by a straight line defining the groundwater component. One estimate of N is based on the formula[3]

$$N = A^{0.2} \qquad (11.3)$$

where N = the time in days

A = the drainage area in square miles

A third procedure is to develop a base flow recession curve using Eq. 11.1 for data from the segment FG, and then back-calculate all base flow to the left of Point F,

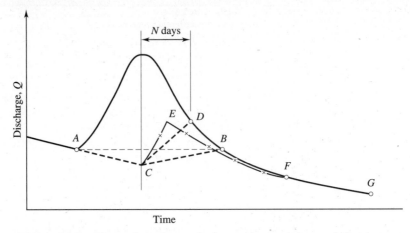

Figure 11.7   Illustration of some hydrograph separation techniques.

where the computed curve begins to deviate from the actual hydrograph, marking the end of direct runoff. The curve is projected backward arbitrarily to some Point E below the inflection point and its shape from C to E is arbitrarily assigned. A fourth widely used method is to draw a line between A and F, and a fifth common method is to project the line AC along the slope to the left of A, and then connect Points C and

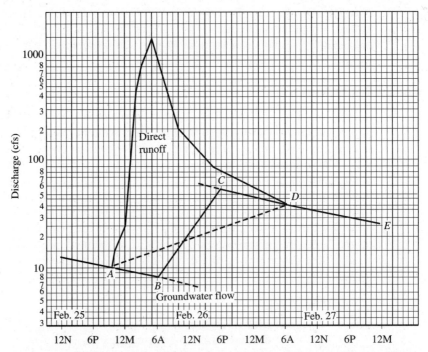

Figure 11.8   Illustration of base flow separation: hydrograph for the Uharie River near Trinity, North Carolina, February 25, 1939. (U.S. Department of Agriculture, Soil Conservation Service).

B. All these methods are approximate since the separation of hydrographs is partly a subjective procedure.

Figure 11.8 illustrates two graphical separation techniques to determine surface runoff and groundwater flow components. Line AD represents the simple procedure of connecting the point of the beginning of direct runoff with the first point on the groundwater recession curve (an advantage over the horizontal line technique because the time base of direct runoff is much shorter). Curve ABCD is constructed from the extension of the base flow recession curve.

11.5 HYDROGRAPH TIME RELATIONSHIPS

Wave travel time is defined as the time required for direct runoff originating at the most remote point in the channel to reach the outlet. The last drop of direct runoff to pass the outlet conceptually travels over the water surface and reaches the outlet at the speed of a small surface wave, rather than at a speed equal to the average velocity of flow. The wave travel time is faster than the average velocity and varies with channel shape and other factors. For a rectangular channel, the ratio is approximately 5/3 (see Section 13.1 for other wave velocities). The time base (Fig. 11.4) of a hydrograph is considered to be the time from which the concentration curve begins until the direct-runoff component reaches zero. An equation for time base may take the form

$$T_b = t_s + t_r \qquad (11.4)$$

where T_b = the time base of the direct runoff hydrograph
t_s = the duration of runoff-producing rain
t_r = the excess rainfall release time

Watershed lag time, illustrated in Fig. 11.4, is defined as the time from the center of mass of effective rainfall to the center of mass of direct runoff.[11] Other definitions and several equations relating lag time to watershed characteristics are provided in Section 11.7 and subsequent chapters.

Because of its importance in unit hydrograph theory, the excess-rainfall release time is introduced. This is defined as the time required for the last, most remote drop of excess rain that fell on the watershed to pass the outlet, signalling the cessation of direct runoff. It is easily determined as the time interval between the end of rain and the end of direct runoff. Only that part of the outflow which classifies as direct runoff (excess rain) is considered in determining the release time. Watershed outflow normally continues after cessation of direct runoff, in the form of interflow and base flow. Release time is very similar by definition to wave travel time and time of concentration (Section 11.6).

A foundational assumption of unit hydrograph theory[12] is that the watershed excess release time is a constant, regardless of the storm duration, and is related to basin factors rather than meteorological characteristics. The excess release time is also conceptually identical with the time base of an instantaneous unit hydrograph (IUH). This is the runoff hydrograph from 1.0 in. of excess rain applied uniformly over the watershed in an instant of time (see Chapter 12). Both wave travel time and excess-rainfall release time are often used synonymously with time of concentration.

11.6 TIME OF CONCENTRATION

The most common definition of time of concentration originates from consideration of overland flow. If a uniform rain is applied to a tract, the portions nearest the outlet contribute runoff at the outlet almost immediately. As rain continues, the depth of excess on the surface grows and discharge rates increase throughout. Runoff contributions from various points upstream arrive at later times, adding themselves to continuing runoff from nearer points, until flow eventually arrives from all points on the watershed, "concentrating" at the outlet. Thus, concentration time is the time required, with uniform rain, for 100 percent of a tract of land to contribute to the direct runoff at the outlet.[9]

As a second popular definition, the concentration time is often equated with either the excess-rainfall release time or the wave travel time because the time for runoff to arrive at the outlet from the most remote point after rain ceases is assumed to be indicative of the time required for 100 percent contribution from all points during any uniform storm having sufficient duration. The latter definition is often preferred because few storm durations exceed the time of concentration, making determination of t_c possible only by examining excess rain recession.

Because time of concentration is conceptually the time required for 100 percent of the watershed to contribute, it is also often defined as the time from the end of excess rainfall to the inflection point on the hydrograph recession limb (e.g., see Fig. 12.2). The reasoning used in this definition is that direct runoff ceases at the point of inflection.

For a small tract of land experiencing uniform rain, the entire area contributes at approximately the same time that the runoff reaches an equilibrium. This gives rise to yet another definition of time of concentration. If rain abruptly ceased, the direct runoff would continue only as long as the excess-rainfall release time t_r. On the basis of the second definition, excess release time and time of concentration can be considered equivalent.

Numerous equations relating time of concentration to watershed parameters have been developed. Table 11.1 summarizes several popular versions. Other variations are presented in Chapters 12, 15, 16, and 25.

11.7 BASIN LAG TIME

The relative timing of rainfall and runoff must be known if drainage areas having subbasins are to be modeled or if continuous simulation is desired. A basic measure of timing is basin lag, which locates the hydrograph's position relative to the causative storm pattern. It is most often defined as the difference in time between the center of mass of effective rainfall and the center of mass of runoff produced. Other definitions are also used. Two of these are (1) the time interval from the maximum rainfall rate to the peak rate of runoff and (2) the time from the center of mass of effective rainfall to the peak rate of flow. Time lag is characterized by the ratio of a flow length to a mean velocity of flow and is thus a property that is influenced by the shape of the drainage area, the slope of the main channel, channel roughness and geometry, and the storm pattern.

TABLE 11.1   SUMMARY OF TIME OF CONCENTRATION FORMULAS

Method and date	Formula for t_c (min)	Remarks
Kirpich (1940)	$t_c = 0.0078L^{0.77}S^{-0.385}$ L = length of channel/ditch from headwater to outlet, ft S = average watershed slope, ft/ft	Developed from SCS data for seven rural basins in Tennessee with well-defined channel and steep slopes (3% to 10%); for overland flow on concrete or asphalt surfaces multiply t_c by 0.4; for concrete channels multiply by 0.2; no adjustments for overland flow on bare soil or flow in roadside ditches.
USBR Design of Small Dams (1973)	$t_c = 60(11.9L^3/H)^{0.385}$ L = length of longest watercourse, mi H = elevation difference between divide and outlet, ft	Essentially the Kirpich formula; developed from small mountainous basins in California (U.S. Bureau of Reclamation, 1973, pp. 67–71).[14]
Izzard (1946)[15]	$t_c = \dfrac{41.025(0.0007i + c)L^{0.33}}{S^{0.333}i^{0.667}}$ i = rainfall intensity, in/h c = retardance coefficient L = length of flow path, ft S = slope of flow path, ft/ft	Developed in laboratory experiments by Bureau of Public Roads for overland flow on roadway and turf surfaces; values of the retardance coefficient range from 0.0070 for very smooth pavement to 0.012 for concrete pavement to 0.06 for dense turf; solution requires iteration; product i times L should be ≤ 500.
Federal Aviation Administration (1970)[16]	$t_c = 1.8(1.1 - C)L^{0.50}/S^{0.333}$ C = rational method runoff coefficient L = length of overland flow, ft S = surface slope, %	Developed from air field drainage data assembled by the Corps of Engineers; method is intended for use on airfield drainage problems, but has been used frequently for overland flow in urban basins.
Kinematic Wave Formulas Morgali and Linsley (1965)[17] Aron and Erborge (1973)[18]	$t_c = \dfrac{0.94L^{0.6}n^{0.6}}{(i^{0.4}S^{0.3})}$ L = length of overland flow, ft n = Manning roughness coefficient i = rainfall intensity in/h S = average overland slope ft/ft	Overland flow equation developed from kinematic wave analysis of surface runoff from developed surfaces; method requires iteration since both i (rainfall intensity) and t_c are unknown; superposition of intensity–duration–frequency curve gives direct graphical solution for t_c.
SCS Lag Equation (1972)[19]	$t_c = \dfrac{1.67\,L^{0.8}[(1000/CN) - 9]^{0.7}}{1900\,S^{0.5}}$ L = hydraulic length of watershed (longest flow path), ft CN = SCS runoff curve number S = average watershed slope, %	Equation developed by SCS from agricultural watershed data; it has been adapted to small urban basins under 2000 acres; found generally good where area is completely paved; for mixed areas it tends to overestimate; adjustment factors are applied to correct for channel improvement and impervious area; the equation assumes that $t_c = 1.67 \times$ basin lag.
SCS Average Velocity Charts (1975, 1986)[20]	$t_c = \dfrac{1}{60} \Sigma \dfrac{L}{V}$ L = length of flow path, ft V = average velocity in feet per second from Fig. 3-1 of TR 55 for various surfaces	Overland flow charts in Ref. 20 provide average velocity as function of watercourse slope and surface cover.

Source: After Ref. 13.

Various studies have been conducted for the purpose of developing relations descriptive of time lag. Most prominent of these was the work by Snyder on large natural watersheds.[21] His original equation has been widely used and modified in various ways by other investigators. Eagleson has proposed an equation for lag time on sewered drainage areas having a minimum size of 141 acres.[22] An early investigation (1936) on small drainage areas (2–4 acres) was conducted by Horner in his classical work on urban drainage in St. Louis, Missouri.[23] Horner's work was inconclusive in that it did not yield a defined procedure, but he did conclude that the comparatively wide range in the lag time at each location led to the inference that the lag was a variable, its value being determined more by rainfall characteristics than by characteristics of the drainage area.

Snyder's study based on data from the Appalachian Mountain region produced the following equation for lag time:[21]

$$t_l = C_t (L_{ca} L)^{0.3} \tag{11.5}$$

where t_l = the lag time (hr) between the center of mass of the rainfall excess for a specified type of storm and the peak rate of flow

L_{ca} = the distance along the main stream from the base to a point nearest the center of gravity of the basin (mi)

L = length of the main stream channel (mi) from the base outlet to the upstream end of the stream and including the additional distance to the watershed divide

C_t = a coefficient representing variations of types and locations of streams

For the area studied, the constant C_t was found to vary from 1.8 to 2.2, with somewhat lower values for basins with steeper slopes. The constant is considered to include the effects of slope and storage. The value of t_l is assumed to be constant for a given drainage area, but allowance is made for the use of different values of lag for different types of storms. The relation is considered applicable to drainage areas ranging in size from 10 to 10,000 mi².

In a study of sewered areas ranging in size from 0.22 to 7.51 mi², Eagleson[22] developed the equation

$$t_l = \frac{\bar{L}}{(1.5/n)\bar{R}^{2/3}\bar{S}^{1/2}} \tag{11.6}$$

where t_l = lag time, the center of mass of rainfall excess to the peak discharge (sec)

$\bar{L}$ = the mean travel distance (ft), which is equal to the length of that portion of the sewer which flows full

n = the weighted Manning's coefficient for the main sewer

$\bar{R}$ = the weighted hydraulic radius of the main sewer flowing full

$\bar{S}$ = the weighted physical slope of the main sewer

Eagleson's equation directly includes the effects of channel geometry and slope, as well as basin shape, and thus represents a refinement of the Snyder approach. It also indirectly includes the important effect of the storm pattern.

Linsley and Ackerman give examples of application of the following modified form of Snyder's equation.[24]

$$t_l = C_t \frac{(LL_{ca})^a}{\sqrt{s}} \tag{11.7}$$

where s is a weighted slope of the channel and the other variables are as defined previously.

Other investigators have represented time lag by equations of the form

$$t_l = K \frac{L}{\sqrt{s}} \tag{11.8}$$

Numerous other derivations of relations for watershed lag times can be found in standard hydrologic texts and periodical literature. Others are included with some of the synthetic unit hydrograph discussions in Chapter 12.

■ Summary

Understanding the structure of hydrographs is important to many design and water supply applications. The hydrograph represents the portion of the hydrologic cycle that engineers most often need in order to determine rates of flow in streams for setting bridge lengths and elevations, designing flood protection measures, and establishing areal extent of flooding. Similarly, the volume of drainage into a reservoir or past a water supply diversion is determined from the area under the hydrograph. Accurate estimates of these volumes are important to design of dams, reservoirs, pipelines, and numerous other structures.

After grasping the fundamentals of hydrograph components, including the time relationships presented in this chapter, the reader should be well prepared for the quantitative developments of hydrograph theory and applications presented throughout Chapters 12 through 16 and in Part Five.

PROBLEMS

11.1. Refer to Fig. 11.1. Replot this hydrograph and use two different techniques to separate the base flow.

11.2. Obtain streamflow data for a water course of interest. Plot the hydrograph for a major runoff event and separate the base flow.

11.3. For the event of Problem 11.2, tabulate the precipitation causing the surface runoff and determine the duration of runoff–producing rain. Estimate the time of concentration and use Eq. 11.4 to estimate the time base of the hydrograph. Compare this with the time base computed from the hydrograph.

11.4. Tabulated below are total hourly discharge rates at a cross section of a stream. The drainage area above the section is 1.0 acre.
   a. Plot the hydrograph on rectangular coordinate paper and label the rising limb (concentration curve), the crest segment, and the recession limb.

    b. Determine the hour of cessation of the direct runoff using a semilog plot of Q versus time.

    c. Use the base flow portion of your semilog plot to determine the groundwater recession constant K.

    d. Carefully construct and label base flow separation curves on the graph of Part a, using two different methods.

Time (hr)	Q (cfs)	Time (hr)	Q (cfs)
0	102	8	210
1	100	9	150
2	98	10	105
3	220	11	75
4	512	12	60
5	630	13	54
6	460	14	48.5
7	330	15	43.5

11.5. On a neat sketch of a typical total runoff hydrograph, show or dimension the (a) storm hyetograph, (b) beginning of direct runoff, (c) cessation time of direct runoff, (d) base flow separation assuming that *additional* contributions to base flow are negligible during the period of rise, and (e) crest segment of the hydrograph.

11.6. For an urban watershed assigned by your instructor, obtain measures of the watershed area, length, and slope, and compare estimates of the time of concentration using the Kirpich, USBR, FAA, and SCS Lag equations in Table 11.1.

REFERENCES

1. American Society of Civil Engineers, *Hydrology Handbook,* Manuals of Engineering Practice, No. 28. New York: ASCE, 1957.

2. Donn G. DeCoursey, "A Runoff Hydrograph Equation," U.S. Department of Agriculture, Agricultural Research Service, Feb. 1966, pp. 41–116.

3. R. K. Linsley, M. A. Kohler, and J. L. H. Paulhus, *Applied Hydrology.* New York: McGraw-Hill, 1949.

4. R. E. Horton, "Erosional Development of Streams and Their Drainage Basins: Hydrophysical Approach to Quantitative Morphology," *Bull. Geol. Soc. Am.* **56**(1945).

5. R. E. Horton, "An Approach Toward a Physical Interpretation of Infiltration Capacity," *Proc. Soil Sci. Soc. Am.* **5**, 399–417(1940).

6. J. R. Philip, "An Infiltration Equation with Physical Significance," *Soil Sci.* **77**(1954).

7. R. E. Horton, *Surface Runoff Phenomena.* Ann Arbor, MI: Edwards Bros., 1935.

8. R. J. M. DeWiest, *Geohydrology.* New York: Wiley, 1965.

9. "Hydrology," in *Engineering Handbook,* Sec. 4, U.S. Department of Agriculture, Soil Conservation Service, 1972.

10. B. S. Barnes, "Discussion of Analysis of Runoff Characteristics by O. H. Meyer," *Trans. ASCE* **105**(1940).

11. A. B. Taylor and H. E. Schwartz, "Unit Hydrograph Lag and Peak Flow Related to Basin Characteristics," *Trans. Am. Geophys. Union* **33**(1952).

12. L. K. Sherman, "Streamflow from Rainfall by the Unit-Graph Method," *Eng. News-Rec.* **108**(1932).

13. D. F. Kilber, "Desk-top methods for urban stormwater calculation," Ch. 4 in *Urban Stormwater Hydrology,* Water Resources Monograph No. 7, American Geophysical Union, Washington, D. C., 1982.

14. U.S. Bureau of Reclamation, *Design of Small Dams,* 2nd ed., Washington, D.C., 1973.

15. C. F. Izzard, "Hydraulics of Runoff from Developed Surfaces," Proceedings, 26th Annual Meeting of the Highway Research Board, 26, pp. 129–146, December 1946.

16. Federal Aviation Administration, "Circular on Airport Drainage," Report A/C 050-5320-5B, Washington, D.C., 1970.

17. J. R. Morgali, and R. K. Linsley, "Computer Analysis of Overland Flow," *J. Hyd. Div., Am. Soc. Civ. Eng.,* 91, no. HY3, May 1965.

18. G. Aron, and C. E. Egborge, "A Practical Feasibility Study of Flood Peak Abatement in Urban Areas," U.S. Army Corps of Engineers, Sacramento, Calif., March 1973.

19. Soil Conservation Service, "National Engineering Handbook, Sec. 4, Hydrology," U.S. Dept. of Agriculture, U.S. GPO, Washington, D.C., 1972.

20. Soil Conservation Service, "Urban Hydrology for Small Watersheds," Technical Release 55, Washington, D.C., 1975 (updated, 1986).

21. F. F. Snyder, "Synthetic Unit Graphs," *Trans. Am. Geophys. Union* **19,** 447–454(1938).

22. Peter S. Eagleson, "Characteristics of Unit Hydrographs for Sewered Areas," paper presented before the ASCE, Los Angeles, CA, 1959, unpublished.

23. W. W. Horner, and F. L. Flynt, "Relation Between Rainfall and Runoff from Small Urban Areas," *Trans. ASCE* **62**(101), 140–205(Oct. 1956).

24. R. K. Linsley, Jr., and W. C. Ackerman, "Method of Predicting the Runoff from Rainfall," *Trans. ASCE* **107**(1942).

Chapter .12

Unit Hydrographs

■ Prologue

The purpose of this chapter is to:

- Define *unit hydrographs* and show their utility in hydrologic studies and design.
- Develop fully the current methods of obtaining, analyzing, and synthesizing unit hydrographs.
- Present methods for converting unit hydrographs for one storm duration to other storm durations.

Ways to predict flood peak discharges and discharge hydrographs from rainfall events have been studied intensively since the early 1930s. One approach receiving considerable use is called the *unit hydrograph method*.

12.1 UNIT HYDROGRAPH DEFINITION

The concept of a unit hydrograph was first introduced by Sherman[1,2] in 1932. He defined a unit graph as follows:[2]

> If a given one-day rainfall produces a 1-in. depth of runoff over the given drainage area, the hydrograph showing the rates at which the runoff occurred can be considered a unit graph for that watershed.

Thus, a unit hydrograph is the hydrograph of *direct runoff* (excludes base flow) for any storm that produces exactly 1.0 inch of net rain (the total runoff after abstractions). Such a storm would not be expected to occur, but Sherman's assumption is that the ordinates of a unit hydrograph are $1.0/P$ times the ordinates of the direct runoff hydrograph for an *equal-duration* storm with P inches of net rain.

The term "unit" has to do with the net rain amount of 1.0 inch and does not mean to imply that the duration of rain that produced the hydrograph is one unit, whether an hour, day, or any other measure of time. The storm duration, X, that produced the unit hydrograph must be specified because a watershed has a different unit hydrograph for each possible storm duration. An *X-hour unit hydrograph* is defined as a direct runoff hydrograph having a 1.0-in. volume and resulting from an *X*-hour storm having a net rain rate of $1/X$ in./hr. A 2-hr unit hydrograph would have a 1.0-in. volume produced by a 2-hr storm, and a 1-day unit hydrograph would be produced by a storm having 1.0 in. of excess rain uniformly produced during a 24-hr period. The value *X* is often a fraction. Figure 12.1 illustrates a 2-hr, 12-hr, and 24-hr unit hydrograph for a given watershed.

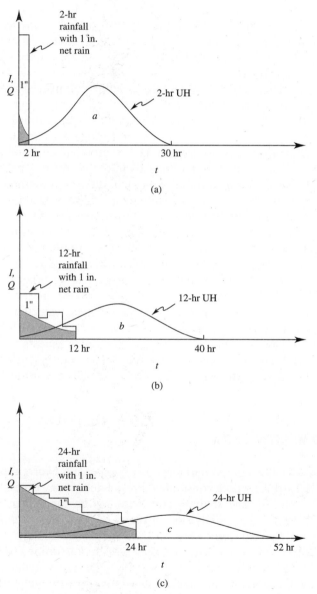

Figure 12.1 Illustration of 2-hr, 12-hr, and 24-hr unit hydrographs for the same watershed (Note: a = b = c = 1″ × A).

By Sherman's assumption, application of an *X*-hour unit graph to design rainfall excess amounts other than 1 in. is accomplished simply by multiplying the rainfall excess amount by the unit graph ordinates, since the runoff ordinates for a given duration are assumed to be directly proportional to rainfall excess. A 3-hr storm producing 2.0 in. of net rain would have runoff rates 2 times the values of the 3-hr unit hydrograph. One-half inch in 3 hr would produce flows half the magnitude of the 3-hr unit hydrograph. This principle of proportional flows is expanded in Section 12.3 and applies only to equal duration storms.

Implicit in deriving the unit hydrograph is the assumption that rainfall is distributed in the same temporal and spatial pattern for all storms. This is generally not true; consequently, variations in ordinates for different storms of equal duration can be expected.

This chapter is organized to define unit hydrographs first, then present methods of deriving unit hydrographs from actual rainfall and runoff records (Section 12.2). After familiarizing the reader with the origin of unit hydrographs, Section 12.3 presents methods of applying unit hydrographs to generate direct runoff hydrographs for any storm with durations that are multiple integers of the U.H. duration.

The construction of unit hydrographs for storms with other than integer multiples of the derived duration is facilitated by a method known as the *S-hydrograph* developed by Morgan and Hulinghorst.[3] The procedure, as explained in Section 12.4, employs a unit hydrograph to form an S-hydrograph resulting from a continuous applied rainfall. The need to alter duration of a unit hydrograph led to studies of the shortest possible storm duration—the instantaneous unit rainfall. The concept of *instantaneous unit hydrograph* (IUH) is traced to Clark[6] and can also be used (Section 12.5) is constructing unit hydrographs for other than the derived duration.

The previous discussion assumes that the analyst has runoff and rainfall data for deriving a unit hydrograph for the subject watershed. The application of unit hydrograph theory to ungauged watersheds received early attention by Snyder[4] and also by Taylor and Schwartz,[5] who tried to relate aspects of the unit hydrograph to watershed characteristics. As a result, a full set of synthetic unit-hydrograph methods emerged. A number of these are presented in Section 12.6.

12.2 DERIVATION OF UNIT HYDROGRAPHS FROM STREAMFLOW DATA

Data collection preparatory to deriving a unit hydrograph for a gauged watershed can be extremely time consuming. Fortunately, many watersheds have available records of streamflow and rainfall, and these can be supplemented with office records of the Water Resources Division of the U.S. Geological Survey.[7] Rainfall records may be secured from *Climatological Data*[8] published for each state in the United States by the National Oceanic and Atmospheric Administration (NOAA). Hourly rainfall records for recording rainfall stations are published as a *Summary of Hourly Observations* for the location. Summaries are listed for approximately 300 first-order situations in the United States.

To develop a unit hydrograph, it is desirable to acquire as many rainfall records as possible within the study area to ensure that the amount and distribution of rainfall over the watershed is accurately known. Preliminary selection of storms to use in deriving a unit hydrograph for a watershed should be restricted to the following:

1. Storms occurring individually, that is, simple storm structure.
2. Storms having uniform distribution of rainfall throughout the period of rainfall excess.
3. Storms having uniform spatial distribution over the entire watershed.

These restrictions place both upper and lower limits on size of the watershed to be employed. An upper limit of watershed size of approximately 1000 mi^2 is overcautious, although general storms over such areas are not unrealistic and some studies of areas up to 2000 mi^2 have used the unit-hydrograph technique. The lower limit of watershed extent depends on numerous other factors and cannot be precisely defined. A general rule of thumb is to assume about 1000 acres. Fortunately, other hydrologic techniques help resolve unit hydrographs for watersheds outside this range.

The preliminary screening of suitable storms for unit-hydrograph formation should meet more restrictive criteria before further analysis:

1. Duration of rainfall event should be approximately 10–30 percent of the drainage area lag time.
2. Direct runoff for the selected storm should range from 0.5 to 1.75 in.
3. A suitable number of storms with the same duration should be analyzed to obtain an average of the ordinates (approximately five events). Modifications may be made to adjust different unit hydrographs to a single duration by means of S-hydrographs or IUH procedures.
4. Direct runoff ordinates for each hydrograph should be reduced so that each event represents 1 in. of direct runoff.
5. The final unit hydrograph of a specific duration for the watershed is obtained by averaging ordinates of selected events and adjusting the result to obtain 1 in. of direct runoff.

Constructing the unit hydrograph in this way produces the integrated effect of runoff resulting from a representative set of equal duration storms. Extreme rainfall intensity is not reflected in the determination. If intense storms are needed, a study of records should be made to ascertain their influence upon the discharge hydrograph by comparing peaks obtained utilizing the derived unit hydrograph and actual hydrographs from intense storms.

Essential steps in developing a unit hydrograph for an isolated storm are:

1. Analyze the streamflow hydrograph to permit separation of surface runoff from groundwater flow, accomplished by the methods developed in Section 11.4.
2. Measure the total volume of surface runoff (direct runoff) from the storm producing the original hydrograph. This is the area under the hydrograph after groundwater base flow has been removed.
3. Divide the ordinates of the direct runoff hydrograph by total direct runoff volume in inches, and plot these results versus time as a unit graph for the basin.

4. Finally, the effective duration of the runoff-producing rain for this unit graph must be found from the hyetograph (time history of rainfall intensity) of the storm event used.

Procedures other than those listed are required for complex storms or in developing synthetic unit graphs when data are limited. Unit hydrographs can also be transposed from one basin to another under certain circumstances. An example illustrates the derivation of a unit hydrograph.

EXAMPLE 12.1 _____

Using the total direct runoff hydrograph given in Fig. 12.2, derive a unit hydrograph for the 1715 ac drainage area.

Solution

1. Separate the base or groundwater flow to get the total direct runoff hydrograph. A common method is to draw a straight line *AC* that begins when

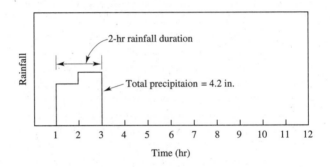

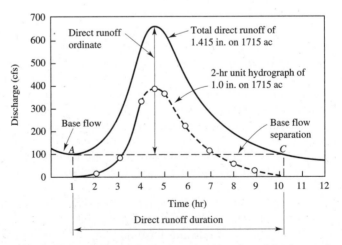

Figure 12.2   Illustration of the derivation of a unit hydrograph from an isolated storm.

the hydrograph starts an appreciable rise and ends where the recession curve intersects the base flow curve. The important point here is to be consistent in methodology from storm to storm.

2. The depth of direct runoff over the watershed is calculated using

$$\frac{\Sigma \ (DR \times \Delta t)}{area} = \frac{2447 \ \text{cfs-hr}}{1715 \ \text{ac}} = 1.4'' \qquad (12.1)$$

where DR is the average height of the direct runoff ordinate during a chosen time period Δt (in this case $\Delta t = 1.0$ hr). The values of DR determined from Fig. 12.2 are listed in Table 12.1.

3. Compute ordinates of the unit hydrograph by using

$$\frac{Q_s}{V_s} = \frac{Q_u}{1} \qquad (12.2)$$

where Q_s = the magnitude of a hydrograph ordinate of direct runoff having a volume equal to V_s (in.) at some instant of time after start of runoff

Q_u = the ordinate of the unit hydrograph having a volume of 1 in. at some instant of time

In this example the values are obtained by dividing the direct runoff ordinates by 1.415. Table 12.1 outlines the computation of the unit-hydrograph ordinates.

4. Determine the duration of effective rainfall (rainfall that actually produces surface runoff). As stated previously, the unit hydrograph storm duration

TABLE 12.1   DETERMINATION OF A 2-HR UNIT HYDROGRAPH
FROM AN ISOLATED STORM

(1) Time (hr)	(2) Runoff (cfs)	(3) Base flow (cfs)	(4) Direct runoff, (2)–(3) (cfs)	(5) 2-hr unit hydrograph ordinate, (4) ÷ 1.415 (cfs)
1	110	110	0	0
2	122	110	12	8.5
3	230	110	120	84.8
4	578	110	468	331
4.7	666	110	556	393
5	645	110	535	379
6	434	110	324	229
7	293	110	183	129
8	202	110	92	65.0
9	160	110	50	35.3
10	117	110	7	4.9
10.5	105	105	0	0
11	90	90	0	0
12	80	80	0	0

should not exceed about 25 percent of the drainage area lag time, but violates this rule for the example. From Fig. 12.2, the rain duration is 2 hr.

5. Using the values from Table 12.1, plot the unit hydrograph shown in Fig. 12.2.   ∎∎

12.3 UNIT HYDROGRAPH APPLICATIONS BY LAGGING METHODS

Once an X-hr unit hydrograph has been derived from streamflow data (or synthesized from basin parameters, Section 12.6) it can be used to estimate the direct runoff hydrograph shape and duration for virtually any rain event. Applications of the X-hr UH to other storms begins with *lagging procedures,* used for storms having durations that are integer multiples of the derived duration. Applications to storms with fractional multiples of X, known as *S-hydrograph* and *IUH* procedures, are discussed in Sections 12.4 and 12.5.

Because unit hydrographs are applicable to effective (net) rain, the process of applying UH theory to a storm begins by first abstracting the watershed losses from the precipitation hyetograph, resulting in an effective rain hyetograph. Any of the procedures detailed in Chapter 4 can be applied. The remainder of this discussion assumes that the analyst has already abstracted watershed losses from the storm.

If the duration of another storm is an integer multiple of X, the storm is treated as a series of end-to-end X-hour storms. First, the hydrographs from each X increment of rain are determined from the X-hour unit hydrograph. The ordinates are then added at corresponding times to determine the total hydrograph.

EXAMPLE 12.2 ──

Discharge rates for the 2-hr unit hydrograph shown in Fig. 12.3 are:

Time (hr)	0	1	2	3	4	5	6
Q (cfs)	0	100	250	200	100	50	0

Develop hourly ordinates of the total hydrograph resulting from a 4-hr design storm having the following excess amounts:

Hour	1	2	3	4
Excess (in.)	0.5	0.5	1.0	1.0

Solution.   The 4-hr duration of the design storm is an integer multiple of the unit hydrograph duration. Thus, the total hydrograph can be found by adding the contributions of two 2-hr increments of end-to-end rain, as shown in Fig. 12.3c. The first 2-hr storm segment has 1.0 in. of net rain and thus reproduces a unit hydrograph. The second 2-hr storm segment has 2.0 in. of net rain (in 2 hr); thus its ordinates are twice those of a 2-hr unit hydrograph. The total hydrograph,

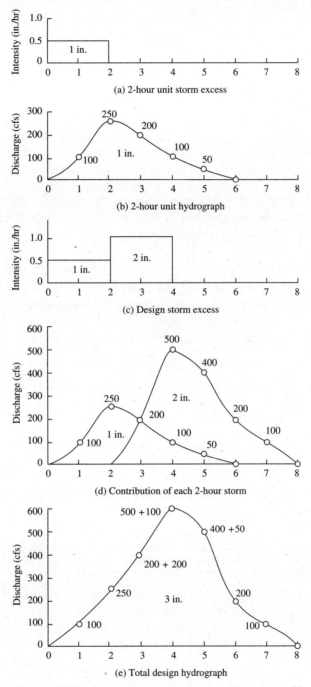

Figure 12.3   Example 12.2 derivation of total runoff hydrograph using a 2-hr unit hydrograph.

Fig. 12.3e, is found by summing the two contributions at corresponding times. Note in Fig. 12.3d that runoff from the second storm begins when the second rain begins, not at the beginning of the first storm. ■■

This method of "lagging" is based on the assumption that linear response of the watershed is not influenced by previous storms—that is, one can superimpose hydrographs offset in time and the flows will be directly additive. The simplest way to develop composite direct runoff hydrographs for multiple-hour storms is in a spreadsheet. Care must be taken, however, in visually confirming, as in Example 12.2, that the start and end points of runoff from each contributing X-hr increment of rain are properly selected. A common error is to lag each additional contributing hydrograph by Δt, the time interval between readings, rather than X, the associated duration with the given unit hydrograph. Also, the multiplier for the UH ordinates must be the net rain occurring in X hours, not the rain occurring in the time increment Δt. Example 12.3 illustrates these points.

EXAMPLE 12.3 _____

Using the derived 2-hr unit hydrograph in Table 12.1, determine the direct runoff hydrograph for a 4-hr. storm having the following excess rain amounts:

Hour	1	2	3	4
Excess rain, in.	0.7	0.7	1.2	1.2

Solution

1. Tabulate the unit hydrograph at intervals of the selected time interval, Δt, as shown in Table 12.2.

TABLE 12.2   UNIT HYDROGRAPH APPLICATION OF EXAMPLE 12.3

Time (hr)	Effective rainfall (in.)	Unit hydrograph (cfs)	Contrib. of first 2-hr rain UH × 1.4"	Contrib. of second 2-hr rain UH × 2.4"	Total outflow hydrograph (cfs)
0	—	0	0	—	0
1	0.7	8.5	11.9	—	11.9
2	0.7	84.8	119	0	119
3	1.2	331	463	20.4	483
4	1.2	379	531	203	734
5		229	321	794	1115
6		129	181	910	1091
7		65	91	550	641
8		35.3	49.4	310	359
9		4.9	6.9	156	163
10		0	0	84.7	84.7
11		—	—	11.8	11.8
12		—	—	0	0

2. Determine the correct UH multiplier for each X-hr interval. Because X is 2 hrs for this example, the first two hours of the storm produce a total net rain of 1.4 inches. Similarly, the last two hours of the storm produce 2.4 inches of net rain.

3. Determine the correct start and end times for each of the two hydrographs and tabulate the contribution of the 1.4-inch and 2.8-inch rains at the appropriate lag times. Because the second X-hr storm started at $t = 3$ hrs, runoff for this storm cannot begin until $t = 3$ hrs as shown in Table 12.2.

4. Add the contributions at each time to determine the total runoff hydrographs for the 4-hr storm.

5. Check the tabular solution by plotting each of the two hydrographs and sum the ordinates at each t, as shown in Fig. 12.4. ■■

In addition to using a given X-hr UH for determining the runoff hydrograph for a given storm, lagging of the X-hr UH can be used to develop other duration unit hydrographs. The procedure is the same as applying the X-hr UH to 1.0 in. of net rain in Y hours. As earlier, Y must be an integer multiple of X. For example, if a 1-hr unit hydrograph is available for a given watershed, a unit hydrograph resulting from a 2-hr storm is obtained by plotting two 1-hr unit hydrographs, with the second unit hydrograph lagged 1 hr, adding ordinates, and dividing by 2. This is demonstrated in Fig. 12.5, where the dashed line represents the resulting 2-hr unit hydrograph. Thus the 1 in. of rainfall contained in the original 1-hr duration has been distributed over a 2-hr period.

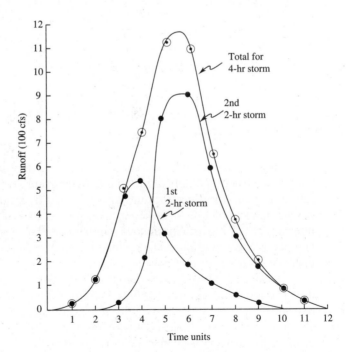

Figure 12.4 Synthesized hydrograph for Example 12.3 derived by the unit hydrograph method.

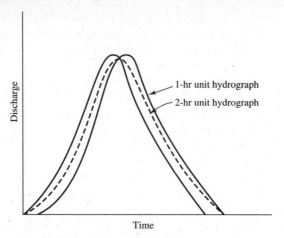

Figure 12.5   Unit hydrograph lagging procedure to develop another unit hydrograph.

Modifications of the original unit-hydrograph duration can be made so that two 1-hr unit hydrographs are used to form a 2-hr unit hydrograph; two 2-hr unit hydrographs result in a 4-hr diagram, and so on. Care must be taken not to mix durations in the lagging procedure, since errors are introduced; a 1-hr and a 2-hr unit hydrograph do not represent a 3-hr unit hydrograph. Lagging procedure is therefore restricted to multiples of the original duration according to the expression

$$D^1 = nD \qquad (12.3)$$

where D^1 = the possible durations of the unit hydrograph by lagging methods
$\qquad D$ = the original duration of any given unit hydrograph
$\qquad n$ = 1, 2, 3, . . .

12.4 S-HYDROGRAPH METHOD

The S-hydrograph method overcomes restrictions imposed by the lagging method and allows construction of any duration unit hydrograph. By observing the lagging system just described, it is apparent that for a 1-hr unit hydrograph, the 1-in. rainfall excess has an intensity of 1 in./hr, whereas the 2-hr unit hydrograph is produced by a rainfall intensity of 0.5 in./hr. Continuous lagging of either one of these unit hydrographs is comparable to a continuously applied rainfall at either 0.5 in./hr or 1 in./hr intensity, depending on which unit hydrograph is chosen.

As an example, using the 1-hr unit hydrograph, continous lagging represents the direct runoff from a constant rainfall of 1 in./hr as shown in Fig. 12.6a. The cumulative addition of the initial unit hydrograph ordinates at time intervals equal to the unit storm duration results in an S-hydrograph (see Fig. 12.7). Graphically, construction of an S-hydrograph is readily accomplished with a pair of dividers. The maximum

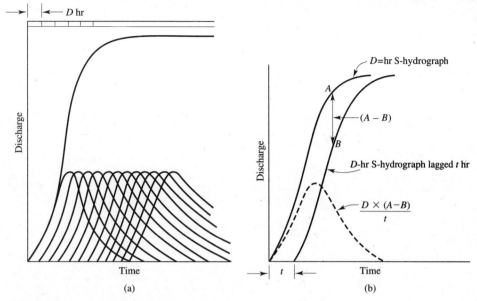

Figure 12.6 S-hydrograph method.

discharge of the S-hydrograph occurs at a time equal to D hours less than the time base of the initial unit hydrograph as shown in Fig. 12.6a.

To construct a pictorial 2-hr unit hydrograph, simply lag the first S-hydrograph by a second S-hydrograph a time interval equal to the desired duration. The difference in S-hydrograph ordinates must then be divided by 2. Any duration t unit hydrograph may be obtained in the same manner once another duration D unit hydrograph is known. Simply form a D-hr S-hydrograph; lag this S-hydrograph t hr, and multiply the difference in S-hydrograph ordinates by D/t. Accuracy of the graphical procedure depends on the scales chosen to plot the hydrographs. Tabular solution of the S-hydrograph method is also employed, but hydrograph tabulations must be at intervals of the original unit-hydrograph duration.

EXAMPLE 12.4 _____

Given the following 2-hr unit hydrograph, use S-hydrograph procedures to construct a 3-hr unit hydrograph.

Time (hr)	0	1	2	3	4	5	6
Q (cfs)	0	100	250	200	100	50	0

Solution. The 2-hr unit hydrograph is the runoff from a 2-hr storm of 0.5 in./hr. The S-hydrograph is formed from a net rain rate of 0.5 in./hr lasting indefinitely as shown in Fig. 12.6a. Its ordinates are found by adding the 2-hr unit-hydrograph (UH) runoff rates from each contributing 2-hr block of rain:

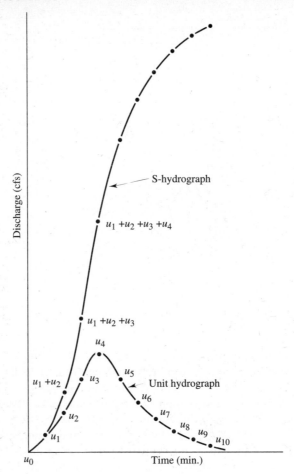

Figure 12.7   S-hydrograph.

Time (hr)	1st 2-hr	2nd 2-hr	3rd 2-hr	4th	S-hydrograph
0	0				0
1	100				100
2	250	0			250
3	200	100			300
4	100	250	0		350
5	50	200	100		350
6	0	100	250	0	350
7		50	200	100	350
8		0	100	250	350

To find a 3-hr hydrograph, the S-curve is lagged by 3 hr and subtracted as shown in Fig. 12.6b. This results in a hydrograph from a 3-hr storm of 0.5 in./hr, or 1.5 in. total. Thus the ordinates need to be divided by 1.5 to produce the 3-hr unit hydrograph:

Time (hr)	S-hydrograph	Lagged S-hydrograph	Difference	3-hr Unit hydrograph
0	0		0	0
1	100		100	67
2	250		250	167
3	300	0	300	200
4	350	100	250	167
5	350	250	100	67
6	350	300	50	33
7	350	350	0	0

■■

12.5 THE INSTANTANEOUS UNIT HYDROGRAPH

The unit-hydrograph method of estimating a runoff hydrograph can be used for storms of extremely short duration. For example, if the duration of a storm is 1 min and a unit volume of surface runoff occurs, the resulting hydrograph is the 1-min unit hydrograph. The hydrograph of runoff for any 1-min storm of constant intensity can be computed from the 1-min unit hydrograph by multiplying the ordinates of the 1-min unit hydrograph by the appropriate rain depth. A storm lasting for many minutes can be described as a sequence of 1-min storms (Fig. 12.8). The runoff hydrograph from each 1-min storm in this sequence can be obtained as in the preceding example. By superimposing the runoff hydrograph from each of the 1-min storms, the runoff hydrograph for the complete storm can be obtained.

From the unit hydrograph for any duration of uniform rain, the unit hydrograph for any other duration can be obtained. As the duration becomes shorter, the resulting unit hydrograph approaches an instantaneous unit hydrograph. The instantaneous unit hydrograph (IUH) is the hydrograph of runoff that would result if 1 in. of water were spread uniformly over an area in an instant and then allowed to run off.[9]

To develop an IUH, any I in./hr S-hydrograph must first be obtained. The resulting S-curve is lagged by the interval Δt to develop a Δt-hour unit hydrograph. The resulting Δt-hour unit graph becomes an IUH when Δt is set to 0.0 in the limit.

If a continuing I in./hr excess storm produces the original and lagged S-hydrographs of Fig. 12.6b, the Δt-hour unit hydrograph is the difference between the two curves, divided by the amount of excess rain depth in Δt hours, or

$$Q_t(\Delta t\text{-hr UH}) = \frac{Q_A - Q_B}{I \, \Delta t} \qquad (12.4)$$

The $Q_A - Q_B$ differences are divided by $I \, \Delta t$ to convert from a storm with $I \, \Delta t$ inches in Δt hours to one with 1.0 in. in Δt hours, which is the definition of a Δt-hour unit graph.

As Δt approaches zero, Eq. 12.4 becomes

$$Q_t(\text{IUH}) = \frac{1}{I} \frac{dQ}{dt} \qquad (12.5)$$

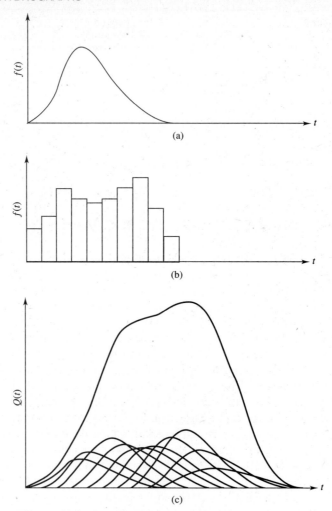

Figure 12.8 Unit-hydrograph description of the runoff process. (a) Unit hydrograph; (b) a sequence of 1-min storms; (c) superposition of runoff hydrographs for each of the 1-min storms. (After Schaake.[9])

which shows that the flow at time t is proportional to the slope of the S-hydrograph at time t. In applications, the slope is approximated by $\Delta Q/\Delta t$, and the IUH ordinates can be estimated from pairs of closely spaced points of the S-hydrograph.

If an IUH is supplied, the above process can be reversed, and any X-hour unit graph can be found by averaging IUH flows at X-hr intervals, or

$$Q_t(X\text{-hr UH}) \cong \tfrac{1}{2}(\text{IUH}_t + \text{IUH}_{t-x}) \qquad (12.6)$$

Use of this approximate equation is allowed for small X values and permits direct calculation of a unit graph from an IUH, bypassing the normal S-hydrograph procedure.

EXAMPLE 12.5

Given the following 1.0 in./hr S-hydrograph, determine the IUH, and then use it to estimate a 1-hr UH.

Time (hr)	0	0.5	1.0	1.5	2.0	2.5	3.0	3.5	4.0
S-curve (cfs)	0	50	200	450	500	650	700	750	800

Solution.   The IUH is found from Eq. 12.5. The slope at time t is approximated by $(Q_{t+0.5} - Q_{t-0.5})/\Delta t$

Time	S-Curve	IUH $\cong \Delta Q/\Delta t$
0	0	0
0.5	50	200
1	200	400
1.5	450	300
2	500	200
2.5	650	200
3	700	100
3.5	750	100
4	800	50
4.5	800	0
5	800	0

The 1-hr UH is obtained from Eq. 12.6, using readings at 1-hr intervals:

Time	IUH$_t$	IUH$_{t-1}$	1-hr UH
0	0	0	0
1	400	0	200
2	200	400	300
3	100	200	150
4	50	100	75
5	0	50	25
6	0	0	0

◾◾

The reader should verify that the 1-hr UH obtained through use of the IUH is approximately the same as that obtained by lagging the S-hydrograph 1 hr, subtracting, and converting the difference to a 1-hr UH.

The ordinates of the IUH represent the relative effect of antecedent rainfall intensities on the runoff rate at any instant of time. By plotting the IUH with time increasing to the left rather than to the right (see Fig. 12.9), and then superimposing this plot over the rainfall hyetograph (plotted with time increasing to the right as in Fig. 12.9), the relative weight given to antecedent rainfall intensities (as a function of time into the past) is easily observed. In other words, the runoff rate at any time is

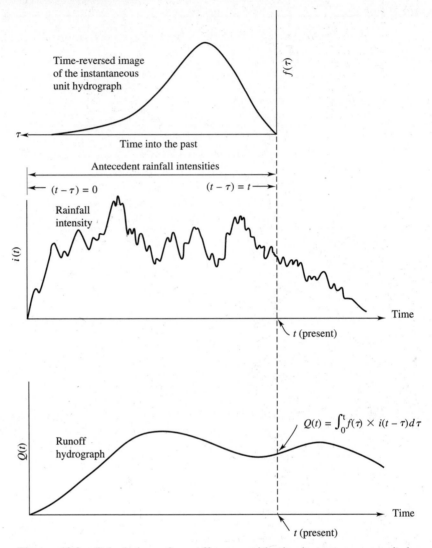

Figure 12.9   Calculation of runoff rates with the instantaneous unit hydrograph. The runoff rate at any time is a weighted average of the antecedent rainfall intensities. The time-reversed image of the instantaneous unit hydrograph represents the weighting function. (After Schaake.[9])

computed as a weighted average of the previous rainfall intensities. Therefore, the computed runoff hydrograph is the weighted, moving average of the rainfall pattern and the weighting function is the time-reversed image of the unit hydrograph.[9]

Stated mathematically, the runoff rate at any time is given by the *convolution integral*

$$Q(t) = \int_0^t f(\tau)i(t - \tau)d\tau \qquad (12.7)$$

where $Q(t)$ = the surface runoff rate at time t
             $f(\tau)$ = the ordinate of the IUH at time τ
     $i(t - \tau)$ = the rainfall intensity (after abstraction of the appropriate infiltration losses) at time $t - \tau$

The variable τ represents time into the past so that time $t - \tau$ occurs before time t. The limits on the integral allow τ to vary between a past and present time (i.e., $\tau = 0$, $t - \tau = 0$). The integral gives a continuous weighting of previous rainfall intensities by the ordinates of the IUH.

12.6 SYNTHETIC UNIT HYDROGRAPHS

As previously noted, the linear characteristics exhibited by unit hydrographs for a watershed are a distinct advantage in constructing more complex storm discharge hydrographs. Generally, however, basic streamflow and rainfall data are not available to allow construction of a unit hydrograph except for relatively few watersheds; therefore, techniques have evolved that allow generation of *synthetic unit hydrographs*.

Gamma Distribution

The shapes of hydrographs often closely match a two-parameter gamma function, given by

$$f(x) = \frac{x^\alpha e^{-x/\beta}}{\beta^{\alpha+1}\Gamma(\alpha + 1)} \tag{12.8}$$

where $0 < x < \infty$. The parameter α is a dimensionless shape factor (must be greater than -1), and β is a positive scale factor having the same units as x and controlling the base length. The product of α and β gives the value x corresponding to the apex, or maximum value of $f(x)$. For $\alpha > 1$, the distribution has a single apex and plots similar to hydrograph shapes, as shown in Fig. 12.10. The distribution mean is $\beta(\alpha + 1)$, and variance is $\beta^2(\alpha + 1)$.

Many of the synthetic unit hydrograph procedures result in only three to five points on the hydrograph, through which a smooth curve must be fitted. In addition to the requirement that the curve passes through all the points, the area under the hydrograph must equal the runoff volume from one unit of rainfall excess over the watershed. This latter requirement is often left unchecked and can result in considerable errors in performing calculations through the use of ordinates of a hydrograph that do not represent a "unit" of runoff.

The most useful feature of the gamma distribution function (explained in greater detail later) is that it guarantees a unit area under the curve. It can conveniently be used to synthesize an entire hydrograph if the calculated peak flow rate Q_p and its associated time t_p are known. This uses a procedure developed by Aron and White.[10]

If time t is substituted for x in Eq. 12.8, the time to peak t_p is $\alpha\beta$. At this point, the function $f(t)$ equals the peak flow rate Q_p, or

$$Q_p = \frac{C_v A \alpha^{\alpha+1}}{t_p e^\alpha \Gamma(\alpha + 1)} = \frac{C_v A}{t_p}\phi(\alpha) \tag{12.9}$$

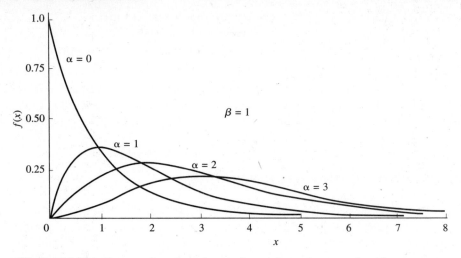

Figure 12.10   Gamma function shapes for various shape and scale parameter values.

where $C_v A$ is the unit volume of runoff from a basin with area A. The conversion factor $C_v = 1.008$ is selected to make $\phi(\alpha)$ dimensionless.

The function $\phi(\alpha)$ is shown by Aron and White to be related to α by[11]

$$\alpha = 0.045 + 0.5\phi + 5.6\phi^2 + 0.3\phi^3 \tag{12.10}$$

Collins shows that this can be approximated reasonably well in the range $1 < \alpha < 8$ by[12]

$$\alpha = 0.5\phi + 5.9\phi^2 \tag{12.11}$$

Combining this with Eq. 12.9 gives

$$\alpha = 0.5\frac{Q_p t_p}{C_v A} + 5.9\left(\frac{Q_p t_p}{C_v A}\right)^2 \tag{12.12}$$

To fit a unit graph using Eqs. 12.9 and 12.12, the peak flow rate and time must be estimated. Several of the methods described subsequently allow this. Next, $\phi(\alpha)$ is found from Eq. 12.9, and α from Eq. 12.10 or 12.11. The unit hydrograph can now be constructed by calculating Q at any convenient multiple, a, of t_p. Substituting at_p for x in Eq. 12.8 gives the flow at $t = at_p$ as

$$Q_{at_p} = Q_p a^\alpha e^{(1-a)\alpha} \tag{12.13}$$

which can be solved for all the flow rates of the hydrograph.

EXAMPLE 12.6 _____

The peak flow rate for the unit hydrograph of a 36,000-acre watershed is 1720 cfs and occurs 12 hr following the initiation of runoff. Use Eq. 12.8 to synthesize the rest of the hydrograph.

Solution. From Eq. 12.9,

$$\phi(\alpha) = \frac{1720(12)}{1.008(36{,}000)} = 0.57$$

From Eq. 12.10 (and 12.11),

$$\alpha = 2.2$$

The hydrograph is then found from Eq. 12.13:

$$Q_{at_p} = 1720a^{2.2}e^{2.2(1-a)}$$

Solving for a few points, we obtain the following values:

10.0	$t = at_p$ (hr)	Q (cfs)
0	0	0
0.5	6	1125
1.0	12	1720
2.0	24	876
5.0	60	9
10.0	120	0

Sufficient intermediate points should be generated to define the entire shape of the hydrograph. ■■

Snyder's Method

One technique employed by the Corps of Engineers[13] and many others is based on methods developed by Snyder[4] and expanded by Taylor and Schwartz.[5] It allows computation of lag time, time base, unit-hydrograph duration, peak discharge, and hydrograph time widths at 50 and 75 percent of peak flow. By using these seven points, a sketch of the unit hydrograph is obtained, Fig. 12.11, and checked to see if it contains 1 in. of direct runoff.

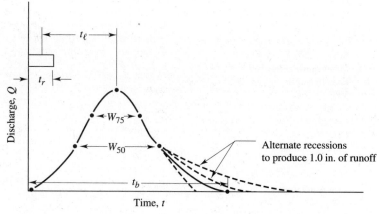

Figure 12.11 Snyder's synthetic unit hydrograph.

Time to Peak   Snyder's method of synthesizing a unit hydrograph assumes that the peak flow rate occurs at the watershed lag, estimated from Eq. 11.5. Its location is established as shown on Fig. 12.11. The lag time and peak discharge rate are both correlated with various physiographic watershed characteristics. For the lag time, the variables L and L_{ca} for Eq. 11.5 are estimated from map measurements, and C_t is developed for the locale, using Snyder's estimates or other sources. Table 12.3 summarizes a variety of C_t values for various regions.

It is assumed that lag time is a constant for a particular watershed—that is, uninfluenced by variations in rainfall intensities or similar factors. The use of L_{ca} accounts for the watershed shape, and C_t takes care of wide variations in topography, from plains to mountainous regions.

Steeper slopes tend to generate lower values of C_t, with extremes of 0.4 noted in Southern California and 8.0 along the Gulf of Mexico and Rocky Mountains. When snowpack accumulations influence peak discharge, values of C_t will be between one sixth to one third of Snyder's values.

Time Base   The time base of a synthetic unit hydrograph (see Fig. 12.11) by Snyder's method is

$$t_b = 3 + \frac{t_l}{8} \tag{12.14}$$

where t_b = the base time of the synthetic unit hydrograph (days)
t_l = the lag time (hr)

TABLE 12.3   TYPICAL SNYDER'S COEFFICIENTS FOR U.S. LOCALITIES

Location	Range of C_t	Average C_t	Range of C_p	Average C_p
Appalachian Highlands	1.8–2.2	2.0	0.4–0.8	0.6
West Iowa	0.2–0.6	0.4	0.7–1.0	0.8
Southern California	—	0.4	—	0.9
Ohio	0.6–0.8	0.7	0.6–0.7	0.6
Eastern Gulf of Mexico	—	8.0	—	0.6
Central Texas	0.4–2.3	1.1	0.3–1.2	0.8
North and Mid-Atlantic states	—	$0.6/\sqrt{S}^a$	—	—
Sewered urban areas	0.2–0.5	0.3	0.1–0.6	0.3
Mountainous watersheds	—	1.2	—	—
Foothills areas	—	0.7	—	—
Valley areas	—	0.4	—	—
Eastern Nebraska	0.4–1.0	0.8	0.5–1.0	0.8
Corps of Engineers training course	0.4–8.0	0.3–0.9	—	—
Great Plains	0.8–2.0	1.3	—	—
Rocky Mountains	1.5–8.8	5.4	—	—
SW desert	0.7–1.9	1.4	—	—
NW coast and Cascades	2.0–4.4	3.1	—	—
21 urban basins	0.3–0.9	0.6	—	—
Storm sewered areas	0.2–0.3	0.2	—	—

^{a}Channel slope S.

Equation 12.14 gives reasonable estimates for large watersheds but will produce excessively large values for smaller areas. A general rule of thumb for small areas is to use three to five times the time to peak as a base value when sketching a unit hydrograph. In any event, the time base should be adjusted as shown in Fig. 12.11 until the area under the unit hydrograph is 1.0″.

Duration   The duration of rainfall excess for Snyder's synthetic unit-hydrograph development is a function of lag time

$$t_r = \frac{t_l}{5.5} \tag{12.15}$$

where t_r = duration of the unit rainfall excess (hr)
t_l = the lag time from the centroid of unit rainfall excess to the peak of the unit hydrograph

This synthetic technique always results in an initial unit-hydrograph duration equal to $t_l/5.5$. However, since changes in lag time occur with changes in duration of the unit hydrograph, the following equation was developed to allow lag time and peak discharge adjustments for other unit-hydrograph durations.

$$t_{lR} = t_l + 0.25(t_R - t_r) \tag{12.16}$$

where t_{lR} = the adjusted lag time (hr)
t_l = the original lag time (hr)
t_R = the desired unit-hydrograph duration (hr)
t_r = the original unit-hydrograph duration = $t_l/5.5$ (hr)

Peak Discharge   If one assumes that a given duration rainfall produces 1 in. of direct runoff, the outflow volume is some relatively constant percentage of inflow volume. A simplified approximation of outflow volume is $t_l \times Q_P$, and the equation for peak discharge can be written

$$Q_P = \frac{640C_PA}{t_{lR}} \tag{12.17}$$

where Q_P = peak discharge (cfs)
C_P = the coefficient accounting for flood wave and storage conditions; it is a function of lag time, duration of runoff producing rain, effective area contributing to peak flow, and drainage area
A = watershed size (mi²)
t_{lR} = the lag time (hr)

Thus peak discharge can be calculated given lag time and coefficient of peak discharge C_P. Values for C_P range from 0.4 to 0.8 and generally indicate retention or storage capacity of the watershed. Larger values of C_P are generally associated with smaller values of C_t. Typical values are tabulated in Table 12.3.

Hydrograph Construction   From Eqs. 11.5, 12.14, 12.15, and 12.17 plot three points for the unit hydrograph and sketch a synthetic unit hydrograph, remembering that total direct runoff amounts to 1 in. An analysis by the Corps of Engineers (see Fig. 12.12) gives additional assistance in plotting time widths for points on the hydrograph located at 50 and 75 percent of peak discharge.[13] As a general rule of thumb, the time width at W_{50} and W_{75} ordinates should be proportioned each side of the peak in a ratio of 1:2 with the short time side on the left of the synthetic unit-hydrograph peak. As noted earlier, for smaller watersheds, Eq. 12.14 gives unrealistic values for the base time. If this occurs, a value can be estimated by multiplying total time to the peak by a value of from 3 to 5. This ratio can be modified based on the amount and time rate of depletion of storage water within the watershed boundaries.

The envelope curves in Fig. 12.12 are defined by

$$W_{50} = 830/(Q_p/A)^{1.1} \qquad (12.18)$$

$$W_{75} = 470/(Q_p/A)^{1.1} \qquad (12.19)$$

The seven points formed through the use of these equations can be plotted and a smooth curve drawn. To assure a unit hydrograph, the curve shape and ordinates should be adjusted until the area beneath the curve is equivalent to one unit of direct runoff depth over the watershed area. This can be done by hand-fitting and planimetering or by curve-fitting.

Hudlow and Clark[14] used least-squares regression techniques to fit a Pearson type III (gamma) probability density function (refer to Chapter 26) through the seven Snyder unit-hydrograph points. This function has an area of 1.0 and a shape similar

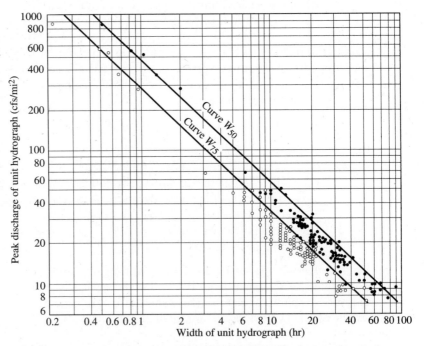

Figure 12.12   Unit hydrograph width at 50 and 75 percent of peak flow.
●, observed value of W_{50}. ○, observed value of W_{75}.

to that of natural hydrographs. The shape is given by

$$Q_t = Q_p \left(\frac{t}{t_p}\right)^a e^{-(t-t_p)/b} \qquad (12.20)$$

where a and b are shape and scale parameters. Hudlow and Clark present a trial-and-error solution to the least-squares normal equations, using Newton's method, to develop estimates of a and b.

The application of Snyder's synthetic unit-hydrograph method to areas other than the original study area should be preceded by a reevaluation of coefficients C_t and C_P in Eqs. 11.5 and 12.17. This analysis can be accomplished by the use of unit hydrographs in the region under study which have the proper lag time–rainfall duration ratio; that is, $t_r = t_l/5.5$. If another rainfall duration is selected, variations of C_t and C_P can be expected.

SCS Method

A method developed by the Soil Conservation Service for constructing synthetic unit hydrographs is based on a dimensionless hydrograph (Fig. 12.13). This dimensionless graph is the result of an analysis of a large number of natural unit hydrographs from a wide range in size and geographic locations. The method requires only the

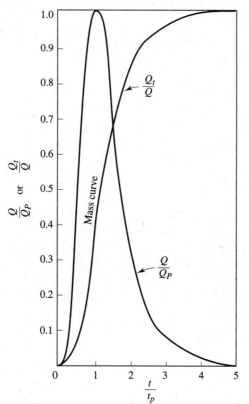

Figure 12.13 Dimensionless unit hydrograph and mass curve. (After Mockus.[15])

determination of the time to peak and the peak discharge as follows:

$$t_p = \frac{D}{2} + t_l \qquad (12.21)$$

where t_p = the time from the beginning of rainfall to peak discharge (hr)
D = the duration of rainfall (hr)
t_l = the lag time from the centroid of rainfall to peak discharge (hr)

The ratios corresponding to Fig. 12.13 are listed in Table 12.4. The peak flow for the hydrograph is developed by approximating the unit hydrograph as a triangular shape with base time of $\frac{8}{3}t_p$ and unit area. The reader should verify that this produces

$$Q_P = \frac{484A}{t_p} \qquad (12.22)$$

where Q_P = peak discharge (cfs)
A = drainage area (mi^2)
t_P = the time to peak (hr)

The time base of $\frac{8}{3}t_p$ is based on empirical values for average rural experimental watersheds and should be reduced (causing increased peak flow) for steep conditions or increased (causing decreased peak flow) for flat conditions. The resulting coefficient in Eq. 12.22 ranges from nearly 600 for steep mountainous conditions to 300 for flat swampy conditions.

A relation of t_l to size of watershed can be used to estimate lag time. Typical relations from two geographic regions are

$$t_l = 1.44A^{0.6} \text{ Texas} \qquad (12.23a)$$
$$t_l = 0.54A^{0.6} \text{ Ohio} \qquad (12.23b)$$

TABLE 12.4   COORDINATES OF SCS DIMENSIONLESS UNIT HYDROGRAPH OF FIGURE 12.13

t/t_p	Q/Q_P	t/t_p	Q/Q_P
0	0	1.4	0.75
0.1	0.015	1.5	0.66
0.2	0.075	1.6	0.56
0.3	0.16	1.8	0.42
0.4	0.28	2.0	0.32
0.5	0.43	2.2	0.24
0.6	0.60	2.4	0.18
0.7	0.77	2.6	0.13
0.8	0.89	2.8	0.098
0.9	0.97	3.0	0.075
1.0	1.00	3.5	0.036
1.1	0.98	4.0	0.018
1.2	0.92	4.5	0.009
1.3	0.84	5.0	0.004

The average lag is $0.6t_c$, where t_c is the time of concentration, defined by SCS as either the time for runoff to travel from the furthermost point in the watershed (called the upland method) or the time from the end of excess rain to the inflection of the unit hydrograph. For the first case,

$$t_c = 1.7t_p - D \qquad (12.24)$$

The dimensionless unit hydrograph, Fig. 12.13, has a point of inflection at approximately $1.7t_p$. If the lag time of $0.6t_c$ is assumed, Eqs. 12.21 and 12.24 give

$$D = 0.2t_p \qquad (12.25)$$

or
$$D = 0.133t_c \qquad (12.26)$$

A small variation in D is permissible, but it should not exceed $0.25t_p$ or $0.17t_c$. Once the $0.133t_c$-hour unit hydrograph is developed, unit hydrographs for other durations can be developed using S-hydrograph or IUH procedures.

By finding a value of t_l, a synthetic unit hydrograph of chosen duration D is obtained from Fig. 12.13.

Another equation used by the SCS is

$$t_l = \frac{l^{0.8}(S + 1)^{0.7}}{1900Y^{0.5}} \qquad (12.27)$$

where t_l = the lag time (hr)
l = length to divide in feet
Y = average watershed slope in percent
S = the potential maximum retention (in.) = $(1000/CN) - 10$, where CN is a curve number described in Chapter 4

The lag from Eq. 12.27 is adjusted for imperviousness or improved watercourses, or both, if the watershed is in an urban area. The multiple to be applied to the lag is

$$M = 1 - P(-6.8 \times 10^{-3} + 3.4 \times 10^{-4}CN - 4.3 \times 10^{-7} CN^2$$
$$-2.2 \times 10^{-8}CN^3) \qquad (12.28)$$

where CN is the curve number for urbanized conditions, and P can be either the percentage impervious or the percentage of the main watercourse that is hydraulically improved from natural conditions. If part of the area is impervious and portions of the channel are improved, two values of M are determined, and both are multiplied by the lag.

EXAMPLE 12.7

For a drainage area of 70 mi^2 having a lag time of $8\frac{1}{2}$ hr, derive a unit hydrograph of duration 2 hr. Use the SCS dimensionless unit hydrograph.

Solution

1. Using Eq. 12.21 we obtain

$$t_P = \tfrac{2}{2} + 8\tfrac{1}{2} = 9\tfrac{1}{2} \text{ hr}$$

2. From Eq. 12.22

$$Q_p = \frac{484 \times 70}{9.5}$$

$$Q_p = 3560 \text{ cfs occurring at } t = 9\frac{1}{2}\text{ hr}$$

3. Using Fig. 12.13, we find the following:
   a. The peak flow occurs at $t/t_p = 1$ or at $t = 9\frac{1}{2}$ hr.
   b. The time base of the hydrograph $= 5t_p$ or 47.5 hr.
   c. The hydrograph ordinates are:
      1. At $t/t_p = 0.5$, $Q/Q_P = 0.43$; thus at $t = 4.75$ hr, $Q = 1531$ cfs.
      2. At $t/t_p = 2$, $Q/Q_P = 0.32$; thus at $t = 19$ hr, $Q = 1139$ cfs.
      3. At $t/t_p = 3$, $Q/Q_P = 0.07$; thus at $t = 28.5$ hr, $Q = 249$ cfs.
4. Check $D/t_p = 0.21$; OK. ■■

Gray's Method

Another method of generating synthetic unit hydrographs has been developed by Gray.[16] An approximate upper limit of watershed size for application of this method to the geographic areas of central Iowa, Missouri, Illinois, and Wisconsin is 94 mi^2. The method is based on dimensionalizing the incomplete gamma distribution and results in a dimensionless graph of the form

$$Q_{t/P_R} = \frac{25.0(\gamma')^q}{\Gamma(q)} (e^{-\gamma't/P_R}) \left(\frac{t}{P_R}\right)^{q-1} \tag{12.29}$$

where $Q_t/P_R =$ percent flow in 0.25 P_R at any given t/P_R value
q and $\gamma =$ shape and scale parameters, respectively
$\Gamma =$ the gamma function of q, equal to $(q - 1)!*$
$e =$ the base of natural logarithms
$P_R =$ the period of rise (min)
$t =$ time (min)

The relation for γ' is defined as $\gamma' = \gamma P_R$ and $q = 1 + \gamma'$.

This form of the dimensionless unit hydrograph (Fig. 12.14) allows computation of the discharge ordinates for the unit hydrograph at times equal to $\frac{1}{4}$ intervals of the period of rise P_R, that is, the time from the beginning of rainfall to the time of peak discharge of the unit hydrograph.

Correlations with physiographic characteristics of the watershed can be developed to get the values of both P_R and γ'.

As an example, the storage factor P_R/γ' has been linked with watershed parameters $L/\sqrt{S_c}$, where L is the length of the main channel of the watershed in miles measured from the outlet to the uppermost part of the watershed (Fig. 12.15); S_c is defined as an average slope in percent obtained by plotting the main channel profile

---

*If q is not an integer $\Gamma(q) = \Gamma(N + Z) = (N - 1 + Z)(N - 2 + Z) \cdots (1 + Z)/\Gamma(1 + Z)$, where N equals the integer. For any q, the function is approximated by

$$\Gamma(q) = q^q e^{-q} \sqrt{\frac{2\pi}{q}} \left(1 + \frac{1}{12q} + \frac{1}{288q^2} - \frac{139}{51,480q^3} - \frac{571}{2,488,320q^4} + \cdots \right)$$

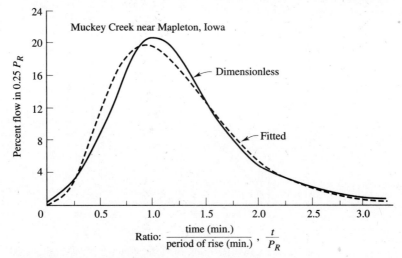

Figure 12.14 Dimensionless graph and fitted two-parameter gamma distribution for Watershed 5. (After Gray.[16])

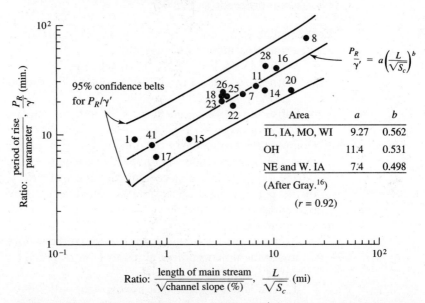

Figure 12.15 Relation of storage factor, P_R/γ', and watershed parameter, $L\sqrt{S_c}$, for watersheds in Nebraska, Iowa, Missouri, Illinois, and Wisconsin. (After Gray.[16])

and drawing a straight line through the outlet elevation such that the positive and negative areas between the stream profile and the straight line are equal. The storage factor P_R/γ' can also be correlated with the period of rise P_R as shown in Fig. 12.16. These two correlations allow solution of Eq. 12.29 and produce a synthetic unit hydrograph of duration $P_R/4$ for an ungauged area.

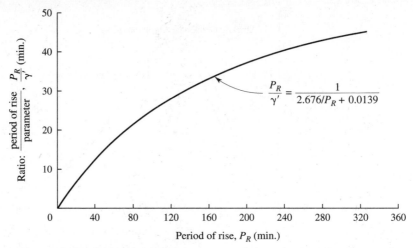

Figure 12.16   Relation of storage factor P_R/γ' and period of rise P_R. (After Gray.[16])

The solution proceeds as follows:

1. Determine L, S_c, and A for the ungauged watershed.
2. Determine parameters P_R, γ', and q.
   a. With $L/\sqrt{S_c}$, use Fig. 12.15 to select P_R/γ'.
   b. With P_R/γ', use Fig. 12.16 to obtain P_R. Compute γ' as the ratio $P_R/(P_R/\gamma')$.
   c. Substitute γ' obtained in Step 2b into the equation $q = 1 + \gamma'$, and solve for q.
3. Compute the ordinates for the dimensionless graph using Eq. 12.29. Compute the percent flow in $0.25P_R$ for values of $t/P_R = 0.125$, 0.375, 0.625, . . . , and every succeeding increment of $t/P_R = 0.250$ until the sum of the percent flows approximates 100 percent. Also compute the peak percentage by substituting $t/P_R = 1$.
4. Compute the unit hydrograph.
   a. Compute the necessary factor to convert the volume of the direct runoff under the dimensionless graph to 1 in. of precipitation excess over the entire watershed.
   1. The volume of the unit hydrograph $= V$

$$V = 1 \text{ in.} \times A \text{ mi}^2 \times 640 \,\frac{\text{acre}}{\text{mi}^2}$$

$$\times \frac{1}{12 \text{ in./ft}} \times 43,560 \,\frac{\text{ft}^2}{\text{acre}}$$

   2. The volume of the dimensionless graph $= V_D$

$$V_D = \Sigma q_i \times 0.25 \times P_R \times 60 \,\frac{\text{sec}}{\text{min}}$$

   3. Solve for Σq_i by equating V and V_D, since they must be equal.

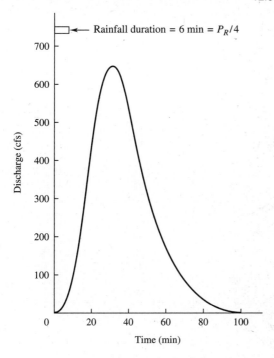

Figure 12.17 Derived hydrograph of Example 12.8 using Gray's method. Note that Gray's method results in a unit hydrograph for a $P_R/4$-hr storm.

b. Convert the dimensionless graph ordinates to the unit-hydrograph ordinates

$$Q_u = \frac{\text{percent flow in } 0.25P_R}{100} \sum q_i$$

c. Translate time base of dimensionless graph to absolute time units by multiplying $t/P_R \times P_R$ for each computed point. Remember that runoff does not commence until the centroid of rainfall, or at a time $P_R/8$.

An example problem demonstrates the solution of Gray's method for a Missouri watershed. A plot representing the derived hydrograph is shown in Fig. 12.17. Direct runoff commences at the centroid of rainfall. Thus it is necessary to add $D/2$ or $P_R/8$ to Column 2, Table 12.5, to obtain the proper times of the unit hydrograph ordinates, shows in Column 6.

EXAMPLE 12.8

For the given data, use Gray's method to construct a unit hydrograph for the Green Acre watershed, where drainage area $= 0.62 \text{ mi}^2$, length $= 0.98 \text{ mi}$, and $S_c = 1.45\%$.

Procedure

1. a. Figure 12.15; $L/\sqrt{S} = 0.813 \text{ mi}$; $P_R/\gamma' = 8.25 \text{ min}$.
    b. Figure 12.16; $P_R/\gamma' = 8.25 \text{ min}$; $P_R = 24.9 \text{ min}$.
    c. $q = 1 + \gamma' = 4.02$; $\gamma' = 3.02$.
2. a. Tabulate percent flow in $0.25P_R$ for $t/P_R = 0.250$:

TABLE 12.5   TABULATION FOR EXAMPLE 12.8

(1) t/P_R	(2) Time (min)	(3) Percent flow in 0.25P_R	(4) Cumulated flow	(5) UH (cfs)	(6) Actual time (min)
0.000	0	0	0	0	0
0.125	3.1	0.45	0.45	17.3	6.1
0.375	9.3	5.80	6.2	224	12.3
0.625	15.6	12.70	18.9	490	18.6
0.875	21.8	16.35	35.3	631	24.8
1.000	24.9	16.85	—	651	27.9
1.125	28.0	16.25	51.5	628	31.0
1.375	34.2	14.20	65.7	548	37.2
1.625	40.4	11.10	76.8	428	43.4
1.875	46.7	7.97	84.8	308	49.7
2.125	52.9	5.55	90.4	214	55.9
2.375	59.3	3.56	93.9	138	62.3
2.625	65.5	2.28	96.2	88.0	68.5
2.875	71.7	1.41	97.6	54.4	74.7
3.125	78.0	0.86	98.5	33.3	81.0
3.375	84.2	0.50	99.0	19.3	87.2

3. a. 1. $V = 1 \times 0.62 \times 640 \times 43,560/12 = 14.4 \times 10^5 \text{ ft}^3$.
  2. $V_D = 0.25 \times 24.9 \times 60 \times \Sigma\, q_i = 373.5\, \Sigma\, q_i$
  3. $\Sigma\, q_i = 3860$.
  b. Column 5 is tabulated by multiplying 3860 times values in Column 3 divided by 100.
  c. Column 2 is obtained by multiplying 24.9 times values in Column 1.
  d. Column 3 comes from solution of Eq. 12.29.   ■■

Espey 10-Minute Synthetic Unit Hydrograph

A regional analysis of 19 urban watersheds was conducted by Espey and Altman[17] and resulted in a set of regression equations that provide seven points of a 10-min hydrograph. The entire hydrograph is developed by fitting a smooth curve through the points using eye-fitting or curve-fitting procedures. In either case, a unit area is necessary.

The equations for time to peak (minutes), peak discharge (cfs), time base (minutes), and width at 50 and 75 percent of the peak flow rate are

$$T_p = 3.1L^{0.23}S^{-0.25}I^{-0.18}\phi^{1.57} \tag{12.30}$$

$$Q_P = 31.62 \times 10^3 A^{0.96}T_p^{-1.07} \tag{12.31}$$

$$T_B = 125.89 \times 10^3 AQ_p^{-0.95} \tag{12.32}$$

$$W_{50} = 16.22 \times 10^3 A^{0.93}Q_p^{-0.92} \tag{12.33}$$

$$W_{75} = 3.24 \times 10^3 A^{0.79}Q_p^{-0.78} \tag{12.34}$$

where L = total distance (ft) along the main channel from the point being considered to the upstream watershed boundary

S = main channel slope (ft/ft) defined by $H/0.8L$, where H is the difference in elevation between the point on the channel bottom at a distance of $0.2L$ downstream from the upstream watershed boundary and a point on the channel bottom at the downstream point being considered

I = impervious area within the watershed (%)

ϕ = a dimensionless watershed conveyance factor

A = watershed drainage area (mi^2)

T_p = time of rise of the unit hydrograph (min)

Q_P = peak flow of the unit hydrograph (cfs)

T_B = time base of the unit hydrograph (min)

W_{50} = width of the hydrograph at 50% of Q_P (min)

W_{75} = width of the unit hydrograph at 75% of Q_P (min)

The coefficients of determination (explained in Chapter 26) for the five equations ranged from 80 to 94 percent. The watershed conveyance factor is found from Fig. 12.18. The W_{50} and W_{75} widths are normally drawn with one-third of the calculated width placed to the left of the peak and two-thirds to the right.

Clark's IUH Time–Area Method

A synthetic unit hydrograph that utilizes an instantaneous unit hydrograph (IUH) was developed in 1945 by Clark.[6] It has been widely used, is often called the time–area method, and has appeared in several computer programs for hydrograph analysis (see Chapters 24 and 25).

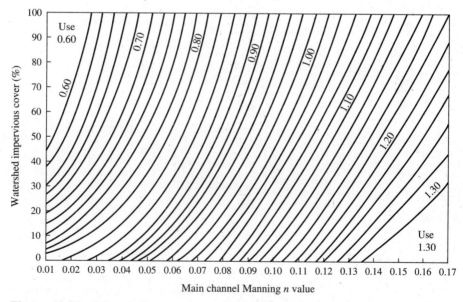

Figure 12.18 Watershed conveyance factor ϕ as a function of percent watershed impervious cover I and weighted main channel Manning n value, for Espey method.

The technique recognizes that the discharge at any point in time is a function of the translation and storage characteristics of the watershed. The translation is obtained by estimating the overland and channel travel time of runoff, which is then combined with an estimate of the delay caused by the storage effects of a watershed.

The translation of excess rainfall from its point of falling to the watershed mouth is accomplished using the time–area curve for the watershed. This is a histogram of incremental runoff versus time, constructed as shown in Fig. 12.19. The dashed lines in Fig. 12.19a subdivide the basin into several areas. Each line identifies the locus of points having equal travel times to the outlet. The isochrones are drawn equal "times" apart, and sufficient zones are selected to fully define the time–area relation.

The time–area graph of Fig. 12.19b is a form of unit hydrograph. The area beneath the curve integrates to 1.0 unit of rain depth over the total area A, and it has a translation hydrograph shape if sufficient subareas are delineated.

If one unit of net rain is placed on the watershed at $t = 0$, the runoff from A_1 would pass the outlet during the first Δt period at an average rate of A_1 units of runoff per unit of time. The volume discharged would be A_1 units of area times one unit of rain. After all areas contribute, one unit of rainfall over the entire area would have passed the outlet.

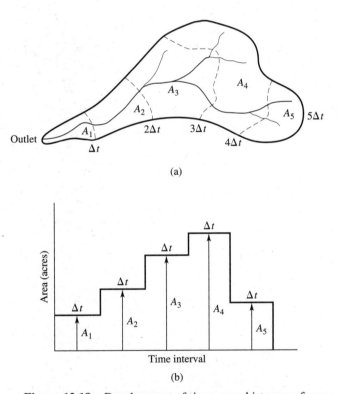

(a)

(b)

Figure 12.19   Development of time–area histogram for use with Clark's method: (a) isochrones spaced Δt apart (shown as dashed lines) and (b) time–area histogram.

The impact of watershed storage on the translation hydrograph is incorporated by routing the time–area histogram through a hypothetical linear reservoir located at the watershed outlet, having a retardance coefficient K equivalent to that of the watershed. For the simplest form of reservoir, the storage S_t at time t is linearly related to the outflow Q_t at time t, or

$$S_t = KQ_t \qquad (12.35)$$

where K is a constant of proportionality called the storage coefficient. It has time units and is often approximated by the lag time of the watershed.

From continuity, the inflow, storage, and outflow for the reservoir are related by

$$I_t - Q_t = \frac{dS_t}{dt} = K\frac{dQ}{dt} \qquad (12.36)$$

If the differential is discretized to $\Delta Q/\Delta t$, and if Q_2 and Q_1 are the flows at t and $t - 1$, then Eq. 12.36 becomes

$$\bar{I}_{\Delta t} - \bar{Q}_{\Delta t} = K\frac{Q_2 - Q_1}{\Delta t} \qquad (12.37)$$

Because $\bar{Q} = (Q_1 + Q_2)/2$, the flow at the end of any Δt is

$$Q_2 = C_0\bar{I} + C_1 Q_1 \qquad (12.38)$$

where

$$C_0 = \frac{2\,\Delta t}{2K + \Delta t} \qquad (12.39)$$

and

$$C_1 = \frac{2K - \Delta t}{2K + \Delta t} \qquad (12.40)$$

The IUH is found from Eq. 12.38 by solving for Q_2 at the end of each successive time interval.

EXAMPLE 12.9 _____

Given the following 15-min time–area curve, find the IUH for the 1000-acre watershed. Then determine the 15-min synthetic unit hydrograph. The storage coefficient K is 30 min.

Time interval (min)	Area between isochrones (acres)
0–15	100
15–30	300
30–45	500
45–60	100

Solution.   From Eqs. 12.39 and 12.40, $C_0 = 0.4$ and $C_1 = 0.6$. Routing is most easily accomplished in a tableau as follows:

Time (hr)	$\bar{I}$ (acre-in./Δt)	$\bar{I}$ (cfs)	$C_0\bar{I} + C_1 Q_1$	IUH (cfs)
0				
	100	400	160 + 0	0
0.25				
	300	1200	480 + 96	80
0.50				
	500	2000	800 + 346	368
0.75				
	100	400	160 + 688	861
1.00				
	0	0	0 + 509	997
1.25				
	0	0	0 + 305	679
1.50				
	0	0	0 + 407	⋮
⋮	⋮	⋮	⋮	

The IUH has a characteristically long recession due to the magnitude of K for this example. Note that after 1.25 hr, the flow becomes 0.6 times the previous flow and continues to decay at this rate indefinitely. As discussed in Section 11.5, the time base of the IUH should equal the excess-runoff release time, which is one definition of time of concentration. Clark's method often produces prolonged runoff because of this shortcoming.

The 15-min unit hydrograph is found using Eq. 12.6, or

$$Q_t(\text{15-min UH}) = \tfrac{1}{2}(\text{IUH}_t + \text{IUH}_{t-15})$$

This results in:

Time (hr)	IUH (cfs)	15-min UH (cfs)
0	0	0
0.25	160	80
0.50	576	368
0.75	1146	861
1.00	848	997
1.25	509	679
1.50	⋮	⋮
⋮		

■■

If the watershed lag time is not available, the K value can also be estimated by recognizing that $Q_t = K\,dQ/dt$ when the inflow is zero in Eq. 12.36. This occurs at approximately the inflection point on the recession of Fig. 12.2, when inflow to the channel ceases. If hydrograph data are available, the estimate of the K value is the ratio of the flow rate to the slope of the hydrograph at this particular point on the hydrograph.

Nash's Synthetic IUH

One of the earliest formulations of the IUH was developed by Nash.[18] Instead of characterizing runoff as translation followed by storage in a single linear reservoir as Clark did, Nash viewed the watershed as a series of n identical linear storage reservoirs, each having the same storage coefficient K. The first instantly ($t = 0$) receives a volume equal to a full inch of net rain from the entire watershed. This water then passes through reservoirs 1, 2, 3, . . . , n, with each providing an additional diffusion effect on the original 1-in. rain.

The number of reservoirs, n, is uniquely related to the reservoir storage coefficient K and the watershed lag time. Once the IUH is developed, it can be used to synthesize any other hydrograph by application of the convolution integral, Eq. 12.7, or from the approximate methods discussed in Section 12.3.

The derivation of Nash's equation for IUH begins from continuity at the first reservoir:

$$I_t - Q_{1t} = \frac{ds}{dt}\bigg|_t \tag{12.41}$$

where Q_{1t} is the outflow from reservoir 1 at time t. Substituting $S_t = KQ_{1t}$ at time $t > 0$ (for an IUH, I_t is zero after $t = 0$), we obtain

$$-Q_{1t} = \frac{ds}{dt}\bigg|_{t>0} = K\frac{dQ_{1t}}{dt}\bigg|_{t>0} \tag{12.42}$$

which can be written

$$\frac{dQ_{1t}}{Q_{1t}} = -\frac{1}{K}\,dt \tag{12.43}$$

Integration from $t = 0^+$ to time t gives

$$\ln Q_{1t} - \ln Q_{1t}\big|_{t=0^+} = -\frac{t}{K} \tag{12.44}$$

which reduces by taking antilogarithms to

$$\frac{Q_{1t}}{Q_{1t}\big|_{t=0^+}} = e^{-t/K} \tag{12.45}$$

Because $Q_{1t} = S_t/K$ and $S_{t=0} = 1$ in., then

$$Q_{1t} = \frac{1}{K}e^{-t/K} \tag{12.46}$$

where Q_{1t} has units of depth per unit of time. Equation 12.46 is an exponential decay function having an initial value of $1/K$ at $t = 0$. This monotonically decreasing outflow from reservoir 1 becomes inflow to the second reservoir.

The second reservoir is initially empty. The continuity relation

$$Q_{1t} - Q_{2t} = K\frac{dQ_{2t}}{dt} \tag{12.47}$$

is solved, giving

$$Q_{2t} = \frac{1}{K^2} t e^{-t/K} \tag{12.48}$$

This equation has a full hydrograph shape, beginning with zero flow at time zero, peaking at the maximum of the function, and eventually receding to zero.

Similarly derived, the hydrograph flowing from nth reservoir has the form

$$Q_{nt} = \frac{1}{(n-1)!\, K^n} t^{n-1} e^{-t/K} \tag{12.49}$$

which is the two-parameter gamma function,

$$Q_{nt} = \frac{1}{\Gamma(n) K^n} t^{n-1} e^{-t/K} \tag{12.50}$$

Because the outflow from the nth reservoir was caused by 1 in. of excess rain falling instantaneously, Eq. 12.50 describes an IUH.

Estimation of K and n   Values of K and n are needed for application of Nash's IUH. By integration, the centroid of the distribution (Eq. 12.50) occurs at $t = nK$. From classical calculus maximization, the peak flow occurs at $t = K^n(n-1)$. The second moment of the IUH about $t = 0$ is $n(n+1)K^2$. Trial combinations of n and K can be used to develop the IUH from Eq. 12.50, and the moments of the plotted distribution can be estimated to verify the products nK and $n(n+1)K^2$. If the IUH is discretized into $m\ \Delta t$ increments, the moments are approximated by

$$\text{First moment} \cong \sum_{}^{m} t_i Q_i\, \Delta t \tag{12.51}$$

and

$$\text{Second moment} \cong \sum_{}^{m} t_i^2 Q_i\, \Delta t \tag{12.52}$$

Another less tedious approach is to use the definition of lag time as the time from centroid of rain to the centroid of the hydrograph. For an IUH, this is the same as the centroidal distance. Thus, if the lag time can be determined from equations such as those in Section 11.6, the product nK can be established, reducing the number of trials.

Some investigators have attempted to relate Nash's K and n parameters to basin and storm characteristics using regression techniques. Rao et al.[19] developed relations for urban areas greater than 5 mi^2:

$$K = \frac{0.575 A^{0.389} D^{0.222}}{(1+I)^{0.622} P_{\text{net}}^{0.106}} \tag{12.53}$$

and

$$t_l = \frac{0.831 A^{0.458} D^{0.371}}{(1+I)^{1.622} P_{\text{net}}^{0.267}} \tag{12.54}$$

where $nK = t_l$ and A is area in square miles, D is net rain duration in hours, P_{net} is the net (effective) rain depth in inches, and I is the ratio of impervious to total area.

Colorado Urban Hydrograph Procedure (CUHP)

Synthetic unit hydrographs can be tailored for regional use. As an example, the Colorado urban hydrograph procedure[20] provides 5-min synthetic unit hydrographs for use in the Denver metropolitan area. It is based on Snyder's method and is considered applicable to watersheds in the size range from 90 acres to 10 mi^2, with "regular" shapes (length $\simeq$ 4 times width). It was developed in the early 1980s and modified in 1984 to reflect refinements from early applications. In its pre-1984 form, the Snyder/CUHP C_t and C_p values for use in Eqs. 11.5 and 12.17 are

$$C_t = 7.81/P_a^{0.78} \qquad (12.55)$$

and
$$C_p = 0.89C_t^{0.46} \qquad (12.56)$$

where P_a is the percent impervious. These regression equations were developed for $P_a > 30$ percent, using data for 96 storms over 19 urban watersheds. The given equations apply to normal watershed conditions and need to be adjusted for steep, flat, or sewered basins. If an urban area is fully sewered, the calculated C_t from Eq. 12.55 is decreased 10 percent. If sparsely sewered, a 10 percent increase is made. If the average slope S of the lower 80 percent of the main water course is flat (less than 0.01 ft/ft) the C_t value becomes

$$C_t = 3.12/P_a^{0.78}S^{0.2} \qquad (12.57)$$

and for steep areas ($S > 0.025$ ft/ft),

$$C_t = 3.75/P_a^{0.78}S^{0.2} \qquad (12.58)$$

The C_p coefficient is determined from Eq. 12.56 and adjusted to 10 percent up or down for fully or sparsely sewered conditions, respectively.

For the CUHP applications, Eqs. 12.18 and 12.19 become

$$W_{50} = 500A/Q_p \qquad (12.59)$$
$$W_{75} = 260A/Q_p \qquad (12.60)$$

where A and Q_p have units of square miles and cubic feet per second. For plotting W_{50}, the smaller of 35 percent or $0.6T_p$ is placed left of the peak. For W_{75}, 45 percent is placed to the left, or $0.424T_p$ if $0.6T_p$ was used for W_{50}. T_p is the time from beginning of runoff of the unit rainfall to the peak time.

Several investigators have suggested that Snyder's C_t and slope S are correlated.[4,8] The original CUHP procedure was altered to recognize this relation, making the adjustments in C_t unnecessary. For the modified version, Eqs. 12.57 and 12.58 are bypassed, and the time to peak rather than lag time is used in Snyder's Eq. 11.5, where

$$t_p = C_t(LL_{ca}/\sqrt{S})^{0.48} \qquad (12.61)$$

The revised time coefficient C_t is obtained from Fig. 12.20a, and the peak coefficient is found from

$$C_p = PC_tA^{0.15} \qquad (12.62)$$

where S = weighted average slope of basin along the stream to the upstream basin boundary (ft/ft)

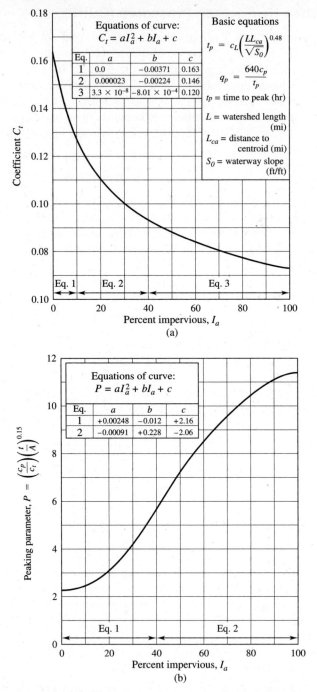

Figure 12.20 Snyder's C_p and C_t coefficients for urban areas, for use with CUHP: (a) relation between C_t and imperviousness and (b) relation between peaking parameter and imperviousness.

P = coefficient, depending on imperviousness, from Fig. 12.20b

A = drainage area (mi^2)

L = length of the main stream channel (mi) from the outlet to the divide

L_{ca} = length along the main channel (mi) from the outlet to a channel point nearest the watershed centroid

The peak rate and time of the unit hydrograph can be developed using Eqs. 12.17 and 12.61. After four additional points defined by Eqs. 12.59 and 12.60 are plotted, the rest of the 5-min hydrograph can be fitted to provide a total area representing 1.0 in. of direct runoff. A hand fit is applied, or the mathematical curve-fitting techniques described early in Section 12.5 can be used if a gamma (or any other) distribution is considered appropriate.

For small watersheds (less than 90 acres), the time to peak is

$$t_p = 0.39 \frac{(P_a^2 - 0.36P_a + 0.07)}{P_a^2 - 0.49P_a + 0.14}(T_c - 6P_a) \qquad (12.63)$$

where T_c is the time of concentration in minutes, and P_a is the percent impervious.

■ Summary

Unit hydrograph methods allow the hydrologist to estimate runoff volumes and rates for virtually any storm. By far, the greatest number of problems in practice are evaluated using unit hydrograph procedures. Most of the current computer models use unit hydrograph procedures as described in Chapters 23, 24, and 25. These models are simply computer programs that perform the unit hydrograph syntheses and convolution steps described in this chapter. Any software user should understand the origin, applicability, and parameter estimation procedures for each unit hydrograph method selected. The most successful uses of the computer models will result from a thorough familiarity with the processes described in this chapter.

PROBLEMS

12.1. Given the following storm pattern and assuming a triangular unit hydrograph for one time unit, determine the composite hydrograph.

Storm pattern				
Time unit	1	2	3	4
Rainfall	1	1	4	2

Unit hydrograph base length = 6 time units; time of rise = 2 time units; and maximum ordinate = $\frac{1}{3}$ rainfall unit height.

12.2. Given a rainfall duration of 1 time unit, an effective precipitation of 1.5 in., and the following hydrograph, determine (a) the unit hydrograph and (b) the composite hydrograph for the given storm sequence.

Hydrograph for 1.5 in. net rain in 1 time unit													
Time units	1	2	3	4	4.5	5	6	7	8	9	10	11	12	13
Flow (cfs)	100	98	220	512	620	585	460	330	210	150	105	75	60	54

Storm sequence				
Time units	1	2	3	4
Precipitation (in.)	0.4	1.1	2.0	1.5

12.3. Solve Problem 12.2 if the storm sequence is as follows:

Storm sequence			
Time units	1	2	3
Precipitation (in.)	0.3	1.4	0.9

12.4. Using U.S. Geological Survey records, or other data, select a streamflow hydrograph for a large, preferably single-peaked runoff event. Separate the base flow and determine a unit hydrograph for the area.

12.5. For the unit hydrograph of Problem 12.1, construct an S-hydrograph.

12.6. For the unit hydrograph computed in Problem 12.2, construct an S-hydrograph.

12.7. Use the S-hydrograph of Problem 12.6 to find a 3 time-unit unit hydrograph.

12.8. Given a watershed of 100 mi^2, assume that $C_t = 1.8$, the length of main stream channel is 18 mi, and the length to a point nearest the centroid is 10 mi. Use Snyder's method to find (a) the time lag, (b) the duration of the synthetic unit hydrograph, and (c) the peak discharge of the unit hydrograph.

12.9. Apply Snyder's method to the determination of a synthetic unit hydrograph for a drainage area of your choice.

12.10. Use Fig. 12.13 to determine a 2-hr unit hydrograph if the drainage area is 60 mi^2 and Eq. 12.23a is applicable.

12.11. Solve Problem 12.10 using Eq. 12.23b.

12.12. Assuming a Nebraska location, use Gray's method to determine a unit hydrograph: drainage area = 1.0 mi^2, length = 0.6 mi, S_c = 1.3 percent.

12.13. A drainage area in Nebraska contains 30 mi^2. The length of the main channel is 10 mi and the representative watershed slope is 2.5 percent. Use Gray's method to determine a unit hydrograph.

12.14. Discharge rates for a flood hydrograph passing the point of concentration for a 600-acre drainage basin are given in the table below. The flood was produced by a uniform rainfall rate of 2.75 in./hr, which started at 9 A.M., abruptly ended at 11 A.M. and resulted in 5.00 in. of direct surface runoff. The base flow (derived from influent seepage) prior to, during, and after the storm was 100 cfs.

Time	8 A.M.	9	10	11	12	1 P.M.	2	3	4	5	6	7
Measured discharge	100	100	300	500	700	800	600	400	300	200	100	100

a. At what times did direct runoff begin and cease?

b. Determine the ϕ index (in./hr) for the basin.

c. Derive the 2-hr unit-hydrograph ordinates (cfs) for each time listed.

d. Estimate the time of concentration (excess release time) for the basin.

e. At what time would direct surface runoff cease if the rainfall of 2.75 in./hr had begun at 9 A.M. and had lasted for 8 hr rather than 2?

f. Determine the peak discharge rate (cfs) and the direct runoff (in.) for a uniform rainfall of 2.75 in./hr and a duration of 8 hr.

12.15. Measured total hourly discharge rates (cfs) from a 3.10-mi^2 drainage basin are listed in the accompanying table. The hydrograph was produced by a rainstorm having a uniform intensity of 2.60 in./hr starting at 9 A.M. and abruptly ending at 11 A.M. The base flow from 8 A.M. to 3 P.M. was a constant 100 cfs. The volume of direct runoff, determined as the area under the direct surface runoff hydrograph, is 1000 cfs-hr.

Time	8 A.M.	9	10	11	12	1 P.M.	2	3
Measured discharge	100	100	300	600	400	200	100	100

a. At what time did the direct runoff begin?

b. Determine the net rain (in.) corresponding to the volume of the direct surface runoff of 1000 cfs-hr.

c. Determine the ϕ index for the basin.

d. Derive a 2-hr unit hydrograph for the basin by tabulating time in hours and discharge in cfs.

e. What is the excess release time of the basin?

f. For the same basin, use the derived 2-hr unit hydrograph to determine the direct runoff rate (cfs) at 4 P.M. on a day when excess (net) rainfall began at 1 P.M. and continued at a net intensity of 2 in./hr for 4 hr, ceasing abruptly at 5 P.M.

12.16. A 5-hr unit hydrograph for a 4250-acre basin is shown in the accompanying sketch. The given hydrograph actually appeared as a direct runoff hydrograph from the basin, caused by rain falling at an intensity of 0.30 in./hr for a duration of 5 hr, beginning at $t = 0$.

a. Determine the excess release time of the basin.

b. Determine the ϕ index for the basin.

c. What percentage of the drainage basin was contributing to direct runoff 4 hr after rain began ($t = 4$)?

d. Use your response to part c to determine Q_P, as shown in the sketch. Do not scale Q_P from the drawing.

e. Note that rain continued to fall between $t = 3$ and $t = 5$. Why did the hydrograph form a plateau between $t = 3$ and $t = 5$, rather than continue to rise during those 2 hours?

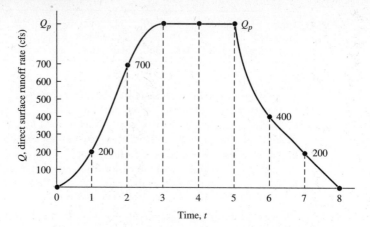

f. Use the given 5-hr unit hydrograph to determine the direct runoff rate (cfs) at 7 P.M. on a day when rain fell at an intensity of 0.60 in./hr from 1 P.M. to 11 P.M.

12.17. The 2-hr unit hydrograph for a basin is given by the following table:

Time (hr)	0	1	2	3	4	5	6	7	8	9
Q (cfs)	0	60	200	300	200	120	60	30	10	0

a. Determine the hourly discharge values (cfs) from the basin for a net rain of 5 in./hr and a rainfall duration of 2 hr.

b. Determine the direct runoff (in.) for the storm of part a. What is the direct runoff for a net rain of 0.5 in./hr and a duration of 2 hr?

c. Rain falls on the basin at a rate of 4.5 in./hr for a 2-hr period and abruptly increases to a rate of 6.5 in./hr for a second 2-hr period. Convert these actual intensities to net rain intensities using a ϕ index of 0.5 in./hr. Construct a table that properly lags and amplifies the 2-hr unit hydrograph, and determine the hourly ordinates (cfs) of direct runoff for the storm. The derived direct runoff hydrograph should begin and end with zero discharge values.

12.18. Given the following 2-hr unit hydrograph for a drainage basin, determine hourly ordinates of the 4-hr unit hydrograph:

Time (hr)	0	1	2	3	4	5	6
Q (cfs)	0	50	300	400	200	50	0

12.19. Use the following 4-hr unit hydrograph for a basin to determine the peak discharge rate (cfs) resulting from a net rain of 3.0 in./hr for a 4-hr duration followed immediately by 2.0 in./hr for a 4-hr duration.

Time (hr)	0	2	4	6	8	10
Q (cfs)	0	200	300	100	50	0

12.20. Compare the time from the peak to the end of runoff for the SCS triangular unit hydrograph with the time of concentration, t_c. Discuss.

12.21. Prove that the area under the rising limb of the SCS basic dimensionless hydrograph equals that of the triangular unit hydrograph, that is, 37.5 percent of the total.

12.22. By calculus, show that the maximum value of $f(x)$ in Eq. 12.15 occurs when $x = \alpha\beta$, for $\alpha \geq 1$. Also solve for the centroidal distance by taking the first moment about the y axis.

12.23. According to the rational method (see Chapter 15) of estimating peak flow from small areas, the peak rate for a storm with uniform continuing intensity is equal to the net rain rate and occurs at the time of concentration. For what conditions, if any, would Eqs. 12.63 and 12.17 result in agreement of the peak magnitude and time, estimated by CUHP, with those of the rational method? Discuss.

12.24. Describe two methods that could be used to construct a 2-hr unit hydrograph using a 1-hr unit hydrograph for a basin.

12.25. Measured total hourly discharge rates (cfs) from a 2.48-mi^2 drainage basin are tabulated below. The hydrograph was produced by a rainstorm having a uniform intensity of 2.60 in./hr starting at 9 A.M. and abruptly ending at 11 A.M. The base flow from 8 A.M. to 3 P.M. was a constant 100 cfs.

Time	8 A.M.	9	10	11	12	1 P.M.	2	3
Discharge (cfs)	100	100	300	450	300	150	100	100

a. At what time did direct runoff begin?
b. Determine the gross and net rain depths (inches).
c. Derive a 2-hr unit hydrograph for the basin by tabulating time in hours and discharge in cubic feet per second.
d. Derive a 4-hr unit hydrograph for the basin.
e. Derive a 1-hr unit hydrograph for the basin.

12.26. Given below is a 3-hr unit hydrograph for a watershed. The ϕ-index is 1.5 in./hr. Desired is the DRH for an 18-hr storm having six successive 3-hr rainfall rates of 2.5, 3.5, 1.5, 4.0, 6.5, 2.5 in./hr.

Time (hr)	0	1	2	3	4	5	6	7	8	9	10	11	12	13	14
Q(IUH)	0	10	40	60	80	100	90	70	60	50	40	30	20	10	0

12.27. Use the following 2-hr unit hydrograph to determine the peak direct-runoff discharge rate (cfs) resulting from a net rain of 2.0 in./hr for 5 hr.

Time (hr)	0	1	2	3	4	5	6	7
Q(cfs)	0	50	200	300	200	150	100	0

12.28. The ordinate for a 5-hr unit hydrograph is 300 cfs at a time 4 hr after the beginning of net rainfall. A storm with a uniform intensity of 3 in./hr and a duration of 5 hr occurs over the basin. What is the runoff rate after 4 hr if the ϕ index is 0.5 in./hr?

12.29. Given below is an IUH for a watershed. Use the IUH to find hourly DRH rates for a net rain of 4 in. in a 2-hr period.

Time (hr)	0	1	2	3	4	5	6	7	8	9	10
Q(IUH)	0	10	40	50	60	80	100	80	20	10	0

12.30. A 2-hr unit hydrograph for a basin is shown in the sketch.
   a. Determine the peak discharge (cfs) for a net rain of 5.00 in./hr and a duration of 2 hr.
   b. What is the total direct surface runoff (in inches) for the storm described in part a?
   c. A different storm with a net rain of 0.50 in./hr lasts for 4 hr. What is the discharge at 8 P.M. if the rainfall started at 4 P.M.?

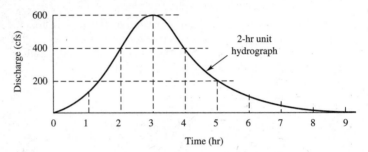

12.31. Recorded flow rates for a net rain of 1.92 inches in 12 hours are shown in the table. If the base flow is 375 cfs throughout the storm, determine the 12-hr unit hydrograph, and convert it to a 6-hr unit hydrograph. Then apply the 6-hr unit hydrograph to determine the total hydrograph (including 400 cfs base flow) for a 24-hr storm having four 6-hr blocks of net rain at rates of 0.7, 3.8, 10.8, and 1.8 in. per hour.

Time in hours	Observed flow (cfs)
0	375
6	825
12	2200
18	3650
24	3900
30	3200
36	2375
42	1725
48	1250
54	900
60	650
66	490
72	410
78	375

12.32. Starting with a triangular-shaped unit hydrograph with a base length of $2.67t_p$ and a height of q_p, derive Eq. 12.22, $q_p = 484A/t_p$. State the units of each term used in the derivation.

12.33. The SCS synthetic unit hydrograph is derived by computing the peak discharge rate (cubic feet per second) from $q_p = 484A/t_p$. In the derivation, it was actually assumed that q_p in./hr $= 0.75V/t_p$, where V is the volume of direct runoff (inches), t_p is the time to peak flow (hours), and A is the basin area (square miles). Derive the first equation from the second.

12.34. Which of the techniques for synthesizing a unit hydrograph requires the least computational effort in developing the entire unit hydrograph? Which probably requires the most?

REFERENCES

1. W. D. Mitchell, "Unit Hydrographs in Illinois," Illinois Waterways Division, 1948.
2. L. K. Sherman, "Stream-Flow from Rainfall by the Unit-Graph Method," *Eng. News-Rec.* **108,** 501–505(Apr. 1932).
3. Rand Morgan and D. W. Hulinghorst, "Unit Hydrographs for Gauged and Ungauged Watersheds," U.S. Engineers Office, Binghamton, NY, July 1939.
4. F. F. Snyder, "Synthetic Unit Graphs," *Trans. Am. Geophys. Union* **19,** 447–454(1938).
5. A. B. Taylor and H. E. Schwartz, "Unit Hydrograph Lag and Peak Flow Related to Basin Characteristics," *Trans. Am. Geophys. Union* **33,** 235–246(1952).
6. C. O. Clark, "Storage and the Unit Hydrograph," *ASCE Trans.* **110,** 1419–1446(1945).
7. *Water Supply Papers,* U.S. Geological Survey, Water Resources Division. Washington, D.C.: U.S. Government Printing Office, 1966–1970.
8. *Hourly Precipitation Data,* National Oceanic and Atmospheric Administration. Washington, D.C.: U.S. Government Printing Office, 1971.
9. John C. Schaake, Jr., "Synthesis of the Inlet Hydrograph," Tech. Rept. No. 3, Department of Sanitary Engineering and Water Resources, The Johns Hopkins University, Baltimore, MD, 1965.
10. G. Aron and E. White, "Fitting a Gamma Distribution over a Synthetic Unit Hydrograph," *Water Resources Bull.* **18**(1) (Feb. 1982).
11. G. Aron and E. White, "Reply to Discussion," *Water Resources Bull.* **19**(2) (Apr. 1983).
12. M. Collins, "Discussion—Fitting a Gamma Distribution over a Synthetic Unit Hydrograph," *Water Resources Bull.* **19**(2) (Apr. 1983).
13. "Flood-Hydrograph Analysis and Computations," U.S. Army Corps of Engineers, *Engineering and Design Manuals,* Em1110-2-1405. Washington, D.C.: U.S. Government Printing Office, Aug. 1959.
14. M. D. Hudlow and R. A. Clark, "Hydrograph Synthesis by Digital Computer," *Proc. ASCE J. Hyd. Div.* (May 1969).
15. V. Mockus, "Use of Storm and Watershed Characteristics in Synthetic Hydrograph Analysis and Application," U.S. Department of Agriculture, Soil Conservation Service, 1957.
16. D. M. Gray, "Synthetic Unit Hydrographs for Small Drainage Areas," *Proc. ASCE J. Hyd. Div.* **87**(HY4) (July 1961).
17. W. H. Espey and D. G. Altman, "Nomographs for Ten-minute Unit Hydrographs for Small Urban Watersheds," Environmental Protection Agency, Rept. EPA-600/9-78-035, Washington, D.C., 1978.
18. J. E. Nash, "The Form of the Instantaneous Unit Hydrograph," IASH Publ. No. 45, Vol. 3, 1957.
19. R. A. Rao, J. W. Delleur, and B. Sarma, "Conceptual Hydrologic Models for Urbanizing Basins," *Proc. ASCE J. Hyd. Div.* (HY7) (July 1972).
20. University of Colorado at Denver, "First Short Course on Urban Storm Water Modeling Using Colorado Urban Hydrograph Procedures," Department of Civil Engineering, June 1985.

Chapter 13

Hydrograph Routing

■ Prologue

The purpose of this chapter is to:

- Present techniques for determining the effect of streams and reservoirs on hydrograph shapes as the hydrographs move downstream through the systems.
- Distinguish between the two major classifications of hydrograph routing techniques.
- Familiarize the reader with procedures for determining when to apply each of the various routing methods.

Flood forecasting, reservoir design, watershed simulation, and comprehensive water resources planning generally utilize some form of routing technique. Routing is used to predict the temporal and spatial variations of a flood wave as it traverses a river reach or reservoir. Routing techniques may be classified into two categories— *hydrologic routing* and *hydraulic routing*.

Hydrologic routing employs the equation of continuity with either a linear or curvilinear relation between storage and discharge within a river or reservoir. Hydraulic routing, on the other hand, uses both the equation of continuity and the equation of motion, customarily the momentum equation. This particular form utilizes the partial differential equations for unsteady flow in open channels. It more adequately describes the dynamics of flow than does the hydrologic routing technique.

Applications of hydrologic routing techniques to problems of flood prediction, evaluations of flood control measures, and assessments of the effects of urbanization are numerous. Most flood warning systems instituted by NOAA and the Corps of Engineers incorporate this technique to predict flood stages in advance of a severe storm. It is the method most frequently used to size spillways for small, intermediate, and large dams. Hydrologic river and reservoir routing and hydraulic river routing techniques are presented in separate sections of this chapter.

13.1 HYDROLOGIC RIVER ROUTING

The first reference to routing a flood hydrograph from one river station to another was by Graeff in 1883.[1] The technique was based on the use of wave velocity and a rating curve of stage versus discharge. Hydrologic river routing techniques are all founded upon the equation of continuity

$$I - O = \frac{dS}{dt} \tag{13.1}$$

where
I = the inflow rate to the reach
O = the outflow rate from the reach
dS/dt = the rate of change of storage within the reach

Three of the most popular hydrologic river routing techniques are described in subsequent paragraphs.

Muskingum Method

Storage in a stable river reach can be expected to depend primarily on the discharge into and out of a reach and on hydraulic characteristics of the channel section. The storage within the reach at a given time can be expressed as[2]

$$S = \frac{b}{a}[XI^{m/n} + (1 - X)O^{m/n}] \tag{13.2}$$

Constants a and n reflect the stage discharge characteristics of control sections at each end of the reach, and b and m mirror the stage-volume characteristics of the section. The factor X defines the relative weights given to inflow and outflow for the reach.

The Muskingum method assumes that $m/n = 1$ and lets $b/a = K$, resulting in

$$S = K[XI + (1 - X)O] \tag{13.3}$$

where K = the storage time constant for the reach
x = a weighting factor that varies between 0 and 0.5.

Application of this equation has shown that K is usually reasonably close to the wave travel time through the reach and X averages about 0.2.

Behavior of the flood wave due to changes in the value of the weighting factor X is readily apparent from examination of Fig. 13.1. The resulting downstream flood wave is commonly described by the amount of translation—that is, the time lag—and by the amount of attenuation or reduction in peak discharge. As can be noted from Fig. 13.1, the value $X = 0.5$ results in a pure translation of the flood wave.

Application of Eqs. 13.1 and 13.3 to a river reach is a straightforward procedure if K and X are known. The routing procedure begins by dividing time into a number of equal increments, Δt, and expressing Eq. 13.1 in finite difference form, using subscripts 1 and 2 to denote the beginning and ending times for Δt. This gives

$$\frac{I_1 + I_2}{2} - \frac{O_1 + O_2}{2} = \frac{S_2 - S_1}{\Delta t} \tag{13.4}$$

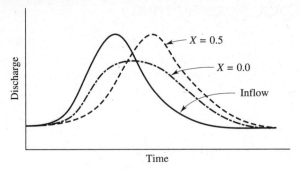

Figure 13.1 Effect of weighting factor.

The routing time interval Δt is normally assigned any convenient value between the limits of $K/3$ and K.

The storage change in the river reach during the routing interval from Eq. 13.3 is

$$S_2 - S_1 = K[X(I_2 - I_1) + (1 - X)(O_2 - O_1)] \qquad (13.5)$$

and substituting this into Eq. 13.4 results in the Muskingum routing equation

$$O_2 = C_0 I_2 + C_1 I_1 + C_2 O_1 \qquad (13.6)$$

in which

$$C_0 = \frac{-KX + 0.5\,\Delta t}{K - KX + 0.5\,\Delta t} \qquad (13.7)$$

$$C_1 = \frac{KX + 0.5\,\Delta t}{K - KX + 0.5\,\Delta t} \qquad (13.8)$$

$$C_2 = \frac{K - KX - 0.5\,\Delta t}{K - KX + 0.5\,\Delta t} \qquad (13.9)$$

Note that K and Δt must have the same time units and also that the three coefficients sum to 1.0.

Theoretical stability of the numerical method is accomplished if Δt falls between the limits $2KX$ and $2K(1 - X)$. The theoretical value of K is the time required for an elemental (kinematic) wave to traverse the reach. It is approximately the time interval between inflow and outflow peaks, if data are available. If not, the wave velocity can be estimated for various channel shapes as a function of average velocity V for any representative flow rate Q. Velocity for steady uniform flow can be estimated by either the Manning or Chézy equation: The approximate wave velocities for different channel shapes are given in Table 13.1.

TABLE 13.1 KINEMATIC WAVE VELOCITIES FOR VARIOUS CHANNEL SHAPES

Channel shape	Manning equation	Chézy equation
Wide rectangular	$\frac{5}{3}V$	$\frac{3}{2}V$
Triangular	$\frac{4}{3}V$	$\frac{5}{4}V$
Wide parabolic	$\frac{11}{9}V$	$\frac{7}{6}V$

Since I_1 and I_2 are known for every time increment, routing is accomplished by solving Eq. 13.6 for successive time increments using each O_2 as O_1 for the next time increment. Example 13.1 illustrates this row-by-row computation.

EXAMPLE 13.1

Perform the flood routing for a reach of river given $X = 0.2$ and $K = 2$ days. The inflow hydrograph with $\Delta t = 1$ day is shown in Table 13.2, column 1. Assume equal inflow and outflow rates on the 16th.

Solution. If $\Delta t = 1$ day and $X = 0.2$ and $K = 2$ days, then Eqs. 13.7 to 13.9 give $C_0 = 0.0477$, $C_1 = 0.428$, and $C_2 = 0.524$. Row-by-row computation is given in Table 13.2.   ■■

Determination of Muskingum K and X   Values of K and X for Muskingum routing are commonly estimated using K equal to the travel time in the reach and an average value of $X = 0.2$. If inflow and outflow hydrograph records are available for one or more floods, the routing process is easily reversed to provide better values of K and X for the reach. To illustrate the latter method, instantaneous values of S versus

TABLE 13.2

Date	(1) Inflow	(2) $C_0 I_2$	(3) $C_1 I_1$	(4) $C_2 Q_1$	(5) Computed outflow
3-16	4,260	—	—	—	4,260
17	7,646	364	1,823	2,232	4,419
18	11,167	532	3,272	2,315	6,119
19	16,730	798	4,779	3,206	8,783
20	21,590	1,029	7,160	4,602	12,791
21	20,950	999	9,240	6,702	16,941
22	26,570	1,267	8,966	8,877	19,110
23	46,000	2,194	11,371	10,013	23,578
24	59,960	2,860	19,688	12,355	34,903
25	57,740	2,754	25,662	18,289	46,705
26	47,890	2,284	24,712	24,473	51,469
27	34,460	1,643	20,496	26,970	49,109
28	21,660	1,033	14,748	25,733	41,514
29	34,680	1,654	9,270	21,753	32,677
30	45,180	2,155	14,843	17,122	34,120
31	49,140	2,343	19,337	17,879	39,559
4-1	41,290	1,969	21,031	20,729	43,729
2	33,830	1,613	17,672	22,914	42,199
3	20,510	978	14,479	22,112	37,569
4	14,720	702	8,778	19,686	29,166
5	11,436	545	6,300	15,283	22,128
6	9,294	443	4,894	11,595	16,932
7	7,831	373	3,977	8,872	13,222
8	6,228	297	3,351	6,928	10,576
9	6,083	290	2,665	5,542	8,497

$XI + (1 - X)O$ are first graphed for several selected values of X as shown in Example 13.2. Because S and $XI + (1 - X)O$ are assumed to be linearly related via Eq. 13.3, the accepted value of X is that which gives the best linear plot (the narrowest loop). After plotting, the value for K is determined as the reciprocal of the slope through the narrowest loop, since from Eq. 13.3

$$K = \frac{S}{XI + (1 - X)O} \tag{13.10}$$

Instantaneous values of S for the graphs in Example 13.2 were determined by solving for S_2 in Eq. 13.4 for successive time increments. A value of $S_1 = 0$ was used for the initial increment, but the value is arbitrary since only the slope and not the intercept of Eq. 13.3 is desired. The S_2 values are plotted against average weighted discharges, $X\bar{I} + (1 - X)\bar{O}$ in Table 13.3. A preferable method would be to plot S_2 values against corresponding values of instantaneous (rather than average) values of $XI_2 + (1 - X)O_2$, using recorded values of inflow and outflow (not provided).

TABLE 13.3

Date	$\bar{I} = \dfrac{I_1 + I_2}{2}$ (cfs)	$\bar{O} = \dfrac{O_1 + O_2}{2}$ (cfs)	S_2^a (10^3cfs-days)	Weighted discharge (cfs) $X\bar{I} + (1 - X)\bar{O}$		
				X = 0.1	X = 0.2	X = 0.3
3-16	5,870	4,180	1.7	4,350[b]	4,520	4,690
17	9,310	6,970	4.0	7,200	7,440	7,670
18	12,900	7,560	9.4	8,090	8,630	9,160
19	20,500	14,200	15.7	14,800	15,500	16,100
20	21,000	18,300	18.4	18,600	18,800	19,100
21	23,400	18,500	23.3	19,000	19,500	20,000
22	32,500	21,300	34.5	22,400	23,500	24,700
23	55,400	29,300	60.6	31,900	34,500	37,100
24	62,700	39,700	83.6	42,000	44,300	46,600
25	52,600	48,700	97.5	49,100	49,500	50,000
3-26	43,200	53,300	87.4	52,300	51,300	50,300
27	25,200	48,700	73.9	46,400	44,000	41,700
28	22,800	37,100	59.6	35,700	34,200	32,800
29	41,200	35,800	65.0	36,300	36,900	37,400
30	50,400	35,800	79.6	37,300	38,700	40,200
31	45,300	35,800	89.1	36,800	37,700	38,600
4-1	38,800	42,700	85.2	42,300	41,900	41,500
2	27,000	44,100	68.0	42,400	40,800	39,000
3	16,200	35,400	48.9	33,500	31,600	29,600
4	12,400	25,200	36.1	23,900	22,600	21,400
5	10,200	16,400	29.9	15,800	15,200	14,500
6	8,080	11,500	26.5	11,200	10,800	10,500
7	6,010	9,380	23.1	9,040	8,710	8,370
8	5,050	7,860	20.3	7,300	7,300	7,020

[a] Note: $S_2 \cong S_1 + \bar{I} \Delta t - \bar{O} \Delta t$ [see Eq. 13.4].
[b] Example: $4350 = 0.1(5870) + (1 - 0.1)(4180)$.

EXAMPLE 13.2

Given inflow and outflow hydrographs on the Muckwamp River, determine K and X for the river reach. (See Table 13.3.)

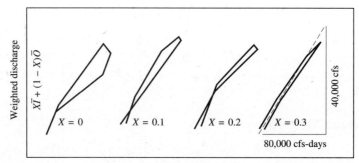

Solution. Selecting the narrowest loop gives $X = 0.3$; $K = 80,000$ cfs-days/40,000 cfs = 2.0 days. These values could now be used to route other floods through the reach as in Example 13.1.

Inherent in this procedure is the postulate that the water surface in the reach is a uniform unbroken surface profile between upstream and downstream ends of the section. Additionally, it is presupposed that K and X are constant throughout the range of flows. If significant departures from these restrictions are present, it may be necessary to work with shorter reaches of the river or to employ a more sophisticated approach. ■■

Muskingum Crest Segment Routing Sometimes it is desirable to solve for a single outflow rate or route only a portion of an inflow hydrograph by the Muskingum method (e.g., the crest segment when only the peak outflow is desired). This is easily accomplished by successively numbering the inflow rates as $I_1, I_2, I_3, \ldots, I_n,$ $I_{n+1}, \ldots$, and rewriting Eq. 13.6 as

$$O_n = C_0 I_n + C_1 I_{n-1} + C_2 O_{n-1} \tag{13.11}$$

where O_n is the outflow rate at any time n. The outflow O_{n-1} is next eliminated from Eq. 13.11 by making the substitution

$$O_{n-1} = C_0 I_{n-1} + C_1 I_{n-2} + C_2 O_{n-2} \tag{13.12}$$

By repeated substitutions for the right-side outflow term $O_{n-2}, O_{n-3}, \ldots$ can each be eliminated and O_n can be expressed as a function only of the first n inflow rates or, finally,

$$O_n = K_1 I_n + K_2 I_{n-1} + K_3 I_{n-2} + \cdots + K_n I_1 \tag{13.13}$$

where $K_1 = C_0$
$K_2 = C_0 C_2 + C_1$
$K_3 = K_2 C_2$
$K_i = K_{i-1} C_2$ for $i > 2$

Using data from Example 13.1 to find the outflow rate on 3-26, we obtain

$$K_1 = C_0 = 0.0477$$
$$K_2 = C_0 C_2 + C_1 = 0.0477(0.524) + 0.428 = 0.4530$$
$$K_3 = K_2 C_2 = 0.4530(0.524) = 0.2374$$
$$\vdots$$
$$K_{11} = K_{10} C_2 = 0.0013$$

Thus the outflow on 3-26 is calculated as $O_{11} = 0.0477 \times (47,890) + 0.4530 \times (57,740) + \cdots + 0.0013(4260) = 51,469$ cfs.

SCS Convex Method

The U.S. Soil Conservation Service (SCS) developed a coefficient channel routing technique, similar to the Muskingum method, in their *National Engineering Handbook*.[3] It has had widespread application in planning and design and can be used successfully even when limited storage data for the reach are available. Until 1983, the procedure was used for all streamflow hydrograph routing in TR-20, the SCS storm event simulation computer program described in Chapter 24. Newer versions of TR-20 use the att-kin method described in Section 13.3.

Analysis of Fig. 13.2 produces the working equation for the convex routing method. Because the areas under both curves are equal, and because the peak outflow

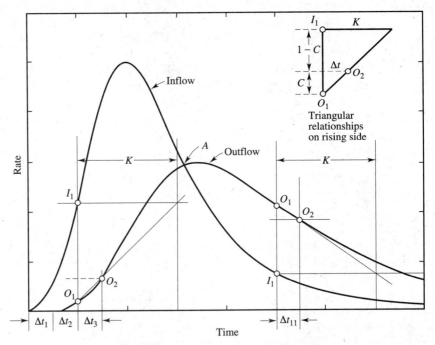

Figure 13.2   Geometric relations used in the SCS convex routing method. (After U.S. Soil Conservation Service.[3])

is less than (and occurs later than) the peak inflow, the curves cross at some point A, resulting in the fact that the value O_2 will always fall between I_1 and O_1. At any time, the vertical distance of O_2 above O_1 (or below O_1 on the right of A) is a fraction C_t of the difference $I_1 - O_1$ as shown in the inset of Fig. 13.2. By proportionate vertical distances

$$O_2 = O_1 + C_t(I_1 - O_1) \tag{13.14}$$

This could be used to route the entire inflow hydrograph if C_t could be established. From Eq. 13.14,

$$C_t = \frac{O_2 - O_1}{I_1 - O_1} \tag{13.15}$$

Because Δt is one limb of the triangle in the inset to Fig. 13.2,

$$\frac{\Delta t}{K} = \frac{O_2 - O_1}{I_1 - O_1} \tag{13.16}$$

where the constant K is the horizontal time from O_1 to the intersection of the line passing through O_1 and O_2. Thus C_t is a function of both Δt and K, or

$$C_t = \frac{\Delta t}{K} \tag{13.17}$$

The technique is valid only if C_t falls between 0.0 and 1.0, and more importantly, if O_2 is always between O_1 and I_1 as indicated in Fig. 13.2. The former can be controlled by selection of Δt, and the latter requirement is satisfied by the mathematical theory arising from analysis of convex sets, which explains the use of this term in naming the routing method.

Determination of K and C_t   Proof that K from Fig. 13.2 is a constant is left to the reader. It is a storage parameter with time units and can be approximated by the K from the Muskingum method. Similarly, C_t is approximately twice the Muskingum X. The reach length divided by wave velocity (estimated from Table 13.1) will provide another estimate of K, or actual measurements of reach travel time can be used. Equation 13.17 can then be solved for C_t. The value recommended by the SCS, in the absence of other estimates, is

$$C_t = \frac{V}{V + 1.7} \tag{13.18}$$

where V is the velocity for a representative steady discharge, and $V + 1.7$ approximates the celerity (speed) of a kinematic wave traveling through the reach. The units of V in Eq. 13.18 are feet per second (fps).

Routing by Eq. 13.14 is easily accomplished after the C_t value is estimated. The interval Δt should be one-fifth or less of the time to peak of the inflow hydrograph to assure a sufficient number of calculated outflow rates to define the hydrograph. As with all routing methods, the time interval should be selected so that one point falls at or near the peak and other locations of rapid change.

Unlike other routing methods, the convex method equation for O_2 is independent of I_2. Thus the procedure can be used to forecast outflow from a reach without knowing the concurrent inflow. This provides a method for early calculation and warning for floods. Flow recorders can be linked through microprocessors to warning systems that calculate downstream flood potentials at least one full routing-time interval ahead of the flood.

The procedure can be reversed to find the inflow hydrograph for a given outflow hydrograph, or it can route a cumulative mass curve of inflow to the reach instead of the hydrograph itself.

Muskingum–Cunge Method

Several attempts to overcome the limitations of the Muskingum method have not been totally successful because of computational complexity or difficulties in physically interpreting the routing parameters.[4,5] The Muskingum parameters are best derived from streamflow measurements and are not easily related to channel characteristics.

Cunge[6] blended the accuracy of the diffusion wave method (see Section 13.3) with the simplicity of the Muskingum method, resulting in one of the most recommended techniques for general use. It is classified as a hydrologic method, yet it gives results comparable with hydraulic methods.

Cunge showed that the finite-difference form of the Muskingum equation becomes the diffusion wave equation if the parameters for both methods are appropriately related. From Eqs. 13.1 and 13.3, the Muskingum equation is

$$K\frac{d}{dt}[XI + (1 - X)O] = \bar{I} - \bar{O} \tag{13.19}$$

Substituting Q_i for I and Q_{i+1} for O, and rewriting in finite-difference form, we obtain

$$\frac{K}{\Delta t}[XQ_i^{t+1} + (1 - X)Q_{i+1}^{t+1} - XQ_i^t - (1 - X)Q_{i+1}^t]$$
$$= \tfrac{1}{2}(Q_i^{t+1} - Q_{i+1}^{t+1} + Q_i^t - Q_{i+1}^t) \tag{13.20}$$

If K is set equal to $\Delta x/c$, Eq. 13.20 is also the finite-difference form of

$$\frac{\partial Q}{\partial t} + c\frac{\partial Q}{\partial x} = 0 \tag{13.21}$$

which is called the *kinematic wave equation* (see Eq. 13.59) and can be derived by combining the continuity and momentum (or friction) equations.[7] The variable Δx denotes an increment of distance along the stream axis and c is the wave speed.

The equation to be used for routing is obtained from Eq. 13.20 by solving for the unknown flow rate,

$$Q_{i+1}^{t+1} = C_0 Q_i^{t+1} + C_1 Q_i^t + C_2 Q_{i+1}^t \tag{13.22}$$

where

$$C_0 = \frac{\Delta t/K - 2X}{2(1 - x) + \Delta t/K} \tag{13.23}$$

$$C_1 = \frac{\Delta t/K + 2X}{2(1 - x) + \Delta t/K} \tag{13.24}$$

$$C_2 = \frac{2(1 - x) - c\,\Delta t/\Delta x}{2(1 - x) + \Delta t/K} \tag{13.25}$$

Because $K = \Delta x/c$, it represents the time for a wave to travel the routing reach length Δx, moving at velocity c. Cunge shows that the velocity c is the celerity of a kinematic wave previously described (Table 13.1).

When $X = 0.5$ and $c\,\Delta t/\Delta x = 1.0$, the routing equation produces translation without attenuation. When $\Delta x = 0$ (zero reach length), no translation or attenuation occurs.

If previous flood data are available, the routing parameter c can be extracted by reversing the routing calculations. Estimates of the parameters can also be obtained from flow and channel measurements.

The value of X for use in Cunge's formulation is

$$X = \frac{1}{2}\left(1 - \frac{q_0}{S_0 c\,\Delta x}\right) \tag{13.26}$$

where S_0 = channel bottom slope (dimensionless)
$\quad\quad q_0$ = discharge per unit width (cfs/ft), normally determined for the peak rate

The value of celerity c can be estimated as a function of the average velocity V by

$$c = mV \tag{13.27}$$

where V is the average velocity Q/A, and m is about $\frac{5}{3}$ for wide natural channels. The coefficient m comes from the uniform flow equation

$$Q = bA^m \tag{13.28}$$

which reduces, by taking partial derivatives, to

$$\frac{\partial Q}{\partial A} = m\frac{Q}{A} = mV \tag{13.29}$$

Substituting this into the continuity equation

$$\frac{\partial Q}{\partial x} + \frac{\partial A}{\partial t} = 0 \tag{13.30}$$

gives Eq. 13.21 if $c = mV$. If discharge data are available, m can be estimated from Eq. 13.28. Values for common shape channels are given in Table 13.1.

The routing can now be done using either constant m and c parameters (i.e., using a single average velocity) or variable parameters (using each new velocity V). Equation 13.27 is solved for c, the value X is derived from Eq. 13.26, and Eqs. 13.23 to 13.25 are solved using $K = \Delta x/c$.

When using this method, the values of Δx and Δt should be selected to assure that the flood wave details are properly routed. Nominally, the time to peak of inflow is broken into 5 or 10 time increments Δt. To give both temporal and spatial resolution, the total reach length L can be divided into several increments of Δx length, and outflow from each is treated as inflow to the next.

EXAMPLE 13.3

Use the Muskingum–Cunge method to route the hydrograph from Example 13.2. Use $S_0 = 0.0001$, $\Delta x = 545$ mi, flow cross-sectional area at $Q = 59,960$ is 5996 ft^2, width at $Q = 59,960$ is 60 ft, and $\Delta t = 1.0$ day (as in Example 13.2).

Solution.   From the inflow, the peak rate of 59,960 cfs gives

$$q_0 = \frac{Q_p}{T_p} = \frac{59,960}{60} = 1000 \text{ cfs/ft}$$

$$V_p = \frac{Q_p}{A_p} = \frac{59,960}{5996} = 10 \text{ fps}$$

$$c = \tfrac{5}{3}V_p = 16.7 \text{ fps}$$

From Eq. 13.26

$$X = \frac{1}{2}\left[1 - \frac{1000}{0.0001(16.7)545(5280)}\right] = 0.4$$

$$K = \frac{\Delta x}{c} = \frac{545(5280)}{16.7} = 172,800 \text{ sec}$$

and
$$C_0 = -0.1765$$
$$C_1 = 0.7647$$
$$C_2 = 0.2941$$

The routing for a portion of the hydrograph is as follows:

Date, t	$C_0 Q^{t+1}_{\text{inflow}}$	$C_1 Q^{t}_{\text{inflow}}$	$C_2 Q^{t}_{\text{outflow}}$	Q^{t+1}_{outflow}
3-16	−1350	3260	0	1910 (3-17)
17	−1970	5850	560	4440
18	−2950	8540	1310	6900
19	−3810	12,790	2030	11,010
20	−3700	16,510	3240	16,050
21	−4690	16,020	4720	16,050
22	−8120	20,320	4720	16,920
23	−10,580	35,180	4980	29,580
24	−10,190	45,850	8700	44,360
25	−8450	44,150	13,050	48,750
26	−6080	36,620	14,340	44,880
27	−3820	26,350	13,200	35,730
3-28	−6120	16,560	10,510	20,950 (3-29)

Note that the peak outflow of 48,750 cfs on March 26 occurs on the same date as in Example 13.2 but has experienced slightly greater attenuation from the Muskingum–Cunge example.

The value C_1 is always positive, and negative values of C_2 are not particularly troublesome. Although C_0 is negative in this example, this condition should be avoided in practice. As seen from Eq. 13.23, negative values of C_0 are avoided when

$$\frac{\Delta t}{K} > 2X \quad \blacksquare\blacksquare \tag{13.31}$$

Other Methods

Other hydrologic river routing procedures have been developed, including the working R&D method, straddle-stagger method, Tatum method, and multiple storage method. They all appear as options in HEC-1, the U.S. Army Corps of Engineer's event simulation and routing model described in Chapter 24.

13.2 HYDROLOGIC RESERVOIR ROUTING

The *storage indication* method of routing a hydrograph through a reservoir is also called the *modified Puls* method.[8] A flood wave passing through a storage reservoir is both delayed and attenuated as it enters and spreads over the pool surface. Water stored in the reservoir is gradually released as pipe flow through turbines or outlet works, called *principal* spillways, or in extreme floods, over an *emergency* spillway.

Flow over an ungated emergency spillway weir section can be described from energy, momentum, and continuity considerations by the form

$$O = CYH^x \tag{13.32}$$

where O = the outflow rate (cfs)
Y = the length of the spillway crest (ft)
H = deepest reservoir depth above the spillway crest (ft)
C = the discharge coefficient for the weir or section, theoretically 3.0
x = exponent, theoretically $\frac{3}{2}$

Flow through a free outlet discharge pipe is similarly described by Eq. 13.32

where Y = the cross-sectional area of the discharge pipe (ft^2)
H = head above the free outlet elevation (ft)
C = the pipe discharge coefficient, theoretically $\sqrt{2g}$
x = exponent, theoretically $\frac{1}{2}$

Flow equations for other outlet conditions are available in hydraulics textbooks. Storage values for various pool elevations in a reservoir are readily determined from computations of volumes confined between various pool areas measured from topographic maps. Since storage and outflow both depend only on pool elevation, the resulting storage-elevation curve and the outflow-elevation relation (Eq. 13.32) can easily be combined to form a storage-outflow graph. Storage in a reservoir depends only on the outflow, contrasted to the dependence on the inflow and outflow in river routing (Eq. 13.3).

For convenience, S is often defined as the "surcharge storage" or the storage above the emergency spillway crest. Normally the overflow rate is zero when S is zero. If the graphed storage-outflow relation is found to be linear, and if the slope of the line is defined as K, then

$$S = KO \tag{13.33}$$

and the reservoir is called a *linear reservoir*. Routing through a linear reservoir is a special case of Muskingum river routing shown in Fig. 13.1 using $x = 0.0$ in Eq. 13.3. Note also that the outflow rate in Fig. 13.1 is increasing only while the inflow exceeds the outflow. This observation is consistent with the assumptions that the inflow immediately goes into storage over the entire pool surface and that the outflow depends only on this storage.

Routing through a linear reservoir is easily accomplished by first dividing time into a number of equal increments and then substituting $S_2 = KO_2$ into Eq. 13.4 and solving for O_2, which is the only remaining unknown for each time increment.

To route an emergency flood through a *nonlinear* reservoir, the storage-outflow relation and the continuity equation, Eq. 13.4, are combined to determine the outflow and storage at the end of each time increment Δt. Equation 13.4 can be rewritten as

$$I_n + I_{n+1} + \left(\frac{2S_n}{\Delta t} - O_n\right) = \frac{2S_{n+1}}{\Delta t} + O_{n+1} \tag{13.34}$$

in which the only unknown for any time increment is the term on the right side. Pairs of trial values of S_{n+1} and O_{n+1} could be generated that satisfy Eq. 13.34 and checked in the storage-outflow curve for confirmation. Rather than resort to this trial procedure, a value of Δt is selected and points on the storage outflow curve are replotted as the "storage indication" curve shown in Fig. 13.3. This graph allows a *direct* determination of the outflow O_{n+1} once a value of the ordinate $2S_{n+1}/\Delta t + O_{n+1}$ has been calculated from Eq. 13.34. The second unknown, S_{n+1}, can be read from the S-O curve (which could also be plotted on the graph in Fig. 13.3) or found from Eq. 13.34.

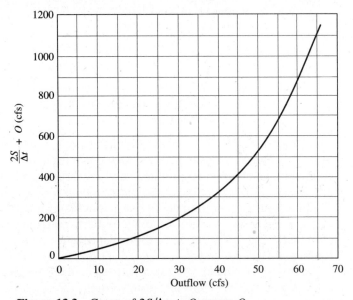

Figure 13.3   Curve of $2S/\Delta t + O$ versus O.

This row-by-row numerical integration of Eq. 13.34 with Fig. 13.3 is illustrated using $\Delta t = 1$ hr in Example 13.4.

EXAMPLE 13.4

Given the triangular-shaped inflow hydrograph and the $2S/\Delta t + O$ curve of Fig. 13.3 find the outflow hydrograph for the reservoir assuming it to be completely full at the beginning of the storm. (See Table 13.4.)

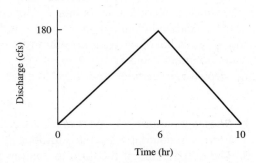

In selecting a routing period Δt, generally at least five points on the rising limb of the inflow hydrograph are employed in the calculations. An increased number of points on the rising limb, that is, a small Δt, improves the accuracy, since as $\Delta t \to 0$ the numerical integration approaches the true limit of the function being integrated, in this case dS/dt.

Column 3 in Table 13.4 comes from the given inflow hydrograph, column 4 is simply the addition of $I_n + I_{n+1}$, and Columns 5 and 7 are initially zero, since in this problem the reservoir is assumed full at the commencement of inflow. Therefore, there is no available storage.

TABLE 13.4   ROUTING TABLE

(1)	(2)	(3)	(4)	+	(5)	=	(6)	(7)	(8)
Time (hr)	n	I_n (cfs)	$I_n + I_{n+1}$ (cfs)		$\dfrac{2S_n}{\Delta t} - O_n$ (cfs)		$\dfrac{2S_{n+1}}{\Delta t} + O_{n+1}$ (cfs)	O_{n+1} (cfs)	S_{n+1} (cfs–hr)
0	1	0	30		0		30	5	12.5
1	2	30	90		20		110	18	46
2	3	60	150		74		224	32	96
3	4	90	210		160		370	43	164
4	5	120	270		284		554	52	250
5	6	150	330		450		780	58	361
6	7	180	315		664		979	63	458
7	8	135	225		853		1078	65	506
8	9	90	135		948		1085	65	510
9	10	45	45		953		998	64	467
10	11	0	0		870		870	62	404
11	12	0	0		746		746	58	344
12	13	0	0		630		630	54	288

The starting value for $n = 1$ in column 6 is computed as the sum of columns 4 and 5 from Eq. 13.34

$$(I_1 + I_2) + \left(\frac{2S_1}{\Delta t} - O_1\right) = \frac{2S_2}{\Delta t} + O_2$$

$$30 + 0 = \frac{2S_2}{\Delta t} + O_2$$

Entering the ordinate of Fig. 13.3 with the value 30 from column 6 gives a value for O_2 of 5 cfs, which is recorded in column 7. The corresponding end-of-time-interval storage, S_2, is calculated from columns 6 and 7 and recorded in column 8. Moving to the second row, a value of the term in column 5 can now be found for $n = 2$ using S_2 and O_2 from columns 7 and 8.

The stepwise procedure used to get outflow figures for all n can be summarized as

1. Entries in columns 1 and 3 are known from the given inflow hydrograph.
2. Entries in column 4 are the additions of $I_n + I_{n+1}$ in column 3.
3. The initial value of the term in column 5 is zero, though it could also be based on any arbitrary starting storage value, and columns 4 and 5 are added to produce the value in column 6.
4. The $2S/\Delta t + O$ versus O plot is entered with known values of $2S/\Delta t + O$ to find values of O for column 7.
5. Columns 6 and 7 are solved for S_{n+1}, which is recorded in column 8.
6. Advance to the next row and calculate the next value for column 5 using the values in the preceding row for O and S from columns 7 and 8.
7. Add the value in column 5 to the advanced sum in column 4 and enter the result in column 6 for the new period under consideration.
8. The new outflow for column 7 is again found from the relation of $2S/\Delta t + O$ as in Fig. 13.3.
9. The corresponding new storage in column 8 is found by solving from columns 6 and 7.
10. Steps 6 through 9 are repeated until the entire outflow hydrograph is generated. ■■

13.3 HYDRAULIC RIVER ROUTING

As noted previously, hydraulic routing employs both the equation of continuity and the equation of motion. Closed-form solutions to the complete hydraulic routing equations do not exist. Thus the application of these techniques demands computer operations. Numerous approaches are available for numerical integration of the equations.

Hydraulic routing techniques are helpful in solving river routing problems, overland flow, or sheet flow. Hydraulic routing proceeds from the simultaneous solution of expressions of continuity and momentum. The general forms of the combination for rivers are called the *spatially varied unsteady flow equations*.

These equations also apply to sheet flow or overland flow and include terms for laterally incoming rainfall. They can be simplified and used to resolve river routing problems.[9] For completeness of presentation, a general form of the spatially varied unsteady flow equations will be presented first.

Equation of Continuity

The equation of continuity states that inflow minus outflow equals the change in storage. To relate this concept to a river section under a condition of rainfall or lateral inflow, consider an element of length Δx and *unit* width into the page as shown in Fig. 13.4.

The total inflow is

$$\rho\left(V - \frac{\partial V}{\partial x}\frac{\Delta x}{2}\right)\left(y - \frac{\partial y}{\partial x}\frac{\Delta x}{2}\right)\Delta t + \rho\int_{x}^{x+\Delta x}\int_{t}^{t+\Delta t}i(x, t)\,dt\,dx \qquad (13.35)$$

The total outflow is

$$\rho\left(V + \frac{\partial V}{\partial x}\frac{\Delta x}{2}\right)\left(y + \frac{\partial y}{\partial x}\frac{\Delta x}{2}\right)\Delta t \qquad (13.36)$$

The change in storage is

$$\rho\frac{\partial y}{\partial t}\Delta x\,\Delta t \qquad (13.37)$$

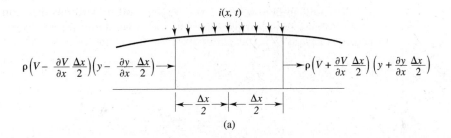

(a)

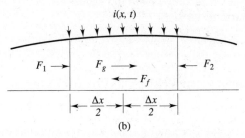

(b)

Figure 13.4 Continuity and momentum elements (where ρ = the density of water, V = the average velocity, y = the depth, i = the lateral inflow per elemental Δx, and S = the slope of the river bottom).

Consequently, continuity gives

$$-\rho\left(y\frac{\partial V}{\partial x}\Delta x + V\frac{\partial y}{\partial x}\Delta x\right)\Delta t + \rho\bar{i}\,\Delta x\,\Delta t - \rho\frac{\partial y}{\partial t}\Delta x\,\Delta t = 0 \qquad (13.38)$$

where $\bar{i}$ is the average lateral inflow resulting from rainfall over Δx and Δt.

The continuity equation of unsteady flow with lateral inflow is obtained by simplifying Eq. 13.38

$$y\frac{\partial V}{\partial x} + V\frac{\partial y}{\partial x} + \frac{\partial y}{\partial t} = \bar{i} \qquad (13.39)$$

For other than a unit width, Eq. 13.39 takes the form

$$A\frac{\partial V}{\partial x} + V\frac{\partial A}{\partial x} + \frac{\partial A}{\partial t} - \bar{i} = 0 \qquad (13.40)$$

where A is the width times the depth, y.

Momentum Equation

In accordance with Newton's second law of motion, the change of momentum per unit of time on a body is equal to the resultant of all external forces acting on the body. The following derivation of the momentum equation of spatially varied unsteady flow is presented subject to the following assumptions: (1) the flow is unidirectional and velocity uniform across the flow section; (2) the pressure is hydrostatic; (3) the slope of the river bottom is relatively small; (4) the Manning formula may be used to evaluate the friction loss due to shear at the channel wall; (5) lateral inflow enters the stream with no velocity component in the direction of flow; and (6) the value of $\bar{i}$ represents the spatial and time variations of lateral inflow.

Forces acting on an element of length Δx and unit width are shown in Fig. 13.4b. The forces F_1 and F_2 represent hydrostatic forces on the element and are expressed as

$$F_1 = \gamma\left[\bar{y}A - \frac{\partial(\bar{y}A)}{\partial x}\frac{\Delta x}{2}\right] \qquad (13.41)$$

$$F_2 = \gamma\left[\bar{y}A + \frac{\partial(\bar{y}A)}{\partial x}\frac{\Delta x}{2}\right] \qquad (13.42)$$

where $\bar{y}$ is the distance from the water surface to the centroid of the area. The resultant hydrostatic force is $F_1 - F_2$, or

$$F_p = -\gamma\frac{\partial(\bar{y}A)}{\partial x}\Delta x \qquad (13.43)$$

By assuming a small slope for the river bottom, the gravitational force is given by

$$F_g = \gamma AS\,\Delta x \qquad (13.44)$$

The frictional force along the bottom is equal to the friction slope S_f multiplied by the weight of water in an element Δx.

$$F_f = \gamma AS_f\,\Delta x \qquad (13.45)$$

The rate of change of momentum in the length Δx may be expressed as

$$\frac{d(mV)}{dt} = m\frac{dV}{dt} + V\frac{dm}{dt} \tag{13.46}$$

in which m is the mass of fluid.

If it is assumed that the incoming lateral inflow enters the moving fluid with no velocity component in the direction of flow and $\bar{i}$ represents the spatial and time variations of the lateral inflow, the rate of change of momentum for the element can be expressed as

$$\frac{d(mV)}{dt} = \rho A\,\Delta x\,\frac{dV}{dt} + \rho V\bar{i}\,\Delta x \tag{13.47}$$

where dV/dt represents

$$\frac{dV}{dt} = \frac{\partial V}{\partial t} + V\frac{\partial V}{\partial x} \tag{13.48}$$

The rate of change of momentum is therefore

$$\rho A\,\Delta x\!\left(\frac{\partial V}{\partial t} + V\frac{\partial V}{\partial x}\right) + \rho V\bar{i}\,\Delta x \tag{13.49}$$

Equating the rate of change of momentum to all external forces acting on the element results in

$$\frac{\partial V}{\partial t} + V\frac{\partial V}{\partial x} + \frac{g}{A}\frac{\partial(\bar{y}A)}{\partial x} + \frac{V\bar{i}}{A} = g(S - S_f) \tag{13.50}$$

Now for a unit width element, the relation simplifies to

$$\frac{\partial V}{\partial t} + V\frac{\partial V}{\partial x} + g\frac{\partial y}{\partial x} + \frac{V}{y}\bar{i} - g(S - S_f) = 0 \tag{13.51}$$

Equations 13.50 and 13.51 form a set of simultaneous expressions that can be solved for V and y subject to the appropriate boundary conditions.

Kinematic, Diffusion, and Dynamic Waves

For the case with zero lateral inflow, i, Eq. 13.51 can be solved for the friction slope

$$S_f = S - \frac{\partial y}{\partial x} - \frac{V}{g}\frac{\partial V}{\partial x} - \frac{1}{g}\frac{\partial V}{\partial t} \tag{13.52}$$

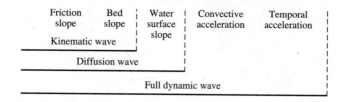

The three types of analysis of unsteady flow routing in open channels are classified as *kinematic, diffusion* (also called *noninertia*), and *dynamic* wave analyses, depending on which terms in Eq. 13.52 are retained. The three techniques differ not only by including different terms of Eq. 13.52 but also in the assumptions regarding flow conditions for satisfying the momentum equation. Table 13.5 shows these assumptions.

TABLE 13.5 ASSUMPTIONS USED IN VARIOUS HYDRAULIC ROUTING METHODS

Method	Common flow condition	Water surface profile
Kinematic	Steady	Uniform
Diffusion	Steady	Nonuniform
Full dynamic	Unsteady	Nonuniform

Steady flow is defined as flow that does not change with time, and uniform flow is flow with a water surface paralleling the bed slope. For steady uniform flow, the rating curve (stage-discharge curve) is a single curve without hysteresis loops. Steady nonuniform flow has constant discharge but varying water surface slope such as that found at the entrance to a reservoir or at the approach of a waterfall.

One way of selecting the applicable method is to examine the rating curve and assess whether it is the same for rising and falling stages. The choice of routing equation depends on whether the difference is small (kinematic), relatively large (dynamic), or somewhere in-between (diffusion).

The kinematic wave method assumes that the inertia terms of Eq. 13.52 are negligible and that the friction slope equals the bed slope S. Momentum conservation is approximated by assuming steady uniform flow, and routing is accomplished by combining the continuity equation with any form of friction loss equation. Typically, either the Manning equation or Chézy equation is used. The Chézy equation is

$$V = C\sqrt{RS} \qquad (13.53)$$

and the Manning equation is

$$V = \frac{1.486}{n} R^{2/3} S^{1/2} \qquad (13.54)$$

where C and n are friction coefficients, S is the friction slope, and R is the hydraulic radius (area divided by wetted perimeter). Both give velocity in feet per second if area and wetted perimeter are input using square feet and feet units.

Either of these equations can be substituted into the kinematic portion of Eq. 13.52, equating slope of energy grade line with bed slope to account for momentum. The continuity equation, Eq. 13.39, for this case reduces to

$$\frac{\partial Q}{\partial x} + \frac{\partial A}{\partial t} = 0 \qquad (13.55)$$

The Manning or Chézy equation has the form

$$Q = bA^m \qquad (13.56)$$

which, after taking derivatives, is

$$\frac{\partial Q}{\partial A} = bmA^{m-1} = m\frac{Q}{A} = mV \tag{13.57}$$

Multiplying Eq. 13.55 by dQ/dA gives

$$mV\frac{\partial Q}{\partial x} + \frac{\partial Q}{\partial t} = 0 \tag{13.58}$$

or, if $c = mV$, the kinematic routing equation is

$$c\frac{\partial Q}{\partial x} + \frac{\partial Q}{\partial t} = 0 \tag{13.59}$$

which can be solved using several numerical schemes.[10] One given by Li[11] solves the nonlinear combination of Eqs. 13.56 and 13.59 using Newton's method.

The celerity c in Eq. 13.59 is equal to mV, or dQ/dA. For a wide, rectangular channel subject to Manning friction, $m = \frac{5}{3}$; for Chézy friction, $m = \frac{3}{2}$. Values for other shapes are given in Table 13.1. The celerity is also given by

$$c = \frac{dQ}{dA} = \frac{1}{B}\frac{dQ}{dy} \tag{13.60}$$

where B is the top width of the channel. Thus celerity is related to the slope of the rating curve, which varies with stage. Most applications assume that c is constant and equal to mV.

When the effects of the water surface slope cannot be ignored, the profile converts from uniform to nonuniform and the slope term $\partial y/\partial x$ in Eq. 13.52 cannot be ignored. This form has had numerous applications and is particularly useful when slopes are relatively flatter than those appropriate to kinematic wave assumptions.

The diffusion wave equation, with water slope $\partial y/\partial x$ retained, is

$$\frac{\partial Q}{\partial t} + c\frac{\partial Q}{\partial x} = d\frac{\partial^2 Q}{\partial x^2} \tag{13.61}$$

The left-hand side is the kinematic wave equation and the right accounts for the diffusion effect of nonuniform water surface profiles. The hydraulic diffusion d is given by

$$d = \frac{q}{2S} \tag{13.62}$$

where q is the flow per unit width of channel, and S is the bed slope. This term reveals why kinematic wave analysis is valid when bed slopes are steep (resulting in small d) or when the channel is extremely wide (resulting in small q). For flat bed slopes, the hydraulic diffusion coefficient is particularly important.

Numerical solutions of Eq. 13.61 are presented by several investigators.[12-14] These normally involve substitution of the relation

$$\frac{\partial Q}{\partial x} = -B\frac{\partial y}{\partial t} \tag{13.63}$$

into Eq. 13.61, where B is the channel top width. The Manning or Chézy equation is used for the friction slope, where

$$S_f = \frac{Q^2}{K^2} \qquad (13.64)$$

in which the conveyance K is $Q/\sqrt{S_f}$ from either equation. Diffusion waves apply to a wider range of problems than kinematic formulations, but their use may not be warranted because it requires about the same effort as dynamic routing.

The third type of wave analysis accounts for all terms in Eq. 13.52, including the nonuniform, unsteady, and inertia components. It is referred to as the "dynamic" or "full dynamic" formulation. Dynamic wave solutions are far more complicated but are often necessary for analysis of flow along very flat slopes, flow into large reservoirs, highly unsteady dam-break flood waves, or reversing (e.g., tidal) flows. These conditions are often encountered on coastal plains. As a general rule, full dynamic wave analysis becomes necessary when

$$S > \frac{15}{T_p} \sqrt{\frac{D}{g}} \qquad (13.65)$$

where S = bed slope
   T_p = time to peak (sec) of the inflow hydrograph
   D = average flow depth (ft)
   g = gravitational acceleration

SCS Att-Kin TR-20 Method

In 1983, the SCS replaced the convex method (Section 13.1) with the modified *att-kin* (*att*enuation-*kin*ematic) method as the agency's preferred channel routing method.[15] The 1964 SCS TR-20 (Chapter 24) single-event simulation model used the convex method but was subsequently modified to route by the att-kin method.

The procedure is a blend of the storage indication and kinematic wave methods. Figure 13.5 shows the two-step process of simulating attenuation first by means of

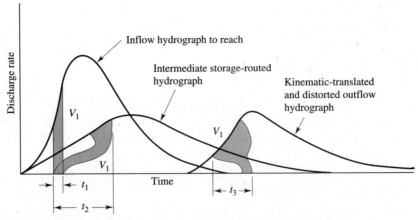

Figure 13.5   Routing principles used in the SCS att-kin method.

storage routing and then translating the wave in time by the kinematic wave method to account for the fact that routed flow rates not only decrease in magnitude but also require time to traverse the length of the routing reach. The storage routing portion provides attenuation with instantaneous translation, and the kinematic wave routing provides translation and distortion but does not attenuate the peak. Both are needed to produce the desired effect. The previously mentioned full dynamic equations simultaneously account for both effects but are difficult to solve.

Figure 13.5 helps to visualize the process. The same volume V_1 of water flowing into the reach during time t_1 would flow out of a hypothetical storage reservoir during interval t_2. This same volume would translate and distort downstream by kinematic action, flowing out of the reach during interval t_3.

Through its theoretical development and selection of routing coefficients, the att-kin method equations satisfy the physical propagation and timing of the peak flow rate first. Conservation of mass is also assured (areas under the three hydrographs of Fig. 13.5 are equal).

The actual process routes the inflow hydrograph through storage, then translates the peak flow rate, without attenuation, to its final time location in the outflow hydrograph. The location in time of the peak outflow is assumed equal to that corresponding to the maximum storage in the reach during passage of the flood.

Because celerity changes with storage, the other flows of the storage-routed hydrograph are translated, each by a different celerity, to their respective final times and values.

The storage indication routing is accomplished by substitution of the relation

$$Q = KS^m \tag{13.66}$$

into the continuity equation, Eq. 13.30, where S is the storage and K and m are coefficients. Kinematic routing solves the unsteady flow equation (13.59) with

$$Q = bA^m \tag{13.67}$$

where b and m are input coefficients, and A is cross-sectional area. If L is the length of routing reach, and if the cross-sectional area throughout L is relatively constant, the storage is given by

$$S = LA \tag{13.68}$$

These equations are combined in an iterative fashion to assure that the peak flow resulting from the kinematic routing equals the peak resulting from storage routing, and simultaneously ensuring that the time of the peak outflow occurs at the time of maximum storage in the reach, or

$$Q_p = KS_p^m \tag{13.69}$$

Input to the method requires selection of a reach length and estimates of b and m for use in Eq. 13.59. As discussed for Table 13.1 and Eq. 13.27, m can be shown to be a factor relating average velocity (under bankfull conditions) with wave celerity, or

$$m = \frac{c}{V} \tag{13.70}$$

The larger m becomes, the shorter the travel time, A value of $m < 1.0$ would incorrectly make the celerity slower than the average flow velocity. Studies by SCS resulted in a recommendation of $\frac{5}{3}$ for general use. Significant errors resulted for m values greater than 2.0. Equation 13.67 is appropriate for cross sections having a single channel with regular shape. Complex cross sections are more diffcult to evaluate, but m values can be developed from a rating table for the stream.[15]

As the coefficient b decreases, attenuation of the peak flow increases due to reduced velocity and increased storage in the reach. The value b can be estimated by plotting Q and A on log-log paper and fitting the linear form of Eq. 13.67 (refer to Table 26.4). The slope would be m and the intercept at $A = 1$ would be b. The SCS has also developed nomographs for estimating b and m.[15]

As a general guideline, the reach length L should be increased to a value that results in a kinematic wave travel time c greater than the selected time increment, or

$$L_R \geq c\,\Delta t$$

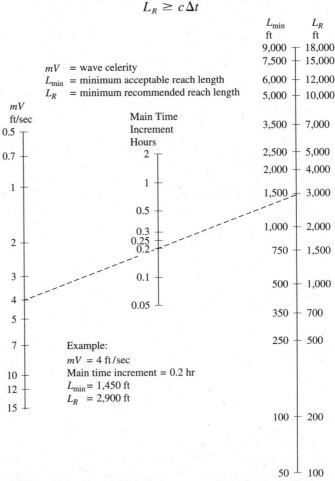

Figure 13.6   SCS nomograph for determining reach length for att-kin method of routing. (After U.S. Soil Conservation Service, "Computer Program for Project Formulation," Technical Release 20, Revised, Appendix G, 1983.)

where L_R is the recommended length, $c = mV$, and Δt is the time increment. The minimum recommended L_R is that giving a wave travel time equal to about half the time increment. This lower value may result in analytical difficulty when lengthy inflow hydrographs or steep streams are encountered. It also results in the peak outflow time being rounded up to the full time increment Δt. If several reaches were routed, this incremental time error would accumulate. Thus a reach length between $c\,\Delta t$ and $c\,\Delta t/2$ is acceptable, but a length greater than $c\,\Delta t$ is recommended. Figure 13.6 provides the range of minimum acceptable and minimum recommended routing reach lengths.

Numerous applications of hydraulic routing techniques appear in the literature; each generally structured for a specific situation. The need to perform hydraulic routing is frequently undertaken in conjunction with a simulation study as will be further discussed in Chapter 23. Material presented here and in Chapter 23 is by no means exhaustive but rather is presented so that an interested student can understand the structuring processes of hydrologic modeling.

■ Summary

Computer software for hydrograph synthesis and routing is available from numerous public and private vendors. Widely used federal agency routines are detailed in Chapter 24. Virtually every computer model of streamflow processes includes one or more hydrograph routing procedures. Some, such as HEC-1 (Chapter 24), provide several choices. In this chapter, the most widely applied hydrologic and hydraulic routing procedures are presented. Comparisons of the models can be found in the literature.[16, 17] At several locations in the chapter, suggestions are given regarding when each of the methods should be applied. The reader who will engage in river or reservoir routing is encouraged to develop the comparison requested in Problem 13.25 before leaving this chapter.

PROBLEMS

13.1. Discuss the main differences between hydrologic and hydraulic routing techniques.

13.2. The Muskingum river routing equation, $O_2 = C_0 I_2 + C_1 I_1 + C_2 O_1$, was derived by substituting the storage equation $S_i = K[XI_i + (1 - X)O_i]$, where $S_1 = K[XI_1 + (1 - X)O_1]$ and $S_2 = K[XI_2 + (1 - X)O_2]$ into the continuity equation $\bar{I} = \bar{O} + \Delta S/\Delta t$ and combining like terms. In these equations, I_1, O_1, and S_1 are the inflow, outflow, and storage at the beginning of the time period; and I_2, O_2, and S_2 are the corresponding values at the end of the time period. The terms $\bar{I}$ and $\bar{O}$ are the average inflow and outflow during the time period, and ΔS is the change in storage. Perform the described derivation and verify the equations for C_0, C_1, and C_2.

13.3. If the Muskingum K value is 12 hr for a reach of a river, and if the X value is 0.2, what would be a reasonable value of Δt for routing purposes?

13.4. A river reach has a storage relation given by $S_i = aI_i + bO_i$. Derive a routing equation for O_2 analogous to the Muskingum equation (13.6). Give equations for the coefficients of I_1, O_1, and I_2.

13.5. List the steps (starting with a measured inflow and outflow hydrograph for a river reach) necessary to determine the Muskingum K and X values. If the inflow and outflow are recorded in cubic feet per second, state the units that would result for K and X if your list of steps is followed.

13.6. Given the following inflow hydrograph:

Hour	Inflow (cfs)	Outflow (cfs)
6 A.M.	100	100
Noon	300	
6 P.M.	680	
Midnight	500	
6 A.M.	400	
Noon	310	
6 P.M.	230	
Midnight	100	

Assume that the outflow hydrograph at a section 3-mi downstream is desired.
  a. Compute the outflow hydrograph by the Muskingum method using values of $K = 11$ hr and $X = 0.13$.
  b. Plot the inflow and outflow hydrographs on a single graph.
  c. Repeat steps (a) and (b) using $X = 0.00$.

13.7. Given the following values of measured discharges at both ends of a 30-mi river reach:
  a. Determine the Muskingum K and X values for this reach.
  b. Holding K constant (at your determined value), use the given inflow hydrograph to determine and plot three outflow hydrographs for values of X equal to the computed value, 0.5, and 0.0. Plot the actual outflow and numerically compare the root mean square of residuals when each of the three calculated hydrographs is compared with the measured outflow.

Time	Inflow (cfs)	Outflow (cfs)
6 A.M.	10	10
Noon	30	12.9
6 P.M.	68	26.5
Midnight	50	43.1
6 A.M.	40	44.9
Noon	31	41.3
6 P.M.	23	35.3
Midnight	10	27.7
6 A.M.	10	19.4
Noon	10	15.1
6 P.M.	10	12.7
Midnight	10	11.5
6 A.M.	10	10.8

13.8. Select a stream in your geographic region that has runoff records. Use the Muskingum method of routing to find K and X.

13.9. Precipitation began at noon on June 14 and caused a flood hydrograph in a stream. As the hydrograph passed, the following measured streamflow data at cross sections A and B were obtained:

Time June 14–17	Inflow, Section A (cfs)	Outflow, Section B (cfs)
6 A.M.	10	10
Noon	10	10
6 P.M.	30	13
Midnight	70	26
6 A.M.	50	43
Noon	40	45
6 P.M.	30	41
Midnight	20	35
6 A.M.	10	28
Noon	10	19
6 P.M.	10	15
Midnight	10	13
6 A.M.	10	11
Noon	10	10

a. Determine the Muskingum K and X values for the river reach.

b. Determine the hydrograph at Section B if a different storm produced the following hydrograph at Section A:

Time	Inflow (cfs)	Time	Inflow (cfs)
6 A.M.	100	Noon	400
Noon	100	6 P.M.	300
6 P.M.	200	Midnight	200
Midnight	500	6 A.M.	100
6 A.M.	600	Noon	100

13.10. The outflow rate (cfs) and storage (cfs-hr) for an emergency spillway of a certain reservoir are linearly related by $O = S/3$, where the number 3 has units of hours. Use this and the continuity equations $S_2 + O_2 \Delta t/2 = \bar{I} \Delta t + S_1 - O_1 \Delta t/2$ to determine the peak outflow rate from the reservoir for the following inflow event:

Time (hr)	I (cfs)	S (cfs-hr)	O (cfs)
0	0	0	0
2	400		
4	600		
6	200		
8	0		

13.11. A simple reservoir has a linear storage-indication curve defined by the equation

$$O = \frac{O}{2} + \frac{S}{\Delta t}$$

where Δt is equal to 1.0 hr. If S at 8 A.M. is 0 cfs-hr, use the continuity equation to route the following hydrograph through the reservoir:

Time	8 A.M.	9 A.M.	10 A.M.	11 A.M.	Noon	1 P.M.
I (cfs)	0	200	400	200	0	0

13.12. For a vertical-walled reservoir with a surface area A show how the two routing equations (13.32 and 13.34) could be written to contain only O_2, S_2, and known values (computed from O_1, S_1, and so on). Eliminate H from all the equations. How could these two equations be solved for the two unknowns?

13.13. Given: Vertical-walled reservoir, surface area = 1000 acres; emergency spillway width = 97.1 ft (ideal spillway); H = water surface elevation (ft) above the spillway crest; and initial inflow and outflow are both 100 cfs.

   a. In acre-ft and cfs-days, determine the values for reservoir storage S corresponding to the following values of H: 0, 0.5, 1, 1.5, 2, 3, 4 ft.

   b. Determine the values of the emergency spillway Q corresponding to the depths named in part a.

   c. Carefully plot and label the discharge-storage curve (cfs versus cfs-days) and the storage-indication curve (cfs versus cfs, Fig. 13.3) on rectangular coordinate graph paper.

   d. Determine the outflow rates over the spillway at the ends of successive days corresponding to the following inflow rates (instantaneous rates at the ends of successive days): 100, 400, 1200, 1500, 1100, 700, 400, 300, 200, 100, 100, 100. Use a routing table similar to the one used in Example 13.4 and continue the rotating procedure until the outflow drops below 10 cfs.

   e. Plot the inflow and outflow hydrographs on a single graph. Where should these curves cross?

13.14. Route the given inflow hydrograph through the reservoir by assuming the initial water level is at the emergency spillway level (1160 ft) and that the principal spillway is plugged with debris. The reservoir has a 500-ft-wide ideal emergency spillway (C = 3.0) located at the 1160-ft elevation. Storage-area-elevation data are

Elevation (ft)	Area of pool (ft² × 10⁶)	Storage (ft³ × 10⁶)
1110	0	0
1120	0.85	4.25
1140	3.75	50.25
1158	9.8	172.15
1160	10.8	192.75
1162	11.8	215.35
1164	12.8	239.95
1166	13.8	266.55
1168	14.85	295.20
1180	25.0	528.55

The inflow hydrograph data are

Time (hr)	I (cfs)
0.0	0
0.5	3,630
1.0	10,920
1.5	10,720
2.0	5,030
2.5	1,600
3.0	460
3.5	100
4.0	10
4.5	0

13.15. Given the following data, route a storm hydrograph through a full reservoir and plot on the same graph the inflow and resulting outflow hydrograph for the Green Acre watershed. The bottom of the rectangular spillway is placed at elevation 980.0. Given: Area $= 0.64$ mi^2; length $= 1.10$ mi; $L_{ca} = 0.53$ mi; $C_t = 2.00$; $C_p = 0.62$; outflow $= CYH^{3/2}$; $C = 3.5$, $L = 10$ ft; and the following storage-elevation curve table.

a. Find the 15-min unit hydrograph by Snyder's method.
b. Find the 30-min unit hydrograph.
c. Find the hydrograph that results from 1.8 in. of rain for the first 30 min and 0.63. in. for the next 15 min.
d. Develop a $2S/\Delta t + O$ versus O curve using a routing period of 15 min and the outflow and storage curves provided.
e. Route the storm hydrograph through the reservoir assuming it is full to the bottom of the spillway elevation 980.
f. Indicate maximum height of water in the reservoir.
g. At what elevation should the top of the dam be placed to obtain 5 ft of freeboard?

Elevation (ft)	Incremental storage 10^4(ft^3)	Total storage 10^4(ft^3)
960		0
	40	
970		40
	210	
980		250
	590	
990		840
	1240	
1000		2080

13.16. Repeat Problem 13.15 with the reservoir initially empty.

13.17. A flood hydrograph is to be routed by the Muskingum method through a 10-mi reach with $K = 2$ hr. Into how many subreaches must the 10-mi river reach be divided in order to use $\Delta t = 0.5$ hr and still satisfy the stability criteria $K/3 \leq \Delta t \leq K$?

13.18.   Repeat Problem 13.6a by dividing the 3-mi reach into two subreaches with equal K values of 5.5 hr. Compare the results.

13.19.   Discuss the problems associated with the use of a reservoir routing technique such as the storage-indication method in routing a flood through a river reach.

13.20.   Verify Eq. 13.51.

13.21.   Precipitation began at noon on June 14 and caused a flood hydrograph in a stream. As the storm passed, the following streamflow data at cross sections A and B were obtained:

Time June 14–17	Inflow Section A (cfs)	Outflow Section B (cfs)
6 A.M.	10	10
Noon	10	10
6 P.M.	30	13
Midnight	70	26
6 A.M.	50	43
Noon	40	45
6 P.M.	30	41
Midnight	20	35
6 A.M.	10	28
Noon	10	19
6 P.M.	10	15
Midnight	10	13
6 A.M.	10	11
Noon	10	10

a. Determine the Muskingum K and X values for the river reach.

b. Determine the hydrograph at Section B if a different storm produced the following hydrograph at Section A (continue computations until outflow falls below 101 cfs):

Time	Inflow (cfs)	Time (cont.)	Inflow (cfs)
6 A.M.	100	Noon	400
Noon	100	6 P.M.	300
6 P.M.	200	Midnight	200
Midnight	500	6 A.M.	100
6 A.M.	600	Noon	100

13.22.   If the Muskingum K value is 12 hr for a reach of a river, and if the X value is 0.2, what would be a reasonable value of Δt for routing purposes?

13.23. Given the following values of measured discharges at both ends of a 30-mi river reach:

Time	Inflow (cfs)	Outflow (cfs)
6 A.M.	10	10
Noon	30	12.9
6 P.M.	68	26.5
Midnight	50	43.1
6 A.M.	40	44.9
Noon	31	41.3
6 P.M.	23	35.3
Midnight	10	27.7
6 A.M.	10	19.4
Noon	10	15.1
6 P.M.	10	12.7
Midnight	10	11.5
6 A.M.	10	10.8

a. Determine the Muskingum K and X values for this reach.
b. Holding K constant (at your determined value), use the given inflow hydrograph to determine and plot three outflow hydrographs for values of X equal to the computed value, 0.5, and 0.0.

13.24. Given the following inflow hydrograph:

Time	Inflow (cfs)	Outflow (cfs)
6 A.M.	10	10
Noon	30	
6 P.M.	68	
Midnight	50	
6 A.M.	40	
Noon	31	
6 P.M.	23	
Midnight	10	

Assume that the outflow hydrograph at a section 3-mi downstream is desired.
a. Compute the outflow hydrograph by the Muskingum method using values of $K = 11$ hr and $X = 0.13$.
b. Plot the inflow and outflow hydrographs on a single graph.
c. Repeat Steps (a) and (b) using $X = 0.00$.

13.25. Carefully review the chapter and consult one other hydrology textbook for all references to applicability of each of the routing procedures presented. Compile the results into a list or table. This table should be retained and consulted frequently.

REFERENCES

1. Graeff, "Traité d'hydraulique." Paris, 1883, pp. 438–443.
2. V. T. Chow, *Open Channel Hydraulics.* New York: McGraw-Hill, 1959.
3. U. S. Soil Conservation Service, *National Engineering Handbook,* Notice NEH 4-102. Washington, D.C.: U.S. Government Printing Office, August 1972.
4. S. Hayami, *On the Propagation of Flood Waves,* Bulletin 1. Kyoto, Japan: Disaster Prevention Institute, 1951.
5. T. N. Keefer and R. S. McQuivey, "Multiple Linearization Flow Routing Model," *Proc. ASCE J. Hyd. Div.* **100**(HY7) (July 1974).
6. J. A. Cunge, "On the Subject of a Flood Propagation Method," *J. Hyd. Res. IAHR* **7**(2), 205–230(1967).
7. D. L. Brakensiek, "Kinematic Flood Routing," *Trans. ASCE* **10**(3) (1967).
8. U. S. Department of the Interior, "Water Studies," *Bureau of Reclamation Manual,* Vol. IV, Sec. 6.10. Washington, D.C.: U. S. Government Printing Office, 1947.
9. T. E. Harbaugh, "Numerical Techniques for Spatially Varied Unsteady Flow," University of Missouri Water Resources Center, Rept. No. 3, 1967.
10. R. K. Price, "Comparison of Four Numerical Methods for Flood Routing," *Proc. ASCE J. Hyd. Div.* **100**(HY7) (July 1974).
11. R. M. Li, D. B. Simons, and M. A. Stevens, "Nonlinear Kinematic Wave Approximation for Water Routing," *Water Resources Res. AGU* **II**(2) (Apr. 1975).
12. J. A. Harder and L. V. Armacost, "Wave Propagation in Rivers," HEL Ser. 8, No. 1. Hydraulic Engineering Laboratory, College of Engineering, University of California, Berkeley, June 1966.
13. D. J. Gunaratnam and F. E. Perkins, "Numerical Solution of Unsteady Flows in Open Channels" Hydrodynamics Rept. 127, MIT, Cambridge, MA, July 1970.
14. G. DiSilvio, "Flood Wave Modification Along Channels," *Proc. ASCE J. Hyd. Div.* **95**(HY7) (1969).
15. U.S. Department of Agriculture, Soil Conservation Service, "Simplified Dam-Breach Routing Procedure," Tech. Release 66, Mar. 1979.
16. Streldoff, T., et al., "Comparative Analysis of Flood Routing Methods," Research Doc. 24, Hydrologic Engineering Center, U. S. Army Corps of Engineers, Davis, CA, 1980.
17. D. L. Gunlach and W. A. Thomas, "Guidelines for Calculating and Routing a Dam-Break Flood," Research Note 5, Hydrologic Engineering Center, U. S. Army Corps of Engineers, Davis, CA, 1977.

Chapter 14

Snow Hydrology

■ **Prologue**

The purpose of this chapter is to:

- Indicate the importance of snowmelt to water supply and management in cold regions.
- Describe methods for measuring snowfall and describing its water-producing capabilities.
- Describe the physics of snowmelt.
- Present models for estimating snowmelt under various conditions of temperature, relative humidity, wind speed, topography, ground cover, and snowpack.

14.1 INTRODUCTION

In many regions, snow is the dominant source of water supply. Mountainous areas in the West are prime examples. Goodell has indicated that about 90 percent of the yearly water supply in the high elevations of the Colorado Rockies is derived from snowfall.[1] Equally high proportions are also likely in the Sierras of California and numerous regions in the Northwest. A significant but lesser share of the annual water yield in the Northeast and Lake states also originates as snow. It is important that the hydrologist understand the nature and distribution of snowfall and the mechanisms involved in the snowmelt process.

Snowmelt usually begins in the spring. The runoff derived is normally out of phase with the periods of greatest water need; therefore, various control schemes such as storage reservoirs have been developed to minimize this problem. An additional point of significance is that some of the greatest floods result from combined large-scale rainstorms and snowmelt. Streamflow forecasting is highly dependent on adequate knowledge of the extent and characteristics of snow fields within the watershed. The water yield from snowfall can be increased by minimizing the vaporization of snow and melt water. Timing the yield can be managed within limits by controlling the rate of snowmelt. Early results can be be obtained by speeding the melt process, whereas the snowmelt period can be extended or delayed by retarding it. The annual snowfall distribution in the United States is shown in Fig. 14.1.

Figure 14.1 Mean annual snowfall in the United States in inches, 1899–1938. (U.S. Weather Bureau.)

An adequate understanding of meteorological factors is as much a prerequisite in considering the snowmelt process as it is in dealing with evapotranspiration. The atmosphere supplies moisture for both snowfall and condensation of water vapor on the snowpack, regulates the exchange of energy within a watershed, and is a controlling factor in snowmelt rates.

As in the rainfall–runoff process, geographic, geologic, topographic, and vegetative factors also are operative in the snow accumulation–snowmelt runoff process.

For rainfall–runoff relations, point rainfall measures are used in estimating areal and time distributions over the basin. A similar approach is taken in snow hydrology although the point-areal relations are usually more complex. Mathematical equations can be used to determine the various components of snowmelt at a given location. Adequate measures of depth and other snowpack properties can also be obtained at specific locations. The use of these measurements in estimating amount and distribution in area and time of snow over large watershed areas is a much less rigorous procedure. Usually, average conditions related to particular areal subdivisions over time are used as the foundation for basin-wide hydrologic estimates. Such procedures are often in the category of index methods (Section 14.6).

Snowmelt routines have been incorporated in numerous hydrologic models, some of which also include water quality dimensions. A good accounting of the fundamentals of the snowmelt process and of contemporary snowmelt modeling approaches may be found in Refs. 2–15 listed at the end of this chapter.

14.2 SNOW ACCUMULATION AND RUNOFF

Under the usual conditions encountered in regions with heavy snowfall, the runoff from the snowpack is the last occurrence in a series of events beginning when the snowfall reaches the ground. The time interval from the start to the end of the process might vary from as little as a day or less to several months or more. Newly fallen snow has a density of about 10 percent (the percentage of snow volume its water equivalent would occupy), but as the snow depth enlarges, settling and compaction increase the density.[15]

The temperature in a deep layer of accumulated snow is often well below freezing after prolonged cold periods. When milder weather sets in, melting occurs first at the snowpack surface. This initial meltwater moves only slightly below the surface and again freezes through contact with colder underlying snow. During the refreezing process, the heat of fusion released from meltwater raises the snowpack temperature. Heat is also transferred to the snowpack from overlying air and the ground. During persistent warm periods, the temperature of the entire snowpack continually rises and finally reaches 32°F. With continued melting, water begins flowing down through the pack. The initial melt component is retained on snow crystals in capillary films. Once the liquid water-holding capacity of the snow is reached, the snow is said to be *ripe*. Throughout the foregoing process, pack density increases due to the refreezing of meltwater and buildup of capillary films. After the water-holding capacity is reached, the density remains relatively constant with continued melt. Meltwater that exceeds the water-holding capacity will continue to move down through the snowpack until the ground is finally reached. At this point runoff can occur. Three situations that may exist at the ground interface when meltwater reaches it are described by Horton.[16]

First, consider the case where the melt rate is less than the infiltration capacity of the soil. In addition, downward capillary pull of the soil coupled with gravity exceeds the same pull of the snow less gravity. The meltwater directly enters the soil and a slush layer is not formed.

The second case occurs when a soil's infiltration capacity is greater than the melt rate, but the net capillary pull of the snowpack exceeds that of the soil aided by gravity. Capillary water builds up in the overlying snow until equilibrium is reached at which upward and downward forces balance. A slush layer forms and provides a supply of water that infiltrates the soil as rapidly as it enters the slush layer.

The final situation is one in which the melt rate exceeds the infiltration capacity. A slush layer forms and water infiltrates the soil at the infiltration capacity rate. Excess water acts in a manner analogous to surface runoff but at a much decreased overland flow rate.

As warm weather continues, the melt process is maintained and accelerated until the snowcover is dissipated.

14.3   SNOW MEASUREMENTS AND SURVEYS

Snow measurements are obtained through the use of standard and recording rain gauges, seasonal storage precipitation gauges, snow boards, and snow stakes. Rain gauges are usually equipped with shields to reduce the effect of wind.[3] Snow boards are about 16 in. square, laid on the snow so that new snowfall which accumulates between observation periods will be found above them. Care must be taken to assure that adverse wind effects or other conditions do not produce an erroneous sample at the gauging location. Snow stakes are calibrated wooden posts driven into the ground for periodic observation of the snow depth or inserted into the snowpack to determine its depth.

Direct measurements of snow depth at a single station are generally not very useful in making estimates of the distributon over large areas, since the measured depth may be highly unrepresentative because of drifting or blowing. To circumvent this problem, snow-surveying procedures have been developed. Such surveys provide information on the snow depth, water equivalent, density, and quality at various points along a snow course. All these measures are of direct use to a hydrologist.

The water equivalent is the depth of water that would weigh the same amount as that of the sample. In this way snow can be described in terms of inches of water. Density is the percentage of snow volume that would be occupied by its water equivalent. The quality of the snow relates to the ice content of the snowpack and is expressed as a decimal fraction. It is the ratio of the weight of the ice content to the total weight. Snow quality is usually about 0.95 except during periods of rapid melt, when it may drop to 0.70–0.80 or less. The thermal quality of snow, Q_t, is the ratio of heat required to produce a particular amount of water from the snow, to the quantity of heat needed to produce the same amount of melt from pure ice at 32°F. Values of Q_t may exceed 100 percent at subfreezing temperatures. The density of dry snow is approximately 10 percent but there is considerable variability between samples. With aging, the density of snow increases to values on the order of 50 percent or greater.

A snow course includes a series of sampling locations, normally not fewer than 10 in number.[14] The various stations are spaced about 50–100 ft apart in a geometric pattern designed in advance. Points are permanently marked so that the same locations will be surveyed each year—very important if snow course memoranda are to be correlated with areal snowcover and depth, expected runoff potential, or other significant factors. Survey data are obtained directly by foresters and others, by aerial photographs and observations, and by automatic recording stations that telemeter information to a central processing location.

In the western United States the Soil Conservation Service coordinates many snow surveys. Various states, federal agencies, and private enterprises are also engaged in this type of activity. Sources of snow survey data are summarized in Ref. 14.

14.4 POINT AND AREAL SNOW CHARACTERISTICS

The estimation of areal snow depth and water equivalent from point measurement data is highly important in hydrologic forecasting.

Estimates of Areal Distribution of Snowfall

Normally, taking arithmetic averages or using Thiessen polygons does not provide reliable results for estimating areal snow distribution from point gaugings. This is because orographic and topographic effects are often pronounced, and gauging networks frequently are not dense enough to permit the straightforward use of normal averaging techniques. However, regional orographic effects are relatively constant from year to year and storm to storm for tracts that are small when compared with the areal extent of general storms occurring in the region.[2] This circumstance permits many useful approaches in estimating the areal snow distribution once the basic pattern has been found for a region.

One method used to estimate basin precipitation from point observations assumes that the ratio of station precipitation to basin precipitation is approximately constant for a storm or storms. This can be stated as[2]

$$\frac{P_b}{P_a} = \frac{N_b}{N_a} \tag{14.1}$$

or

$$P_b = \frac{P_a N_b}{N_a} \tag{14.2}$$

where P_b = the basin precipitation
P_a = the observed precipitation at a point or group of stations
N_b = the annual precipitation for the basin
N_a = the normal annual precipitation for the control station or stations

The normal annual precipitation is determined from a map (carefully prepared if it is to be representative) displaying the mean annual isohyets for the region. The precipitation is determined by planimetering areas between the isohyets. If the number of stations used and their distribution adequately depict the basin, Eq. 14.2 can provide a good approximation. For stations not uniformly distributed, weighting coefficients based on the percentage of the basin area portrayed by a gauge are sometimes used in determining N_a for the group.

Another system used in estimating areal snowfall is the isopercental method. In this approach, the storm or annual station precipitation is expressed as a percentage of the normal annual total. Isopercental lines are drawn and can be superimposed on a normal annual precipitation map (NAP) to produce new isohyets representing the storm of interest. A NAP map indicates the general nature of the basin's topographic effects, while the isopercental map shows the deviations from this pattern. The advantage of this method over preparing an isohyetal map directly is that relatively consistent storm pattern features of the NAP can be taken into consideration as well as observed individual storm variations.

Estimates of Basin-Wide Water Equivalent

A hydrologist must be concerned not only with the amount and areal distribution of snowfall, but also with estimating the water equivalent of this snowpack over the basin, since in the final analysis it is this factor that determines runoff. Basin water equivalent may be given as an index or reported in a quantitative manner such as inches depth for the watershed.

The customary procedure for determining the basin water equivalent is to take observed data from snow course stations and to provide an index of basin conditions. Various procedures employ averages, weighted averages, and other approaches to accomplish this.[2] The important point to remember is that the usefulness of any index is based on how well it represents the overall basin conditions, not on how favorably it describes a particular point value. Indexes do not actually provide a quantitative evaluation of the property they cover. Instead, they give relative changes in the factor. By introducing additional data, however, an index can be used in a prediction equation. For example, if the basin water equivalent can be estimated by subtracting the runoff and loss components from the precipitation input, the index can be correlated with actual basin water equivalent in a quantitative manner.

Areal Snowcover

Estimates of the areal distribution of snowfall are very helpful in making hydrologic forecasts. A knowledge of actual areal extent of snowcover on the ground at any given time is also applied in hydrograph synthesis and in making seasonal volumetric forecasts of the runoff. Observations of snowcover are generally obtained by ground and air reconnaissance and photography. Between snowcover surveys, approximations of the extent of the snowcover are based on available hydrometeorological data. Snowcover depletion patterns within a given basin are normally relatively uniform from year to year; thus snowcover indexes can often be developed from data gathered at a few representative stations.

14.5 THE SNOWMELT PROCESS

The snowmelt process converts ice content into water within the snowpack. Rates differ widely due to variations in causative factors to be discussed later. These divergencies are not as strikingly apparent when considering drainage from the snowpack, however, since the pack itself tends to filter out these non-uniformities so that the drainage exhibits a more consistent rate.

Energy Sources for Snowmelt

The heat necessary to induce snowmelt is derived from short- and long-wave radiation, condensation of vapor, convection, air and ground conduction, and rainfall. The most important of these sources are convection, vapor condensation, and radiation. Rainfall ranks about fourth in importance while conduction is usually a negligible source.

Energy Budget Considerations

If snowmelt is considered as a heat transfer process, an energy budget equation can be written to determine the heat equivalent of the snowmelt. Such an equation is of the form[2]

$$H_m = H_{rl} + H_{rs} + H_c + H_e + H_g + H_p + H_q \qquad (14.3)$$

where H_m = the heat equivalent of snowmelt
   H_{rl} = net long-wave radiation exchange between the snowpack and surroundings
   H_{rs} = the absorbed solar radiation
   H_c = the heat transferred from the air by convection
   H_e = the latent heat of vaporization derived from condensation
   H_g = the heat conduction from the ground
   H_p = the rainfall heat content
   H_q = the internal energy change in the snowpack

In this equation H_{rs}, H_g, and H_p are all positive; H_{rl} is usually negative in the open; H_e and H_q may take on positive or negative values; and H_c is normally positive. The actual amount of melt from a snowpack for a given total heat energy is a function of the snowpack's thermal quality. The heat energy required to produce a centimeter of water from pure ice at 32°F is 80 langleys (g-cal/cm²). Therefore, 203.2 langleys are needed to get 1 in. of runoff from a snowpack of 100 percent thermal quality. If the term H_m represents the combined total heat input in langleys, an equation for snowmelt M in inches is[2]

$$M = \frac{H_m}{203.2Q_t} \qquad (14.4)$$

where Q_t is the thermal quality of the snowpack. For subfreezing snowpacks, Q_t exceeds 1; for ripe snowpacks having some water content, Q_t is less than one. A typical value for these conditions is reported to be 0.97.[6] Figure 14.2 gives a graphical solution to this equation for several values of Q_t.

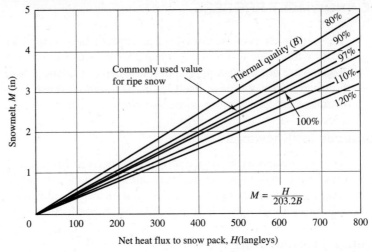

Figure 14.2 Snowmelt resulting from thermal energy. (After U.S. Army Corps of Engineers.[2])

EXAMPLE 14.1

Given a snowpack with thermal quality of 0.90, determine the snowmelt in inches if the total input is 137 langleys.

Solution. Use Eq. 14.4

$$M = H_m/203.2Q_t$$
$$M = 137/203.2 \times 0.90$$
$$M = 0.75 \text{ in.} \quad \blacksquare\blacksquare$$

Turbulent Exchange

The quantity of heat transferred to a snowpack by convection and condensation is commonly determined from turbulent exchange equations. Such an approach has been widely used, since measurements of temperature and vapor pressure must be made in the turbulent zone where vertical water vapor, temperature, and wind velocity gradients are controlled by the action of eddies. In the following two subsections several practical equations for estimating condensation and convection melt are given. Here a combined theoretical equation is presented to acquaint the reader with the theory of turbulent exchange.

The basic turbulent exchange equation can be written[2]

$$Q = A\frac{dq}{dz} \tag{14.5}$$

where Q = the time rate of flow of a specified property of the air such as water vapor through a unit horizontal area

dq/dz = the vertical gradient of the property

q = the property
z = the elevation
A = an exchange coefficient

Property q must be unaffected by the vertical transport. Properties pertinent to this discussion are the air temperature, water vapor, and wind velocity. Theoretically, the potential temperature should be used, but air temperatures measured at normal distances above the snowpack do not cause serious errors. The potential temperature of dry air is that which the air would take if brought adiabatically from its actual pressure to a standard pressure.

Gradients of the various properties of importance here follow a power law distribution where conditions of atmospheric stability exist.[17] This qualification is characteristic of the atmosphere's state over snowfields.[2] Logarithmic profiles are more nearly representative of neutral or unstable atmospheric conditions. The power law provides that the ratio of values of a property determined at two levels above the snow is equivalent to the ratio of the levels raised to some power. Thus,

$$\frac{q_2}{q_1} = \left(\frac{z_2}{z_1}\right)^{1/n} \tag{14.6}$$

where q = the value of the property
z = the elevation (with subscripts denoting the level)
n = the power law exponent

If z_1 is made equal to 1, q_1 assumed to be the property value at this height, and the subscript dropped for the second level, Eq. 14.6 becomes

$$q = q_1 z^{1/n} \tag{14.7}$$

The magnitude of q is taken as the difference in values of q measured at the level z and the snow surface. For example, if $T = 38°F$ at height z, and temperature is the property of interest, then $q = (38 - 32) = 6°F$. The gradient of the property dq/dz can be obtained by differentiating Eq. 14.7:

$$\frac{dq}{dz} = \left(\frac{q_1}{n}\right) z^{(1-n)/n} \tag{14.8}$$

If this expression is substituted in the basic turbulent exchange equation (14.5), the following relation is obtained:

$$Q = A\left(\frac{q_1}{n}\right) z^{(1-n)/n} \tag{14.9}$$

Thus, eddy exchange of the property at a specified elevation z is determined from observations of the property at unity level. The exchange coefficient is also related to elevation z. For equilibrium conditions up to the usual levels of measurement of moisture and temperature, gradients of these variables are such that the eddy transfer of moisture and heat is constant with height. Then the exchange coefficient A must be inversely related to the property gradient, or

$$\frac{A}{A_1} = \frac{(dq/dz)_1}{dq/dz} \qquad (14.10)$$

Substitution in Eq. 14.8 for dq/dz gives the following result:

$$A = A_1 z^{(n-1)/n} \qquad (14.11)$$

since $(dq/dz)_1 = q_1/n$ for $z = 1$. Now, if the value of A from Eq. 14.11 is inserted in Eq. 14.9,

$$Q = A_1 \left(\frac{q_1}{n}\right) \qquad (14.12)$$

The exchange coefficient at an observation level has been shown to be directly proportional to the wind velocity measured at that elevation.[18] Therefore, it may be written that

$$A_1 = k v_1 \qquad (14.13)$$

where $v_1 = $ the wind velocity at Level one
$\qquad k = $ a constant of proportionality

Substituting for A_1 in Eq. 14.12 gives

$$Q = \left(\frac{k}{n}\right) q_1 v_1 \qquad (14.14)$$

Using the power law equation (14.7), we find that

$$q_1 = q z^{-1/n} \qquad (14.15)$$

and
$$v_1 = v z^{-1/n} \qquad (14.16)$$

After making these substitutions in Eq. 14.14 and denoting the observation level of v as z_b, and that of q as z_a (since these may be different), Eq. 14.14 becomes

$$Q = \left(\frac{k}{n}\right)(z_a z_b)^{-1/n} q_a v_b \qquad (14.17)$$

This is a generalized turbulent exchange equation. Consideration is now given to developing specific theoretical equations for condensation and convection melt.

First, consider the case of the condensation melt. The property to be transported in this case is water vapor, and since the exchange coefficient expresses the transfer of an air mass, it is necessary to determine the moisture content of the air mass. This can be accomplished by using the specific humidity, which gives the weight of the water vapor contained in a unit weight of moist air. Equation 14.18 can be used to calculate specific humidity:

$$q = \frac{0.622e}{p_a} \qquad (14.18)$$

where $e = $ the vapor pressure
$\qquad p_a = $ the total pressure of the moist air

Inserting this expression in Eq. 14.17 for q_a yields[2]

$$Q_e = \left(\frac{k}{n}\right)(z_a z_b)^{-1/n}\left(\frac{0.622}{p}\right)e_a v_b \tag{14.19}$$

where Q_e is the moisture transfer to the snow surface per unit time.

The vapor pressure property e_a is the difference in vapor pressures between the level z_a and the snow surface. The value of e_a may thus be either positive (condensation) or negative (evaporation). In addition to the condensate on the snow surface, release of the latent heat of vaporization (about 600 cal/g) from the condensate will melt 7.5 g of snow per gram of condensate if the thermal quality is 100 percent. Multiplying moisture transfer Q_e by 8.5 (melt plus condensate per gram of condensate), the time rate of condensation melt (M_e) becomes

$$M_e = 8.5\left(\frac{k}{n}\right)(z_a z_b)^{-1/n}\left(\frac{0.622}{p}\right)e_a v_b \tag{14.20}$$

The proportionality constant k is a complex function related to the air density and other factors. Since the density of air is a function of elevation, k also varies with height. The constant may be made independent of density, and therefore of elevation, by introducing a factor to compensate directly for the density–elevation relation. Atmospheric pressure serves to accomplish this, and the equation can be adjusted by multiplying by the ratio p/p_0, where p is the pressure at the snowfield elevation and p_0 is the sea level pressure. Introducing this ratio in Eq. 14.20 and a new constant k_1, which is related to sea level pressure, gives[2]

$$M_e = 8.5\left(\frac{k_1}{n}\right)(z_a z_b)^{-1/n}\left(\frac{0.622}{p_0}\right)e_a v_b \tag{14.21}$$

For convection melt, the property of importance in Eq. 14.17 is air temperature. To convert air temperature into thermal units, the specific heat of air c_p must be introduced. Putting these values in Eq. 14.17, heat transfer by eddy exchange H_c converts to[2]

$$H_c = \left(\frac{k}{n}\right)(z_a z_b)^{-1/n} c_p T_a v_b \tag{14.22}$$

Since the latent heat of fusion is 80 cal/g, convective snowmelt M_c in grams in the cgs system is given by $H_c/80$, or

$$M_c = \left(\frac{1}{80}\right)\left(\frac{k}{n}\right)(z_a z_b)^{-1/n} c_p T_a v_b \tag{14.23}$$

Introducing the elevation density correction p/p_0, we obtain

$$M_c = \left(\frac{1}{80}\right)\left(\frac{k_1}{n}\right)(z_a z_b)^{-1/n}\left(\frac{p}{p_0}\right)c_p T_a v_b \tag{14.24}$$

Equations 14.21 and 14.24 can be combined into a single convection-condensation melt M_{ce} equation of the form[2]

$$M_{ce} = \frac{k_1}{n}(z_a z_b)^{-1/n}\left(\frac{1}{80}\frac{p}{p_0}c_p T_a + 8.5\frac{0.622}{p_0}e_a\right)v_b \tag{14.25}$$

This is a generalized theoretical equation for snowmelt that results from the turbulent transfer of water vapor and heat to the snowpack. It is assumed that the exchange coefficients for heat and water vapor are equal. Their evaluation is accomplished by experimentation.[2]

A combined physical equation of the general nature of Eq. 14.25 has been developed by Light.[19] Widely used, its individual convection and condensation melt components are discussed in following sections. The combined form of the Light equation is

$$D = \frac{\rho k_0^2}{80 \ln(a/z_0) \ln(b/z_0)} U\left[c_p T + (e - 6.11) \frac{423}{p}\right] \qquad (14.26)$$

where
D = the effective snowmelt (cm/sec)
ρ = air density
k_0 = von Kármán's coefficient = 0.38
z_0 = the roughness parameter = 0.25
a, b = the levels at which the wind velocity, temperature, and vapor pressure are measured, respectively
U = the wind velocity
c_p = the specific heat of air
T = the air temperature
e = the vapor pressure of the air
p = the atmospheric pressure

Convection

Heat for snowmelt is transferred from the atmosphere to the snowpack by convection. The amount of snowmelt by this process is related to temperature and wind velocity. The following equation can be used to estimate the 6-hr depth of snowmelt in inches by convection:[20]

$$D = KV(T - 32) \qquad (14.27)$$

where
V = the mean wind velocity (mph)
T = the air temperature (°F)

On the basis of the theory of air turbulence and heat transfer (turbulent exchange), a theoretical value for the exchange coefficient K of $0.00184 \times 10^{-0.0000156h}$ has been given by Light.[19] In this relation h, the elevation in feet, is used to reflect the change in barometric pressure due to the difference in altitude. The expression is said to represent conditions for an open, level snowfield where measurements of wind and temperature are made at heights of 50 and 10 ft, respectively, above the snow. Values of the expression $10^{-0.0000156h}$ vary from 1.0 at sea level to 0.70 at 10,000 ft of elevation. The actual values of K are normally less than the theoretical figure due to such factors as forest cover. Empirical 6-hr K values have been reported in the literature.[20]

Anderson and Crawford[15] give an expression for the hourly snowmelt due to convection as

$$M = \frac{cV(T_a - 32)}{Q_t} \qquad (14.28)$$

where M = the hourly melt (in.)
V = the wind velocity (mi/hr)
T_a = the surface air temperature (°F)
Q_t = the snow quality
c = a turbulent exchange coefficient determined empirically

Temperature measurements are at 4 ft, with wind gauged at 15 ft. The corresponding value of c is reported as 0.0002.

Condensation

Heat given off by condensing water vapor in a snowpack is often the most important heat source, particularly when temperatures are in the higher ranges (50–60°F). To melt a pound of ice at 32°F, a thermal input of 144 Btu is required. A pound of moisture originating from the condensation process at 32°F produces about 1073 Btu. On this basis, 1 in. of condensate produces approximately 7.5 in. of water from the snow. A total yield of around 8.5 in. of snowmelt including the condensate is thus derived.

A water vapor supply at the snow surface is formed by the turbulent exchange process; consequently, a mass transfer equation similar to those presented for evaporation studies fits the melt process. An equation for hourly snowmelt from condensation takes the form[15]

$$M = \frac{bV}{Q_t}(e_a - 6.11) \qquad (14.29)$$

where b = an empirical constant
e_a = the vapor pressure of the air (mb), the numerical value
6.11 = the saturation vapor pressure (mb) over ice at 32°F (e_a must exceed 6.11)

Also, M, Q_t, and V are as previously defined. The constant b has a value of 0.001 for temperature and wind measurements at 4 and 15 ft, respectively.[15]

A similar expression but for 6-hr snowmelt (D) is given as

$$D = K_1 V(e_a - 6.11) \qquad (14.30)$$

where the theoretical value of K_1 is said by Light to equal 0.00578 if wind and temperature data are obtained at the 50- and 10-ft levels, respectively, and the snowfield is level and open.[19] Actual figures based on observation are generally lower than this due mainly to forest influences. A value of 0.0032 has been reported by Wilson[20] for three study basins in Wyoming. For condensation melt to occur, the dew-point temperature must exceed 32°F. When it drops below that level, evaporation occurs at the snow surface. An equation for snow evaporation takes the form

$$E = \frac{kV(e_a - e_s)}{Q_t} \qquad (14.31)$$

where E = the hourly evaporation in inches
e_s = the saturation vapor pressure over the snow
k = an empirical constant

Also, V, e_a, and Q_t are as defined before.[15] In the expression $k = 0.0001$, temperature and wind measurements are taken as for Eq. 14.30, and the temperature of the air is assumed equal to that of the snow surface for temperatures below 32°F.

Radiation Melt

The net amount of short- and long-wave radiation received by a snowpack can be a very important source of heat energy for snowmelt. Under clear skies, the most significant variables in radiation melt are insolation, reflectivity or albedo of the snow, and air temperature. Humidity effects, while existent, are usually not important. When cloud cover exists, striking changes in the amount of radiation from an open snowfield are in evidence. The general nature of these effects is illustrated in Fig. 14.3.[2] Combined short- and long-wave radiation exchange as a function of cloud height and cover is represented. Radiation melt is shown to be more significant in the spring than in the winter. It should also be noted that winter radiation melt tends to increase with cloud cover and decreasing cloud height as a result of the more dominant role played by long-wave radiation during that period.

Forest canopies also exhibit important characteristics in regulating radiative heat exchange. These effects differ somewhat from those exhibited by the cloud cover, especially where short-wave radiation is concerned. Clouds and trees both limit insolation, but clouds are very reflective, while a large amount of the intercepted insolation is absorbed by the forest. Consequently, the forest is warmed and part of the incident energy directly transferred to snow in the form of long-wave radiation; an additional fraction is transferred indirectly by air also heated by the forest.

Figure 14.4 illustrates some effects of forest canopy on radiation snowmelt. The figure typifies average conditions for a coniferous cover in the middle latitudes.[2] In winter, the maximum radiation melt is associated with complete forest cover, and in spring the greatest radiation melt occurs in the open. Generalizations should not be drawn from these curves, which indicate relative seasonal effects of forest cover on radiation melt for the conditions described. Another factor affecting radiation melt is the land slope and its aspect (orientation). Radiation received by north-facing slopes is less than that for south-exposure inclines in the northern hemisphere, for example.

Solar energy provides an important source of heat for snowmelt. Above the earth's atmosphere, the thermal equivalent of solar radiation normal to the radiation path is 1.97 langleys/min (1 langley is approximately 3.97×10^{-3} Btu/cm²). The actual amount of radiation reaching the snowpack is modified by many factors such as the degree of cloudiness, topography, and vegetal cover. The importance of vegetal cover in influencing snowmelt, long recognized, has prompted many forest management schemes to regulate snowmelt.[1,14,21,24]

Two basic laws are applicable to radiation. Planck's law states that the temperature of a blackbody is related to the spectral distribution of energy that it radiates. Integration of Planck's law for all wavelengths produces Stefan's law,

$$R_a = \sigma T^4 \tag{14.32}$$

where R_a = the total radiation
   σ = Stefan's constant $[0.813 \times 10^{-10}$ langley/(min-K^{-4})]
   T = the temperature (K)

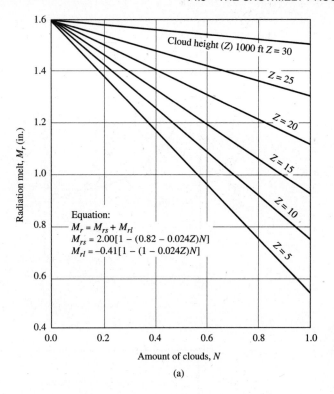

Equation:
$$M_r = M_{rs} + M_{rl}$$
$$M_{rs} = 2.00[1 - (0.82 - 0.024Z)N]$$
$$M_{rl} = -0.41[1 - (1 - 0.024Z)N]$$

(a)

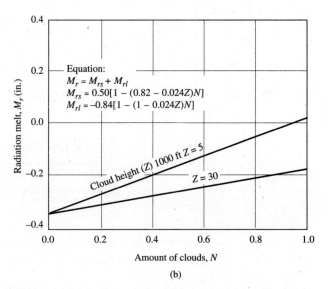

Equation:
$$M_r = M_{rs} + M_{rl}$$
$$M_{rs} = 0.50[1 - (0.82 - 0.024Z)N]$$
$$M_{rl} = -0.84[1 - (1 - 0.024Z)N]$$

(b)

Figure 14.3   Daily radiation melt in the open with cloudy skies: (a) during spring, May 20; and (b) during winter, February 15. (After U.S. Army Corps of Engineers.[2])

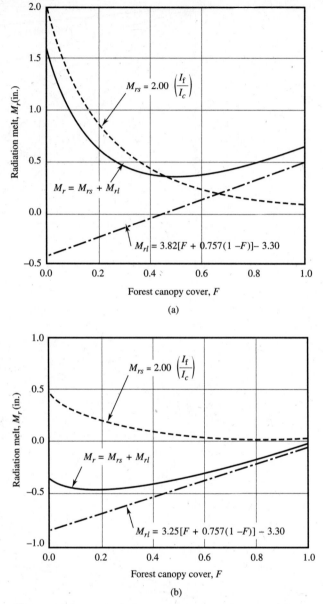

Figure 14.4 Daily radiation melt in the forest with clear skies: (a) during spring, May 20; and (b) during winter, February 15. (After U.S. Army Corps of Engineers.[2])

Because snow radiates as a blackbody, the amount of radiation is related to its temperature (Planck's law), and total energy radiated is according to Stefan's law. Long-wave radiation by a snowpack is determined in a complex fashion through the interactions of temperature, forest cover, and cloud conditions.

Direct solar short-wave radiation received at the snow surface is not all transferred to sensible heat. Part of the radiation is reflected and thus lost for melt purposes. Short-wave reflection is known as *albedo* and ranges from about 40 percent for melting snow late in the season to approximately 80 percent for newly fallen snow. Values as high as 90 percent have also been reported in several cases.[22] This property of the snowpack to reflect large fractions of the insolation explains why the covers persist and air temperatures remain low during clear, sunny, winter periods.

That portion of short-wave radiation not reflected and available for snowmelt may become long-wave radiation or be conducted within the snowpack. Some heat may also be absorbed by the ground with no resultant melt if the ground is frozen. An expression for hourly short-wave radiation snowmelt is given as[2]

$$M = \frac{H_m}{203.2Q_t} \tag{14.33}$$

where H_m = the net absorbed radiation (langleys)
203.2 = a conversion factor for changing langleys to inches of water

When the snow quality is 1, long-wave radiation is exchanged between the snowcover and its surroundings. Snowmelt from net positive long-wave radiation follows Eq. 14.33. If the net long-wave radiation is negative (back radiation), there is an equivalent heat loss from the snowpack.

An approximate method of estimating 12-hr snowmelt D_{12} (periods midnight to noon, noon to midnight) from direct solar radiation has been given by Wilson.[20] The relation is of the form

$$D_{12} = D_0(1 - 0.75m) \tag{14.34}$$

where D_0 = the snowmelt occurring in a half-day in clear weather
m = the degree of cloudiness (0 for clear weather, 1.0 for complete overcast)

Suggested values for D_0 are 0.35 in. (March), 0.42 in. (April), 0.48 in. (May), and 0.53 in. (June) within latitudes 40–48°.[20]

Rainfall

Heat derived from rainfall is generally small, since during those periods when rainfall occurs on a snowpack, the temperature of the rain is probably quite low. Nevertheless, at higher temperatures, rainfall may constitute a significant heat source; it affects the aging process of the snow and frequently is very important in this respect. An equation for hourly snowmelt from rainfall is[15]

$$M = \frac{P(T_w - 32)}{144Q_t} \tag{14.35}$$

where P = the rainfall (in.)
T_w = the web-bulb temperature assumed to be that of the rain
This equation is based on the relation between heat required to melt ice (144 Btu per pound of ice) and the amount of heat given up by a pound of water when its temperature is decreased by one degree.

Daily snowmelt by rainfall estimates are given by

$$M_d = 0.007 P_d (T_a - 32) \qquad (14.36)$$

where M_d = the daily snowmelt (in.)
         P_d = the daily rainfall (in.)
         T_a = the mean daily air temperature (°F) of saturated air taken at the 10-ft level[23]

Conduction

Major sources of heat energy to the snowpack are radiation, convection, and condensation. Under usual conditions, the reliable determination of hourly or daily melt quantities can be founded on these heat sources plus rainfall if it occurs. An additional source of heat, negligible in daily melt computations but perhaps significant over an entire melt season, is ground conduction.

Ground conduction melt is the result of upward transfer of heat from ground to snowpack due to thermal energy that was stored in the ground during the preceding summer and early fall. This heat source can produce meltwater during winter and early spring periods when snowmelt at the surface does not normally occur. Heat transfer by ground conduction can be expressed by the relation[2]

$$H_q = K \frac{dT}{dz} \qquad (14.37)$$

where K = the thermal conductivity of soil
         dT/dz = the temperature gradient perpendicular to soil surface

The snowmelt from ground conduction is generally exceedingly small. Wilson notes that after about 30 days of continuous snowcover, heat transferred from the ground to the snow is insignificant.[20] The amount of snowmelt from ground conduction during a snowmelt season has been estimated at approximately 0.02 in./day.[23] Ground conduction does act to provide moisture to the soil; thus, when other favorable conditions for snowmelt occur, a more rapid development of runoff can be expected.

This section has emphasized the physics of snowmelt. The manner in which heat can be provided to initiate the melt process was discussed. Equations 14.27–14.31 and 14.33–14.37 inclusive can be used to estimate the melt at a given point. The task of computing runoff from snowmelt in a basin cannot be approached in such a simple fashion, since there are many complex factors operative. The remainder of this chapter is devoted to the general subject of runoff from snowmelt investigations. Figure 14.5 illustrates hourly variation in the principal heat fluxes to a snowpack for a cloudy day.

EXAMPLE 14.2 _____

During a completely cloudy April period of 12 hr, the following averages existed for a ripe snowpack located at 10,000 ft above sea level at a latitude of 44° N: air temperature 50° F; mean wind velocity, 10 mph; relative humidity, 65%; average rainfall intensity, 0.03 in./hr for 12 hr; wet bulb psychrometer reading, 48° F. Estimate the snowmelt in in. of water for convection, condensation, radiation, and warm rain for the 12 hr period.

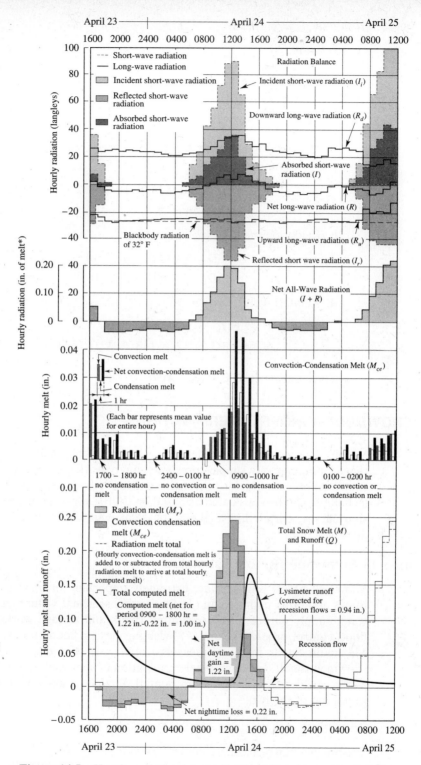

Figure 14.5 Hourly variation in principal heat fluxes to a snowpack for a cloudy day. (After U.S. Army Corps of Enginers.[2])

Solution

a. Convection melt, 6 hr

$$D = KV(T - 32)$$
$$D = 2 \times 0.7 \times 0.00184 \times 10 \times (50 - 32) = 0.50 \text{ in.} \qquad (14.27)$$

b. Condensation melt, 6 hr

$$D = K_1 V(e_a - 6.11)$$
$$D = 2 \times 0.00578 \times 10 \times (12.19 \times 0.65 - 6.11)$$
$$= 0.21 \text{ in.} \qquad (14.30)$$

c. Radiation melt, 12 hr

$$D_{12} = D_0(1 - 0.75 \text{ m})$$
$$D_{12} = 0.42 \times (1 - 0.75 \times 1) = 0.11 \qquad (14.34)$$

d. Rainfall melt, hourly

$$M = P(T_w - 32)/144Q_t$$
$$M = [0.03 \times 12 \times (48 - 32)]/(144 \times 0.97) = 0.04 \qquad (14.35)$$

Thus, total melt is 0.86 in. ■■

14.6 SNOWMELT RUNOFF DETERMINATIONS

Various approaches to runoff determination from snowmelt have been followed. They range from relatively simple correlation analyses that completely ignore the physical snowmelt process to relatively sophisticated methods using physical equations. Most techniques can be considered as based on degree-day correlations, analyses of recession curves, correlation analyses, physical equations, or various indexes. Each is discussed in turn.

Purposes of Snowmelt Runoff Estimates

Snowmelt runoff estimates are extremely important for many regions of the United States and other countries in (1) forecasting seasonal water yields for a diversity of water supply purposes, (2) regulating rivers and storage works, (3) implementing flood control programs, and (4) selecting design floods for particular watersheds. Maximum floods in many areas are often due to a combination of rainfall and snowmelt runoff. In effect, the determination of snowmelt runoff has the same utility as the calculation of runoff from rainfall. In some areas it will, in fact, be the more important of the two.

Snowpack Condition

The manner in which runoff from either rainfall or snowmelt is affected by conditions prevalent within the snowpack is of primary interest to a hydrologist. Various views on storage characteristics of a snowpack have been advanced. These range from the concept that a snowpack can retain large amounts of liquid water to the hypothesis that snowpack storage is negligible. There is no universally applicable relation, and it becomes important to base any runoff considerations on a knowledge of the character of a snowpack at the time of study. Winter runoff is related to a snowpack's condition, whereas in the spring, once active melt begins, little or no delay in the transport of melt or rainfall through the snowpack occurs.

For drainage basins in mountainous areas, snowpack storage effects may be approximated by subdividing the watershed into relatively uniform areas. Normally, this will be accomplished by using elevation zones. Snowpack at the lowest levels may be conditioned to transmit readily rain or meltwater, whereas in higher elevations a liquid water deficit may prevail. At uppermost elevations, the snowpack may be very dry and cold and thus in a condition for the optimum storage of water. The storage potential of the watershed zones must be based on representative measurements of the snow depth, density, temperature, water equivalent, moisture content, and snowpack character. The snowpack character relates to the physical structure of the pack. Unfortunately, adequate measures of all these factors are not always available or easily obtained. Estimates of changes between sampling periods are usually indexed to readily observed meteorologic variables.

The formulation of snowpack storage and time delay characteristics can be fashioned by assuming a homogeneous pack. In this case, storage is related directly to the liquid water deficit and cold content of the pack. Time delay is a function of the inflow rate. It is considered that the snowpack storage potential must be entirely satisfied before runoff begins. In reality this is not the case, but the assumption permits an analysis to be made. As melt proceeds, the storage potential of any snowpack diminishes.

Storage of a snowpack before runoff commences is considered to be the sum of the equivalent water requirement to raise the temperature of the snowpack to 0°C (cold content W_c) and the liquid water-holding capacity of the snowpack. If the cold content is given in inches of water needed to bring a snowpack temperature to 0°C, it may be represented by[2]

$$W_c = \frac{W_0 T_s}{160} \qquad (14.38)$$

where T_s = the mean snowpack temperature below 0°C
W_0 = the initial water equivalent of the snowpack in inches for an assumed specific heat of ice of 0.5

The time t_c in hours needed to raise the snowpack temperature to 0°C is thus given by

$$t_c = \frac{W_0 T_s}{160(i + m)} \qquad (14.39)$$

where i is the rainfall intensity (in./hr) and m is the rate of melt (in./hr). Storage required to meet the liquid water deficit of the snowpack is given by

$$S_f = \frac{f_p}{100}(W_0 + W_c) \qquad (14.40)$$

where S_f = the amount of water stored (in.)
f_p = the percent deficiency in liquid water of the snowpack

The time in hours t_f needed to fill the storage S_f is given by

$$t_f = \frac{f_p(W_0 + W_c)}{100(i + m)} \tag{14.41}$$

It has been specified that the total storage potential S_p to be met prior to the runoff is given as

$$S_p = W_c + S_f \tag{14.42}$$

This is also known as "permanent" storage, since it is not available to the runoff until the snowpack has finally melted. An additional storage component transitory storage S_t is that water stored in the snowpack while moving through it to become runoff. Until initiation of runoff, the transitory storage in inches can be expressed as

$$S_t = \frac{D(i + m)}{V} \tag{14.43}$$

where D = the depth of the snowpack (ft)
$\quad\quad\quad V$ = the rate of transmission through the snowpack (ft/hr)

The delay time of water in passing through the snowpack t_t is thus

$$t_t = \frac{D}{V} \tag{14.44}$$

for t_t in hours. Assuming that W_c is very small compared with W_0, the depth of the snowpack is given by

$$D = \frac{W_0}{\rho_s} \tag{14.45}$$

with ρ_s the density of the snowpack. Then

$$t_t = \frac{W_0}{\rho_s V} \tag{14.46}$$

Before the runoff commences, the total water S stored in the snowpack, in inches, is given by

$$S = W_c + S_f + S_t \tag{14.47}$$

which can also be written

$$S = W_0\left(\frac{T_s}{160} + \frac{f_p}{100} + \frac{i + m}{\rho_s V}\right) \tag{14.48}$$

The total time in hours that passes before runoff is produced is thus[2]

$$t = t_c + t_f + t_t \tag{14.49}$$

or $\quad\quad\quad\quad\quad\quad t = W_0\left[\frac{T_s}{160(i + m)} + \frac{f_p}{100(i + m)} + \frac{1}{\rho_s V}\right] \tag{14.50}$

After establishing the active runoff from the snowpack, the only significant term in Eq. 14.49 is t_t, and this is usually small compared with the overall basin lag and

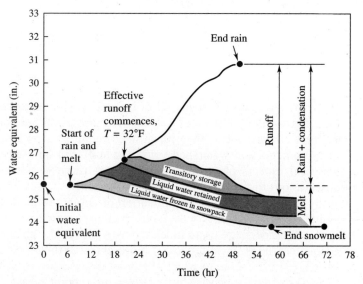

Figure 14.6 Water balance in a snowpack during rainfall.

can be neglected. With increased snowmelt and runoff, additional increments of water previously withheld by snow blockage to drainage outlets and other factors are released. Adequate quantification of this cannot be accomplished at present.[24] A deep snowpack, say 15 ft, having a mean temperature of $-5°C$, could store about 4 in. of liquid water before the onset of runoff. Figure 14.6 illustrates the nature of the water balance in a snowpack during a rainstorm.

EXAMPLE 14.3

A core sample of a snowpack produces the following information: air temperature, 68°F; relative humidity, 20 percent; snowpack density, 0.2; snowpack depth, 10 ft; snowpack temperature, 22°F.

    a. What is the vapor pressure of the air?
    b. Will condensation on the snowpack occur, based on the vapor pressure?
    c. What is the cold content of one sq ft of surface area of the snowpack?
    d. Is the snowpack ripe?

Solution

    a. From Appendix Table A.2, $e_a = 23.3$ mb
        for a relative humidity of 20 percent, $e_a = 4.66$ mb
    b. Condensation will not occur since the air is unsaturated.
    c. Cold content

$$W_c = W_0 T_s / 160$$

$$\text{Temperature} = 22°F = -5.6°C$$

$$W_c = 0.20 \times 120 \times 5.6/160 = 0.84 \text{ in.}$$

d. The snowpack is not ripe since its temperature is below freezing. ■■

Indexes

Hydrologic indexes are made up of hydrologic or meteorologic variables to describe their functioning. The index variable is more easily measured or handier than the element it represents. When mean fixed relations are known to exist between point measurements and watershed values, indexes can be used to record both areal and temporal aspects of basin values. Indexes serve to permit (1) readily obtainable observations to depict hydrologic variables or processes which themselves cannot be easily measured, and (2) simplification of computational methods by allowing individual observations or groups of observations to replace watershed values in time and space. The adequacy of an index is based on (1) the ability of the index to describe adequately the physical process it represents, (2) the random variability of the observation, (3) the degree to which the point observation is typical of actual conditions, and (4) the nature of variability between the point measurement and basin means.[2] Indexes may be equations or simple coefficients, and variable or constant.

The types of data required to make comprehensive thermal budget studies are normally unavailable in whole or part for watersheds other than experimental ones. As a result, a hydrologist must make the best use of information at hand. The most commonly available data are daily maximum and minimum temperatures, humidity, and wind velocity. Less prevalent are continuous measurements of these data, and few stations record solar radiation or the duration of sunshine. Hourly cloudiness data can sometimes be obtained from local airport weather stations.

A completely general index for reliably describing snowmelt–runoff relations for all basins has not been established. Most indexes include coefficients valid only for specific topographic, meteorologic, hydrologic, and seasonal conditions and are therefore limited in applicability to other watersheds. Table 14.1 shows some types of indexes that have been used successfully in snowmelt investigations.

TABLE 14.1   SOME INDEXES USED TO DESCRIBE THERMAL BUDGET VARIABLES

Thermal budget component	Index
Absorbed short-wave radiation	Duration of sunshine data Diurnal temperature range
Long-wave radiation[a]	Air temperature for heavy forested areas For open areas long-wave radiation should be estimated
Convective heat exchange	$(T_a - T_b)V$, where T_a is air temperature, T_b the snow surface temperature or base temperature, and V the wind speed
Heat of condensation	$(e_a - e_s)V$, where e_a and e_s are vapor pressures of air and snow surface or a base value, and V is wind speed

[a] Figure 14.7 illustrates an approximate linear relation between melt and long-wave radiation used by the U.S. Army Corps of Engineers for index purposes.

The snowmelt runoff equation stated in terms of thermal budget indexes is

$$Y = a + \sum_{i=1}^{n} b_i X_i \qquad (14.51)$$

where Y = the snowmelt runoff
   a = a regression constant
   b_i = the regression coefficients
   X_i = individual indexes

Various indexes usable to represent the terms of Eq. 14.51 are selected and a standard regression analysis performed to determine a and b_i. It should be noted that every term in the heat budget equation is not always significant for a particular analysis, and thus the number of X_i will vary for different basins and conditions. A final melt equation

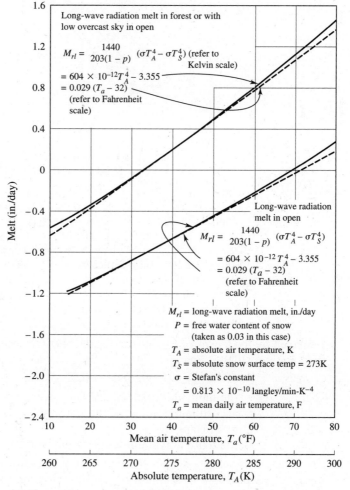

Figure 14.7   Long-wave radiation melt, with linear approximation. (After U.S. Army Corps of Engineers.[2])

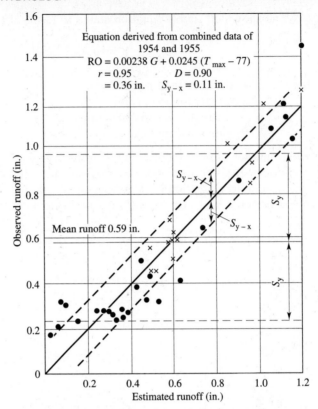

Figure 14.8 Observed versus estimated runoff for ($\times$) 1954 and ($\bullet$) 1955. RO = the daily generated snowmelt runoff (in.) depth over a snow-covered area; G = the daily net all-wave radiation absorbed by snow in the open (langleys); T_{max} = the daily maximum temperature for Boise (°F); r = the coefficient of correlation; D = the coefficient of determination; S_y = the standard deviation of observed runoff (in.); S_{y-x} = the standard error of the stimated runoff (in.). (After U.S. Army Corps of Engineers.[2])

developed by the Corps of Engineers[2] for the partly forested Boise River basin above Twin Springs, Idaho, was

$$Q = 0.00238G + 0.0245(T_{max} - 77) \tag{14.52}$$

where Q = the daily snowmelt runoff (in.) over the snow-covered area

G = an estimated value of the daily all-wave radiation exchange in the open (langleys)

T_{max} = the daily maximum temperature at Boise (°F)

The equation is said to predict the daily snowmelt runoff values within 0.11 in. of observed values about 67 percent of the time. Figure 14.8 illustrates this relation.

In attempting to develop suitable indexes for snowmelt, a hydrologist should seek the approach most closely resembling the thermal budget of the area, within the limitations of available data.

Temperature Indexes

The atmospheric temperature is an extremely useful parameter in snowmelt determination. It reflects the extent of radiation and the vapor pressure of the air; it is also sensitive to air motion. Frequently, it is the only adequate meteorologic variable regularly on hand, so widespread use has been made of degree-day relations in snowmelt computations.

A *degree day* is defined as a deviation of 1° from a given datum temperature consistently over a 24-hr period. In snowmelt computations, the reference temperature is usually 32°F. If the mean daily temperature is 43°F, for example, this is equivalent to 11 degree days above 32°F. If the temperature does not drop below freezing during the 24-hr period, there will be 24 degree hr for each degree departure above 32°F. In this example there would be 264 degree hr for the day of observation.

Various ways of estimating the mean temperature have enabled investigators to take several approaches. One method is simply to average the maximum and minimum daily temperatures. Bases other than 32°F are also used. Regardless of the particular attack employed, a degree hour or degree day is an index to the amount of heat present for snowmelt or other purposes and has proved useful in point-snowmelt and runoff from snowmelt determinations.

The standard practice in developing snowmelt relations on the basis of temperature is to correlate degree days or degree hours with the snowmelt or basin runoff. In some cases, other factors are introduced to define forest cover effects and/or other influences. Another approach often used is to calculate a degree-day factor—the ratio of runoff or snowmelt to accumulated degree days that produced the runoff or melt.

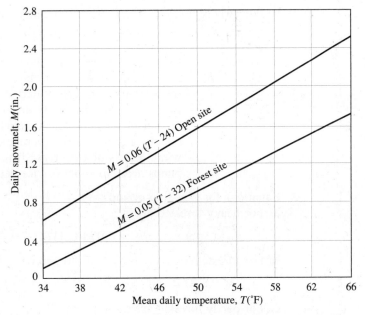

Figure 14.9 Mean temperature index. The equations are applicable only for the range of temperatures shown in the diagram. (After U.S. Army Corps of Engineers.[2])

Unfortunately, the degree-day factor has been found to vary seasonally and between basins; therefore, single representative values should be used with caution. Point-degree-day factors for snow-covered basins range from 0.015 to 0.20 in. per degree per day when melting occurs. Gartska states that an average point value of 0.05 can be used to represent spring snowmelt, provided that caution is used. Linsley and others state that basin mean degree-day factors are usually between 0.06 and 0.15 in./degree day under conditions of continuous snowcover and at melting temperatures. Figure 14.9 illustrates temperature index equations for springtime snowmelt for clear and forested areas.[2]

Generalized Basin Snowmelt Equations

Extensive studies by the U.S. Army Corps of Engineers at various laboratories in the West have produced several general equations for snowmelt during (1) rain-free periods and (2) periods of rain.[24] When rain is falling, heat transfer by convection and condensation is of prime importance. Solar radiation is slight, and long-wave radiation can readily be determined from theoretical considerations. When rain-free periods prevail, both solar and terrestrial radiation become significant and may require direct evaluation. Convection and condensation are usually less critical during rainless intervals. The equations are summarized as follows:[2]

1. Equations for periods with rainfall.
   a. For open (cover below 10 percent) or partly forested (cover from 10 to 60 percent) watersheds,

$$M = (0.029 + 0.0084kv + 0.007P_r)(T_a - 32) + 0.09 \qquad (14.53)$$

   b. For heavily forested areas (over 80 percent cover),

$$M = (0.074 + 0.007P_r)(T_a - 32) + 0.05 \qquad (14.54)$$

   where M = the daily snowmelt (in./day)
   P_r = the rainfall intensity (in./day)
   T_a = the temperature of saturated air at 10-ft level (°F)
   v = the average wind velocity at 50-ft level (mph)
   k = the basin constant, which includes forest and topographic effects, and represents average exposure of the area to wind. Values of k decrease from about 1.0 for clear plains areas to about 0.2 for dense forests

2. Equations for rain-free periods.
   a. For heavy forested areas,

$$M = 0.074(0.53T_a' + 0.47T_d') \qquad (14.55)$$

   b. For forested areas (cover of 60–80 percent),

$$M = k(0.0084v)(0.22T_a' + 0.78T_d') + 0.029T_a \qquad (14.56)$$

   c. For partly forested areas,

$$M = k'(1 - F)(0.0040I_i)(1 - a)$$
$$+ k(0.0084v)(0.22T_a' + 0.78T_d') + F(0.029\ T_a') \qquad (14.57)$$

d. For open areas,

$$M = k'(0.00508I_i)(1 - a) + (1 - N)(0.0212T_a' - 0.84)$$
$$+ N(0.029T_c') + k(0.0084v)(0.22T_a' + 0.78T_d') \qquad (14.58)$$

where M, v, k = as previously described

T_a' = the difference between the 10-ft air and the snow surface (°F) temperatures

T_d' = the difference between the 10-ft dew-point and snow-surface temperatures (°F)

I_i = the observed or estimated insolation (langleys)

a = the observed or estimated mean snow surface albedo

k' = the basin short-wave radiation melt factor (varies from 0.9 to 1.1), which is related to mean exposure of open areas compared to an unshielded horizontal surface

F = the mean basin forest-canopy cover (decimal fraction)

T_c' = the difference between the cloud-base and snow-surface temperatures (°F)

N = the estimated cloud cover (decimal fraction)

Note that the use of equations of the type given must be related to the areal extent of the snowcover if realistic values are to be obtained. Present methods of determining this are not totally adequate.

EXAMPLE 14.4

a. Use Eq. 14.53 to estimate the snowmelt at an elevation of 3000 ft in a partly forested area if the rainfall intensity is 0.3 in./day, the wind velocity is 20 mph, and the temperature of the saturated air is 42°F.

b. Rework your solution for a dense forest cover and a saturated air temperature of 53°F.

Solution

a. $M = (0.029 + 0.0084kv + 0.007P_r)(T_a - 32) + 0.09$

$M = (0.029 + 0.0084 \times 0.5 \times 20 + 0.007 \times 0.3)(42 - 32) + 0.09$

$M = 1.24$ in./day

b. $M = (0.074 + 0.007P_r)(T_a - 32) + 0.05$

$M = (0.074 + 0.007 \times 0.3)(53 - 32) + 0.05$

$M = 1.65$ in./day ■■

The Water Budget

The water budget can be used to estimate the snowmelt runoff from a watershed.[2] Such an approach has particular merit for areas where hydrometeorologic records are short. Difficulty with the method is the usual lack of satisfactory data to quantify the various components properly. A hydrologic budget equation for the earth's surface (Eq. 1.1) can be written

$$R = P - L - \Delta S \qquad (14.59)$$

where P = the gross precipitation
R = the runoff
L = the losses
ΔS = the change in storage

For snowmelt computations this equation is modified somewhat.

Gross precipitation for a given period P is now defined as the sum of precipitations in the form of snow P_s and rain P_r, or

$$P = P_r + P_s \qquad (14.60)$$

This may also be written as

$$P = P_r + P_s = P_n + L_i \qquad (14.61)$$

where P_n = net precipitation
L_i = interception loss

A further refinement yields

$$P = P_{rn} + L_{ri} + P_{sn} + L_{si} \qquad (14.62)$$

where P_{rn}, P_{sn} = net rainfall and snowfall, respectively
L_{ri}, L_{si} = the rain and snow interception, respectively

Figure 14.10 indicates the nature of snow interception by forested areas. Additional information on interception can be found in Chapter 3.

The total loss L is

$$L = L_{si} + L_{ri} + L_e + Q_{sm} \qquad (14.63)$$

where L_e = the evapotranspiration loss
Q_{sm} = the change in available soil moisture

The storage term ΔS is then given as

$$\Delta S = (W_2 - W_1) + Q_g \qquad (14.64)$$

where W_2, W_1 = the final and initial water equivalents of the snowpack, respectively
Q_g = the ground and channel storage

Inserting values for P, L, and ΔS from Eqs. 14.62–14.64 in Eq. 14.59 gives

$$R = P_{rn} + L_{ri} + P_{sn} + L_{si} - L_{si} - L_{ri} - L_e - Q_{sm} - (W_2 - W_1) - Q_g$$
$$(14.65)$$

and canceling positive and negative values of L_{ri} and L_{si} produces

$$R = P_{rn} + P_{sn} - (W_2 - W_1) - Q_{sm} - Q_g - L_e \qquad (14.66)$$

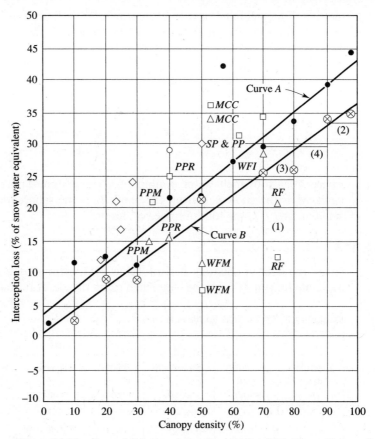

Figure 14.10 Snowfall interception loss. (After U.S. Army Corps of Engineers.[2])

The expression $P_{sn} - (W_2 - W_1)$ represents the snowmelt M; therefore,

$$R = P_{rn} + M - Q_{sm} - Q_g - L_e \qquad (14.67)$$

If reliable estimates of the terms in Eq. 14.67 can be secured, the basin discharge R is computable.

Elevation-Band Procedure

Runoff from snowmelt on a watershed can be estimated from calculations of excess water made available on a series of contributing areas (bands) at various elevations in the watershed. The practice is as follows: divide the watershed into several subareas or bands; estimate the quantity of snowmelt, rainfall, and losses generated on each band during a prescribed interval of time; and use the weighted sum of these contributions to provide an estimate of the excess water available for runoff. For each band, it is assumed that snowmelt, rainfall, and losses are uniform over the band. The subareas are considered to be either snow-covered or snow-free and melting or not melting. For each band, snowmelt is computed using equations of the type presented earlier, rainfall is estimated based on expectations or historic information, and losses are estimated as described in Chapters 3 through 5. Once these estimates have been made for each band, the following equation serves to provide a weighted value of excess water available for runoff from the basin.

$$M = \frac{\sum_{i=1}^{n} [(P_i + M_i - L_i)]A_i}{\sum_{i=1}^{n} A_i} \qquad (14.68)$$

where M = snowmelt water available for runoff (cm/day), P_i is the rainfall on the band, M_i is the snowmelt from the band, L_i is the subarea loss, A_i is the size of the subarea, and n is the total number of bands.

EXAMPLE 14.5

Given the data in columns 1–5 of Table 14.2, estimate the amount of excess water available for runoff from the watershed using the elevation-band method (Eq. 14.68).

TABLE 14.2   DATA FOR EXAMPLE 14.5

(1)	(2)	(3)	(4)	(5)	(6)	(7)
Elevation band no.	Subarea size sq km	Snowmelt cm/d	Rainfall cm/d	Losses cm/d	(3) + (4) + (5)	(6) × (2)
1	230	0.02	0.90	0.40	1.32	303.60
2	224	0.40	1.10	0.50	2.00	448.00
3	289	0.60	1.80	0.70	3.10	895.90
4	213	0.70	1.90	0.60	3.20	681.60
5	193	0.35	2.20	0.30	2.85	550.05
6	167	0.00	2.40	0.10	2.50	417.50
Totals	1316					3296.65

Solution.   The solution for the numerator is the sum of the products given in column 7 of spreadsheet Table 14.2; the solution for the denominator is the sum of the subareas given in column 2 of the table.

The excess water available for runoff = 3269.65/1316 = 2.51 cm/day.

■■

Hydrograph Recessions

Recession curves have been discussed in Chapter 11 and take the general form

$$Q = Q_0 e^{-kt} \qquad (14.69)$$

where Q = the discharge at time t
Q_0 = the initial rate of flow
k = a recession constant

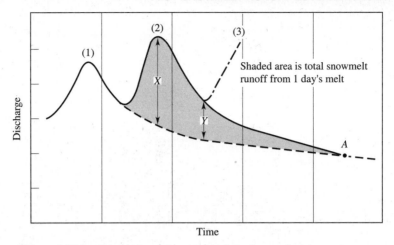

Figure 14.11 Separation of a snowmelt hydrograph.

Studies of daily streamflow by hydrographs permit evaluation of the amount of runoff derived from snowmelt. The technique used is essentially one of separation of the daily hydrographs. Figure 14.11 (not to scale and oversimplified) illustrates the procedure. Assume that the first, second, and succeeding peaks, respectively, fit snowmelt days. If the ultimate recession curve is extended backward in time, at a point A the recession curve from Hydrograph 2 will intersect it. The area between recessions from Hydrograph 1 and Hydrograph 2 (shown cross-hatched) is the melt attributed to Day 1. In like manner, a series of snowmelt hydrographs can be studied to determine their individual melt components. By observing such hydrograph features as the height to peak X, the height to trough Y, and the form of the recession, volumetric and rate forecasts of snowmelt runoff can be made. A more comprehensive treatment of this subject can be found in Ref. 25.

Hydrograph Synthesis

Syntheses of runoff hydrographs associated with snow hydrology are of two types. The first is a short-term forecast. The second kind is the development of flow distribution for a complete melt season or a particular rain-on-snow occurrence. Short-term forecasting is very helpful in preparing plans to operate reservoirs or other flow controls, while the synthesis of particular storm period hydrographs is basic to calculating design floods. To forecast a few days in advance, only the present state of a snowfield and streamflow need be known to make an estimate. For long-term forecasting, it is necessary to have the reliable prediction of various meteorological parameters in addition to a knowledge of the initial conditions. Known historic parameters can be used for reconstructing historic flows whereas assumed or generated parameters satisfy design flood syntheses. Figure 14.12 displays some common hydrometeorologic data.

In snowmelt hydrograph syntheses, several factors (not of great concern where only rainfall exists) must be carefully considered. First, a drainage basin with snowcover cannot be accepted as a homogeneous system, since the areal extent of the

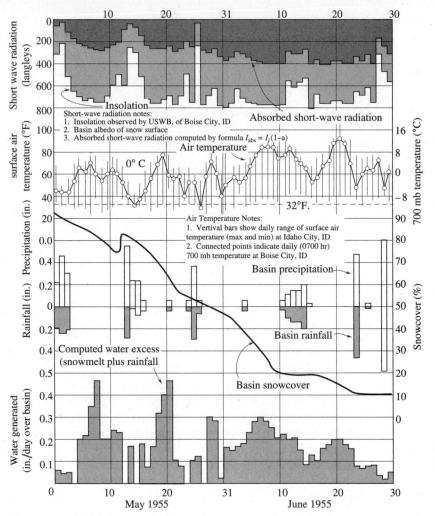

Figure 14.12 Hydrometeorologic data and computation of water generated. (After U.S. Army Corps of Engineers.[2])

blanket is highly important. Where only snowmelt flows are developed, the contributing area need not be the entire drainage—only that portion with snowcover. If rainfall occurs during the snowcover period, contributions can come from bare areas while other expanses may produce combined runoff. The nature of losses in such cases may differ greatly for nonsnow overlayed and covered locations.

The altitude is an exceedingly pertinent factor in the hydrology of tracts subjected to snowfall. Rates of snowmelt decrease with elevation due to a general reduction in temperature with height. Orographic effects and the temperature-elevation relations tend to raise the amount of precipitation with altitude. Greater snowcover depth occurs because of increased precipitation and reduced melt rates. As a result, the basin-wide melt and cover-area increase with height as the snowline is approached, then diminish with elevation over the higher places normally completely snow covered until late in the season. A snowpack exhibits another important trait in relation to rainstorms. In the spring, relatively little runoff occurs from snow-free regions compared with that from a snowfield for moderate rainfalls. During very cold weather, the situation during heavy rains is often reversed, since a dry snowpack can retain significant amounts of water.

Two basic approaches introduce elevation effects into procedures for hydrograph synthesis.[2] The first divides the basin into a series of elevation zones where the snow depth, precipitation losses, and melt are assumed uniform. A second method considers the watershed as a unit, so adjustments are made to account for the areal extent of the snowcover, varying melt rates, precipitation, and other factors.

To synthesize a snowmelt hydrograph, information on the precipitation losses, snowmelt, and time distribution of the runoff are needed. Snowmelt is generally estimated by index methods for forecasting, but in design flood synthesis the heat budget approach is the most used. Precipitation is determined from gaugings and historic or generated data. Losses are defined in two ways where snowmelt is involved. For rain-on-snow hydrographs all the water is considered a loss if delayed very long in reaching a stream. This is basically the concept of direct runoff employed in rainstorm hydrograph analysis. For hydrographs derived principally from snowmelt, only that part of the water which becomes evapotranspiration, or deep percolation, or permanently retained in the snowpack is considered to be lost. Assessing the time distribution of runoff from snow-covered areas is commonly done with unit hydrographs or storage routing techniques. For rain-on-snow events, normal rainfall-type unit hydrographs are applied; for the distribution of strictly snowmelt excess, special long-tailed unit graphs are employed. Storage routing techniques are widely exercised to synthesize spring snowmelt hydrographs, perhaps dividing them into several components and different representative storage times.

The time distribution of snowmelt runoff differs from that of rainstorms due mainly to large contrasts in the rates of runoff generation. For flood flows associated with rainfall only, direct runoff is the prime concern, and time distribution of base flow is only approximated. Big errors in estimates of base flow are not generally of any practical significance where major rainstorm floods occur. In rainstorm flows, infiltrated water is treated as part of the base flow component and little effort is directed toward determining its time distribution when it appears as runoff. In using the unit-hydrograph approach to estimate snowmelt hydrographs, it is customary to separate the surface and subsurface components and route them independently.

Storage routing has been used extensively for routing floods through reservoirs or river reaches. It is also applicable in preparing runoff hydrographs. In snowmelt runoff estimates, the rainfall and meltwater are treated as inputs to be routed through the basin, using storage times selected from the hydrologic characteristics of the watershed. Two basic hydrologic routing approaches are related to the assumption of (1) reservoir-type storage or (2) storage that is a function of inflow and outflow. These methods were treated in depth in Chapter 13.

Storage routing techniques that separate runoff into surface and groundwater components, assign different empirically derived storage times to each, and then route them separately have been employed.[26] An additional system uses a multiple storage,

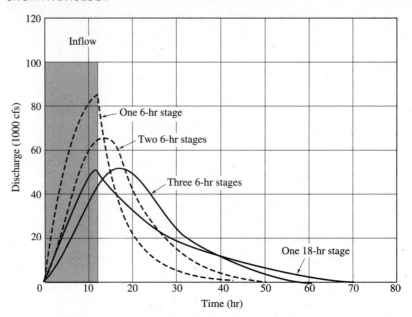

Figure 14.13 Example of multiple-stage reservoir-type storage routing. This figure illustrates the use of multiple-storage reservoir-type storage routing for evaluating time distribution of runoff in a manner analogous to unit hydrographs. (After U.S. Army Corps of Engineers.[2])

reservoir-type storage scheduling.[2] In this method inflow is routed through two or more stages of storage successively. Figure 14.13 illustrates such an approach. Any desired travel time can be obtained by properly selecting the storage time and the number of stages. Retention times between steps may also be varied to reflect various basin hydrologic characteristics. Clark has suggested that the use of single-stage routing after translating input in time permits computations to be simplified.[27]

The most practiced method for synthesizing snowmelt runoff hydrographs has been the unit hydrograph. The character of snowmelt unit graphs differs primarily in time base length from that of rainstorm unit plots. As discussed in Chapter 12, rainstorm unit hydrographs often are derived from single isolated storm events. In snowmelt runoff, rates of water excess are small and approximately continuous. As a result, the use of S-hydrographs is indicated.[2]

The S-hydrograph method has considerable utility, since it allows (1) adjustments to the derived unit hydrograph for nonuniform generation rates, (2) adjusting the observed time period to a desired interval, (3) ready adjustments of the area under the unit hydrograph to unit volume, (4) averaging several hydrographs to get a unit hydrograph, and (5) a particular unit hydrograph to be separated into two unit graphs of unequal generation periods (a particularly useful technique in snow hydrology). Figure 14.14 shows an example of the S-hydrograph method in adjusting for nonuniform generation rates of water excess.

Once a percentage S-hydrograph is derived, a unit hydrograph of any desired period can be obtained as indicated in Fig. 14.15. Ordinates of the S-hydrograph are

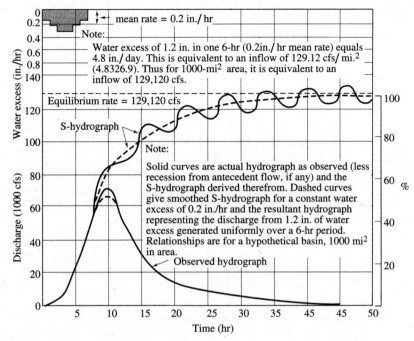

Figure 14.14 S-hydrograph–unit-hydrograph relations. (After U.S. Army Corps of Engineers.[2])

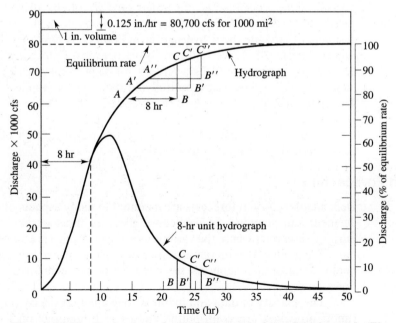

Figure 14.15 Derivation of unit hydrograph from S-hydrograph. The ordinates of the 8-hr UHG are determined as shown by incremental differences taken 8 hr apart. (After U.S. Army Corps of Engineers.[2])

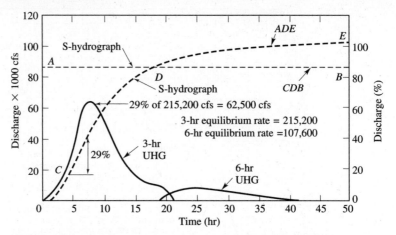

Figure 14.16   Derivation of unit hydrographs having different periods from a divided S-hydrograph. S-hydrograph *CDB* is used to derive the 3-hr UHG; S-hydrograph *ADE* is used to derive the 6-hr UHG. (After U.S. Army Corps of Engineers.[2])

relative to the equilibrium rate associated with a specified water excess for the selected period. The choice of a "best period" for a snowmelt unit hydrograph creates some problems. The rising portion is best defined by a relatively short period, whereas the long tail is correlated with a long interval. A method of choosing two points has been devised, one to represent the early periods of runoff and one to represent the later periods.[2] The procedure is illustrated on Fig. 14.16, although the recession time chosen is unrealistically shortened for purposes of explanation. Assume that it is desirable to separate the S-hydrograph at Point *D*. A horizontal line drawn through *D* cuts the original S-curve into two S-hydrographs, *CDB* and *ADE*. By choosing any desired time periods, unit hydrographs for each S-curve can be obtained by taking incremental differences and multiplying them by the applicable conversion factors. The figure describes the procedure covering a 3-hr period for Curve *CDB* and a 6-hr period for Curve *ADE*.

Figures 14.17 and 14.18 illustrate the reconstitution of several hydrographs using methods discussed in this section.

Runoff Forecasting

Dependable seasonal forecasts are essential to many aspects of water resources management and planning. In particular, they are indispensable to planning multiple-purpose reservoirs where the various uses generally are in competition and sometimes not compatible. The increased emphasis being placed on water management for quality control also supports the need for better techniques in predicting seasonal flows.

The sparsity of data and inability to evaluate properly the controlling factors in runoff processes are contributing causes to limitations on forecasting methods in many cases. Nevertheless, satisfactory forecasts can be made if proper attention is given to the significant variables. When factors of importance are reliably determined

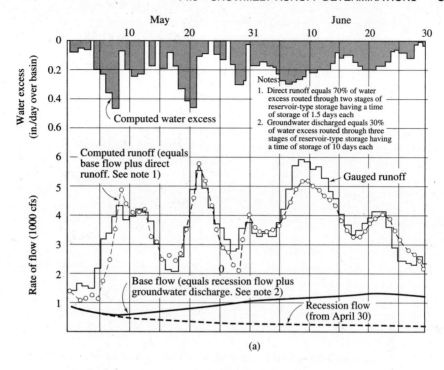

(a)

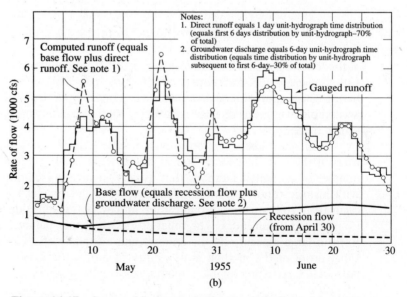

(b)

Figure 14.17   Seasonal hydrograph of mean daily flows: (a) storage-routing method and (b) unit-hydrograph method. (After U.S. Army Corps of Engineers.[2])

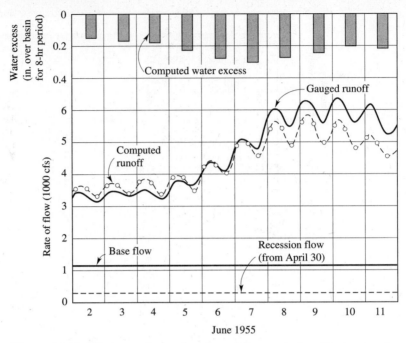

Figure 14.18   Hydrograph showing diurnal flow variation. Discharge values were computed at 8-hr invervals as indicated. All meltwater was considered to be generated during the 8-hr period 1000–1800 hr. (After U.S. Army Corps of Engineers.[2])

at the time of the forecast, most errors result from conditions that set in after observations were made. Subsequent events may have little effect on forecasts in some basins where approximate divergencies can be fairly well quantified. Note also that an acceptable magnitude of error will vary with the circumstances for which the prediction is made.

Principal forecasting tools are the water budget, index methods, and combinations of the two. The water budget has already been discussed in detail, as have a number of snowmelt methods that usually involve correlation of specific variables with historic runoff records. Indexes represent factors such as precipitation, water equivalent, soil moisture, groundwater, and evapotranspiration. Figure 14.19 plots an evapotranspiration index. The reliability of the water balance and index methods is fixed in large measure by the amount and adequacy of available hydrologic data.[28] Final proof of the method is its ability to model the simulated system accurately. Adequate forecasting by index methods normally requires records for approximately 25 years.

■ Summary

In many regions of the United States and other nations, water derived from snowmelt is the major source of supply. An understanding of snowmelt processes is an important adjunct to the management of watersheds in these cold regions. Water supply

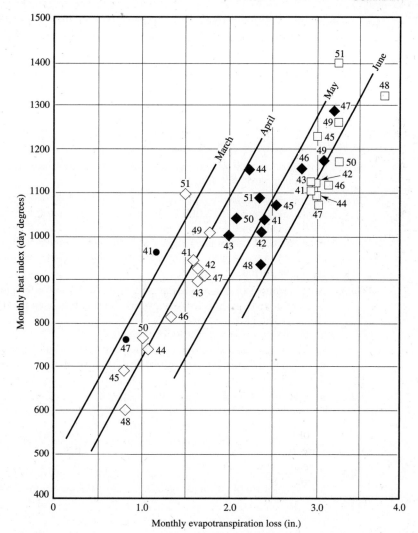

Figure 14.19 Loss by evapotranspiration. (●) March; (◇) April; (◆) May; (□) June.
Notes: 1. Plotted points show the basin evapotranspiration losses computed by Thornthwaite's method, for given monthly day degrees at Detroit. 2. Monthly day degrees are the sum of degrees of the daily maximum temperatures above base 32°F. 3. Lines show the most probable basin evapotranspiration for given day degrees for indicated months. 4. Loss computations are for an 11-year period 1941–1951; points for which losses are zero are not plotted. (After U.S. Army Corps of Engineers.[2])

models for cold regions must address snow accumulation, snowmelt, and the translation of snowmelt into streamflow responses.

To estimate the quantity and timing of water supplies that may be produced from snowpacks on a watershed, it is necessary to survey the snowpack, calculate melt rates from contributing portions of the snowpack, estimate losses from the melt, and route the excess meltwater through the relevant stream channel system.

A number of theoretical and empirical models for snowmelt estimation under various conditions of temperature, ground cover, snowpack condition, and other factors have been discussed in this chapter.[29-37] These models can be applied to real-time water supply management problems (reservoir operation, for example) as well as to long-term water supply forecasting for a watershed of interest.

PROBLEMS

14.1. Given a snowpack with thermal quality of 0.87, determine the snowmelt in inches if the total input is 135 langleys.

14.2. During a partly cloudy April period of 12 hr, the following averages existed for a ripe snowpack located at 10,000 ft above sea level at a latitude of 44°N: air temperature 50°F; mean wind velocity, 8 mph; relative humidity, 65 percent; average rainfall intensity, 0.04 in./hr for 12 hr; wet bulb psychrometer reading, 48°F. Estimate the snowmelt in in. of water for convection, condensation, radiation, and warm rain for the 12-hr period.

14.3. A core sample of a snowpack produces the following information: air temperature, 68°F; relative humidity, 25%; snowpack density, 0.2; snowpack depth, 8 ft; snowpack temperature, 22°F.
   a. What is the vapor pressure of the air?
   b. Will condensation on the snowpack occur, based on the vapor pressure?
   c. What is the cold content of one sq ft of surface area of the snowpack?
   d. Is the snowpack ripe?

14.4. **a.** Use Eq. 14.53 to estimate the snowmelt at an elevation of 3000 ft in a partly forested area if the rainfall intensity is 0.2 in./day, the wind velocity is 15 mph, and the temperature of the saturated air is 44°F.
   b. Rework your solution for a dense forest cover and a saturated air temperature of 57°F.

14.5. Given the data in Columns 1–5 of Table 14.2, estimate the amount of excess water available for runoff from the watershed using the elevation-band method (Eq. 14.68).

14.6. Are the snowmelt effects of condensation, convection, radiation, warm rain, and conduction additive? Answer by analyzing the conditions that produce large amounts of snowmelt by each process and examine the conditions to determine if the effects are additive.

REFERENCES

1. B. C. Goodell, "Snowpack Management for Optimum Water Benefits," Conference Preprint 379, ASCE Water Resources Engineering Conference, Denver, CO, May 1966.
2. Corps of Engineers, "Snow Hydrology," NTIS PB 151 660, North Pacific Division, Corps of Engineers, Portland, OR, June 1956.
3. Corps of Engineers, "Runoff Evaluation and Streamflow Simulation by Computer," Tech. Rep., North Pacific Division, Corps of Engineers, Portland, OR, 1971.
4. E. A. Anderson, "National Weather Service River Forecast System—Snow Accumulation and Ablation Model," NOAA Tech. Memo NWS HYDRO-17, U.S. Department of Commerce, Washington, D.C., 1973.

5. E. A. Anderson, "A Point Energy and Mass Balance Model of a Snow Cover," NOAA Tech. Rep. NWS 19, U.S. Department of Commerce, Washington, D.C., Feb. 1976.

6. P. S. Eagleson, *Dynamic Hydrology*. New York: McGraw-Hill, 1970.

7. G. Fleming, *Computer Simulation Techniques in Hydrology*. New York: American Elsevier, 1975.

8. D. M. Gray (ed.), *Handbook on the Principles of Hydrology*. Port Washington, NY: Water Information Center, 1970.

9. Hydrologic Engineering Center, "Storage, Treatment, Overflow, Runoff Model, STORM," User's Manual, Generalized Computer Program 723-S8-L7520, HEC, Corps of Engineers, Davis, CA, Aug. 1977.

10. R. C. Johanson, J. C. Imhoff, and H. H. Davis, "User's Manual for Hydrological Simulation Program—Fortran (HSPF)," EPA-600/9-80-015, Environmental Protection Agency, Athens, GA, Apr. 1980.

11. R. K. Linsley, M. A. Kohler, and J. L. H. Paulhus, *Hydrology for Engineers*. New York: McGraw-Hill, 1975.

12. L. A. Roesner, H. M. Nichandros, R. P. Shubinski, A. D. Feldman, J. W. Abbott, and A. O. Friedland, "A Model for Evaluating Runoff Quality in Metropolitan Master Planning," ASCE Urban Water Resources Research Program Tech. Memo No. 23 (NTIS PB 234 312), ASCE, New York, Apr. 1974.

13. W. C. Huber, J. P. Heaney, S. J. Nix, R. E. Dickinson, and D. J. Polmann, "Storm Water Management Model User's Manual, Version III," EPA-600/2-84-109a (NTIS PB84-198423), Environmental Protection Agency, Cincinnati, OH, Nov. 1981.

14. Ven Te Chow (ed.), *Handbook of Applied Hydrology*. New York: McGraw-Hill, 1964.

15. E. A. Anderson and N. H. Crawford, "The Synthesis of Continuous Snowmelt Runoff Hydrographs on a Digital Computer," Department of Civil Engineering, Stanford University, Stanford, CA, Tech. Rep. No. 36, June 1964.

16. R. E. Horton, "Phenomena of the Contact Zone Between the Ground Surface and a Layer of Melting Snow," Transactions of Meetings of International Commission of Snow and Glaciers, Edinburgh, Sept. 1936, International Association of Hydrology Bulletin 23.

17. The Johns Hopkins University, Laboratory of Climatology, "Micrometeorology of the Surface Layer of the Atmosphere; the Flux of Momentum, Heat, and Water Vapor, Final Report," *Publ. Climatology* **7**(2) (1954).

18. H. U. Sverdrup, "The Eddy Conductivity of the Air over a Smooth Snow Field," *Geofys. Publik.* **11**(7) (1936).

19. P. Light, "Analysis of High Rates of Snow Melting," *Trans. Am. Geophys. Union* **22,** Part 1, pp. 195–235(1941).

20. W. T. Wilson, "An Outline of the Thermodynamics of Snowmelt," *Trans. Am. Geophys. Union* **22,** Part 1 (1941).

21. J. Kittredge, *Forest Influences*. New York: McGraw-Hill, 1948.

22. H. T. Mantis et al., "Review of the Properties of Snow and Ice," U.S. Army Corps of Engineers, Snow, Ice, and Permafrost Research Establishment, SIPRE Rept. 4, July 1951.

23. U.S. Army Corps of Engineers, "Runoff from Snowmelt," Engineering and Design Manuals, EM 1110-2-1406, Jan. 1960.

24. S. S. Butler, *Engineering Hydrology*. Englewood Cliffs, NJ: Prentice-Hall, 1957.

25. W. U. Gartska, L. D. Love, B. C. Goodell, and F. A. Bertle, "Factors Affecting Snowmelt and Streamflow," U.S. Bureau of Reclamation and U.S. Forest Service, 1958.

26. A. L. Zimmerman, "Reconstruction of the Snow-Melt Hydrograph in the Payette River Basin," Proceedings of the Western Snow Conference, Portland, OR, Apr. 1955.

27. C. O. Clark, "Storage and the Unit Hydrograph," *Trans. ASCE* **110,** 1419–1446(1945).

28. M. C. Quick, "River Flood Flows: Forecasts and Probabilities," *Proc. ASCE J. Hyd. Div.* **91**(HY3) (May 1965).

29. V. M. Ponce, *Engineering Hydrology: Principles and Practices,* Englewood Cliffs, New Jersey: Prentice Hall, 1989.

30. D. M. Gray and D. H. Male, *Handbook of Snow,* Toronto: Pergamon Press, 1981.

31. J. R. Meiman, "Snow Accumulation Related to Elevation, Aspect and Forest Canopy," Proceedings, Workshop and Seminar in Snow Hydrology, Ottawa: Queen Printers of Canada, pp. 35–47, 1970.

32. J. O. Rhea and L. O. Grant, "Topographic Influences on Snowfall Patterns in Mountainous Terrain," in *Advanced Concepts and Techniques in the Study of Snow and Ice Resources,* National Academy of Sciences, Washington, D.C., 1974.

33. U.S. Army Corps of Engineers, North Pacific Division, "Program Description and User Manual for SSARR Model, Streamflow Synthesis and Reservoir Regulation," Portland, Oregon, Draft, April 1986.

34. E. A. Anderson, "Development and Testing of Snow Pack Energy Balance Equations," *Water Resources Research,* **3**(1): 19–38, 1967.

35. R. L. Laramie and J. C. Schaake, Jr., "Simulation of the Continuous Snowmelt Process," Technical Report No. 143, Cambridge, MA: MIT Department of Civil Engineering, Ralph M. Parsons Laboratory, 1972.

36. G. A. McKay, "Problems of Measuring and Evaluating Snow Cover," in Snow Hydrology, Proceedings of a Workshop Seminar, University of New Brunswick, 1968.

37. E. L. Peck, "Snow Measurement Predicament," *Water Resources Research,* **8**(1): pp. 244–248, 1972.

Chapter 15

Urban and Small Watershed Hydrology

■ Prologue

The purpose of this chapter is to:

- Review the empirical and analytical methods available for predicting peak runoff rates from urban and small watersheds.
- Introduce the popular *rational method* for predicting peak discharge rates from small urban watersheds.
- Present a representative sample of equations and charts used for estimating peak rates of runoff for urban and small watersheds.
- Illustrate how some of the procedures used in urban stormwater software (Chapter 25) were developed and what assumptions apply.
- Provide the reader with an understanding of the applicability of the methods described.

Effective disposal of storm water is essential. Drainage systems have changed from primitive ditches to complex networks of curbs, gutters, and surface and underground conduits. Along with the increasing complexity of these systems has come the need for a more thorough understanding of basic hydrologic processes. Simple rules of thumb and crude empirical formulas are no longer adequate.

Methods presented in Chapters 12 and 25 allow the generation of complete hydrographs from actual or hypothetical storms occurring over urban areas. This chapter presents techniques that provide the magnitude of the peak flow rate for a given storm or a given frequency. Many applications allow use of the peak in analysis and design, and the procedures are much less time consuming than unit-hydrograph or urban modeling techniques.

15.1 INTRODUCTION

Methods used in estimating quantities of storm water runoff from urban drainage areas and other small watersheds may be classified as the rule-of-thumb approach, the empirical approach, and analytical approaches that describe physical processes.

The Rule-of-Thumb Approach

An early inquiry into urban rainfall–runoff processes was precipitated by the storm of June 20, 1857, on the Savoy Street sewer in London. One inch of rain fell in 75 min, producing a maximum flow of 0.34 ft³/sec-acre. Based on information then available, the distinguished engineers Bidder, Hawksley, and Bazalgette concluded that 0.25 in. of rainfall would contribute about 0.125 in. to the sewer, and 0.40 in. would yield approximately 0.25 in. At this time, a general English rule of thumb was that about half of the rainfall would appear as runoff from urban surfaces. Such early guidelines were forerunners of modern urban hydrologic models.

Empirical Lumped-Parameter Approach

Following the early rules of thumb, empirical formulas became the principal mechanism for determining quantities of runoff. Most second-generation approaches were lumped-parameter approaches. They are characterized by (1) consideration of the entire drainage area as a single unit, (2) estimation of flow at only the most downstream point, and (3) the assumption that rainfall is uniformly distributed over the drainage area. The foremost example of this approach is the *rational method,* which was based on an experiment involving 4 years of rainfall data using nonrecording rain gauges and 1 year of runoff data estimated from high water marks on pairs of white-washed sticks. Five open ditches were used for flow determination.

The rational method is described by the statement $Q = CIA$, where Q equals the peak runoff rate in cfs, C is a runoff coefficient, I is the rainfall rate in in./hr, and A is the drainage area in acres (see Section 15.2). The rational method has been used for over half a century with little change in its original form. It is a standard method of urban storm drainage design today. Persistence in the use of this formula can be attributed in part to its simplicity. The present analytical effort in urban hydrology should bring about some change in design concepts, but new techniques should not sacrifice the practicing professional's desire for easy-to-apply procedures.

A second example of the lumped-parameter approach is the unit-hydrograph method developed by Leroy K. Sherman in 1932. In Chapter 12 it was stated that the X-hour unit hydrograph is the hydrograph of 1 in. of runoff from a drainage area produced by a uniform rainfall lasting X units of time. Once determined, the unit graph can be used to construct the hydrograph for a storm of any magnitude and duration. Originally, the unit-hydrograph concept was applied mainly to stream or river basins, but it is now used for urban and small watershed networks as well. The concept of the instantaneous unit hydrograph has been a major factor in its expansion to use on small drainage areas. The instantaneous unit hydrograph operates on an effective precipitation applied in zero time. This is theoretical, but the assumption makes the hydrograph independent of the duration of effective precipitation and eliminates one variable from hydrograph analysis. A number of models using an instantaneous unit hydrograph or an approximation of it have been reported in the literature.[1–3] One of the pioneers was J. E. Nash.[4]

Physical Process Approach

This approach is characterized by an attempt to quantify all pertinent physical phenomena from the input (rainfall) to the output (runoff).[5–8] This usually involves the following steps: (1) determine a design storm; (2) deduct losses from the design storm to arrive at an excess rainfall rate: (3) determine the flow to a gutter or some defined channel by overland flow equations; (4) route these gutter or small channel flows to the main channel; (5) route the flow through the principal conveyance system (pipe, canal, or stream); and (6) determine the outflow hydrograph. The result obtained is affected by the accuracy of calculating losses and hydraulic phenomena and the validity of the simplifying assumptions. If errors are small and noncumulative, the prediction of the runoff is valid.

In the past, most microscopic procedures dealt solely with individual storm events. With the advent of modern computers, the trend has been more toward the continuous simulation of hydrologic processes.[3, 9, 10]

Simulation

Analyses of many urban and other small watersheds are effected through the use of simulation models. As discussed in Chapter 21, hydrologic systems may be simulated by physical models, analog models, or digital models. Digital models such as EPA's storm water management model (SWMM) are widely used for small watershed studies and for the hydrologic design of storm drainage systems. Chapter 23 describes the development of and uses for continuous streamflow simulation models.

Quantity–Quality Models

Evaluation of the water quality produced in the runoff from small watersheds is becoming increasingly more important.[11, 12] Unfortunately, data on this aspect are very limited and usually not adequate for continuous simulation. Variations in runoff quality result from changes in season and geographical area. Urban and rural practices such as lawn watering, irrigation, car washing, and others affect the ultimate destination of fertilizers and other chemicals.[13–15] A better understanding of the complex nature of physical, chemical, and biological processes that affect the water quality is needed, and a number of operational models have been developed.[16, 17]

The state of the art in modeling urban and small watershed runoff quantity and quality has progressed from simple rules of thumb to complex simulation models incorporating all fundamental hydrologic processes. The mechanics of all these processes are not completely known and some empiricism remains. In fact, small watershed modeling is still part art and part science, but as more data become available and modeling technology improves, a greater degree of sophistication and reliability will result. Perhaps the greatest underlying need is for more data characterizing the water cycle on urban areas. Information on both the quantity and quality of urban runoff is in short supply.

15.2 PEAK FLOW FORMULAS FOR URBAN WATERSHEDS

Numerous methods are available for estimating the peak rates of runoff required for design applications in small urban and rural watersheds. Some incorporate the rainfall–runoff process, whereas others are completely empiric or correlative in that they predict peak runoff rates by correlating the flow rates with simple drainage basin characteristics such as area or slope.

Both categories of peak flow determination have had wide application; however, two relatively major difficulties are normally encountered in applying the techniques. First, the rainfall-runoff formulas, such as the *rational formula,* are difficult to apply unless the return periods for rainfall and runoff are assumed to be equal. Also, estimates of coefficients required by these formulas are subjective and have received considerable criticism. The empiric and correlative methods are limited in application because they are derived from localized data and are not valid when extrapolated to other regions.

The most fundamental peak flow formulas and empiric-correlative methods, due to their simplicity, persist in dominating the urban design scene, and several of the most popular forms are briefly described to acquaint the reader with methods and assumptions. Urban runoff simulation techniques are described in Chapter 25.

Rational Formula

The rational formula for estimating peak runoff rates was introduced in the United States by Emil Kuichling in 1889.[18] Since then it has become the most widely used method for designing drainage facilities for small urban and rural watersheds. Peak flow is found from

$$Q_p = CIA \tag{15.1}$$

where Q_p = the peak runoff rate (cfs)
C = the runoff coefficient (assumed to be dimensionless)
I = the *average* rainfall intensity (in./hr), for a storm with a duration equal to a critical period of time t_c
t_c = the time of concentration (see Chapter 11)
A = the size of the drainage area (acres)
CI = the average net rain intensity (in./hr) for a storm with duration = t_c

The runoff coefficient can be assumed to be dimensionless because 1.0 acre-in./hr is equivalent to 1.008 ft³/sec. Typical C values for storms of 5–10-year return periods are provided in Table 15.1.

The rationale for the method lies in the concept that application of a steady, uniform rainfall intensity will cause runoff to reach its maximum rate when all parts of the watershed are contributing to the outflow at the point of design. That condition is met after the elapsed time t_c, the time of concentration, which usually is taken as the time for a wave to flow from the most remote part of the watershed. At this time, the runoff rate matches the net rain rate.

Figure 15.1 graphically illustrates the relation. The IDF curve is the rainfall intensity–duration–frequency relation for the area and the peak intensity of the runoff is $Q/A = q$, which is proportional to the value of I defined at t_c. The constant of proportionality is thus the runoff coefficient, $C = (Q/A)/I$. Note that Q/A is a point value and that the relation, as it stands, yields nothing of the nature of the rest of the hydrograph.

The definition chosen for t_c can adversely affect a design using the rational formula. If the average channel velocity is used to estimate the travel time from the most remote part of the watershed (a common assumption), the resulting design

TABLE 15.1 TYPICAL *C* COEFFICIENTS FOR 5-
TO 10-YEAR FREQUENCY DESIGN

Description of area	Runoff coefficients
Business	
Downtown areas	0.70–0.95
Neighborhood areas	0.50–0.70
Residential	
Single-family areas	0.30–0.50
Multiunits, detached	0.40–0.60
Multiunits, attached	0.60–0.75
Residential (suburban)	0.25–0.40
Apartment dwelling areas	0.50–0.70
Industrial	
Light areas	0.50–0.80
Heavy areas	0.60–0.90
Parks, cemeteries	0.10–0.25
Playgrounds	0.20–0.35
Railroad yard areas	0.20–0.40
Unimproved areas	0.10–0.30
Streets	
Asphaltic	0.70–0.95
Concrete	0.80–0.95
Brick	0.70–0.85
Drives and walks	0.75–0.85
Roofs	0.75–0.95
Lawns; sandy soil:	
Flat, 2%	0.05–0.10
Average, 2–7%	0.10–0.15
Steep, 7%	0.15–0.20
Lawns; heavy soil:	
Flat, 2%	0.13–0.17
Average, 2–7%	0.18–0.22
Steep, 7%	0.25–0.35

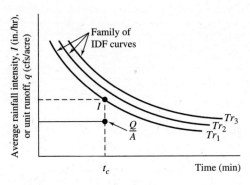

Figure 15.1 Rainfall–runoff relation for the rational method.

discharge could be less than that which might actually occur during the life of the project. The reason is that wave travel time through the watershed is faster than average discharge velocity (see Section 13.1). As a result of using the slower velocity V, the peak time (t_c) is overestimated, the resulting intensity I from IDF curves is too small, and the rational flow rate Q is underestimated.

Rational Method Applications   Most applications of the rational formula in determining peak flow rates utilize the following steps:

1. Estimate the time of concentration of the drainage area.
2. Estimate the runoff coefficient, Table 15.1.
3. Select a return period T_r and find the intensity of rain that will be equaled or exceeded, on the average, once every T_r years. To produce equilibrium flows, this design storm must have a locally derived IDF curve such as Fig. 27.13 or Fig. 15.2 using a rainfall duration equal to the time of concentration.
4. Determine the desired peak flow Q_p from Eq. 15.1.
5. Some design situations produce larger peak flows if design storm intensities for durations less than t_c are used. Substituting intensities for durations less than t_c is justified only if the contributing area term in Eq. 15.1 is also reduced to accommodate the shortened storm duration.

One of the principal assumptions of the rational method is that the predicted peak discharge has the same return period as the rainfall IDF relation used in the

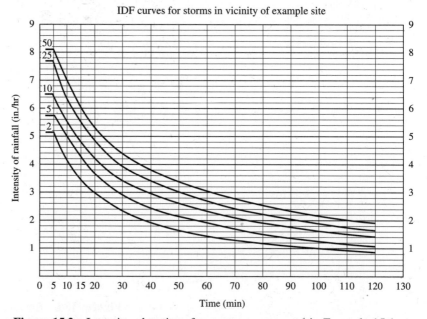

Figure 15.2   Intensity–duration–frequency curves used in Example 15.1.

prediction. Another assumption, and one that has received close scrutiny by investigators,[19,20] is the constancy of the runoff coefficient during the progress of individual storms and also from storm to storm. The coefficient is usually selected from a list based on the degree of imperviousness and infiltration capacity of the drainage surface. Because $C = I_{net}/I$, the coefficient must vary if it is to account for antecedent moisture, nonuniform rainfall, and the numerous conditions that cause abstractions and attenuation of flood-producing rainfalls. In practice, a composite, weighted average runoff coefficient is computed for the various surface conditions. Times of concentration are determined from the hydraulic characteristics of the principal flow path, which typically is divided into two parts, overland flow and flow in defined channels; the times of flow in each segment are added to obtain t_c.

Another assumption with the rational method is that the equation is most applicable to antecedent moisture conditions that exist for frequent storms, in the range of the 2- to 10-yr recurrence interval, representative of storms traditionally used for design of residential storm drain systems. Because more severe, less frequent storms often have wetter antecedent moisture conditions, the rational coefficient is increased by multiplying it by a frequency factor. The commonly used multipliers for less frequent storms are:

Return period (yrs)	Multiplier
2–10	1.0
25	1.1
50	1.2
100	1.25

EXAMPLE 15.1

Use the rational method to find the 10-year and 50-year design runoff rates for the area shown in Fig. 15.3. The IDF rainfall curves shown in Fig. 15.2 are applicable.

Solution

1. Time of concentration:

$$t_c = t_1 + t_2 = 15 + 5 = 20 \text{ min}$$

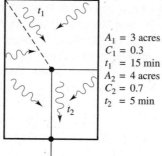

$A_1 = 3$ acres
$C_1 = 0.3$
$t_1 = 15$ min
$A_2 = 4$ acres
$C_2 = 0.7$
$t_2 = 5$ min

Figure 15.3 Hypothetical drainage system for Example 15.1.

2. Runoff coefficient:

$$\overline{C} = [(3 \times 0.3) + (4 \times 0.7)]/(3 + 4) = 0.53 \text{ for 10-yr event}$$
$$\overline{C} = 1.2(0.53) = 0.64 \text{ for 50-yr event}$$

3. Rainfall intensity—from Fig. 15.2:

$$I_{10} = 4.2 \text{ in./hr}$$
$$I_{50} = 5.3 \text{ in./hr}$$

4. Design peak runoff:

$$Q_{10} = CIA = 0.53 \times 4.2 \times 7 = 16 \text{ cfs}$$
$$Q_{50} = CIA = 0.64 \times 5.3 \times 7 = 24 \text{ cfs} \quad \blacksquare\blacksquare$$

Rational Method Discussion   The runoff coefficient in the rational formula is dependent on the soil type, antecedent moisture condition, recurrence interval, land use, slope, amount of urban development, rainfall intensity, surface and channel roughness, and duration of storm. Tables and graphs generally allow determination of C from only two or three of these factors. Nomographs and regression equations can provide relations among more factors. One such relation, applicable only in the region for which it was derived, is[21]

$$C = 7.2(10^{-7})CN^3 \, T^{0.05}[(0.01CN)^{0.6}]^{-S^{0.2}}(0.001CN^{1.48})^{0.15-0.1I}[(P + 1)/2]^{0.7}$$

$$(15.2)$$

where CN = SCS curve number (Chapter 4)

T = recurrence interval (years)

S = average land slope (%)

I = rain intensity (in./hr)

P = percent imperviousness

The rational formula is a simple model to express a complex hydrologic system. Yet the method continues to be used in practice with results implying acceptance by designers, officials, and the public. The method is easy to apply and gives consistent results. From the standpoint of planning, for example, the method demonstrates in clear terms the effects of development: runoff from developed surfaces increases because times of concentration decrease and runoff coefficients increase.

For storm drainage systems, the designer is normally asked to estimate the peak flow rate that might be equalled or exceeded at least once in a given number of years (described as the *frequency* — see Section 10.4). For designs using the rational formula, the frequency of the peak runoff event is assumed equal to the frequency of the rain event (an event being defined as some rain depth in a given duration). Studies have explored this assumption.[22] Figure 15.4 shows cumulative log–normal probability functions (Chapter 26) fitted to observations of rainfall and runoff on a 47-acre area in Baltimore, Maryland, with an average surface imperviousness of 0.44. The data are partial series fitted independently to the observed rainfall sequence and the observed runoff sequence. Thus the largest runoff does not necessarily correspond to the largest ranked rainfall, and a similar lack of correspondence between any runoff

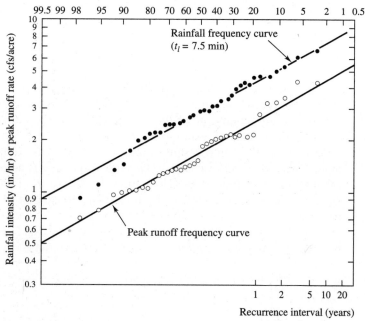

Figure 15.4   Distributions of recorded rainfall and runoff. (After Schaake.[22])

and the rainfall that produced it holds for the ranked position of the observations in the arrays of the two separate sequences. In Fig. 15.4, the 5-year rainfall frequency of 6.5 in./hr corresponds to a runoff frequency of 4.0 cfs/acre; the ratio indicates a runoff coefficient of approximately 0.6. Although the two sequences are each closely log–normal, they tend to converge, which suggests that the runoff coefficient increases slightly with more intense, less frequent storms. In the design range, however, the results tend to support the assumption of the rational method that the recurrence interval of the runoff equals the recurrence interval of the rainfall. It should be noted that the rainfall distributions in Figs. 15.1 and 15.4 have similar properties. All IDF curves are drawn through the average rainfall intensities derived from many different storms of record; any single IDF curve does not represent the progress of a single storm. For lack of historical runoff data, the designer turns to the rational method to construct from the rainfall history what amounts to a *runoff* intensity–duration–frequency relation.

   The most critical (highest peak) runoff event is often assumed to be caused by a storm having a duration equal to the time of concentration of the watershed. If the rainfall IDF curve is steep in the design range, several durations should be tested for the given frequency to assure that no other storm of equal probability produces a higher peak runoff rate. Most applications of the rational method do not include this test because the assumption that the peak occurs at t_c is commensurate with the other inherent assumptions.

The rational method is used in the design of urban storm drainage systems serving areas up to six hundred acres in size. For areas larger than 1 mi², hydrograph or other techniques are generally warranted. Considerable judgment is required in selecting both the runoff coefficients and times of concentration. A common procedure is to select coefficients and assume that they remain constant throughout the storm. As the design proceeds from point to point downstream, a composite weighted C factor is computed for the drainage area above each point. The time of concentration is composed of an inlet time (the overland and any channel flow times to the first inlet) plus the accumulated time of flow in the system to the point of design.

Figure 15.5 is an example of a design aid for predicting overland flow times. Calculation of flow time in storm drains can readily be estimated knowing the type of pipe, slope, size, and discharge.[23] Generally, the pipe is assumed to flow full for this calculation. (See Fig. 15.6.) Nomographs also are available to solve the Manning equation for flow in ditches and gutters. The estimation of inlet time is frequently based solely on judgment; reported values vary from 5 to 30 min. Densely developed areas with impervious tracts immediately adjacent to the inlet might be assigned inlet periods of 5 min, but a minimum value of 10–20 min is more usual.

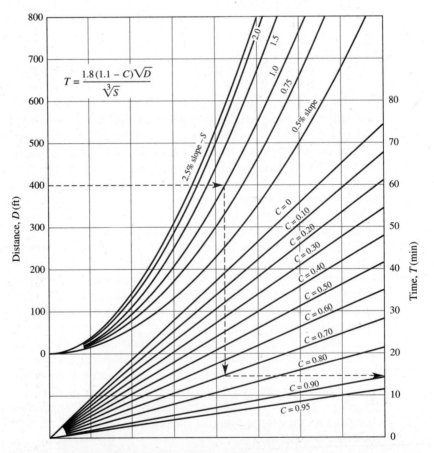

Figure 15.5 Surface flow time curves. (After Federal Aviation Agency.[23])

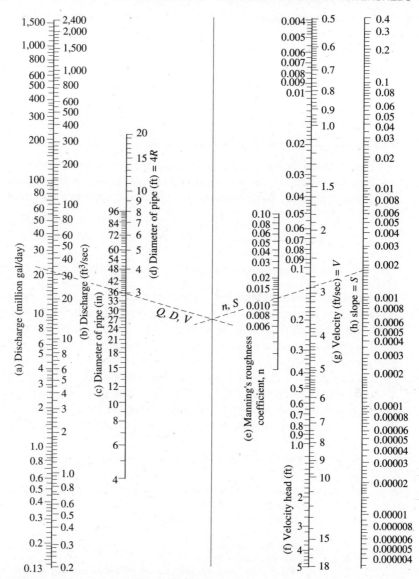

Figure 15.6 Flow in pipes (Manning's formula). (After Ref. 24.)

Most designers applying this method do not use the time of concentration in its strictest sense; rather, the largest sum of inlet time plus travel time in the storm drain system is taken as the time of concentration. Caution is required in applying the method. Peak discharge is not the summation of the individual discharges, because peaks from subareas occur at different times. The runoff from subareas should be rechecked for each area under consideration. The average intensity I is that for the time of concentration of the total area drained. While I decreases as the design proceeds downstream, the size of the contributing area increases and Q normally increases continuously. It should be noted that the design at each point downstream is a new solution of the rational method. The only direct relation from point to point derives from the means for determining an increment of time to be added for a new time of concentration. The effect is to provide an equal level of protection (i.e., an equal frequency of surcharging) at all points in the system. Example 15.2 is reproduced from standard design references to illustrate the application of the rational method to an urban system.[24]

EXAMPLE 15.2

Based on the storm sewer arrangement of Fig. 15.7a, determine the outfall discharge. Assume that $C = 0.3$ for residential areas and $C = 0.6$ for business tracts. Use a 5-year frequency rainfall from Fig. 15.7b and assume a minimum 20-min inlet time.

Solution. The principal factors in the design are listed in Table 15.2. Additional columns can be provided to list elevations of manhole inverts, sewer inverts, and ground elevations. This information is helpful in checking designs and for subsequent use in drawing final design plans. (See Table 15.3.) ■■

Modified Rational Method The rational method is truly "rational" in that the peak flow rate is simply set equal to the net rain rate after sufficient time occurs for the entire watershed to contribute runoff. This results for any storm equalling or exceeding the time of concentration of the watershed. The net rain rate is a fraction of the gross rain rate, I, and the fraction is C. Incorporating A in the equation simply transforms runoff from inches per hour to volume per hour. This trait allows the conceptual extension of the rational method to problems in which the runoff hydrograph is required. This is particularly applicable in designs, such as detention basin design, requiring volume of runoff as well as peak flow rates.

TABLE 15.2   DEFINITION OF COLUMN HEADINGS IN TABLE 15.3

Column	Comment
1	Line being investigated
2, 3	Inlet or manhole being investigated
4	Length of the line
5	Subarea of the inlet
6	Accumulated subareas
7	Value of the concentration time for the area draining into the inlet
8	Travel time in the pipe line
9	Weighted C for the area being drained
10	Rainfall intensity based on time of concentration and a 5-year frequency curve
11	Unit runoff $q = CI$
12	Accumulated runoff that must be carried by line
13	Slope of line
14	Size of pipe
15	Pipe capacity
16	Velocity in full pipe
17	Actual velocity in pipe

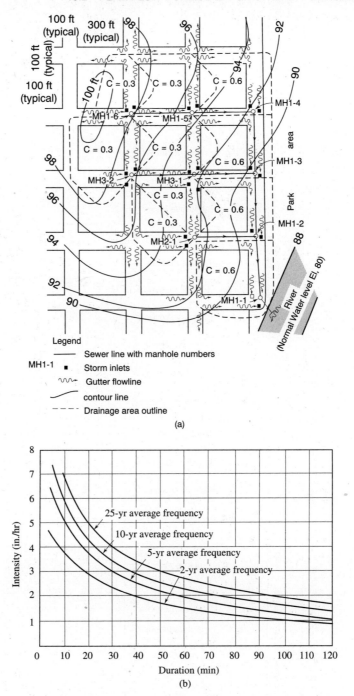

Figure 15.7 Sample storm drainage problem: (a) typical storm sewer design plan and (b) intensity–duration–frequency rainfall curves for Davenport, Iowa. (After Ref. 24.)

TABLE 15.3 TYPICAL STORM SEWER COMPUTATIONS FOR THE RATIONAL METHOD

	Manhole number				Area (acres)		Flow time (min)										Design flow							Sewer invert		Ground elevation	
Line	From	To	Length (ft)	Increment	Total	To upper end	In section	Average runoff coefficient	Rainfall (in./hr)	Runoff (cfs/acre)	Total runoff (cfs)	Slope of sewer (%)	Diameter (in.)	Capacity, full	Velocity, full	Velocity, (fps)	Velocity head (ft)	Depth of flow (in.)	Total energy (ft)	Manhole losses (ft)	Manhole invert drop (ft)	Fall in sewer (ft)	Upper end	Lower end	Upper end	Lower end
(1)	(2)	(3)	(4)	(5)	(6)	(7)	(8)	(9)	(10)	(11)	(12)	(13)	(14)	(15)	(16)	(17)	(18)	(19)	(20)	(21)	(22)	(23)	(24)	(25)	(26)	(27)
1	1–6	1–5	400	2.64	2.64	20.0	1.4	0.30	3.7	1.11	2.93	0.85	12	3.3	4.0	4.6		9			. . .	3.40	93.00	89.60	98.4	94.9
1	1–5	1–4	400	3.61	6.25	21.4	1.2	0.30	3.6	1.08	6.75	0.75	18	9.2	5.1	5.6		11			0.40	3.00	89.20	86.20	94.9	91.8
1	1–4	1–3	400	3.88	10.13	22.6	1.2	0.42	3.4	1.43	14.50	0.45	24	15.2	4.8	5.6		18			0.40	1.80	85.80	84.00	91.8	89.7
3	3–2	3–1	400	5.55	5.55	20	1.1	0.30	3.7	1.11	6.16	1.00	15	6.4	5.1	5.9		12			. . .	4.00	91.00	87.00	96.2	92.3
3	3–1	1–3	400	6.43	11.98	21.1	1.1	0.30	3.6	1.08	12.92	0.60	24	17.5	5.5	6.1		15			0.60	2.40	86.40	84.00	92.3	89.7
1	1–3	1–2	400	3.92	26.03	23.8	1.1	0.39	3.3	1.29	33.60	0.30	36	37.0	5.1	5.9		26			0.80	1.20	83.20	82.00	89.7	89.5
2	2–1	1–2	400	2.52	2.52	20	1.4	0.30	3.7	1.11	2.80	0.90	12	3.2	4.1	4.7		9			. . .	3.60	87.50	83.90	92.7	89.5
1	1–2	1–1	400	3.86	32.41	24.9	1.1	0.41	3.2	1.31	42.50	0.24	42	50.0	5.2	5.9		29			0.40	0.96	81.60	80.64	89.5	88.5
1	1–1	Out-fall	125	5.44	37.85	26.0	. . .	0.44	3.2	1.41	53.20	0.30	42	56.0	5.7	6.6		33			0.10	0.38	80.54	80.16	88.5	. . .

Source: Reference 24.

322

In the *modified rational method,* a full hydrograph is developed rather than simply estimating the peak flow rate, using the following reasoning. If the storm duration exceeds the time of concentration, the runoff rate would rise to the rational formula peak value, then stay constant until net rain ceases. At that point, runoff rates would decrease to zero as excess rain is released from the basin. If the rainfall-excess release time (see Chapter 11) is equal to the time of concentration, the hydrograph would have an approximate trapezoid shape rising to the peak at $t = t_c$, remaining flat until $t =$ the rain duration, D, and then falling along a straight line until $t = D + t_c$. Many software packages for urban hydrology incorporate the modified rational method for hydrograph analysis. The method is approximate and should not be applied to watersheds over 50 acres in size.

SCS TR-55 Method

The U.S. Soil Conservation Service developed procedures for estimating runoff volume and peak rates of discharge from urban areas.[25] They are known collectively as *TR-55* and individually as the *graphical method, chart method,* and *tabular method.* The three methods adjust rural procedures in NEH-4[26] to urban conditions by increasing the curve number CN for impervious areas and reducing the lag time t_l for imperviousness and channel improvements. Allowances are also made for various watershed shapes, slopes, and times of concentration. The SCS designed the first two methods to be used for estimating peak flows, and the third for synthesizing complete hydrographs. The tabular method and chart method (used for small watersheds up to 2000 acres) were revised in 1986,[27] but are described here to help explain the evolution of the methods. All three were developed for use with 24-hr storms. Use with other storm durations is not advised.

The graphical method was developed for homogeneous watersheds, up to 20 mi^2 in size, on which the land use and soil type may be represented by the *runoff curve number.* As shown in Chapter 4, the runoff curve number is simply a third variable in a graph of rainfall versus runoff.

The SCS peak discharge graph shown in Fig. 15.8 is limited to applications where only the peak flow rate is desired for 24 hr, Type-II storm distributions (see Chapter 16). A Type-II storm distribution is typical of the 24-hr thunderstorm experienced in all states except the Pacific Coast states. Figure 15.8 was developed from numerous applications of the SCS TR-20 event simulation model described in Chapter 24. To apply Fig. 15.8, the watershed time of concentration in hours is entered into the graph to produce the peak discharge rate in cfs/mi^2 of watershed per inch of net rain during the 24-hr period. The 24-hr net rain is estimated from the 24-hr gross amount using the SCS curve number approach described in Chapter 4.

The effect of urbanization can be estimated using Fig. 15.9. Once the composite curve number (CN) has been estimated for the previous area, a modified curve number is determined by entering Fig. 15.9 with the value of the percent impervious area on the modified watershed, reading vertically to the curve corresponding to the CN for the pervious watershed, and then reading horizontally to determine the modified composite runoff curve number that would be used in determining the net rain depth for the urbanized watershed.

Use of the 1975 graphical method is restricted by the assumptions of the tabular method. The method is a composite of results for one case of the tabular method. This

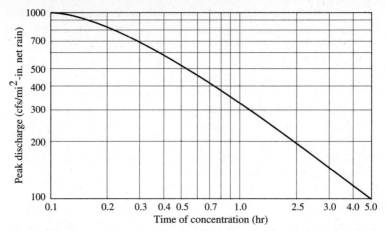

Figure 15.8 Peak discharge (cfs/mi²/in.) of runoff versus time of concentration t_c for 24-hr, Type-II storm distribution. (After U.S. Soil Conservation Service.[25])

restricts its applications to runoff volumes greater than about 1.5 in. (if the curve number is less than 60). Time of concentration should range between 0.1 and 2.0 hr, and the initial abstraction should not exceed about 25 percent of the precipitation.

The chart method allows determination of peak flows for 24-hr Type-II storms over watersheds having a fixed length/width relation and no ponding areas. Three charts are used for flat, moderate, or steep slopes of approximately 1, 4, or 16 percent. Tables of adjustments for intermediate slopes are provided in the technical release.

Several microcomputer software packages for urban hydrology have been developed.[28] Over two-thirds are based on SCS procedures, but caution should be applied

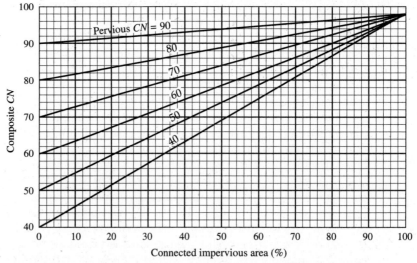

Figure 15.9 Percentage of impervious areas versus composite *CN*s for given pervious area *CN*s. (After U.S. Soil Conservation Service.[25])

in assuming that the commercial programs fully imitate TR-55 or other SCS hand-book methods. An ideal TR-55 package would include all three methods, would carry SCS endorsement, would state all assumptions and limitations, and would incorporate all SCS adjustments for peak coefficient, percent imperviousness, percentage of channel improved, ponding or swampy areas, length/width ratio variations, and slope. Its use should also be cautioned for other than 24-hr storms having a Type-II SCS distribution. Packages not adhering to these limitations would not be qualified as TR-55 procedures.

A significant problem in some of the commercial software packages is the use of a triangular-shaped unit hydrograph for convolution to produce hydrographs for storms of various durations. The SCS used a triangular shape to conceptualize the peak flow rate of a curvilinear unit hydrograph, but has never endorsed use of other than either the curvilinear shape discussed in Section 12.5 or the tabulated hydrographs given in the TR-55 manual. For further reading, the SCS published a guide[29] for the use of the 1975 TR-55 intended to clarify procedures in the original technical release.

Prevailing SCS TR-55 Method

The 1986 edition of TR-55,[27] rather than the 1975 version, is recommended for use. It incorporates several years of results of research and experiences with the original edition. The revisions include the following:

1. Three additional rain distributions (see Fig. 16.17).
2. Expansion of the chapter on urban runoff curve numbers.
3. A procedure for calculating travel times of sheet flow.
4. Deletion of the chart method.
5. Modifications to the graphical peak discharge method and tabular hydrograph method.
6. TR-55 computer program.

Rather than relying totally on Fig 15.9, the new TR-55 uses Table 15.4 and Fig. 15.10 to provide urban runoff curve numbers for certain instances indicated in the table.

For the new graphical method, an urban curve number and the 24-hr design rain depth are estimated, then an initial abstraction I_a is determined from the SCS runoff equation (Chapter 4) or from Table 15.5. The peak flow is found from linear interpolation of the curves in Figs. 15.11, 15.12, 15.13, or 15.14, depending on the rainfall distribution type (Fig. 16.17). If the computed I_a/P ratio falls outside the curves, the nearest curve should be used. If the watershed contains a percentage of ponds or swampy areas, the peak flow is multiplied by a reduction coefficient from Table 15.6.

EXAMPLE 15.3 ――――――――――――――――――――――――――――――――――――――

A 1280-acre urban Tennessee watershed has a 6.0-hr time of concentration, $CN = 75$ from Table 15.4, and 5 percent of the area is ponded. The 25-year, 24-hr rain is 6.0 in. Find the 25-year peak discharge.

TABLE 15.4 RUNOFF CURVE NUMBERS FOR URBAN AREAS (see Sec. 4.9 for other values)

Cover description		Curve numbers for hydrologic soil group[a]			
Cover type and hydrologic condition	Average percent impervious area[b]	A	B	C	D
Fully developed urban areas (vegetation established)					
Open space (lawns, parks, golf courses, cemeteries, etc.)[c]					
Poor condition (grass cover <50%)		68	79	86	89
Fair condition (grass cover 50–75%)		49	69	79	84
Good condition (grass cover > 75%)		39	61	74	80
Impervious areas					
Paved parking lots, roofs, driveways, etc. (excluding right-of-way)		98	98	98	98
Streets and roads					
Paved; curbs and storm sewers (excluding right-of-way)		98	98	98	98
Paved; open ditches (including right-of-way)		83	89	92	93
Gravel (including right-of-way)		76	85	89	91
Dirt (including right-of-way)		72	82	87	89
Western desert urban areas					
Natural desert landscaping (pervious areas only)[d]		63	77	85	88
Artificial desert landscaping (impervious weed barrier, desert shrub with 1–2-in. sand or gravel mulch and basin borders)		96	96	96	96
Urban districts					
Commercial and business	85	89	92	94	95
Industrial	72	81	88	91	93
Residential districts by average lot size					
$\frac{1}{8}$ acre or less (town houses)	65	77	85	90	92
$\frac{1}{4}$ acre	38	61	75	83	87
$\frac{1}{3}$ acre	30	57	72	81	86
$\frac{1}{2}$ acre	25	54	70	80	85
1 acre	20	51	68	79	84
2 acres	12	46	65	77	82
Developing urban areas					
Newly graded areas (pervious areas only, no vegetation)[e]		77	86	91	94
Idle lands (CNs are determined using cover types similar to those in Table 4.7).					

[a] Average runoff condition, and $I_a = 0.2S$.

[b] The average percent impervious area shown was used to develop the composite CNs. Other assumptions are as follows: impervious areas are directly connected to the drainage system, impervious areas have a CN of 98, and pervious areas are considered equivalent to open space in good hydrologic condition. CNs for other combinations of conditions may be computed using Fig. 15.9 or 15.10.

[c] CNs shown are equivalent to those of pasture. Composite CNs may be computed for other combinations of open space cover type.

[d] Composite CNs for natural desert landscaping should be computed using Fig. 15.9 or 15.10 based on the impervious area percentage (CN = 98) and the pervious area CN. The pervious area CNs are assumed equivalent to desert shrub in poor hydrologic condition.

[e] Composite CNs to use for the design of temporary measures during grading and construction should be computed using Fig. 15.9 or 15.10 based on the degree of development (impervious area percentage) and the CNs for the newly graded pervious areas.

Source: U.S. Soil Conservation Service.[27]

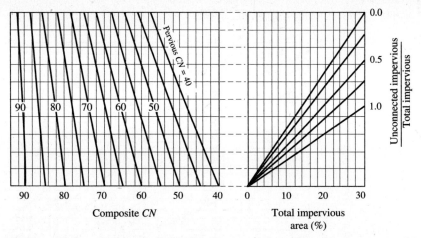

Figure 15.10 Graph of 1986 TR-55 composite *CN* with unconnected impervious area, or total impervious area, less than 30 percent. (After U.S. Soil Conservation Service.[27])

TABLE 15.5 I_a VALUES FOR RUNOFF CURVE NUMBERS

Curve number	I_a (in.)	Curve number	I_a (in.)
40	3.000	70	0.857
41	2.878	71	0.817
42	2.762	72	0.778
43	2.651	73	0.740
44	2.545	74	0.703
45	2.444	75	0.667
46	2.348	76	0.632
47	2.255	77	0.597
48	2.167	78	0.564
49	2.082	79	0.532
50	2.000	80	0.500
51	1.922	81	0.469
52	1.846	82	0.439
53	1.774	83	0.410
54	1.704	84	0.381
55	1.636	85	0.353
56	1.571	86	0.326
57	1.509	87	0.299
58	1.448	88	0.273
59	1.390	89	0.247
60	1.333	90	0.222
61	1.279	91	0.198
62	1.226	92	0.174
63	1.175	93	0.151
64	1.125	94	0.128
65	1.077	95	0.105
66	1.030	96	0.083
67	0.985	97	0.062
68	0.941	98	0.041
69	0.899		

Source: U.S. Soil Conservation Service.

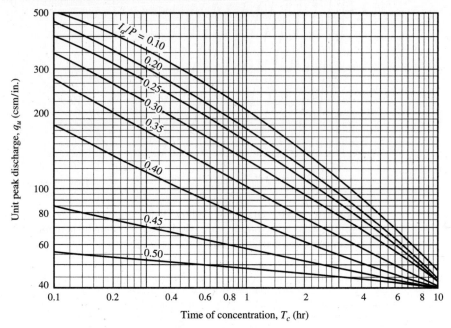

Figure 15.11   Unit peak discharge (q_u) for SCS Type-I rainfall distribution. (After U.S. Soil Conservation Service.)

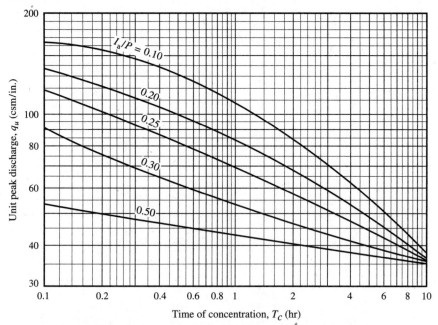

Figure 15.12   Unit peak discharge (q_u) for SCS Type-IA rainfall distribution. (After U.S. Soil Conservation Service.)

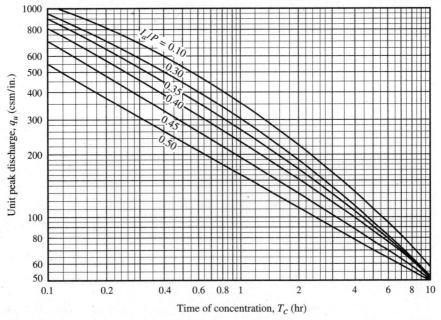

Figure 15.13 Unit peak discharge (q_u) for SCS Type-II rainfall distribution. (After U.S. Soil Conservation Service.)

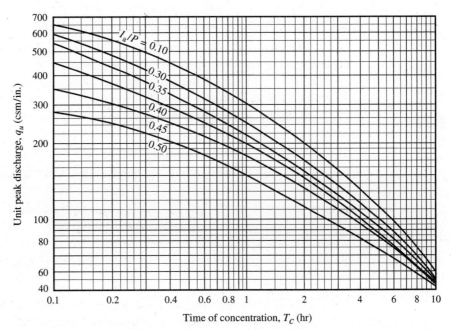

Figure 15.14 Unit peak discharge (q_u) for SCS Type-III rainfall distribution. (After U.S. Soil Conservation Service.)

TABLE 15.6 ADJUSTMENT FACTOR (F_p) FOR POND AND SWAMP AREAS THAT ARE SPREAD THROUGHOUT THE WATERSHED

Percentage of pond and swamp areas	F_p
0	1.00
0.2	0.97
1.0	0.87
3.0	0.75
5.0	0.72

Source: U.S. Soil Conservation Service.

Solution. From Fig. 16.17, the Type-II storm applies to Tennessee. From Table 15.5, $I_a = 0.667$. Thus $I_a/P = 0.11$. From Fig. 15.13, $q_u = 96$ csm/in. From Chapter 4, the runoff from 6.0 in. is 3.28 in. Since 5 percent of the area is ponded, the peak flow is adjusted using Table 15.6, giving $F_p = 0.72$. Thus

$$Q = (96 \text{ csm/in.})(3.28 \text{ in.})(2.0 \text{ mi}^2)(0.72) = 453 \text{ cfs} \quad \blacksquare\blacksquare$$

The graphical method provides peak discharges only. If a hydrograph is needed or watershed subdivision is required, the *tabular* method[27] should be used. The event simulation model TR-20 should be used if the watershed is very complex or a higher degree of accuracy is required (see Chapter 24).

Assumptions of the graphical method include:

The method should be used only if the weighted *CN* is greater than 40.

The T_c values with the method may range from 0.1 to 10 hr.

The watershed must be hydrologically homogeneous, that is, describable by one *CN*. Land use, soils, and cover must be distributed uniformly throughout the watershed.

The watershed may have only one main stream or, if more than one, the branches must have nearly equal times of concentration.

The method cannot perform channel or reservoir routing.

The F_p factor can be applied only for ponds or swamps that are not on the flow path.

Accuracy of peak discharge estimated by this method will be reduced if I_a/P values are used that are outside the range given.

When this method is used to develop estimates of peak discharge for present and developed conditions of a watershed, use the same procedure for estimating T_c.

Both the graphical and tabular methods are derived from TR-20 output. The use of T_c permits them to be used for any size watershed within the scope of the curves or tables. The tabular method can be used for a heterogeneous watershed that is divided into a number of homogeneous subwatersheds. Hydrographs for the subwatersheds can be routed and added.

The tabular method is described in the technical release and is not detailed here. In using the method, the following steps are employed:

1. Subdivided the watershed into areas that are relatively homogeneous and have convenient routing reaches.
2. Determine drainage area of each subarea in square miles.
3. Estimate T_c for each subarea in hours. The procedure for estimating T_c is outlined in TR-55.
4. Find the travel time for each routing reach in hours.
5. Develop a weighted CN for each subarea.
6. Select an appropriate rainfall distribution according to Fig. 16.17.
7. Determine the 24-hr rainfall for the selected frequency (Chapter 16).
8. Calculate total runoff in inches computed from CN and rainfall (Chapter 4).
9. Find I_a for each subarea from Table 15.5.
10. Using the ratio of I_a/P and T_c for each subarea, select one of the hydrographs tabulated in TR-55.
11. Multiply the hydrograph ordinates (csm/in.) by the area (mi^2) and runoff (in.) of each respective subarea.
12. Route and combine the hydrographs.

The SCS recommends that TR-20, rather than the tabular method, be used if any of the following conditions apply:

Travel time is greater than 3 hr.

T_c is greater than 2 hr.

Drainage areas of individual subareas differ by a factor of 5 or more.

The TR-55 procedures have been incorporated by SCS in a computer program. Copies are available from the U.S. National Technical Information Service.

15.3 PEAK FLOW FORMULAS FOR SMALL RURAL WATERSHEDS

SCS TP-149 Method

TR-55 is the SCS procedure for urban watersheds, TR-20 is the unit-hydrograph procedure for larger agricultural watersheds (see Chapter 24), and TP-149 was developed to allow estimation of peak flow rates from small (5-2000 acres) agricultural watersheds.[30] It consists of a series of 42 charts from which the peak discharge of a 24-hr rainfall can be determined.

Input to the procedure is the drainage area, average watershed slope, storm distribution type (I or II), watershed composite curve number, and depth of rainfall. Figures 15.15 and 15.16 illustrate the numerous charts in the TP. Shown are type-I and type-II curves for moderately sloped watersheds, with $CN = 70$ for both. Similar charts are available for the combinations given in Table 15.7. Applications of TP-149 to watersheds having curve numbers other than the 5-unit increments of Table 15.7, or for slopes other than 1, 4, or 16 percent, can be accomplished by arithmetic or logarithmic interpolation between adjacent chart values.

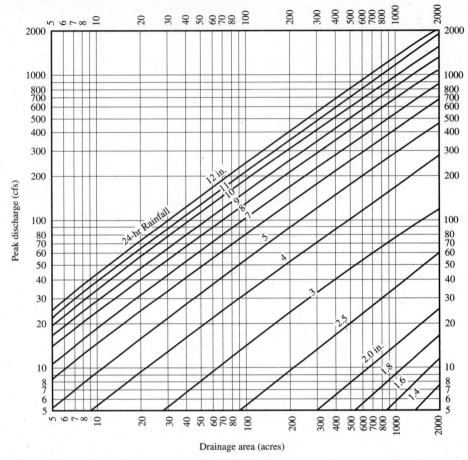

Figure 15.15   TP-149 peak rates of discharge for small watersheds, Type-I storms: 24-hr rainfall, moderate slopes, and $CN = 70$. (After U.S. Soil Conservation Service, "A Method of Estimating Volume and Rate of Runoff in Small Watersheds," U.S. Department of Agriculture, Jan. 1968.)

TABLE 15.7   CHARTS AVAILABLE IN TP-149 FOR PEAK FLOW RATES OF SMALL WATERSHEDS

Storm distribution type	Slope type	Slope range (%)	Curve number, CN
I, II	Flat, 1%	0–3	60, 65, 70, 75, 80, 85, 90
I, II	Moderate, 4%	3–8	60, 65, 70, 75, 80, 85, 90
I, II	Steep, 16%	8–30	60, 65, 70, 75, 80, 85, 90

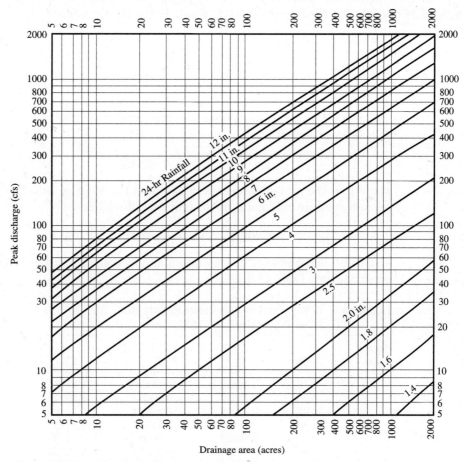

Figure 15.16 TP-149 peak rates of discharge for small watersheds, Type-II storms: 24-hr rainfall, moderate slopes, $CN = 70$. (After U.S. Soil Conservation Service, "A Method of Estimating Volume and Rate of Runoff in Small Watersheds," U.S. Department of Agriculture, Jan. 1968.)

EXAMPLE 15.4 _____

Compare the peak flow rates from Type-I and Type-II storms using Figs. 15.15 and 15.16. Assume that only storm type changes and all other conditions are equal.

Solution.   A 4-in. rain over 200 acres on a watershed with $CN = 70$ results in $Q_p = 52$ cfs for a Type-I storm (Fig. 15.15) and $Q_p = 97$ cfs for Type II (Fig. 15.16). Thus the storm distribution type makes a significant difference in results of peak flow estimation using SCS techniques. ∎∎

Federal Highway Administration SCS Peak Flow Design Method

The Federal Highway Administration (FHWA) lists in their Hydrologic Engineering Circular No. 19, Hydrology (1995 Ed.) a procedure for estimating peak flow rates for homogeneous, small-to-medium sized watersheds having times of concentration between 0.1 and 10 hours. It employs an SCS regression equation that has coefficients determined from data on different rainfall distribution types and ratios of the initial abstraction I_a (see Chapter 4) and total precipitation, P. The peak discharge in metric units is calculated from

$$q_p = q_u\, AQ \tag{15.3}$$

where q_p is the peak discharge in m^3/sec, A is the drainage area in sq. km., Q is the net rain depth in cm, and q_u is the unit peak discharge from

$$\log q_u = C_0 + C_1 \log t_c + C_2 \log^2 t_c \tag{15.4}$$

in which t_c is the time of concentration in hours, and the regression coefficients are obtained from Table 15.8.

TABLE 15.8   COEFFICIENTS FOR FHWA HEC-19 SCS PEAK DISCHARGE METHOD

Rainfall type	I_a/P	C_0	C_1	C_2
I	0.10	2.30550	−0.51429	−0.11750
	0.20	2.23537	−0.50387	−0.08929
	0.25	2.18219	−0.48488	−0.06589
	0.30	2.10624	−0.45695	−0.02835
	0.35	2.00303	−0.40769	0.01983
	0.40	1.87733	−0.32274	0.05754
	0.45	1.76312	−0.15644	0.00453
	0.50	1.67889	−0.06930	0.0
IA	0.10	2.03250	−0.31583	−0.13748
	0.20	1.91978	−0.28215	−0.07020
	0.25	1.83842	−0.25543	−0.02597
	0.30	1.72657	−0.19826	0.02633
	0.50	1.63417	−0.09100	0.0
II	0.10	2.55323	−0.61512	−0.16403
	0.30	2.46532	−0.62257	−0.11657
	0.35	2.41896	−0.61594	−0.08820
	0.40	2.36409	−0.59857	−0.05621
	0.45	2.29238	−0.57005	−0.02281
	0.50	2.20282	−0.51599	−0.01259
III	0.10	2.47317	−0.51848	−0.17083
	0.30	2.39628	−0.51202	−0.13245
	0.35	2.35477	−0.49735	−0.11985
	0.40	2.30726	−0.46541	−0.11094
	0.45	2.24876	−0.41314	−0.11508
	0.50	2.17772	−0.36803	−0.09525

Source: After U.S. Federal Highway Administration, Hec-19, Hydrology, FHWA-IP-95, 1995.

The procedure has the following limitations:

- use with homogeneous watersheds (*CN*s from zone to zone should not differ by 5)
- *CN* should be greater than 50
- t_c should be between 0.1 and 10 hours
- I_a/P should be between 0.1 and 0.5
- t_c should be about same for any of the main channels, if watershed has more than one main channel
- no channel or reservoir routing is allowed
- no storage facility on main channel
- watershed area in storage ponds and lakes should be less than 5 percent

Synthetic Unit-Hydrograph Peak Rate Formulas

Peak flow rates from small watersheds can also be determined using the synthetic unit-hydrograph techniques described in Chapter 12. A storm having a duration defined by Eq. 12.22 will produce, according to Snyder's method of synthesizing unit hydrographs, a peak discharge for 1.0 in. of net rain given by Eq. 12.17, or

$$Q_p = \frac{640C_p A}{t_{lR}} \tag{15.5}$$

Similarly, the peak flow rate resulting from a storm with duration D given by Eq. 12.22 or 12.23 is, according to the SCS method for constructing synthetic unit hydrographs, equal to

$$Q_p = \frac{484A}{t_p} \tag{15.6}$$

where t_p is the time from the beginning of the effective rain to the time of the peak runoff rate, which by definition is the watershed lag time plus half the storm duration. Both of Eqs. 15.5 and 15.6 apply to 1.0 in. of net rain occurring in the duration D. Either can be multiplied by P_{net} for other storm depths with equal durations. Peak flows for storms with durations other than D would need to be determined by unit-hydrograph methods.

Discharge–Area and Regression Formulas

A multitude of peak flow formulas relating the discharge rate to drainage area have been proposed and applied. Gray[31] lists 35 such formulas, and Maidment[32] compares many others. Most of these empiric equations are derived using pairs of measurements of drainage area and peak flow rates in a regression equation having the form

$$Q = CA^m \tag{15.7}$$

where Q = the peak discharge associated with a given return period
A = the drainage area
C, m = regression constants

Popular discharge-area formulas in the form of Eq. 15.7 include the Meyers equation[33]

$$Q = 10{,}000A^{0.5} \tag{15.8}$$

where A = the drainage area, which must be 4 mi^2 or more
           Q = the ultimate maximum flood flow (cfs)

This example gives only one flow rate of unknown frequency and is chosen only to illustrate the form of flood flow equations. A program of determining flood magnitudes for a range of frequencies on a state-by-state basis has been completed by the USGS using the multiple regression techniques discussed in Chapters 26 and 27 and illustrated for Virginia in Problem 27.25. Similar formulas are available from the USGS for other states. Software containing all the USGS regression equations for the United States is available from the U.S. Geological Survey and Federal Highway Administration as part of the HYDRAIN software package.

U.S. Geological Survey Index–Flood Method

The U.S. Geological Survey *index–flood* method described in Section 27.4 is a graphical regional correlation of the recurrence interval with peak discharge rates. The steps involved in the derivation of a regional flood index curve are outlined in Section 27.4. The first step in applying the technique to a watershed is to determine the mean annual flood, defined as the flood magnitude having a return period of 2.33 years. Mean annual floods for ungauged watersheds are found from regression equations similar in form to Eq. 15.7. For example, the USGS report[34] on flood magnitudes and frequencies in Nebraska gives, in cfs,

$$Q_{2.33} = CA^{0.7} \qquad\qquad (15.9)$$

where A = the contributing drainage area in mi^2
           C = a regional coefficient obtained from Fig. 15.17

Once the mean annual flood magnitude is obtained, other annual flood magnitudes can easily be determined from the appropriate index–flood curve (see Fig. 26.4c). The use of such curves in urban hydrology is limited because the USGS data network for the index-flood method seldom includes watersheds smaller than 10 mi^2. The USGS regression equations, described later, are applicable for watersheds in the $1-10$ mi^2 range and larger.

Cyprus Creek Formula

Extremely flat areas pose particular difficulties to the hydrologist, including estimates of infiltration, runoff volume, and peak runoff rates. Flooding in these areas tends to be shallow and widespread. Flow velocities are low, and water stands on the surface for relatively long periods of time. These areas are often distinguished by networks of straight drainage channels that have been constructed to store and eventually discharge the excess rain.

The SCS developed a procedure[35] to calculate the instantaneous peak flow from flatland areas based on first calculating the capacity of canals that would be needed to limit flat-area flooding for the design storm to a duration that would prevent excess crop damage, and then to apply a multiplier to this rate to obtain the instantaneous peak for the design of drainage structures. The procedure is illustrated in Fig. 15.18.[35]

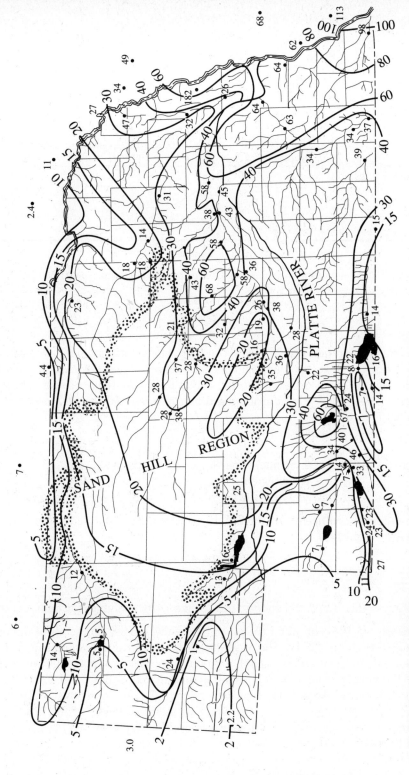

Figure 15.17 Variation of the mean annual flood coefficient *C* in Nebraska, exclusive of the main stems of the Platte Rivers. Dotted band represents the boundary of the Sand Hill Region. (After Furness.[34])

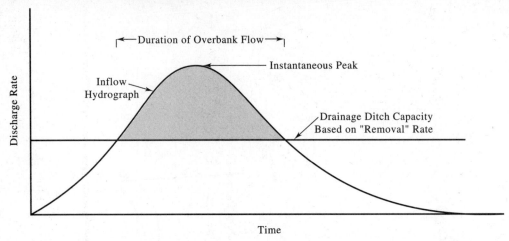

Figure 15.18   Illustration of relation between Cyprus Creek "removal" rate and peak instantaneous flow. (After Soil Conservation Service.[35])

The selected duration was 24 hours, considered to be the maximum allowable time for inundation of crops. An equation, called the *Cyprus Creek formula*, was developed to determine the canal design flow rate, called the *24-hr removal rate*. The equation, based on rainfall depth, contributing drainage area, and the SCS composite curve number is

$$Q = CA^{5/6} \qquad (15.10)$$

where Q = required channel capacity for 24-hr removal (cfs)
         C = drainage coefficient
         A = drainage area (sq mi)

The drainage coefficient, C, for Eq. 15.10 is found from an equation developed by Stephens and Mills[36]

$$C = 16.39 + (14.75\, Q_{scs}) \qquad (15.11)$$

where

Q_{scs} = the SCS direct runoff (in.) for the 24-hr design event from Fig. 4.14.

Once Eq. 15.10 is solved for the given frequency, the instantaneous peak flow rate is obtained from Fig. 15.19. The procedure is limited to drainage areas from 1 to about 200 square miles. It is suggested that ratios of the peak instantaneous rate to the 24-hr canal removal rate be limited to values greater than or equal to 1.0. For flatland areas that have part of the area in storm sewers, the SCS recommends that the peak flows from Fig. 15.19 be increased by the amounts indicated in Fig. 15.20. The SCS further recommends restricting use of this procedure to watersheds that have slopes that are less than 0.002. For steeper slope watersheds, other methods such as TR-20, TR-55, TP 149, or regression equations are recommended.

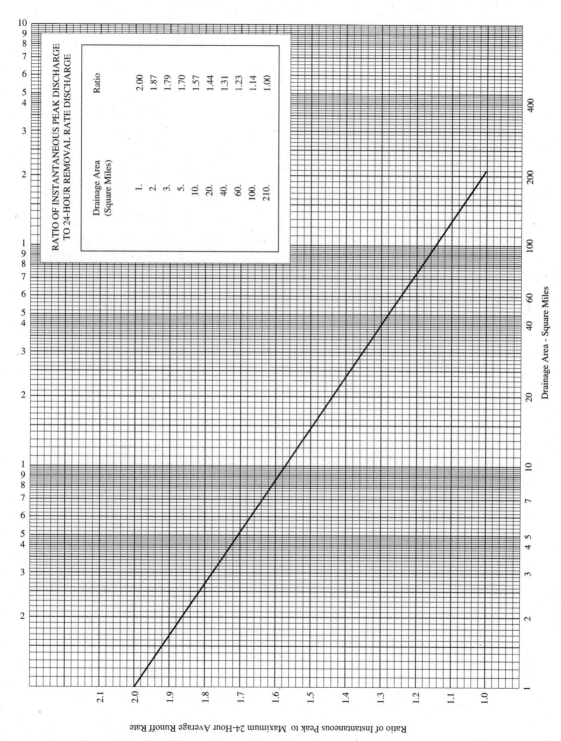

Figure 15.19 Ratio of Cyprus Creek 24-hr removal rate discharge to peak instantaneous flow rate. (After Soil Conservation Service.[36])

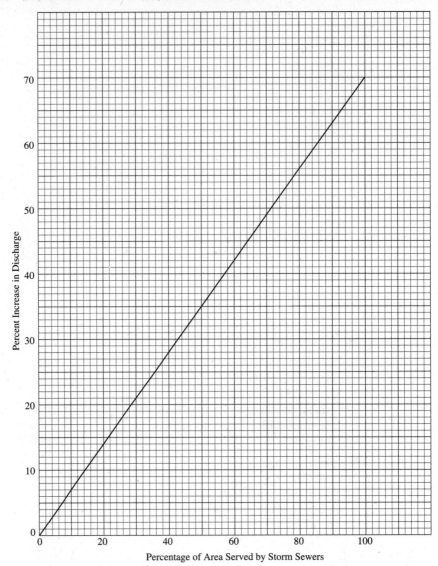

Figure 15.20   Effect of urban storm sewers on peak discharge for urban areas. (After U.S. Geological Survey.[47])

EXAMPLE 15.5 _____

Use the Cyprus Creek method to determine the peak 50-yr flow rate from a 1.0 sq mi drainage area that has a CN = 80, is 50 percent storm sewered, and has a 50-yr, 24-hr rainfall depth of 12.0 inches.

Solution.   From Fig. 4.14, the direct runoff for 12 inches of rain is 9.45 in. The drainage coefficient, C, is found from Eq. 15.11,

$$C = 16.39 + (14.75)(9.45) = 155.8$$

The 24-hr removal rate is found from Eq. 15.10,

$$Q = 155.8(1.0)^{5/6} = 155.8 \text{ cfs}$$

From Fig. 15.19 the ratio of instantaneous rate to removal rate is 2.0, giving a design flow rate of 311.6 cfs if no storm sewers existed. From Fig. 15.20, it is found that the unsewered area discharge should be increased by 35 percent for a watershed with 50 percent storm sewers. The final design flow is 1.35 × 311.6 = 420.7 cfs.   ■■

U.S. Geological Survey Regression Equations for Urban Areas

The U.S. Geological Survey, in cooperation with the Federal Highway Administration, conducted a nationwide study of flood magnitude and frequency in urban watersheds.[37] The investigation involved 269 gauged basins at 56 cities in 31 states, including Hawaii. The locations are shown in Fig. 15.21. Basin sizes ranged from 0.2 to 100 mi^2.

Multiple linear regression (see Chapter 27) of a variety of independent parameters was conducted to develop peak flow equations that could be applied to small, ungauged urban watersheds throughout the United States. Similar USGS regression equations for large rural basins are described in Chapter 27.

The simplest form of the developed regression equations involves the three most significant variables identified. These were contributing area A (mi^2), basin development factor BDF (dimensionless), and the corresponding peak flow RQ_i (cfs) for the ith frequency from an identical rural basin in the same region as the urban watershed. The latter variable accounts for regional variations, and estimates can be developed from any of the applicable USGS flood frequency reports (see Section 27.4). The three-parameter equations for the 2-, 5-, 10-, 25-, 50-, 100-, and 500-year flows are given as[37]

$$Q_2 = 13.2A^{0.21}(13 - BDF)^{-0.43}RQ_2^{0.73} \tag{15.12}$$

$$Q_5 = 10.6A^{0.17}(13 - BDF)^{-0.39}RQ_5^{0.78} \tag{15.13}$$

$$Q_{10} = 9.51A^{0.16}(13 - BDF)^{-0.36}RQ_{10}^{0.79} \tag{15.14}$$

$$Q_{25} = 8.68A^{0.15}(13 - BDF)^{-0.34}RQ_{25}^{0.80} \tag{15.15}$$

$$Q_{50} = 8.04A^{0.15}(13 - BDF)^{-0.32}RQ_{50}^{0.81} \tag{15.16}$$

$$Q_{100} = 7.70A^{0.15}(13 - BDF)^{-0.32}RQ_{100}^{0.82} \tag{15.17}$$

$$Q_{500} = 7.47A^{0.16}(13 - BDF)^{-0.30}RQ_{500}^{0.82} \tag{15.18}$$

These were developed from data at 199 of the 269 original sites. The other sites were deleted because of the presence of detention storage or missing data. All these equations have coefficients of determination above 0.90.

Figure 15.22 shows the correspondence of estimated and observed values used in developing Eq. 15.15. Forty percent of the values fall within one standard deviation of the regression line. Graphs for other recurrence intervals are similar to the 10-year graph shown in Fig. 15.22.

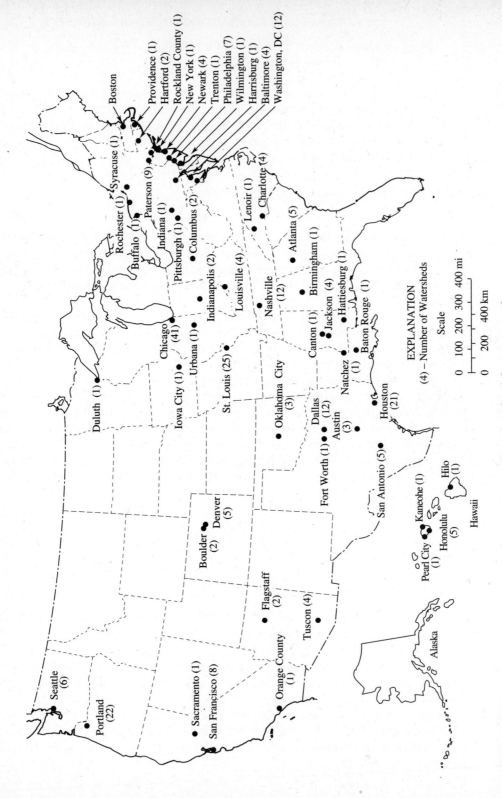

Figure 15.21 Location of runoff stations used in nationwide urban flood-frequency study. (After Sauer et al.[37])

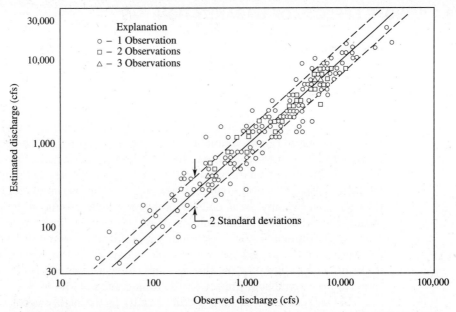

Figure 15.22 Comparison of observed and estimated 10-year urban peak discharges by Eq. 15.15. (After Sauer et al.[37])

The basin development factor *BDF* is an index of the prevalence of four drainage elements: storm sewers, channel improvements, impervious channel linings, and curb-and-gutter streets. A value of zero indicates the development elements are not prevalent. A maximum value of 12 indicates full development. Each of the four elements are given up to three units of *BDF*. This is accomplished by dividing the basin area into thirds and giving each third up to four *BDF* points (one for each element) depending on whether more than 50 percent of the drainage features are developed.

EXAMPLE 15.6 ───

Use the USGS index-flood method (Eq. 15.9 and Figs. 15.17 and 27.4c) to develop an estimate of the 50-year flood for a 100-mi^2 rural watershed at Lincoln, Nebraska. Then use the regional USGS regression equations to estimate the 50-year flood if the upper third of the watershed has 60 percent of storm sewers, 30 percent of channel improvements, 74 percent of impervious channel linings, and 82 percent of curb-and-gutter streets.

Solution. From Fig. 15.17, the *C* coefficient for Lincoln is 63. Equation (15.9) gives $Q_{2.33} = 1580$ cfs. Assuming that Fig. 15.28c applies, $Q_{50}/Q_{2.33} = 2.5$. Solving gives $Q_{50} = 3960$ cfs for the undeveloped watershed. Because three of the four development factors are more than half complete in one-third of the watershed, $BDF = 3$. From Eq. 15.16,

$$Q_{50} = 8.04(100)^{0.15}(10)^{-0.32}(3960)^{0.81}$$

or $Q_{50} = 6300$ cfs for the developed watershed. ■■

15.4 RUNOFF EFFECTS OF URBANIZATION

In Chapter 10, basin characteristics affecting runoff were discussed. The effects of slope, area size, soil and rock structure, and other factors were illustrated. From these discussions it is easy to understand that modifications of the land surface have varying effects on the runoff characteristics of a given drainage area.

If a heavily forested area with its thick layer of mulch is converted to cropland or pasture, the soil is disturbed and the overlying absorptive cover is changed. The result is increased runoff volume and a change in the timing of flows. When lowlands or marshes are surface-drained, the flooding characteristics of these areas are modified. The drains serve to remove the water at an accelerated rate, thus increasing the peak flows and runoff volumes. Inasmuch as there is usually a significant linkage between low, swampy areas and the underlying underground system, this relation is changed as well. The rapid removal of water from the drained area decreases the time—and consequently the opportunity for infiltration—and the net effect is usually a lowering of the underlying water table. Changes in the vegetal cover affect the infiltration capacities of soils, and land-use changes that modify the nature of vegetation can have significant impact on the timing and volume of flows.

Urbanization of the land usually results in the highly accelerated removal of storm water with corresponding increases in the volume and peak rate of runoff. In many cases, infiltration might be all but eliminated and a very high percentage of the storm rainfall becomes runoff. On the other hand, by increasing an area's storage capacity and delaying the outflow, it is possible to increase the timing and delay the peak rate of runoff. For example, a shopping center parking lot can be graded and its drains sized to permit several inches of ponding during intense storms. This delays the downstream arrival of flows from the area and significantly reduces the hydrograph peaks. By understanding the effects of land-use change on the hydrology of an area, it is possible to put this knowledge to beneficial use. Several aspects of this are discussed in Refs. 38–46.

The principal effects of land-use change have been classified by Leopold as follows:[47] (1) changes in peak flow characteristics, (2) changes in total runoff, (3) changes in water quality, and (4) changes in hydrologic amenities (the appearance or impression a watercourse and its environment leaves with the observer). A summary of the major hydrologic effects of changes in the forested cover, agricultural intensity, or wetland content is presented in Table 15.9.[32]

Change in Runoff Characteristics

Land-use changes can increase or decrease the volume of runoff and the maximal rate and timing of flow from a given area. The most influential factors affecting flow volume are the infiltration rate and surface storage. Changes in interception and other factors are usually of negligible importance.[48] The peak rate of flow is related to the flow volume. This can be exemplified by the unit-hydrograph principle, which states that with other things constant, the peak flow rate varies directly with the volume of flow. This relation is illustrated by Fig. 15.23. From this, it can be seen that land-use practices that decrease flow volume also decrease the peak rate of flow, and vice versa.

TABLE 15.9   SUMMARY OF THE MAJOR HYDROLOGIC EFFECTS OF LAND-USE CHANGE

Land-use change	Component affected	Principal hydrologic process involved	Geographic scale and likely magnitude of effect
Afforestation (deforestation has converse effect except where disturbance caused by forest clearance may be of overriding importance)	Annual flow	Increased interception in wet periods Increased transpiration in dry periods through increased water availability to deep root systems	Basin scale; magnitude proportional to forest cover, world average is 34 mm year^{-1} reduction for 10% increase in forest cover
	Seasonal flow	Increased interception and increased dry period transpiration will increase soil moisture deficits and reduce dry season flow	Basin scale; can be of sufficient magnitude to stop dry season flows
		Drainage activities associated with planting may increase dry season flows through intial dewatering and also through long-term effects of the drainage system	Basin scale; drainage activities will increase dry season flows
		Cloud water (mist or fog) deposition will augment dry season flows	High-altitude basins only; increased cloud water deposition may have a significant effect on dry season flows
	Floods	Interception reduces floods by removing a proportion of the storm rainfall and by allowing buildup of soil moisture storage	Basin scale; effect is generally small but greatest for small storm events
		Management activities: cultivation, drainage, road construction, all increase floods	Basin scale; increased floods for all sizes of storm events
	Water quality	Leaching of nutrients is less from forests through reduced surface runoff and reduced fertilizer applications	Basin scale; variable but leaching can be an order of magnitude less than from agricultural land
		Deposition of most atmospheric pollutants is higher to forests because of reduced aerodynamic resistance	Basin scale; leads to acidification of catchments and runoff

Source: After Maidment, D. R., *Handbook of Hydrology,* McGraw-Hill, Inc., 1993. Reproduced with permission of the publisher.[32]

TABLE 15.9   (continued)

Land-use change	Component affected	Principal hydrologic process involved	Geographic scale and likely magnitude of effect
	Erosion	High infiltration rates in natural, mixed forests reduce surface runoff and erosion	Basin scale; reduces erosion
		Slope stability is enhanced by reduced soil pore water pressure and binding of forest roots	Basin scale; reduces erosion
		Windthrow of trees and weight of tree crop reduce slope stability	Basin scale; increases erosion
		Soil erosion, through splash detachment, is increased from forests without an understory of shrubs or grass	Basin scale; increases erosion
		Management activities: cultivation, drainage, road construction, felling, all increase erosion	Basin scale; management activities are often more important than the direct effect of the forest
	Climate	Increased evaporation and reduced sensible heat fluxes from forests affect climate	Micro, meso, and global scale; forests generally cool and humidify the atmosphere; a 2°C increase in regional temperature is predicted for Amazonia if deforestation continues
Agricultural intensification	Water quantity	Alternation of transpiration rates affects runoff	Basin scale; effect is marginal
		Timing of storm runoff altered through land drainage	Basin scale; significant effect
	Water quality: fertilizers	Application of inorganic fertilizers	Basin scale; increased nutrient concentrations in surface and groundwaters
	Pesticides	Application of nonselective and persistent pesticides poses health risks to humans and animal life	Basin, regional, and global scale; effects can be long lasting

Source: After Maidment.[32]

TABLE 15.9   (continued)

Land-use change	Component affected	Principal hydrologic process involved	Geographic scale and likely magnitude of effect
	Farm wastes	Inadequate disposal of farm organic and inorganic water pollutes surface and groundwater bodies	Basin scale; effect on groundwater and surface waters
	Erosion	Cultivation without proper soil conservation measures and uncontrolled grazing on steep slopes increases erosion	Basin scale; effects are very site-dependent
Draining wetlands	Seasonal flow	Upland peat bogs, groundwater fens, and African dambos have little effect in maintaning dry season flows	Basin scale; drainage or removal of wetland will not reduce, and may increase dry season flows

Source: After Maidment.[32]

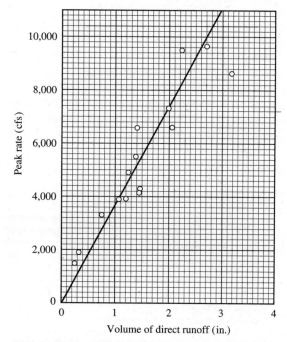

Figure 15.23   Typical peak-volume relation of annual floods at Eagle Creek, Indianapolis, Indiana. $A = 170$ mi^2 (After Mockus.[48])

Some effects of land-use treatment on the direct runoff are shown in Table 15.10. The overall effect of any measure is a function of the extent to which that measure can be or is applied in a watershed or the way in which the measure is used. For example, contour furrows can be made to have a large or small effect on the direct runoff depending on the dimensions of the furrows. Changing the vegetative cover from spring oats to spring wheat would probably have little effect on runoff produced, while a change from oats to a permanent meadow could have a significant effect.[48]

TABLE 15.10 EFFECTS OF SOME LAND-USE AND TREATMENT MEASURES ON THE DIRECT RUNOFF

| | Reduction in direct runoff volume is due to | |
| | Increasing infiltration rates[a] | Increasing surface storage |
Measure		
Land-use change that increases plant or root density[b]	X	
Increasing mulch or litter	X	
Contouring		X
Countour furrowing		X
Level terracing		X
Graded terracing		X

[a] Assuming soils not frozen.
[b] Example: Row crop to grass for hay; poor pasture to good pasture.
Source: After Mockus.[48]

In addition to land-use effects on the volume and rate of flow, lag effects (delay between upstream production of flow and its arrival at a downstream location) are also noticed. Land-use and treatment measures can modify lag by (1) increasing or decreasing the infiltration (reducing or increasing the surface runoff), or (2) by increasing or decreasing the flow distance or flow velocity. Some of these effects are shown in Table 15.11.

TABLE 15.11 RELATIVE EFFECTS OF LAND-USE AND TREATMENT MEASURES ON TYPES OF LAG

| | Increasing infiltration with resultant effect on subsurface flow[a] | | Effect of increasing surface flow distance or decreasing velocity | |
Measure	Small watersheds	Large watersheds	Small watersheds	Large watersheds
Land-use changes that increase plant or root density[b]	Can be large	Can be large	Not usually considered	
Increasing mulch or litter	Can be large	Can be large	Not usually considered	
Contouring	Can be large	Usually negligible	Can be large	Negligible
Contour furrowing	Can be large	Can be large	Not usually considered	
Level terracing	Can be large	Can be large	Not usually considered	
Graded terracing	Usually negligible	Usually negligible	Can be large	Negligible

[a] Assuming soils not frozen.
[b] Example: Row crop to grass; poor pasture to good pasture.
Source: After Mockus.[48]

The effects of urbanization require special mention, as they often have a pronounced impact on the characteristics of an area's hydrology. Urbanization generally increases the volume of the runoff and peak rate of flow and decreases the watershed's time lag. Figure 15.24 illustrates the effects on lag time and hydrograph peak for hypothetical unit hydrographs.[47] The runoff volume is determined mainly by infiltration and the nature of surface storage. The land slope, the soil type, the nature of the vegetative cover, and the degree of imperviousness of the watershed are all important factors. Figure 15.25 illustrates the combined effects of increased imperviousness and sewerage on the mean annual flood for a 1-mi² drainage area. An often overlooked but potentially important effect of increased runoff is the accompanying reduction in groundwater recharge. Where urban areas are expansive, local groundwater supplies can be seriously reduced.

In general, the peak rate of runoff will increase more rapidly than the volume of runoff as urbanization occurs. This is because of the increase in the rate of overland flow to stream channels and the resultant decrease in concentration time of the basin. Water flows more quickly from streets and roofs than from naturally vegetated areas, and conveyances such as storm sewers and lined open channels increase the flow velocities and thus decrease the lag time. The reduction in time lag (or concentration time) of the basin is extremely important as it affects the frequency or return period for a given level of flow. For example, the storm that was found to be the "50-year" storm on a basin having a 6-hr lag, will no longer be the "50-year" storm if the lag is reduced to 3 hr by urbanization. A study of Fig. 15.26 illustrates this point.

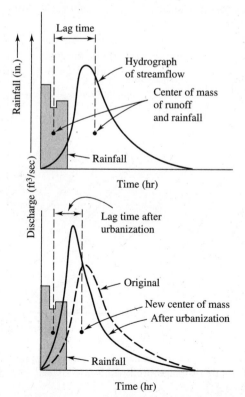

Figure 15.24 Hypothetical unit hydrographs relating the runoff to rainfall, with definitions of significant parameters. (U.S. Geological Survey Circular 554.)

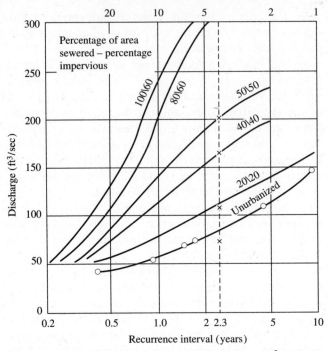

Figure 15.25 Flood frequency curves for a 1-mi² basin in various states of urbanization. (U.S. Geological Survey Circular 554.)

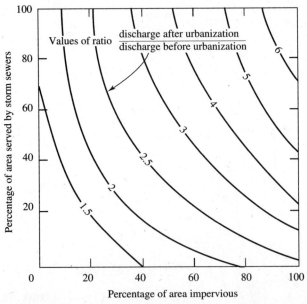

Figure 15.26 Effect of urbanization on the mean annual flood for a 1-mi² drainage area. (U.S. Geological Survey Circular 554.)

Water Quality

Changes in water quality due to land-use practices can be either positive or negative. Table 15.9 describes some of the water quality effects of forestation and agricultural intensification. The principal effect of land-use change on water quality is the introduction of waste materials such as nutrients, road salts, various chemicals, and oil and gasoline products. An especially important water quality problem is the rapid increase in sediment load in streams owing to the exposure of bare soil to storm runoff during and after periods of development. Urbanization has caused increases in sediment yield on the order of 100–250 times that of rural areas. Such increases result from the denuding of sites and the upsetting of balances of natural drainage networks to the flows they must carry. Streams tend to construct and maintain channels that exceed the bank-full stage at a recurrence interval of about 2 years. If the number of flows above bank-full stage is increased due to urbanization, or other causes, the banks and bed of erodible channels will not remain stable but will be enlarged through erosion.

Application

A knowledge of the manner in which land-use changes and land treatments can modify the runoff process is extremely important. Various proposed changes can be simulated and their effect evaluated before decisions to implement these practices are made. Designs can be improved and features incorporated into traditional design practices that will save funds, reduce adverse environmental impacts, and even enhance the quality of life. New uses for excess flows such as recreational ponds, artificial recharge, and urban irrigation can be found. By considering the total water management instead of only the fast removal of storm water runoff, many positive impacts are obtainable. Table 15.12 summarizes some measures for modifying the runoff process.

■ Summary

Many small watershed drainage structures can be completely designed using procedures presented in this chapter to develop estimates of the peak flow rate for a given frequency. Literally hundreds of thousands of culverts, bridges, bridge footings, river training structures, scour control measures, stream bank erosion protection structures, storm sewers, and open channels have been designed by first calculating the peak flow rate and then using hydraulic procedures to determine the appropriate size of structure, based on allowable depth and velocity for the given flow rate and site conditions. Chapter 16 describes the application of these and other hydrologic procedures in designing hydraulic structures, both large and small.

TABLE 15.12   MEASURES FOR REDUCING AND DELAYING URBAN STORM RUNOFF

Area	Reducing runoff	Delaying runoff
Large flat roof	Cistern storage Rooftop gardens Pool storage or fountain storage Sod roof cover	Ponding on roof by constricted downspouts Increasing roof roughness 1. Rippled roof 2. Graveled roof
Parking lots	Porous pavement 1. Gravel parking lots 2. Porous or punctured asphalt Concrete vaults and cisterns beneath parking lots in high- value areas Vegetated ponding areas around parking lots Gravel trenches	Grassay strips on parking lots Grassed waterways draining parking lot Ponding and detention measures for impervious areas 1. Rippled pavement 2. Depressions 3. Basins
Residential	Cisterns for individual homes or groups of homes Gravel driveways (porous) Contoured landscape Groundwater recharge 1. Perforated pipe 2. Gravel (sand) 3. Trench 4. Porous pipe 5. Dry wells Vegetated depressions	Reservoir or detention basin Planting a high delaying grass (high roughness) Gravel driveways Grassy gutters or channels Increased length of travel of runoff by means of gutters and diversions
General	Gravel alleys Porous sidewalks Mulched planters	Gravel alleys

Source: After U.S. Department of Agriculture, Soil Conservation Service, 1972.

PROBLEMS

15.1.   In using the rational formula, $Q = CIA$, for the design of any structure, what do the terms Q, C, I, and A represent? In selecting a T-year design storm intensity, why are the rainfall duration and time of concentration equated? Answer by noting the effect of selecting duration less than, and greater than, the time of concentration.

15.2.   A 53,200-acre area has a ϕ index of 0.10 in./hr. A storm with a constant rainfall rate of 0.7 in./hr lasts for 6 hr.
   a. What is the rational formula peak discharge in cfs if time of concentration is 4 hr?
   b. What is the runoff rate (cfs) at the end of the fifth hour after the rainfall begins?

15.3.   The time of concentration for a 6-acre parking lot is 20 min. Which of the following storms gives the greatest peak rate of runoff by the rational formula ($Q = CIA$) if $C = 0.6$? State any assumptions.
   a. 4 in./hr for 10-min duration.
   b. 1 in./hr for 40-min duration.

15.4.   A 10.0-mi^2 drainage basin has a time of concentration of 100 min and a constant ϕ index of 0.25 in./hr. If rain falls uniformly over the basin at a rate of 2.75 in./hr for a duration of 200 min, sketch the approximate hydrograph and determine the maximum discharge rate (cfs) at the basin outlet.

15.5. A 10.0-mi^2 drainage basin with a time of concentration of 100 min receives rainfall at a rate of 2.75 in./hr for a period of 200 min. If the runoff coefficient is 0.8, determine the discharge rate (cfs) from the basin 130 min after the rain began.

15.6. In using the rational formula ($Q = CIA$) for peak discharge rates, what does the product CI represent? What is the meaning of I in terms of the design discharge and rainfall duration?

15.7. The 4-hr unit hydrograph for a 5600-acre watershed is

Time (hr)	0	2	4	6	8	10	12
Q (cfs)	0	400	1000	800	400	200	0

The local 10-year IDF curve is linear with the equation $I = 5.6 - 0.2D$, where I is rain intensity in in./hr and D is the rain duration in hours. Use the unit-hydrograph procedures of Chapter 12 to determine the peak 10-year flow rate for the watershed. Compare this with the rational formula estimate of the 10-year peak. ($\phi = 1.0$ in./hr.) Use t_c as the time from the cessation of rain to the cessation of runoff.

15.8. Rework Example 15.2 based on a $C = 0.2$ and $C = 0.4$. Compare and discuss the effect of C on the discharge at the outfall.

15.9. A watershed has area A. Starting with a triangular-shaped unit hydrograph with a base length of $2.67t_p$ and a height of Q_p, derive Eq. 15.6 (see also Eq. 12.25), $Q_p = 484A/t_p$. State and carry units of each term used in the derivation.

15.10. Use both the rational and the SCS peak flow equation (15.5) to determine the 5-year design discharge (cfs) for a storm drain that receives runoff from a 300-acre area having a length of 1200 ft, a Manning n value of 0.050, a slope of 0.001 ft/ft, and a runoff coefficient of 0.60. Use a lag time of $0.6t_c$ and the Kirpich formula for t_c; also, use the Fig. 15.7 IDF curves for predicting needed precipitation depths.

15.11. Using the SCS dimensionless unit hydrograph described in Chapter 12, determine the peak discharge for a net storm of 10 in. in 2 hr on a 400-acre basin with a time to peak of 4 hr and a lag time of 3 hr. Compare with Eq. 12.17.

15.12. A 10.00-mi^2 watershed with a 100-min time of concentration receives rainfall at a rate of 2.75 in./hr for a period of 200 min.
a. Determine the peak discharge (cfs) from the watershed if $C = 0.4$.
b. Estimate the discharge rate (cfs) 150 min after the beginning of rainfall.
c. Estimate the discharge rate from the watershed 40 min after the beginning of rainfall.

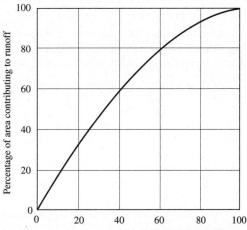

Time (min) after beginning of rainfall

15.13. A storm gutter receives drainage from both sides. On the left it drains a rectangular 600-acre area of $t_c = 60$ min. On the right it drains a relatively steep 300-acre area of $t_c = 10$ min. The ϕ index on both sides is 0.5 in./hr. Use the intensity-duration-frequency curves in Fig. 15.7 to determine the peak discharge (cfs) with a 25-year recurrence interval for (a) the 600-acre area alone, (b) the 300-acre area alone, and (c) the combined area assuming that the proportion of the 600-acre area contributing to runoff at any time t after rain begins is $t/60$.

15.14. A drainage basin has a time of concentration of 8 hr and produces a peak Q of 4032 cfs for a 10-hr storm with a net intensity of 2 in./hr. Determine the peak flow rate and the time base (duration) of the direct surface runoff for a net rain of 4 in./hr lasting (a) 12 hr, (b) 8 hr, and (c) 4 hr. State any assumptions used.

15.15. A 1.0-mi^2 parking lot has a runoff coefficient of 0.8 and a time of concentration of 40 min. For the following three rainstorms, determine the peak discharge (cfs) by the rational method: (a) 4.0 in./hr for 10 min, (b) 1.0 in./hr for 40 min, and (c) 0.5 in./hr for 60 min. State any assumption regarding area contributing after various rainfall durations.

15.16. The concentration time varies with discharge but is relatively constant for large discharges. From this statement, why do engineers feel confident in using the rational formula?

15.17. Determine the 50-year flood for a 20-mi^2 basin at the northwest corner of Nebraska. Use the index–flood method and assume that Fig. 26.4 applies.

15.18. Determine the entire frequency curve for the basin in Problem 15.17 and plot it on probability paper.

15.19. Use the index–flood method to determine the 10- and 50-year peaks for a 6400-acre drainage basin near Lincoln, Nebraska. Assume that Fig. 26.4 applies.

15.20. For the drainage basin in Problem 15.19 determine the probability that the 20-year peak will be equaled or exceeded at least once (a) next year and (b) in a 4-yr. period. Refer to Section 26.1.

15.21. For a 100-mi^2 drainage basin near Lincoln, Nebraska, use the index–flood method to determine the probability that next year's flood will equal or exceed 3000 cfs.

15.22. Use Fig. 26.4 to determine the return period (years) of the mean annual flood for that region. How does this compare with the theoretical value for a Gumbel distribution? How does it compare with a normal distribution? Refer to Section 26.6.

15.23. Use the Cyprus Creek method to determine the 25-yr peak discharge for the watershed described in Example 15.3. Assume that the watershed is nearly flat.

15.24. You are asked to determine the magnitude of the 50-year flood for a small, rural drainage basin (near your town) that has no streamflow records. State the names of at least two techniques that would provide estimates of the desired value.

15.25. The drainage areas, channel lengths, and relevant elevations (underlined) for several subbasins of the Oak Creek Watershed at Lincoln, Nebraska, are shown in Fig. 24.8. The watershed has a SCS curve number of $CN = 75$ which may be used to determine the direct runoff for any storm. Assume that IDF curves in Fig. 27.13 apply at Lincoln. Treat the entire watershed as a single basin and determine the 50-year flood magnitude at Point 8 using:
a. The rational method.
b. The SCS peak flow graph, Fig. 15.8.
c. Snyder's method of synthetic unit hydrographs, Eq. 15.5.
d. The USGS index–flood method. Figure 26.4 applies.

15.26. Repeat Problem 15.25 with Subarea I excluded. Compare the results with Problem 15.25 and comment on the effectiveness at Point 8 for the 50-year event of the Branched Oak Reservoir at Point 9. (This reservoir will easily store the 100-year flood from Area I.)

15.27. Repeat Problem 15.25 for Subarea A.

15.28. Repeat Problem 15.25 for Subarea I.

15.29. Describe completely how the magnitude of the 30-year flood for a watershed is determined by the USGS index–flood method.

15.30. A rural watershed with a composite CN of 70 is being urbanized. Eventually, 36 percent of the area will be impervious. Determine the increase in runoff that can be expected for a 6.2-in. rain.

15.31 Using the peak flow for the SCS dimensionless unit hydrograph in Ch. 12, determine the peak discharge for a net storm of 10 in. in 2 hr on a 400-acre basin with a time to peak of 4 hr and a lag time of 3 hr.

15.32. A timber railroad bridge in Nebraska at Milepost 271.32 on the railroad system shown in the sketch is to be replaced with a new concrete structure. The 50- and 100-year flood magnitudes are needed to establish the low chord and embankment elevations, respectively. Determine the design flow rates using the SCS TP-149 method. The bridge drains the zone marked, about 45 acres. The moderately sloped basin lies in a Type-II storm region, the curve number is 70, and the 24-hr 50- and 100-year rainfall depths are 8.6″ and 9.4″ respectively.

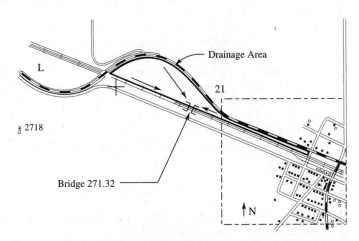

15.33. Repeat Problem 15.32 using the FHWA HEC-19 peak flow SCS design method. The time of concentration is 0.2 hrs. Values of I_a can be determined from the relationships in Fig. 4.14. Provide the answers in both metric and English units.

REFERENCES

1. J. Amorocho and W. E. Hart, "A Critique of Current Methods in Hydrologic Systems Investigations," *Trans. Am. Geophys. Union* **45**(2), 307–321(June 1964).
2. K. P. Singh, "Nonlinear Instantaneous Unit-Hydrograph Theory," *ASCE J. Hyd. Div.* **90**(HY2), Part I, 313-347(Mar. 1964).

3. W. T. Sittner, C. E. Schauss, and J. C. Monro, "Continuous Hydrograph Synthesis with an API-Type Hydrologic Model," *Water Resources Res.* **5**(5), 1007–1022(1969).

4. J. E. Nash, "The Form of the Instantaneous Unit Hydrograph," *Int. Assoc. Sci. Hyd.* **3**(45), 114–121(1957).

5. D. R. Dawdy and T. O'Donnel, "Mathematical Models of Catchment Behavior," *Proc. ASCE J. Hyd. Div.* **91**(HY4), 124–127(July 1965).

6. S. L. S. Jacoby, "A Mathematical Model for Nonlinear Hydrologic Systems," *J. Geophy. Res.* **71**(20), 4811–4824(Oct. 1966).

7. R. Prasad, "A Nonlinear Hydrologic System Response Model," *Proc. ASCE J. Hyd. Div.* **93**(HY4)(1967).

8. A. L. Tholin and C. T. Keifer, "Hydrology of Urban Runoff," *J. ASCE* **85,** 47–106(Mar. 1959).

9. N. H. Crawford and R. K. Linsley, Jr., "Digital Simulation in Hydrology: Stanford Watershed Model IV," Department of Civil Engineering, Stanford University, Stanford, CA, Tech. Rep. No. 39, July 1966.

10. John C. Schaake, Jr., "Synthesis of the Inlet Hydrograph," Tech. Rep. 3, Storm Drainage Research Project, Johns Hopkins University, Baltimore, MD, June 1965.

11. American Public Works Association, "Water Pollution Aspects of Urban Runoff," Federal Water Pollution Control Administration, 1969.

12. American Society of Civil Engineers, First Year Report, "Urban Water Resources Research," Sept. 1968.

13. W. Viessman, Jr., "Modeling of Water Quality Inputs from Urbanized Areas," Urban Water Resources Research, Study by ASCE Urban Hydrology Research Council, Sept. 1968, pp. A79–A103.

14. S. R. Weible, R. B. Weidner, A. G. Christianson, and R. J. Anderson, "Characterization, Treatment, and Disposal of Urban Storm Water," in *Proceedings of the Third International Conference, International Association on Water Pollution Research* (S. H. Jenkins, ed.). Elmsford, NY: Pergamon Press, 1969.

15. S. R. Weible, R. B. Weidner, J. M. Cohan, and A. G. Christianson, "Pesticides and Other Contaminants in Rainfall and Runoff," *J. Am. Water Works Assoc.* **58**(8), 1675(Aug. 1966).

16. Division of Water Resources, Department of Civil Engineering, University of Cincinnati, Cincinnati, OH, "Urban Runoff Characteristics," Water Pollution Control Research Series, EPA, 1970.

17. Metcalf and Eddy, Inc., University of Florida, Gainesville, Water Resources Engineers, Inc., "Storm Water Management Model," Environmental Protection Agency, Vol. 1, 1971.

18. E. Kuichling, "The Relation Between the Rainfall and the Discharge of Sewers in Populous Districts," *Trans. ASCE,* **20**(1889).

19. W. W. Horner, "Modern Procedure in District Sewer Design," *Eng. News* **64,** 326(1910).

20. W. W. Horner and F. L. Flynt, "Relation Between Rainfall and Runoff from Small Urban Areas," *Trans. ASCE* **20**(140), (1936).

21. R. L. Rossmiller, "The Runoff Coefficient in the Rational Formula," Engineering Research Institute, Iowa State University, Feb. 1981.

22. J. C. Schaake, Jr., J. C. Geyer, and J. W. Knapp, "Experimental Examination of the Rational Method," *Proc. ASCE J. Hyd. Div.* **93**(HY6) (Nov. 1967).

23. Federal Aviation Agency, Department of Transportation, "Airport Drainage," Advisory Circular, A/C 150-5320-5B. Washington, D.C.: U.S. Government Printing Office, 1970.

24. "Design and Construction of Sanitary Storm Sewers," *ASCE Manuals and Reports on Engineering Practice,* No. 37, 1970.

25. U.S. Soil Conservation Service, "Urban Hydrology for Small Watersheds," Tech. Release No. 55, Jan. 1975.

26. U.S. Soil Conservation Service, "Hydrology," *National Engineering Handbook.* Sec. 4, Washington, D.C.: U.S. Government Printing Office. 1972.

27. U.S. Soil Conservation Service, "Urban Hydrology for Small Watersheds," Tech. Release 55 (2nd ed.), June 1986.

28. G. L. Lewis, "A Shopper's Guide to Urban Stormwater Micro Software," in Proceedings of the ASCE Hydraulics Division Specialty Conference, Orlando, FL, Aug. 1985.

29. U.S. Soil Conservation Service, "Guide for the Use of Technical Release No. 55," Albany, NY, Dec. 1977.

30. U.S. Soil Conservation Service, "A Method for Estimating Volume and Rate of Runoff in Small Watersheds," Tech. Paper 149, rev., Washington, D.C., 1973.

31. D. M. Gray (ed.), *Handbook on the Principles of Hydrology.* Port Washington, NY: Water Information Center, 1973.

32. Maidment, D. R. (ed.), *Handbook of Hydrology.* New York: McGraw-Hill, Inc., 1993.

33. C. S. Jarvis, "Floods," in *Hydrology* (O. E. Meinzer, ed.), Chap. 11-G. New York: McGraw-Hill, 1942.

34. L. W. Furness, "Floods in Nebraska, Magnitude and Frequency," U.S. Geological Survey Rep. to Nebraska Department of Roads and Irrigation, Apr. 1955.

35. Soil Conservation Service, "Guide to Determine Instantaneous Peak Flow for Flatland Areas," Texas SCS Engineering Technical Note No. 210-18-TX8 Hydrology, U.S. Department of Agriculture, February 1985.

36. Stephens and Mills, "Using the Cypress Creek Formula to Estimate Runoff Rates in the Southern Coastal Plain and Adjacent Flatwoods Land Resource Areas," U.S. Agricultural Research Service Report ARS 41–95, U.S. Department of Agriculture, 1970.

37. V. B. Sauer, W. D. Thomas, V. A. Stricker, and K. V. Wilson, "Flood Characteristics of Urban Watersheds in the United States," U.S. Geological Survey Water-Supply Paper 2207. Washington, D.C.: U.S. Government Printing Office, 1983.

38. D. G. Anderson, "Effects of Urban Development on Floods in Northern Virginia," U.S. Geological Survey Open-File Report, 1968.

39. W. H. Espey, C. W. Morgan, and F. D. Masch, "Study of Some Effects of Urbanization on Storm Runoff from a Small Watershed," Texas Water Development Board, Rep. No. 23, 1966.

40. P. N. Felton and H. W. Lull, "Suburban Hydrology Can Improve Watershed Conditions," *Public Works* **94** (1963).

41. E. E. Harris and S. E. Tantz, "Effect of Urban Growth on Streamflow Regimen of Permanente Creek, Santa Clara County, California," U.S. Geological Survey Water Supply Paper 1591-B, 1964.

42. L. D. James, "Using a Computer to Estimate the Effects of Urban Development of Flood Peaks," *Water Resources Res.* **1**(2), (1965).

43. F. J. Keller, "The Effect of Urban Growth on Sediment Discharge," Northwest Branch Anacostia River Basin, Maryland, in "Short Papers in Geology and Hydrology," U.S. Geological Survey Professional Paper 450-C, 1962.

44. H. W. Lull and W. E. Sopper, "Hydrologic Effects from Urbanization of Forested Watersheds in the Northeast," Upper Darby, PA, Northeastern Forest Experimental Station, 1966.

45. M. G. Wolman and P. A. Schick, "Effects of Construction on Fluvial Sediment, Urban and Suburban Areas of Maryland," *Water Resources Res.* **3**(2), (1967).

46. R. L. Bras and F. E. Perkins, "Effects of Urbanization on Catchment Response," *Proc. ASCE J. Hyd. Div.* **101**(HY3) (Mar. 1975).

47. Luna B. Leopold, "Hydrology for Urban Land Planning," USGS Circular No. 554. Washington, D.C.: U.S. Government Printing Office, 1968.

48. Victor Mockus, "Hydrologic Effect of Land Use and Treatment," Chap. 12, *SCS National Engineering Handbook,* Sec. 4, "Hydrology." Washington, D.C.: U.S. Department of Agriculture, Soil Conservation Service, 1972.

Chapter 16

Hydrologic Design

■ Prologue

The purpose of this chapter is to:

- Introduce the hydrologist to procedures used in the United States for designing structures for safe and effective passage of flood flows.
- Give sufficient information for the designer to select the applicable criteria for designing hydraulic structures.
- Provide a discussion of design storm hyetographs and provide methods for selecting the duration, depth, and distribution of precipitation for design.
- Demonstrate how design floods can be developed without using precipitation data.
- Discuss particular design methods including airport drainage, urban storm sewer design, and flood control reservoir design.
- Describe the U.S. Federal Emergency Management Agency (FEMA) flood plain management system and present the hydrologic fundamentals of flood plain analysis.

Readers are encouraged to review the material in Chapters 26 and 27 prior to studying design procedures presented in this chapter.

Predicting peak discharge rates or synthesizing complete discharge hydrographs for use in designing minor and major structures are two of the more challenging aspects of engineering hydrology. Minor types of hydraulic structures range from small crossroad culverts, levees, drainage ditches, urban storm drain systems, and airport drainage structures to the spillway appurtenances of small dams. When lumped together with major structures, all require varying amounts of hydrologic design information. Generally, a hydrologist is required to provide peak rates of discharge for a design frequency, a stage height at a design frequency or a complete discharge hydrograph for a design storm. Other information such as sedimentation rates, low-flow frequency analysis, groundwater analysis, and reservoir yield studies are often conducted as part of a design project.

Most designs involving hydrologic analyses use a design flood that simulates some severe future event or imitates some historical event. If streamflow records are unavailable, design flood hydrographs are synthesized from available storm records using the rainfall–runoff procedures of Chapters 2, 12, and 15. Only in rare cases are streamflow records adequate for complex designs, particularly in small watersheds. Regional analyses and the empiric-correlative methods discussed in Chapter 15 are useful for determining peak flow rates at ungauged sites. Methods presented in Chapter 12 and in this chapter are used for developing entire hydrographs necessary for many engineering designs.

Hydrologic methods for designing minor and major structures are described in this chapter. Included are discussions of data needs, frequency levels, methods for synthesizing design storms, and hazard assessments for floodplains and dams.

16.1 HYDROLOGIC DESIGN PROCEDURES

Procedures for estimating design flood flows (interest can be in either the peak flow rate or the entire hydrograph) include methods that examine historical or projected flood flows to arrive at a suitable estimate (*flow-based* methods), and methods that evaluate the storms that produce floods, and then convert the storms to flood flow rates (*precipitation-based* methods). In each case, the analysis can be based on selecting a design frequency and determining the associated flood (call *frequency-based* methods), developing designs for a range of flood frequencies and narrowing the final choice on the basis of long-term costs and benefits (called *risk-based* methods), or designing on the basis of an estimate of the probable maximum storm or maximum flood that could occur at the site (called *critical-event* methods).

Minor Structure Design   Minor structure design is largely based on frequency-based or sometimes risk-based methods. Several steps in the hydrologic approach to *minor* structure design are common to most design handbooks and adopted techniques. The general steps (each is illustrated subsequently) are:

1. Determine the duration of the critical storm, usually equated to the time of concentration of the watershed.
2. Choose the design frequency.
3. Obtain the storm depth based on the selected frequency and duration.
4. Compute the net direct runoff (several methods were presented in Chapter 4).
5. Select the time distribution of the rainfall excess.
6. Synthesize the unit hydrograph for the watershed (see Chapter 12).
7. Apply the derived rainfall excess pattern to the synthetic unit hydrograph to get the runoff hydrograph.
8. Establish the frequency of the calculated flood (usually assumed equal to the design storm frequency).

Major Structure Design   Hydrologic design aspects of *major* structures are considerably more complex than those of a small dam, crossroad culvert, or urban drainage system. A design storm hydrograph for a large dam still is required but it is put to greater use. The design storm hydrograph is routed to determine the adequacy of spillways and outlets operated in conjunction with reservoir storage. The economic selection of the spillway size from the various possibilities dictates the final design and is a function of the degree of protection provided for downstream life and property, project economy, agency policy and construction standards, and reservoir operational requirements. Major structure design is largely based on critical event methods presented in Section 16.5.

Water Resource System Design   Most information and techniques presented in this chapter are directed toward the flood protection aspect of small and large structures. Needless to say, a major structure is designed for more than just flood protection; it is multipurpose and may provide storage for irrigation, power, water supply, navigation, and low-flow augmentation. The proper allocation of storage to these uses requires an understanding of the entire streamflow history in terms of the frequency of occurrence of low flows and average monthly, seasonal, and yearly flows, as well as the historical and design floods. Material is presented in Part Five to provide a hydrologist with the tools to develop complete streamflow histories for a complex multipurpose system involving various combinations of minor and major structures, water development projects, and management practices.

Flow-Based Methods

For design locations where records of stream flows are available, or where flows from another basin can be transposed to the design location, a design flood magnitude can be estimated directly from the stream flows by any of the following methods:

1. Frequency analysis of flood flows at the design location or from a similar basin in the region.
2. Use of regional flood frequency equations, normally developed from regression analysis (see Chapter 26) of gauged flood data.
3. Examination of the stream and floodplain for signs of highest historical floods and estimation of the flow rates using measurements of the cross-section and slope of the stream.

Precipitation-Based Methods

Where stream-gauging records are unavailable or inadequate for streamflow estimation, design floods can be estimated by evaluating the precipitation that would produce the flood, and then converting the precipitation into runoff by any of the rainfall-runoff methods described in Chapters 10–15 or 21–27. Typical methods include:

1. Design using the greatest storm of record at the site, by converting the precipitation to runoff.
2. Transposition of a severe historical storm from another similar watershed in the region.

3. Frequency analysis of precipitation and conversion of design storm to runoff.

4. Use of a theoretical probable maximum precipitation (PMP), or fraction of PMP, based on meteorological analyses.

Because the flood flow rate is desired in all cases, the flow-based methods are preferred over conversion of precipitation to runoff. Due to the relatively longer period of time and greater number of locations at which precipitation amounts have been recorded, precipitation-based methods are used in the majority of designs, especially with small and very large basins. Flow-based methods are typically used in the midrange of basin sizes.

Frequency-Based Methods

Regardless of whether flow or precipitation data are used, designs most often proceed by selecting a minimum acceptable recurrence interval and using procedures from Chapter 27 to determine the corresponding worst condition storm or flood that could be equalled or exceeded during the selected recurrence interval. Criteria for selecting design recurrence intervals are summarized in Section 16.3. Results from frequency analysis of flood flow data normally provide reliable estimates of 2-, 5-, 10-, and 25-year flows. Extrapolation beyond the range of the period of flow records is allowed, but is less reliable.

Risk-Based Methods

Recent trends in design of minor (and major) structures are toward the use of *economic risk analyses* rather than frequency-based designs. The risk method selects the structure size as that which minimizes total expected costs. These are made up of the structure costs plus the potential flood losses associated with the particular structure. The procedure is illustrated in Fig. 16.1. The total expected cost curve is the sum of

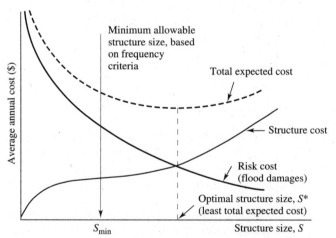

Figure 16.1 Principles of economic risk analysis for structure size selection. (U.S. Federal Highway Administration, Hydraulic Engineering Circular No. 17).

the other two curves. Risk costs (flood damages, structure damages, road and bridge losses, traffic interruptions) and structure costs are estimated for each of several sizes. The optimal size is that with the smallest sum. Structures selected by risk analysis are normally constrained to sizes equal to or larger than those resulting from traditional frequency-based methods.

Critical Event Methods

Because of the high risk to lives or property below major structures, their design generally includes provisions for a flood caused by a combination of the most severe meteorologic and hydrologic conditions that are possible. Instead of designing for some frequency or least expected total cost, flood handling facilities for the structures are sized to safely store or pass the most critical storm or flood possible. Methods for designing by critical event techniques include:

1. Estimating the probable maximum precipitation (PMP) and determining the associated flood flow rates and volumes by transforming the precipitation to runoff.
2. Determining the probable maximum flood (PMF) by determining the PMP and converting it to a flood by application of a rainfall-runoff model, including snowmelt runoff if pertinent.
3. Examining the flood plain and stream to identify palaeo-flood evidences such as high-water marks, boulder marks on trees or banks, debris lines, historical accounts by local residents, or geologic or geomorphologic evidences.
4. In some cases, the critical event method involves estimating the magnitude of the 500-yr event by various frequency or approximate methods. Often, such as in mapping floodplains, the 500-yr flood is estimated as a multiple of the 100-yr event, ranging from 1.5 to 2.5. Due to lack of longer-term records, frequency-based estimates are seldom attempted for recurrence intervals exceeding 500 years.

16.2 DATA FOR HYDROLOGIC DESIGN

The design of any structure requires a certain amount of data, even if only a field estimate of the drainage area and a description of terrain type and cover. The following material identifies some general data types and sources.

Physiographic Data

The hydrologic study for any structure requires a reliable topographic map. United States Geological Survey topographic maps usually are available. The mapping of the United States is almost complete with 15-minute quadrangles, and many of these areas are mapped by 7.5-minute quadrangles. County maps and aerial photos can also be used to advantage in making preliminary studies of the watershed.

Based on an area map, a careful investigation of the watershed's drainage behavior must be made. Additional information can be obtained from USGS maps that depict predominant rock formations. Soil types and the infiltration and erosive characteristics of soils can be secured from U.S. Soil Conservation districts or university extension divisions.

The drainage areas contributing to large dams require stricter analysis of an area's hydrology than is necessary in designing minor structures. The possibility of a uniformly intense rainfall over the entire basin is an unrealistic assumption for large watersheds. The influence of temporal and spatial variations of the rainfall should thus be considered. For major dams, the estimated "worst possible" rainfall values are generally converted to a design discharge hydrograph, which is then used in reservoir routing calculations to proportion reservoir and spillway size, surcharge storage, and any additional outlets needed to maintain power requirements or sustained downstream flow for navigation, irrigation, or water supply. The basic concern in hydrologic design of a large dam is to protect downstream interests using a realistic estimate for the design storm hydrograph.

Topographic map detail necessarily shifts with the type and purpose of the structure being designed. Field reconnaissance always increases the understanding of an area's hydrology no matter how insignificant the structure might be.

Hydrologic Data

One difficulty in hydrologic design is that of getting adequate data for the region under study. Considerable data can be acquired from previously published reports issued by governmental agencies and/or universities. The following is a list of federal agencies that publish hydrologic data:

> Agricultural Research Service
> Soil Conservation Service
> Forest Service
> U.S. Army Corps of Engineers
> National Oceanic and Atmospheric Administration
> Bureau of Reclamation
> Department of Transportation
> U.S. Geological Survey, Topographic Division
> U.S. Geological Survey, Water Resources Division

Additional data often can be procured from departments of state governments, interstate commissions, and regional and local agencies.

Meteorologic Data

The National Weather Service, couched in the National Oceanic and Atmospheric Administration, is the primary source of meteorologic data published in a variety of forms, including their Hydrometeorologic Report (HMR) series. Figure 16.2 shows the applicable reports for various geographic and topographic regions of the

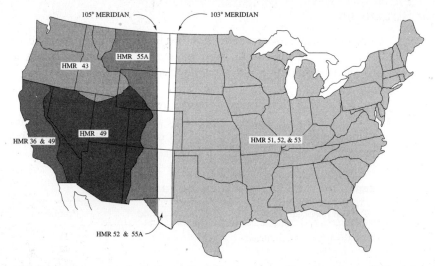

Figure 16.2   Hydrometeorological report series coverage of conterminous United States. (U.S. Bureau of Reclamation.)

United States.[1] Numerous other federal state, and local agencies collect and analyze precipitation information—especially those who design, inspect, or regulate large hydraulic structures.

Current practice for estimating design storm hyetographs requires knowledge of the meteorologic characteristics of storms in the region, maximum amount of precipitable moisture in the atmosphere, causes of precipitation, frequencies of total storm depths for various durations of storms, and influence of snowmelt for storms over the region. In some areas such as foothill regions of major mountain chains, topography has a very distinct impact on precipitation.

16.3 HYDROLOGIC DESIGN—FREQUENCY CRITERIA

Selection of frequency for the design of a structure is most often based on potential damage to property, danger to life, and economic losses such as interruption of commerce. A standard of practice involves selection of a frequency, then designing for the worst condition expected to occur for that frequency. Where danger to life is involved, a great amount of controversy exists over appropriate design standards.

All projects involve some risks to property and life, but where direct danger to human life is absent, the design can proceed through selection of an accepted frequency level and design of the least cost structure that provides this protection. As an alternative to least structure cost, economic risk analysis can be used. For this method, the final design frequency is optimized rather than preselected. Structure sizes that would accommodate storms for several frequencies are tested, and the one with the least total expected cost is used. These costs include not only the actual construction costs but also the flood damage risks and costs due to interruption of services and commerce. Either annual or present worth economic analyses can be used.

Minor Structures

The design frequencies shown in Table 16.1 are typical of levels generally encountered in minor structure design. An example of variations that do occur is the design frequency of a culvert, which under cases of excessive backwater could effectively halt traffic.

The Soil Conservation Service recommends the use of a 25-year frequency for minor urban drainage design if there is no potential loss of life or risk of extensive damage such as first-floor elevations of homes. A 100-year frequency is commonly recommended when extensive property damage may occur.[2]

TABLE 16.1   MINOR STRUCTURE DESIGN FREQUENCIES

Type of minor structure		Return period, T_r	Frequency = $1/T_r$
Highway crossroad drainage[a]			
0–400	ADT[a]	10 yr	0.10
400–1700	ADT	10–25 yr	0.10–0.04
1700–5000	ADT	25 yr	0.04
5000–	ADT	50 yr	0.02
Airfields		5 yr	0.20
Railroads		25–50 yr	0.04–0.02
Storm drainage		2–10 yr	0.50–0.10
Levees		2–50 yr	0.50–0.02
Drainage ditches		5–50 yr	0.20–0.02

[a] ADT = average daily traffic. (After Ref. 3).

Large Dams

Dams require hydrologic analysis during the design of the original structure and during periodic safety evaluations. Significant economic and human losses are possible when large quantities of water are rapidly released from storage.

Initial heights of retarded water behind the dam, disregarding the total volume of stored water, can produce destructive flood waves for a considerable distance downstream. Based on two criteria, the Task Force on Spillway Design Floods recommended the classification of large dams as listed in Table 16.2. The type of construction has not been included in this grouping, although it affects the extent of failure resulting from overtopping.

Many of the federal agencies have adopted definitions for hydraulic elements of dams. The following list is used by the Soil Conservation Service:

A *spillway* is an open or closed channel, or both, used to convey excess water from a reservoir. It may contain gates, either manually or automatically controlled, to regulate the discharge of excess water.

The *principal spillway* is the ungated spillway designed to convey the water from the retarding pool at release rates established for the structure.

The *emergency spillway* of a dam is the spillway designed to convey water in excess of that impounded for flood control or other beneficial purposes.

TABLE 16.2 DESIGN CRITERIA FOR LARGE DAMS

Category (1)	Impoundment danger potential		Failure damage potential[a]		Spillway design flood (6)
	Storage (acre-ft)[b] (2)	Height (ft) (3)	Loss of life (4)	Damage (5)	
Major; failure cannot be tolerated	>50,000	>60	Considerable	Excessive or as matter of policy	Probable maximum; most severe flood considered reasonably possible on the basin
Intermediate	1000–50,000	40–100	Possible but small	Within financial capability of owner	Standard project; based on most severe storm or meteorological conditions considered reasonably character-istic of the specific region
Minor	<1000	<50	None	Of same magni-tude as cost of the dam	Frequency basis; 50–100-year recurrence interval

[a] Based on consideration of height of dam above tailwater, stoarage volume, and length of damage reach, present and future potential population, and economic development of floodplain.
[b] Storage at design spillway pool level.
Source: After Snyder.[3]

The *retarding pool* is the reservoir space allotted to the temporary im-poundment of floodwater. Its upper limit is the elevation of the crest of the emergency spillway.

Retarding storage is the volume in the retarding pool.

The *sediment pool* is the reservoir space allotted to the accumulation of incoming sediment during the life of the structure.

Sediment storage is the volume allocated to total sediment accumulation.

Sediment pool elevation is the elevation of the surface of the anticipated sediment accumulation at the dam.

An *earth spillway* is an unvegetated open channel spillway in earth mate-rials.

A *vegetated spillway* is a vegetated open channel spillway constructed of earth materials.

A *ramp spillway* is a vegetated spillway constructed on the downstream face of an earth dam.

The *control section* in an open channel spillway is that section where accelerated flow passes through critical depth.

The *inlet channel* of an emergency spillway is the channel upstream from the control section.

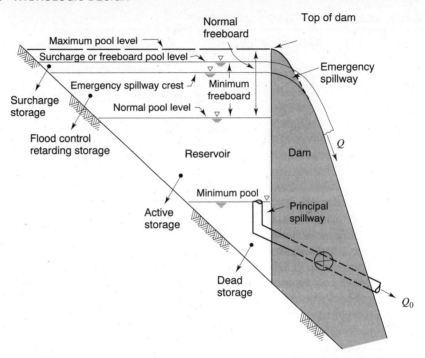

Figure 16.3 Multipurpose reservoir pool levels and storage zones.

The *exit channel* of an emergency spillway is that portion of the channel downstream from the control section which conducts the flow safely to a point where it may be released without jeopardizing the integrity of the structure.

The *emergency spillway hydrograph* is that hydrograph used to establish the minimum design dimensions of the emergency spillway.

The *freeboard hydrograph* is the hydrograph used to establish the minimum elevation of the top of the dam.

Several of these features are illustrated in Fig. 16.3.

Small Dams

Small dams customarily are designed using two or more levels of frequency to provide an emergency spillway and ensure an adequate allowable freeboard. Figure 16.3 shows a typical small dam with normal freeboard (NF) and minimal freeboard (MF). The freeboard values for earth dams with riprap protection on the upstream slope, shown in Table 16.3, are based on wave runup caused by storm winds with 100-mph wind velocities. Minimal freeboard pertains to wind velocities of 50 mph. The fetch is defined as the perpendicular distance from the structure to the windward shore. If smooth concrete rather than riprap is used on the upstream face, the freeboard values shown should be increased 50 percent.[4]

TABLE 16.3 USBR RECOMMENDED NORMAL AND MINIMUM FREEBOARD VALUES, FT

Fetch (mi)	NF	MF
<1	4	3
1	5	4
2.5	6	5
5	8	6
10	10	7

Source: After Ref. 4.

The U.S. Soil Conservation Service design criteria for principal spillways of small dams are given in Table 16.4. The SCS Technical Release No. 60 should be consulted for full interpretation of this table.[5] Design frequency requirements are selected to fit the planned or foreseeable use of the structures. The SCS classifies structures into three groups:[6]

Class a. Structures located in rural or agricultural areas where failure might damage farm buildings, agricultural land, or township or country roads.

Class b. Structures located in predominantly rural or agricultural areas where failure might damage isolated homes, main highways or minor railroads, or cause interruption of use or service of relatively important public utilities.

Class c. Structures located where failure might cause loss of life, serious damage of homes, industrial and commercial buildings, important public utilities, main highways, or railroads.

The physical size of a small dam can range to over 100 ft in height but generally is restricted to structures retarding less than 25,000 acre-ft of storage at the emergency spillway crest. Small dams generally receive special attention if they are constructed in populated areas where dam failure could cause the loss of life. Many flood deaths have been caused by dam or levee failure. When this possibility exists, the design storm for small dams is established by use of the probable maximum precipitation, PMP. The PMP is generally defined as the reasonable maximization of the meteorological factors that operate to produce a maximum storm. Other definitions have been proposed,[7] including:

1. The PMP is the maximum amount and duration of precipitation that can be expected to occur on a drainage basin.
2. The PMP is the flood that may be expected from the most severe combination of critical meteorologic and hydrologic conditions that are reasonably possible in the region. The PMP has a low, but unknown, probability of occurrence. It is neither the maximum observed depth at the design location nor region nor a value that is completely immune to exceedance.

TABLE 16.4 SCS DESIGN CRITERIA FOR PRINCIPAL SPILLWAYS OF SMALL DAMS

Class of dam	Purpose of dam	$V_s H_e^1$	Existing or planned upstream dams	Precipitation data for maximum frequency[2] of use of emergency spillway type: Earth	Vegetated
(a)	Single[3] irrigation only	Less than 30,000	None	0.5DL[4]	0.5DL
		Greater than 30,000	None	0.75DL	0.75DL
	Single or multiple[5]	Less than 30,000	None	P_{50}	P_{25}^6
		Greater than 30,000	None	$0.5(P_{50} + P_{100})$	$0.5(P_{25} + P_{50})$
		All	Any[7]	P_{100}	P_{50}
(b)	Single or multiple	All	None or any	P_{100}	P_{50}
(c)	Single or multiple	All	None or any	P_{100}	P_{100}

[1] Product of reservoir storage volume V_s (acre-feet) times effective height of dam H_e (feet).
[2] Precipitation depths for indicated return periods (years).
[3] Applies to irrigation dams on ephemeral streams in areas where mean annual rainfall is less than 25 in.
[4] DL = design life (years).
[5] Class (a) dams involving industrial or municipal water are to use minimum criteria equivalent to that of Class (b).
[6] In the case of a ramp spillway, the minimum criteria should be increased from P_{25} to P_{100}.
[7] Applies when the failure of the upstream dam may endanger the lower dam.
Source: Soil Conservation Service.

Estimates of PMP are based on an investigation by the U.S. Weather Bureau conducted to establish the maximum possible amount of precipitable water that could be achieved throughout the United States.[8,9] Figure 16.4 provides estimates of precipitable water over watersheds between sea level and 8,000 ft. Figure 16.5 extends the estimates above 8,000 ft.[1] Point values of PMP for the same locale may vary with duration of storm causing the precipitation. Figure 16.6 provides PMP estimates for 6-hr storms. These and similar published charts for other durations are helpful in selecting the PMP for any region in the United States.

The design frequencies for principal spillways for small SCS Class a, b, or c dams are provided in Table 16.5. These are based on 6-hr rainfall depths for (1) the 100-year frequency (Fig. 16.7) and (2) the PMP (Fig. 16.6). Design storm depths for all watersheds having a time of concentration less than 6-hr are established in Table 16.5. For those watersheds with greater time of concentration, adjustments are made to the 6-hr storm depth to account for the greater amounts of direct runoff in a longer period of time. These adjustments are discussed in Section 16.4.

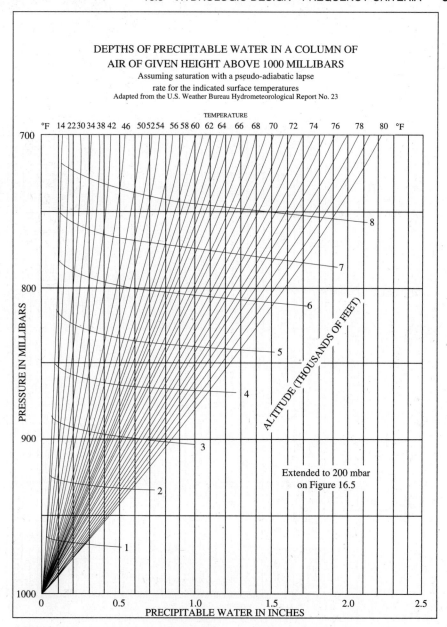

Figure 16.4 Diagram for precipitable water determination from 1,000 to 700 millibars. (U.S. Bureau of Reclamation.)

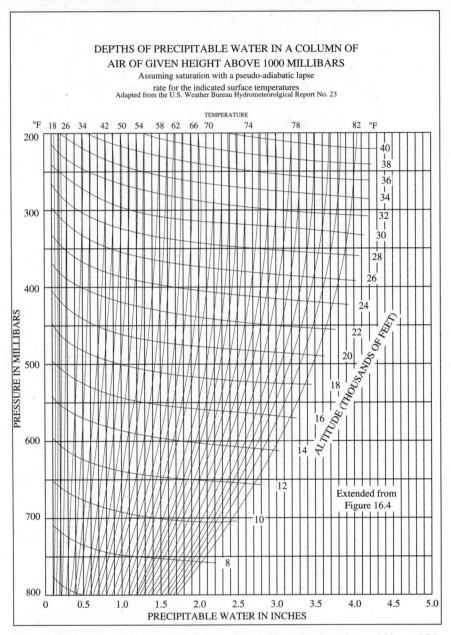

Figure 16.5 Diagram for precipitable water determination from 800 to 200 millibars. From [28]. 103-D-1908. (U.S. Bureau of Reclamation.)

Major Structures

Three general terms described in Table 16.2 are employed to designate design floods for major structures: (1) the probable maximum flood, (2) the standard project flood, and (3) the frequency-based flood. The concept of *flood* in this section is described by an entire discharge hydrograph that is generally synthesized from rainfall estimates. Corresponding to the three flood designations are the storm values, that is, the depth of rainfall, referred to in terms of (1) the probable maximum precipitation, PMP; (2) the standard project storm, SPS; and (3) the frequency-based storm. Design of the reservoir system components is commonly based on one of these representative terms. Descriptions of each are given in the next section on design storms.

16.4 DESIGN STORMS

Once the design frequency has been established, the next step in a structure design is the determination of six storm parameters: the storm duration, the duration of rainfall excess, the point depth, any areal depth adjustment, the storm intensity and time distribution, and the areal distribution pattern.

Duration

The length of storm used by the SCS in designing emergency and freeboard hydrographs for small dams is of 6-hr duration or t_c, whichever is greater. Often, the minor structure being designed cannot be justified economically on the basis of this length of storm. For many minor structures, particularly urban drainage structures, a design flood hydrograph is based on a storm duration equal to the time of concentration of the watershed. This procedure uses the rational method of Chapter 15 or the synthetic unit hydrographs of Chapter 12 along with a critical storm pattern produced by arranging the rainfall excess pattern into the most critical sequence. The SCS uses 24-hr durations for all urban watershed studies.

Durations of approximately 6 hr or less are satisfactory for small watersheds, but the lengths of storms in large areas require storm depths for periods of up to 10 days. Frequency-based values are available for durations of from 2 to 10 days for locations within the United States.[10] Similar data are also available for other selected areas outside the United States. Generally, however, design criteria for large dams require estimates of storm depths that do not have frequency levels assigned.

Duration of Rainfall Excess

Initial rainfall during most storms infiltrates or is otherwise abstracted, and the duration of excess rain T_0 is less than the actual rain duration by an amount equal to the time that initial abstractions occur. Excess rain duration T_0 can be estimated for a 6-hr storm as a function of the curve number CN and precipitation P from

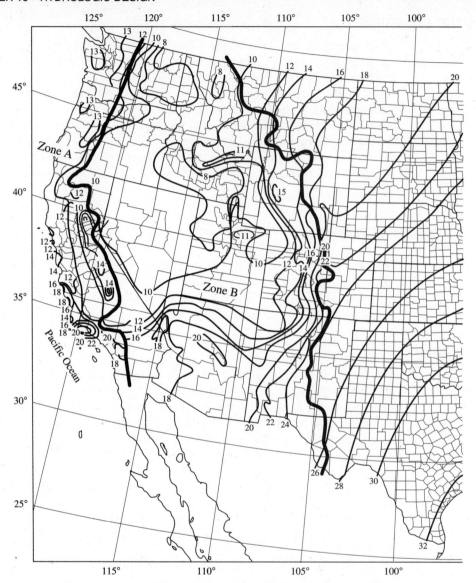

Figure 16.6   The 10-mi^2 or less PMP for 6-hr duration (in.). (U.S. Weather Bureau, NOAA.)

Fig. 16.8. This family of curves was developed by the SCS,[11] where P is the storm depth and CN is a loss parameter defined in Chapter 4. A CN of 100 represents zero losses so that $T_0 = 6$ hr for $CN = 100$. Table 16.6 is used to find the duration of excess rain for any storm duration greater than 6 hr. The rainfall ratio is the abstraction P^* lost before runoff (Table 16.7) divided by the total precipitation amount P. The time ratio from Table 16.6 is multiplied by the rainfall duration to obtain T_0.

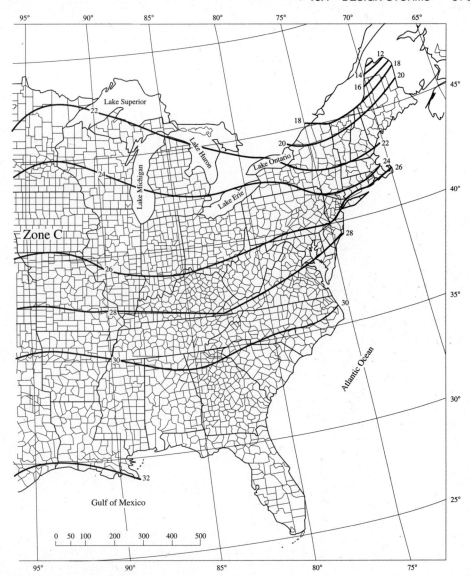

Figure 16.6 Continued

Depth

The probable maximum precipitation or frequency-based 100-year 6-hr storm depths at any point can be determined from Figs. 16.6 and 16.7. A convenient means of obtaining storm depths for durations other than 6 hr is to use a table or graph of multipliers for various durations. The U.S. Bureau of Reclamation[6] applies the multipliers in Table 16.8 to the PMP from Fig. 16.6 to determine other duration PMP depths for areas west of the 105° meridian. Similar USBR data east of the meridian are not available.

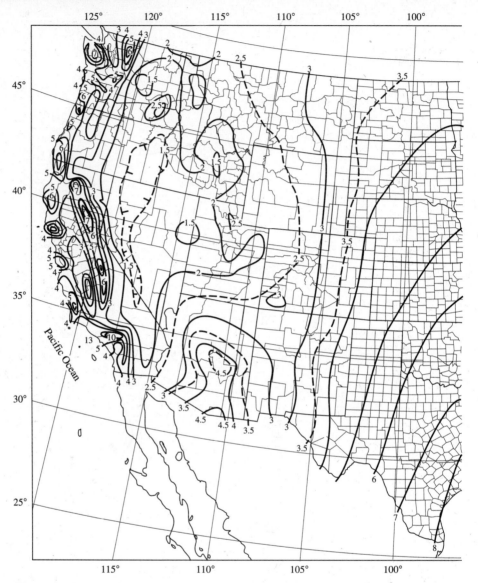

Figure 16.7 The 100-year frequency 6-hr precipitation (in.) for 10-mi^2 or less. (U.S. Weather Bureau, NOAA.)

The Tennessee Valley Authority (TVA) has regional regulatory and resource development authorities for much of the Tennessee River basin. Over the years they have developed dam design criteria, including their own definition of PMP. They recommend use of Table 16.9 for adjusting 6-hr storm depths to other duration storms.[14]

The U.S. Soil Conservation Service curve[12] of Fig. 16.9 is available for use in adjusting the PMP and 100-year 6-hr point rainfall depths from Figs. 16.6 and 16.7.

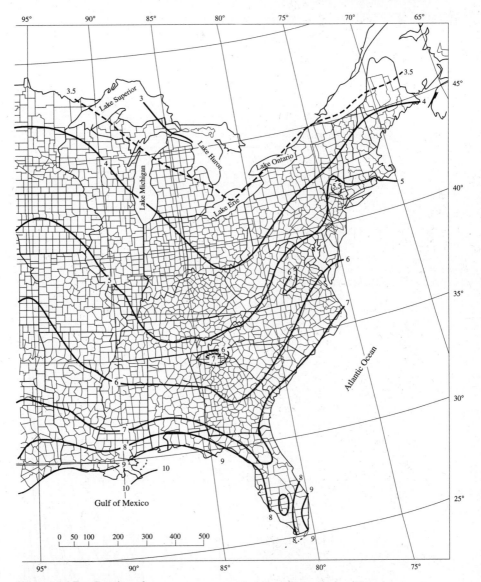

Figure 16.7 Continued

Taken together, Figs. 16.6, 16.7, and 16.9 allow the determination for minor structure design of all values in Table 16.5 for any storm duration.

IDF Relationships

For short-duration storms over small areas, the most convenient method of determining storm depth is to acquire the intensity-duration-frequency curve for the locale, enter the graph with the selected duration and frequency, and convert the resulting intensity to depth of rain over the selected duration. Area adjustments in the depth

TABLE 16.5   SCS DESIGN CRITERIA FOR EMERGENCY SPILLWAYS OF SMALL DAMS

Class of dam	$V_e H_e^1$	Existing or planned upstream dams	Precipitation Data[2] for	
			Emergency spillway hydrograph	Freeboard hydrograph
(a)[3]	Less than 30,000	None	P_{100}	$P_{100} + 0.12(\text{PMP} - P_{100})$
	Greater than 30,000	None	$P_{100} + 0.06(\text{PMP} - P_{100})$	$P_{100} + 0.26(\text{PMP} - P_{100})$
	All	Any[4]	$P_{100} + 0.12(\text{PMP} - P_{100})$	$P_{100} + 0.40(\text{PMP} - P_{100})$
(b)	All	None or any	$P_{100} + 0.12(\text{PMP} - P_{100})$	$P_{100} + 0.40(\text{PMP} - P_{100})$
(c)	All	None or any	$P_{100} + 0.26(\text{PMP} - P_{100})$	PMP

[1] Product of reservoir storage volume V_s (acre-feet) times effective height of dam H_e (feet).
[2] Precipitation depths for either 100-yr return period (P_{100}) or PMP.
[3] Class (a) dams involving industrial or municipal water are to use minimum criteria equivalent to that of Class (b).
[4] Applies when the failure of the upstream dam may endanger the lower dam.
Source: Soil Conservation Service.

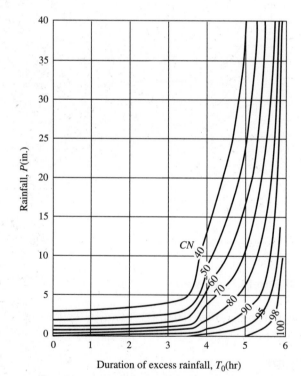

Figure 16.8 Duration of excess rainfall for SCS 6-hr design storms. (After Ref. 12.)

TABLE 16.6   RAINFALL AND TIME RATIOS FOR DETERMINING T_0 WHEN
STORM DURATION IS GREATER THAN 6 hr

Rainfall ratio	Time ratio	Rainfall ratio	Time ratio	Rainfall ratio	Time ratio	Rainfall ratio	Time ratio
0	1.00	0.070	0.852	0.140	0.746	0.210	0.684
0.002	0.995	0.072	0.848	0.142	0.744	0.212	0.682
0.004	0.990	0.074	0.844	0.144	0.742	0.214	0.680
0.006	0.985	0.076	0.841	0.146	0.740	0.216	0.679
0.008	0.981	0.078	0.837	0.148	0.739	0.218	0.677
0.010	0.976	0.080	0.833	0.150	0.737	0.220	0.675
0.012	0.971	0.082	0.830	0.152	0.735	0.222	0.673
0.014	0.967	0.084	0.827	0.154	0.733	0.224	0.672
0.016	0.962	0.086	0.824	0.156	0.732	0.226	0.670
0.018	0.957	0.088	0.821	0.158	0.730	0.228	0.668
0.020	0.952	0.090	0.818	0.160	0.728	0.230	0.667
0.022	0.948	0.092	0.815	0.162	0.726	0.232	0.666
0.024	0.943	0.094	0.812	0.164	0.724	0.234	0.666
0.026	0.938	0.096	0.809	0.166	0.723	0.236	0.665
0.028	0.933	0.098	0.806	0.168	0.721	0.238	0.665
0.030	0.929	0.100	0.803	0.170	0.719	0.240	0.664
0.032	0.924	0.102	0.800	0.172	0.717		
0.034	0.919	0.104	0.797	0.174	0.716	(change in	
0.036	0.915	0.106	0.794	0.176	0.714	tabulation	
0.038	0.911	0.108	0.791	0.178	0.712	increment)	
0.040	0.908	0.110	0.788	0.180	0.710	0.250	0.662
0.042	0.904	0.112	0.785	0.182	0.709	0.300	0.651
0.044	0.900	0.114	0.782	0.184	0.707	0.350	0.640
0.046	0.896	0.116	0.779	0.186	0.705	0.400	0.628
0.048	0.893	0.118	0.776	0.188	0.703	0.450	0.617
0.050	0.889	0.120	0.773	0.190	0.702	0.500	0.606
0.052	0.885	0.122	0.770	0.192	0.700	0.550	0.595
0.054	0.882	0.124	0.767	0.194	0.698	0.600	0.583
0.056	0.878	0.126	0.764	0.196	0.696	0.650	0.542
0.058	0.874	0.128	0.761	0.198	0.695	0.700	0.500
0.060	0.870	0.130	0.758	0.200	0.693	0.750	0.447
0.062	0.867	0.132	0.755	0.202	0.691	0.800	0.386
0.064	0.863	0.134	0.751	0.204	0.689	0.850	0.310
0.066	0.859	0.136	0.749	0.206	0.687	0.900	0.220
0.068	0.856	0.138	0.747	0.208	0.686	0.950	0.116

Source: After Ref. 12.

may be necessary for basins larger than about one square mile. The IDF curves are available from several sources, including NOAA, the National Weather Service, and more often from the city, county or parish, or state engineer. One such set of curves was provided in Fig. 2.8 for use in the vicinity of Baltimore, Maryland. Equations that describe the shapes of IDF curves have been developed for a number of major U.S. cities.[15] For small structure designs, the distribution of the selected design storm from an IDF curve is often assumed to be uniform. Alternatively, other distributions described subsequently may be applied.

TABLE 16.7   RAINFALL PRIOR TO EXCESS RAINFALL

CN	P*(in.)	CN	P*(in.)	CN	P*(in.)	CN	P*(in.)	CN	P*(in.)
100	0	86	0.33	72	0.78	58	1.45	44	2.54
99	0.02	85	0.35	71	0.82	57	1.51	43	2.64
98	0.04	84	0.38	70	0.86	56	1.57	42	2.76
97	0.06	83	0.41	69	0.90	55	1.64	41	2.88
96	0.08	82	0.44	68	0.94	54	1.70	40	3.00
95	0.11	81	0.47	67	0.98	53	1.77	39	3.12
94	0.13	80	0.50	66	1.03	52	1.85	38	3.26
93	0.15	79	0.53	65	1.08	51	1.92	37	3.40
92	0.17	78	0.56	64	1.12	50	2.00	36	3.56
91	0.20	77	0.60	63	1.17	49	2.08	35	3.72
90	0.22	76	0.63	62	1.23	48	2.16	34	3.88
89	0.25	75	0.67	61	1.28	47	2.26	33	4.06
88	0.27	74	0.70	60	1.33	46	2.34	32	4.24
87	0.30	73	0.74	59	1.39	45	2.44	31	4.44

Source: After Ref. 12.

TABLE 16.8   CONSTANTS FOR EXTENDING 6-hr
PMP DESIGN STORMS IN AREAS
WEST OF THE 105° MERIDIAN TO
LONGER DURATION PERIODS

Duration (hr)	Constant[a]	Duration (hr)	Constant[a]
8	1.16	22	1.74
10	1.31	24	1.80
12	1.43	30	1.95
14	1.50	36	2.10
16	1.56	42	2.25
18	1.62	48	2.38
20	1.68		

[a] Multiply 6-hr point rainfall from Fig. 16.6 by the indicated constant.
Source: After Ref. 6.

TABLE 16.9   TVA RATIOS FOR ADJUSTING
6-HR STORM DEPTHS FOR
OTHER DURATIONS

Duration	1	2	3	6	12	24
Ratio	0.51	0.68	0.80	1.00	1.13	1.24

Areal Adjustment

The rainfall depths shown in Figs. 16.6 and 16.7 were derived from frequency analyses (Chapter 27) of *point* measurements and are considered to be applicable only for areas up to 10 mi^2. For larger watersheds the areal depths are less; adjustment must be made to account for smaller rainfall depths over larger areas.

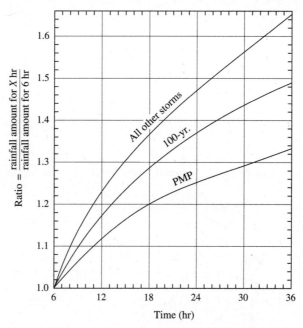

Figure 16.9 Relative increase in rainfall amount for storm durations over 6 hr for SCS dam designs. (After Ref. 12.)

The U.S. Weather Bureau[13] developed Fig. 16.10 as a guide in reducing point depths to areal depths for areas up to 400 mi². For small watersheds, the SCS applies the ratios from Fig. 16.11 to 6-hr map values from Figs. 16.6 and 16.7. Any PMP value from Fig. 16.6 for major designs is modified according to Fig. 16.12. This curve is used by the U.S. Bureau of Reclamation in areas west of the 105° meridian. For drainage areas up to 100 sq mi, the TVA recommends use of Fig. 16.13 for adjusting the expected rainfall over 1 sq mi (approximately equal to the point rainfall) to larger areas.

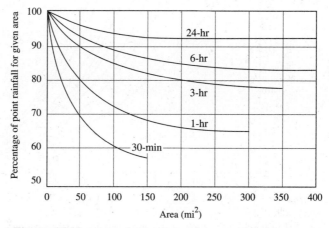

Figure 16.10 Area-depth curves for use with duration frequency values. (After Ref. 13.)

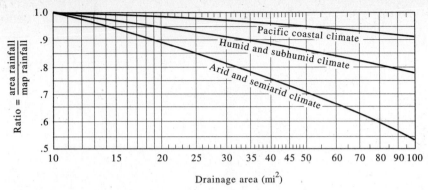

Figure 16.11 Rainfall ratios for 10–100 mi² for SCS dams. (After Ref. 12.)

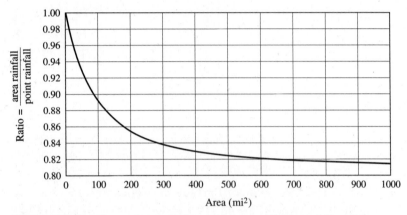

Figure 16.12 Conversion ratio from 6-hr point PMP rainfall to 6-hr area rainfall for area west of 105° meridian. (After Ref. 6.)

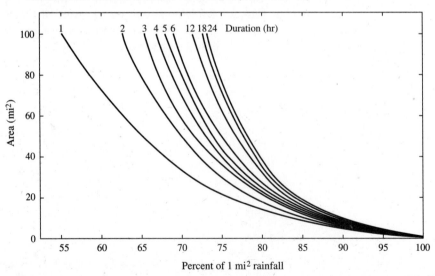

Figure 16.13 TVA graph for adjusting 1 sq mi rain depth to rain depth for basin areas up to 100 sq mi. (After Ref. 14.)

Time Distribution

After the storm depth and duration have been established, the designer must select a representative hyetograph. The choice will significantly affect the shape and peak value of the resulting runoff hydrograph. Any decision must be based on either the worst-possible storm pattern or on an analysis of recorded storm distribution patterns.

Huff[11] divided recorded storm distribution patterns from small midwestern watersheds into four equal probability groups from the most severe (first quartile) to the mildest (fourth quartile). The median curve for the first quartile storms is given in Fig. 16.14, which is used, for example, in the RRL and ILLUDAS simulation models of Chapter 25.

The two rainfall patterns normally investigated are first quartile and second quartile storms. A first quartile distribution has greater portions of rainfall occurring during the early minutes of the storm. Additional information on the most probable storm pattern over areas up to 400 mi^2 has been provided by Huff. Taken together, the two types make up 66 percent of the total number of storms registered on small watersheds in the Midwest. Each has an almost equal chance of occurrence.

Curves for each probability level drawn in Figs. 16.15 and 16.16 can be used to design for several levels of severity. A 90 percent level is the distribution occurring in 10 percent or less of the storms. Eighty percent of the total rainfall occurs in the first 20 percent of storm time for the 10 percent level in the first quartile storm. The passage of intense, prefrontal squall lines, typical of thunderstorms, will produce this particular rainfall distribution. On the other hand, the 90 percent level is more indicative of steady rain or a series of rain showers. The 50 percent or median curve is recommended for most applications.

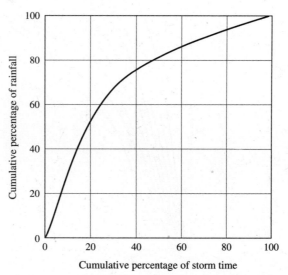

Figure 16.14   Time distribution of storm rainfall, median first quartile curve for point rainfall. (After Huff.[11])

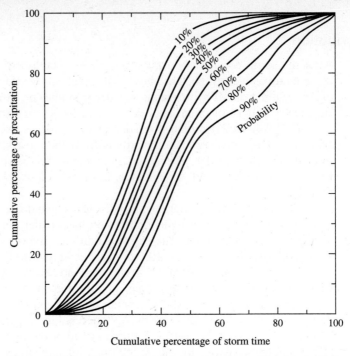

Figure 16.15 Time distribution of second quartile storms. (After Huff.[11])

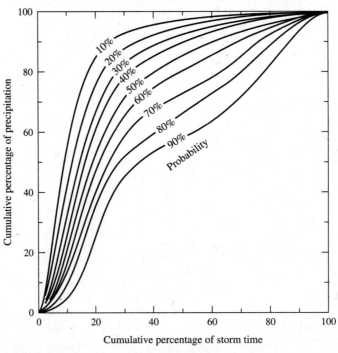

Figure 16.16 Time distribution of first quartile storms. (After Huff.[11])

Figure 16.17 shows the 10, 50, and 90 percent histograms for first quartile storms. Using these storm distributions permits the construction of hyetographs for the design rainfall.

Time distributions for critical storms for small-dam or other minor structure designs are usually assumed to be uniform. The SCS uses a uniform distribution for short-duration storms. Alternatively, Fig. 16.18 is the SCS distribution of the 6-hr storm used in developing emergency spillway and freeboard hydrographs.[12] This curve is very similar to the 50 percent (median) second quartile curve in Fig. 16.15.

The SCS also developed 24-hr storm distributions to represent the critical rainfall and runoff volume for peak discharges from watershed sizes normally studied by their engineers. A set of four rainfall distributions were developed, shown in Fig. 16.19a. They are applicable to the various regions shown in Fig. 16.19b and incorporate brief central periods of intense rain within a longer duration storm. Numerical values for plotting these curves can be found in SCS publications.[16]

Types I and IA represent the Pacific maritime climate with wet winters and dry summers. Type III represents Gulf of Mexico and Atlantic coastal areas where tropical storms bring large 24-hr rainfall amounts. Type II represents the rest of the

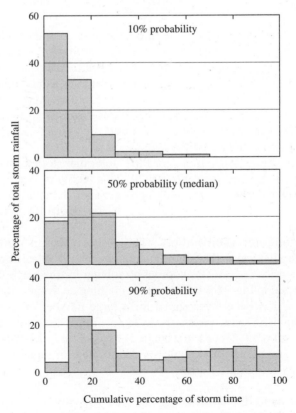

Figure 16.17 Selected histograms for first quartile storms. (After Huff.[11])

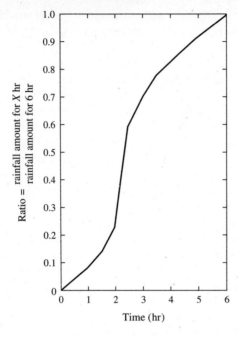

Figure 16.18   A 6-hr design storm distribution for SCS dam design. (After Ref. 12.)

country. For more precise information on boundaries in a state having more than one storm type, contact the respective SCS State Conservation Engineer.

The greatest peak flows from small basins are usually caused by intense, brief rains. These can occur as distinct events or as portions of a longer storm. The 24-hr storm duration is longer than needed to determine peaks from small watersheds but is appropriate for determining runoff volumes. In light of this, the SCS uses them to study peak flows, volumes of runoff, and direct runoff hydrographs from watersheds normally studied by the agency.

Time distributions for PMP and other storms used in major structure design can be constructed from Fig. 16.20. This family of curves is used by the U.S. Bureau of Reclamation[6] in three geographical zones shown in Fig. 16.6. The Corps of Engineers uses a distribution curve similar to Fig. 16.18 for 6-hr SPS analyses.

Triangular Distribution   The simplest design storm distribution is a triangular shape. Because the depth, P, and duration, D, of rain are already established, the peak intensity, i_{max}, is $2P/D$, found by solving for the height of the triangular hyetograph as shown in Fig. 16.21. The only remaining decision is the time to the peak, t_p. The ratio t_p/D has been investigated for a large number of storms at locations in California, Illinois, Massachusetts, New Jersey, and North Carolina. Values range from about 0.3 to 0.5.[17] Once the triangle is constructed, the intensities at regular intervals may be graphically or analytically determined for input to the rainfall-runoff model being used for design.

Blocked IDF Distributions   A frequently used procedure for developing a design storm distribution for short duration storms (up to about 2 hr) is to successively construct blocks of a design storm histogram by using the appropriate intensity-

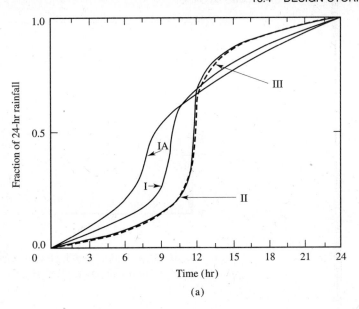

(a)

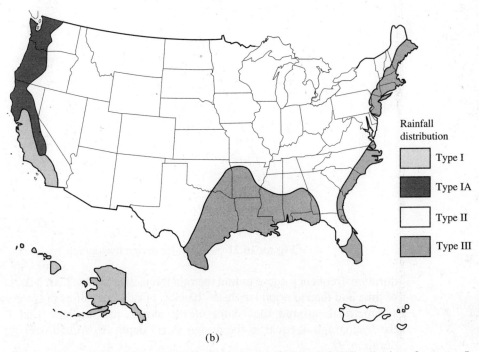

(b)

Figure 16.19   SCS 24-hr rainfall distributions: (a) 24-hr rainfall distributions for zones I, IA, II, and III and (b) approximate boundaries for SCS rainfall distributions. (After Ref. 16.)

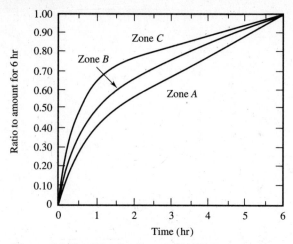

Figure 16.20 Distribution of 6-hr PMP for any area west of the 105° meridian. (After Ref. 6.)

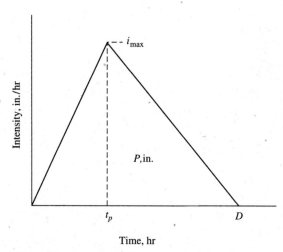

Figure 16.21 Triangular design hyetograph.

duration-frequency curve to find the rain intensities for Δt, $2 \Delta t$, $3 \Delta t$, etc., increments of time and then to organize these "blocks" of rain intensities in some pattern, usually symmetrical, around the center of the storm, making sure that the area under the hyetograph is equal to the design storm depth, P, spread over the design storm duration, D.

To apply the procedure, successive depths of equal-probability storms with durations of Δt, $2\Delta t$, $3\Delta t$, $4\Delta t$, etc., are determined from the IDF curve and tabulated. Next, any of a variety of procedures, such as the *alternating block method*, the *Chicago method*, or the *balanced method*, are available for distributing these blocks and assuring that the total rain depth equals P. Most assume that the highest intensity occurs in the middle of the storm, the second highest occurs next, and so on, working out in both directions from the center block. The balanced method, for example, assumes that a Δt-hr storm with intensity $i_{\Delta t}$ from the IDF curve could occur, with equal probability, during the middle of the D-hr design storm. This intensity is plotted as the middle block of the design storm hyetograph. Next, the rain depth for duration $2\Delta t$ is obtained from the IDF curve. Its distribution is assumed to be a two-bar histogram with the first half matching the intensity of the Δt-hr storm; the second half intensity is calculated by spreading the rest of the rain depth for the 2 Δt-hr duration uniformly over the second Δt interval. The process is repeated for rain depths for storms with durations of $3\Delta t$, $4\Delta t$, . . . , up to D. The goal is to develop a storm hyetograph such that a storm of any duration, centered at the middle of the blocked IDF hyetograph, will have a total rain depth matching the rain depth from the IDF curve for the given duration.

Areal Distribution

Precipitation depths can and do vary from point to point during a storm. Areal variation in design storm depth is normally disregarded except in major structure designs. The usual approach in major structure analysis is to select a design (usually elliptical) or historic (transposed) isohyetal pattern for the PMP or SPS depth and assign precipitation depths to the isohyets in a fashion that gives the desired average depth over the basin. The average depth is determined by the isohyetal method illustrated in Chapter 2.

Four major types of storm patterns are shown in Fig. 16.22 for areas up to 400 mi[2]. These were identified by Huff in his analysis of midwestern storm patterns.[11] The letters H and L represent areas with high and low precipitation depths, respectively. The typical isohyetal pattern for SPS storms has been established as generally elliptical in shape as shown in Fig. 16.23. This pattern is used by the Tennessee Valley

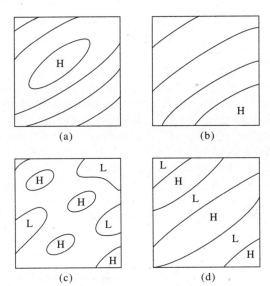

Figure 16.22 Major types of storm patterns: (a) closed elliptical, (b) open elliptical, (c) multicellular; and (d) banded. (After Huff.[11])

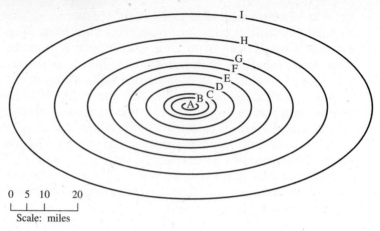

Figure 16.23 Generalized pattern storm.

Isohyet	Area enclosed (mi²)
A	11
B	45
C	114
D	279
E	546
F	903
G	1349
H	2508
I	4458

(After Ref.18.)

Authority (TVA)[18] for areas up to 3000 mi². Variations in the rainfall depth found in a standard project storm will diverge from a maximum at the storm center to a value considerably less than the average depth at the edges of the watershed boundaries. This variation can be determined and incorporated in the design storm.

A slightly modified isohyetal pattern for SPS storms is used by the Corps of Engineers[19] as shown in Fig. 16.24. The percentages shown for isohyets $A, B, \ldots, G$ are multiplied by the 96-hr SPS depth to give an elliptical pattern with the desired average depth. Similar maps for 24-, 48-, or 72-hr storms can be obtained simply by modifying the 96-hr percentages of Fig. 16.24. This is accomplished using the depth-area-duration curves in Fig. 16.24. For example, if a 24-hr storm is used, first note that the A isohyet of Fig. 16.24 encloses an area of 16 mi². From Fig. 16.25 the corresponding SPS percentage for a 24-hr storm is 116 percent rather than the 140 percent value used with a 96-hr storm. Therefore the pattern percentages vary with the selected design storm duration.

An additional aid for constructing design storm distributions over smaller midwestern[18] watersheds (up to 400 mi²) is presented in Table 16.10. The ratio of maximum point rainfall to mean rainfall over the basin is provided and can be used to estimate the maximum depth occurring at a storm center if the mean areal depth is

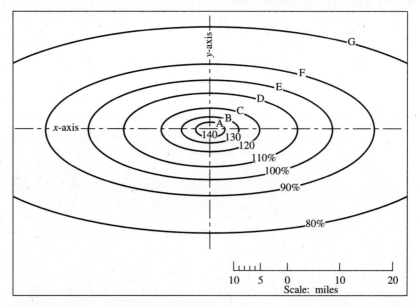

Figure 16.24 Generalized SPS isohyetal pattern for a 96-hr storm. The pattern may be oriented in any direction and may correspond to the depth-area relation represented by a 96-hr storm.

Isohyet	Area (mi²)
A	16
B	100
C	320
D	800
E	1800
F	3700
G	7100

(After Ref. 19.)

known. Ratios for 50-, 100-, and 200-mi² areas are equal to those in Table 16.10 multiplied by 0.91, 0.94, and 0.97, respectively. For uniform rainfall the 95 percent ratios of the table are recommended. With extreme variability the 5 percent ratio applies. The 50 percent ratios approximate average conditions.

16.5 CRITICAL EVENT METHODS

For some structures such as large dams, design of the flood-handling facilities by frequency-based methods is not appropriate. For any fixed frequency, a flood exceeding the design level is possible, and may have catastrophic consequences. Rather than selecting a frequency level for design, the design for large water control structures more often attempts to find the worst flood of history or to calculate the worst possible

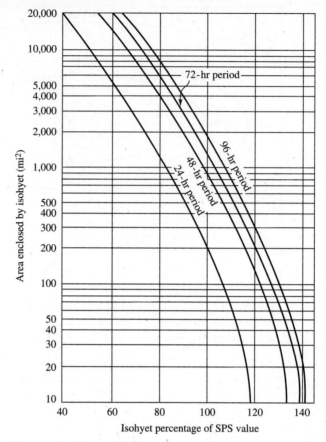

Figure 16.25 SPS depth-area-duration curves by 24-hr storm increments. (After Ref. 19.)

future event, and then design accordingly. These methods include the use of the probable maximum precipitation PMP, probable maximum flood PMF, record high storm depths, record high floods, multiples of frequency-based floods, and paleohydrology.

Probable Maximum Precipitation

Probable maximum precipitation depths for drainage basins in the United States are provided in the respective National Weather Service HMRs[20] identified in Fig. 16.2. The probable maximum storm is defined as the most severe storm considered reasonably possible to occur. The resulting probable maximum flood is customarily obtained by using unit hydrographs and rainfall estimates of the PMP prepared by the National Weather Service[21] (see Figs. 16.6 and 16.26).

TABLE 16.10 RATIO OF MAXIMUM POINT TO MEAN RAINFALL ON 400 mi^2

Rainfall period (hr)	Mean rainfall (in.)							
	0.5	1.0	1.5	2.0	2.5	3.0	4.0	5.0
	5% Probability level ratios (Storms with extreme variation in intensity)							
0.5	5.20	3.00	2.18	1.70	1.41	1.30	1.26	1.22
1	5.50	3.21	2.29	1.80	1.48	1.35	1.30	1.25
2	5.80	3.38	2.44	1.90	1.55	1.41	1.33	1.28
3	6.05	3.54	2.53	1.99	1.61	1.46	1.36	1.30
6		3.77	2.69	2.12	1.72	1.52	1.43	1.35
12		4.01	2.86	2.25	1.83	1.60	1.50	1.40
18		4.14	2.96	2.33	1.90	1.65	1.54	1.43
24		4.27	3.05	2.40	1.96	1.69	1.57	1.45
48		4.55	3.25	2.55	2.08	1.77	1.63	1.50
	50% Probability level ratios (Storms with average time distributions)							
0.5	2.66	2.02	1.57	1.32	1.22	1.16	1.14	1.12
1	3.03	2.15	1.65	1.39	1.27	1.20	1.18	1.16
2	3.46	2.29	1.75	1.46	1.32	1.24	1.21	1.19
3	3.77	2.42	1.85	1.52	1.38	1.28	1.23	1.22
6		2.59	1.98	1.63	1.43	1.33	1.28	1.26
12		2.78	2.12	1.75	1.50	1.39	1.32	1.30
18		2.89	2.20	1.81	1.57	1.43	1.35	1.32
24		3.00	2.28	1.87	1.60	1.47	1.38	1.33
48		3.17	2.44	1.99	1.68	1.53	1.46	1.38
	95% Probability level ratios (Storms with uniform intensities)							
0.5	2.38	1.53	1.28	1.18	1.16	1.13	1.11	1.10
1	2.75	1.72	1.38	1.23	1.20	1.17	1.15	1.14
2	3.15	1.90	1.47	1.28	1.24	1.20	1.18	1.16
3	3.46	2.02	1.53	1.33	1.27	1.22	1.20	1.18
6		2.24	1.67	1.43	1.31	1.27	1.24	1.21
12		2.50	1.78	1.50	1.38	1.31	1.28	1.25
18		2.67	1.89	1.53	1.41	1.33	1.30	1.27
24		2.77	1.92	1.58	1.43	1.35	1.32	1.29
48		3.07	2.04	1.64	1.47	1.40	1.36	1.33

Source: After Huff.[11]

A proposed method to estimate PMP advocated by Hershfield[21] suggests that the 24-hr PMP at a point be computed by the equation

$$PMP_{24} = \bar{P} + KS_n \qquad (16.1)$$

where PMP_{24} = the 24-hr probable maximum precipitation
$\bar{P}$ = the mean of the 24-hr annual maximums over the period of record
K = a constant equal to 15
S_n = the standard deviation of the 24-hr annual maximums

Adjustments to the value of $\bar{P}$ and S_n for the record length are noted by Hershfield. However, for appraisal purposes these adjustments probably will not significantly alter results more than 5–10 percent.

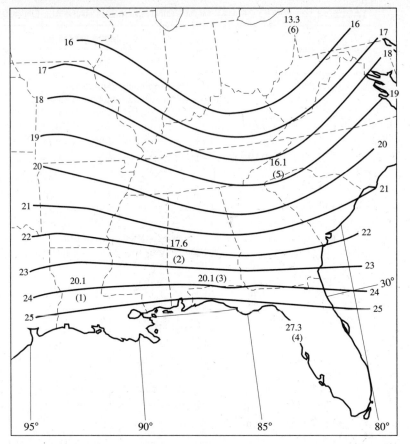

Figure 16.26 Twenty-four-hour 2000-mi² PMP (in.). (1) Alexandria, LA, June 13–17, 1886. (2) Eautaw, AL, April 15–18, 1900. (3) Elba, AL, March 11–16, 1929. (4) Yankeetown, FL, September 3–7, 1950. (5) Altapass, NC, July 13–17, 1916. (6) Jefferson, OH, September 10–13, 1878. (After Ref. 18.)

The U.S. Bureau of Reclamation underwent considerable evaluation of its design criteria for new dams and for safety evaluation of existing dams, following the Teton, Idaho, dam failure in 1976. The policy adopted for modification of existing dams is first to determine whether they will accommodate the peak discharge of the PMF without overtopping. In addition, the dam and appurtenant features must accommodate at least the first 80 percent of the PMF volume without failure. For embankment dams, failure is assumed to occur if overtopping levels are reached.

Recorded Extremes—Creager, and Crippen and Bue Envelope Curves

Where frequency-based methods of PMP/PMF studies are unwarranted, design for critical events can be based on the greatest recorded rain or flood flow for the location. Similarly, tables or curves of flood data can be developed to give the maximum floods of record in the region under study; see Creager flood envelope curves in Fig. 16.27.

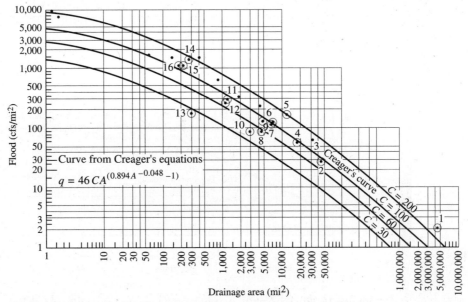

Figure 16.27 Creager envelope curves: ⊙ peak inflow for Harza Projects; · recorded unusual flood discharges. (1) Congo at Inga, Congo. (2) Tigris at Samarra, Iraq. (3) Caroni at Guri, Venezuela. (4) Tigris at Eski Mosul, Iraq. (5) Jhelum at Mangla, Pakistan. (6) Diyala at Derbendi Khan, Iraq. (7) Greater Zab at Bekhme, Iraq. (8) Suriname at Brokopondo, Suriname. (9) Lesser Zab at Doken Dam, Iraq. (10) Pearl River, U.S.A. (11) Cowlitz at Mayfield, U.S.A. (12) Cowlitz at Mossyrock, U.S.A. (13) Karadj, Iran. (14) Agno at Ambuklao, Philippines. (15) Angat, Philippines. (16). Tachien, Formosa. (*Note:* Curves taken from *Hydroelectric Handbook,* by Creager and Justin. New York: Wiley, 1950.)

In cases where estimates of PMP have not been made, volumes of rainfall to be expected can also be approximated from Creager rainfall envelope curves of the world record rainfalls as depicted in Fig. 16.28. Maximum flood flow data for 883 sites up to 25,900 sq km formed the basis for the Crippen and Bue envelope equation given by

$$q_p = 10^{[C_1 + C_2 \log A + C_3 (\log A)^2 + C_4 (\log A)^3]} \tag{16.2}$$

where q_p is the maximum flow (m³/sec), A is the drainage area (sq km) and the coefficients are from Table 16-11 using Figure 16.29.

Standard Project Storm

The standard project storm is another rainfall depth that is used in the design of large dams. This value is usually obtained from a survey of severe storms in the general vicinity of the drainage basin. The storm selected as the SPS may be oriented to produce the maximum amount of runoff for the SPF. Alternatively, severe storms experienced in meteorologically "similar" areas can be transposed over the study area.

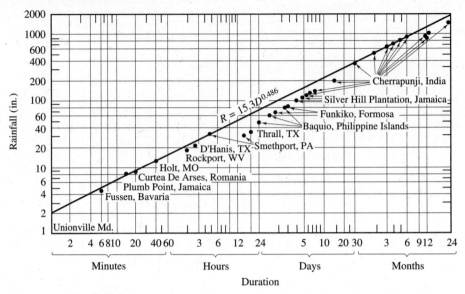

Figure 16.28 Creager curves of world's greatest rainfalls. (After Ref. 18.)

TABLE 16.11 COEFFICIENTS FOR CRIPPEN AND BUE PEAK DISCHARGE ENVELOPE CURVES

Fig. 16.29 Region	Upper limit (sq km)	Coefficients			
		C1	C2	C3	C4
1	26000	3.203865	.8049163	−.0394382	−.0029757
2	7800	3.470923	.7472908	−.0551780	−.0000965
3	26000	3.330746	.8443124	−.0642062	−.0021362
4	26000	3.258400	.8906783	−.0870959	.0022803
5	26000	3.726412	.7964721	−.0899000	.0022744
6	26000	3.500489	.9123848	−.1013380	.0049614
7	26000	3.326333	.8503960	−.0998747	.0042129
8	26000	3.236183	.9193289	−.0947436	.0029486
9	26000	3.503734	.8054884	−.0890172	.0018961
10	2600	3.314692	1.0386350	−.0597463	−.0042542
11	26000	3.231389	.8867450	−.1020535	.0045531
12	18100	3.596209	.8806263	−.0747598	.0000138
13	26000	3.461373	.8519276	−.1094456	.0058948
14	26000	3.073497	.6472710	−.0252243	−.0038285
15	50	3.451746	.9718339	−.0617496	−.0057110
16	2600	3.565536	.9699340	−.0649503	−.0034776
17	26000	3.389030	.9445212	−.0678131	−.0027647
Nationwide	2600	3.743026	.7918884	.0244991	−.0192899

Source: After Crippen, J. R., and C. D. Bue, "Maximum Flood flows in The Conterminous United States," U.S.G.S. Water Supply Paper 1887, 1977.

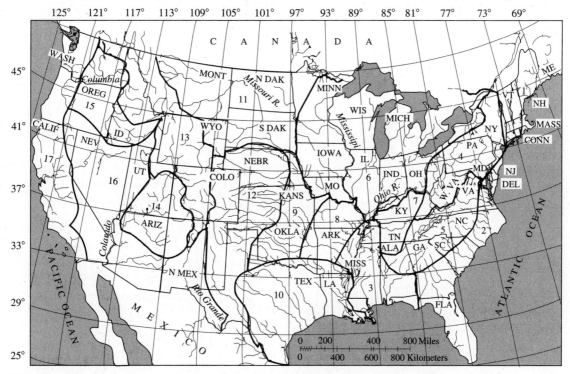

Figure 16.29 Map of the conterminous United States showing Crippen and Bue flood region boundaries. (*Source:* After U.S.G.S. Water Supply Paper 1887).

Transposition limits are based on climatological similarity, synoptic weather patterns, and a knowledge of atmospheric processes. A number of severe storms are selected from the storms of record. The depth–area–duration curves for these are developed and presented in graphical form similar to Fig. 16.25, usually for a particular point location within the basin (referred to as an index point). This position is used as a reference location for further calculations regarding the SPS.

The SPS differs from a PMP estimate and is patterned after a storm of record that causes the most severe rainfall depth–area–duration relation. Appropriate allowances should be made for inclusion of snowmelt in calculating design storm hydrographs from the standard project storm. Generally, the standard project storm rainfall is approximately 50 percent of the PMP. Records of the four or five largest storms should be critically examined to find a suitable composite for use in calculating the SPS. When these data are not available, a reasonable percentage of the PMP can be substituted. For example, when no potential for loss of life exists, the U.S. Bureau of Reclamation[5] divides the 6-hr PMP from Fig. 16.6 by the factors from Figs. 16.30 and 16.31 to obtain a reasonable design storm depth.

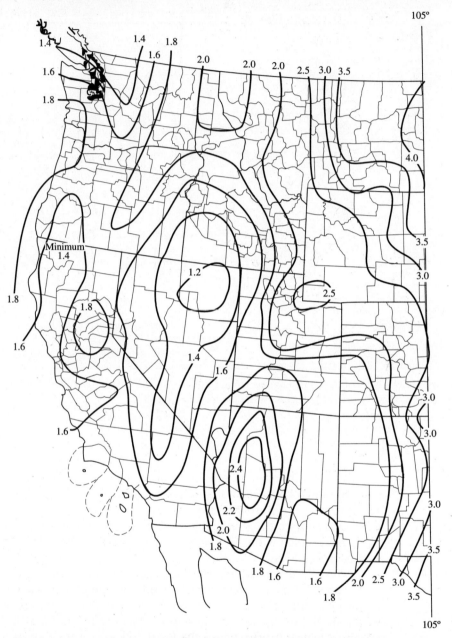

Figure 16.30   Ratio for determining the rainfall applicable for computing the inflow design flood less than the maximum probable for the area west of the 105° meridian. *Note:* Divide the probable maximum precipitation values by the indicated number. (After Ref. 6.)

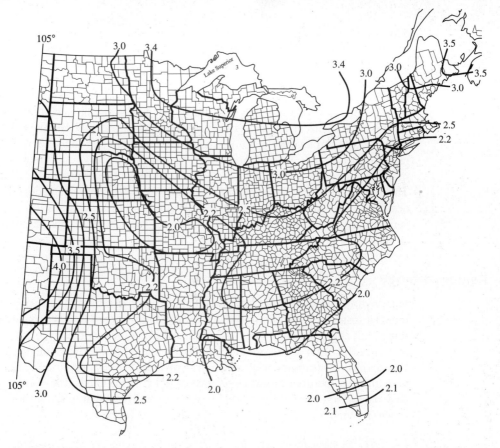

Figure 16.31 Ratio for determining the rainfall applicable for computing the inflow design flood less than maximum probable for the area east of the 105° meridian. *Note:* Divide the probable maximum precipitation by the indicated number. (After Ref. 6.)

Frequency-Based Flood for High-Hazard Dams

The federal agency standard for designing and evaluating spillway capacities for dams of high hazard potential has been the PMF. The use of such a conservative standard has been challenged, especially with concern over the need and costs to modify existing federally licensed dams so that they safely pass the probable maximum flood.

A committee of the National Research Council recommended the continued use of the PMF as the general design standard for proposed high-hazard dams.[22] For existing high-hazard dams the committee recommended that design should take into account estimated flood probabilities, expected project performances, and incremental damages that would result from dam failure for a range of floods up to and including the probable maximum flood.

This latter recommendation of a frequency-based approach signals a significant departure from a long tradition of "no-risk" design based on the PMF. The U.S. Army Corps of Engineers, a long-time supporter of the PMF standard, is investigating the potential of a frequency-based design approach.

Frequency curves (see Chapter 27) can be plotted and used in major and minor structure design for projects along streams for which lengthy records are available.[23] Most often, the location of the dam is not the gauging site, and stream routing techniques can effectively transfer the flood peaks. Regionalized flood frequency data may also be employed for small structures. For example, log–Pearson Type III estimates of peak flows for assigned return periods are readily found from Eq. 27.8 if the mean, standard deviation, and skew of logarithms of annual peaks can be estimated. The regional mean and standard deviation of logarithms often correlate well with drainage area and can be determined from nearby gauged stations. Because large samples are required for the determination of skew coefficients, regional skews such as those in Fig. 16.32 are preferred. Customarily, frequency-based floods are not a part of the design criteria for major structures.

Paleohydrology

Paleohydrology is the study of floods that occurred prior to the time of direct measurement or historical documentation. The rare, high-hazard floods are studied by examining high stage indicators or other indirect, preserved evidences of their occurrence. Such studies infer flood magnitudes from the visible effects of ancient floods on landscapes, vegetation (primarily marks on trees), soils, deposited sediments, or even man-made structures. If such evidence of paleofloods exists, the flood stage and peak flow magnitude can be estimated by direct or indirect hydraulic methods or hydraulic modeling techniques.[24] Types of indicators of paleofloods include:

1. Slackwater sand and silt deposits that result from deposition of suspended sediment transported during extreme floods.
2. Stratigraphic anomalies such as organic layers, tributary alluvium, mud cracks, bed forms such as ripples or dunes, or rounded gravel or cobble deposits.
3. Flood-transported litter such as organic materials, buried trees, compounds found only in buried soils, and archaeological debris or refuse.
4. High water marks from ancient floods such as rock or floating debris scars on tree trunks, floating debris deposits, sediment deposits in elevated caves, and scour lines or other disturbances along canyon or terrace walls.

Paleohydrologic studies can provide important supplemental design flood information, and can most often provide reasonableness checks for the upper limits of the maximum size of floods that have occurred in a river reach.[25, 26]

16.6 AIRPORT DRAINAGE DESIGN

Airport drainage is required to dispose of surface water and minimize the interruption of traffic into and out of the area. The total drainage system has several functions:[28] (1) collect and carry off surface water, (2) remove excess groundwater, (3) lower the

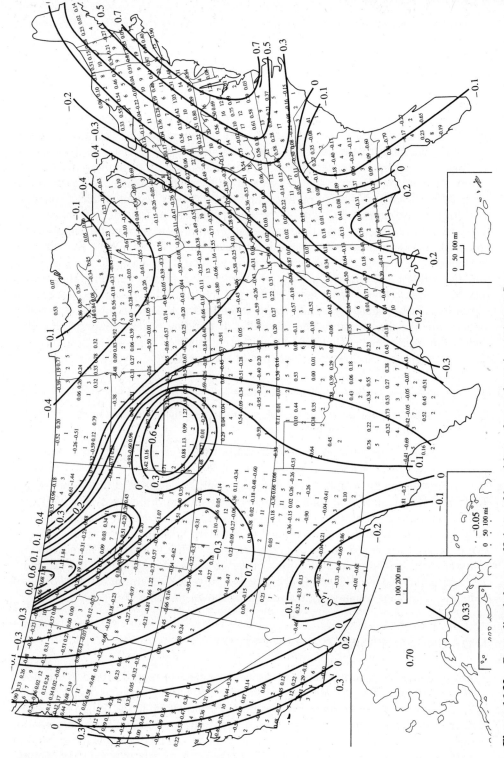

Figure 16.32 Generalized skew coefficients of logarithms of annual maximum streamflow: average skew coefficient by 1° quadrangles. The lower number in each quadrangle is the number of stream gauging stations for which the average shown above it was computed. (After Ref. 23.)

401

water table, and (4) protect all slopes from erosion. Only the problem of collecting and removing the surface water is discussed here.

The design of most airport drains relies on the rational method, although some efforts are being made to evaluate these systems by means of other techniques. Drainage calculations are usually based on collected rainfall data and a 1-ft contour topographic map of the proposed finished site. The entire system of drainage is outlined on this map with the proper identification of subareas, main and lateral storm drains, direction of flow, gradients, inlets, and surface channels. The final design is attained by calculating the most reasonable cost to provide satisfactory drainage.

A step-by-step procedure to design a portion of the surface drainage facilities of an airport is outlined in Example 16.1 as a straightforward application of the rational method. Each subarea size is outlined and a weighted C adopted, based on $C = 0.90$ for the pavement and $C = 0.30$ for the turf areas. The time of concentration in the system is composed of inlet time and duration of travel in the conduit. Figures 15.1 and 15.3 provide estimates for these times based on field research by the Corps of Engineers.[23] The design of the drainage system should be adequate to ensure that ponding will not be excessive in areas adjacent to the runways. The general criteria is that these areas should be at least 75 ft from the bordering pavement. To prevent saturation of the nearby ground, rough calculations to ensure adequate ponding volumes for the design rainfall are desirable. If ponding occurs, routing techniques such as those described in Chapter 13 facilitate calculations of ponding depth from the known storage and outflow characteristics of the system.

EXAMPLE 16.1 ——

Prepare a surface drainage design for the portion of an airfield shown in Fig. 16.33 for the 5-year frequency rainfall in Fig. 16.34.

Solution.   The solution is given in Table 16.12.   ■■

Computation of subarea sizes, values of weighted C, and inlet times are listed in Table 16.13. Calculation of final pipe sizes necessary to drain the system will vary with slope and type of pipe selected. The slope of the pipe is usually controlled by the outlet elevation that must be maintained to allow the system to drain freely. Designs in this example are based on use of a concrete pipe with $n = 0.015$. A nomograph for solution of the discharge as a function of the size and slope is provided in Fig. 16.35. The minimum velocity allowed in the pipe is 2.5 fps to prevent excessive settling of sediment. The final drainage system design is shown in Table 16.14.

16.7 DESIGN OF URBAN STORM DRAIN SYSTEMS

Storm drainage from urban areas starts as sheet flow on paved areas. These flows combine with flows from rooftops and hydrographs from pervious zones forming inflow to surface swales or open street gutters. For floods up to the design frequency, surface runoff enters storm sewers through street inlets or outlets from detention or retention storage areas. Most modern storm drain design incorporates these types of

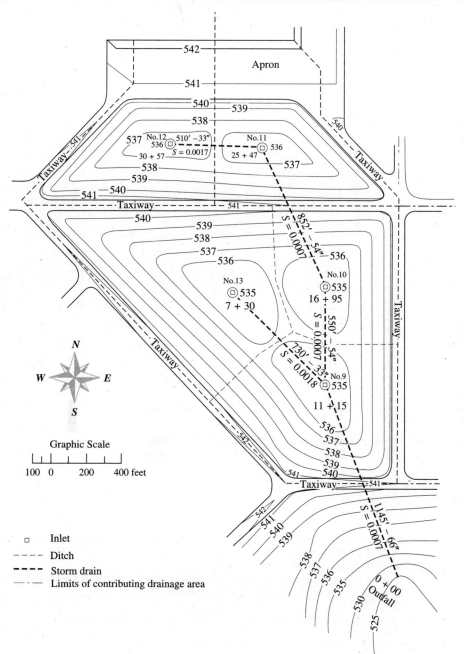

Figure 16.33 Portion of airport showing drainage design. (After Ref. 28.)

IDF curves for storms in vicinity of example site

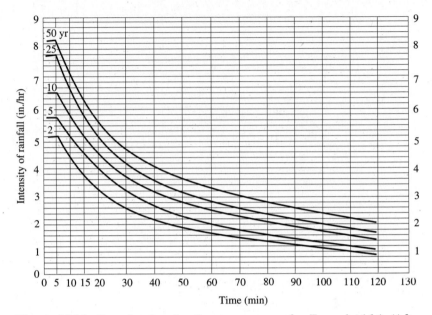

Figure 16.34 Intensity-duration-frequency curve for Example 16.1 (After Ref. 28.)

TABLE 16.12 DESCRIPTIONS OF THE COMPUTATIONS IN TABLE 16.14

Column	Comment
1	Inlet being investigated
2	Line segment
3	Length of line
4	Inlet time
5	Flow time in line obtained by dividing the length of the line by the velocity of the drain
6	Time of concentration
7	Weighted value of C
8	Rainfall intensity based on the time of concentration as duration and 5 years as the frequency of design
9	Acreage of subarea immediately tributary to the inlet
10	$Q = CIA$
11	Accumulated runoff that must be carried by the next line being computed
12	Velocity of flow through line A/A of the nomograph Fig. 15.3. Note that these correspond to full pipe flow. Velocities for actual flow rates can also be used
13	Pipe size from Fig. 16.35.
14	Slope of the line
15	Pipe capacity that must be larger than the estimated flow
16	Invert elevation of the pipeline
17	Pertinent remarks relative to the design

TABLE 16.13 DESIGN DATA FOR DRAINAGE EXAMPLE IN TABLE 16.14

Inlet number	Tributary area to inlets (acres)					Distance remote point to inlet (ft)			Time for overland flow (min)		
	Pavement	Turf	Both	Subtotal	C^a	Pavement	Turf	Total	Pavement	Turf	Total
12	4.78	9.91	14.69	14.69	0.49	100	790	890	4	37	41
11	5.48	9.24	14.72	29.41	0.53	90	750	840	4	36	40
10	1.02	10.95	11.97	41.38	0.35	65	565	630	3.5	31.3	34.8
13	1.99	19.51	21.50	21.50	0.35	110	1140	1250	4.3	44.3	48.6
9	1.46	14.59	16.05	78.93	0.35	85	612	697	3.9	32.4	36.3
Totals	14.73	64.20	78.93	78.93							

^{a}Weighted C based on $C = 0.9$ for pavement and $C = 0.3$ for turf.
Source: Reference 28.

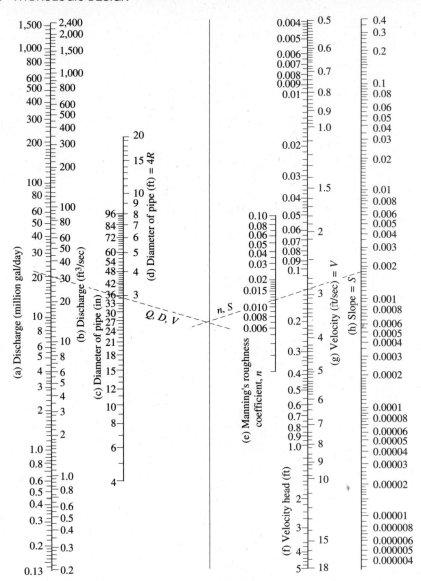

Figure 16.35 Flow in pipes (Manning's formula). (After Ref. 29.)

facilities to keep runoff from minor storms off streets and parking lots. For major storms, good system designs incorporate facilities to handle the overflows at storm sewer inlets (flows in excess of the design flows), as well as regional stormwater detention and retention facilities for 25-, 50-, or even 100-yr events. Figure 16.36 is from a child's book to illustrate the underground world; it provides an excellent vignette of urban storm drainage, water, electrical, and other systems.[27] The calculation of individual hydrographs to the inlet of each section is influenced by the infiltration capacity of the pervious areas, overland flow delays, depression storage,

TABLE 16.14 DRAINAGE SYSTEM DESIGN DATA

Inlet (1)	Line segment (2)	Length of segment (ft) (3)	Inlet time (min) (4)	Flow time (min) (5)	Time of concentration (min) (6)	Runoff coefficient C (7)	Rainfall intensity (in./hr) (8)	Tributary area A (acres) (9)	Runoff Q (cfs) (10)	Accumulated runoff (cfs) (11)	Velocity of drain (ft/sec)[a] (12)	Size of pipe (in.)[b] (13)	Slope of pipe (ft/ft) (14)	Capacity of pipe (cfs) (15)	Invert elevation (16)	Remarks (17)
12	12–11	510	41	2.7	41	0.49	2.40	14.69	17.28	17.28	3.18	33	0.0017	18.90	530.65	(n = 0.015)
11	11–10	852	40	5.0	43.7	0.53	2.31	14.72	18.02	35.30	2.84	54	0.0007	45.00	528.03	See note below
10	10–9	550	34.8	3.3	48.7	0.35	2.15	11.97	9.01	44.31	2.84	54	0.0007	45.00	527.44	See note below
13	13–9	730	48.6	3.7	48.6	0.35	2.16	21.50	16.25	16.25	3.27	33	0.0018	19.40	530.11	
9	9–out	1145	36.3	5.9	52.3	0.35	2.03	16.05	11.40	71.96	3.24	66	0.0007	77.00	526.05	
Out															525.25	

[a] Minimum velocity is 2.5 fps.

[b] Minimum pipe size is 12-in. diameter for maintenance purposes.

Note: The time of concentration for inlet 11 is 43.7 min(41 + 2.7 = 43.7), which is the most time-remote point for the inlet. Also, the time of concentration for inlet 10 is 48.8 min(41 + 2.7 + 5.0 = 48.7).

Source: Reference 28.

Figure 16.36   Children's artist rendering of urban underground system, including storm drains. From UNDERGROUND, Copyright © 1976 by David Macaulay. Reprinted with permission of Houghton Mifflin Co. All rights reserved.

detention in gutters, house drains, catchbasins, and the storm sewer systems, and interception in extensively landscaped locations.

Two items normally accounted for in urban storm drain design are:

1. *Infiltration*. The ability of the soil to infiltrate water depends on many characteristics of the soil as noted in Chapter 3. The range of values given in the following table is typical of various bare soils after 1 hr of continuous rainfall.

TYPICAL INFILTRATION RATES

Soil group	Infiltration (in./hr)
High (sandy, open-structured)	0.50–1.00
Intermediate (loam)	0.10–0.50
Low (clay, close-structured)	0.01–0.10

The influence of grass cover increases these values 3 to 7.5 times.

2. *Retention.* This is usually assumed to be 0.10 in. for pervious surfaces such as lawns and normal urban pervious surfaces.

Development of hydrologic parameters for design of storm sewer pipes, street gutters, or detention basins is by the rational method (or modified rational method—see Chapter 25) when peak flow rates and approximate hydrographs are adequate, or unit hydrograph and kinematic wave hydrograph synthesis methods when greater detail is needed. The latter usually involve use of public domain or vendor-developed stormwater design software. The hydrologic aspects of computerized hydrologic design tools are detailed in Chapter 25. In addition to the material presented in this text, descriptions of uses of the rational method, modified rational method, ILLUDAS, TR-55, SWMM, DR3M, and other tools in designing urban storm drainage facilities are addressed in numerous urban drainage design texts and handbooks. Additionally, many state departments of transportation or city and county engineer's offices have developed locally applicable drainage design manuals. As well, the American Society of Civil Engineers has developed a "model" drainage design manual for local adaptation, available by contacting ASCE in New York. Finally, the discussion of urban models in Chapter 25 includes a useful "shopper's guide" to urban drainage analysis and design software.

16.8 FLOODPLAIN ANALYSIS

Due to heavy monetary and other floodplain losses over the years, the federal government has been conducting studies of floodplains of the nation's waterways and methods of protecting life and property and preventing overdevelopment that causes increased water levels and more widespread flooding. Hydrology is a key ingredient in these studies for identifying potential flow rates, studying effects of dams, open channels, and other water control structures on hydrographs and determining volumes of floodwaters that will need to be safely stored and conveyed by the waterways and floodplains.

U.S. National Flood Insurance Program (NFIP)

In 1968, the U.S. Department of Housing and Urban Development (HUD), later called the Federal Emergency Management Agency (FEMA), initiated the NFIP to identify flood hazard areas and to provide occupants of floodplains with mapping of the flood-prone areas and access to low-cost flood insurance. The NFIP requires local governments to adopt and implement flood management programs that prevent developments in excess of national standards.

Since the inception of the National Flood Insurance Program, flood hazard areas have been mapped in over 18,000 communities in the United States. The program cost over $1.0 billion to complete and has since converted to a maintenance effort of updating and expanding the maps as developments occur. Each of these studies has required either approximate or detailed evaluation of peak flow rates for a range of recurrence intervals. The 100-year discharge, called the *base flood,* has been determined in all cases. The portion of the floodplain occupied by the base flood has been mapped, allowing communities to determine whether a property is in the 100-yr floodplain, and in many cases, what water surface elevation would be experienced at the property during the base flood.

Figure 16.37 illustrates the typical NFIP mapping and floodplain management procedure. Surveyed valley and channel cross-sections are used in determining the 100-yr flow depth, allowing the hydrologist to delineate the lateral extent of flooding during the 100-yr flood. Then a *floodway* width is generally determined as that portion of the floodplain that is reserved in order to discharge the 100-year flood without cumulatively increasing the water surface more than 1.0 ft. This procedure is illustrated in Fig. 16.38. The floodway is most often centered over the main stream channel, but can be offset or even split into several zones.

Development within the floodway is allowed only if compensated by relocating the floodway or mitigating the water surface increase due to the development. The *flood fringe* is that portion of the floodplain outside the floodway in which development is allowed, up to a point of full encroachment by buildings, roadbeds, berms, and so forth. As much as seven to ten percent of the total land area of the United States lies within the 100-year floodplain. The largest areas of floodplain are in the southern parts of the country, and the most populated floodplains are along the north Atlantic coast, the Great Lakes region, and in California.

The floodplain mapping effort produced a large amount of data and analyses useful to design hydrologists. The products of the program include:

1. The 10-, 50-, 100-, and 500-year frequency discharge for streams.
2. The 10-, 50-, 100-, and 500-year flood elevations for riverine, coastal, and lacustrine floodplains.

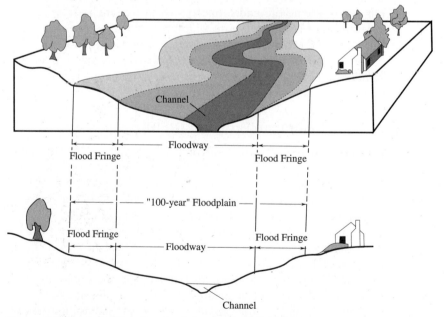

Figure 16.37 Definition sketch of floodplain delineations.

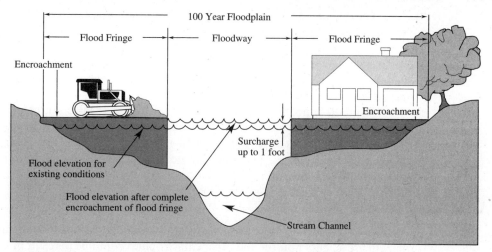

Figure 16.38 Procedure for determining the floodway width.

3. The 100- and 500-year mapped floodplain delineations at scales ranging from 1:4800 to 1:24,000.
4. The 100-year floodway data and mapping.
5. Coastal high hazard area mapping (areas subject to significant wave hazards).
6. Floodway flow velocities.
7. Insurance risk zones.

This information is provided in the form of three products:

1. *Flood Insurance Study Reports* provide general program and community background information, tabulated flood discharge data, tabulated floodway data including velocity, floodway width, and surcharge information, tabulated flood insurance zone data, and profiles of the 10-, 50-, 100-, and 500-year flood elevation versus stream distances for riverine flooding.
2. *Flood Insurance Rate Maps* (FIRM maps) provide delineations of the 100- and 500-year floodplains, base flood elevations, coastal high hazard areas, and insurance risk zones on a planimetric base at a scale between 1:4800 and 1:24,000.
3. *Flood Boundary Hazard Maps* provide delineations of the 100- and 500-year floodplains, locations of surveyed floodplain and channel cross sections used in hydraulic analyses, and delineations of the 100-year floodway on a planimetric or topographic base at a scale between 1:4800 and 1:24,000.

Hydrology for Floodplain Studies

Flood flow frequency estimates for gauged locations in NFIP studies are based on log-Pearson Type III (see Chapter 27) analysis of streamflow records. Annual peak flows and historical data are fitted according to procedures recommended by FEMA.

For ungauged locations, flood flow frequency estimates are developed through regional frequency analysis or through rainfall–runoff modeling. Equations published by the U.S. Geological Survey relate peak discharges of various frequencies to various drainage basin characteristics such as size, slope, elevation, shape, and land use. These equations are developed using multiple regression techniques (see Chapter 26) at gauged sites throughout the region.

Rainfall–runoff modeling techniques (Chapter 24) use synthetic rainfall hyetographs. Storm-event models, such as the Corps HEC-1 and SCS TR-20 packages, employ design storms of particular frequencies and then mathematically simulate the physical runoff process. The resulting peak discharge is assumed to have the same frequency as the rainfall.

U.S. Flood Hazards

Despite considerable effort and expenditure in identification of floodplains and flood hazard areas, dam failures and other catastrophies continue to result in severe damage to life, property, and the environment. Floods from hurricanes, intense rainstorms, and rapid snowmelt or structure failure have all contributed to the loss of life. A tabulation of events causing more than 100 deaths in the United States is provided in Table 16.15. As indicated, the majority are hurricane related, principally concentrated in the east-coast and Gulf of Mexico regions as shown in Fig. 16.39.

Monetary losses from floods are also large. Table 16.16 shows a number of past U.S. floods producing over $50 million in flood damages each, given in 1966 dollars. Collectively, these floods have produced flood damages in billions of dollars, distributed through the years as shown in Fig. 16.40.

The Federal Insurance Administration evaluated the floodplain areas in the communities mapped by FEMA. By using demographic and economic information, projections of future property at risk of flooding could be made. Results suggest that billions of investments in flood damageable property have occurred in floodplains. Table 16.17 lists the breakdown, by state, of estimated 1990 development value that will be in harm's way.

Dam Break Hazards

Table 16.18 lists the outflow rates, peak depth, and storage at the time of failure for 18 significant dam failures in the United States. The death rate for dam failures is related to the polpulation at risk (PAR). This term describes those people who would need to take some action to avoid the rising water.

Figures 16.41 and 16.42 show the losses as functions of PAR for low (less than 1.5 hr) and high (greater than 1.5 hr) advance warning times, respectively. The high-warning-time losses are significantly less. This strongly supports the incorporation of early warning and flood delay features in the design of any structure. Data used in plotting Figs. 16.41 and 16.42 are given in Table 16.19.

Table 16.20 provides a typical time line required for alerting downstream residents of a severe storm and potential dam failure. The values given are hypothetical, and apply to an assumed 15-mi reach between the storm center and the populated area.

TABLE 16.15   FLOODS CAUSING 100 OR MORE DEATHS IN THE
UNITED STATES

Year	Stream or place	Lives lost	Cause
1831	Barataria Isle, LA	150	Hurricane tidal flood
1856	Isle Derniere, LA	320	Hurricane tidal flood
1874	Connecticut River tributary	143	Dam failure
1875	Indianola, TX	176	Hurricane tidal flood
1886	Sabine, TX	150	Hurricane tidal flood
1889	Johnstown, PA	2100	Dam failure
1893	Vic. Grand Isle, LA	2000	Hurricane tidal flood
1899	Puerto Rico	3000	Hurricane tide and waves
1900	Galveston, TX	6000+	Hurricane tidal flood
1903	Central States	100+	Rainfall–river floods
1903	Heppner, OR	247	Rainfall–river floods
1906	Gulf coast	151	Hurricane tidal flood
1909	Gulf coast—New Orleans	700	Hurricane tidal flood
1913	Miami, Muskingham, and Ohio Rivers	467	Rainfall–river floods
1913	Brazos River, TX	177	Rainfall–river floods
1915	Louisiana and Texas Gulf coast	550	Hurricane tidal flood
1919	Louisiana and Texas Gulf coast	284	Hurricane tidal flood
1921	Upper Arkansas River	120	Rainfall–river flood
1926	Miami and Clewiston, FL	350	Hurricane tidal and river flood
1927	Lower Mississippi River	100+	Rainfall–river flood
1927	Vermont	120	Rainfall–river flood
1928	Puerto Rico	300	Hurricane tide and waves
1928	Lake Okeechobee, FL	2400	Hurricane tidal flood
1928	San Francisco, CA	350	Dam failure
1932	Puerto Rico	225	Hurricane tide and waves
1935	Florida Keys	400	Hurricane tidal flood
1935	Republican River, KS, NE	110	Rainfall–river flood
1936	Northeastern United States	107	Rainfall, snowmelt–river floods
1937	Ohio River	137	Rainfall–river flood
1938	New England coast	200	Hurricane tidal and river flood
1955	Northeastern United States	115	Hurricane rainfall–river floods
1957	West coast, LA	556	Hurricane tide and river floods
1960	Puerto Rico	107	Hurricane rainfall–river floods
1972	Buffalo Creek, WV	125	Dam disaster
1972	Rapid Creek, SD	245	Rainfall
1976	Big Thompson, CO	139	Rainfall

Figure 16.39 Number by state of major hurricanes in the United States, 1899–1989. (*Source:* National Hurricane Center, National Weather Service, NOAA.)

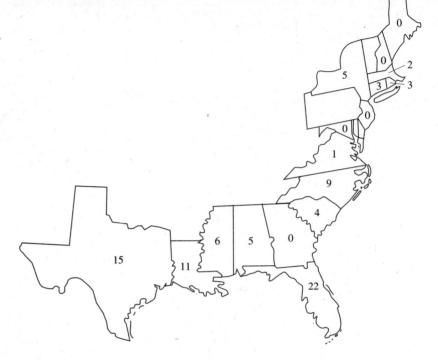

Figure 16.40 Average annual flood damages in the U.S., 1916–85. (*Source:* National Weather Service, NOAA.)

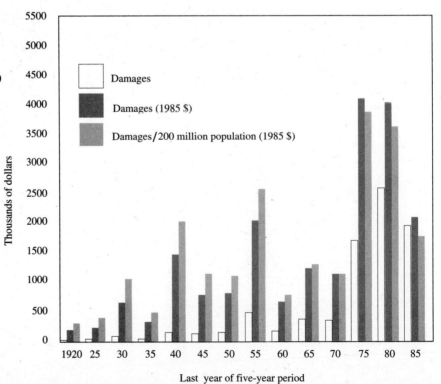

TABLE 16.16 FLOODS RESULTING IN DAMAGES EXCEEDING $50 MILLION IN THE UNITED STATES

Year	Stream or place	Damage ($ millions) Contemporary dollars	Damage ($ millions) 1966 Dollars	Cause
1844	Upper Mississippi River	N.A.[a]	1,161	Rainfall–river flood
1889	Johnstown, PA	20	84	Dam failure
1900	Galveston, TX	25	100	Hurricane tidal floods
1903	Passaic and Delaware Rivers	25	273	Rainfall and dam failure
1903	Missouri River basin	50	N.A.	Rainfall–river flood
1913	Ohio River basin	150	516	Rainfall–river flood
1913	Brazos and Colorado Rivers, TX	128	349	Hurricane rainfall–river floods
1921	Arkansas River	13	64	Rainfall–river flood
1926	Miami and Clewiston, FL	70	130	Hurricane tidal and river floods
1926	Illinois River	N.A.	51	Rainfall–river floods
1927	New England	50	178	Rainfall–river flood
1927	Lower Mississippi	284	N.A.	Rainfall–river flood
1928	Puerto Rico	50	90	Hurricane tide and waves
1935	Susquehanna and Delaware Rivers	36	185	Rainfall–river flood
1936	Northeastern United States	221	374	Rainfall–river flood
1936	Ohio River basin	150	371	Rainfall–snowmelt flood
1937	Ohio River basin	418	996	Rainfall–river flood
1938	New England streams	125	376	Hurricane tidal and river floods
1938	California streams	100	294	Rainfall–river floods
1942	Mid–Atlantic coastal streams	28	103	Rainfall–river floods
1943	Central states	172	N.A.	Rainfall–river floods
1944	South Florida	63	117	Hurricane tidal and river floods
1944	Missouri River basin	52	N.A.	Rainfall–river floods
1945	Hudson River basin	24	75	Rainfall–river floods
1945	South Florida	54	98	Hurricane tidal and river floods
1945	Ohio River basin	34	61	Rainfall–river flood
1947	South Florida	60	88	Hurricane tidal and river floods
1947	Missouri River basin	178	N.A.	Rainfall–river floods
1948	Columbia River basin	102	226	Rainfall–river floods
1950	San Joaquin River, CA	32	57	Rainfall–river floods
1951	Kansas River basin	883	N.A.	Rainfall–river floods
1952	Missouri River basin	180	N.A.	Snowmelt floods
1952	Upper Mississippi River	198	N.A.	Rainfall–river floods
1954	New England streams	180	216	Hurricane tidal floods
1955	Northeastern United States	684	879	Hurricane tidal and river floods
1955	California and Oregon streams	271	405	Rainfall–river floods
1957	Ohio River basin	65	72	Rainfall–river floods
1957	Texas rivers	144	188	Rainfall–river floods
1959	Ohio River basin	114	120	Rainfall–river floods
1960	South Florida	78	86	Hurricane tidal and river floods
1961	Texas coast	300	336	Hurricane tidal floods
1964	Florida	325	342	Hurricane tidal and river floods
1964	Ohio River basin	106	112	Rainfall–river floods
1964	California streams	173	183	Rainfall–river floods
1964	Columbia River—North Pacific	289	311	Rainfall–river floods
1965	South Florida	139	144	Hurricane tidal and river floods
1965	Upper Mississippi River	158	162	Rainfall–snowmelt river flood
1965	Platte River, CO, NE	191	N.A.	Rainfall–river floods
1965	Arkansas River, CO, KS	61	65	Rainfall–river floods
1965	New Orleans and vicinity	322	338	Hurricane tidal flood

[a]N.A. = not available.

Source: U.S. Water Resources Council, 1968.

TABLE 16.17 ESTIMATED PROPERTY VALUE AT RISK FROM FLOODING, RANKED IN DECREASING ORDER, BASED ON 1990 COSTS

Rank	State	Property value, X $1000
1	California	163,323,192
2	Florida	131,548,814
3	Texas	72,376,950
4	Louisiana	45,402,322
5	Arizona	45,094,183
6	New Jersey	38,945,265
7	New York	32,005,900
8	Illinois	26,880,755
9	Massachusetts	23,813,115
10	Pennsylvania	18,888,390
11	Virginia	17,441,420
12	Maryland	16,330,448
13	Washington	16,245,009
14	Ohio	15,273,147
15	Michigan	13,449,078
16	North Carolina	12,993,067
17	Wisconsin	12,181,725
18	Georgia	11,832,494
19	Connecticut	11,717,290
20	Missouri	11,654,861
21	Indiana	10,786,741
22	Minnesota	10,655,164
23	Nebraska	10,360,574
24	Oklahoma	9,501,778
25	Alabama	9,274,903
26	South Carolina	9,220,305
27	Tennessee	8,037,425
28	Colorado	7,137,757
29	Oregon	6,861,790
30	Mississippi	6,134,073
31	New Mexico	5,519,278
32	Kansas	5,279,194
33	Iowa	5,261,678
34	Rhode Island	4,312,117
35	Kentucky	4,170,637
36	North Dakota	3,924,872
37	Utah	3,812,936
38	Nevada	3,437,813
39	Arkansas	3,005,150
40	Delaware	2,954,467
41	Maine	2,416,322
42	West Virginia	2,098,262
43	New Hampshire	1,991,453
44	South Dakota	1,430,610
45	Idaho	1,391,498
46	Hawaii	1,323,905
47	Vermont	1,091,099
48	Wyoming	1,081,460
49	Montana	881,661
50	Alaska	647,818

Source: B. R. Mrazik, "Status of Floodplain Hazard Evaluation Under the National Flood Insurance Program," Federal Emergency Management Agency, Washington, DC, 1986.

TABLE 16.18 FLOOD PEAKS FROM ACTUAL DAM FAILURES

Name of dam, location, and year of failure	Peak outflow (ft^2/sec)	Depth (ft)	Storage at time of failure (acre-ft)
Johnstown, PA, 1889	250,000	72	11,500
Hatchtown, UT, 1914	179,000	52	13,600
Schaeffer, CO, 1921	164,000	90	3,600
Apishapa, CO, 1923	242,000	105	18,500
Castlewood, CO, 1933	126,000	70	5,000
Fred Burr, MT, 1948	23,100	33.5	610
Oros, Brazil, 1960	410,000	116	570,000
Little Deer Creek, UT, 1963	47,000	70	1,000
Baldwin Hills, CA, 1963	35,000	59	700
Swift, MT, 1964	881,000	157	34,300
Lower Two Medicine, MT, 1964	63,500	36	20,930
Hell Hole, CA, 1964	260,000	100	24,800
Buffalo Creek, WV, 1972	50,000	46	404
Teton, ID, 1976	2,300,000	275	251,700
Sandy Run, PA, 1977	15,300	28	46
Laurel Run, PA, 1977	37,000	42	310
Kelly Barnes, PA, 1977	24,000	34	630
Florida Power & Light, FL, 1979	110,000	28	110,000
Lawn Lake, CO, 1982	18,000	26	674

Source: U.S. Bureau of Reclamation.

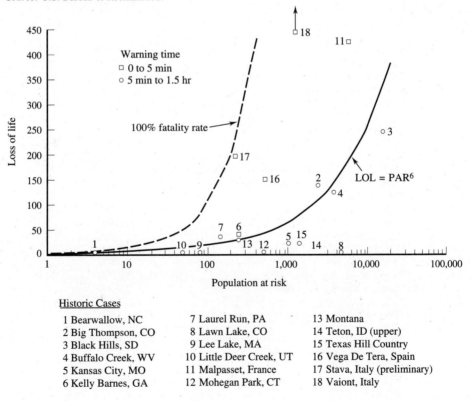

Historic Cases

1 Bearwallow, NC	7 Laurel Run, PA	13 Montana
2 Big Thompson, CO	8 Lawn Lake, CO	14 Teton, ID (upper)
3 Black Hills, SD	9 Lee Lake, MA	15 Texas Hill Country
4 Buffalo Creek, WV	10 Little Deer Creek, UT	16 Vega De Tera, Spain
5 Kansas City, MO	11 Malpasset, France	17 Stava, Italy (preliminary)
6 Kelly Barnes, GA	12 Mohegan Park, CT	18 Vaiont, Italy

Figure 16.41 Loss of life for several dam failures when warning time was less than 1.5 hr. (U.S. Bureau of Reclamation.)

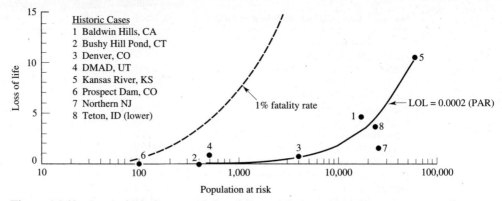

Figure 16.42 Loss of life for several dam failures when warning time was greater than 1.5 hr. (U.S. Bureau of Reclamation.)

TABLE 16.19 FLOODING CASE STUDIES

Location[a]	PAR	Loss of life	Hours of warning
Bearwallow, NC, 1976	4	4	0
Big Thompson, CO, 1976*	2,500	139	<1.0
Black Hills, SD, 1972**	17,000	245	<1.0
Buffalo Creek, WV, 1972	4,000	125	<1.0
Kansas City, MO, 1977*	1,000	25	<1.0
Kelly Barnes, PA, 1977	250	39	<0.5
Laurel Run, PA, 1977	150	40	0
Lawn Lake, CO, 1982	5,000	3	<1.5
Lee Lake, MA, 1968	80	2	<1.0
Little Deer Creek, UT, 1963	50	1	<1.0
Malpasset, France, 1959	6,000	421	0
Mohegan Park, CT, 1963	500	6	0
Montana, 1964 (Swift and Lower Two Medicine Dams)	250	27	<1.5
Teton, ID, 1976 (dam through Wilford)	2,000	7	<1.5
Texas Hill Country, 1978*	1,500	25	<1.5
Vega De Tera, Spain, 1959	500	150	0
Baldwin Hills, CA, 1963	16,500	5	1.5
Bushy Hill Pond, CT, 1982	400	0	2–3
Denver, CO, 1965*	3,000	1	3.0
DMAD, UT, 1983	500	1	1–12
Kansas River, KS, 1951*	58,000	11	>1.5
Northern New Jersey, 1984*	25,000	2	>2.0
Prospect Dam, CO, 1980	100	0	>5.0
Teton, ID, 1976 (Rexburg to American Falls)	23,000	4	>1.5

[a]One asterisk (*) indicates a flash flood; all others are dam failures. Two asterisks (**) indicate both flood and a dam failure contributed to loss of life.

Source: U.S. Bureau of Reclamation.

TABLE 16.20   EXAMPLE OF A TIME LINE FOR A HYPOTHETICAL THUNDERSTORM EVENT

Time (hr)	Event
0	Start of thunderstorm precipitation
1.0	NWS radar indicates heavy, sustained precipitation centered 15 mi (24 km) upstream
	Reservoir level = 1678 ft (dam creast = 1694); inflow = 500 ft³/sec; outflow = 500 ft³/sec
1.5	NWS issues flash flood warning; project office personnel leave for the dam
2.0	Project personnel arrive at the dam: reservoir level = 1678.5 ft; inflow = 15,000 ft³/sec; outflow = 5465 ft³/sec
	High-water warning issued by project personnel to sheriff, to warn and evacuate campers and anglers
2.5	Reservoir level = 1680.6 ft; inflow = 20,000 ft³/sec; outflow = 5800 ft³/sec
3.0	Reservoir level = 1688.0 ft; inflow = 12,000 ft³/sec; outflow = 7810 ft³/sec
	Warning issued to authorities by project personnel to evacuate areas threatened by a dam failure
3.4	Public warning initiated by authorities
3.5	Reservoir level = 1692 ft; inflow = 10,000 ft³/sec; outflow = 9474 ft³/sec
	Begin overtopping
3.7	Dam failure begins
4.5	Maximum dam failure: outflow = 300,000 ft³/sec is reached

Source: U.S. Bureau of Reclamation.

■ Summary

Much of the material presented in hydrology textbooks is introduced to define hydrologic principles that should be understood by engineers before attempting to design hydraulic structures. This chapter compiles these principles into a set of basic tenets for design of major and minor hydraulic structures.

For designs using streamflow records, the analyst need only select the basis of design (frequency-based, risk-based, critical-event design), and then couple the procedures described here with those in Chapters 15, 26, and 27. Design analyses that require transformation of precipitation data to streamflows can proceed by integrating procedures developed in Chapters 12, 13, and 14 and the design storm methods presented here into the models documented in Chapters 22, 23, 24, and 25. The design storm depth, duration, areal coverage, time distribution, and orientation described here are necessary to successfully complete the rainfall-runoff simulations.

PROBLEMS

16.1. Calculate a 100-year minor structure design storm hyetograph for Subarea I in the Oak Creek watershed, Fig. 24.8.

16.2. Calculate a design storm hyetograph for a 100-year frequency level that has a 6-hr duration. Assume that the storm occurred near Springfield, Illinois, and compute the rainfall excess distribution likely to occur only 10 percent of the time. Compare this with the median (50 percent) distribution.

16.3. Determine the U.S. Bureau of Reclamation minimum (50 mph) and normal (100 mph) freeboard values for a small dam with a smooth concrete upstream face. The fetch for the reservoir is 5 mi.

16.4. Needed is a design inflow hydrograph to a reservoir at a site where no records of streamflow are available. List the general steps you would take as a hydrologist in developing the entire design inflow hydrograph.

16.5. Calculate a major structure design storm for Subarea I in the Oak Creek watershed, Fig. 24.8. Assume that the storm will have a uniform areal distribution.

16.6. Construct a major structure design storm pattern for the entire Oak Creek watershed, Fig. 24.8, using Figs. 16.24 and 16.25. Describe how you would orient the pattern to create the most severe conditions at Point 8.

16.7. Show how the histograms in Fig. 16.17 were developed from the curves in Fig. 16.16. Also, construct a 50 percent probability histogram similar to the one in Fig. 16.17 for second quartile storms.

16.8. Given the following watershed data, route a storm hydrograph through a reservoir at the watershed outlet and plot the resulting outflow hydrograph. Show all the work in a neat and logical order. Area = 3.75 mi^2; length = 5.80 mi; elevation of the bottom of the spillway = 1160.0 ft; spillway width B = 500 ft; spillway coefficient C = 3.0. Required:

a. A 45-min unit hydrograph for this watershed by Gray's method.

b. The storm (inflow) hydrograph resulting from a storm that produced 6.0 in. of rainfall excess during the first 45-min period and 4.0 in. during the last 45-min interval.

c. Route the storm hydrograph through the reservoir. Assume that at the start of rainfall, the elevation of the water surface in the reervoir is 1158.0 ft.

d. Elevation for the top of the dam. Discuss.

e. Spillway design adequacy. Discuss.

Elevation (ft)	Distance (ft)	Area × 10^{-6} ft^2 (of respective contours)
1,110	0	0.0
1,120	1,400	0.85
1,140	4,600	3.75
1,160	8,000	10.80
1,180	11,500	25.00
1,200	14,700	
1,220	17,400	
1,240	20,000	
1,260	22,600	
1,280	24,800	
1,300	26,600	
1,320	28,000	
1,340	29,200	
1,360	30,000	
1,376	30,624	

16.9. Repeat Problem 16.8 for a Class b structure in an area selected by the instructor.

16.10. Repeat Problem 16.9 for a Class a structure.

16.11. The 2-hr unit hydrograph for a 5600-acre watershed is:

Time (hr)	0	1	2	3	4	5	6	7
Q (cfs)	0	400	1000	800	300	200	100	0

The local 10-year IDF curve is linear with the equation I = 5.6 − 0.2D, where I is rain intensity in inches per hour and D is the rain duration in hours. Use unit-hydrograph concepts to determine the direct runoff hydrograph for a 10-year design storm. (ϕ = 0.6 in./hr.)

16.12. Except for soil type, two drainage basins are otherwise identical. For the same storm, which basin would produce the most direct runoff, one containing SCS Group A soils or one containing Group D soils?

16.13. Compare the TVA storm depth adjustments in Table 16.9 with the USBR values from Table 16.8 for durations of 6, 12, and 24 hr. Discuss.

16.14. Compare the TVA storm depth adjustments in Fig. 16.13 with the SCS values from Fig. 16.11 for areas of 20, 40, 60, 80, and 100 sq mi. Discuss.

16.15. Using the IDF curves in Fig. 2.8 for Baltimore, MD, develop a triangular 10-yr design storm hyetograph for a 300-acre watershed having a 1.2-hr time of concentration. Use $t_p/D = 0.4$.

16.16. Repeat Problem 16.15 but develop a blocked IDF 10-yr design storm hyetograph using the balanced method. Discuss the differences between the resulting triangular and blocked hyetographs, especially noting the different effects they would have on runoff hydrographs.

For Problems 16.17–16.21, refer also to Chapter 26.

16.17. What return period (years) must an engineer use in the design of a bridge opening if it is acceptable to have only a 19 percent risk that flooding will occur at least once in two consecutive years?

16.18. A temporary spillway for a dam has a capacity of 3000 cfs. Any discharge equaling or exceeding 3000 cfs will result in damage or destruction of the spillway structure. If the frequency of a 3000-cfs flood is 0.10, what is the probability that the capacity of the spillway will be equaled or exceeded at least once in a 3-year period required for construction of a permanent spillway?

16.19. A building site near a small natural stream was flooded 20 times in the last 50 years.
   a. During the next 3 years, what is the risk to any construction on the site?
   b. What is the probability of flooding next year?
   c. What is the probability of flooding at least in one of the next 3 years?
   d. What is the probability of three consecutive years of safe construction?

16.20. The 75-year record of peak annual earthquake magnitudes at Horseshoe, Arizona, reveals that the values follow an extreme-value distribution with a mean of 5.2 and a standard deviation of 2.0 on the Richter scale. Determine the probability of completing the construction of a nuclear power plant without having an earthquake magnitude exceeding 12.0 during the 10-year construction period.

16.21. Historical records of power failures in a buried power cable near a stream reveal that power failure occurs for a variety of reasons on the average of once every 5 years. Records also show that whenever the stream floods the chances of a power failure are increased to 40 percent. The river reaches flood stage once every 10 years.
   a. Demonstrate that a flood in the stream and a power failure in the cable are dependent events.
   b. Are the two events mutually exclusive? Why?
   c. Find the probability of the joint occurrence of a flood and a power failure in any year.
   d. Find the probability of the occurrence of either event in any year.
   e. Find the probability of a power failure in both of two consecutive years.

REFERENCES

1. U.S. Bureau of Reclamation, *Flood Hydrology Manual, A Water Resources Technical Publication,* U.S. Department of the Interior, Denver Office, 1989.
2. U.S. Soil Conservation Service, "Earth Dams," Engineering Memorandum 27 (rev.), Sec. A, Washington D.C., March 1965.
3. F. F. Snyder, "Hydrology of Spillway Design: Large Structures—Adequate Data," *Proc. ASCE J. Hyd. Div.* **90**(HY3), 239–259(May 1964).
4. *Design of Small Dams,* Bureau of Reclamation, U.S. Department of the Interior. Washington, D.C.: U.S. Government Printing Office, 1965.
5. U.S. Soil Conservation Service, "Earth Dams and Reservoirs," *Technical Release No. 60,* U.S. Department of Agriculture, Oct. 1985.
6. H. O. Ogrosky, "Hydrology of Spillway Design: Small Structures—Limited Data," *Proc. ASCE J. Hyd. Div.* **90**(HY3), 295–310(May 1964).
7. Federal Emergency Management Agency, Proceedings, Probable Maximum Precipitation and Probable Maximum Flood Workshop, Proceedings, FEMA Workshop, Berkeley Springs, W. Virginia, May, 1990.
8. "Generalized Estimates of Maximum Probable Precipitation Over the United States East of the 105th Meridian," U.S. Weather Bureau Hydrometeorological Report No. 23, June 1947.
9. "Generalized Estimates of Probable Maximum Precipitation for the United States West of the 105th Meridian for Areas to 400 Square Miles and Durations to 24 Hours," U.S. Weather Bureau Tech. Paper 38, 1960.
10. "Two-to-Ten-Day Precipitation for Return Periods of 2 to 100 Years in the Contiguous United States," U.S. Department of Commerce, Tech. Paper No. 49, 1964.
11. F. A. Huff, "Time Distribution of Rainfall in Heavy Storms," *Water Resources Res.* **3**(4), 1007–1019(1967).
12. "Hydrology," Suppl. A to Sec. 4, *Engineering Handbook,* U.S. Department of Agriculture, Soil Conservation Service, 1968.
13. "Rainfall Frequency Atlas of the United States for Durations from 30 Minutes to 24 Hours and Returns Periods from 1 to 100 Years," U.S. Weather Bureau Tech. Paper 40, 1963.
14. National Oceanic Atmospheric Administration, *Hydrometeorological Report No. 56,* "Probable Maximum and TVA Precipitation Estimates with Areal Distributions for Tennessee River Drainages Less Than 3000 Square Miles in Area," 1986.
15. David Kibler, (ed.), *Water Resources Monograph 7, Urban Storm Water Hydrology,* American Geophysical Union, 1982.
16. Soil Conservation Service, "Urban Hydrology for Small Watersheds," U.S. Department of Agriculture, Technical Release 55, June 1986.
17. B. C. Yen, and V. T. Chow, "Design Hyetographs for Small Drainage Structures," Proceedings, ASCE, JHD, V. 106, HY6, 1980.
18. "Probable Maximum TVA Precipitation for Tennessee River Basins Up to 3000 Square Miles in Area and Durations to 72 Hours," U.S. Department of Commerce, ESSA, Weather Bureau Hydrometeorological Rep. No. 45, 1969.
19. "Standard Project Flood Determinations," Civil Engineer Bulletin No. 52-8, EM 1110-2-1411, U.S. Department of the Army, Office of the Chief of Engineers, Washington, D.C., revised 1965, pp. 1–19.
20. U.S. National Weather Service, "Hydrometeorological Reports," Washington, D.C..
21. D. M. Hershfield, "Estimating the Probable Maximum Precipitation," *Proc. ASCE J. Hyd. Div.* **87**(HY5), 99–116(1961).

22. National Research Council, "Safety of Dams—Floods and Earthquake Criteria," Committee on Safety Criteria for Dams, National Academy Press, 1985.

23. "Guidelines for Determining Flood Flow Frequencies," U.S. Water Resources Council, Hydrology Committee Bulletin 17B, Revised, Washington, D.C., Sept., 1981.

24. V. R. Baker, "Paleoflood Hydrology and Extraordinary Flood Events," J. Hydrology, v. 96, 1987.

25. R. M. Jarrett, "Paleohydrologic techniques used to define the spatial occurrence of floods," in *Geomorphology, 3:* Elsevier Science Publishers B. V., Amsterdam, 1990.

26. V. R. Baker, "Magnitude and Frequency of Palaeofloods," *Floods: Hydrological, Sedimentological and Geomorphological Implications,* John Wiley & Sons Ltd, 1989.

27. D. Macaulay, *Underground,* Houghton Mifflin Company, Boston, 1976.

28. "Airport Drainage," Federal Aviation Agency, Advisory Circular, AC No. AC150/5320-5A. Washington, D.C.: U.S. Government Printing Office, 1966.

29. "Design and Construction of Sanitary Storm Sewers," ASCE Manuals and Reports on Engineering Practice, No. 37.

PART FOUR

GROUNDWATER HYDROLOGY

Chapter 17

Groundwater, Soils, and Geology

■ **Prologue**

The purpose of this chapter is to:

- Introduce the subject of *groundwater hydrology*.
- Describe how soil and geologic properties affect groundwater storage and movement.
- Define types of aquifers.
- Indicate the relationship between surface water and groundwater systems.

17.1 INTRODUCTION

The amount of water stored below ground in the United States exceeds by a significant amount all aboveground storage in streams, rivers, reservoirs, and lakes including the Great Lakes.[1] This enormous reservoir sustains streamflow during precipitation-free periods and constitutes the major source of fresh water for many arid localities. Figure 17.1 indicates the distribution and nature of primary groundwater areas of the United States.

The quantification of the volume and rate of flow of groundwater in various regions is a difficult task because volumes and flow rates are determined to a considerable extent by the geology of the region. The character and arrangement of rocks and soils are important factors, and these are often highly variable within a groundwater reservoir. An additional difficulty is the inability to measure directly many critical geologic and hydraulic reservoir characteristics.

The difficulties associated with determining the quantitative aspects of groundwater resources are paled, however, by those associated with their quality. And in many localities, it is the quality dimension that is most critical. More and more evidence is being uncovered indicating that many aquifers have been contaminated, at least locally, by the improper disposal of chemical and other wastes, by leachates from solid waste disposal sites, and from infiltrating storm water discharges. As a result, protection of groundwater quality has become a national policy, and in many

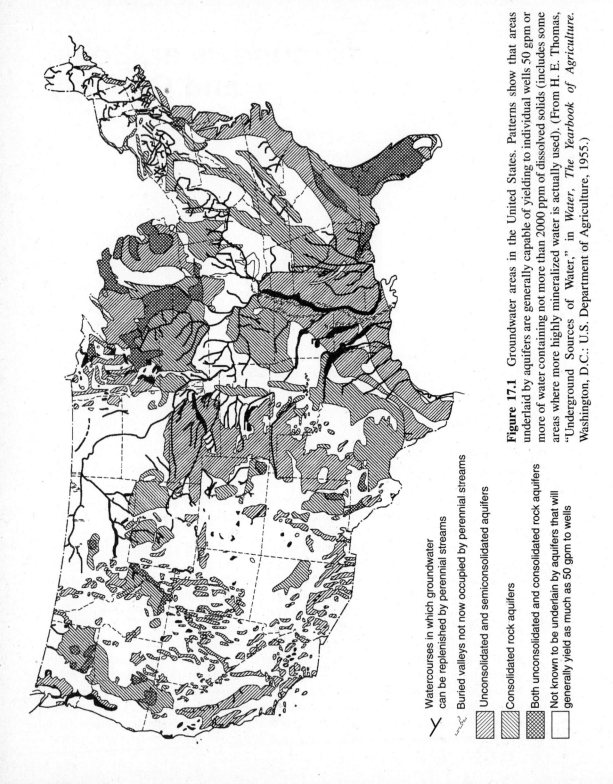

Figure 17.1 Groundwater areas in the United States. Patterns show that areas underlaid by aquifers are generally capable of yielding to individual wells 50 gpm or more of water containing not more than 2000 ppm of dissolved solids (includes some areas where more highly mineralized water is actually used). (From H. E. Thomas, "Underground Sources of Water," in *Water, The Yearbook of Agriculture*, Washington, D.C.: U.S. Department of Agriculture, 1955.)

Watercourses in which groundwater can be replenished by perennial streams

Buried valleys not now occupied by perennial streams

Unconsolidated and semiconsolidated aquifers

Consolidated rock aquifers

Both unconsolidated and consolidated rock aquifers

Not known to be underlain by aquifers that will generally yield as much as 50 gpm to wells locations it has become more important than overdrafts of groundwater supplies. Today, the hydrologist must be concerned with both the quality and quantity aspects of groundwater. Furthermore, there is emerging an increasing specialization in groundwater quality modeling. This latter type of modeling is generally beyond the scope of this text but information on this topic may be found in Refs. 2–6.

17.2   GROUNDWATER FLOW—GENERAL PROPERTIES

Understanding the movement of groundwater requires a knowledge of the time and space dependencies of the flow, the nature of the porous medium and fluid, and the boundaries of the flow system.

Groundwater flows are usually three-dimensional. Unfortunately, the solution of such problems by analytic methods is complex unless the system is symmetric.[7,8] In other cases, space dependency in one of the coordinate directions may be so slight that assumption of two-dimensional flow is satisfactory. Many problems of practical importance fall into this class. Sometimes one-dimensional flow can be assumed, thus further simplifying the solution.

Fluid properties such as velocity, pressure, temperature, density, and viscosity often vary in time and space. When time dependency occurs, the issue is termed an *unsteady flow problem* and solutions are usually difficult. On the other hand, situations where space dependency alone exists are *steady flow problems*. Only homogeneous (single-phase) fluids are considered here. For a discussion of multiple phase flow, Refs. 5 and 8 are recommended.

Boundaries to groundwater flow systems may be fixed geologic structures or free water surfaces that are dependent for their position on the state of the flow. A hydrologist must be able to define these boundaries mathematically if the groundwater flow problems are to be solved.

Porous media through which groundwaters flow may be classified as isotropic, anisotropic, heterogeneous, homogeneous, or several possible combinations of these. An *isotropic* medium has uniform properties in all directions from a given point. *Anisotropic* media have one or more properties that depend on a given direction. For example, permeability of the medium might be greater along a horizontal plane than along a vertical one. *Heterogeneous* media have nonuniform properties of anisotropy or isotropy, while *homogeneous* media are uniform in their characteristics.

17.3   SUBSURFACE DISTRIBUTION OF WATER

Groundwater distribution may generally be categorized into zones of aeration and saturation. The saturated zone is one in which all voids are filled with water under hydrostatic pressure. In the zone of aeration, the interstices are filled partly with air, partly with water. The saturated zone is commonly called the *ground water zone*. The zone of aeration may ideally be subdivided into several subzones. Todd classifies these as follows:

1. *Soil water zone.* A soil water zone begins at the ground surface and extends downward through the major root band. Its total depth is variable and dependent on soil type and vegetation. The zone is unsaturated except during periods of heavy infiltration. Three categories of water classification may be encountered in this region: hygroscopic water, which is adsorbed from the air; capillary water, held by surface tension; and gravitational water, which is excess soil water draining through the soil.

2. *Intermediate zone.* This belt extends from the bottom of the soil-water zone to the top of the capillary fringe and may change from nonexistence to several hundred feet in thickness. The zone is essentially a connecting link between a near-ground surface region and the near-water-table region through which infiltrating fluids must pass.

3. *Capillary zone.* A capillary zone extends from the water table (Fig. 17.2) to a height determined by the capillary rise that can be generated in the soil. The capillary band thickness is a function of soil texture and may fluctuate not only from region to region but also within a local area.

4. *Saturated zone.* In the saturated zone, groundwater fills the pore spaces completely and porosity is therefore a direct measure of storage volume. Part of this water (specific retention) cannot be removed by pumping or drainage because of molecular and surface tension forces. Specific retention is the ratio of volume of water retained against gravity drainage to gross volume of the soil.

Water that can be drained from a soil by gravity is known as the *specific yield.* It is expressed as the ratio of the volume of water that can be drained by gravity to the gross volume of the soil. Values of specific yield depend on the soil particle size, shape and distribution of pores, and degree of compaction of the soil. Average values for alluvial aquifers range from 10 to 20 percent. Meinzer and others have developed procedures for determining the specific yield.[12]

17.4 GEOLOGIC CONSIDERATIONS

The determination of groundwater volumes and flow rates requires a thorough knowledge of the geology of a groundwater basin. In bedrock areas, hydrologic characteristics of the rocks, that is, their location, size, orientation, and ability to store or transmit water, must be known. In unconsolidated rock areas, basins often contain hundreds to thousands of feet of semiconsolidated to unconsolidated fill deposits that originated from the erosion of headwater areas. Such fills often contain extensive quantities of stored water. The characteristics of these basin fills must be evaluated.

A knowledge of the distribution and nature of geohydrologic units such as *aquifers, aquifuges,* and *aquicludes* is essential to proper planning for development or management of groundwater supplies. In addition, bedrock basin boundaries must be located and an evaluation made of their leakage characteristics.

An aquifer is a water-bearing stratum or formation that is capable of transmitting water in quantities sufficient to permit development. Aquifers may be considered

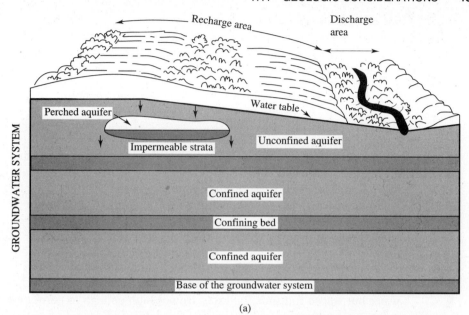

Figure 17.2 Definition sketches of groundwater systems and mechanisms for recharge and withdrawal: (a) aquifer notation[10] and (b) components of the hydrologic cycle affecting groundwater.[11]

as falling into two categories, confined and unconfined, depending on whether a water table or free surface exists under atmospheric pressure. Storage volume within an aquifer is changed whenever water is recharged to, or discharged from, an aquifer. In the case of an unconfined aquifer this may easily be determined as

$$\Delta S = S_y \, \Delta V \qquad (17.1)$$

where ΔS = the change in storage volume
        S_y = the average specific yield of the aquifer
       ΔV = the volume of the aquifer lying between the original water table and the
           water table at some later specific time

For saturated, confined aquifers, pressure changes produce only slight modifications in the storage volume. In this case, the weight of the overburden is supported partly by hydrostatic pressure and somewhat by solid material in the aquifer. When hydrostatic pressure in a confined aquifer is reduced by pumping or other means, the load on the aquifer increases, causing its compression, with the result that some water is forced out. Decreasing the hydrostatic pressure also causes a small expansion, which in turn produces an additional release of water. For confined aquifers, water yield is expressed in terms of a *storage coefficient* S_c, defined as the volume of water an aquifer takes in or releases per unit surface area of aquifer per unit change in head normal to the surface. Figure 17.2 illustrates the classifications of aquifers.

In addition to water-bearing strata exhibiting satisfactory rates of yield, there are also non-water-bearing and impermeable strata that may contain large quantities of water but whose transmission rates are not high enough to permit effective development. An aquifuge is a formation impermeable and devoid of water; an aquiclude is an impervious stratum.

In the following three chapters, the mechanics of groundwater flow and the elements of groundwater modeling will be introduced. The techniques presented all depend on a knowledge of the physical system to be modeled. Before a numerical model can be developed, a conceptual framework must be devised. This framework must take into account the region's topography and geology; the types of aquifers, their thickness, lateral extent, boundaries, lithological variations, and characteristics; the nature and extent of recharge and discharge areas, their rates of discharge and recharge; and the elevation of the water table.[7]

Topography

To understand how a groundwater system operates, it is essential to know something about the region's surface. A topographic map should be compiled showing all surface water bodies, including streams, lakes, and artificial channels and/or ponds, as well as land surface contours. Furthermore, an inventory of pumping wells, observation wells, and exploration wells should be made for purposes such as identifying types of soils and rocks, pinpointing discharge locations and rates, and determining water table elevations.

Subsurface Geology

The geologic structure of a groundwater basin governs the occurrence and movement of the groundwater within it. Specifically, the number and types of water-bearing formations, their vertical dimensions, interconnections, hydraulic properties, and outcrop patterns must be understood before the system can be analyzed.[6] Once the subsurface conditions have been identified, contour maps of the upper and lower boundaries of aquifers, water table contour maps, and maps of aquifer characteristics can be prepared. Well-drillers logs, experimental test wells, and other geophysical exploration methods can be used to obtain the needed geologic data.[5-9, 13, 14]

17.5 FLUCTUATIONS IN GROUNDWATER LEVEL

Any circumstance that alters the pressure imposed on underground water will also cause a variation in the groundwater level. Seasonal factors, changes in stream and river stages, evapotranspiration, atmospheric pressure changes, winds, tides, external loads, various forms of withdrawal and recharge, and earthquakes all may produce fluctuations in the water table level or piezometric surface, depending on whether the aquifer is free or confined.[9] It is important that an engineer concerned with the development and utilization of groundwater supplies be aware of these factors. The engineer should also be able to evaluate their importance relative to operation of a specific groundwater basin.

17.6 GROUNDWATER–SURFACE WATER RELATIONS

Notwithstanding that water resource development has often been based on the predominant use of either surface water or groundwater, it must be emphasized that these two components of the total water resource are interdependent. Changes in one component can have far-reaching effects on the other. Coordinated development and management of the combined resource are critical. Linkage between surface waters and groundwaters should be investigated in all regional studies so that adverse effects can be noted if they exist and opportunities for joint management understood.

In Part Three it was shown how surface stream flows are sustained by the groundwater resource, and it was also pointed out that groundwaters are replenished by infiltration derived from precipitation on the earth's surface.

Underground reservoirs are often extensive and can serve to store water for a multitude of uses. If withdrawals from these reservoirs consistently exceed recharge, *mining* occurs and ultimate depletion of the resource results. By properly coordinating the use of surface water and groundwater supplies, optimum regional water resource development seems most likely to be assured. Several studies directed toward this coordinated use have been initiated.[15, 16]

■ Summary

The importance of groundwater to the health and well-being of humans is well documented. Groundwater is a major source of freshwater for public consumption, industrial uses, and the irrigation of crops. For example, more than half of the freshwater used in Florida for all purposes comes from groundwater sources, and about 90 percent of that state's population depends on groundwater for its potable water supply. The need to husband this resource is clear. Quantity and quality dimensions are both important.

Groundwater protection and management practices must be based on an understanding of groundwater sources, the manner in which groundwater is distributed below the earth's surface, geologic, topographic, and soil characteristics of the region, and the interconnections between groundwater and surface water sources.

REFERENCES

1. J. G. Ferris, "Ground Water," *Mech. Eng.* (Jan. 1960).
2. The Conservation Foundation, "Groundwater Protection," Final Report of the National Groundwater Policy Forum, Washington, D.C., 1987.
3. E. F. Wood, Raymond A. Ferrara, William G. Gray, and George F. Pinder, *Groundwater Contamination from Hazardous Wastes.* Englewood Cliffs, NJ: Prentice-Hall, 1984.
4. V. I. Pye, R. Patrick, and J. Quarles, *Groundwater Contamination in the United States.* Philadelphia: University of Pennsylvania Press, 1983.
5. R. A. Freeze and J. A. Cherry, *Groundwater.* Englwood Cliffs, NJ: Prentice-Hall, 1979.
6. D. R. Maidment (ed.), *Handbook of Hydrology,* New York: McGraw-Hill, 1993.
7. J. Boonstra and N. A. de Ridder, "Numerical Modeling of Groundwater Basins," International Institute for Land Reclamation and Improvement, The Netherlands, 1981.
8. R. J. M. DeWiest, *Geohydrology.* New York: Wiley, 1965.
9. D. K. Todd, *Groundwater Hydrology.* New York: Wiley, 1960.
10. R. C. Heath, "Groundwater Regions of the United States," Geological Survey Water Supply Paper No. 2242. Washington, D.C.: U.S. Government Printing Office, 1984.
11. U.S. Environmental Protection Agency, "The Report to Congress: Waste Disposal Practices and Their Effects on Groundwater," Executive summary. U.S. EPA, PB 265–364, 1977.
12. O. E. Meinzer, "The Occurrence of Groundwater in the United States," U.S. Geological Survey, Water-Supply Paper No. 489, 1923.
13. H. H. Cooper, Jr. and C. E. Jacob, "A Generalized Graphical Method for Evaluating Formation Constants and Summarizing Well-Field History," *Trans. Am. Geophys. Union* **27,** 526–534(1946).
14. O. E. Meinzer, "Outline of Methods for Estimating Groundwater Supplies," U.S. Geological Survey, Water-Supply Paper 638-C, Washington, D.C., 1932.
15. Nathan Buras, "Conjunctive Operation of Dams and Aquifers," *Proc. ASCE J. Hyd. Div.* **89**(HY6) (Nov. 1963).
16. F. B. Clendenen, "A Comprehensive Plan for the Conjunctive Utilization of a Surface Reservoir with Underground Storage for Basin-Wide Water Supply Development: Solano Project California," Doctor of Eng. thesis, University of California, Berkeley, 1959.

Chapter 18

Mechanics of Flow

■ **Prologue**

The purpose of this chapter is to:

- Present the principles of groundwater flow.
- Describe soil properties that affect groundwater storage and movement.
- Describe the relevant hydrodynamic equations.
- Relate the mechanics of groundwater flow to modeling regional groundwater systems and calculating flows to wells and other groundwater collection devices.

18.1 HYDROSTATICS

Water located in pore spaces of a saturated medium is under pressure (called *pore pressure*), which can be determined by inserting a piezometer in the medium at a point of interest. If Location A (Fig. 18.1) is considered, it can be seen that pore pressure is given by

$$p = h_a \gamma \tag{18.1}$$

where p = the pore pressure (gauge pressure)
h_a = the head measured from the point to the water table
γ = the specific weight of water

Pore pressure is considered positive or negative, depending on whether the pressure head is measured above (positive) or below (negative) the point under consideration. If an arbitrary datum is established, the total head or piezometric head above the datum is

$$P_p = z + h \tag{18.2}$$

where P_p is known as the piezometric potential. In Fig. 18.1 this is equal to $h_a + z_a$ for Point A in the saturated zone and $z_b - h_b$ for Point B in the unsaturated zone. The

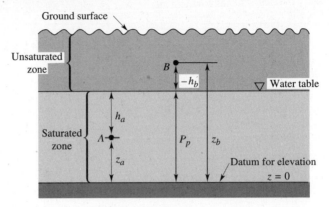

Figure 18.1   Definition sketch showing hydrostatic pressures in a porous medium.

term h_a is the pore pressure of A while $-h_b$ denotes tension or vacuum (negative pore pressure) at B.

18.2 GROUNDWATER FLOW

Analogies can be drawn between flow in pipes under pressure and in fully saturated confined aquifers. The flow of groundwater with a free surface is also similar to that in an open channel. A major difference is the geometry of a groundwater system flow channel as compared with common hydraulic pipe flow or channel systems. The problem can easily be recognized by envisioning a discharging cross section composed of a number of small openings, each with its own geometry, orientation, and size so that the flow velocity issuing from each pore varies in both magnitude and direction. Difficulties in analyzing such systems are apparent. Computations are usually based on macroscopic averages of fluid and medium properties over a given cross-sectional area.

Unknown quantities to be determined in groundwater flow problems are density, pressure, and velocity if constant temperature conditions are assumed to exist.[1-14] In general, water is considered incompressible, so the number of working variables is reduced. An exception to this is discussed later relative to the storage coefficient for a confined aquifer. Primary emphasis here will be placed on the flow of water in a saturated porous medium.

18.3 DARCY'S LAW

Darcy's law for fluid flow through a horizontal permeable bed is stated as[1]

$$Q = -KA \frac{dh}{dx} \tag{18.3}$$

where A = the total cross-sectional area including the space occupied by the porous material
K = the hydraulic conductivity of the material
Q = the flow across the control area A

In Eq. 18.3

$$h = z + \frac{p}{\gamma} + C \tag{18.4}$$

where h = the piezometric head
z = the elevation above a datum
p = the hydrostatic pressure
C = an arbitrary constant

If the specific discharge $q = Q/A$ is substituted in Eq. 18.3,

$$q = -K \frac{d}{dx}\left(z + \frac{p}{\gamma}\right) \tag{18.5}$$

Note that q also equals the porosity n multiplied by the pore velocity V_p. Darcy's law is widely used in groundwater flow problems. Several applications are illustrated in later sections.

Darcy's law is limited in applicability to cases where the Reynolds number is on the order of 1. For Reynolds numbers less than 1, Darcy's law may be considered valid. Deviations from Darcy's law have been shown to occur at Reynolds numbers as low as 2, depending on such factors as grain size and shape. The Reynolds number N_R is defined herein as

$$N_R = \frac{\rho q d}{\mu} \tag{18.6}$$

where q = the specific discharge
d = the mean grain diameter
ρ = fluid density
μ = dynamic viscosity

For many conditions of practical importance (zones lying adjacent to collecting devices are an exception), Darcy's law has been found to apply.

Of special interest is the fact that the Darcy equation is analogous to Ohm's law

$$i = \left(\frac{1}{R}\right)E \tag{18.7}$$

where i = the current
R = the resistance
E = the voltage

Current and velocity are analogous, as are K and $1/R$, and E and dh/dx. The similarity of the two equations is the basis for electric analog models of groundwater flow systems.[2,3]

EXAMPLE 18.1

Water temperature in an aquifer is 60°F and the rate of water movement = 1.2 ft/day. The average particle diameter in the porous medium is 0.08 in. Find the Reynolds number and indicate whether Darcy's law is applicable.

Solution.   Equation 18.6 gives the Reynolds number as

$$N_R = \frac{\rho q d}{\mu}$$

This may also be written as

$$N_R = \frac{q d}{v}$$

From Table A.2 in Appendix A, v is found to be 1.21×10^{-5} ft²/sec. Converting the velocity q into units of ft/sec gives $q = 1.2/86,400 = 1.39 \times 10^{-5}$. The mean grain diameter in ft $= 0.08/12 = 0.0067$. Substituting these values in the equation, we obtain

$$N_R = \frac{1.39 \times 10^{-5} \times 0.0067}{1.21 \times 10^{-5}}$$

$$= 0.0077$$

Since $N_R < 1.0$, Darcy's law does apply.   ■■

18.4 PERMEABILITY

The hydraulic conductivity K is an important parameter that is often separated into two components, one related to the medium, the other to the fluid. The product

$$k = C d^2 \tag{18.8}$$

called the *specific* or *intrinsic permeability,* is a function of the medium only. In Eq. 18.8, d represents the mean grain diameter of the particles and C is a constant shape factor associated with packing, size distribution, and other factors.[2,4] By using this definition, hydraulic conductivity, also known as the *coefficient of permeability,* can be written

$$K = \frac{k \gamma}{\mu} \tag{18.9}$$

Dimensions of intrinsic permeability are L^2. Since values of k given as ft² or cm² are extremely small, a unit of measure known as the *darcy* has been widely adopted.

$$1 \text{ darcy} = 0.987 \times 10^{-8} \text{ cm}^2 \quad \text{or} \quad 1.062 \times 10^{-11} \text{ ft}^2$$

Several ways of expressing hydraulic conductivity are reported in the literature. The U.S. Geological Survey has defined the standard coefficient of permeability K_s as the number of gallons per day of water passing through 1 ft² of medium under a

TABLE 18.1 SOME VALUES OF THE STANDARD COEFFICIENT OF
PERMEABILITY AND INTRINSIC PERMEABILITY FOR SEVERAL
CLASSES OF MATERIALS

Material	Approximate range K_s (gpd/ft^2)	Approximate range k (darcys)
Clean gravel	10^6–10^4	10^5–10^3
Clean sands; mixtures of clean gravels and sands	10^4–10	10^3–1
Very fine sands; silts; mixtures of sands, silts, clays; stratified clays	10–10^{-3}	1–10^{-4}
Unweathered clays	10^{-3}–10^{-4}	10^{-4}–10^{-5}

TABLE 18.2 CONVERSION FACTORS FOR PERMEABILITY AND HYDRAULIC
CONDUCTIVITY UNITS[1]

	Permeability, k			Hydraulic conductivity, K		
	cm^2	ft^2	darcy	m/s	ft/s	U.S. gal/day/ft^2
cm^2	1	1.08×10^{-3}	1.01×10^8	9.80×10^2	3.22×10^3	1.85×10^9
ft^2	9.29×10^2	1	9.42×10^{10}	9.11×10^5	2.99×10^6	1.71×10^{12}
darcy	9.87×10^{-9}	1.06×10^{-11}	1	9.66×10^{-6}	3.17×10^{-5}	1.82×10^1
m/s	1.02×10^{-3}	1.10×10^{-6}	1.04×10^5	1	3.28	2.12×10^6
ft/s	3.11×10^{-4}	3.35×10^{-7}	3.15×10^4	3.05×10^{-1}	1	6.46×10^5
U.S. gal/day/ft^2	5.42×10^{-10}	5.83×10^{-13}	5.49×10^{-2}	4.72×10^{-7}	1.55×10^{-6}	1

[1] R. Allan Freeze and John A. Cherry, *Groundwater*, 1979, p. 29. Reprinted by permission of Prentice-Hall, Englewood Cliffs, New Jersey.
Note: To obtain k in ft^2, for example, multiply k in cm^2 by 1.08×10^{-3}

unit-hydraulic gradient at a temperature of 60°F. Another measure, called the *field coefficient of permeability K_f*, is defined as

$$K_f = K_s \left(\frac{\mu_{60}}{\mu_f} \right) \tag{18.10}$$

where μ_{60} = the dynamic viscosity of water at 60°F
μ_f = the dynamic viscosity at the prevailing field temperature

Since the temperature effect on the coefficient of permeability is small over the range of temperatures normally encountered for groundwater in practice, the correction specified by Eq. 18.10 is seldom used.[12]

Table 18.1 gives the value of intrinsic permeability and the standard coefficient of permeability for several classes of materials. Table 18.2 gives a set of conversion factors for the units of k and K commonly used. Because considerable variation in these parameters within the materials classifications shown in Table 18.1 can occur, it is important that thorough geologic surveys be conducted to support groundwater modeling efforts.

For many groundwater analyses, it is convenient to use the coefficient of transmissivity

$$T = K_f b \tag{18.11}$$

where K_f = the field hydraulic conductivity
         b = the saturated depth of the aquifer

The coefficient of transmissivity is widely used in the water well industry. If K is expressed in gal/day/ft^2, then T has units of gal/day/ft. The range of values of T can be found by multiplying the pertinent K values from Table 18.1 by the range of expected aquifer thicknesses. It has been determined that aquifers worth considering for water supply development have values of T greater than about 100,000 gal/day/ft (0.015 m^2/s).[12]

EXAMPLE 18.2

Laboratory tests of an aquifer material give a standard coefficient of permeability $K_s = 3.78 \times 10^2$ gpd/ft^2. If the prevailing field temperature is 50°F, find the field coefficient of permeability K_f.

Solution.   Using Eq. 18.10 we obtain

$$K_f = K_s\left(\frac{\mu_{60}}{\mu_f}\right)$$

From Table A.2, Appendix A, the kinematic viscosity at 60°F = 1.21×10^{-5} ft^2/sec and at 50°F it is 1.41×10^{-5} ft^2/sec. For constant density,

$$K_f = \frac{3.78 \times 10^2 \times 1.21 \times 10^{-5}}{1.41 \times 10^{-5}}$$

and

$$K_f = 3.24 \times 10^2 \text{ gpd/ft}^2   ■■$$

18.5 VELOCITY POTENTIAL

Potential theory is directly applicable to groundwater flow computations. The *velocity potential* ϕ is a scalar function of time and space. The potential is defined by

$$\phi(x, y, z) = -K\left(z + \frac{p}{\gamma}\right) + C \tag{18.12}$$

where C is an arbitrary constant. By definition, its derivative with respect to any given direction is the velocity of flow in that direction. Thus we can write

$$u = \frac{\partial \phi}{\partial x} \quad v = \frac{\partial \phi}{\partial y} \quad w = \frac{\partial \phi}{\partial z} \tag{18.13}$$

where u, v, and w are the velocities in the x, y, and z directions, respectively, and K is assumed constant. In vector notation this becomes

$$V = \text{grad } \phi = \nabla\phi \tag{18.14}$$

with V the combined velocity vector and

$$\text{grad } \phi = \frac{\partial \phi}{\partial x}\mathbf{i} + \frac{\partial \phi}{\partial y}\mathbf{j} + \frac{\partial \phi}{\partial z}\mathbf{k} = \nabla\phi \tag{18.15}$$

18.6 HYDRODYNAMIC EQUATIONS

The determination of values for the variables u, v, w, and h is the target of most groundwater flow problems. The first three variables are the specific discharge components in the x, y, and z directions, respectively, while h is the total head at a specified point in the flow domain. To effect a solution, four equations involving these variables are needed. These are the equations of motion in each direction plus the continuity equation.

The equations of motion are based on Newton's second law,

$$F = ma \tag{18.16}$$

where F = the force
m = the mass
a = the acceleration

Considering forces acting on a fluid element, accelerations in the three coordinate directions may be determined according to Eq. 18.16. If frictionless flow is assumed (reasonable for many cases of flow in porous media), the body forces plus the surface force (pressure) must be equivalent to the total force in each direction. In the manner of Harr,[5] the following equations (Euler's equations) in the three coordinate directions are obtained:

$$\frac{\partial u}{\partial t} + u\frac{\partial u}{\partial x} + v\frac{\partial u}{\partial y} + w\frac{\partial u}{\partial z} = X - \frac{1}{\rho}\frac{\partial p}{\partial x} \tag{18.17}$$

$$\frac{\partial v}{\partial t} + u\frac{\partial v}{\partial x} + v\frac{\partial v}{\partial y} + w\frac{\partial v}{\partial z} = Y - \frac{1}{\rho}\frac{\partial p}{\partial y} \tag{18.18}$$

$$\frac{\partial w}{\partial t} + u\frac{\partial w}{\partial x} + v\frac{\partial w}{\partial y} + w\frac{\partial w}{\partial z} = Z - \frac{1}{\rho}\frac{\partial p}{\partial z} - g \tag{18.19}$$

where X, Y, Z, and g are body forces per unit mass. For steady flow [u, v, w, and h, $\neq f(t)$], the first terms on the left-hand side of each equation vanish. With laminar groundwater flow in the range of validity of Darcy's law, velocities are small (often on the order of 5 ft/yr to 5 ft/day).[2] Thus for steady laminar flow, Eqs. 18.17–18.19 reduce to

$$X = \frac{1}{\rho}\frac{\partial p}{\partial x} \quad Y = \frac{1}{\rho}\frac{\partial p}{\partial y} \quad Z = \frac{1}{\rho}\frac{\partial p}{\partial z} + g \tag{18.20}$$

In most groundwater flow problems the velocity head is negligible; thus p may be given as $\rho g(h - z)$. Then Eq. 18.20 becomes

$$X = g\frac{\partial h}{\partial x} \quad Y = g\frac{\partial h}{\partial y} \quad Z = g\frac{\partial h}{\partial z} \tag{18.21}$$

Remembering that Darcy's law defines $\partial h/\partial x = -u/K$, and so on, it follows that

$$X = -\frac{gu}{K} \quad Y = -\frac{gv}{K} \quad Z = -\frac{gw}{K} \tag{18.22}$$

For steady laminar flow, the body forces are linear functions of velocity and Eqs. 18.17–18.19 may be written

$$g\frac{\partial h}{\partial x} = -g\frac{u}{K} \tag{18.23}$$

$$g\frac{\partial h}{\partial y} = -g\frac{v}{K} \tag{18.24}$$

$$g\frac{\partial h}{\partial z} = -g\frac{w}{K} \tag{18.25}$$

where

$$u = -K\frac{\partial h}{\partial x} \quad v = -K\frac{\partial h}{\partial y} \quad w = -K\frac{\partial h}{\partial z} \tag{18.26}$$

This demonstrates that the equations of motion fit Darcy's law for steady laminar flow.

The continuity equation may be stated as[6]

$$\frac{\partial \rho}{\partial t} + \frac{\partial(\rho u)}{\partial x} + \frac{\partial(\rho v)}{\partial y} + \frac{\partial(\rho w)}{\partial z} = 0 \tag{18.27}$$

This equation is valid for a compressible fluid with time-dependent properties. In steady compressible flow the first term becomes zero, and for steady incompressible flow the equation becomes

$$\frac{\partial u}{\partial x} + \frac{\partial v}{\partial y} + \frac{\partial w}{\partial z} = 0 \tag{18.28}$$

Now since $u = \partial\phi/\partial x$, and so on, Eq. 18.28 becomes

$$\nabla^2\phi = \frac{\partial^2\phi}{\partial x^2} + \frac{\partial^2\phi}{\partial y^2} + \frac{\partial^2\phi}{\partial z^2} = 0 \tag{18.29}$$

which is known as the *Laplace equation*. With steady-state laminar flow, groundwater motion is completely described by the continuity equation subject to appropriate boundary conditions.

If the hydraulic conductivity K is constant, Eq. 18.29 can be written

$$\nabla^2 h = 0 \tag{18.30}$$

the expression of steady incompressible flow in a homogeneous isotropic porous medium.

For unsteady flow, the compressibility of both aquifer and water are pertinent. Consider a small element of porous medium that has a volume $\Delta x\,\Delta y\,\Delta z$. Then the term in a continuity equation representing a change in storage is defined by

$$\frac{\partial(\rho n\,\Delta x\,\Delta y\,\Delta z)}{\partial t} \tag{18.31}$$

Presupposing that compressive forces are predominant in the vertical (z) direction, we can neglect lateral changes. Thus in terms of the element described, only Δz is considered variable. A storage expression written as the sum of three terms involving partial derivatives of the variables Δz, ρ, and porosity n is[4]

$$\frac{\partial(\rho n \, \Delta x \, \Delta y \, \Delta z)}{\partial t} = \left(n\rho \frac{\partial(\Delta z)}{\partial t} + \rho \, \Delta z \frac{\partial n}{\partial t} + n \, \Delta z \frac{\partial \rho}{\partial t} \right) \Delta x \, \Delta y \qquad (18.32)$$

The three elements on the right can be expressed in terms of pore pressure p, the aquifer compressibility α, and the fluid compressibility β.[2,4]

Fluid compressibility is defined as the reciprocal of its bulk modulus of elasticity. It is given by[2]

$$\beta = -\frac{\partial V/V}{\partial p} \qquad (18.33)$$

where V = the volume
        p = the pore pressure

If the piezometric surface of a confined aquifer is lowered a distance of one unit, the amount of water released from a column of aquifer of unit horizontal cross-sectional area is defined as the storage coefficient S. This is analogous to the specific yield S_y of an unconfined aquifer. Obviously, in Eq. 18.33 S is equivalent to ∂V. Furthermore, if the aquifer column is of height b, $V = b$. The change in pressure ∂p is equivalent to the negative product of the change in head (one unit) and specific weight of water. Making these substitutions in Eq. 18.33 we find that

$$\beta = \frac{S}{\gamma b} \qquad (18.34)$$

Now if the aquifer material is considered elastic, that is, if Δz and n can be modified, the volume change can be expressed in terms of alteration in the density of the material due to the difference in packing. Thus

$$\frac{\partial V}{V} = -\frac{\partial \rho}{\rho} \qquad (18.35)$$

Introducing Eqs. 18.34 and 18.35 into Eq. 18.33 gives

$$\partial \rho = \frac{\rho S}{b\gamma} \, \partial p \qquad (18.36)$$

Next, substituting this expression for $\partial \rho$ in Eq. 18.27 we obtain

$$\frac{\partial(\rho u)}{\partial x} + \frac{\partial(\rho v)}{\partial y} + \frac{\partial(\rho w)}{\partial z} = -\frac{\rho S}{b\gamma} \frac{\partial p}{\partial t} \qquad (18.37)$$

The left-hand side of this equation can be expanded to

$$\rho \left(\frac{\partial u}{\partial x} + \frac{\partial v}{\partial y} + \frac{\partial w}{\partial z} \right) + \left(u \frac{\partial \rho}{\partial x} + v \frac{\partial \rho}{\partial y} + w \frac{\partial \rho}{\partial z} \right) \qquad (18.38)$$

The second term is normally very small compared with the first and can be neglected. The validity of this assumption improves as the flow angle decreases. By using

Eq. 18.38 and the foregoing assumption, Eq. 18.37 becomes

$$\frac{\partial u}{\partial x} + \frac{\partial v}{\partial y} + \frac{\partial w}{\partial z} = -\frac{S}{b\gamma}\frac{\partial p}{\partial t} \tag{18.39}$$

or if isotropic conditions prevail,

$$K\nabla^2 h = \frac{S}{b\gamma}\frac{\partial p}{\partial t} \tag{18.40}$$

since from Eq. 18.26 $u = -K\,\partial h/\partial x$, and so on. Inserting γh for p and the transmissivity T for Kb produces

$$\nabla^2 h = \frac{S}{T}\frac{\partial h}{\partial t} \tag{18.41}$$

which is the general equation for unsteady flow in a confined aquifer of constant thickness b.

The storage coefficient S and the transmissivity are commonly called the *formation constants* of a confined aquifer. For an unconfined aquifer Eq. 18.41 reverts to

$$\nabla^2 h = \frac{S}{Kb}\frac{\partial h}{\partial t} \tag{18.42}$$

since b is a function of the change in head. The unsteady flow equation for an unconfined aquifer is nonlinear in form. The solution of such an equation is discussed by Jacob.[7] Where variations in saturated thickness of unconfined aquifers are minor, Eq. 18.41 may be used as an approximation.[2]

For unconfined aquifers, the right-hand side of Eq. 18.42 is often negligible so that the equation

$$\nabla^2 h = 0 \tag{18.30}$$

is frequently valid for both steady and unsteady flow.

18.7 FLOWLINES AND EQUIPOTENTIAL LINES

Many problems of practical interest in groundwater hydrology can be considered two-dimensional flow problems. The equation of continuity for steady incompressible flow in an isotropic medium then becomes

$$\frac{\partial u}{\partial x} + \frac{\partial v}{\partial y} = 0 \tag{18.43}$$

$$\nabla^2 h = \frac{\partial^2 h}{\partial x^2} + \frac{\partial^2 h}{\partial y^2} = 0 \tag{18.44}$$

and

$$\nabla^2 \phi = \frac{\partial^2 \phi}{\partial x^2} + \frac{\partial^2 \phi}{\partial y^2} = 0 \tag{18.45}$$

The Laplace equation is satisfied by two conjugate harmonic functions ϕ and ψ.[2,5] Curves $\phi(x, y) =$ constant are orthogonal to the curves $\psi(x, y) =$ constant. The function $\phi(x, y)$ is the velocity potential, the function $\psi(x, y)$ is known as the *stream function* and is defined by

$$u = \frac{\partial \psi}{\partial y} \quad v = -\frac{\partial \psi}{\partial x} \tag{18.46}$$

Substituting Eq. 18.46 into Eq. 18.43 yields

$$\frac{\partial^2 \psi}{\partial x \, \partial y} - \frac{\partial^2 \psi}{\partial y \, \partial x} = 0 \tag{18.47}$$

It has already been shown that

$$u = \frac{\partial \phi}{\partial x} \quad v = \frac{\partial \phi}{\partial y}$$

so we can write

$$\frac{\partial \phi}{\partial x} = \frac{\partial \psi}{\partial y} \quad \frac{\partial \phi}{\partial y} = -\frac{\partial \psi}{\partial x} \tag{18.48}$$

These are known as the *Cauchy-Riemann equations*. The stream function satisfies both the equation of continuity and the equations of Cauchy-Riemann. It can also be shown that the Laplace equation is satisfied and therefore[4,5]

$$\nabla^2 \psi = \frac{\partial^2 \psi}{\partial x^2} + \frac{\partial^2 \psi}{\partial y^2} = 0 \tag{18.49}$$

Refer now to Fig. 18.2. If V is a velocity vector tangent to a particle flow path 3–4, then it can be decomposed into two components u and v.[6] By geometry of the figure

$$\frac{v}{u} = \frac{dy}{dx} = \tan \alpha \tag{18.50}$$

and thus

$$v \, dx - u \, dy = 0 \tag{18.51}$$

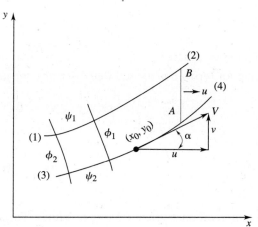

Figure 18.2   Definition sketch for a stream function.

If Eqs. 18.46 are substituted into Eq. 18.50, then

$$\frac{\partial \psi}{\partial x} dx + \frac{\partial \psi}{\partial y} dy = 0 \qquad (18.52)$$

The total differential $d\psi$ is equal to zero, and ψ must be a constant. A series of curves $\psi(x, y)$ equal to a succession of constants can be drawn and will be tangent at all points to the velocity vectors. These curves trace the flow path of a fluid particle and are known as *streamlines* or *flowlines*. An important property of the stream function is demonstrated with the aid of Fig. 18.2. Consider the flow crossing a vertical section AB between streamlines defined as ψ_1 and ψ_2. If the discharge across the section is designated as Q, it is apparent that

$$Q = \int_{\Psi_2}^{\Psi_1} u \, dy \qquad (18.53)$$

or

$$Q = \int_{\Psi_2}^{\Psi_1} d\psi \qquad (18.54)$$

and

$$Q = \psi_1 - \psi_2 \qquad (18.55)$$

Equation 18.55 illustrates the important property that flow between two streamlines is constant. Streamline spacing reveals the relative magnitudes of flow velocities between them. Higher values are associated with narrower spacings, and vice versa.

The curves in Fig. 18.2 designated as ϕ_1 and ϕ_2, called *equipotential lines,* are determined by velocity potentials $\phi(x, y) = $ constant. These curves intersect the flowlines at right angles, illustrated in the following way. The total differential $d\phi$ is given by

$$d\phi = \frac{\partial \phi}{\partial x} dx + \frac{\partial \phi}{\partial y} dy \qquad (18.56)$$

Substituting for terms $\partial \phi / \partial x$ and $\partial \phi / \partial y$ their equivalents u and v gives us

$$u \, dx + v \, dy = 0 \qquad (18.57)$$

and

$$\frac{dy}{dx} = -\frac{u}{v} \qquad (18.58)$$

Thus equipotential lines are normal to flowlines. The system of flowlines and equipotential lines forms a flow net.

One significant point of difference between ϕ and ψ functions is that equipotential lines exist only when the flow is irrotational. For two-dimensional flow the condition of irrotationality is said to exist when the z component of vorticity ζ_z is zero, or

$$\zeta_z = \left(\frac{\partial v}{\partial x} - \frac{\partial u}{\partial y}\right) = 0 \tag{18.59}$$

Proof of this is given by Eskinazi.[6] Substituting for u and v in Eq. 18.59 in terms of ϕ, we obtain

$$\frac{\partial^2 \phi}{\partial x\, \partial y} - \frac{\partial^2 \phi}{\partial y\, \partial x} = 0 \tag{18.60}$$

This indicates that when the velocity potential exists, the criterion for irrotationality is satisfied.

Once either streamlines or equipotential lines in a flow domain are determined, the other is automatically known because of the relations in Eq. 18.48. Thus

$$\psi = \int \left(\frac{\partial \phi}{\partial x}\, dy - \frac{\partial \phi}{\partial y}\, dx\right) \tag{18.61a}$$

and

$$\phi = \int \left(\frac{\partial \psi}{\partial y}\, dx - \frac{\partial \psi}{\partial x}\, dy\right) \tag{18.61b}$$

It is enough then to determine only one of the functions, since the other can be obtained using relations Eqs. 18.61a and 18.61b. The complex potential given by

$$w = \phi + i\psi \tag{18.62}$$

where i, the square root of -1, is widely used in analytic flow net analyses.[4,5] Of special importance is the fact that

$$\nabla^2 w = \nabla^2 \phi + i\nabla^2 \psi = 0 \tag{18.63}$$

satisfies the conditions of continuity and irrotationality simultaneously.

Equations presented in this section have been limited to the case of two-dimensional flow. Extension to three dimensions would be obtained in a similar fashion.

18.8 BOUNDARY CONDITIONS

To solve groundwater flow problems it is necessary that appropriate boundary conditions be specified. Some of the more commonly encountered ones are described in this section; more comprehensive discussions are found elsewhere.[8,9]

Boundary conditions discussed can be categorized as follows: impervious boundaries, surfaces of seepage, constant head boundaries, and lines of seepage (free surfaces).

Impervious boundaries may be artificial objects such as concrete dams, rock strata, or soil strata that are highly impervious. In Fig. 18.3 the impervious boundary

AB represents such a limit. Since flow cannot cross an impervious boundary, velocity components normal to it vanish and the impervious boundary is a streamline. In other words, at the boundary, ψ = constant.

Next look at the upstream face of the earth dam *BC*. At any point of elevation *y* along *BC* the pressure can be assumed hydrostatic, or

$$p = \gamma(h - y) \tag{18.64}$$

The definition of a velocity potential states that

$$\phi = -K\left(\frac{p}{\gamma} + y\right) + C \tag{18.65}$$

Substituting for pressure in Eq. 18.65 yields

$$\phi = -K\left[\frac{\gamma(h - y)}{\gamma} + y\right] + C \tag{18.66}$$

and

$$\phi = -Kh + C \tag{18.67}$$

Thus for a constant reservoir level *h* and an isotropic medium,

$$\phi = \text{constant}$$

and surface *BC*, often termed a *reservoir boundary,* is an equipotential line.

The free surface or line of seepage *CD* in Fig. 18.3 is seen to be a boundary between the saturated and unsaturated zones. Since flow does not occur across this boundary, it is obviously also a streamline. Pressure along this free surface must be constant, and therefore along *CD*

$$\phi + Ky = \text{constant} \tag{18.68}$$

This is a linear relation in ϕ, and therefore equal vertical falls along *CD* must be associated with successive equipotential drops. One important groundwater flow problem is to determine the location of the line of seepage.

The surface of seepage *DE* of Fig. 18.3 represents the location at which water seeps through the downstream face of the dam and trickles toward point *E*. The pressure along *DE* is atmospheric. The surface of seepage is neither a flowline nor an equipotential line.

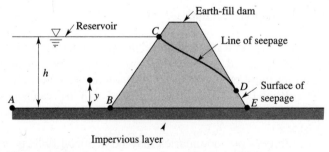

Figure 18.3   Some common boundary conditions.

18.9 FLOW NETS

Flow nets, or graphical representations of families of streamlines and equipotential lines, are widely used in groundwater studies to determine quantities, rates, and directions of flow. The use of flow nets is limited to steady incompressible flow at constant viscosity and density for homogeneous media or for regions that can be compartmentalized into homogeneous segments. Darcy's law must be applicable to the flow conditions.

The manner in which a flow net can be used in problem solving is best explained with the aid of Fig. 18.4. This diagram shows a portion of a flow net constructed so that each unit bounded by a pair of streamlines and equipotential lines is approximately square. The reason for this will be clear later.

A flow net can be determined exactly if functions ϕ and ψ are known beforehand. This is often not the case, and as a result, graphically constructed flow nets are widely used. The preparation of a flow net requires application of the concept of square elements and adherence to boundary conditions. Graphical flow nets are usually difficult for a beginner to create, but with reasonable practice an acceptable net can be drawn. Various mechanical methods for graphical flow net construction are presented in the literature and are not discussed here.[5,9]

After a flow net has been constructed, it can be analyzed using the geometry of the net and by applying Darcy's law.

Remembering that $h = p/\gamma + z$, we find that Fig. 18.4 shows that the hydraulic gradient G_h between two equipotential lines is given by

$$G_h = \frac{\Delta h}{\Delta s} \qquad (18.69)$$

Then by applying Darcy's law, in the manner of Todd,[2] the flow increment between adjacent streamlines is

$$\Delta q = K \, \Delta m \left(\frac{\Delta h}{\Delta s}\right) \qquad (18.70)$$

where Δm represents the cross-sectional area for a net of unit width normal to the plane of the diagram. If the flow net is constructed in an orthogonal manner and

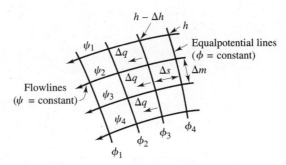

Figure 18.4 Segment of an orthogonal flow net.

composed of approximately square elements,

$$\Delta m \approx \Delta s \quad \text{and} \quad \Delta q = K \, \Delta h \qquad (18.71)$$

Now if there are n equipotential drops between the equipotential lines, it is evident that

$$\Delta h = \frac{h}{n}$$

where h is the total head loss over the n spaces. If the flow is divided into m sections by the flowlines, then the discharge per unit width of the medium is

$$Q = \sum_{i=1}^{m} \Delta q = \frac{Kmh}{n} \qquad (18.72)$$

When the medium's hydraulic conductivity is known, the discharge can be computed using Eq. 18.72 and a knowledge of flow net geometry.

Where the flow net has a free surface or line of seepage, the entrance and exit conditions given in Fig. 18.3 are useful. A more comprehensive discussion of these conditions is given in Ref. 10.

Some trouble arises in flow net construction at locations where the velocity becomes infinite or vanishes. Such points are known as *singular points* and according to DeWiest may be placed in three separate categories.[4] In the first classification flowlines and equipotential lines do not intersect at right angles. Such a situation often occurs when a boundary coincides with a flowline; Point A in Fig. 18.5 is an example.

The second classification has a discontinuity along the boundary that abruptly changes the slope of the streamline. In Fig. 18.6 Points A, B, and C represent such discontinuities. At Points A and C the velocity is infinite, while at Point B it is zero. If the angle of discontinuity measured in a counterclockwise direction inside the flow field is less than 180°, the velocity is zero; if larger than 180°, it is infinite. The angle at A is 270°, for example.

The third category includes the case where a source or sink exists in the flow net. Under these circumstances the velocity is infinite, since squares of the flow net

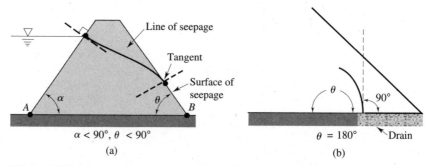

Figure 18.5 Some entrance and exit conditions for the line of seepage. (After Casagrande.[10])

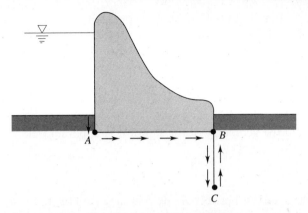

Figure 18.6   Flowline slope discontinuities.

approach zero size as the source or sink is approached. Wells and recharge wells represent sinks and sources in a practical sense and are discussed later.

18.10 VARIABLE HYDRAULIC CONDUCTIVITY

It is common for flow within a porous medium of one hydraulic conductivity to enter another region with a different hydraulic conductivity. When such a boundary is crossed, flowlines are refracted. The change in direction that occurs can be determined as a function of the two permeabilities involved in the manner of Todd and DeWiest.[2,4] Figure 18.7 illustrates this.

Consider two soils of permeabilities K_1 and K_2 which are separated by the boundary LR shown in Fig. 18.7. The directions of the flowlines before and after crossing the boundary are defined by angles θ_1 and θ_2.

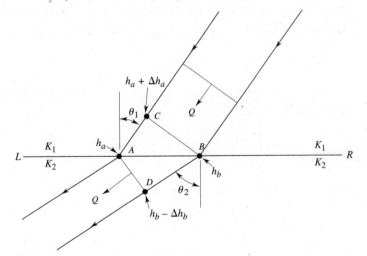

Figure 18.7   Flowline refraction.

For continuity to be preserved, the velocity components in media K_1 and K_2, which are normal to the boundary, must be equal, since the cross-sectional area at the boundary is AB for a unit depth. Using Darcy's law and noting the equipotential drops h_a and h_b, we find

$$K_1 \frac{\Delta h_a}{AC} \cos \theta_1 = K_2 = \frac{\Delta h_b}{BD} \cos \theta_2 \qquad (18.73)$$

From the geometry of the figure it is apparent that

$$AC = AB \sin \theta_1$$
$$BD = AB \sin \theta_2$$

The head loss between A and B is shown on the figure to be equal to both Δh_a and Δh_b, and since there can be only a single value,

$$\Delta h_a = \Delta h_b$$

Introducing these expressions in Eq. 18.73 produces

$$\frac{K_1}{\tan \theta_1} = \frac{K_2}{\tan \theta_2} \qquad (18.74)$$

For refracted flow in a saturated porous medium, the ratio of the tangents of angles formed by the intersection of flowlines with normals to the boundary is given by the ratio of hydraulic conductivities. As a result of refraction, the flow net on the K_2 side of the boundary will no longer be squares if the equipotential line spacing DB is maintained. To adjust the net on the K_2 side, the relation

$$\frac{\Delta h_b}{\Delta h_a} = \frac{K_1}{K_2} \qquad (18.75)$$

can be used where $\Delta h_b \neq \Delta h_a$.

Equipotential lines are also refracted in crossing permeability boundaries. The relation for this is

$$\frac{K_1}{K_2} = \frac{\tan \alpha_2}{\tan \alpha_1} \qquad (18.76)$$

where α is the angle between the equipotential line and a normal to the boundary of permeability.[4]

18.11 ANISOTROPY

In many cases hydraulic conductivity is dependent on the direction of flow within a given layer of soil. This condition is said to be anisotropic. Sedimentary deposits often fit this aspect, with flow occurring more readily along the plane of deposition than across it. Where the permeability within a plane is uniform but very small across it as compared to that along the plane, a flow net can still be used after proper adjustments are made. A discussion of this is given elsewhere.[4,5,11,12] Nonhomogeneous aquifers require special consideration but may sometimes be analyzed by using representative or average parameters. A detailed study is outside the scope of this book.[3-5,12]

18.12 DUPUIT'S THEORY

Groundwater flow problems in which one boundary is a free surface can be analyzed on the basis of Dupuit's theory of unconfined flow. This theory is founded on two assumptions made by Dupuit in 1863.[13] First, if the line of seepage is only slightly inclined, streamlines may be considered horizontal and, correspondingly, equipotential lines will be essentially vertical. Second, slopes of the line of seepage and the hydraulic gradient are equal. When field conditions are known to be satisfactorily represented by these assumptions, the results obtained according to Dupuit's theory compare very favorably with those arrived at by more rigorous techniques.

Figure 18.8 is useful in translating the foregoing assumptions into a mathematical statement. Consider an element given in the figure which has a base area $dx\, dy$ and a vertical height h. Writing the continuity equation in the x direction and considering steady flow to be the case,

$$\text{inflow}_{x_0} = \text{velocity}_{x_0} \times \text{area}_{x_0} \tag{18.77}$$

The velocity at $x = 0$ is given by Darcy's law as

$$u_{x_0} = -K\frac{\partial h}{\partial x} \tag{18.78}$$

Thus the discharge across the element at $x = 0$ is

$$Q_0 = -K\frac{\partial h}{\partial x}h\, dy \tag{18.79}$$

The outflow at $x = dx$ is obtained by a Taylor's series expansion as

$$Q_{dx} = -K\frac{\partial h}{\partial x}h\, dy + dx\frac{\partial}{\partial x}\left(-K\frac{\partial h}{\partial x}h\, dy\right) + \cdots \tag{18.80}$$

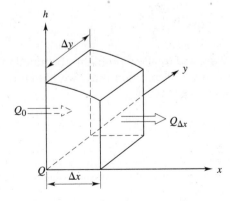

Figure 18.8 Definition sketch for development of Dupuit's equation.

Subtracting the outflow from the inflow if K is considered constant, we obtain

$$I_x - O_x = K\,dx\,dy\,\frac{\partial}{\partial x}\left(h\frac{\partial h}{\partial x}\right) \tag{18.81}$$

or

$$I_x - O_x = \frac{K\,dx\,dy}{2}\,\frac{\partial}{\partial x}\left(\frac{\partial h^2}{\partial x}\right) \tag{18.82}$$

where dx and dy are considered fixed lengths. A similar consideration in the y direction yields

$$I_y - O_y = \frac{K\,dx\,dy}{2}\,\frac{\partial}{\partial y}\left(\frac{\partial h^2}{\partial y}\right) \tag{18.83}$$

Assuming that there is no movement in the vertical direction, these are the only components of the inflow and outflow. Furthermore, still dealing with steady flow, the change in storage must be zero. As a result,

$$\frac{K\,dx\,dy}{2}\,\frac{\partial}{\partial x}\left(\frac{\partial h^2}{\partial x}\right) + \frac{K\,dx\,dy}{2}\,\frac{\partial}{\partial y}\left(\frac{\partial h^2}{\partial y}\right) = 0 \tag{18.84}$$

and since $(K\,dx\,dy)/2$ is constant, this reduces to

$$\frac{\partial^2 h^2}{\partial x^2} + \frac{\partial^2 h^2}{\partial y^2} = 0 \tag{18.85}$$

or

$$\nabla^2 h^2 = 0 \tag{18.86}$$

Consequently, according to Dupuit's assumptions, Laplace's equation for the function h^2 must be satisfied.[14]

In the particular case where recharge is occurring as a result of infiltrated water reaching the water table, a simple adjustment may be made to Eq. 18.85. If the recharge intensity (dimensionally LT^{-1}) is specified as R, then the total recharge to the element of Fig. 18.8 is $R\,dx\,dy$ and the continuity equation for steady flow becomes

$$K\frac{dx\,dy}{2}\left(\frac{\partial^2 h^2}{\partial x^2} + \frac{\partial^2 h^2}{\partial y^2}\right) + R\,dx\,dy = 0 \tag{18.87}$$

or more simply,

$$\nabla^2 h^2 + \frac{2}{K}R = 0 \tag{18.88}$$

Now, applying Dupuit's theory to the flow problem illustrated in Fig. 18.9, and assuming one-dimensional flow in the x direction only, we obtain the discharge per unit width of the aquifer given by Darcy's law:

$$Q = -Kh\frac{dh}{dx} \tag{18.89}$$

In this instance h is the height of the line of seepage at any position x along the impervious boundary. For the one-dimensional example considered here, Eq. 18.85

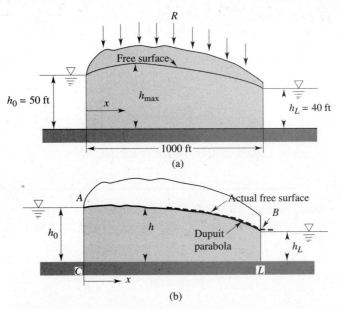

Figure 18.9 Steady flow in a porous medium between two water bodies: (a) free surface with infiltration and (b) free surface without infiltration.

becomes

$$\frac{d^2h^2}{dx^2} = 0 \qquad (18.90)$$

Upon integration,

$$h^2 = ax + b \qquad (18.91)$$

where a and b are constants.

Then for boundary conditions at $x = 0$, $h = h_0$,

$$b = h_0^2 \qquad (18.92)$$

Differentiation of Eq. 18.91 yields

$$2h \frac{dh}{dx} = a \qquad (18.93)$$

Also from Darcy's equation, $h\, dh/dx = -Q/K$. Making this substitution, we obtain

$$a = \frac{-2Q}{K} \qquad (18.94)$$

and inserting the values of the constants in Eq. 18.91, we obtain

$$h^2 = -2\frac{Q}{K}x + h_0^2 \qquad (18.95)$$

This is the equation of a free surface. It is a parabola (often called *Dupuit's parabola*). If the existence of a surface of seepage at B is ignored, and noting that at $x = L$, $h = h_L$, we find that Eq. 18.95 becomes

$$h_L^2 = -\frac{2QL}{K} + h_0^2 \tag{18.96}$$

or

$$Q = \frac{K}{2L}(h_0^2 - h_L^2) \tag{18.97}$$

which is known as the *Dupuit equation*.

EXAMPLE 18.3 _____

Refer to Fig. 18.9a. Given the dimensions shown and a recharge intensity R of 0.01 ft/day, find the discharge at $x = 1000$ ft using Dupuit's equation. Assume that $K = 8$.

Solution.   Note that

$$\frac{dQ}{dx} = R$$

or

$$Q = Rx + C$$

At $x = 0$,

$$Q = Q_0$$

therefore,

$$Q = Rx + Q_0$$

Also,

$$Q = -Kh\frac{dh}{dx}$$

$$-Kh\frac{dh}{dx} = Rx + Q_0$$

Integrating yields

$$\left.\frac{-Kh^2}{2}\right|_{h_0}^{h_L} = \left.\frac{Rx^2}{2}\right|_0^L + \left.Q_0 x\right|_0^L$$

and inserting the limits, we obtain

$$\frac{-K(h_L^2 - h_0^2)}{2} = \frac{RL^2}{2} + Q_0 L$$

$$Q_0 = \frac{K(h_0^2 - h_L^2)}{2L} - \frac{RL}{2}$$

Then since $Q = Rx + Q_0$,

$$Q = R\left(x - \frac{L}{2}\right) + \frac{K(h_0^2 - h_L^2)}{2L}$$

$$R = 0.01 \times 7.5 = 0.075 \text{ gpd/ft}^2$$

$$Q = 0.075(1000 - 500) + \frac{8(50^2 - 40^2)}{2000}$$

$$= 0.075 \times 500 + \frac{8 \times 900}{2000}$$

$$= 37.5 + 3.6$$

$$= 41.1 \text{ gpd/ft}^2 \quad \blacksquare\blacksquare$$

■ Summary

Understanding the movement of groundwater requires a knowledge of the time and space dependency of the flow, nature of the porous medium and fluid, and the boundaries of the flow system. In particular, groundwater development and management depend on understanding the storage properties of the associated soils and rocks and the ability of these subsurface materials to transmit water. Fundamental to the mechanics of groundwater flow is Darcy's law (Eq. 18.3). Using this equation along with a knowledge of the hydraulic conductivity K, estimates of flow can be had. The hydrodynamic equations presented in this chapter serve as models for a variety of groundwater flow calculations. Applications are given in Chapters 19 and 20.

PROBLEMS

18.1. What is the Reynolds number for flow in a soil when the water temperature is 55°F, the velocity is 0.5 ft/day, and the mean grain diameter is 0.08 in.?

18.2. The water temperature in an aquifer is 60°F, the velocity is 1.0 ft/day. The average particle diameter of the soil is 0.06 in. Find the Reynolds number and indicate whether Darcy's law applies.

18.3. Rework Problem 18.2 assuming the temperature is 65°F and the velocity is 0.8 ft/day.

18.4. A laboratory test of a soil gives a standard coefficient of permeability of 3.8×10^2 gpd/ft^2. If the prevailing field temperature is 60°F, find the field coefficient of permeability.

18.5. Rework Problem 18.4 assuming K_s is 3.8×10^2 gpd/ft^2 and the temperature is 65°F.

18.6. Given the well and flow net data in the following figure, find the discharge using a flow net solution. The well is fully penetrating; $K = 2.87 \times 10^{-4}$ ft/sec, $a = 180$ ft, $b = 43$ ft, and $c = 50$ ft.

18.7. Rework Problem 18.6 assuming $K = 8.2 \times 10^{-5}$ m/sec, $a = 85$ m, $b = 21$ m, and $c = 26$ m.

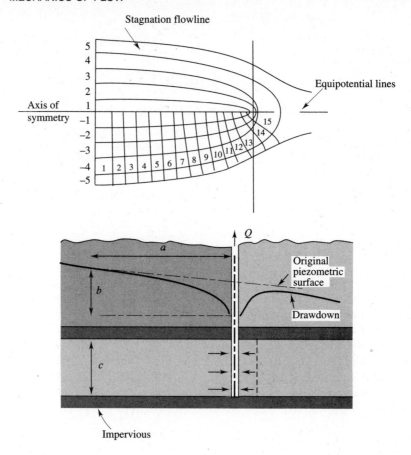

18.8. Rework Problem 18.6 assuming $K = 8.4 \times 10^{-5}$ m/sec, $a = 100$ m, $b = 22$ m, and $c = 35$ m.

18.9. A stratum of clean sand and gravel 15 ft deep has a coefficient of permeability of $K = 3.25 \times 10^{-3}$ ft/sec, and is supplied with water from a channel that penetrates to the bottom of the stratum. If the water surface in an infiltration gallery is 2 ft above the bottom of the stratum, and its distance to the channel is 50 ft, what is the flow into a foot of gallery? Use Eq. 18.97.

REFERENCES

1. Henri Darcy, *Les fontaines publiques de la ville de Dijon.* Paris: V. Dalmont, 1856.
2. D. K. Todd, *Groundwater Hydrology.* New York: Wiley, 1960.
3. William C. Walton, *Groundwater Resource Evaluation.* New York: McGraw-Hill, 1970.
4. R. J. M. DeWiest, *Geohydrology.* New York: Wiley, 1965.
5. M. E. Harr, *Groundwater and Seepage.* New York: McGraw-Hill, 1962.
6. Salamon Eskinazi, *Principles of Fluid Mechanics.* Boston: Allyn and Bacon, 1962.
7. C. E. Jacob, "Flow of Groundwater," in *Engineering Hydraulics* (Hunter Rouse, ed.) New York: Wiley, 1950.

8. M. Muskat, *The Flow of Homogeneous Fluids Through Porous Media.* Ann Arbor, MI: J. W. Edwards, 1946.

9. D. W. Taylor, *Fundamentals of Soil Mechanics.* New York: Wiley, 1948.

10. A. Casagrande, "Seepage Through Dams," in *Contributions to Soil Mechanics, 1925–1940.* Boston: Boston Society of Civil Engineers, 1940.

11. M. S. Hantush, "Modification of the Theory of Leaky Aquifers," *J. Geophys. Res.* **65,** 3713–3725(1960).

12. R. A. Freeze and J. A. Cherry, *Groundwater.* Englewood Cliffs, NJ: Prentice-Hall, 1979.

13. Jules Dupuit, *Etudes théoriques et pratiques sur le mouvement des eau dans les canaux de couverts et à travers les terrains perméables.* 2nd ed. Paris: Dunod, 1863.

14. P. Ya. Polubarinova-Kochina, *Theory of Groundwater Movement.* Princeton, NJ: Princeton University Press, 1962.

Chapter 19

Wells and Collection Devices

■ **Prologue**

The purpose of this chapter is to:

- Present methods for calculating confined and unconfined steady radial flow toward a well.
- Describe procedures for dealing with unsteady groundwater flow conditions.
- Describe a method for estimating flow to an infiltration gallery.

Groundwater is collected primarily by wells, although infiltration galleries are sometimes used where the circumstances are appropriate.[1] Outflows from natural springs are also amenable to collection, but once these waters exit the ground, they become surface flows and are handled as such. Wells are holes or shafts, usually vertical, excavated in the earth for the purpose of bringing groundwater to the surface. Infiltration galleries are horizontal conduits for intercepting and collecting groundwater by gravity flow. Problems of groundwater flow to wells and infiltration galleries can be solved by applying Darcy's law.

19.1 FLOW TO WELLS

A well system can be considered as composed of three elements—the well structure, pump, and discharge piping.[2] The well itself contains an open section through which water enters and a casing to transport the flow to the ground surface. The open section is usually a perforated casing or slotted metal screen permitting water to enter and at the same time preventing collapse of the hole. Occasionally, gravel is placed at the bottom of the well casing around the screen.

When a well is pumped, water is removed from the aquifer immediately adjacent to the screen. Flow then becomes established at locations some distance from the well in order to replenish this withdrawal. Because of flow resistance offered by the soil, a head loss results and the piezometric surface adjacent to the well is depressed, producing a cone of depression (Fig. 19.1), which spreads until equilibrium is reached and steady-state conditions are established.

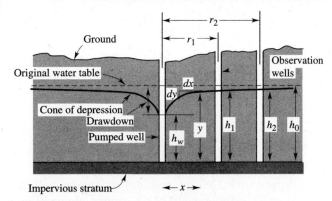

Figure 19.1   Well in an unconfined aquifer.

The hydraulic characteristics of an aquifer (which are described by the storage coefficient and aquifer permeability) can be determined by laboratory or field tests. The three most commonly used field methods are the application of tracers, the use of field permeameters, and aquifer performance tests.[3] A discussion of aquifer performance tests is given here along with the development of flow equations for wells.[2,4,5]

Aquifer performance tests may be either equilibrium or nonequilibrium tests. In an equilibrium test the cone of depression must be stabilized for a flow equation to be derived. For a nonequilibrium test the derivation includes a condition that steady-state conditions have not been reached. Adolph Thiem published the first performance tests based on equilibrium conditions in 1906.[6]

19.2 STEADY UNCONFINED RADIAL FLOW TOWARD A WELL

The basic equilibrium equation for an unconfined aquifer can be derived using the notation of Fig. 19.1. Here flow is assumed to be radial; the original water table is considered to be horizontal; the well is presumed to fully penetrate the aquifer of infinite areal extent; and steady-state conditions must prevail. Then flow toward the well at any distance x away must equal the product of the cylindrical element of area at that section and the flow velocity. With Darcy's law this becomes

$$Q = 2\pi x y K_f \frac{dy}{dx} \tag{19.1}$$

where $2\pi xy$ = the area through any cylindrical shell, in ft^2 with the well as its axis

K_f = the hydraulic conductivity (ft/sec)

dy/dx = the water table gradient at any distance x

Q = the well discharge (ft^3/sec)

Integrating over the limits specified, we find that

$$\int_{r1}^{r2} Q\, \frac{dx}{x} = 2\pi K_f \int_{h1}^{h2} y\, dy \tag{19.2}$$

$$Q \ln \frac{r_2}{r_1} = \frac{2\pi K_f(h_2^2 - h_1^2)}{2} \tag{19.3}$$

and

$$Q = \frac{\pi K_f(h_2^2 - h_1^2)}{\ln(r_2/r_1)} \tag{19.4}$$

Converting K_f to the field units of gpd/ft^2, Q to gpm, and ln to log, we can rewrite Eq. 19.4 as

$$K_f = \frac{1055Q \log(r_2/r_1)}{h_2^2 - h_1^2} \tag{19.5}$$

If the drawdown in the well does not exceed one half of the original aquifer thickness h_0, reasonable estimates of Q or K_f can be obtained by using Eq. 19.4 or 19.5, even if the height h_1 is measured at the well periphery where $r_1 = r_w$, the radius of the well boring.

EXAMPLE 19.1 _____

An 18-in. well fully penetrates an unconfined aquifer of 100-ft depth. Two observation wells located 100 and 235 ft from the pumped well are known to have drawdowns of 22.2 and 21 ft, respectively. If the flow is steady and $K_f = 1320$ gpd/ft^2, what would be the discharge?

Solution. Equation 19.4 is applicable, and for the given units this is

$$Q = \frac{K(h_2^2 - h_1^2)}{1055 \log(r_2/r_1)}$$

$$\log(r_2/r_1) = \log(235/100) = 0.37107$$

$$h_2 = 100 - 21 = 79 \text{ ft}$$

$$h_1 = 100 - 22.2 = 77.8 \text{ ft}$$

$$Q = \frac{1320(79^2 - 77.8^2)}{1055 \times 0.37107}$$

$$= 634.44 \text{ gpm} \quad \blacksquare\blacksquare$$

19.3 STEADY CONFINED RADIAL FLOW TOWARD A WELL

The basic equilibrium equation for a confined aquifer can be obtained in a similar manner, using the notation of Fig. 19.2. The same assumptions apply. Mathematically, the flow in ft^3/sec is found from

$$Q = 2\pi xm K_f \frac{dy}{dx} \tag{19.6}$$

Integrating, we obtain

$$Q = 2\pi K_f m \frac{h_2 - h_1}{\ln(r_2/r_1)} \tag{19.7}$$

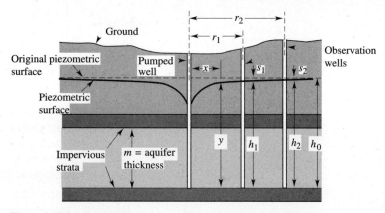

Figure 19.2   Radial flow to a well in a confined aquifer.

The coefficient of permeability may be determined by rearranging Eq. 19.7 to the form

$$K_f = \frac{528Q \, \log(r_2/r_1)}{m(h_2 - h_1)} \tag{19.8}$$

where Q = gpm
             K_f = the permeability (gpd/ft^2)
             r, h = ft

EXAMPLE 19.2

Determine the permeability of an artesian aquifer being pumped by a fully penetrating well. The aquifer is 90 ft thick and composed of medium sand. The steady-state pumping rate is 850 gpm. The drawdown of an observation well 50 ft away is 10 ft; in a second observation well 500 ft away it is 1 ft.

Solution

$$
\begin{aligned}
K_f &= \frac{528Q \, \log(r_2/r_1)}{m(h_2 - h_1)} \\
&= \frac{528 \times 850 \times \log 10}{90 \times (10 - 1)} \\
&= 554 \text{ gpd/ft}^2 \quad \blacksquare\blacksquare
\end{aligned}
$$

19.4 WELL IN A UNIFORM FLOW FIELD

For a steady-state well in a uniform flow field where the original piezometric surface is not horizontal, a somewhat different situation from that previously assumed prevails. Consider the artesian aquifer shown in Fig. 19.3. The heretofore assumed circular area of influence becomes distorted in this case. A solution is possible by applying potential theory, by using graphical means, or, if the slope of the piezometric surface is very slight, Eq. 19.7 may be employed without serious error.

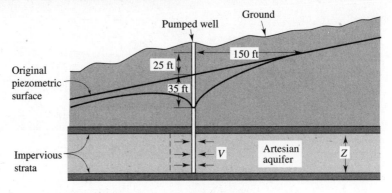

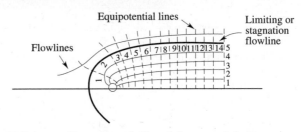

Figure 19.3   Well in a uniform flow field and flow net definition.

Figure 19.3 provides a graphical solution to a uniform flow field problem. First, an ortholgonal flow net consisting of flowlines and equipotential lines must be constructed. This should be done so that the completed flow net will be composed of a number of elements that approach little squares in shape. Once the net is complete, if can be analyzed by considering the net geometry and using Darcy's law in the manner of Todd.[3]

EXAMPLE 19.3 _____

Find the discharge to the well of Fig. 19.3 by using an applicable flow net. Consider the aquifer to be 35 ft thick, $K_f = 3.65 \times 10^{-4}$ fps, and other dimensions as shown.

Solution.   Using Eq. 18.72, we find that

$$q = \frac{Kmh}{n}$$

where $h = 35 + 25 = 60$ ft

$m = 2 \times 5 = 10$

$n = 14$

$q = \dfrac{3.65 \times 10^{-4} \times 60 \times 10}{14}$

$= 0.0156$ cfs per unit thickness of the aquifer

The total discharge Q is thus

$$Q = 0.0156 \times 35 = 0.55 \text{ cfs or } 245 \text{ gpm} \quad \blacksquare\blacksquare$$

19.5 WELL FIELDS

When more than one unit in a well field is pumped, there is a composite effect on the free water surface. This consequence is illustrated by Fig. 19.4 in which the cones of depression are seen to overlap. The drawdown at a given location is equal to the sum of the individual drawdowns.

If, within a particular well field, pumping rates of the pumped wells are known, the composite drawdown at a point can be determined. In like manner, if the drawdown at one point is known, the well flows can be calculated.

If the drawdown at a given point is designated as m, and subscripts $1, 2, \ldots, n$ are used to relate this drawdown to a particular well (e.g., m_1 refers to the drawdown for W_1), for the total drawdown m_T at some location[3]

$$m_T = \sum_{i=1}^{n} = m_i \tag{19.9}$$

The number of wells, their rate of pumping, and well-field geometry and characteristics determine the total drawdown at a specified location.

Again considering Eq. 19.4, we obtain

$$h_0^2 - h^2 = \frac{Q}{\pi K} \ln \frac{r_0}{r} \tag{19.10}$$

It can be seen that the drawdown for a well pumped at rate Q can be computed if h_0, r_0, and r are known. It follows then from Eq. 19.9 that for n pumped wells in an unconfined aquifer

$$h_0^2 - h^2 = \sum_{i=1}^{n} \frac{Q_i}{\pi K} \ln \frac{r_{0i}}{r_i} \tag{19.11}$$

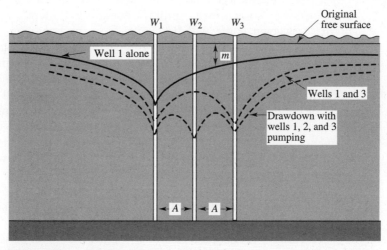

Figure 19.4   Combined effect of pumping several wells at equal rates.

where h_0 = the original height of the water table

h = the combined effect height of the water table after pumping n wells

Q_i = the flow rate of the ith well

r_{0i} = distance of the ith well to a location at which the drawdown is considered negligible

r_i = the distance from well i to the point at which the drawdown is being investigated

Todd indicates that values of r_0 used in practice often range from 500 to 1000 ft.[3] The impact of this assumption is softened because Q in Eq. 19.10 is not very sensitive to r_0. Equation 19.11 should be used only where drawdowns are relatively small.

For flow in a confined aquifer the expression for combined drawdown becomes

$$h_0 - h = \sum_{i=1}^{n} \frac{Q_i}{2\pi Km} \ln \frac{r_{0i}}{r_i} \qquad (19.12)$$

Equations for well flow covering a variety of particular well-field patterns are reported in the literature.[3, 7] Those given here are applicable for steady flow in a homogeneous isotropic medium.

19.6 THE METHOD OF IMAGES

Some groundwater flow problems subjected to boundary conditions negating the direct use of radial flow equations can be transformed into infinite systems fitting these equations by applying the method of images.[2, 8, 9]

When a stream is located near a pumped well and the stream and aquifer are interconnected, the drawdown curve of a pumped well may be affected as shown in Fig. 19.5. Another boundary condition often affecting the drawdown of a well is an impervious formation that limits the extent of the aquifer. The cone of depression of a pumped well is not affected until the boundary is intersected. After that, the shape of the drawdown curve will be changed by the boundary. Boundary effects can frequently be evaluated by means of "image wells." The boundary condition is replaced by either a recharging or a discharging well that is pumped or recharged at a rate equivalent to that of the pumped well. That is, in an infinite aquifer, drawdowns of the real and image wells would be identical. The image well is located at a distance from the boundary equal to that of the real well but on the opposite side (Fig. 19.5). Streams are replaced by recharge wells while impermeable boundaries are supplanted by pumped image wells. Computations for the case of a well and impervious boundary directly follow the procedures outlined under the section on well fields. For the well and stream system, the recharge image well is considered to have a negative discharge. The heads are then added according to this sign convention.

The procedure for combining drawdown curves of real and image wells to obtain an actual drawdown curve is illustrated graphically for the example shown in Fig. 19.5. More detailed information on other cases can be found elsewhere.[9, 10]

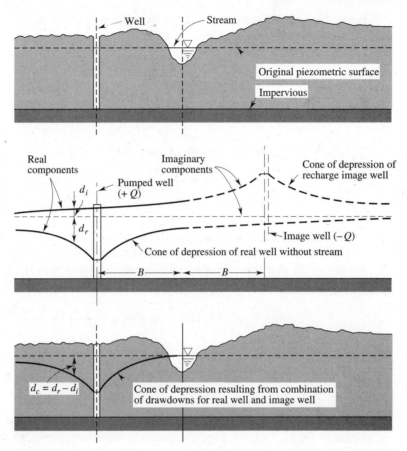

Figure 19.5 Drawdown in a pumping well whose aquifer is connected to a stream.

19.7 UNSTEADY FLOW

When a new well is first pumped, a large portion of the discharge comes directly from the storage volume released as the cone of depression develops. Under these circumstances the equilibrium equations overestimate permeability and therefore the yield of the well. When steady-state conditions are not encountered—as is usually the situation in practice—a nonequilibrium equation must be used. Two approaches can be taken, the rather rigorous method of C. V. Theis or a simplified procedure such as that proposed by Jacob.[11, 12]

In 1935 Theis published a nonequilibrium approach that takes into consideration time and storage characteristics of the aquifer.[11] His method uses an analogy between heat transfer described by the Biot–Fourier law and groundwater flow to a well. The method provides a solution to Eq. 18.41 for given initial and boundary conditions. Application of the method is appropriate for confined aquifers of constant thickness. For use under conditions of unconfined flow, vertical components of flow must be negligible, and changes in aquifer storage through water expansion and aquifer compression must also be negligible relative to the gravity drainage of pores as the water table drops as a result of pumping.[13]

Theis states that the drawdown (s) in an observation well located at a distance r from the pumped well is given by

$$s = \frac{Q}{4\pi T} \int_u^\infty \frac{e^{-u}}{u} \, du \tag{19.13}$$

in which Q = (constant) pumping rate (L^3T^{-1} units), T = aquifer transmissivity (L^2T^{-1} units), and u is a dimensionless variable defined by

$$u = r^2 \frac{S_c}{4tT} \tag{19.14}$$

where r is the radial distance from the pumping well to an observation well, S_c is the aquifer storativity (dimensionless), and t is time. The integral in Eq. 19.13 is usually known as the *well function* of u and is commonly written as $W(u)$. It may be evaluated from the infinite series

$$W(u) = -0.577216 - \ln u + u - \frac{u^2}{2 \times 2!} + \frac{u^3}{3 \times 3!} \cdots \tag{19.15}$$

Using this notation, Eq. 19.13 can be written as

$$s = \frac{QW(u)}{4\pi T} \tag{19.16}$$

The basic assumptions employed in the Theis equation are essentially the same as those in Eq. 19.7 except for the nonsteady-state condition. Some values of the well function are given in Table 19.1.

In American practice, Eqs. 19.13 and 19.14 commonly appear in the following form,

$$s = \frac{114.6Q}{T} \int_u^\infty \frac{e^{-u}}{u} \, du \tag{19.17}$$

$$u = \frac{1.87r^2S_c}{Tt} \tag{19.18}$$

where T is given in units of gpd/ft, Q has units of gpm, and t is the time in days since the start of pumping.

Equations 19.13 and 19.14 can be solved by comparing a log–log plot of u versus $W(u)$ known as a *type curve,* with a log–log plot of the observed data r^2/t versus s. In plotting type curves, $W(u)$ and s are ordinates, u and r^2/t are abscissas. The two curves are superimposed and moved about until segments coincide. In this operation the axes must remain parallel. A coincident point is then selected on the matched curves and both plots marked. The type curve then yields values of u and $W(u)$ for the desired point. Corresponding values of s and r^2/t are determined from a plot of the observed

TABLE 19.1   VALUES OF $W(u)$ FOR VARIOUS VALUES OF u

u	1.0	2.0	3.0	4.0	5.0	6.0	7.0	8.0	9.0
$\times 1$	0.219	0.049	0.013	0.0038	0.0011	0.00036	0.00012	0.000038	0.000012
$\times 10^{-1}$	1.82	1.22	0.91	0.70	0.56	0.45	0.37	0.31	0.26
$\times 10^{-2}$	4.04	3.35	2.96	2.68	2.47	2.30	2.15	2.03	1.92
$\times 10^{-3}$	6.33	5.64	5.23	4.95	4.73	4.54	4.39	4.26	4.14
$\times 10^{-4}$	8.63	7.94	7.53	7.25	7.02	6.84	6.69	6.55	6.44
$\times 10^{-5}$	10.94	10.24	9.84	9.55	9.33	9.14	8.99	8.86	8.74
$\times 10^{-6}$	13.24	12.55	12.14	11.85	11.63	11.45	11.29	11.16	11.04
$\times 10^{-7}$	15.54	14.85	14.44	14.15	13.93	13.75	13.60	13.46	13.34
$\times 10^{-8}$	17.84	17.15	16.74	16.46	16.23	16.05	15.90	15.76	15.65
$\times 10^{-9}$	20.15	19.45	19.05	18.76	18.54	18.35	18.20	18.07	17.95
$\times 10^{-10}$	22.45	21.76	21.35	21.06	20.84	20.66	20.50	20.37	20.25
$\times 10^{-11}$	24.75	24.06	23.65	23.36	23.14	22.96	22.81	22.67	22.55
$\times 10^{-12}$	27.05	26.36	25.96	25.67	25.44	25.26	25.11	24.97	24.86
$\times 10^{-13}$	29.36	28.66	28.26	27.97	27.75	27.56	27.41	27.28	27.16
$\times 10^{-14}$	31.66	30.97	30.56	30.27	30.05	29.87	29.71	29.58	29.46
$\times 10^{-15}$	33.96	33.27	32.86	32.58	32.35	32.17	32.02	31.88	31.76

Source: After L. K. Wenzel, "Methods for Determining Permeability of Water Bearing Materials with Special Reference to Discharging Well Methods," U. S. Geological Survey, Water-Supply Paper 887, Washington, DC, 1942.

data. Inserting these values in Eqs. 19.13 and 19.14 and rearranging, values for transmissibility T and storage coefficient S_c can be found.

Often this procedure can be shortened and simplified. When r is small and t large, Jacob found that values of u are generally small.[12] Thus terms in the series of Eq. 19.15 beyond the second one become negligible and the expression for T becomes

$$T = \frac{264Q(\log t_2 - \log t_1)}{h_0 - h} \tag{19.19}$$

which can be further reduced to

$$T = \frac{264Q}{\Delta h} \tag{19.20}$$

where Δh = drawdown per log cycle of time $[(h_0 - h)/(\log t_2 - \log t_1)]$
Q = well discharge (gpm)
h_0, h = as defined in Fig. 19.2
T = the transmissibility (gpd/ft)

Field data on drawdown $(h_0 - h)$ versus t are drafted on semilogarthmic paper. The drawdown is plotted on an arithmetic scale, Fig. 19.6. This plot forms a straight line whose slope permits computing formation constants using Eq. 19.20 and

$$S_c = \frac{0.3Tt_0}{r^2} \tag{19.21}$$

with t_0 the time corresponding to zero drawdown. Equation 19.21 is obtained through manipulation of Eq. 19.13.

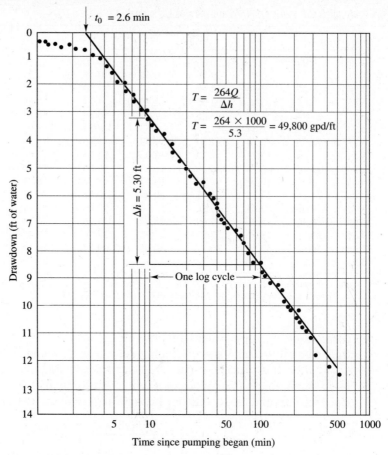

$t_0 = 2.6$ min

$$T = \frac{264Q}{\Delta h}$$

$$T = \frac{264 \times 1000}{5.3} = 49,800 \text{ gpd/ft}$$

$\Delta h = 5.30$ ft

One log cycle

Drawdown (ft of water)

Time since pumping began (min)

Figure 19.6 Pumping test data, Jacob method.

EXAMPLE 19.4 _____

Using the following data, find the formation constants for an aquifer using a graphical solution to the Theis equation. Discharge equals 540 gpm.

Distance from pumped well, r (ft)	r^2/t	Average drawdown, s (ft)
50	1,250	3.04
100	5,000	2.16
150	11,250	1.63
200	20,000	1.28
300	45,000	0.80
400	80,000	0.51
500	125,000	0.33
600	180,000	0.22
700	245,000	0.15
800	320,000	0.10

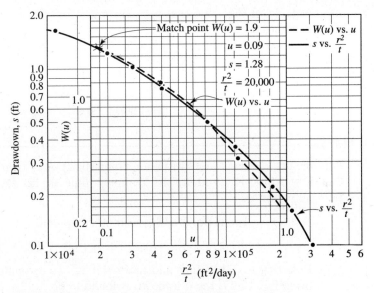

Figure 19.7 Graphical solution to Theis's equation.

Solution.   Plot s versus r^2/t and $W(u)$ versus u as shown in Fig. 19.7. Determine the match point as noted and compute S_c and T using Eqs. 19.7 and 19.8:

$$T = \frac{114.6Q}{s} W(u)$$

$$= \frac{114.6 \times 540}{1.28} \times 1.9 = 91{,}860 \text{ gpd/ft}$$

$$S_c = \frac{uT}{1.87r^2/t}$$

$$= \frac{0.09 \times 91{,}860}{1.87 \times 20{,}000} = 0.22 \quad \blacksquare\blacksquare$$

EXAMPLE 19.5

Using the data given in Fig. 19.6, find the coefficient of transmissibility T and storage coefficient S_c for an aquifer, given $Q = 1000$ gpm and $r = 300$ ft.

Solution.   Find the value of Δh from the graph, 5.3 ft. Then by Eq. 19.20

$$T = \frac{264Q}{\Delta h} = \frac{264 \times 1000}{5.3}$$

$$= 49{,}800 \text{ gpd/ft}$$

Using Eq. 19.21, we find that

$$S_c = \frac{0.3Tt_0}{r^2}$$

Note from Fig. 19.6 that $t_0 = 2.6$ min. Converting to days, we find that this becomes

$$t_0 = 1.81 \times 10^{-3} \text{ days}$$

and

$$S_c = \frac{0.3 \times 49,800 \times 1.81 \times 10^{-3}}{(300)^2}$$

$$= 0.0003 \quad \blacksquare\blacksquare$$

EXAMPLE 19.6 _____

Find the drawdown at an observation point 200 ft away from a pumping well. Given that $T = 3.0 \times 10^4$ gpd/ft, the pumping time is 12 days, $S_c = 3 \times 10^{-4}$, and $Q = 300$ gpm.

Solution. From Eq. 19.18, u can be computed,

$$u = [1.87 \times (200)^2 \times 3 \times 10^{-4}]/[3.0 \times 10^4 \times 12] = 6.23 \times 10^{-5}$$

Referring to Table 19.1 and interpolating, we estimate $W(u)$ to be 9.1. Then, using Eq. 19.17, the drawdown is found to be

$$s = [114.6 \times 9.1 \times 300]/[3.0 \times 10^4] = 10.41 \text{ ft} \quad \blacksquare\blacksquare$$

EXAMPLE 19.7 _____

A well is being pumped at a constant rate of 0.0038 m³/s. Given that $T = 0.0028$ m²/s, $r = 90$ meters, and the storage coefficient $= 0.00098$, find the drawdown in the observation well for a time period of (a) 1,000 sec. and (b) 20 hours.

Solution

a. Using Eq. 19.14, u can be computed as follows,

$$u = [90 \times 90 \times 0.00098]/[4 \times 1000 \times 0.0028]$$
$$u = 0.71$$

Then from Table 19.1, $W(u)$ is found to be 0.36. Applying Eq. 19.16, the drawdown can be determined,

$$s = [0.0038 \times 0.36]/[4 \times \pi \times 0.0028]$$
$$s = 0.039m$$

b. Follow the procedure used in (a)

$$u = [90 \times 90 \times 0.00098]/[4 \times 72,000 \times 0.0028]$$
$$u = 0.0098$$

Then from Table 19.1, $W(u)$ is found to be 4.06 Applying Eq. 19.16, the drawdown can be determined,

$$s = [0.0038 \times 4.06]/[4 \times \pi \times 0.0028]$$
$$s = 0.44m \quad \blacksquare\blacksquare$$

19.8 LEAKY AQUIFERS

The foregoing analyses have dealt with free aquifers or those confined between impervious strata. In reality, many cases exist wherein the confining strata are not completely impervious and water is actually transferred from them to the productive aquifer. The flow regime is altered and computations must include leakage. Since about 1930, leaky aquifers have been the subject of research by investigators such as De Glee, Jacob, Hantush, DeWiest, Walton, Neuman and Witherspoon, and others.[5,14-26] A thorough treatment of their work is beyond the scope of this book; interested readers should consult the indicated references.

19.9 PARTIALLY PENETRATING WELLS

In many actual situations there is only partial penetration of the well. The question then arises as to the applicability of procedures developed previously for full penetration.

Numerous studies of this problem have been conducted.[7,27,28] In 1957 Hantush reported that steady flow to a well just penetrating an infinite leaky aquifer becomes very nearly radial at a distance from the well of about 1.5 times the aquifer thickness.[28] As depth of penetration increases, the approach to radial flow becomes increasingly apparent. Therefore, computation of drawdowns for partially penetrating wells are made using equations for total penetration with relative safety, provided that the distance from the pumped well is greater than 1.5 times the aquifer thickness. At points closer to the well, it is frequently possible to use a flow net or other relations developed for this region.

19.10 FLOW TO AN INFILTRATION GALLERY

An infiltration gallery may be defined as a partially pervious conduit constructed across the path of the local groundwater flow such that all or part of this flow will be intercepted. These galleries are often built in a valley area parallel to a stream so that they can convey the collected flow to some designated location under gravity-flow conditions. Figure 19.8 shows a typical cross section through a gallery with one pervious face.

Computation of discharge to an infiltration gallery with one pervious wall (Fig. 19.8) is accomplished in the manner outlined by Dupuit.[29] Several assumptions must be made to effect the solution. They are that the sine and tangent of the angle of inclination of the water table are interchangeable; that the velocity vectors are everywhere horizontal and uniformly distributed; that the soil is incompressible and isotropic; and that the gallery is of sufficient length that end effects are negligible. While permitting a solution of the problem, these assumptions do limit the utility of the results.

Based on these assumptions, and following the procedure given in Section 18.12, Eq. 18.97 can be used to calculate the discharge per unit width. Using the

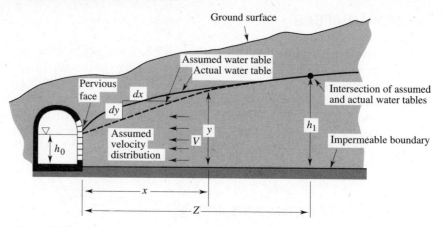

Figure 19.8   Cross-section through an infiltration gallery.

nomenclature of Fig. 19.8, Eq. 18.97 becomes

$$q = \frac{K}{2Z}(h_1^2 - h_0^2)$$

This equation indicates that the computed water table is parabolic. This is often called Dupuit's parabola. Figure 19.8 shows that the computed water table differs from the actual water table in an increasing manner as the gallery face is approached. It is therefore apparent that the computed parabola does not accurately describe the real water table. The differences, however, are small except near the point of outflow, providing the initial assumptions are satisfied. The calculated discharge approximates the true discharge more closely as the ratio of Z/h_1 increases.

EXAMPLE 19.8

A stratum of clean sand and gravel 20 ft deep has a coefficient of permeability of $K = 3.25 \times 10^{-3}$ ft/sec, and is supplied with water from a channel that penetrates to the bottom of the stratum. If the water surface in an infiltration gallery is 3 ft above the bottom of the stratum, and its distance to the channel is 50 ft, what is the flow into a foot of gallery? Use Eq. 18.97.

Solution

$$q = 0.5(3.25 \times 10^{-3})(20 \times 20 - 3 \times 3)/50$$
$$q = 0.012 \text{ cfs, the flow into one foot of gallery} \quad \blacksquare\blacksquare$$

19.11 SALTWATER INTRUSION

The contamination of fresh groundwater by the intrusion of salt water often presents a serious quality problem. Islands and coastal regions are particularly vulnerable. Aquifers located inland sometimes contain highly saline waters as well. Fresh water is lighter than salt water (specific gravity of the latter is about 1.025) and forms a freshwater layer above the underlying salt water. This equilibrium is disturbed when an aquifer is pumped, since salt water replaces the fresh water removed. Under equilibrium conditions, a drawdown of 1 ft in a freshwater table corresponds to a rise of about 40 ft by salt water. Wells subjected to saltwater intrusion obviously have limited pumping rates.

Recharge wells have been drilled in coastal areas to maintain a head sufficient to preclude seawater intrusion, a practice employed effectively in southern California.

19.12   GROUNDWATER BASIN DEVELOPMENT

To use groundwater resources efficiently while simultaneously permitting the maximum development of the resource, equilibrium must be established between withdrawals and replenishments. Economic, legal, political, social, and water quality aspects require full consideration.

Lasting supplies of groundwater will be assured only when long-term withdrawls are balanced by recharge during the corresponding period. The potential of a groundwater basin can be assessed by employing the water budget equation,

$$\sum I - \sum O = \Delta S$$

where the inflow $\sum I$ includes all forms of recharge, the total outflow $\sum O$ includes every kind of discharge, and ΔS represents the change in storage during the accounting period. The most significant forms of recharge and discharge are those listed in Table 19.2.

A groundwater hydrologist must be able to estimate the quantity of water that can be economically and safely produced from a groundwater basin in a specified time period. He or she should also be competent to evaluate the consequences of imposing various rates of withdrawal on an underground supply.

Development of groundwater basins should be based on careful study, since groundwater resources are finite and exhaustible. If the various types of recharge balance the withdrawals from a basin over a period of time, no difficulty will be encountered. Excessive drafts, however, can deplete underground water supplies to a point where economic development is not feasible. The mining of water will ultimately deplete the entire supply.

TABLE 19.2   SOME FORMS OF RECHARGE AND DISCHARGE

Recharge	Discharge
Seepage from streams, ponds, lakes	Seepage to lakes, streams, springs
Subsurface inflows	Subsurface outflows
Infiltrated precipitation	Evapotranspiration
Water recharged artifically	Pumping or other artificial means of collection

■ Summary

The collection of groundwater is accomplished primarily through the construction of wells, and many factors influence the numerical estimation of their performance. Some situations are amenable to solution through the utilization of relatively simple mathematical expressions. Others depend upon sophisticated application of the hydrodynamic equations under various conditions of nonuniformity of aquifer materials and a variety of boundary conditions. The reader is cautioned not to be misled by the simplicity of some of the solutions presented and to observe that many of these relate to special conditions and are not applicable to all groundwater-flow situations.

The rate of movement of water through the ground is of a different magnitude than that through natural or artificial channels or conduits. Typical flow rates range from 5 ft/day to a few feet per year. These low rates of flow exacerbate the impact of contaminant spills on groundwater sources and complicate cleanup since natural flushing from the site may take many years to occur.

The methods described in this chapter for estimating flows to collection devices are based mainly on the principles of fluid flow embodied in Darcy's law. Applications are limited to flows in the laminar range, but under most conditions encountered in the field, Darcy's law applies. Examples of the use of equations describing the mechanics of flow to wells and infiltration galleries were given in this chapter. Both steady-state and unsteady flow conditions were addressed as well.

PROBLEMS

19.1. A 12-in. well fully penetrates a confined aquifer 100 ft thick. The coefficient of permeability is 600 gpd/ft^2. Two test wells located 40 and 120 ft away show a difference in drawdown between them of 9 ft. Find the rate of flow delivered by the well.

19.2. A 12-in. well fully penetrates a confined aquifer 100 ft thick. The coefficient of permeability is 600 gpd/ft^2. Two test wells located 45 and 120 ft away show a difference in drawdown between them of 8 ft. Find the rate of flow delivered by the well.

19.3. Determine the permeability of an artesian aquifer for a fully penetrating well. The aquifer is composed of medium sand and is 100 ft thick. The steady-state pumping rate is 1200 gpm. The drawdown in an observation well 75 ft away is 14 ft, and the drawdown in a second observation well 500 ft away is 1.2 ft. Find K in gallons per day per square foot.

19.4. Consider a confined aquifer with a coefficient of transmissibility T of 680 ft^3/day/ft. At $t = 5$ min, the drawdown $s = 5.6$ ft; at 50 min, $s = 23.1$ ft; and at 100 min, $s = 28.2$ ft. The observation well is 75 ft away from the pumping well. Find the discharge of the well.

19.5. Given the following data: $Q = 59,000$ ft^3/day, $T = 630$ ft^3/day, $t = 30$ days, $r = 1$ ft, and $S_c = 6.4 \times 10^{-4}$. Consider this to be a nonequilibrium problem. Find the drawdown s. Note that for

$$u = 8.0 \times 10^{-9} \qquad W(u) = 18.06$$
$$u = 8.2 \times 10^{-9} \qquad W(u) = 18.04$$
$$u = 8.6 \times 10^{-9} \qquad W(u) = 17.99$$

19.6. Determine the permeability of an artesian aquifer being pumped by a fully penetrating well. The aquifer composed of medium sand is 130 ft thick. The steady-state pumping rate is 1300 gpm. The drawdown in an observation well 65 ft away is 12 ft, and in a second well 500 ft away is 1.2 ft. Find K_f in gpd/ft^2.

19.7. Consider a confined aquifer with a coefficient of transmissibility $T = 700$ ft^3/day-ft. At $t = 5$ min the drawdown $= 5.1$ ft; at 50 min, $s = 20.0$ ft; at 100 min, $s = 26.2$ ft. The observation well is 60 ft from the pumping well. Find the discharge of the well.

19.8. Assume that an aquifer being pumped at a rate of 300 gpm is confined and pumping test data are given as follows. Find the coefficient of transmissibility T and the storage coefficient S. Assume $r = 55$ ft.

Time since pumping started (min)	1.3	2.5	4.2	8.0	11.0	100.0
Drawdown s (ft)	4.6	8.1	9.3	12.0	15.1	29.0

19.9. We are given the following data:

$$Q = 60,000 \text{ ft}^3/\text{day} \qquad t = 30 \text{ days} \quad r = 1 \text{ ft}$$
$$T = 650 \text{ ft}^3/(\text{day})(\text{ft}) \qquad S_c = 6.4 \times 10^{-4}$$

Assume this to be a nonequilibrium problem. Find the drawdown s. Note for

$$u = 8.0 \times 10^{-9} \qquad W(u) = 18.06$$
$$u = 8.2 \times 10^{-9} \qquad W(u) = 18.04$$
$$u = 8.6 \times 10^{-9} \qquad W(u) = 17.99$$

19.10. An 18-in. well fully penetrates an unconfined aquifer 100 ft deep. Two observation wells located 90 and 235 ft from the pumped well are known to have drawdowns of 22.5 and 20.6 ft, respectively. If the flow is steady and $K_f = 1300$ gpd/ft^2, what would be the discharge?

19.11. A confined aquifer 80 ft deep is being pumped under equilibrium conditions at a rate of 700 gpm. The well fully penetrates the aquifer. Water levels in observation wells 150 and 230 ft from the pumped well are 95 and 97 ft, respectively. Find the field coefficient of permeability.

19.12. A well is pumped at the rate of 500 gpm under nonequilibrium conditions. For the data listed, find the formation constants S and T. Use the Theis method.

r^2/t	Average drawdown, h (ft)
1,250	3.24
5,000	2.18
11,250	1.93
20,000	1.28
45,000	0.80
80,000	0.56
125,000	0.38
180,000	0.22
245,000	0.15
320,000	0.10

19.13. We are given a well pumping at a rate of 590 gpm. An observation well is located at $r = 180$ ft. Find S and T using the Jacob method for the following test data.

Drawdown (ft)	Time (min)	Drawdown (ft)	Time (min)
0.43	26	2.00	661
0.94	78	2.06	732
1.08	99	2.12	843
1.20	131	2.15	926
1.34	173	2.20	1034
1.46	218	2.23	1134
1.56	266	2.28	1272
1.63	303	2.30	1351
1.68	331	2.32	1419
1.71	364	2.36	1520
1.85	481	2.38	1611
1.93	573		

19.14. A 24-in. diameter well penetrates the full depth of an unconfined aquifer. The original water table and a bedrock aquifuge were located 50 and 150 ft, respectively, below the land surface. After pumping at a rate of 1700 gpm continuously for 1920 days, equilibrium drawdown conditions were established, and the original water levels in observation wells located 1000 and 100 ft from the center of the pumped well were lowered 10 and 20 ft, respectively.
a. Determine the field permeability (gpd/ft^2) of the aquifer.
b. For the same well, zero drawdown occurred outside a circle with a 10,000-ft radius measured from the center of the pumped well. Inside the circle, the average drawdown in the water table was observed to be 10 ft. Determine the coefficient of storage of the aquifer.

19.15. A well fully penetrates the 100-ft depth of a saturated unconfined aquifer. The drawdown at the well casing is 40 ft when equilibrium conditions are established using a constant discharge of 50 gpm. What is the drawdown when equilibrium is established using a constant discharge of 66 gpm?

19.16. After a long rainless period, the flow in Wahoo Creek decreases by 8 cfs from Memphis downstream 8 mi to Ashland. The stream penetrates an unconfined aquifer, where the water table contours near the creek parallel the west bank and slope to the stream by 0.00020, while on the east side the contours slope away from the stream toward the Lincoln wellfield at 0.00095. Compute the transmissivity of the aquifer knowing $Q = TIL$, where I is the slope and L is the length.

19.17. The time–drawdown data for an observation well located 300 ft from a pumped artesian well (500 gpm) are given in the following table. Find the coefficient of storage (ft^3 of water/ft^3 of aquifer) and the transmissivity (gpd/ft) of the aquifer by the Theis method. Use 3 × 3 cycle log paper.

Time (hr)	Drawdown (ft)	Time (hr)	Drawdown (ft)
1.8	0.27	9.8	1.09
2.1	0.30	12.2	1.25
2.4	0.37	14.7	1.40
3.0	0.42	16.3	1.50
3.7	0.50	18.4	1.60
4.9	0.61	21.0	1.70
7.5	0.84	24.4	1.80

19.18. Over a 100-mi^2 surface area, the average level of the water table for an unconfined aquifer has dropped 10 ft because of the removal of 128,000 area-ft of water from the aquifer. Determine the storage coefficient for the aquifer. The specific yield is 0.2 and the porosity is 0.22.

19.19. Over a 100-mi^2 surface area, the average level of the piezometric surface for a confined aquifer in the Denver area has declined 400 ft as a result of long-term pumping. Determine the amount of the water (acre-ft) pumped from the aquifer. The porosity is 0.3 and the coefficient of storage is 0.0002.

19.20. Find the drawdown at an observation point 250 ft away from a pumping well, given that $T = 3.1 \times 10^4$ gpd/ft, the pumping time is 10 days, $S_c = 3 \times 10^{-4}$, and $Q = 280$ gpm.

19.21. Find the permeability of an artesian aquifer being pumped by a fully penetrating well. The aquifer is 130 ft thick and is composed of medium sand. The steady-state pumping rate is 1300 gpm. The drawdown in an observation well 65 ft away is 12 ft, and in a second well 500 ft away it is 1.2 ft. Find K_f in gpd/ft^2.

19.22. An 18 in. well fully penetrates an unconfined aquifer 100 ft deep. Two observation wells located 90 and 235 ft from the pumped well are known to have drawdowns of 22.5 ft and 20.6 ft respectively. If the flow is steady and $K_f = 1300$ gpd/ft^2, what would be the discharge?

19.23. A well is being pumped at a constant rate of 0.004 m^3/s. Given that $T = 0.0028$ m^2/s, $r = 100$ meters, and the storage coefficient $= 0.001$, find the drawdown in the observation well for a time period of (a) 1 hr, and (b) 24 hours.

19.24. A well is being pumped at a constant rate of 0.003 m^3/s. Given that $T = 0.0028$ m^2/s, the storage coefficient $= 0.001$, and the time since pumping began is 12 hours, find the drawdown in an observation well for a radial distance of (a) 150 m, and (b) 500 m.

REFERENCES

1. J. W. Clark, W. Viessman, Jr., and M. J. Hammer, *Water Supply and Pollution Control,* 2nd ed. New York: Thomas Y. Crowell, 1965.
2. John F. Hoffman, "Field Tests Determine Potential Quantity, Quality of Ground Water Supply," *Heating, Piping, and Air Conditioning* (Aug. 1961).
3. D. K. Todd, *Groundwater Hydrology,* New York: Wiley, 1960.

4. John F. Hoffman, "How Underground Reservoirs Provide Cool Water for Industrial Uses," *Heating, Piping, and Air Conditioning* (Oct. 1960).

5. John F. Hoffman, "Well Location and Design," *Heating, Piping, and Air Conditioning* (Aug. 1963).

6. G. Thiem, *Hydrologische Methodern.* Leipzig: Gebhardt, 1906, p. 56.

7. R. A. Freeze and J. A. Cherry, *Groundwater,* Englewood Cliffs, NJ: Prentice-Hall, 1979.

8. C. E. Jacob, "Flow of Groundwater," in *Engineering Hydraulics* (Hunter Rouse, ed.). New York: Wiley, 1950.

9. C. O. Wisler and E. F. Brater, *Hydrology.* New York: Wiley, 1959.

10. R. J. M. DeWiest, *Geohydrology.* New York: Wiley, 1965.

11. C. V. Theis, "The Relation Between the Lowering of the Piezometric Surface and the Rate and Duration of Discharge of a Well Using Ground Water Storage," *Trans. Am. Geophys. Union* **16,** 519–524(1935).

12. H. H. Cooper, Jr. and C. E. Jacob, "A Generalized Graphical Method for Evaluating Formation Constants and Summarizing Well-Field History," *Trans. Am. Geophys. Union* **27,** 526–534(1946).

13. D. B. McWhorter and D. K. Sunada, *Ground Water Hydrology and Hydraulics.* For Collins, CO: Water Resources Publications, 1977.

14. C. E. Jacob, "Radial Flow in a Leaky Artesian Aquifer," *Trans. Am Geophys. Union* **27,** 198–205(1946).

15. M. S. Hantush, "Plain Potential Flow of Groundwater with Linear Leakage," Ph.D. dissertation, University of Utah, 1949.

16. M. S. Hantush and C. E. Jacob, "Nonsteady Radial Flow in an Infinite Leaky Aquifer and Nonsteady Green's Functions for an Infinite Strip of Leaky Aquifer," *Trans. Am. Geophys. Union* **36,** 95–112(1955).

17. M. S. Hantush and C. E. Jacob, "Flow to an Eccentric Well in a Leaky Circular Aquifer," *J. Geophys. Res.* **65,** 3425–3431(1960).

18. M. S. Hantush, "Analysis of Data from Pumping Tests in Leaky Aquifers," *Trans. Am. Geophys. Union* **37,** 702–714(1956).

19. M. S Hantush, "Modification of the Theory of Leaky Aquifers," *J. Geophys. Res.* **65,** 3713–3725(1960).

20. R. J. M. DeWiest, "On the Theory of Leaky Aquifers," *J. Geophys. Res.* **66,** 4257–4262(1961).

21. R. J. M. DeWiest, "Flow to an Eccentric Well in a Leaky Circular Aquifer with Varied Lateral Replenishment," *Geofis, Pura Aplic.* **54,** 87–102(1963).

22. W. C. Walton, "Leaky Artesian Aquifer Conditions in Illinois," Report of Investigation No. 39, Illinois State Water Survey, 1960.

23. William C. Walton, *Groundwater Resource Evaluation.* New York: McGraw-Hill, 1970.

24. S. P. Neuman and P. A. Witherspoon. 1969a. Theory of flow in a confined two-aquifer system. *Water Resources Res.,* 5, pp. 803–816.

25. S. P. Neuman and P. A. Witherspoon. 1969b. Applicability of current theories of flow in leaky aquifers. *Water Resources Res.,* 5, pp. 817–829.

26. S. P. Neuman and P. A. Witherspoon. 1972. Field determination of the hydraulic properties of leaky multiple-aquifer systems. *Water Resources Res.,* 8, pp. 1284–1298.

27. D. Kirkham, "Exact Theory of Flow into a Partially Penetrating Well," *J. Geophys. Res.* **64,** 1317–1327(1959).

28. M. S. Hantush, "Nonsteady Flow to a Well Partially Penetrating an Infinite Leaky Aquifer," *Proc. Isaqi Sci. Soc.* **1,** 10–19(1957).

29. Jules Dupuit, *Etudes théoriques et pratiques sur le mouvement des eau dans les canaux de couverts et à travers les terrains perméables,* 2nd ed. Paris: Dunod, 1863.

Chapter 20

Modeling Regional Groundwater Systems

■ Prologue

The purpose of this chapter is to:

- Describe the features of large-scale groundwater systems.
- Introduce the principles of finite difference approaches to modeling regional groundwater systems.
- Illustrate the application of groundwater modeling techniques to the Upper Big Blue basin in Nebraska.

The analytical methods described so far have been applicable mainly to the flow of water to individual wells. In this chapter, the concepts of analyzing regional groundwater systems are introduced. Such analyses are requisites for the wise development, management, and operation of expansive groundwater resources.

Given that water quantity and quality aspects must be dealt with jointly in most water resources decision-making processes, regional groundwater models must often be designed to include both of these dimensions.[1-25] The fluid flow aspects of groundwater models are presented in this chapter. Solute transport models are complex and beyond the scope of this book, but a brief introduction to them is given at the end of the chapter. It is important for the reader to understand the importance and role of these water quality-oriented models.

20.1 REGIONAL GROUNDWATER MODELS

Groundwater systems models may be of the analog or the digital (mathematical) variety. The focus of this chapter is on the digital type of model, the type most commonly employed today. Such models are characterized by a set of equations representing the physical processes occurring in an aquifer. These models may be deterministic or probabilistic in nature, but only deterministic models are discussed here. They describe the cause-effect relations stemming from known features of the physical system under study.

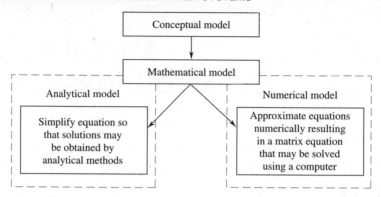

Figure 20.1   Logic diagram for developing a mathematical model. (Courtesy of the National Water Well Association, Worthington, OH.)

Figure 20.1 characterizes the procedure for developing a deterministic mathematical model. A conceptual model is formulated based on a knowledge of the characteristics of the region and an understanding of the mechanics of groundwater flow. The next step is to translate the conceptual model into a mathematical model, usually represented by a partial differential equation or set of equations accompanied by appropriate boundary and initial conditions. Conditions of continuity and conservation of momentum, usually described by Darcy's law, are incorporated in the model. Other model features include artesian or water table condition designation and dimensionality (one-, two-, or three-dimensional). If water quality and/or heat transfer considerations are to be incorporated in the model, additional equations describing conservation of mass for the chemical species involved and conservation of energy are required. Typically used relations are Fick's law for chemical diffusion and Fourier's law for heat transport.

Once the mathematical model has been formulated, it can be applied to the situation at hand. This requires converting the governing equations into forms that facilitate solution. Ordinarily this is achieved through the use of numerical methods such as finite differences or finite elements to represent the applicable partial differential equations. In using a finite difference approach, for example, the region is divided into grid elements and the continuous variables are represented as discrete variables at the nodal points. In this manner, the governing differential equation is replaced by a finite number of algebraic expressions that can be solved in an iterative way. Models of this type find wide application in the estimation of site-specific aquifer behavior. They have proven to be effective under irregular boundary conditions, where there are heterogeneities, and where highly variable pumping or recharge rates are expected.[1] Several types of groundwater models and their applications are summarized in Figure 20.2.

A number of steps must be followed in modeling a targeted groundwater region. Figure 20.3 is illustrative. The first step is to define the boundaries. They may be physical, such as an impervious layer, or arbitrary, such as the choice of a politically,

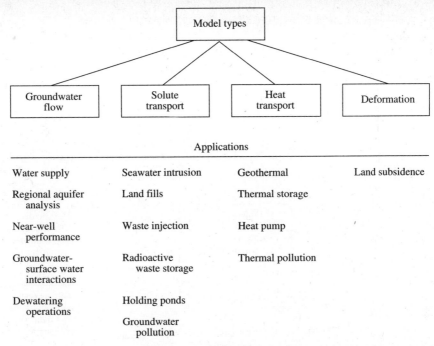

Figure 20.2 Types of groundwater models and typical applications. (Courtesy of the National Water Well Association, Worthington, OH.)

or otherwise, defined subregion. Next, the region is divided into discrete elements by superimposing a rectangular or polygonal grid (see Figure 20.4).

Once the grid is determined, the controlling aquifer parameters (S_c and T) and the initial conditions are set for each grid element. If solute transport is included in the model, additional parameters such as hydrodynamic dispersion properties must also be specified. After all of these specifications have been met, the model can be operated and its output compared with recorded history (history matching). Comparisons of recorded values of head and other features with counterpart model predictions permit parameter adjustments to be made until observed and computed data are considered by the modeler to be in close agreement.

Upon completion of the model's calibration, it can be applied to analyze a variety of management and/or development options. The model's prediction of the outcomes of these alternative strategies can be a valuable aid to decision-making processes. Examples of the types of problems that can be addressed include: the ability of an aquifer to support various levels of use; the impact on an aquifer of varying natural and artificial recharge rates; the effects on underground storage of well location, spacing, and pumping rate; the rate of movement of subsurface contaminants; and saltwater intrusion.

While numerical groundwater models have much to recommend them, caution must be exercised to ensure that they are used and interpreted appropriately. Prickett notes that overkill, inappropriate prediction, and misinterpretation are three ways in which groundwater models can be misused.[5] To avoid these pitfalls, both the modeler

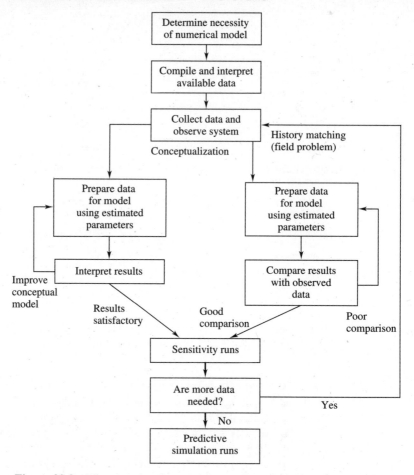

Figure 20.3   Diagram showing model use. (Courtesy of the National Water Well Association, Worthington, OH.)

and user must understand the underlying assumptions upon which the model was founded, its limitations, and its sources of errors. Used wisely, models can be powerful decision-making aids. Used inappropriately, they can lead to erroneous and sometimes damaging proposals.

20.2 FINITE-DIFFERENCE METHODS

Digital simulation requires an adequate mathematical description of the physical processes to be modeled. For groundwater flow this description consists of a partial differential equation and accompanying boundary and initial conditions. The governing equation is integrated to produce a solution that gives the water levels or heads associated with the aquifer being studied at selected points in space and time. The model can simulate years of physical activity in a span of seconds, so that the

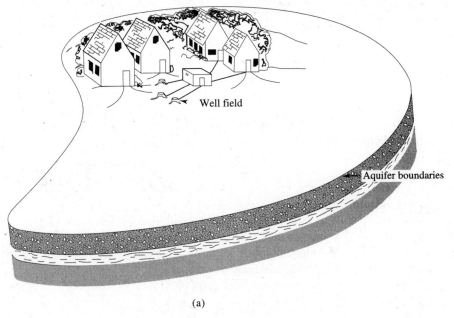

(a)

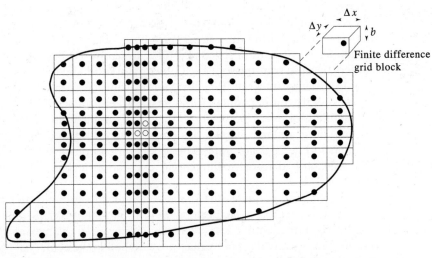

(b)

Figure 20.4 (a) Map view of aquifer showing well field and boundaries. (b) Finite-difference grid for aquifer study, where Δx is the spacing in the x direction, Δy is the spacing in the y direction, and b is the aquifer thickness. Solid dots: block–center nodes; open circles: source–sink nodes. (Courtesy of the National Water Well Association, Worthington, OH.)

consequences of proposed actions can be evaluated before decisions involving construction or social change are implemented. The expectation is that the model runs will lead to wiser and more cost-effective decisions.

The finite-difference method is based on the subdivision of an aquifer into a grid and the analysis of flows associated with zones of the aquifer. The equation that must be solved is derived from continuity considerations and Darcy's law for groundwater motion. This yields the following partial differential equation (a version of Eq. 18.41), describing flow through an areally extensive aquifer. Note that the equation presented here describes the two-dimensional case:

$$\frac{\partial(T\,\partial h/\partial x)}{\partial x} + \frac{\partial(T\,\partial h/\partial y)}{\partial y} = S\frac{\partial h}{\partial t} + W \tag{20.1}$$

where h = total hydraulic head (L),
   x = x direction in a cartesian coordinate system (L),
   y = y direction in a cartesian coordinate system (L),
   S = specific yield of the aquifer (dimensionless),
   T = transmissivity of the aquifer (L^2/T)
   W = source and sink term (L/T)

In the above equation, vertical flow velocities are considered to be negligible everywhere in the aquifer. The following assumptions are implict in the derivation: the flow is two-dimensional; fluid density is constant in time and space; hydraulic conductivity is uniform within the aquifer; flow obeys Darcy's law; and the specific yield of the aquifer is constant in space and time. Equation 20.1 is nonlinear for unconfined aquifers because transmissivity is a function of head and thus the dependent variable.

In order to integrate Eq. 20.1, initial values of head, transmissivity, saturated thickness of the aquifer, and the amounts of water produced by sources and sinks must be identified for every point in the region of the integration. The specific yield and location of geometric boundaries must also be defined. Unfortunately, analytic solutions to Eq. 20.1 are impossible to obtain except for the most trivial cases. It is thus necessary to resort to numerical integration techniques to obtain the desired answers.[8, 10–14]

Application of finite-difference techniques to groundwater flow problems requires that the region of concern be divided into many small subregions or elements (Fig. 20.5). For each of these elements, characteristic values of all the variables in Eq. 20.1 are specified. These values are assigned to the centers of the elements, which are called nodes. The heads in adjacent nodes are related through a finite-difference equation, which is derived from Eq. 20.1. These difference equations can be derived by an appropriate Taylor's series expansion or by mass balance considerations.[8] The resulting algebraic equations can then be solved simultaneously to yield the heads at each node for each time step considered.

It should be understood that the simulation methods presented in this chapter are pointed toward the analysis of regional rather than localized groundwater problems such as the prediction of the drawdown at a particular well. In such cases, the methods discussed in Chapter 19 are usually the most appropriate. Here we are mainly concerned with water level or head changes that might occur over a large area due to prescribed water-use practices.

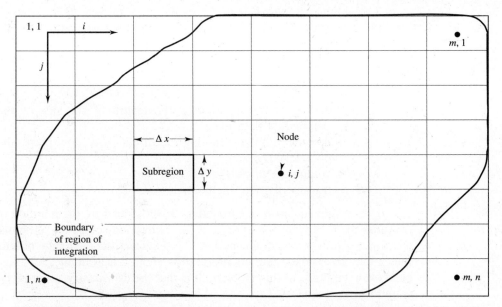

Figure 20.5 Subdivision of a region of integration into computational elements for a finite-difference problem formulation.

Boundary Conditions

In order to integrate Eq. 20.1, the governing boundary conditions must be specified. Two types of boundary condition are discussed here. Others were presented in Chapters 18 and 19.

Where the region of integration is limited by a political or arbitrarily chosen boundary, it is often the policy to employ a constant-gradient boundary condition.[10] In this case, an assumption is made that the gradient of the water table will not change along the boundary even though the water level may rise or fall. Where streams with interconnections to the groundwater system are encountered, stream boundary conditions are employed. Constant-gradient boundaries are expressed mathematically as

$$\frac{\partial h}{\partial s} = g(x, y) \tag{20.2}$$

where $g(x, y)$ = a constant specified at the location x, y throughout the period of
    simulation (dimensionless)
    h = hydraulic head (L)
    s = direction perpendicular to the boundary (L)

Stream boundaries are expressed as

$$h = f(x, y, t) \tag{20.3}$$

where $f(x, y, t)$ = an unknown function of time at the location x, y
    (dimensionless)
    h = hydraulic head (L)

The volumetric rate of flow across the constant-head boundaries described by Eq. 20.2 can be modeled at each time step using the Darcy equation:[10]

$$Q = T \frac{\partial h}{\partial s} \Delta l \tag{20.4}$$

where h = head (L)

Δl = dummy variable denoting the length of the side of the subregion perpendicular to s (L)

s = dummy variable denoting the direction of flow perpendicular to the boundary (L)

Q = volumetric discharge (L³/T)

T = transmissivity at the boundary (L²/T)

Use of this equation at a boundary is illustrated by the notation of Fig. 20.6. Consider the flow from left to right in the x direction across the left-hand side of the elemental region depicted. The node $i - 1, j$ lies outside the region of integration and thus it may be assumed that no information about it is available. An assumption may be made to circumvent this problem. It is that the transmissivity across the boundary is uniform and equal to $T_{i,j}$.

In finite-difference form the head change term in Eq. 18.26 can be stated as

$$\frac{\partial h}{\partial x} = \frac{h_{i,j} - h_{i-1,j}}{\Delta x} \tag{20.5}$$

But the head $h_{i-1,j}$ does not exist, and another approximation is required,

$$h_{i,j} - h_{i-1,j} \simeq h_{i+1,j} - h_{i,j} \tag{20.6}$$

These two expressions are then substituted in Eq. 20.4 to yield

$$Q_{i-1/2,j} \simeq T_{i,j} \frac{h_{i+1,j} - h_{i,j}}{\Delta x} \Delta y \tag{20.7}$$

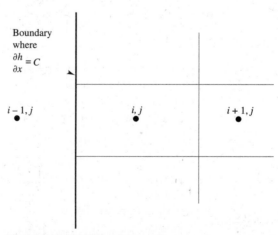

Boundary where $\frac{\partial h}{\partial x} = C$

$i-1,j$

i,j

$i+1,j$

Figure 20.6 Subregions adjacent to a constant-gradient boundary.

At the beginning of each time step, a new volumetric flux is calculated along each constant-gradient boundary. This is accomplished by using the heads and transmissivities computed in the previous time interval.

Surface streams are sometimes treated as constant-head boundaries in groundwater problems. The assumption is adequate where the water level in the surface body is expected to remain unchanged during the time period of the modeling process. In many instances, however, surface flows, and hence heads, are significantly affected by withdrawals or recharges to the interconnected groundwater system. They may then be a limited source of water supply for the groundwater system. To accommodate the surface water-groundwater linkage, a leakage term may be applied.[10] This expression may take the form

$$\text{leakage}_{i,j,k} = -\frac{K_{i,j}}{b_{i,j}}(h_{i,j,k} - h_{i,j,0}) \tag{20.8}$$

where $b_{i,j}$ = thickness of the streambed (L)
$h_{i,j,k}$ = head in the aquifer at node i, j, at time k; $k = 0$ indicates initial conditions (L)
$k_{i,j}$ = hydraulic conductivity at node i, j (L/T).

When Eq. 20.8 is used, the stream is considered to cover the entire area represented by the related node. After each time step the leakage from the stream to the aquifer is calculated and streamflows are depleted accordingly. If the streamflow at a particular node becomes zero, the model can be made to note that the stream is dry and break the hydraulic connection at that point.[10]

Time Steps and Element Dimensions

The success of any finite-difference scheme depends on the incremental values assigned the element dimensions and the time steps. In general, the smaller the dimensions of elements and time increments, the closer the finite-difference approximation to the differential equation. However, as these partitions are made smaller, a price in computational costs and data needs must be paid. Furthermore, oversubdivision may even bring about computational intractability. Thus the object is to select the degree of definition that results in an adequate representation of the system while keeping data and computational costs at a minimum. There are procedures for making such selections, but, except for a brief discussion in the following section, they are not presented here.[10-14]

One-Dimensional Flow Model

To illustrate the finite-difference approach to groundwater problem solving, a one-dimensional conceptualization is discussed. Although most practical-scale models are two- or three-dimensional in character, their development is only an extension of the one-dimensional case. For details of some of the more complex models the reader should consult the appropriate references.[6-8, 10-15] The book by McWhorter and Sunada is easy to read and includes excellent example problems.[8] The treatment of one-dimensional flow taken here follows the approach of that reference.

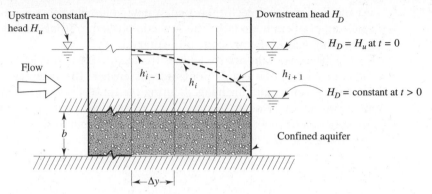

Figure 20.7   Grid notation for a one-dimensional groundwater flow case. (After McWhorter and Sunada.[8])

Let us consider a one-dimensional flow in a confined aquifer system such as that illustrated by Fig. 20.7. It is assumed that the flow is unsteady and that the flowlines are parallel and not time dependent. On this basis, a unit width of the aquifer can be studied and observations made about it can easily be translated to the total system. As shown in the figure, the unit width of the aquifer is Δx. The flow region is overlaid by a grid, and for each grid element, values of hydraulic conductivity K_i, element length y_i, aquifer thickness b_i, storage coefficient S_i, and the initial values of head h_i must be specified. The mass balance for grid element i requires that the inflow ($Q_{i-1 \to i}$) from element $i - 1$ to element i minus the outflow ($Q_{i \to i+1}$) from element i to element $i + 1$ must be balanced by the rate of change in storage which occurs in element i, $\Delta V_i/\Delta t$.

To simplify the problem, let us further consider that the aquifer is of uniform thickness and that it is homogeneous and isotropic. Thus the values of K, S, b, and Δy are constant, and we shall consider that from studies of the aquifer properties, they are also known. Therefore it may be stated that

$$K_1 = K_2 = \cdots = K_m = K$$
$$S_1 = S_2 = \cdots = S_m = S$$
$$b_1 = b_2 = \cdots = b_m = b$$
$$\Delta y_1 = \Delta y_2 = \cdots = \Delta y_m = \Delta y \qquad (20.9)$$

where the subscript m represents the total number of grid elements. Using this notation and Fig. 20.7, we can see that the flow from element $i - 1$ to i is

$$Q_{i-1 \to i} = -KA\frac{h_i^n - h_{i-1}^n}{\Delta y} \qquad (20.10)$$

where i = the element number
          n = the selected time

Equation 20.10 is recognized as Darcy's equation. It is assumed in this representation that the head generating the flow at time n is the difference between the average heads at the two adjacent elements divided by the distance between their centers (nodes). This approximation approaches exactness as Δy diminishes to zero.

The area A appearing in Eq. 20.10 is the cross-sectional area of flow and is obtained as the product of Δx and b. Since we are dealing with a unit width of aquifer, $\Delta x = 1$ and since b is a constant by definition here, Eq. 20.10 may be written

$$Q_{i-1 \rightarrow i} = -T\frac{h_i^n - h_{i-1}^n}{\Delta y} \tag{20.11}$$

where $T = Kb$. A similar expression for the flow from element i to $i + 1$ may be obtained:

$$Q_{i \rightarrow i+1} = -T\frac{h_{i+1}^n - h_i^n}{\Delta y} \tag{20.12}$$

Equations 20.11 and 20.12 represent the inflow and outflow from element i. Considering that continuity conditions must be met, this change in flow across the element must be balanced by the change in storage which occurs during the time step. This is given as

$$\frac{\Delta V_i}{\Delta t} = S \, \Delta y \left(\frac{h_i^{t+\Delta t} - h_i^t}{\Delta t}\right) \tag{20.13}$$

Now inserting these three expressions in the continuity equation (inflow $-$ outflow $=$ change in storage), we get

$$\left(-T\frac{h_i^n - h_{i-1}^n}{\Delta y}\right) - \left(-T\frac{h_{i+1}^n - h_i^n}{\Delta y}\right) = S \, \Delta y \left(\frac{h_i^{t+\Delta t} - h_i^t}{\Delta t}\right) \tag{20.14}$$

By rearrangement, the equation becomes

$$h_{i+1}^n - 2h_i^n + h_{i-1}^n = \frac{S}{T}\frac{(\Delta y)^2}{\Delta t}(h_i^{t+\Delta t} - h_i^t) \tag{20.15}$$

which is known as the explicit or forward difference form of the finite-difference equation if n is designated as the current value of time. If, on the other hand, n is defined as $t + \Delta t$, then the equation is the implicit or backward difference equation. Each of these forms has its own solution techniques.[8] The explicit solution to Eq. 20.15 will be discussed here.

By letting $n = t$ in Eq. 20.1 and rearranging, one obtains

$$h_i^{t+\Delta t} = \frac{T \, \Delta t}{S(\Delta y)^2}(h_{i+1}^t + h_{i-1}^t) + h_i^t\left[1 - \frac{2T \, \Delta t}{S(\Delta y)^2}\right] \tag{20.16}$$

In this case the space derivatives are centered at the beginning of the time step and the single unknown is $h_i^{t+\Delta t}$. Equation 20.16 can be solved explicitly at each element for the head at the next period of time. The solution depends only on a knowledge of the heads in adjacent elements at the beginning of a time step.

It must be recognized that a solution obtained using Eq. 20.16 is only an approximation to the exact solution. The correspondence with the exact solution is related to the choice of Δy and Δt. If the selected values are too large, the difference between the approximate and exact solution can grow as t increases, bringing about an unstable condition. In the one-dimensional homogeneous case discussed here, stability is assured if

$$\frac{T \, \Delta t}{S(\Delta y)^2} < \frac{1}{2} \qquad (20.17)$$

The equation shows that the choice of time and space increments is not independent. Satisfaction of Eq. 20.17 does not guarantee an accurate approximation, however; it only provides for a stable solution.[8]

EXAMPLE 20.1

Refer to the one-dimensional flow problem of Fig. 20.8. Let us assume that the element length is 4 m and that the thickness of the confined aquifer is 2 m. It is further assumed that the head at the left and right sides of the region is 8 m at $t = 0$ and that the head on the right side takes on the value 2 m for all t greater than zero. $K = 0.5$ m/day and $S = 0.02$. As shown in the figure, there are five elements. Using the notation of Eq. 20.16, the initial condition is $h_4^0 = 8.0$ m. Use the explicit method to determine future heads.

Solution

1. First a determination must be made of the time step to use. This may be accomplished using Eq. 20.17.

$$\Delta t < \frac{1}{2} \frac{S(\Delta y)^2}{T} = \frac{1}{2} \frac{(0.02)(4)^2}{1.0} = 0.16 \text{ days}$$

The value of T used in the above expression was obtained using the relation $T = Kb$:

$$T = 0.5 \times 2.0 = 1.0$$

To be on the safe side, we shall choose a time step of 0.1 days, although any value less than 0.16 would have assured stability.

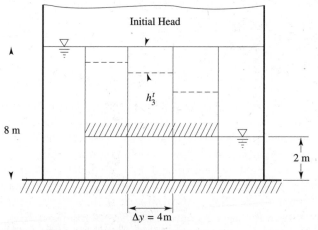

Figure 20.8   Sketch for Example 20.1.

2. For the first time step, $t = 0.1$, we can calculate $h_4^{t+\Delta t}$ and corresponding heads for the other elements using Eq. 20.16. Thus

$$h_4^{t+\Delta t} = \frac{T \Delta t}{S(\Delta y)^2} (h_5^0 + h_3^0) + h_4^0\left[1 - 2\frac{T \Delta t}{S(\Delta y)^2}\right]$$

and substituting numerical values, we get

$$h_4^{0.1} = \frac{1.0(0.1)}{(0.02)(4)^2}(2.0 + 8.0) + 8.0\left[1 - \frac{2(1.0)(0.1)}{(0.02)(4)^2}\right]$$

$$= 3.1 + 3.0 = 6.13 \text{ m}$$

Since h_4^0 and $h_2^0 = 8.0$ m and since $h_1 = 8.0$ by definition, it can easily be shown using Eq. 20.16 that the values of $h_3^{0.1}$ and $h_2^{0.1}$ are not changed from their original level of 8.0 during the first time step.

3. Now consider the second time step, $t + \Delta t = 0.2$ days. For element 4,

$$h_4^{0.2} = \frac{T \Delta t}{S(\Delta y)^2} (h_5^{0.1} + h_3^{0.1}) + h_4^{0.1}\left[1 - \frac{2T \Delta t}{S(\Delta y)^2}\right]$$

$$= 0.31(2.0 + 8.0) + 6.13(0.37) = 5.4 \text{ m}$$

For element 3

$$h_3^{0.2} = \frac{T \Delta t}{S(\Delta y)^2} (h_4^{0.1} + h_2^{0.1}) + h_3^{0.1}\left[1 - \frac{2T \Delta t}{S(\Delta y)^2}\right]$$

$$= 0.31(6.13 + 8.0) + 8.0(0.37) = 7.4 \text{ m}$$

Element 2 does not have a head change until the third time step.

4. The process demonstrated is repeated until the heads have been calculated for the total time period of interest. For this example, they will ultimately reach equilibrium conditions. ■■

This example problem illustrates the mechanics of the finite-difference procedure. Problems of practical scale would require the use of a computer, but the approach would still be the same.

20.3 FINITE-ELEMENT METHODS

The most widely used numerical techniques for solving groundwater flow problems are the finite-difference and finite-element methods. The finite-element method is similar to the finite-difference method in that both approaches lead to a set of N equations in N unknowns that can be solved by relaxation.[6] Nodes in the finite-element method are usually the corner points of an irregular triangular or quadrilateral mesh for two-dimensional applications, while for three-dimensional applications, bricks or tetrahedrons are commonly used. The size and shape of the elements selected are arbitrary. They are chosen to fit the application at hand. They differ from the regular rectangular grid elements used in finite-difference modeling. Elements that are closest to points of flow concentration such as wells are usually smaller than those further removed from such influences. Aquifer parameters such as hydraulic conductivity may be kept constant for a given element but may vary from one to another. To minimize the variational function, its partial derivative with respect to head is evaluated for each node and equated to zero. The procedure results in a set of algebraic equations that can be solved by iteration, matrix solution, or a combination of these methods.[24] Finite-element modelers must understand partial differential equations and the calculus of variations.[6]

The finite-element approach offers some advantages over the finite-difference technique. Often, a smaller nodal grid is required, and this offers economies in computer effort. The finite-element approach can also accommodate one condition that the finite-difference approach is unable to handle.[6] When using the finite-difference method, the principal directions of anisotropy in an anisotropic formation are parallel to the coordinate directions. In cases where two anisotropic formations having different principal directions occur in a flow field, the finite-difference approach cannot produce a solution, whereas the finite-element approach can. The finite-element technique can be used to simulate transient aquifer performance. A detailed discussion of the finite-element technique is beyond the scope of this book, but there are many good references for the interested reader.[6, 16–19, 24]

20.4 MODEL APPLICATIONS

To illustrate how simulation models can be used to provide insights into water management schemes, a model analysis of the Upper Big Blue basin aquifer in Nebraska is presented. The study was conducted by the Conservation and Survey Division of the University of Nebraska under the direction of Huntoon.[10]

The use of groundwater for irrigation in the Upper Big Blue basin was observed to be rapidly increasing and by 1972 about 3.3 wells/mi^2 were in operation. At that time farmers were becoming concerned about the progressive decline of water levels and were seeking guidance about the efficiency of implementing some form of basin-wide water management program. The University of Nebraska designed a model to evaluate the situation and to explore various proposals for recharging the aquifer and for estimating the long-term consequences of several scenarios of water use in the basin.

The study area is shown in Fig. 20.9. Generally the water table is free in the region of interest. Figure 20.10 shows the configuration of the water table as observed in 1953. For modeling purposes, the water-level contours shown were considered to be representative of predevelopment conditions. This assumption was based on the fact that groundwater withdrawals before this time were not extensive. It was also surmised that the contours represented a water table in which an equilibrium existed between natural recharge and discharge in the region. Transmissivities were estimated from drill-hole sample logs recorded in the area. These values are needed for modeling and are also important indices of the potential yield of wells that might be constructed.

As might be suspected, the information of most concern to the local landowners and water planners was the rate of decline of the water table. In particular, it was

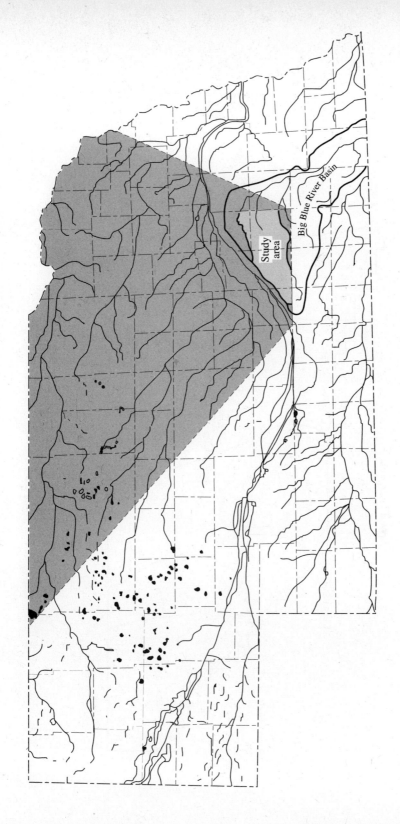

Figure 20.9 Big Blue River basin, Nebraska.

Study area

Big Blue River Basin

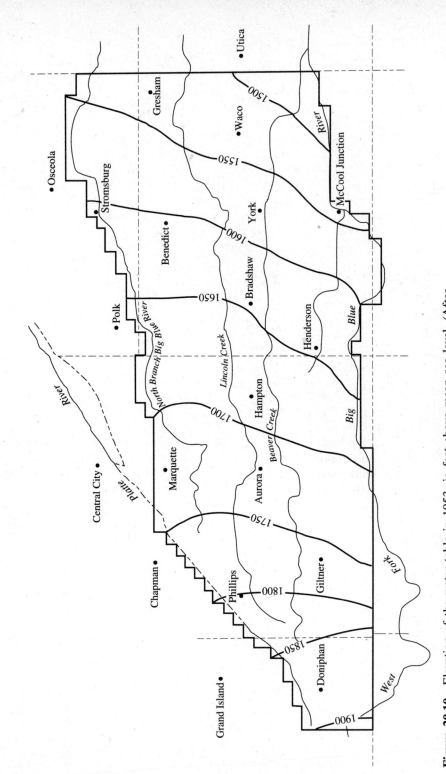

Figure 20.10 Elevation of the water table in 1953, in feet above mean sea level. (After Huntoon.[10])

desired to know how rapidly the groundwater resource would be depleted, where and when water level declines would pose an economic constraint on water use, and what impacts future developments and/or management would have on the rate of decline.

The model developed to explore these features was a two-dimensional representation of flow through an areally extensive aquifer.[10] Equation 20.1, along with the appropriate boundary conditions, constituted the model. The region shown in Fig. 20.9 was divided into a finite-difference grid and, after substitution of the nodal values of T and S, the model was operated to predict water-level changes to the year 2020 for various policies of recharge and for several levels of development. Calibration of the model was accomplished using historic data. The model was operated over the period 1953–1972 using the known distribution of wells and the average net pumpage per well to establish a match between observed and estimated water-level changes. Once this was accomplished, the simulation of future trends proceeded. Figures 20.11 and 20.12 show the correspondence achieved in the matching process.

On the basis of the model studies, it was determined that water levels in the study area would continue to decline even if development was limited to the 1972 level. It was further predicted that some parts of the area would experience severe groundwater shortages by the year 2000. It was found, however, that by employing artificial recharge methods, permanent groundwater supplies could be assured. To assess the effects of artificial recharge, two water delivery systems were modeled. Both of these delivered water from Platte River Valley sources to recharge wells located in the project area. Using these two water delivery systems, three recharge schemes were simulated. The gross effect of introducing the recharge wells was the cancellation of the effects of the proportionate number of pumping wells. Figure 20.13 shows the computed water-level changes at one location under a graduated development plan (projected on the basis of the 1972 rate of development) with no recharge and then with graduated development for each of the three recharge schemes. The continual downward trend in water level with no recharge (curve 1) clearly shows the nature of the problem in the Upper Big Blue basin. The other curves depicting the three artificial recharge options show that stability can be achieved if such an approach is taken.

While the costs of implementing artificial recharge might be excessive, it is apparent that any long-term solution to the declining water table problem, short of reducing use, would require a supplemental source of water.

Operation of the model provided useful insights into the nature of the water table problem and suggested that irrigators should be making some important water management decisions about their future mode of operation.

The modeling of groundwater systems is complex.[10-25] In structuring models such as that just discussed, simplifying assumptions must usually be made. These have to do with aquifer parameters such as transmissivity, specific yield (for unconfined aquifers), and storage coefficient (for confined systems). Furthermore, the boundary conditions are normally approximations of what occurs in the physical system, and assumptions about the uniformity of materials in various subsurface strata are sometimes crude. This does not mean that groundwater models cannot be expected to yield useful results. It does imply that the users of the models must be cautious about how they interpret the output. For example, an areally extensive aquifer model such as that developed by Huntoon for analyzing the Blue River problem can be expected to give

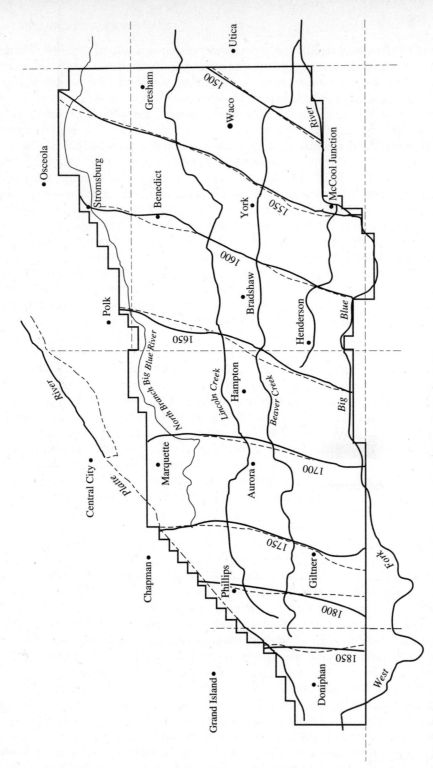

Figure 20.11 Measured and computed water table elevations in 1972. Solid line represents measured data; dashed line represents computed data. (After Huntoon.[10]).

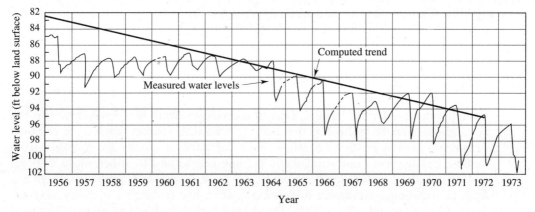

Figure 20.12 Measured and computed water-level trends. (After Huntoon.[10]).

reliable information about water-level trends for various configurations of development. It should not, on the other hand, be considered an accurate predictive tool for monitoring the water-level change at some specific point in the region of concern. This type of information could be derived only from a more detailed modeling of the locality surrounding the point. The information provided by the Blue River model was targeted to show local landowners what the future might hold for several development levels and for several management options. The actual water levels predicted by the model were not of central concern; what was of interest was the determination that unless future development was restricted and supplemental water provided, or unless current uses could be significantly reduced, the outlook in the next 50 years was not good for irrigated farming.

The model thus provided the basis for making some quantitative observations about the future. It also provided insights into the relief that might be expected from artificial recharge. Beyond that, it could be used to model other possible management

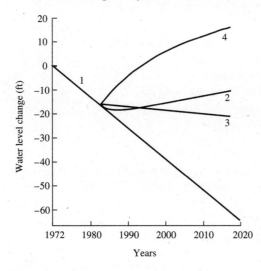

Figure 20.13 Computed water-level changes under a plan of graduated development for conditions of (1) no recharge, (2) recharge under Scheme 1, (3) recharge under Scheme 2, and (4) recharge under Scheme 3. (After Huntoon.[10])

options. A model such as this, carefully used and properly interpreted, can thus add a powerful dimension to decision-making processes.

20.5 GROUNDWATER QUALITY MODELS

Groundwater quality has become a major source of concern in recent years. This has come about from the realization that many groundwater sources that were at one time considered almost pristine have now been degraded in quality by seepages from dumps, leakage from industrial waste holding ponds, and by other waste disposal and/or industrial and agricultural practices. To deal with such problems, there has been an expanding movement to develop quantitative techniques to understand the mechanics of groundwater quality. These models, although not as advanced as their surface water counterparts, are now beginning to play an important role in water quality management.

The subject of groundwater quality modeling is complex and under rapid development. Accordingly, a thorough treatment of the subject is beyond the scope of this book. The importance of this topic cannot be overemphasized, however, and the reader is encouraged to consult the references at the end of the chapter, specifically Refs. 6 and 26–30.

In 1974, Gelhar and Wilson developed a lumped parameter model for dealing with water quality in a stream–aquifer system. The nomenclature and conceptualization of their model are shown in Fig. 20.14.[29] The rationale for using a lumped parameter approach was that when dealing with changes in groundwater quality over long periods of time, temporal rather than spatial variations are most important.

Changes in water table in the Gelhar–Wilson (GW) model are represented by the following equation:

$$p\frac{dh}{dt} = -q + \epsilon + q_r - q_p \tag{20.18}$$

where h = average thickness of the saturated zone
p = average effective porosity
ϵ = natural recharge rate
q = natural outflow from the aquifer
q_r = artificial recharge/unit area
q_p = pumping rate/unit area
t = time

This is just another form of the continuity equation relating inflow, outflow, and the change in storage (left-hand term in Eq. 20.18). The change in concentration of a constituent is given by

$$ph\frac{dc}{dt} + (\epsilon + q_r + \alpha ph)c = \epsilon c_L + q_r c_r \tag{20.19}$$

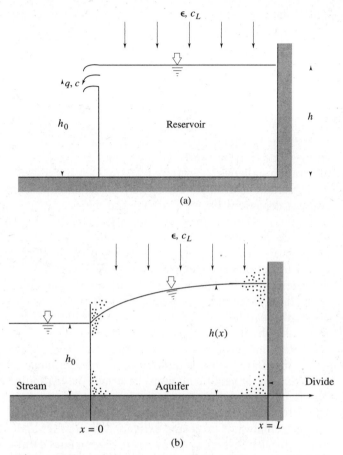

Figure 20.14 Schematic of the Gelhar–Wilson model. (After Novotny and Chesters.[24])

where c = concentration

     c_L = concentration of the natural recharge

     c_r = concentration of the artificial recharge

     c = a first-order rate constant for degradation of the contaminant

       The GW model assumes that dispersion is negligible. This assumption may be made on the basis that the objective of the model is to estimate regional-average concentrations.[29] The model also provides for the determination of hydraulic and solute response times for the system. These are measures of the lag that occurs in the movement of both water and constituent inputs to the system. Gelhar and Wilson assume that the response of an aquifer to a specific input can be likened to that of a well-mixed linear reservoir. Their studies showed that the model's determination of the concentration of constituents leaving an aquifer is representative of the average concentration of the constituent in the aquifer. On this basis, it appears that the model is well suited to estimating the quality of groundwater discharging to a surface stream, providing the aquifer is narrow relative to the length along which discharge occurs.

■ Summary

Groundwater in a regional aquifer system is constantly in motion. The amount stored at any time is affected by artificial and natural recharge, evapotranspiration, flow to springs and surface water courses, and by collection devices such as wells and infiltration galleries.

Natural hydrologic states may be significantly affected by human activities. Aquifer depletions having regional and national economic implications are not uncommon. Depletion of the Ogallala aquifer in the central United States by long-term and extensive water withdrawals for irrigation is a good example. On the other hand, water levels have been made to rise, sometimes inadvertently, by human intervention. Leaky irrigation canals in central Nebraska were at one time responsible for groundwater level rises in some farming locations of a magnitude sufficient to jeopardize use of the land. Once major problems of depletion or over-replenishment occur, they are not easily dealt with. In general, a safe-yield policy for groundwater management has merit and should be considered.[6, 30]

Regional groundwater flow problems are usually modeled by an equation combining Darcy's law and the equation of continuity. The resulting partial differential equation, or set of equations, describes the hydraulic relations within the aquifer. To effect a solution to the governing equation(s), the aquifer's hydraulic features, geometry, and initial and boundary conditions must be determined. Unfortunately, many groundwater problems exist for which exact analytic solutions cannot be obtained. In such cases, it is necessary to rely on numerical methods for modeling. Under such circumstances, an approximate solution is obtained by replacing the basic differential equations with another set of equations that can be solved iteratively on a computer. Both finite difference and finite element methods are applicable.

The finite difference approach described in this chapter replaces the governing partial differential equations with a set of algebraic equations. These can be solved on the computer to produce a set of water table elevations at a finite number of locations in the aquifer.

Once the groundwater model has been calibrated, it can be used to predict the outcomes (impacts) of alternative development and/or management strategies proposed for an aquifer. Such analyses are valuable adjuncts to decision-making processes. Models can, for example, simulate the effects of opening new well fields, analyze changed operating practices for existing well fields, explore schemes for artificial recharge, and predict the impacts of proposed irrigation development plans. Groundwater models can be applied to unconfined aquifers, semiconfined aquifers, confined aquifers, or any combination thereof. They can accommodate large variations in aquifer parameters such as hydraulic conductivity and storage coefficient, and they can be used to analyze unsteady as well as steady flow problems.

PROBLEMS

20.1. Refer to Fig. 20.8. Assume that the element length is 5 m and the thickness of the confined aquifer is 2.5 m. The head at the left and right sides is 8.1 m at $t = 0$, and the head on the right is 2.5 m for all $t > 0$. $K = 0.5$ m/day and $S = 0.02$. Use the explicit method to determine heads at future times.

20.2. Refer to Fig. 20.8. Assume the element length is 10 ft and the thickness of the confined aquifer is 8 ft. The head at the left and right is 21 ft at $t = 0$, and it drops on the right side to 8 ft for all $t > 0$. $K = 1.5$ ft/day and $S = 0.02$. Use the explicit method to calculate future heads.

20.3. Refer to Fig. 20.12. Aside from the trend, what else can you deduce from studying this figure?

20.4. Discuss how you would go about designing a grid for a regional groundwater study. What types of boundary conditions might you specify? Why?

REFERENCES

1. J. W. Mercer and C. R. Faust, *Ground-Water Modeling*. Worthington, OH: National Water Well Association, 1981.
2. C. A. Appel and J. D. Bredehoeft, "Status of Groundwater Modeling in the U.S. Geological Survey," *U.S. Geol. Survey Circular* 737(1976).
3. Y. Bachmat, B. Andres, D. Holta, and S. Sebastian, "Utilization of Numerical Groundwater Models for Water Resource Management," U.S. Environmental Protection Agency Report EPA-600/8-78-012.
4. J. E. Moore, "Contribution of Ground-water Modeling to Planning," *J. Hydrol.* 43(Oct. 1979).
5. T. A. Prickett, "Ground-water Computer Models—State of the Art," *Ground Water* **17**(2), 121–128(1979).
6. R. A. Freeze and J. A. Cherry, *Groundwater*. Englewood Cliffs, NJ: Prentice-Hall, 1979.
7. G. D. Bennett, *Introduction to Ground Water Hydraulics,* book 3, *Applications of Hydraulics*. Washington, D.C.: U.S. Geological Survey, U.S. Government Printing Office, 1976.
8. D. B. McWhorter and D. K. Sunada, *Ground Water Hydrology and Hydraulics*. Fort Collins, CO: Water Resources Publications, 1977.
9. D. K. Todd, *Groundwater Hydrology,* 2d ed. New York: Wiley, 1980.
10. P. W. Huntoon, "Predicted Water-Level Declines for Alternative Groundwater Developments in the Upper Big Blue River Basin, Nebraska," Resource Rep. No. 6, Conservation and Survey Div., University of Nebraska, Lincoln, 1974.
11. D. W. Peacemen and H. H. Rachford, Jr., "The Numerical Solution of Parabolic and Elliptic Differential Equations," *Soc. Indust. Appl. Math. J.* **3**, 28–41(1955).
12. G. F. Pinder and J. D. Bredehoeft, "Application of the Digital Computer for Aquifer Evaluation," *Water Resources Res.* **4**(4), 1069–1093(1968).
13. I. Remson, G. M. Hornberger, and F. J. Molz, *Numerical Methods in Subsurface Hydrology*. New York: Wiley, 1971.
14. T. A. Prickett and C. G. Lonnquist, *Selected Digital Computer Techniques for Groundwater Resource Evaluation,* Illinois State Water Survey Bull. No. 55, 1971.

15. R. R. Marlette and G. L. Lewis, "Digital Simulation of Conjunctive-Use of Groundwater in Dawson County, Nebraska," Civil Engineering Rep., University of Nebraska, Lincoln, 1973.

16. C. S. Desai, and J. F. Abel. 1972. *Introduction to the Finite Element Method.* Van Nostrand Reinhold, New York.

17. G. F. Pinder, and W. G. Gray. 1977. *Finite Element Simulation in Surface and Subsurface Hydrology.* Academic Press, New York, 295 pp.

18. I. Remson, G. M. Hornberger, and F. J. Molz. 1971. *Numerical Methods in Subsurface Hydrology.* Wiley-Interscience, New York.

19. O. C. Zienkiewicz, 1967. *The Finite-Element Method in Structural and Continuum Mechanics.* McGraw-Hill, New York.

20. C. E. Jacob, "Flow of Groundwater," in *Engineering Hydraulics* (Hunter Rouse, ed.). New York: Wiley, 1950.

21. R. A. Young and J. D. Bredehoeft, "Digital Computer Simulation for Solving Management Problems of Conjunctive Groundwater and Surface Water Systems," *Water Resources Res.* **8**(3) (June 1972).

22. H. W. Crooke, "Ground Water Replenishment in Orange County, California," *J. AWWA* (July 1961).

23. P. van der Heijde, Y. Bachmat, J. Bredehoeft, B. Andrews, D. Holtz, and S. Sebastian, *Groundwater Management: The Use of Numerical Models.* Washington, D.C.: American Geophysical Union, 1985.

24. D. R. Maidment (ed.), *Handbook of Hydrology.* New York: McGraw-Hill, 1993.

25. J. Boonstra, and de Ridder, N. A., *Numerical Modeling of Groundwater Basins.* Wageningen, the Netherlands: International Institute for Land Reclamation and Improvement/LRI, 1981.

26. E. F. Wood, Ferrara, R. A., Gray, W. G. and Pinder, G. F., *Groundwater Contamination from Hazardous Wastes.* Englewood Cliffs, N.J.: Prentice-Hall, Inc., 1984.

27. E. K. Nyer, *Groundwater Treatment Technology*, New York: Van Nostrand Reinhold Company, 1985.

28. V. Novotny and G. Chesters, *Handbook of Nonpoint Pollution, Sources and Management.* New York: Van Nostrand Reinhold, 1981.

29. L. W. Gelhar and J. L. Wilson, "Ground Water Quality Modeling," in *Proceedings of the 2nd National Ground Water Quality Symposium.* Washington, D.C.: U.S. Environmental Protection Agency, 1974.

30. W. Viessman, Jr. and M. J. Hammer, *Water Supply and Pollution Control*, 5th edition. New York: HarperCollins College Publishers, 1993.

PART FIVE

HYDROLOGIC MODELING

Chapter 21

Introduction to Hydrologic Modeling

■ Prologue

The purpose of this chapter is to:

- Introduce the types and classes of hydrologic models.
- Illustrate the limitations, alternatives, steps, general components, and data needs of hydrologic simulation models.
- Present a philosophical protocol for performing successful modeling studies.
- Give an overview of groundwater model types.
- Distinguish the need for separate, specific procedures detailed in subsequent Chapters 22, 23, 24, and 25.

Information regarding rates and volumes of flow at any point of interest along a stream is necessary in the analysis and design of many types of water projects. Although many streams have been gauged to provide continuous records of streamflow, planners and engineers are sometimes faced with little or no available streamflow information and must rely on *synthesis* and *simulation* as tools to generate artificial flow sequences for use in rationalizing decisions regarding structure sizes, the effects of land use, flood control measures, water supplies, water quality, and the effects of natural or induced watershed or climatic changes.

The problems of decision making in both the design and operation of large-scale systems of flood control reservoirs, canals, aqueducts, and water supply systems have resulted in a need for mathematical approaches such as simulation and synthesis to investigate the total project. Simulation is defined as the mathematical description of the response of a hydrologic water resource system to a series of events during a selected time period. For example, simulation can mean calculating daily, monthly, or seasonal streamflow based on rainfall; or computing the discharge hydrograph resulting from a known or hypothetical storm; or simply filling in the missing values in a streamflow record.

Simulation is commonly used in generating streamflow hydrographs from rainfall and drainage basin data. The philosophies and overall concepts used in simulation are introduced in this chapter. Chapter 22 summarizes concepts of streamflow synthesis by *stochastic* methods. Chapters 23–25 provide details regarding *deterministic* continuous models, single-event models, urban runoff and storm sewer design models, and water quality models.

Stochastic techniques used to extend records, either rainfall or streamflow, are classified as *synthesis* methods. This procedure relies on the statistical properties of an existing record or regional estimates of these parameters. An overview of synthesis techniques is presented in Chapter 22.

21.1 HYDROLOGIC SIMULATION

In this chapter, *simulation* of all or parts of a surface, groundwater, or combined system implies the use of computers to imitate historical events or predict the future response of the physical system to a specific plan or action. Physical, analog, hybrid, or other models for simulating the behavior of hydraulic and hydrologic systems and system components have had, and will continue to have, application in imitating prototype behavior but are not discussed here.

A few of the numerous event, continuous, and urban runoff computer models for simulating the hydrologic cycle are compared in Table 21.1. As shown in the table, most of the models were developed for, or by, universities or federal agencies. All have moderate-to-extensive input data requirements, and all have from 1 to 10 percent of

TABLE 21.1   DIGITAL SIMULATION MODELS OF HYDROLOGIC PROCESSES

Code name	Model name	Agency or organization	Percentage of inputs by judgment[a]	Date of original development
Continuous streamflow simulation models—Chapter 23				
API	Antecedent Precipitation Index Model	Private	1	1969
USDAHL	1970, 1973, 1974 Revised Watershed Hydrology	ARS	1	1970
SWM-IV	Stanford Watershed Model IV	Stanford University	10	1959
HSPF	Hydrocomp Simulation Program—FORTRAN	EPA	10	1967
NWSRFS	National Weather Service Runoff Forecast System		10	1972
SSARR	Streamflow Synthesis and Reservoir Regulation	Corps	3	1958
PRMS	Precipitation-Runoff Modeling System	USGS	5	1982
SWRRB	Simulator for Water Resources in Rural Basins	USDA	10	1990
Rainfall-runoff event-simulation models—Chapter 24				
HEC-1	HEC-1 Flood Hydrograph Package	Corps	1	1973
TR-20	Computer Program for Project Hydrology	SCS	5	1965
USGS	USGS Rainfall–Runoff Model	USGS	10	1972
HYMO	Hydrologic Model Computer Language	ARS	1	1972
SWMM	Storm Water Management Model	EPA	5	1971
Urban runoff simulation models—Chapter 25				
UCUR	University of Cincinnati Urban Runoff Model	University of Cincinnati	2	1972
STORM	Quantity and Quality of Urban Runoff	Corps	3	1974
MITCAT	MIT Catchment Model	MIT	5	1970
SWMM	Storm Water Management Model	EPA	5	1971
ILLUDAS	Illinois Urban Drainage Area Simulator	Illinois State Survey	1	1972
DR3M	Distributed Routing Rainfall–Runoff Model	USGS	5	1978
PSURM	Pennsylvania State Urban Runoff Model	Pennsylvania State University	5	1979

[a] Judgment percentages are from U.S. Army Waterways Experiment Station.[1]

inputs that are judgment parameters. These are normally validated by repeated trials with the models. The urban runoff models are primarily event simulation models but have been isolated in Table 21.1 because the descriptions of urban models are deferred to Chapter 25.

Several of the major event and continuous streamflow simulation models shown in Table 21.1 are described in Chapters 23 and 24. The Stanford and HEC-1 models are emphasized. For further reference, most models listed in Table 21.1 are briefly described, along with about 100 other models, in the publication "Models and Methods Applicable to Corps of Engineers Studies."[1] Fleming's text presents complete descriptions of the SSARR, SWM, HSP, USDAHL, and other models.[2]

Classification of Simulation Models

In recent decades the science of computer simulation of groundwater and surface water resource systems has passed from scattered academic interests to a practical engineering procedure. The varied nature of developed and applied simulation models has caused a proliferation of categorization attempts. A few of the most descriptive classifications are presented.

Physical vs. Mathematical Models   Physical models include analog technologies and principles of similitude applied to small-scale models. In contrast, mathematical models rely on mathematical statements to represent the system. A laboratory flume may be a 1 : 10 physical model of a stream, while the unit hydrograph theory of Chapter 12 is a mathematical model of the response of a watershed to various effective rain hyetographs.

Continuous vs. Discrete Models   A second classification is achieved by considering physical, analog, and some digital models as *continuous* because the processes occur and are modeled continuously. Many digital simulation models rely on the necessity and advantages of slicing space and time into finite increments, and thus qualify as *discrete* models. A well-known example of the latter is the storage-indication method for routing a flood hydrograph through a reservoir by generating instantaneous reservoir discharge rates at the end points of equally spaced intervals in time.

Dynamic vs. Static Models   Processes that involve changes over time and time-varying interactions can be simulated by *dynamic* models. In contrast, models that examine time-independent processes are frequently called *static*. Few hydrologic simulation models fall into the latter category.

Descriptive vs. Conceptual Models   Descriptive models have had the greatest application and are of particular interest to practicing hydrologists because they are designed to account for observed phenomena through empiricism and the use of basic fundamentals such as continuity or momentum conservation assumptions. Conceptual models, on the other hand, rely heavily on theory to interpret phenomena rather than to represent the physical process. Examples of the latter include models based on probability theory. Recent trends in the use of artificial intelligence and expert systems in water system modeling would classify as conceptual methods.

Lumped vs. Distributed Parameter Models   Models that ignore spatial variations in parameters throughout an entire system are classified as *lumped parameter* models. An example is the use of a unit hydrograph for predicting time distributions of surface runoff for different storms over a homogeneous drainage area. The "lumped parameter" is the X-hour unit hydrograph used for convolution with rain to give the storm hydrograph. The time from end of rain to end of runoff is also a lumped parameter as it is held constant for all storms. *Distributed parameter* models account for behavior variations from point to point throughout the system. Most modern groundwater simulation models are distributed in that they allow variations in storage and transmissivity parameters over a grid or lattice system superimposed over the plan of an aquifer. More recently, surface water systems are being analyzed through use of distributed parameter Geographical Information System (GIS) technologies.

Black-Box vs. Structure-Imitating Models   Both of these models accept input and transform it into output. In the former case, the transformation is accomplished by techniques that have little or no physical basis. The alchemist's purported ability to transform lead into gold or plants into medicine was accomplished in a black-box fashion. In hydrology, black-box models may sometimes transform "plants" into "medicine" even though the reasons for success are not clearly understood. For example, a model that accepts a sequence of numbers, reduces each by 20 percent, and outputs the results might be entirely adequate for predicting the attenuation of a flood wave as it travels through a reach of a given stream. In contrast, a structure-imitating model would be designed to use accepted principles of fluid mechanics and hydraulics to facilitate the transformation.

Stochastic vs. Deterministic Models   Many stochastic processes are approximated by deterministic approaches if they exclude all consideration of random parameters or inputs. For example, the simulation of a reservoir system operating policy for water supply would properly include considerations of uncertainties in natural inflows, yet many water supply systems are designed on a deterministic basis by mass curve analyses, which assume that sequences of historical inflows are repetitive.

Deterministic methods of modeling hydrologic behavior of a watershed have become popular. *Deterministic simulation* describes the behavior of the hydrologic cycle in terms of mathematical relations outlining the interactions of various phases of the hydrologic cycle. Frequently, the models are structured to simulate a streamflow value, hourly or daily, from given rainfall amounts within the watershed boundaries. The model is "verified" or "calibrated" by comparing results of the simulation with existing records. Once the model is adjusted to fit the known period of data, additional periods of streamflow can be generated.

Event-Based vs. Continuous Models   Hydrologic systems can be investigated in greater detail if the time frame of simulation is shortened. Many short-term hydrologic models could be classified as *event-simulation* models as contrasted with *sequential* or *continuous* models. An example of the former is the Corps of Engineers single-event model, HEC-1,[3] and an example of the latter is the Stanford watershed model developed by Crawford and Linsley,[4] which is normally operated to simulate three, four, five, or more years of streamflow. A typical event simulation model might use a time increment of 1 hr or perhaps even 1 min.

Water Budget vs. Predictive Models   Several model classifications have arisen that distinguish between the purposes of the model types. One important comparison is whether the model proposes to predict future conditions using synthesized precipitation and watershed conditions or by verifying historical events. A *water budget model* is defined as a model or set of relationships that confirm the historical balance of inflows, outflows, and changes in storage for a system under study. It is strongly advised that simulation model studies begin by structuring a water budget model, then use the parameters that affirm the balance in any simulation modeling of the watershed. For example, meteorologic data, streamflow records, crop patterns, and water application amounts might be known for a given agricultural watershed. A water budget model would be used to determine the correct evapotranspiration (ET) formula parameters by testing a range of values until a balance in the continuity equation occurs for all time increments. This is often performed on a day-by-day or month-by-month basis. Once the ET parameters are derived from the water budget model, predictive simulations of different crop patterns, meteorologic conditions, or farming practices could be performed with the satisfaction that the relationships in the model corroborate historical water budgets. Because only *primary* hydrologic inputs and outputs are measured (precipitation, temperature, runoff, land use), normal modeling studies require the simultaneous development of water budget parameters for secondary processes such as ET, infiltration, groundwater storage, and temporal and spatial distribution of water applications.

Limitations of Simulation

Because simulation entails a mathematical abstraction of real-world systems, some degree of misrepresentation of system behavior can occur. The extent to which the model and system outputs vary depends on many factors. The test of a developed simulation model consists of verification by demonstrating that the behavior is consistent with the known behavior of the physical system.

Even verified simulation models have limitations in uses for water resources planning and analysis. Simulation models will allow performance assessments of specific schemes but cannot be used efficiently to generate options, particularly optimal plans, for stated objectives. Once a near-optimal plan is formulated by some other technique, a limited number of simulation runs are normally effective for testing and improving the plan by modifying combinations of decision variables using random or systematic sampling techniques. Techniques for generating optimal plans are described in Section 21.3.

Another limitation of simulation models involves changing the operating procedures for potential or existing components of the system being modeled. Programming a computer to handle reservoir storage and release processes, for example, requires large portions to define the operating rules, and considerable reprogramming is required if other operating procedures are to be investigated.

A fourth limitation of simulation models is the potential overreliance on sophisticated output when hydrologic and economic inputs are inadequate. The techniques of operational hydrology can be used to obviate data inadequacies, but these also require input. Controversy over the use of synthetic data centers on the question of whether operational hydrology provides better information than that contained in the input.

Utility of Simulation

Computer simulation of hydrologic processes has several important advantages that should be recognized whenever considering the merits of a simulation approach to a problem that has other possible solutions. One alternative to digital simulation is to build and operate either the prototype system or a physically scaled version. Simulation by physical modeling has been applied successfully to the analysis of many components of systems such as the design of hydraulic structures or the investigation of stream bank stability. However, for the analysis of complex water resource systems comprised of many interacting components, computer simulation often proves to be the only feasible tool.

Another alternative to digital simulation is a hand solution of the governing equations. Simulation models, once formulated, can accomplish identical results in less time. Also, solutions that would be impossible to achieve by hand are frequently achieved by simulation. In addition, the system can be nondestructively tested; proposed modifications of the designs of system elements can be tested for feasibility or compared with alternatives; and many proposals can be studied in a short time period.

An often overlooked advantage of simulation includes the insight gained by gathering, organizing, and processing the data, and by mentally and mathematically formulating the model algorithms that reproduce behavior patterns in the prototype.

Steps in Digital Simulation

A simulation model is a set of equations and algorithms (e.g., operating policies for reservoirs) that describe the real system and imitate the behavior of the system. A fundamental first step in organizing a simulation model involves a detailed analysis of all existing and proposed components of the system and the collection of pertinent data. This step is called the system *identification* or inventory phase. Included items of interest are site locations, reservoir characteristics, rainfall and streamflow histories, water and power demands, and so forth. Typical inventory items required for a simulation study and data needs that are specific to some of the models are detailed in subsequent paragraphs.

The second phase is model *conceptualization,* which often provides feedback to the first phase by defining actual data requirements for the planner and identifying system components that are important to the behavior of the system. This step involves (1) selecting a technique or techniques that are to be used to represent the system elements, (2) formulating the comprehensive mathematics of the techniques, and (3) translating the proposed formulation into a working computer program that interconnects all the subsystems and algorithms.

Following the system identification and conceptualization phases are several steps of the *implementation* phase. These include (1) validating the model, preferably by demonstrating that the model reproduces any available observed behavior for the actual or a similar system; (2) modifying the algorithms as necessary to improve the accuracy of the model; and (3) putting the model to work by carrying out the simulation experiments.

Model Protocol

Five axioms for performing successful model studies, adapted from recommendations by Friedrich,[5] are:

1. Evaluate the data before beginning.
2. Document assumptions.
3. Plan and control the sequence of computer runs.
4. Insist on reasonableness of output.
5. Document, document, document.

Examining and evaluating the basic data are essential. An annotated, bibliographic record of the data sources should be maintained. It is always good advice to program models that output (echo) data values as they are read in. Verification of the numerical values and proper entry of the data can be established from the echo.

Statistics such as the mean, mode, median, range, standard deviation, skewness, kurtosis, and rank order are often helpful in locating entry errors. Checking for inconsistencies can identify errors. Did the runoff occur ahead of the rainfall? Are water levels gradually varied, or are there discontinuities? Do alphabetical characters appear in the data? Will blank values in the data sets be interpreted as missing or zeros? Will zeros in the data sets result in overflows (division by zero)? For hydrograph routing, does the time interval selected fall between the limits recommended for stability and convergence of the numerical method used to solve the differential equations?

Assumptions are also important to the success of a simulation. Assumptions were made by the programmer when developing the model, and additional assumptions are made by users. For example, a program that calculates the standard deviation from an unbiased estimating equation assumes that the sample size must be sufficiently large to validate the estimate. A value of $N = 30$ is often considered minimal. For a TP-149 (Chapter 15) application, is a 24-hr storm being used, as assumed by the method? No computer program should be used prior to reading and understanding the assumptions made by the programmer and becoming aware of the assumptions implicit in the hydrologic process that was programmed.

The low cost of simulation can result in unnecessary runs and may entice users into pursuing output that may not add substantively to the information originally sought. Planning the work (the number and purpose of runs) and working the plan can help avoid this common pitfall. The cost of additional computer time is only a small part of the expense that results from making unwarranted runs.

A study plan can be developed with approximate time and monetary limits to use as a guide during a simulation project. Combining several investigations in a single run is another way to conduct an efficient simulation. Some of the models available allow this. For example, TR-20 (see Chapter 24) allows the generation of flood hydrographs from several storms at once. It is often desirable to generate the 2-, 5-, 10-, 25-, 50-, 100-, and 500-year flood discharge at a single watershed location.

The computer is able to generate far more output than the hydrologist can analyze. Most models incorporate options allowing the user to specify output quantity. In addition to controlling output, a predetermination should be made of which specific analyses will be performed. A tabulation of key output data can be developed to compile and evaluate trends (and make course corrections) after each run. Because deterministic hydrology is about 80 percent accounting, many opportunities exist in simulation for assessing *water budget* balances. If the total recharge to an aquifer is less than the total outflow and withdrawls, but simulated water tables are rising, a check of input and model parameters should be made. Writing important conclusions on the printed output of simulation runs helps document the study and guide revisions in future runs.

Documentation of simulation studies is generally deficient in practice. The record should communicate the findings in a way that provides a later reviewer general understanding of the work plan followed, decisions made, and reasons for each run. The documentation should state assumptions made, provide samples of the input and output, explain input preparation requirements, state how sensitive the results are to parameter changes and assumptions, and document reasons out-of-range parameters were accepted.

Documentation is an ongoing and continual task. It is especially crucial if the model will be employed in regulatory procedures or litigation. A comprehensive documentation process would[6]:

1. Include an outline description of the problem being studied.
2. Identify the equations, techniques, and methods used.
3. Demonstrate the model's validity to this problem.
4. Discuss the code.
5. Include all assumptions used in the code and in preparing the input.
6. List published or known limitations or ranges of the applicability of the model.
7. Characterize the uncertainties in the model; describe sensitivity tests.
8. Describe parameters and data sets used.
9. State the regulatory or legal criteria incorporated in the model.
10. Describe the verification, whether with test data or analytical solutions.
11. Include a narrative description of the results, indicating any unexpected or unusual outcomes.
12. Present any other details deemed relevant.
13. Discuss the model used.
14. Document changes made in the model code.

Components of Hydrologic Simulation Models

Numerous mathematical models have been developed for the purpose of simulating various hydrologic phenomena and systems. A general conceptual model including most of the important components is shown in Fig. 21.1; several others are described subsequently. Imported water in the lower left could be input to reservoir or ground-

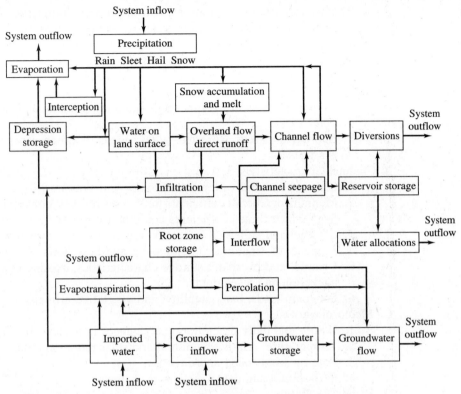

Figure 21.1.   Components of a surface and subsurface water resource system.

water storage or channel flow, or it might be guided directly to water allocations on the far right if either storage or distribution were deemed unnecessary. The routing of channel flow or overland flow could be accomplished by simple lumped parameter techniques, or solutions of the unsteady-state flow equations for discrete segments of the channel could be used. In other words, the selection of techniques and algorithms to represent each component depends on the degree of refinement desired as output and also on knowledge of the system. A distributed parameter approach is justified only when available information is adequate. Components of models are described in Chapters 22–25.

Data Needs for Hydrologic Simulation

The simulation of all or part of a water system requires a data inventory as part of the initial planning process. Most model input data requirements (90 percent or more) are map or field available, or can be empirically determined or obtained from engineering handbooks and equations. A general list of data inventory topics that encompasses most hydrologic–economic modeling needs follows.

A. Basin and Subbasin Characteristics
    1. Lag times, travel times in reaches, times of concentration.
    2. Contributing areas, depressions, mean overland flow distances and slopes.

    3. Design storm abstractions: evapotranspiration, infiltration, depression, and interception losses. Composite curve numbers, infiltration capacities and parameters, ϕ indexes.

    4. Land-use practices, soil types, surface and subsurface divides.

    5. Water-use sites for recreation, irrigation, flood damage reduction, diversions, flow augmentation, and pumping.

    6. Numbering system for junctions, subareas, gauging and precipitation stations.

    7. Imprevious areas, forested areas, areas between isochrones, irrigable acreages.

B. Channel Characteristics

    1. Channel bed and valley floor profiles and slopes.

    2. Manning or Chézy coefficients for various reaches, or hydraulic or field data from which these coefficients could be estimated.

    3. Channel and valley cross-sectional data for each river reach.

    4. Seepage information; channel losses and base flows.

    5. Channel and overbank storage characteristics, existing or proposed channelization and levee data.

    6. Sediment loads, bank stability, and vegetative growth.

C. Meteorologic Data

    1. Hourly and daily precipitation for gauges in or near the watershed.

    2. Temperature, relative humidity, and solar radiation data.

    3. Data on wind speed and direction.

    4. Evaporation pan data.

D. Water Use Data

    1. Flows returned to streams from treatment plants or industries.

    2. Diversions from streams and reservoirs.

    3. Transbasin diversions from and to the basin.

    4. Stream and ditch geometric properties and seepage characteristics.

    5. Irrigated acreages and irrigation practices, including water use efficiencies.

    6. Crop types and water consumption requirements.

    7. Past conservation practices such as terracing, installation of irrigation return pits, and conservation tillage.

    8. Stock watering practices.

    9. Presence and types of phreatophytes in stream valleys and along ditch banks.

E. Streamflow Data

    1. Hourly, daily, monthly, annual streamflow data at all gauging stations, including statistical analyses.

    2. Flood frequency data and curves at gauging stations, or regional curves for ungauged sites, preferably on an annual and seasonal basis.

    3. Flow duration data and curves at gauging stations (also any synthesized data for ungauged areas).

    4. Rating curves; stage–discharge, velocity–discharge, depth-discharge curves for certain reaches.

5. Flooded area curves.
6. Stage versus area flooded.
7. Stage versus frequency curves.
8. Stage versus flood damage curves, preferably on a seasonal basis.
9. Hydraulic radius versus discharge curves.
10. Streamflows at ungauged sites as fractions of gauged values.
11. Return flows as fractions of water-use allocations diverted for consumptive use.
12. Seasonal distributions of allocations to users.
13. Minimal streamflow to be maintained at each site.
14. Mass curves and storage–yield analyses at gauged sites.

F. Design Floods and Flood Routing
   1. Design storm and flood determination; temporal and spatial distribution and intensity.
   2. Maximum regional storms and floods.
   3. Selection and verification of flood routing techniques to be used and necessary routing parameters.
   4. Base flow estimates during design floods.
   5. Available records of historic floods.

G. Reservoir Information
   1. List of potential sites and location data.
   2. Elevation–storage curves.
   3. Elevation–area curves.
   4. Normal, minimal, other pool levels.
   5. Evaporation and seepage loss data or estimates.
   6. Sediment, dead storage requirements.
   7. Reservoir economic life.
   8. Flood control operating policies and rule curves.
   9. Outflow characteristics, weir and outlet equations, controls.
   10. Reservoir-based recreation benefit functions.
   11. Costs versus reservoir storage capacities.
   12. Purposes of each reservoir and beneficiaries and benefits.

Nonmodeling Assessments

After researching the available data and information, the need for simulation can be assessed. If a decision is made to proceed, the appropriate simulation model can be selected, a sequence planned, and data prepared.

Transformation of raw data into usable form does not always require a simulation model. Much of the usual information needed for water resources assessments can be prepared by hand or by using analytical procedures available in microcomputer format. Typical nonmodeling analyses include the following:

1. Identify water-user groups and all basin sites for hydropower production, reservoir-based recreation, irrigation, flood damage reduction from

reservoir capacity, industrial and municipal water supply, diversions, and flow augmentation.

2. Compile annual and seasonal streamflows and flood records at each gauged site for the period of record at each site.

3. Determine the fraction of the allocation to each consumptive use that is assumed to return to the stream at each user site in the basin.

4. Perform frequency analyses of annual streamflow and flood values at each gauge site in the basin.

5. Determine for each reservoir site the evaporation and seepage losses.

6. Select mean probabilities to be used in the firm and secondary yield analyses.

7. Develop flood peak probability distributions at each potential flood damage center or reach in the basin.

8. Determine the fraction of water to be allocated during each period to each water-use site in the basin.

9. Determine existing and proposed hydropower plant capacities and load factors.

10. Identify any minimal allowable streamflows to be maintained for flow augmentation at each flow augmentation reach in the basin.

11. Specify any maximal or minimal constraints on any of the annual or seasonal water allocations, storage capacities, or target yields.

12. Specify any constraints on maximal or minimal dead storage, active storage, flood control storage, or total storage capacities at any or all of the reservoir sites in the basin.

13. Determine annual capital, operation, maintenance, and replacement (OMR) costs at each reservoir site as functions of a range of total reservoir capacities or scales of development.

14. Determine benefits as functions of energy produced.

15. Determine annual capital and OMR costs at each hydropower production site as functions of various plant capacities.

16. Determine benefit–loss functions for a variety of allocations to domestic, commercial, industrial, and diversion uses.

17. Determine short-run losses as functions of deviations (both deficit and surplus) in planned or target allocations to user sites.

18. Develop benefit functions at each irrigation site in the basin. This analysis requires information on the area of land that can be irrigated per unit of water allocated, the quantities of each crop that can be produced per unit area of land, the total fixed and variable costs of producing each crop, and the unit prices that will clear the market of any quantity of each crop.

19. Develop flood-damage-reduction benefit functions at each potential flood damage site. This analysis requires records of historical and/or simulated floods, channel storage capacities, and flood control reservoir operating policies.

20. Develop reservoir-based recreation benefit functions at each recreation site in the basin.

21.2 GROUNDWATER SIMULATION

Digital simulation models are used in a different manner to study the storage and movement of water in a porous medium. Distributed rather than lumped parameter models are used to imitate observed events and to evaluate future trends in the development and management of groundwater systems. The equations describing the flow of water in a porous medium were derived in Chapter 18 and modeling of regional systems was discussed in Chapter 20. This section deals primarily with techniques used in solving the hydrodynamic equations of motion and continuity, followed by brief discussions of (1) typical input requirements, (2) techniques of calibrating and verifying the models, and (3) the sensitivity of groundwater models to parameter changes. An example of the calibration and application of a groundwater model is also provided.

Model Types

Groundwater studies involve the adaptation of a particular code to the problem at hand. Several popular public domain computer codes for solving various types of groundwater flow problems are listed in Table 21.2. The codes become models when the system being studied is described to the code by inputting the system geometry and known internal operandi (aquifer and flow field parameters, initial and boundary conditions, and water use and flow stresses applied in time to all or parts of the system). Codes have emerged in four general categories: *groundwater flow* codes, *solute transport* codes, *particle tracking* codes, and *aquifer test data analysis* programs.[10]

Groundwater flow codes provide the user with the distribution of heads in an aquifer that would result from a simulated set of distributed recharge-discharge stresses at cells or line segments. From Darcy's law, the flow passing any two points can be calculated from the head differential. The codes are used to model both confined and unconfined aquifers. Each can be structured to model regional flow, or flow in proximity of a single well or wellfield. Steady-state and transient conditions can be evaluated. Boundaries can be barriers, full or partially penetrating streams and lakes, leaky zones, or constant head or constant gradient perimeters. By application of Darcy's law, the seepage velocities of groundwater can be determined after solving for the head differentials.

When groundwater seepage velocities are known, the advection, dispersion, and changes in concentration of solutes can be modeled. Solute transport models build on groundwater flow models by the addition of advection, dispersion, and/or chemical reaction equations. If the chemical, dispersion, or dilution concentration changes due to groundwater flow are not important, particle tracking codes model transport by advection and provide an easier method than solute transport models to track the path and travel times of solutes that move under the influence of head differentials. Aquifer test data programs provide users with computer solutions to many of the hand calculations (Chapter 17) needed to graph and interpret aquifer test data for determining aquifer and well parameters.

TABLE 21.2 GROUNDWATER MODELING CODES

Acronym for code	Description	Source	Year
	Groundwater flow models		
PLASM	Two-dimensional finite difference	Ill. SWS	1971
MODFLOW	Three-dimensional finite difference	USGS	1988
AQUIFEM-1	Two- and three-dimensional finite element	MIT	1979
GWFLOW	Package of 7 analytical solutions	IGWMC	1975
GWSIM-II	Storage and movement model	TDWR	1981
GWFL3D	Three-dimensional finite difference	TDWR	1991
MODRET	Seepage from retention ponds	USGS	1992
	Solute transport models		
SUTRA	Dissolved substance transport model	USGS	1980
RANDOMWALK	Two-dimensional transient model	Ill. SWS	1981
MT3D	Three-dimensional solute transport	EPA	1990
AT123D	Analytical solution package	DOE	1981
MOC	Two-dimensional solute transport	USGS	1978
HST3D	3-D heat and solute transport model	USGS	1992
	Particle tracking models		
FLOWPATH	Two-dimensional steady state	SSG	1990
PATH3D	Three-dimensional transient solutions	Wisc GS	1989
MODPATH	Three-dimensional transient solutions	USGS	1991
WHPA	Analytical solution package	EPA	1990
	Aquifer test analyses		
TECTYPE	Pump and slug test by curve matching	SSG	1988
PUMPTEST	Pumping and slug test	IGWMC	1980
THCVFIT	Pumping and slug test	IGWMC	1989
TGUESS	Specific capacity determination	IGWMC	1990

Note: IGWMC = International Groundwater Modeling Center; Ill. SWS = Illinois State Water Survey; SSG = Scientific Software Group; EPA = Environmental Protection Agency; USGS = U.S. Geological Survey; Wisc. GS = Wisconsin Geological Survey; MIT = Massachusetts Institute of Technology; TDWR = Texas Department of Water Resources; DOE = Department of Energy.

Solution Techniques

With few exceptions, the hydrodynamic equations for groundwater flow have no analytical solutions, and groundwater modeling relies on *finite-difference* and *finite-element* methods to provide approximate solutions to a wide variety of groundwater problems. The choice of method is normally driven by the system to be modeled. Other numerical methods include *boundary integral* methods, *integrated finite difference* methods, and *analytic element* methods.

These solutions, as with streamflow simulation models, are facilitated by first subdividing the region to be modeled into subareas. Groundwater system subdivision depends more on geometric criteria and less on topographic criteria in the sense that the region is overlaid by a regular or semiregular pattern of node points at which (or between which) specific measures of aquifer and water system parameters are input and other parameters are calculated. Approximate solutions of simultaneous linear and nonlinear equations are found by making initial estimates of the solution values, testing the estimates in the equations of motion and continuity, adjusting the values, and finally accepting minor violations in the basic principles or making further adjustments of the parameters in an orderly and converging fashion.

The orderly solution of finite difference analogs of the steady-state or unsteady-state partial differential equation of motion for flow of groundwater in a confined aquifer or an unconfined aquifer is obtained by *relaxation* methods. An early relaxation solution of the equation is discussed by Jacob.[7] For two-dimensional problems, the iterative alternating-direction-implicit (ADI) method developed by Peaceman and Rachford[8] is often adopted.

Prickett and Lonnquist[9] used the ADI technique to calculate fluctuations in water table elevations at all nodes in an aquifer model by proceeding through time in small increments from a known initial state. Their model is computationally efficient and readily applied and is particularly attractive for use with problems involving time variables and numerous nodes. The primary aquifer parameters are the permeability and storage coefficient, which, if assumed constant over the aquifer plan, result in a homogeneous and isotropic condition. For those familiar with relaxation methods, the Gauss–Seidel and the successive over-relaxation (SOR) methods have had application in solving difference equations.

Data Requirements

Input to groundwater system models may be classified as spatial and temporal. Spatial input includes initial or projected water table maps, saturated thickness data over the region, land surface contour maps, transmissivity maps, regional variations in storage coefficients, locations and types of wells and canals, locations and types of aquifer boundaries both lateral and vertical, a node coordinate system, actual or net pumpage rates, percolation and recharge rates for precipitation and other applied waters, logs of drilled wells, geologic stratigraphy, and soil types and cropping patterns.

Time-dependent data requirements for aquifer models involve principally the formulation of time schedules, using a range of time increments for such variables as pumping rates, precipitation hyetographs, canal and streamflow hydrographs, groundwater evapotranspiration rates, and development variables such as the timing of added wells or other system components. Because each temporal schedule can apply only to a particular subset of node positions, the time-dependent requirements are also spatial.

In addition to the listed input parameters, aquifer models require reliable estimates of the percentages of waters in the land phase that actually percolate to the aquifer being modeled. These estimates can be based on knowledge of the physical processes involved in unsaturated flow through porous medium but are most often obtained as judgment parameters that are modified during the calibration phase of the simulation. Simply stated, the lateral movement and the changes in piezometer or water table levels are easily modeled if the node-by-node stresses (withdrawal rates or recharge rates) are known. The latter parameters are governed by the complex movement of water in the unsaturated soil zone and by the random precipitation and consumptive use patterns of the region. The art of modeling groundwater systems lies in the ability to evaluate these parameters.

Calibration

Groundwater model calibration removes some of the guesswork involved in parameter determination. Several combinations of parameters, based on available knowledge of the physical system, are tested in the model during a period for which records are available. Simulated results are then compared with historical events. After structuring the model, calibration is achieved by operating the model during the study period by imposing historical precipitation amounts, canal diversions, evaporation and evapotranspiration rates, streamflows and stream levels, pumping rates during known periods, and other stresses on the aquifer. Calibration is achieved after the flow, storage, and other parameters have been adjusted within reasonable limits to produce the best imitation of recorded events.

Case Example

A typical finite-difference study involving surface water and groundwater modeling in central Nebraska was performed by Marlette and Lewis.[11] The region involved is shown in Fig. 21.2. In addition to the surface irrigation system represented by the several canals and laterals, over 1200 wells withdraw water from the aquifer between the Platte River and the Gothenburg and Dawson County canals. The aquifer recharge and withdrawal amounts as percentages of precipitation, snowfall, pumped water, delivered canal water, evaporation, and evapotranspiration were estimated using a mix of judgment and physical process evaluations. The resulting set that produced the best comparison with recorded events at the six observation wells shown in Fig. 21.2 is summarized in Table 21.3. Samples of the comparison between recorded and simulated water levels in the Dawson County study during a 2-year calibration period are shown in Figs. 21.3 and 21.4.

The Prickett and Lonnquist model was applied in the Dawson County study. The storage coefficient for this unconfined aquifer was established by calibration trials as 0.25 and the adopted permeability was 61 m/day. Other trials were made using various combinations of S and K, with S ranging from 0.10 to 0.30 and with K ranging between 41 and 102 m/day. As with most unconfined aquifer models, water table elevations were most sensitive to fluctuations in the storage coefficient. Figure 21.5 is

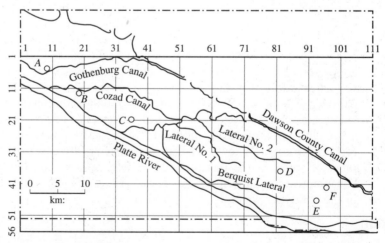

Figure 21.2.   Grid coordinates for Dawson County, Nebraska, aquifer model. ○ = observation well.

TABLE 21.3 ADOPTED RECHARGE CRITERIA FOR WATER ALLOCATIONS
OVER THE DAWSON COUNTY AQUIFER

System component	Allocation and applied amounts	Aquifer recharge/withdrawal as a percentage of applied amount
Rainfall	Recorded depth if daily amount exceeded 0.25 cm at all nodes	30
Snowfall	25% of recorded depths at all nodes	30
Pumped water	Average rate of 50 l/sec at all well nodes during irrigation seasons	50
Delivered canal water	Recorded daily rates, applied to land surface one node laterally uphill and two nodes downhill from canal	30
Evaporation	Observed daily lake evaporation depth at all marsh and water surface nodes	100
Evapotranspiration	125% of daily lake evaporation, applied at all alfalfa nodes	15

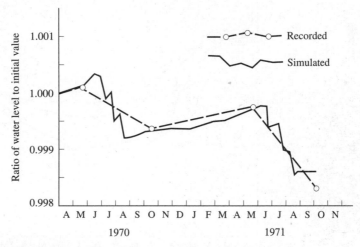

Figure 21.3. Simulated and recorded water levels at observation
well D in Fig. 21.2. $I = 82$; $j = 37$.

a typical summary of the calibration results at a single observation well located at Position F in Fig. 21.2.

After verification, the Dawson County model was applied to investigate the short-term influence of several management schemes. Included among the schemes were investigations involving the complete removal or shutdown of the surface water canals, and other tests in which isolated canal contributions to recharge were determined by operating the model with single canals and comparing results with water table fluctuations for identical conditions with all canals removed. Many other applications of the model are possible. This particular study revealed that recharge from

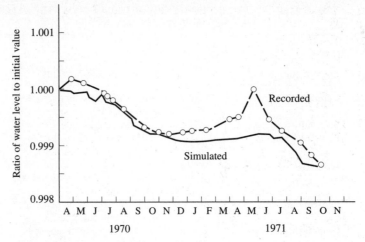

Figure 21.4. Simulated and recorded water levels at observation well F in Fig. 21.2. $I = 97; j = 42$.

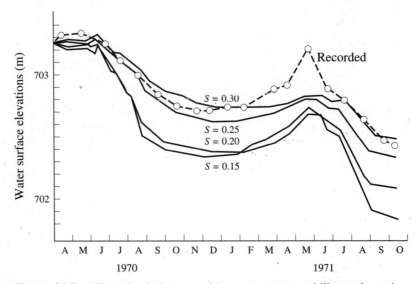

Figure 21.5. Water-level changes with constant permeability and varying storage coefficients. Well F, Fig. 21.2; $K = 61\text{m/day}$; $I = 97; J = 42$.

the existing canal system contributes to the water balance of the aquifer but is not the dominant factor in the short run. The natural recharge from precipitation and from the Platte River account for the long-term water table stability in the region.

21.3 HYDROLOGIC SIMULATION PROTOCOL

The use of hydrologic simulation as a tool in the decision-making process is not new but is of a different, more sophisticated and more encompassing form. A model is a representation of an actual or proposed system that permits the evaluation and manipulation of many years of prototype behavior. This is the feature that makes the use of these tools so attractive and holds such potential for the analysis of even the largest, most complex systems. It is also the principal feature that makes this approach so well suited to water resources system planning and analysis.

Apart from the use of conventional hand methods and some elementary models, planning has traditionally been a practice of judgment. This is changing, however, as quantitative tools are developed that permit the analysis of large numbers of alternatives and plans. Judgment, an essential element of the process, is not ruled out but is strengthened through new insights that were not available to those in the planning profession a few years ago.

Planners are continually required to anticipate the future and ask "What if?" and "What's best?" questions. Quantitative planning techniques, such as simulation can provide detailed information about more planning alternatives for less cost than any other approach available. Development of these tools has occurred principally at universities and federal agencies.

Combined Use of Simulation and Optimization Models

An important second type of quantitative planning tool should be mentioned at this point. *Screening* models are designed to utilize limited system information to select a best plan among many alternatives for a specified objective or set of objectives. Hence screening models, or optimization models as they are often called, are oriented toward *plan formulation* in contrast to the *plan evaluation* function of simulation models. Simulation models are suited to detailed analyses of specific alternatives and yield reliable information on which to base final designs or operating policies. Screening models address the question, "If our goal is . . . , how can we best proceed?" Simulation models, on the other hand, ask, "If our plan is to . . . , will the plan work and what will the system look like after we are finished?" Used together to take advantage of the special merits of each, these two tools become a powerful adjunct to traditional planning technologies. Complete descriptions of techniques and case studies of screening, optimization, and simulation models are presented in Refs. 12–15.

Final design values should be determined by assigning the optimization results to the system elements and operating a detailed simulation model over time using a sequence of known or synthesized precipitation amounts and/or streamflows, while at the same time accumulating benefits over time for flood control, reservoir and streamside recreation, water yields, streamflow augmentation, sediment and erosion control, and any other factors not considered or approximated in the preliminary screening model. Several simulation runs with slight adjustments in decision variables should result in a plan that best meets the objectives and is a significant improvement in plans generated by conventional methods.

Results of optimization models will provide readily obtained and useful information for initiating more refined simulation analyses in order to test the most promising measures and arrive at final plans for the design, construction, and operation of a water resource system. Even though optimization models will provide information for decisions regarding both the development and management of a system, the use of postoptimization simulation is recommended, primarily because of the simplifying assumptions often required in preliminary screening. Detailed simulation is unsuited for preliminary screening of development alternatives owing to time and cost limitations. Unless a new generation of computers evolves, current time and size limitations do not allow screening by simulating all alternatives unless substantial sacrifices in realism are made. For the present, preliminary screening followed by detailed simulation appears to be the most effective means for arriving at optimal water development and management plans.

21.4 CORPS OF ENGINEERS SIMULATION MODELS

In 1964, the U.S. Army Corps of Engineers developed a specialty branch located at the Hydrologic Engineering Center (HEC) in Davis, California. The facility provides a center for applying academic research results to practical needs of the Corps field offices. In addition, the center provides training and technical assistance to government agencies in advanced hydrology, hydraulics, and reservoir operations.

Over the years, a large number of analytical tools were developed at HEC. Table 21.4 summarizes the computer programs in categories of hydrology, river/reservoir hydraulics, reservoir operations, stochastic hydrology, river/reservoir water

TABLE 21.4 HEC WATER RESOURCE COMPUTER PROGRAMS

Name	Date of latest version	Purpose
	Hydrology Models	
HEC-1, Flood Hydrograph Package	September 1980	Simulates the precipitation runoff process in any complex river basin. Optimizes parameters. Computes expected annual flood damage. Optimizes size of flood control system components.
Basin Rainfall and Snowmelt Computation	July 1966	Computes area-average precipitation and snowmelt for many subbasins of a river basin using gauge data and weightings (included in HEC-1).
Unit Graph and Hydrograph Computation	July 1966	Computes subbasin interception/infiltration, unit hydrographs base flow, and runoff hydrograph (included in HEC-1).
Unit Graph Loss Rate Optimization	August 1966	Estimates best-fit values for unit graph and loss rate parameters from given precipitation and subbasin runoff (included in HEC-1).
Hydrograph Combining and Routing	August 1966	Combines runoff from subbasins at confluences and routes hydrographs through a river network using hydrologic routing methods (included in HEC-1).

TABLE 21.4   (*Continued*)

Name	Date of latest version	Purpose
Streamflow Routing Optimization	July 1966	Estimates best-fit values for hydrologic streamflow routing parameters with given upstream, downstream, and local inflow hydrographs (included in HEC-1).
Interior Drainage Flood Routing	November 1978	Computes seepage, gravity and pressure flow, pumping and overtopping discharges for pond areas behind levees or other flow obstructions. Main river elevation and ponding area elevation−area-capacity data are used in computing discharges.
Storage, Treatment, Overflow, Runoff Model ("STORM")	July 1976	Simulates the precipitation runoff process for a single, usually urban, basin for many years of hourly precipitation data. Simulates quality of urban runoff and dry weather sewage flow. Evaluates quantity and quality of overflow for combinations of sewage treatment plant storage and treatment rate.
River/reservoir hydraulics		
HEC-2, Water Surface Profiles	August 1979	Computes water surface profiles for steady, gradually varied flow in rivers and tributaries using natural or artificial cross sections. Flow may be sub- or super-critical. Analyzes allowable encroachment for a given rise in water surface.
Gradually Varied Unsteady Flow Profiles	January 1976	Simulates one-dimensional, unsteady, free surface flows in a branching river network. Natural and artificial cross sections may be used. Uses an explicit centered difference computational scheme.
Geometric Elements from Cross Section Coordinates ("GEDA")	June 1976	Computes tables of hydraulic elements for use by the Gradually Varied Unsteady Flow Profiles or other programs. Interpolates values for area, top width, n value, and hydraulic radius at evenly spaced locations along a reach.

TABLE 21.4 *(Continued)*

Name	Date of latest version	Purpose
River reservoir hydraulics (continued)		
Stream Hydraulics Package ("SHP")	October 1978	Performs dynamic streamflow routing in a complex river network using full St. Venant, kinematic wave, modified Puls, or Muskingum routines. Can optimize the storage-discharge function for modified Puls given inflow and outflow hydrographs from a reach.
Spillway Rating and Flood Routing	October 1966	Computes a spillway rating curve for concrete ogee or broadcrested weir spillway with or without discharge from a conduit or sluice. Routes the spillway design flood using a gated or uncontrolled spillway to determine maximum water surface elevation.
Spillway Rating—Partial Tainter Gate Opening	July 1966	Computes discharge rating curve for ogee-type weirs with Tainter (radial-type) gates at any size opening.
Spillway Gate Regulation Curve	February 1966	Computes gate regulation schedule curves for a reservoir knowing the area–capacity curves, induced surcharge envelope curve, and the slope of the recession portion of an inflow hydrograph.
Reservoir Area–Capacity Tables by Conic Method	July 1966	Computes surface area and storage volume for various reservoir water surface elevations using a conic method.
Reservoir operation		
HEC-3, Reservoir System Analysis for Conservation	July 1973	Simulates operation of reservoirs for a complex system of reservoirs and purposes for multiyear sequences of *monthly* streamflows without streamflow routing. Demands for hydropower, low-flow augmentation, and various water supplies are satisfied at the reservoir or at downstream control points within constraints for reservoir storage, release capacities, balance between reservoirs, and for various hydrothermal power systems.

TABLE 21.4 *(Continued)*

Name	Date of latest version	Purpose
	Reservoir operation (continued)	
HEC-5, Simulation of Flood Control and Conservation Systems	June 1979	Simulates operation of reservoirs for flood control, low-flow augmentation, hydropower and water supply throughout a complex river system. Reservoir operation may be accomplished at variable time intervals during a multiyear simulation. Downstream flood routing is performed for flood control operation and expected annual damages may be computed.
Reservoir Yield	August 1966	Simulates the operation of a single reservoir and one downstream control point for hydropower, water supply, and water quality. Operates on a monthly time interval without streamflow routing.
	Stochastic hydrology	
HEC-4, Monthly Streamflow Simulation	January 1971	Analyzes monthly streamflows at a number of interrelated stations to determine their statistical characteristics. Can generate multiyear synthetic monthly streamflow series using the log–Pearson Type III distribution representation of the historical data.
Flood Flow Frequency Analysis	June 1976	Computes log–Pearson Type III frequency statistics of annual maximum flood peaks according to the United States Water Resource Council Guidelines, Bulletin 17a. Program automatically adjusts for zero flood years, incomplete records, low and high outliers, and historical information. Computes expected probability.
Regional Frequency Computation	July 1972	Performs log–Pearson Type III frequency computations of annual maximum hydrologic events at several stations. Used in regional frequency analysis preserving intercorrelation between stations and regional skew. Computes statistics of flow at each station for several durations if desired.

TABLE 21.4 (*Continued*)

Name	Date of latest version	Purpose
River/reservoir water quality		
Water Quality for River-Reservoir Systems ("WQRRS")	October 1978	Simulates multiparameter water quality in rivers (horizontal segments) and in reservoirs (vertical segments) for steady-state or dynamic flow conditions. Complete aquatic-biologic system is simulated to obtain resulting water quality.
Statistical and Graphical Analysis of Water Quality Data	October 1978	Performs statistical analyses of water quality time series and station results obtained from Water Quality for River-Reservoir Systems or other models. Plots up to 11 parameters versus time at desired stations.
HEC-5Q, Reservoir System Operation including Water Quality Control	September 1979	Simulates the operation of a single reservoir in the same manner as HEC-5 but also includes operation for control of water temperature in the reservoir and at downstream control points.
Heat Exchange Program	December 1972	Computes equilibrium temperatures and surface heat exchange coefficients for use in estimating net heat exchange between a water surface and the atmosphere. Requires daily meteorologic data at locations of interest.
Thermal Simulation Program	June 1973	Determines the annual temperature cycle of a water body by heat balance analysis of inflows, outflows, and surface heat transfer.
Reservoir Temperature Stratification	September 1969	Simulates monthly temperature variations in horizontal strata within a reservoir using inflow, outflow, and meteorologic data, reservoir and dam configuration, and downstream release requirements.
River/reservoir sedimentation		
HEC-6, Scour and Deposition in Rivers and Reservoirs	March 1976	Simulates one-dimensional sediment transport, scour, and deposition in a river system that may have reservoirs. Steady-state water surface profiles are

TABLE 21.4 (*Continued*)

Name	Date of latest version	Purpose
River/reservoir sedimentation (continued)		
		computed; and at each cross section, discharge, inflowing sediment load, gradation of bed material and armoring are considered in computing the scour or deposition. Dredging may be analyzed.
Suspended Sediment Yield	March 1968	Computes annual suspended sediment load corresponding to observed daily water discharge and suspended sediment loads.
Reservoir Delta Sedimentation	July 1967	Computes the expected ultimate profile of sediment deposits forming the delta at a reservoir inflow point.
Deposit of Suspended Sediment	June 1967	Computes the distribution and location of sediments in a reservoir. Sediment inflow, trap efficiency, and size distribution of passing sediments are calculated. Sedimentation is calculated in main body of reservoir as well as in tributary arms.
Groundwater		
Finite Element Solution of Steady State Potential Flow Problems	July 1970	Computes groundwater or seepage flows for steady, two-dimensional or axisymmetric flow through heterogeneous, anisotropic porous media of virtually any geometry.
Flood damage		
Expected Annual Flood Damage Computation ("EAD")	June 1977	Computes flood damage for the economic evaluation of flood control and flood-plain management plans. Damages may be computed for a specific event, expected annual, or equivalent annual for a given interest rate and time period. Analyzes multiple damage reaches and types of damage for several flood management plans.
Interactive Nonstructural Analysis Package	February 1980	Analyzes flood damage reduction benefits for various nonstructural measures such as flood proofing, raising structures, and relocating structures.

TABLE 21.4   (*Continued*)

Name	Date of latest version	Purpose
Geographic information analysis		
Resource Information and Analysis ("RIA")	September 1978	Analyzes geographic data stored in grid cell data banks. Analyzes, tabulates, and displays (printer plots) map analyses for locational attractiveness, impact assessment, distance determination, and coincidence of geographic features.
Hydrologic Parameters ("HYDPAR")	November 1978	Computes interception/infiltration and unit-hydrograph parameters from grid cell data for use in HEC-1 watershed model. Computes parameters for sub-basins in a river basin and stores results for access by HEC-1.
Damage Calculation ("DAMCAL")	November 1978	Computes stage versus damage functions for index locations along the floodplain. Uses grid cell data bank representation of land use and topography and aggregates damages to index location for a range of water surface elevations.

Source: A. D. Feldman, "HEC Models for Water Resources Systems Simulation: Theory and Experience," in *Advances in Hydroscience,* Vol. 12. Orlando, FL: Academic Press, 1981.

quality, river/reservoir sedimentation, groundwater, flood damage, and geographic information analysis. Figure 21.6 shows the relations among, and potential use of, three of the earliest HEC models. This figure illustrates the concept that many of the HEC models can be combined in a single investigation to provide useful tools for total system analysis. A listing of private outlets for the software and vendors who provide technical support can be obtained from the HEC.

HEC-HMS Hydrologic Modeling System   The Hydrologic Engineering Center is developing next-generation computer models to replace those in current use. This package, called the HEC-HMS, is targeted to include "object-oriented" versus traditional "procedural-oriented" technology. A new graphical user interface allows the user to edit, execute, and view model data in a window environment. The new approach is designed to provide a logical way to express problems, breaking them down into individual understandable entities and defining interactions among the entities. The technology is described by Pabst.[16]

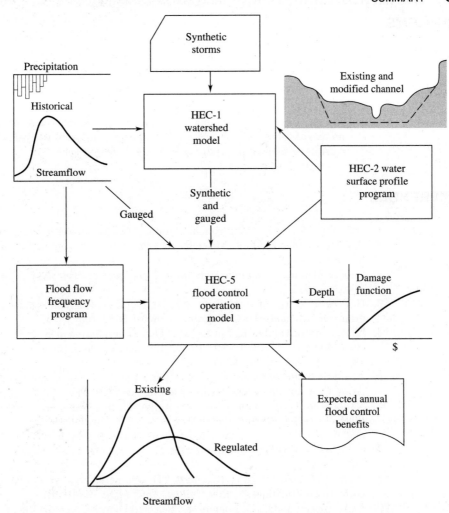

Figure 21.6. Example of the use of three HEC models to study flood control alternatives. (After A. D. Feldman, "HEC Models for Water Resources System Simulation: Theory and Experience," *Advances in Hydroscience,* Vol.12. Orlando, FL: Academic Press, 1981.)

■ Summary

The models introduced in this chapter have enjoyed growing use in practice for almost two decades. Without exception, model users report that the protocol for problem identification, model selection and conceptualization, model calibration and verification, and model documentation are fundamentals that should be mastered early and sustained throughout the process. Any use of the procedures presented in the next five chapters should be preceded by a review of the principles in this chapter.

PROBLEMS

21.1. Simulation and synthesis are treated separately in Chapters 22 and 23. List the most distinguishing characteristics of each method and give an example of each.

21.2. List at least three reasons many of the developed models of the rainfall–runoff process might not be used by hydrologists.

21.3. You are asked to determine a design inflow hydrograph to a reservoir at a site where no records of streamflow are available. List general steps you would take as a hydrologist in developing the entire design inflow hydrograph.

REFERENCES

1. U.S. Army Waterways Experiment Station, "Models and Methods Applicable to Corps of Engineers Urban Studies," Miscellaneous Paper H-74-8, National Technical Information Service, Aug. 1974.

2. George Fleming, *Computer Simulation Techniques in Hydrology*. New York: American Elsevier, 1975.

3. U.S. Army Corps of Engineers, "HEC-1 Flood Hydrograph Package," Users and Programmers Manuals, HEC Program 723-X6-L2010, Jan. 1973.

4. N. H. Crawford and R. K. Linsley, Jr., "Digital Simulation in Hydrology: Stanford Watershed Model IV," Department of Civil Engineering, Stanford University, Stanford, CA, Tech. Rep. No. 39, July 1966.

5. A. J. Friedrich, "Management of Computer Use in Solving Engineering Problems," U.S. Army Corps of Engineers, Hydrologic Engineering Center, Davis, CA, 1979.

6. National Research Council, *Ground Water Models—Scientific and Regulatory Applications*. Water Science and Technology Board, Commission on Physical Sciences, Mathematics, and Resources, National Academy Press, Washington, D.C., 1990.

7. C. E. Jacob, "Flow of Groundwater," in *Engineering Hydraulics* (Hunter Rouse, ed.). New York: Wiley, 1950.

8. D. W. Peaceman and N. H. Rachford, "The Numerical Solution of Parabolic and Elliptic Differential Equations," *J. Soc. Indust. Appl. Math.* **3**, (1955).

9. T. A. Prickett and C. G. Lonnquist, "Selected Digital Computer Techniques for Groundwater Resource Evaluation," Illinois State Water Survey Bull. No. 55, 1971.

10. D. R. Maidment, (ed.), *Handbook of Hydrology*. New York: McGraw-Hill, 1993.

11. R. R. Marlette and G. L. Lewis, "Digital Simulation of Conjunctive-Use of Groundwater in Dawson County, Nebraska," Civil Engineering Report, University of Nebraska, Lincoln, 1973.

12. W. K. Johnson, "Use of Systems Analysis in Water Resource Planning," *Proc. ASCE J. Hyd. Div.* (1974).

13. R. deNeufville and D. H. Marks, *Systems Planning and Design Case Studies in Modeling Optimization and Evaluation*. Englewood Cliffs, NJ: Prentice-Hall, 1974.

14. D. P. Loucks, "Stochastic Methods for Analyzing River Basin Systems." Cornell University Water Resources and Marine Sciences Center, Ithaca, NY, Aug. 1969.

15. A. Maass, (ed.), *Design of Water Resources Systems*. Cambridge, MA: Harvard University Press, 1962.

16. A. F. Pabst, "Next Generation HEC Catchment Modeling," Proceedings, ASCE Hydraulics Division Symposium on Engineering Hydrology, San Francisco, CA, July 25–30, 1993.

Chapter 22

Hydrologic Time Series Analysis

■ Prologue

The purpose of this chapter is to:

- Show how time series analysis is used for generating synthetic hydrologic records.
- Give definitions of terms used to describe the stochastic aspects of hydrologic series.
- Introduce fundamentals of streamflow synthesis including mass curve analysis, random generation of sequences, serial-dependent sequences, and sequences having prescribed frequency distributions.

Time-series analysis of hydrologic variables has become a practical methodology for generating synthetic sequences of precipitation or steamflow values that can be used for a range of applications from filling in missing data in a gauged record to extending monthly streamflow records[1], and from analyzing long-term reliability of yields of watersheds[2] or reservoirs[3,4] to forecasting floods or snowmelt runoff quantities from synthetic precipitation sequences.[5] These *synthetic hydrology* techniques augment the simulation tools described in Chapter 21. Both have experienced widespread use by hydrologists and engineers.[6] Synthesis involves the generation of a sequence of values for some hydrologic variable (daily, monthly, seasonal, or annual). The techniques are most often applied to produce streamflow sequences for use in reservoir design or operation studies but can also be used to generate rainfall sequences that can subsequently be input to simulation models.

If historical flows could be considered to be representative of all possible future variations that some project will experience during its lifetime, there would be little need for synthetic hydrology. The historical record is seldom adequate for predicting future events with certainty, however. The exact historical pattern is unlikely to recur, sequences of dry years (or wet years) may not have been as severe as they may become, and the single historical record gives the planner limited knowledge of the magnitude of risks involved.

Synthesis enables hydrologists to deal with data inadequacies, particularly if record lengths are not sufficiently extensive. Short historical records of hydrologic variables such as streamflow are extended to longer sequences using hydrologic synthesis and other techniques of the broad science known as *operational hydrology*.[7] These new, synthetic sequences either preserve the statistical character of the historical records or follow a prescribed probability distribution, or both. When coupled with computer simulation techniques, the techniques provide hydrologists with improved design and analysis capabilities.

The methods described in this chapter are based on probability and statistics. The material presented in Chapter 26 should be reviewed prior to studying this chapter.

22.1 SYNTHETIC HYDROLOGY

Hydrologic synthesis techniques are classified as (1) *historical repetition methods,* such as mass curve analyses, which assume that historical records will repeat themsevles in as many end-to-end repetitions as required to bracket the planning period; (2) *random generation techniques,* such as Monte Carlo techniques, which assume that the historical records are a number of random, independent events, any of which could occur within a defined probability distribution; and (3) *persistence methods,* such as Markov generation techniques, which assume that flows in sequence are dependent and that the next flow in sequence is influenced by some subset of the previous flows. Historical repetition or random generation techniques are normally applied only to annual or seasonal flows. Successive flows for shorter time intervals are usually correlated, necessitating analysis by the Markov generation method.

As with most subfields of hydrology, a number of computer programs for time-series analysis and hydrologic data synthesis have been developed. One of the first, and one of the most widely applied, was the U.S. Army Corps of Engineers model HEC-4 (see Section 21.4) published in 1971.[1] Its use is limited, though, to synthesizing sequences of serially dependent monthly streamflows in a river reach. Other codes[8,9] are available to the hydrologist, however. Additional models and descriptions of theory and applications of time-series analysis of precipitation and streamflow are detailed in a number of available texts and publications.[10-13]

Mass Curve Analysis

One of the earliest and simplest synthesis techniques was devised by Rippl[14] to investigate reservoir storage capacity requirements. His analysis assumes that the future inflows to a reservoir will be a duplicate of the historical record repeated in its entirety as many times end to end as is necessary to span the useful life of the reservoir. Sufficient storage is then selected to hold surplus waters for release during *critical periods* when inflows fall short of demands. Reservoir size selection is easily accomplished from an analysis of peaks and troughs in the mass curve of accumulated synthetic inflow versus time.[15-17] Future flows can be similar, but are unlikely to be identical to past flows. Random generation and Markov modeling techniques produce sequences that are different from, although still representative of, historical flows.

EXAMPLE 22.1

Streamflows past a proposed reservoir site during a 5-year period of record were, respectively, in each year 14,000, 10,000, 6000, 8000, and 12,000 acre-ft. Use Rippl's mass curve method to determine the size of reservoir needed to provide a yield of 9000 acre-ft in each of the next 10 years.

Solution. A 10-year sequence of synthetic flows, using Rippl's assumptions, is shown in Table 22.1. Inflows are set equal to the historical record repeated twice. ∎∎

TABLE 22.1 STREAMFLOWS FOR EXAMPLE 22.1

	Flows (thousands of acre-ft)									
Year	1	2	3	4	5	6	7	8	9	10
Inflow	14	10	6	8	12	14	10	6	8	12
Cumulative inflow	14	24	30	38	50	64	74	80	88	100

When cumulative inflow and cumulative draft are plotted, the maximum deficiency shown in Fig. 22.1 is 4000 acre-ft. Thus a reservoir with a 4000-acre-ft capacity should be placed in the stream. Starting with a full reservoir at the beginning of Year 1, the reader should verify the adequacy of the reservoir by "simulating" a draft of 9000 acre-ft per year for 10 years.

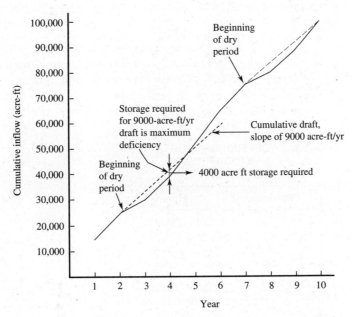

Figure 22.1 Mass curve for Example 22.1: — cumulative inflow; --- cumulative draft.

Random Generation

One method of generating sequences of future flows is a simple random rearrangement of past records. If the stream is ungauged and records are not available, a probability distribution can be selected and a sequence of future flows that follow the distribution and have prescribed statistical moments is generated.

Whenever historical flows are available, a reasonable sequence of future flows can be synthesized by first consulting a table of random numbers, selecting a number, matching this with the rank-in-file number of a past flow, and listing the corresponding flow as the first value in the new sequence. The next random number would be used in a similar fasion to generate the next flow, and so on. Random numbers having no corresponding flows are neglected and the next random number is selected. Table B.3 in Appendix B is a table of uniformly distributed random numbers (each successive number has an equal probability of taking on any of the possible values). To illustrate the use of Table B.3 in the random generation process, the first three years of a synthetic flow sequence could be generated by selecting the 53rd, 74th, and 23rd from the list of past flows. Alternatively, the flows in 1953, 1974, and 1923 could also be selected as the new random sequence.

Most computers have random number generation capabilities in their system libraries. Rather than storing large tables of numbers such as Table B.3, successive random integers are usually generated by the computer.

EXAMPLE 22.2 _____

Annual flows in Crooked Creek were 19,000, 14,000, 21,000, 8000, 11,000, 23,000, 10,000, and 9000 acre-ft, respectively, for years 1, 2, 3, 4, 5, 6, 7, and 8. Generate a 5-year sequence of annual flows, Q_j, by matching five random numbers with year numbers.

Solution. Random integers between 0 and 9 are generated from the computer. The Q_j values in Table 22.2 are selected from the eight given flows by matching the respective year number with the random number. The digit 9 has no corresponding flow in the 8-year sequence, so the next random number, 2, places the 14,000-cfs flow in Year 2 in the first position of the synthetic 5-year sequence. ■■

TABLE 22.2 DEVELOPMENT OF 5-YEAR SYNTHETIC SEQUENCE

j	Random digit	Q_j (acre-ft)
1	9	Skip
2	2	14,000
3	5	11,000
4	8	9,000
5	1	19,000
6	4	8,000

Streamflow synthesis for ungauged streams requires either a regional frequency curve or an equation of the cumulative distribution function. The synthetic sequence can be generated using an assumed CDF shape (see Chapter 26) with regional estimates of distribution parameters such as the mean and standard deviation.

The random generation process for gauged or ungauged basins begins with the selection of a sequence of random numbers p_i from Table B.3 or from digital computers. Each p_i is made fractional by placing a decimal point in front, and the resulting decimal fraction is treated either as $G(x)$ or $F(x)$ (see Chapter 26) in solving for streamflow x.

EXAMPLE 22.3

Generate a 5-year sequence of annual flows in Crooked Creek (see Example 22.2) having a Pearson Type III CDF (see Chapter 26) with a mean of 14,000 acre-ft, a standard deviation of 5000 acre-ft, and a skew coefficient of 0.2.

Solution.   Successive random values of p_i between 0 and 1000 are generated from a random number generator. These are treated as exceedance probabilities $G(Q)$, which in turn are entered into Table B.2 to find the interpolated Pearson Type III frequency factor K (see Table 22.3). Corresponding flows are found from (see Chapter 27)

$$Q_j = \overline{Q} + K_j(s) \qquad (27.5)$$

or

$$Q_j = 14,000 + K_j(5000) \qquad (22.1)$$

■■

TABLE 22.3   DEVELOPMENT OF 5-YEAR PEARSON III SEQUENCE

i	p_I	$G(Q)$ (%)	j	K_i(Table B.2)	Q_i[Eq. (22.1)](acre-ft)
1	123	12.3	1	1.17	19,850
2	61	6.1	2	1.56	21,800
3	529	52.9	3	−0.11	13,450
4	716	71.6	4	−0.62	10,900
5	361	36.1	5	0.36	15,800

22.2 SERIALLY DEPENDENT TIME SERIES ANALYSIS

Markov Generation

The synthesis of sequences of streamflows by random generation ignores the existence of *persistence* present to some extent in most hydrologic sequences. Persistence is the tendency for high flows to be followed by high flows, and low flows to be followed by low flows. Markov models assume that each flow is dependent on one or more of the most recent flows.

Single-period and *multiperiod* Markov models view hydrologic series as chains of serially dependent values, where each value has a deterministic part and a random error part. To illustrate, the *lag-one single-period* Markov chain assumes that the next flow in sequence, Q_{i+1}, is the sum of the mean $\overline{Q}_{j+1}$ plus a dependent fractional part of the deviation of the previous flow Q_i from its mean, $\overline{Q}_j$, plus a random component e. Expressed as an equation, the lag-one Markov model[18] is

$$Q_{i+1} = \overline{Q}_{j+1} + r_j(Q_i - \overline{Q}_j) + e \tag{22.2}$$

where Q_{i+1} = the generated streamflow in period $i + 1$ (day, month, year)
  Q_{j+1} = the mean of observed flows in period $j + 1$
  Q_i = the generated flow in period i
  Q_j = the mean of observed flows in period j
  r_j = the correlation coefficient for the relation of flows for period $j + 1$ to flows for period j
  e = the simulation error due to unexplained variance

Now since e represents the error in the relation, rewriting Eq. 22.2, we obtain

$$Q_{i+1} = \overline{Q}_{j+1} + r_j(Q_i - \overline{Q}_j) + t_i\sigma_{j+1}(1 - r_j^2)^{1/2} \tag{22.3}$$

where t_i = a random number selected from a normal distribution having a zero mean and a unit variance
  σ_{j+1} = the standard deviation of observed flows for the period $j + 1$

To apply Eq. 22.3, it is necessary to calculate the required means, variances, and correlations and then to initiate the synthetic sequence of flows with some value Q_i with $i = 1$. The mean of observed flows is often selected as the initial flow in the random synthetic runoff sequence, but the choice affects all future flows in the sequence. The first 50 flows in the sequence should be discarded as a means of providing a warm-up period to eliminate starting condition bias.[17] Standard normal random deviates, t_i, are generated using the random generation procedures described earlier.

Whenever the single-period Markov generator is used for annual flows, the period $j + 1$ flows and the period j flows have equal means, or $\overline{Q}_{j+1} = \overline{Q}_j$. This assumption is not valid for smaller periods, such as monthly or daily subdivisions. For example, October and November flows seldom have equal means. To reflect different seasonal or monthly means, the *multiperiod* Markov model is used. This requires a double indexing subscript in Eq. 22.3 or

$$Q_{i,j} = \overline{Q}_j + b_j(Q_{i-1,j-1} - \overline{Q}_{j-1}) + t_i\sigma_j(1 - r_j^2)^{1/2} \tag{22.4}$$

where j is the number of seasonal periods in the year and the other terms have been defined except for $b_j = r_j(\sigma_{j+1}/\sigma_j)$ which must be employed, since $\overline{Q}_{j+1} \neq \overline{Q}_j$. This model has been used extensively in streamflow analysis. The single period and multiperiod Markov generation procedures sometimes result in negative flows. These flows must be retained for generating the next flows in sequence and then they may be discarded.

EXAMPLE 22.4

The parameters describing quarterly flows in millions of gallons per day for the Patapsco River are listed in Table 22.4. Apply the information in conjunction with a table of random numbers and create a 2-year record of quarterly flows. The solution is shown in Table 22.5. A flowchart for the process is sketched in Fig. 22.2. ■■

TABLE 22.4 QUARTERLY STREAMFLOW
PARAMETERS, PATAPSCO RIVER

j	b_j	σ_j	r_j	$(1 - r_j^2)^{1/2}$	$\bar{Q}_j$
1	0.66	73	0.57	0.82	68
2	0.43	85	0.41	0.91	137
3	0.14	96	0.43	0.90	183
4	0.91	29	0.36	0.93	107

TABLE 22.5 TABULAR GENERATION OF SYNTHETIC QUARTERLY STREAMFLOWS

i (1)	j (2)	$\bar{Q}_j$ (3)	$\bar{Q}_{j-1}$ (4)	$Q_{i-1,j-1}$ (5)	k (6)	t_j (7)	$b_j(Q_{i-1,j-1} - \bar{Q}_{j-1})$ (8)	$t_i\sigma_j(1 - r_j^2)^{1/2}$ (9)	$Q_{i,j}$ (10) $[(3) + (8) + (9)]$
1	1	68	107	107	0.5374	0.094	0.0	5.6	73.6
2	2	137	68	73.6	0.6338	0.342	2.4	26.5	165.9
3	3	183	137	165.9	0.3530	−0.377	4.0	−30.6	156.4
4	4	107	183	156.4	0.6343	0.343	−24.2	9.3	92.1
5	1	68	107	92.1	0.0263	−1.94	−9.8	−116.4	0.0
6	2	137	68	0.0	0.6455	0.373	−29.2	28.9	136.7
7	3	183	137	136.7	0.8507	1.04	−0.04	84.5	267.5
8	4	107	183	267.5	0.3485	0.39	76.8	10.6	194.5

Column 1. Period index, $i = 1$ to 8 quarters for a 2-year run.

Column 2. Quarterly index, $j = 1$ to 4, two cycles.

Column 3. Mean quarterly flow.

Column 4. Preceding mean quarterly flow.

Column 5. The generated flow for the preceding quarter. Note that the first entry is the same as Column 4; that is, generation begins using the mean flow from the preceding quarter.

Column 6. Uniform random digits from a random number table. Could be produced simply by drawing randomly from a card deck, with replacement.

Column 7. Random normal deviates from Appendix B, Table B.1. If Column 6 is less than 0.5, find z for $(0.5 - k)$ and set $t = -z$; if $k > 0.5$, find z for $(k - 0.5)$ and set $t = z$.

Column 8. Regression component of Eq. (22.4).

Column 9. Random component of Eq. (22.4).

Column 10. Generated flow. Note that a negative flow occurred in Row $i = 5$ and was then set to 0.0. In practice, the negative value should not be cancelled to 0.0 until all subsequent flows have been calculated.

Normal Distributions

The model in Fig. 22.2 is applicable to simulating flows at a single site where the population of flows is normally distributed for each calendar month. The equation as stated preserves the mean, the variance, and the correlation between successive flows of the historic monthly streamflow sequence. The simulation procedure begins by assuming an initial value of Q_i, computing Q_{i+1}, and repeating by replacing Q_i by Q_{i+1}. In extending a record, the initial assumption of Q_i is the last recorded monthly streamflow.

Log–Normal Distributions

Monthly flows frequently do not appear to be from a normally distributed population. Experience has shown, however, that annual flows are generally normal and monthly flows are log–normal or Pearson III. A recursive equation for log–normal simulation

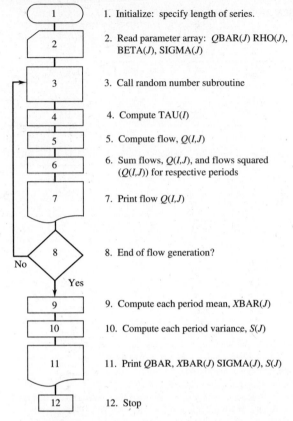

1. Initialize: specify length of series.

2. Read parameter array: $QBAR(J)$ $RHO(J)$, $BETA(J)$, $SIGMA(J)$

3. Call random number subroutine

4. Compute $TAU(I)$

5. Compute flow, $Q(I,J)$

6. Sum flows, $Q(I,J)$, and flows squared $(Q(I,J))$ for respective periods

7. Print flow $Q(I,J)$

8. End of flow generation?

9. Compute each period mean, $XBAR(J)$

10. Compute each period variance, $S(J)$

11. Print $QBAR$, $XBAR(J)$ $SIGMA(J)$, $S(J)$

12. Stop

Figure 22.2   Flowchart for computing normally distributed synthetic streamflows by Markov generation.

presented by Matalas[19] preserves the mean μ_x, standard deviation σ_x, skewness γ_x, and lag-one correlation coefficient r_x for the monthly flows Q. Using $Y = \log(Q - a)$, the recursive equation (22.3) appears as

$$(Y_{i+1} - \mu_y) = r_y(Y_i - \mu_y) + t\sigma_y(1 - r_y^2)^{1/2} \tag{22.5}$$

where Y_{i+1} = the generated log–normal flow in month $i + 1$
   μ_y = the mean of the log–normal flow in month $i + 1$
   r_y = the correlation coefficient for log–normal flow in month $i + 1$ with month i
   σ_y = the standard deviation of observed flows in month i
   t = the normally distributed random number with zero mean and unit variance

Values for a, μ_y, σ_y, and r_y must be found in the following manner to ensure preserving a synthetic sequence that will resemble the historic sequence in terms of μ_x, σ_x, γ_x, and r_x. Relations for μ_x, σ_x^2, γ_x, and r_x are given as

$$\mu_x = a + \exp(\tfrac{1}{2}\sigma_y^2 + \mu_y) \tag{22.6}$$

$$\sigma_x^2 = \exp[2(\sigma_y^2 + \mu_y)] - \exp(\sigma_y^2 + 2\mu_y) \tag{22.7}$$

$$r_x = \frac{\exp(\sigma_y^2 r_y) - 1}{\exp(\sigma_y^2) - 1} \tag{22.8}$$

$$\gamma_x = \frac{\exp(3\sigma_y^2) - 3\exp(\sigma_y^2) + 2}{[\exp(\sigma_y^2) - 1]^{3/2}} \tag{22.9}$$

Since the values of μ_x, σ_x, γ_x, and r_x are known, Eqs. 22.6–22.9 can be solved for μ_y, σ_y, r_y, and a. Equation 22.5 may then be used to step forward in monthly increments finding values of Y_{i+1}. To get the synthetic sequence of streamflow, the value of a is added to the antilog of each value of Y_{i+1}.

Pearson III Distributions

The Pearson III distribution for simulating daily streamflows in reservoir studies has been advocated and is recommended when simulating daily flows for critical flood months.[20] The basic recursion formula is

$$Q_{i+2} = b_1 Q_{i+1} + b_2 Q_i + X_r(1 - r^2) \tag{22.10}$$

where Q_{i+2}, Q_{i+1}, Q_i = the normal standardized variates for successive days
$\quad\quad\quad X_r$ = the random standardized variate
$\quad\quad\quad b_1$, b_2 = regression coefficients
$\quad\quad\quad r$ = the correlation coefficient for the regression equation

This simulation model is a second-order Markov chain; hence each daily flow is assumed to be partially dependent on flows in each of the previous 2 days. Normal standardized variates for successive days are found by first classifying streamflows into months, and determining the logarithmic monthly mean, standard deviation, and skew coefficient.* Next, logarithmic daily flows are standardized by subtracting the mean monthly logarithm from each daily value and dividing by the monthly standard deviation. This standardized variate is transformed to normal by the Pearson Type III approximation

$$t = \frac{6}{g}\left[\left(\frac{g}{2}k + 1\right)^{1/3} - 1\right] + \frac{g}{6} \tag{22.11}$$

where t = the standard normal deviate
$\quad\quad\quad k$ = the Pearson Type III standard deviate
$\quad\quad\quad g$ = the skew coefficient

A typical simulation with this technique first requires calculation of the following statistics from the existing streamflow record:

1. the logarithmic mean, serial correlation coefficient, and skew coefficient of daily flows for each month of the record period,
2. the correlation coefficients between logarithmic flows of successive days, and

---

* In order to avoid zero flows, 0.01 times the monthly mean should be added to each daily flow before taking the logarithm.

3. a regression coefficient of the standard deviation for the daily flow logarithms within each month of record to the logarithm of the monthly total flow.

Given the calculated statistics, the simulation of daily flows could be structured in the following manner:

1. Generate standardized variates from Eq. 22.10.
2. Use the logarithm of the monthly mean flow as an initial estimate of the mean of the logarithmic daily flows for the month.
3. Calculate the standard deviation of flow logarithms by the previously determined regression equation.
4. Apply the inverse of Eq. 22.11,

$$ k = \frac{2}{g} \left[\frac{g}{6} \left(t - \frac{g}{6} \right) + 1 \right]^3 - \frac{2}{g} \tag{22.12} $$

to transform the standardized variates to flows, multiplying by the appropriate standard deviation and adding the mean.
5. Add the difference between the total monthly flow generated and the given monthly flow to the given monthly flow, and repeat the simulation.
6. Multiply daily results of the second simulation by the ratio of the given monthly total to the generated monthly total.

Each simulation technique commonly requires modificiations when applied to individual problems. Methods outlined thus far can be utilized as guides in establishing a procedure to follow in synthesizing flows. Simulation of flows for a given station has been presented with serial correlation, skewness, means, and standard deviations maintained. When generating streamflow sequences for an entire system, the preservation of cross-correlation between stations becomes a significant factor.

Synthetically generated runoff sequences are employed to determine the capacities of reservoirs to satisfy specified demands. Individually generated flow magnitudes are uncertain, as are the synthetically generated sequences of flows. Hydrologists can estimate probabilities of flows by generating several equally likely sequences of flows and then evaluating recurrences of certain values. Herein lies one of the most useful applications of Markov generating techniques.

■ Summary

Hydrologic modeling is often presented as comprising only the deterministic models of the rainfall–runoff process described in Chapters 21, 23, 24, and 25. The fully equipped hydrologist incorporates the synthetic hydrology models described in this chapter in the analysis and design of water resources systems. A growing number of projects are constructed or operated on the basis of synthetic hydrology and time-series analysis each year.

PROBLEMS

22.1. Plot cumulative inflows versus time for the 8-year record in Example 22.2 and determine by mass curve analysis the size of the reservoir needed to provide a yield of 12,000 acre-ft in each of the next 24 years. What is the maximum yield possible?

22.2. Use the annual rainfall from Table 26.2 to generate a 10-year sequence of synthetic annual rain depths for Richmond using Rippl's mass curve assumption.

22.3. Repeat Problem 22.2 using random generation rather than mass curve methods. Use two-digit random numbers from Table B.3 and match these with the last two digits of the year numbers in Problem 26.32.

22.4. Repeat Problem 22.2 using random generation to generate a 10-year synthetic sequence of annual rain depths that has a normal CDF with a mean and standard deviation equal to that of the annual rain data from Problem 26.32.

22.5. Repeat Problem 22.4 assuming that the annual rain depths follow a log–Pearson Type III distribution. For statistics use the mean, standard deviation, and skew of the logarithms of annual rain at Richmond.

22.6. Select a gauged stream in your geographic location and prepare a quarterly model using (a) normal distribution, (b) log–normal distribution, and (c) Pearson Type III distribution.

22.7. Can you convert the simulation problem in Example 22.4 to a log–normal distribution simulation? What difficulties are encountered with the data given in the example?

22.8. Select a month of thunderstorm activity in your region. From published NOAA hourly rainfall data, fit a distribution to the time between storms, and duration of storms, for 20 years of recorded data covering the selected month. Prepare a computer program to randomly generate the times between storms and the durations of storms.

22.9. Flows during 6 years of record were used in synthesizing the mass curve shown on the following page.
    a. Use Rippl's assumption and the graph to determine the missing magnitude of the flow for the 12th year.
    b Determine the reservoir capacity required to allow an annual yield of 2000 acre-ft/yr. Repeat for 500 acre-ft/yr.
    c. Determine the maximum yield possible at the site. How does this value relate statistically to the flows?

22.10. Describe with words and equations how you would develop a table of random precipitation depths that follow a normal distribution and have a mean of 4 in. and a standard deviation of 3 in. Use your method to calculate the first three depths.

22.11. A sequence of uniformly distributed random numbers is given below. Use random generation to generate a 5-year sequence of annual rain depths that will follow a Pearson Type III distribution and will have a mean of 25.8 in., a standard deviation of 4.0 in., and a skew coefficient of −2.20. Random numbers to be used are 20, 01, 90, 03, and 80.

22.12. Total July runoff from a basin is randomly distributed according to a Pearson Type III distribution. The mean is 10,000 acre-ft, the standard deviation is 1000 acre-ft, the skew is −0.6, and the lag-one serial correlation coefficient is 0.50. Start with $Q_1 = 10,000$ and find five more Markov-generated flows if a sequence of randomly selected return periods gives 2, 100, 10, 2, and 50 years.

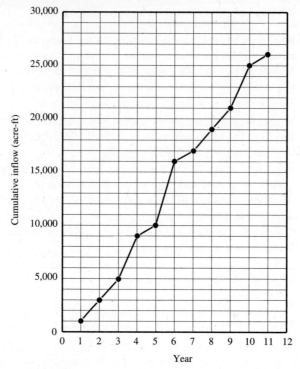

Figure 22.9 Mass curve.

REFERENCES

1. U.S. Army Corps of Engineers, "HEC-4 Monthly Streamflow Simulation," Hydrologic Engineering Center, 1971.
2. R. M. Hirsch, "Synthetic Hydrology and Water Supply Reliability," *Water Resources Research,* v. 15, no. 6, 1979.
3. R. M. Vogel, and J. R. Stedinger, "The Value of Stochastic Streamflow Models in Overyear Reservoir Design Applications," *Water Resources Research,* v. 25, no. 9, 1988.
4. D. K. Frevert, et al., "Use of Stochastic Hydrology in Reservoir Operation," *J. Irrigation and Drainage Engineering, ASCE,* v. 115, no. 3, 1989.
5. J. W. Delleur, et al., "An Evaluation of the Practicality and Complexity of Some Rainfall and Runoff Time Series Models," *Water Resources Research,* v. 12, no. 5, 1976.
6. J. D. Salas, et al., "Applied Modeling of Hydrologic Time Series," Water Resources Publications, Littleton, CO, 1980.
7. G. K. Young, and W. C. Pisano, "Operational Hydrology Using Residuals," *J. Hydraulics Division, ASCE,* v. 94, no. HY4, 1968.
8. U.S. Bureau of Reclamation, "Applied Stochastic Techniques, Personal Computer Version 5.2, User's Manual," Earth Sciences Division, Denver, CO, 1990.
9. J. C. Grygier, and J. R. Stedinger, "SPIGOT, A Synthetic Streamflow Generation Software Package," School of Civil and Environmental Engineering, Cornell University Ithaca, NY, 1990.

10. V. Yevjevich, "Structural Analysis of Hydrologic Time Series," *Hydrology Paper No. 56,* Colorado State University Hydrology Papers, Ft. Collins, CO, 1972.

11. P. J. Brockwell, and R. A. Davis, *Time Series: Theory and Methods,* Springer-Verlag Publishing, New York, 1991 (2nd ed.).

12. M. B. Fiering, *Streamflow Synthesis,* Harvard University Press, Cambridge, MA, 1967.

13. J. W. Delleur, and M. L. Kavvas, "Stochastic Models for Monthly Rainfall Forecasting and Synthetic Generation," *J. Applied Meteorology,* v. 17, no. 10, 1978.

14. W. Rippl, "The Capacity of Storage Reservoirs for Water Supply," *Proc. Inst, Civil Eng. London* **71,** 270(1883).

15. H. E. Hurst, "Long-Term Storage Capacities of Reservoirs," *Trans. ASCE* **116,** 776(1951).

16. H. E. Hurst, R. P. Black, and Y. M. Sunaika, *Long-Term Storage.* London: Constable & Company Ltd., 1965.

17. M. B. Fiering and B. B. Jackson, "Synthetic Streamflows," American Geophysical Union Water Resources Monograph No. 1, 1971.

18. H. A. Thomas and M. B. Fiering, "Mathematical Synthesis of Streamflow Sequences for the Analysis of River Basins by Simulation," in *Design of Water Resources Systems* (A. Maass et al., eds). Cambridge, MA: Harvard University Press, 1962.

19. N. C. Matalas, "Mathematical Assessment of Synthetic Hydrology," *Water Resources Res.* **3**(4), 937–945(1967).

20. L. R. Beard, "Simulation of Daily Streamflow," International Hydrology Symposium, Fort Collins, CO, Sept. 1967.

Chapter 23

Continuous Simulation Models

■ Prologue

The purpose of this chapter is to:

- Introduce and describe computer codes available for performing continuous simulation of surface runoff and streamflow.
- Present in detail how one of the programs—the Stanford watershed model— simulates the miscellaneous components of the hydrologic cycle.
- Show the major similarities and differences of the leading models.
- Provide a detailed case study of how the model parameters are developed from available information.
- Illustrate, by using the case study, the steps involved in calibrating a continuous model and verifying the results.
- Show how well the models are able to replicate gauged streamflows.

Simulation models described in this chapter provide hydrologists with tools for estimating streamflow by *continuously* accounting in time for precipitation, direct runoff, infiltration, evapotranspiration, interflow, deep percolation, base flow, and streamflow. During rain-free intervals between storms, continuous simulation models track the storage of water and its depletion to evaporation, deep percolation, and base flow, until the next rain or snow event occurs.

The models are based on the physical processes described in Chapters 1–14. As such, they classify as deterministic tools. Stochastic modeling procedures described in Chapter 22 can be used to generate synthetic streamflows without simulating the physical processes involved in converting rain and snow into runoff and streamflow. Alternatively, the methods in Chapter 22, or other similar procedures, can be used to synthesize precipitation sequences, which are then input to continuous simulation models and converted to streamflow.

Several of the continuous simulation models identified earlier in Table 21.1 are described here. The Stanford watershed model, Version IV (SWM-IV), is presented in detail as typical of the other models. Many of the others are, in fact, based on

SWM-IV, and several simulate various components of the hydrologic cycle in the same manner. Section 23.2 presents and compares two independent case studies of Stanford model studies, showing how the parameters were determined and how the models were calibrated and applied to the problems being assessed.

23.1 CONTINUOUS STREAMFLOW SIMULATION MODELS

API Model

This model was one of the earliest structured to give a deterministic simulation of a continuous streamflow hydrograph. It was originally tested on watersheds of 68 and 817 mi² and must be calibrated to each watershed to obtain a reliable method of simulating the streamflow.[1] A flow diagram showing the structure is given in Fig. 23.1. Four basic components describe the interrelations pertaining to this model of streamflow in a river: a unit hydrograph, an API (antecedent precipitation index, introduced in Chapter 2 and illustrated in Sec. 10.3), a relation for groundwater recession, and a relation for computing the groundwater flow hydrograph as a function of the direct runoff hydrograph. This model generates both groundwater flow and direct runoff discharge from precipitation values. The API model continues to enjoy widespread popularity and use in simulation modeling.

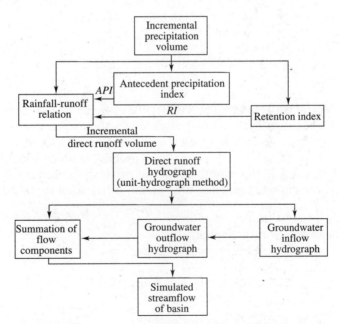

Figure 23.1 Schematic diagram of API-type hydrologic model. (After Sittner et al.[1])

Stanford Watershed Model IV (SWM-IV)

Crawford and Linsley designed this digital computer program to simulate portions (the land phase) of the hydrologic cycle for an entire watershed.[2] The model has undergone much development since its conception and is currently available from the U.S. Environmental Protection Agency under the name HSPF, which is a public domain FORTRAN version (discussed subsequently) of the original program. The SWM-IV has been widely accepted as a tool to synthesize a continuous hydrograph of hourly or daily streamflows at a watershed outlet. A lumped parameter approach is used and data requirements are much less than for alternative distributed models. Hourly and daily precipitation data, daily evaporation data, and a variety of watershed parameters are input.

The relations and linkage of the various components of SWM-IV are shown in Fig. 23.2. Hydrologic fundamentals are used at each point to transform the input data into a hydrograph of streamflow at the basin outlet. Rainfall and evaporation data are first entered into the program. Incoming rainfall is distributed, as shown in Fig. 23.2, among interception, impervious areas such as lakes and streams, and water destined to be infiltrated or to appear in the upper zone as surface runoff or interflow, both of which contribute to the channel inflow. The infiltration and upper zone storage eventually percolate to lower zone storage and to active and inactive groundwater storage. User-assigned parameters govern the rate of water movement between the storage zones shown in Fig. 23.2.

Three zones of moisture regulate soil moisture profiles and groundwater conditions. The rapid runoff response encountered in smaller watersheds is accounted for in the upper zone, while both upper and lower zones control such factors as overland flow, infiltration, and groundwater storage. The lower zone is responsible for longer-term infiltration and groundwater storage that is later released as base flow to the stream. The total streamflow is a combination of overland flow, groundwater flow, and interflow.

Model Structure   The SWM-IV is made up of a sequence of computation routines for each process in the hydrologic cycle (interception, infiltration, routing, and so on). Separate discussions of each component are provided in the following paragraphs. Actual calculations proceed from process to process as illustrated by the arrows in Fig. 23.2. All the moisture that was originally stored in the watershed or was input as precipitation during any time period is balanced in the continuity equation

$$P = E + R + \Delta S \tag{23.1}$$

where P = precipitation
E = evapotranspiration
R = runoff
ΔS = the total change in storage in the upper, lower, and groundwater storage zones

The change in storage for each zone is calculated as the difference between the volumes of inflow and outflow. Furthermore, all hydrologic activity in a time interval is simulated and balanced before the program proceeds to the next time interval. The simulation terminates when no additional data are input.

Interception   Interception is the first of several abstractions modeled by the SWM-IV. All incoming precipitation is intercepted unless the precipitation intensity exceeds the interception rate or if the interception storage fills. Interception rates depend on the precipitation rate and on the watershed cover. Typical values of interception maximums are provided in Table 23.1.

TABLE 23.1   TYPICAL MAXIMUM
              INTERCEPTION RATES

Watershed cover	Interception rate (in./hr)
Grassland	0.10
Moderate forest cover	0.15
Heavy forest cover	0.20

Source: After Crawford and Linsley.[2]

Evapotranspiration   In SWM-IV evapotranspiration (ET) is assumed to occur at the potential rate from interception storage and the "upper" storage zone. The upper zone simulates the depressions and highly permeable surface soils. The lower soil zone simulates the linkage to the groundwater storage zone.

Evapotranspiration from the lower zone is set equal to the *ET opportunity,* defined in Fig. 23.3. ET opportunity is defined as the maximum amount of water available for ET at a particular location during a prescribed time interval. In the modeling logic, ET occurs from several locations (see Fig. 23.2) including the interception storage, upper zone storage, lower zone storage, stream and lake surfaces, and groundwater storage. Evapotranspiration from interception and upper zone storage is set equal to the potential rate, E_p, which is assumed to be the lake evaporation rate calculated as the product of a pan coefficient times the input values of the evaporation pan data. The evaporation of any intercepted water is assumed to occur at a rate equal to the potential evapotranspiration rate and ceases when the interception storage has been depleted.

Evaporation from stream and lake surfaces also occurs at the potential rate. The total volume is governed by the total surface area of streams and lakes (ETL) defined as the ratio of the total stream and lake area in the watershed to the total watershed area. Evapotranspiration from groundwater storage also occurs at the potential rate and is calculated in a similar fashion using a surface area equal to a factor K24EL multiplied by the watershed area. Thus the parameter K24EL represents the fraction of the total watershed area over which evapotranspiration from the groundwater storage will occur. Most investigators set this parameter at a value equal to the fraction of the watershed area covered by phreatophytes. Its value is normally small but can be large, for example, in an agricultural area that has many acres of subirrigated alfalfa.

If interception storage is depleted, the model will attempt to satisfy the potential for ET by drawing from the upper zone storage at the potential rate. Once the upper zone storage is depleted, ET occurs from the lower zone but not at the potential rate; the ET rate from the lower zone is always less than E_p. When interception and the upper zone storage do not satisfy the potential, any excess enters as E_p in Fig. 23.3,

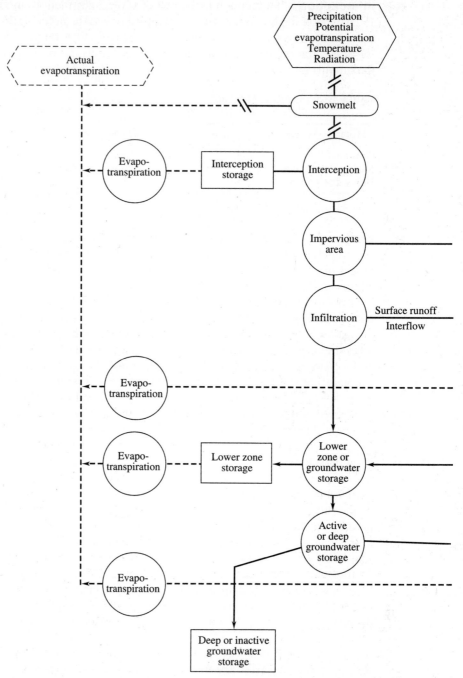

Figure 23.2   Stanford watershed model IV flowchart. (After Crawford and Linsley.[2])

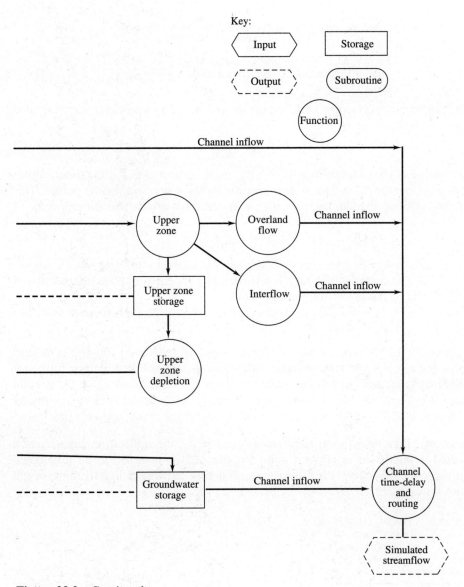

Figure 23.2 Continued

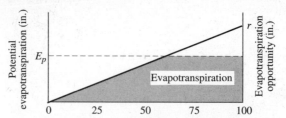

Percentage of area with a daily evapotranspiration opportunity
equal to or less than the indicated value

Figure 23.3   Evapotranspiration relation used in the Stanford watershed model. (After Crawford and Linsley.[2])

and the rate of evapotranspiration from the lower zone is determined from the shaded area, or

$$E = E_p - \frac{E_p^2}{2r} \tag{23.2}$$

The variable r is the evapotranspiration opportunity, defined as the maximum water amount available for ET at a particular location during a prescribed time period. This factor varies from point to point over any watershed from zero to a maximum value of

$$r = K3 \frac{LZS}{LZSN} \tag{23.3}$$

where   LZS = the current soil moisture storage in the lower zone (in.)
LZSN = a nominal storage level, normally set equal to the median value of the lower zone storage (in.)
K3 = an input parameter that is a function of watershed cover as shown in Table 23.2

The ratio LZS/LZSN is known as the lower zone soil moisture ratio and is used to compare the actual lower zone storage with the nominal value at any time. Values of ET opportunity are assumed to vary over a watershed from zero to r along the straight line shown in Fig. 23.3. This assumed linear cumulative distribution of the parameter over an area is also used in evaluating areal distributions of infiltration rates.

Infiltration   Like the evapotranspiration opportunity, the infiltration capacity of a watershed is highly variable from point to point and is assumed to be distributed according to a linear cumulative distribution function shown as a line from the origin to Point b in Fig. 23.4.

TABLE 23.2   TYPICAL LOWER ZONE EVAPOTRANSPIRATION PARAMETERS

Watershed cover	K3
Open land	0.20
Grassland	0.23
Light forest	0.28
Heavy forest	0.30

Source: Crawford and Linsley.[2]

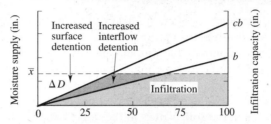

Figure 23.4   Assumed linear areal variation of infiltration capacity over a watershed. (After Crawford and Linsley.[2])

Infiltration into the lower and groundwater storage zones is determined as a function of the moisture supply $\bar{x}$ available for infiltration. Steps to determine infiltration for a given moisture supply $\bar{x}$ are:

1. The net infiltration is determined from the area labeled "infiltration" in Fig. 23.4. This water is assumed to infiltrate into the lower and groundwater storage zones. The area enclosed by the trapezoid is given by the equations in the first row of Table 23.3. If the moisture supply $\bar{x}$ exceeds the maximum infiltration capacity b, the maximum allowed net infiltration is $b/2$, which is the median infiltration capacity.

2. Some of the moisture supply contributes to an increase in the interflow detention during any time increment and is calculated as the region indicated by an arrow in Fig. 23.4. Equations for this area using various ranges in $\bar{x}$ are provided in the second row of Table 23.3. The volume of water in a state of being transported as interflow at any instant is called the interflow detention or detained interflow.

3. Any remaining moisture supplied, ΔD in Fig. 23.4, contributes to increasing the surface detention during the time increment. Equations for this triangular-shaped area are included in Table 23.3 for various values of $\bar{x}$.

TABLE 23.3   EQUATIONS FOR THE SHADED AREAS IN FIG. 23.3

Component	$\bar{x} < b$	$b < \bar{x} < cb$	$\bar{x} > cb$
Net infiltration	$\bar{x} - \dfrac{\bar{x}^2}{2b}$	$\dfrac{b}{2}$	$\dfrac{b}{2}$
Increase in interflow detention	$\dfrac{\bar{x}^2}{2b}\left(1 - \dfrac{1}{c}\right)$	$\bar{x} - \dfrac{b}{2} - \dfrac{\bar{x}^2}{2cb}$	$\dfrac{b}{2}(c - 1)$
Increase in surface detention	$\dfrac{\bar{x}^2}{2cb}$	$\dfrac{\bar{x}^2}{2cb}$	$\bar{x} - \dfrac{cb}{2}$
Percentage of increased detention assigned to interflow	$100\left(1 - \dfrac{1}{c}\right)$	$100\left(1 - \dfrac{\bar{x}^2}{2cb(x - b/2)}\right)$	$100\dfrac{c - 1}{2\bar{x}/b - 1}$

Source: After Crawford and Linsley.[2]

The quantity of net infiltration is controlled largely by the maximum infiltration capacity b, while the parameter c significantly affects hydrograph shapes because the parameter controls the amount of water detained during the time increment. The values of b and c for any time interval depend on the soil moisture ratio, LZS/LZSN, and on the input parameters CB and CC; CB is an index that controls the rate of infiltration and depends on the soil permeability and the volume of moisture that can be stored in the soil. Values in the range from 0.3 to 1.2 are common. The parameter CC is an input value that fixes the level of interflow relative to the overland flow. Values of CC range from 1.0 to 5.0.

If the soil moisture ratio is less than 1.0, the variable b is found from

$$b = \frac{CB}{2^{(4LZS/LZSN)}} \tag{23.4}$$

and when LZS/LZSN is greater than 1.0, the equation for b is

$$b = \frac{CB}{2^{\{4.0+2[(LZS/LZSN)-1.0]\}}} \tag{23.5}$$

These equations were developed by Crawford and Linsley from numerous trials using SWM-IV in many different watersheds. When the soil moisture ratio reaches a value of 2.0, the variable b reaches its minimum value of $\frac{1}{64}$ of CB. The parameter c is determined from

$$c = (CC)2^{(LZS/LZSN)} \tag{23.6}$$

Variations in parameters b and c with changes in LZS/LZSN are shown in Figs. 23.5 and 23.6. Midrange values of CB = 1.0 and CC = 1.0 were used in developing these curves.

Figure 23.7 is a graph of distribution of water among infiltration, interflow, and overland flow for various values of the moisture supply $\bar{x}$. Different values of b and c would produce a different set of curves.

Water stored as overland flow surface detention will either contribute to streamflow or enter the upper zone storage as depicted in Fig. 23.2. The portion that

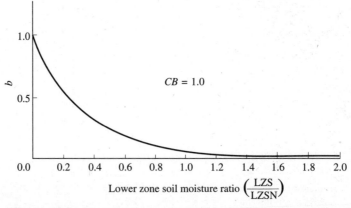

Figure 23.5 Variation in parameter b for various values of the soil moisture ratio. (After Crawford and Linsley.[2])

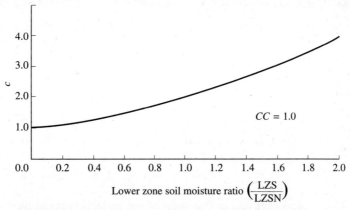

Figure 23.6 Variation in parameter c for various values of the soil moisture ratio. (After Crawford and Linsley.[2])

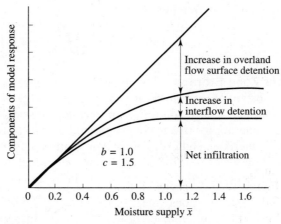

Figure 23.7 Typical SWM-IV response to moisture supply variations. (After Crawford and Linsley.[2])

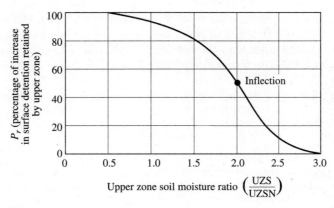

Figure 23.8 Delayed infiltration as a function of upper zone soil moisture ratio. (After Crawford and Linsley.[2])

enters the upper zone storage is called delayed infiltration and is a function of the upper zone soil moisture ratio, UZS/UZSN, as shown in Fig. 23.8. The inflection point occurs at a soil moisture ratio of 2.0. If the ratio is less than 2.0, the percentage retained by the upper zone is given by

$$P_r = 100\left[1.0 - \left(\frac{UZS}{2UZSN}\right)\left(\frac{1.0}{1.0 + UZI1}\right)^{UZI1}\right] \tag{23.7}$$

where UZI1 is determined from

$$UZI1 = 2.0\left[\frac{UZS}{2UZSN} - 1.0\right] + 1.0 \tag{23.8}$$

The curve is defined to the right of the inflection point by

$$P_r = 100\left[\left(\frac{1.0}{1.0 + UZI2}\right)^{UZI2}\right] \tag{23.9}$$

where UZI2 is determined from

$$UZI2 = 2.0\left[\frac{UZS}{UZSN} - 2.0\right] + 1.0 \tag{23.10}$$

Upper Zone Storage The upper storage zone, as shown in Fig. 23.2, receives a large portion of the rain during the first few hours of the storm, while the lower and groundwater storage zones may or may not receive any moisture. The portion of the upper zone storage that is not evaporated or transpired is proportioned to the surface runoff, interflow, and percolation. Percolation (upper zone depletion) from the upper zone to the lower zone in Fig. 23.2 ocurs only when UZS/UZSN exceeds LZS/LZSN. When this occurs, the percolation rate in in./hr is determined from

$$PERC = 0.003(CB)(UZSN)\left(\frac{UZS}{UZSN} - \frac{LZS}{LZSN}\right)^3 \tag{23.11}$$

where CB is an index that controls the rate of infiltration. It ranges from 0.3 to 1.2 depending on the soil permeability and on the volume of moisture that can be stored in the soil. The variables UZS and UZSN are defined as the actual and nominal soil moisture storage amounts in the upper zone. The nominal value of UZSN is approximately a function of watershed topography and cover and is always considered to be much smaller than the nominal LZSN value. The initial estimates of UZSN relative to LZSN are found from Table 23.4.

TABLE 23.4 VALUES OF UZSN AS A FUNCTION OF LZSN FOR INITIAL ESTIMATES IN SIMULATION WITH SWM-IV

Watershed	UZSN
Steep slopes, limited vegetation, low depression storage	0.06LZSN
Moderate slopes, moderate vegetaion, moderate depression storage	0.08LZSN
Heavy vegetal or forest cover, soils subject to cracking, high depression storage, very mild slopes	0.14LZSN

Source: After Crawford and Linesley.[2]

The parameters LZSN and CB must also be estimated at the beginning of a simulation study. The combination that will most satisfactorily reproduce both long- and short-term historical responses to hydrologic inputs can be determined by the following procedure:[2]

1. Assume an initial value for LZSN equal to one quarter of the mean annual rainfall plus 4 in. (used in arid and semiarid regions), or one eighth of the annual mean rainfall plus 4 in. (used in coastal, humid, or subhumid climates).
2. Determine the initial value of UZSN from Table 23.4.
3. Assume a value for CB in the normal range from 0.3 to 1.2.
4. Simulate a period of record using the streamflow, rainfall, and evaporation data and systematically adjust LZSN, UZSN, CB, and other parameters until agreement between synthesized and recorded streamflows is satisfactory. If the annual water budgets do not balance, LZSN is adjusted; CB is adjusted on the basis of comparisons between synthesized and recorded flow rates for individual storms.

Lower Zone Storage and Groundwater   The lower groundwater storage zone in Fig. 23.2 receives water from the net infiltration and from percolation. The percolation rate is determined from Eq. 23.11. The percentage of net infiltration that reaches groundwater storage depends on the soil moisture ratio LZS/LZSN as shown in Fig. 23.9. If this ratio is less than 1.0, the percentage P_g is found from

$$P_g = 100 \left[\frac{LZS}{LZSN} \left(\frac{1.0}{1.0 + LZI} \right)^{LZI} \right] \qquad (23.12)$$

and if LZS/LZSN is greater than 1.0, the percentage is

$$P_g = 100 \left[1.0 - \left(\frac{1.0}{1.0 + LZI} \right)^{LZI} \right] \qquad (23.13)$$

In both equations, the variable LZI is defined as

$$LZI = 1.5 \left[\frac{LZS}{LZSN} - 1.0 \right] + 1.0 \qquad (23.14)$$

Note from Fig. 23.9 that the nominal storage LZSN equals the lower zone storage LZS when half or 50 percent of all the incoming moisture enters groundwater storage.

The outflow from the groundwater storage, GWF, at any time is based on the commonly used linear semilogarithmic plot of base flow discharge versus time. This technique was described in Section 11.4 and illustrated in Fig. 11.8. In modified form the base flow equation is

$$GWF = (LKK4)[1.0 + KV(GWS)](SGW) \qquad (23.15)$$

where LKK4 is defined by

$$LKK4 = 1.0 - (KK24)^{1/96} \qquad (23.16)$$

in which KK24 is the minimum of all the observed daily recession constants (see Section 11.4), where each constant is the ratio of the groundwater discharge rate to the

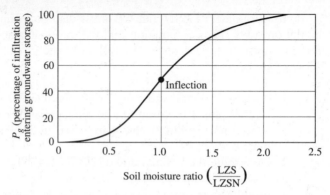

Figure 23.9 Percentage of infiltrated water that reaches groundwater storage. (After Crawford and Linsley.[2])

groundwater discharge rate 24 hr earlier. Thus the recession constant KK24 (K in Eq. 11.1) is determined using $t = 1$ day. The variable GWS in Eq. 23.15 has values that depend on the long-term inflows to groundwater storage. Its value on any given day (e.g., the ith day) is calculated as 97 percent of the previous day's value, adjusted for any inflow to groundwater storage, or $GWS_i = 0.97\,(GWS_{i-1}$ + inflow to groundwater storage during day i).

In Eq. 23.15, SGW is a groundwater storage parameter that reflects the fluctuations in the volume of water stored and ranges from 0.10 to 3.90 in. The term KV in Eq. 23.15 allows for changes that are known to exist in the groundwater recession rates as time passes. When KV is zero, Eq. 23.15 reduces to Eq. 11.1 and the groundwater recession follows the linear semilog relation. If the usual dry season recession rate KK24 is too large for wet periods (when groundwater storages are being recharged by seepage from the streams) the parameter KV is hand-adjusted so that the term $1.0 + KV(GWS)$ will reduce the effective rate to some desired value during recharge periods. Table 23.5 illustrates this computation by showing effective recession rates for various combinations of KK24 and GWS when KV is set equal to 1.0.

The fraction of active or deep groundwater storage that is either lost to deep or inactive groundwater storage (Fig. 23.2) or is diverted as flow across the drainage basin boundary is input as parameter K24L. This fraction is the total inflow to

TABLE 23.5 EFFECTIVE RECESSION RATES FOR VARIOUS COMBINATIONS OF KK24 AND GWS WHEN KV = 1.0

KK24	GWS			
	0.0	0.5	1.0	2.0
0.99	0.99	0.985	0.98	0.97
0.98	0.98	0.970	0.96	0.94
0.97	0.97	0.955	0.94	0.91
0.96	0.96	0.940	0.92	0.88

Source: After Crawford and Linsley.[2]

groundwater and represents all the active groundwater storage that does not contribute to streamflow.

Overland Flow   The overland flow process has been studied by many investigators. A wide range of methods for estimating the velocities and depths of sheet flow over a land surface has been applied and falls in the hydraulic-hydrologic categories of Chapter 13. Hydraulic overland flow methods involve finite-difference and other numerical techniques to solve at various points the partial differential equations of continuity and momentum for unsteady overland flow. The hydrologic methods, including those adopted in SWM-IV, approximate the velocities and depths for unsteady overland flows by a lumped parameter approach that requires much less data than the hydraulic techniques.

Average values of lengths, slopes, and roughnesses of overland flow in the Manning and continuity equations are used in SWM-IV to continuously calculate the surface detention storage D_e. The overland flow discharge rate q is then related to D_e.

As the rain supply rate continues in time, the amount of water detained on the surface increases until an equilibrium depth is established. The amount of surface detention at equilibrium estimated by SWM-IV is

$$D_e = \frac{0.000818i^{0.6}n^{0.6}L^{1.6}}{S^{0.3}} \tag{23.17}$$

where D_e = the surface detention at equilibrium (ft³/ft of overland flow width)
   i = the rain rate (in./hr)
   S = the slope (ft/ft)
   L = the length of overland flow (ft)
   n = Manning's roughness coefficient

The overland flow discharge rate is next determined as a function of detention storage from

$$q = \frac{1.486}{n}S^{1/2}\left(\frac{D}{L}\right)^{5/3}\left[1.0 + 0.6\left(\frac{D}{D_e}\right)^3\right]^{5/3} \tag{23.18}$$

where q = the overland flow discharge rate (cfs per ft of width)
   D = the average detention storage during the time interval

The equation also applies during the recession that occurs after rain ceases, but the ratio D/D_e is assumed to be 1.0. Typical overland flow roughness coefficients are provided in Table 23.6.

The time at which detention storage reaches an equilibrium is determined from

$$t_e = \frac{0.94L^{3/5}n^{3/5}}{i^{2/5}S^{3/10}} \tag{23.19}$$

where t_e is the time to equilibrium (min). Crawford and Linsley show that these equations very accurately reproduce measured overland flow hydrographs.[2]

For each time interval Δt, an end-of-interval surface detention D_2 is calculated from the initial value D_1 plus any water added ΔD (Fig. 23.4) to surface detention storage during the time interval, less any overland flow discharge $\bar{q}$ that escapes from

TABLE 23.6 TYPICAL MANNING EQUATION OVERLAND FLOW
ROUGHNESS PARAMETERS, NN

Watershed cover	Manning's n for overland flow
Smooth asphalt	0.012
Asphalt or concrete paving	0.014
Packed clay	0.03
Light turf	0.20
Dense turf	0.35
Dense shrubbery and forest litter	0.40

Source: After Crawford and Linsley.[2]

detention storage during the time interval. This is simply an expression of continuity, or

$$D_2 = D_1 + \Delta D - \overline{q}\,\Delta t \tag{23.20}$$

The discharge $\overline{q}$ is found from Eq. 23.18 using a value of $D = (D_1 + D_2)/2$. Equations 23.17–23.20 allow the complete determination of overland flow using easily found basin-wide values of the average length, slope, and roughness overland flow.

Interflow   The water temporarily detained as interflow storage is treated in the same fashion as overland flow detention storage. The inflow to interflow detention was defined in Fig. 23.4. The outflow is simulated using a daily recession constant similar to that defined for groundwater discharge. The interflow recession constant IRC is the average ratio of the interflow discharge at any time to the interflow discharge 24 hr earlier. For each 15-min time interval modeled, the outflow from detention storage is

$$\text{INTF} = \text{LIFC4(SRGX)} \tag{23.21}$$

where

$$\text{LIFC4} = 1.0 - (\text{IRC})^{1/96} \tag{23.22}$$

The variable SRGX is the water stored in the interflow detention at any time. Its value continuously changes when the continuity equation is applied to each time interval. The end-of-interval value of SRGX depends, according to continuity, on the value at the beginning of the interval and any inflow to or discharge from the interflow detention during the interval.

Channel Translation and Routing   The Stanford watershed model utilizes a hydrologic watershed routing technique to translate the channel inflow to the watershed outlet. Clark's IUH time–area method described in Section 12.6 is adopted almost as presented in Chapter 12. In place of the net rain hyetograph, the Stanford model views the sum of all channel inflow components as an "inflow" hyetograph. This inflow is then translated in time through the channel to the basin outlet, where it is next routed through an equivalent storage system to account for the attenuation caused by storage in the channel system. Routing through the linear reservoir (linear in the sense that storage is assumed to be directly proportional to the outflow, Eq. 12.35) is accomplished from

$$O_2 = \overline{I} - \text{KS1}(\overline{I} - O_1) \tag{23.23}$$

where O_2 = the outflow rate at the end of the time interval
O_1 = the outflow rate at the beginning
$\bar{I}$ = the average inflow rate during the time interval

Also,

$$KS1 = \frac{K - \Delta t/2}{K + \Delta t/2} \qquad (23.24)$$

Examples of the determination of K and other necessary parameters from watershed data are included in Section 23.2.

Applications of the SWM-IV   Applications of the model typically begin with data for a three- to six-year calibration period for which rainfall and runoff data are available. These data are used to allow successive adjustments of several parameters until the simulated and recorded hydrographs of the streamflow agree. If sufficient data are available, a second period of record may be reserved for use as a control to check the accuracy of the parameters derived from a calibration with the first half of the data.

The Stanford watershed model was originally developed in 1959 and has undergone several modifications since that time. James[3] translated the Crawford and Linsley model from ALGOL to FORTRAN. Several modifications of the FORTRAN version have evolved from a variety of investigations. Included among these are the Kentucky watershed model (KWM),[4,5] the Kentucky self-calibrating version (OPSET),[4] the Ohio State University version, the Texas version,[6] the Hydrocomp Simulation Program (HSP) written in PL/1, and EPA-produced, nonproprietary FORTRAN version of HSP called HSPF, and the National Weather Service runoff forecasting model. Brief descriptions of several of these are included below.

ARS Revised Model of Watershed Hydrology (USDAHL)

Growing interest in the effects of soil types, vegetation, pavements, and farming practices on infiltration and overland flow has resulted in the growth of the USDAHL continuous simulation model. The 1974 version[7] of this model was developed by investigators at the Agricultural Research Service Hydrograph Laboratory.

Input data to the model are relatively extensive. Continuous records of the precipitation, the weekly average temperatures, the weekly average pan-evaporation amounts, and detailed data on soils, vegetation, land use, and cultural practices are required.

The study watershed is initially divided into as many as four distinct land-use or soil-type zones. Fourteen subroutines and a main calling routine compute for each zone the snowmelt, infiltration, overland flow, channel flow, evapotranspiration, groundwater evaporation and movement, groundwater recharge, and return flow.

Evapotranspiration potentials are estimated by applying assigned coefficients to pan-evaporation data. Infiltration for each soil or land-use zone is computed using a modified Holtan equation. Water stored in cracks in dry soils is simulated as a function of soil moisture. Manning's equation and the continuity equation are used to route overland flow. The streamflow is routed by a simultaneous solution of the continuity equation and a storage function. Groundwater movements are calculated by Darcy's equation. The daily snowmelt on each zone is calculated as a function of the temperature at which snowmelt starts, the weighted average vegetative density for the zone, the weekly average air temperature, and the potential snowmelt per day in the zone snowpack. Precipitation falling during a snowmelt day also contributes to the snowmelt equation.

Among other uses, the model has been applied by the Soil Conservation Service in preparing environmental impact statements. Figure 23.10 shows the results of applying the 1974 version to annual runoff from four widely separated and widely diversified ARS experimental watersheds. In addition to the runoff, the model computes the evapotranspiration amounts, soil moisture changes, return flows, and groundwater recharge depths for each of the zones.

Although other modifications are possible, the USDAHL model is specifically designed for relatively small rural watersheds, generally under 20 square miles.

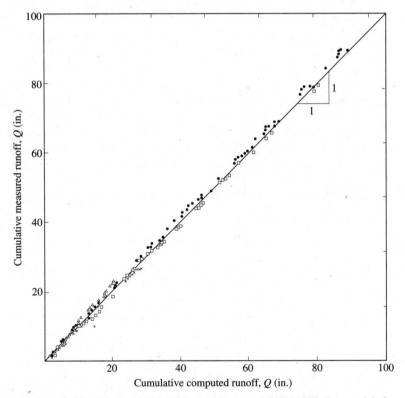

Figure 23.10 Chart showing the accuracy of USDAHL-74 model for estimating the cumulative computed runoff as compared with the cumulative measured runoff at four watersheds. ● W-97, Coshocton, OH: △ W-11, Hastings, NE; □ W-3, Ft. Lauderadale, FL; × W-G, Riesel, TX (After Holtan and Lopez.[7])

National Weather Service River Forecast System (NWSRFS)

Yet another version of the Stanford watershed model was developed by the Hydrologic Research Laboratory staff at the National Weather Service Office of Hydrology.[8] The NWSRFS model was developed for use in forcecasting river flows and stages by the National Weather Service. The model has been applied successfully to several river basins ranging in size from 70 mi^2 in North Carolina to 1000 mi^2 in Oklahoma. River forecasting in large river basins does not require the detail incorporated in SWM for smaller watersheds. For this reason, the NWS model includes two major changes involving the use of a longer time increment, simplified programming, fewer process computations, and a rapid procedure for determining optimal watershed parameters that allow the model to reproduce historical flows accurately.

A 6-hr time increment is used by the model, allowing fewer rainfall inputs and more important, fewer detailed calculations of processes such as overland flow that occur in shorter time periods. Iterations are thus completed more rapidly than with the SWM. As with the OPSET model, the National Weather Service optimization procedure for determining parameter values gives the model a strength not available with the SWM-IV.

Other modifications include hourly computations of infiltration, detention, and upper and lower zone retention and daily computations of percolation of water from the upper soil zone to groundwater storage. Evaporation from stream surfaces and groundwater evapotranspiration are handled separately by the SWM-IV and are jointly computed in the NWS version. As mentioned, overland flow routing is eliminated and is replaced by three types of runoff: surface runoff, interflow, and groundwater flow—representing fast, medium, and slow response.

Input data for model calibration consist of mean daily discharges and instantaneous hydrographs for a few selected runoff events. Rainfall is input as a continuous record of 6-hr basin-wide means determined from areal averaging techniques. Because of the changes in routing increment and in the detail of process simulation, the output from the NWS version is similar in makeup to the SWM-IV output.

COE Streamflow Synthesis and Reservoir Regulation Model (SSARR)

Another widely used continuous streamflow simulation model designed for large basins was developed by the Corps of Engineers.[9] The SSARR model was developed primarily for streamflow and flood forecasting and for reservoir design and operation studies. Prior to the development of NWSRFS, hydrologists at the National Weather Service used the SSARR model. The model has been applied to both rain and snowmelt events.

Applications of the model begin with a subdivision of the drainage basin into homogeneous hydrologic units of a size and character consistent with subdivides, channel confluences, reservoir sites, diversion points, soil types, and other distinguishing features. The streamflows are computed for all significant points throughout the river system.

Rainfall data can be input at any number of stations in the basin. The part that will run off is divided into the base flow, subsurface or interflow, and surface runoff. The division is based on indices and on the intensity of the direct runoff. Each component is simply delayed according to different processes, and all are then combined to produce the final subbasin outflow hydrograph. This subarea runoff is then routed through stream channels and reservoirs to be combined with other subarea hydrographs, all of which become part of the output.

Routings through channels and reservoirs are accomplished by the same technique. This requires an assumption of short stream reaches, and occasional allowances for backwater effects are necessary in the channel routing process. Streamflows are synthesized on the basis of rainfall and snowmelt runoff. Snowmelt can be determined on the basis of the precipitation depth, elevation, air and dew point temperatures, albedo, radiation, and wind speed. Snowmelt options include the temperature index method or the energy budget method.

Input includes the precipitation depths, the watershed-runoff indices for subdividing flow among the three processes, initial reservoir elevations and outflows, drainage areas, bounds on usable storage and allowable discharge from reservoirs, total computation periods, routing intervals, and other special instructions to control plots, prints, and other input-output alternatives.

This model was one of the earliest continuous streamflow simulation models using a lumped parameter representation and has its primary strength in its verified accuracy indicated by tests conducted in several large drainage basins including the Columbia River basin and the Mekong River basin.

Hydrocomp Simulation Program (HSP)

A commercial version of the Stanford water model was developed at Hydrocomp, Inc., named the Hydrocomp Simulation Program.[10] Among several advantages incorporated in HSP are hydraulic reservoir routing techniques and kinematic-wave channel routing techniques. Other major changes include the addition of water quality simulation capabilities. Due to these additions, the model is often referred to as the Hydrocomp water quality model.

The HSP model has been used routinely for several types of hydrologic study including floodplain mapping, water quality studies, storm water and urban flooding studies, urban drainage facility design, and water quality aspects of urban runoff.

The model consists of three computer routines:

1. *Library* allows the use of direct access disk storage to handle input data with efficient data management routines.
2. *Lands* handles the usual SWM lands phase along with added processes in calculating soil moisture budgets, groundwater recharge and discharge, inflow to stream channels, and eutrophication.
3. *Channel* is responsible for assembling and routing all channel inflow through channel networks, lakes, and reservoirs.

The HSP model incorporates a continuous water balance by tracking precipitation through all possible avenues of the hydrologic and water resource system. The groundwater inflow, interflow, and surface runoff are individually simulated, lagged, and combined at appropriate times as the channel inflow. The routing of computed inflow through the channel network utilizes a modified kinematic-wave model. Water quality constituents are related to variable discharge rates and other water parameters so that the coupling of quantity and quality of the runoff is accomplished.

Inputs for simulating water quality include the temperature, radiation, wind, and humidity and observed values of the factors under study which form the basis for calibration. At lest two years of data are preferable for calibration; however, calibration has been achieved with less than one year of data. Other required input includes hourly precipitation in 5-, 15-, or 30-min or greater time increments and potential evapotranspiration; if snow or water quality simulation is desired, the temperature, radiation, wind, and humidity factors are needed.

Outputs from HSP can be obtained for any desired point within the watershed. Included in the output options are values of quality data at outfalls or other points, river stages, reservoir levels, hourly and mean daily discharge rates, stream and lake temperatures, dissolved oxygen and total dissolved solids, algae counts, phosphorus, nitrate, nitrite, ammonia, total nitrogen, phosphate, pH, carbonaceous BOD, coliforms, conservative metals, and the usual daily, monthly, and annual water budgets, snow depths, and end-of-period moisture equivalents.

Typical simulation periods in HSP applications range from 20 to 50 years. Hour-by-hour data are not viewed as an exact sequence of future flows. Rather, the data are used for analysis of the probability of occurrences of ranges in the factors of interest. When used in this manner, the model is functioning with a purpose similar to that of some of the operational hydrology techniques described in Chapter 22.

EPA Hydrocomp Simulation Program—Fortran (HSPF)

Following development of the HSP version of the Stanford model, the U.S. Environmental Protection Agency contracted in 1980 to have public-domain version made available for continuous streamflow simulation and water quality modeling. The original program, written in ALGOL, was converted to FORTRAN 77.[11] The HSPF code is available for PC applications. Substantial portions of water quality modeling algorithms were added to HSP in developing HSPF. The hydrologic cycle processes, however, are essentially the same as in SWM-IV and HSP. One exception is the addition of several routing procedures not previously available.

Parameters that drive the routines in HSPF must be estimated for all the hydrologic processes, making verification of the model difficult because of the numerous combinations of parameter values. Methods of estimating these parameters and calibrating the model are illustrated in the case studies in Section 23.2 (Table 23.8 defines over 35 parameters used in the Stanford model).

USGS Precipitation-Runoff Modeling System (PRMS)

After developing their urban storm-event model, DR3M (see Chapter 25), the U.S. Geological Survey developed several other computer codes to model continuous hydrologic processes. The PRMS performs simulation of daily streamflows for a variety of precipitation, climate, and land use combinations. It is available from the USGS in

PC, minicomputer, or mainframe format.[12] During storms, the model gives output on any prescribed time interval. Between storm periods, the model tracks soil moisture and other storage/depletion zones on a daily basis until the next storm interval. Streamflow is output as mean daily flow rates.

A lumped-parameter approach is utilized in PRMS. The smallest subdivision of the study watershed is a *hydrologic response unit* that is assumed to behave as a homogeneous hydrologic element. The USGS has delineated HRUs in most areas of the United States. Like the DR3M model, the streams, storm sewers, reservoirs, and detention ponds in the watershed are modeled as nodes and interconnecting links. Hydrograph routing in channels and reservoirs is accomplished by kinematic-wave and storage-indication methods, respectively.

ARS Simulator for Water Resources in Rural Basins (SWRRB)

Through a cooperative program with Texas A&M University, the Agricultural Research Service, U.S. Department of Agriculture, developed a continuous daily streamflow simulation program for use in modeling ungauged agricultural areas.[13] This FORTRAN 77 *water budget* model is available from the ARS in PC format. Its development focused on a model that would allow the user to predict impacts of various watershed management practices such as crop rotation, fall plowing, urbanization, conservation tillage, terracing, fallowing, and floodwater detention on monthly and annual water and sediment yields from rural basins.

Sediment yields for each of the subwatersheds, and for the total basin, are calculated using the universal soil loss equation.[14] The method uses watershed hydrology outputs from the rainfall-runoff portion to estimate sediment yield, using inputs of runoff volume, peak flows, soil type and erodibility, crop types, erosion managment factors, and watershed slope and length.

The subbasins are modeled in a lumped-parameter style. Once rainfall or snowfall records are input, physical-process algorithms are linked as shown in Fig. 23.11 for solar radiation, snowmelt, surface runoff, ET, conveyance losses, sediment production from individual storms, evaporation from water surfaces, percolation, and soil moisture accounting. Crop production is also calculated based on crop types, temperature, consumptive use of water, and irrigation practices.

Hydrologic abstractions for each subwatershed are estimated by the SCS curve number (CN) method (Chapter 4, Section 4.9). Soil water budgeting is performed by adding the net moisture input to the soil profile from precipitation after subtracting direct runoff, ET (consumptive use by the crops and evaporation from soils), percolation, and return flow (groundwater flow back to the streams). Direct surface runoff is set equal to the net rain from the CN method. For input to the sediment yield component, a peak flow rate for individual storms is estimated using a modified rational method (Chapter 15). Snowmelt is calculated by the degree-day method described in Chapter 14.

Expert Systems

Recent trends in systems analysis are leading to development of *expert system* (ES) techniques that rely on *artificial intelligence* for use in planning and design of water resources projects. A computer can be given information obtained from extensive

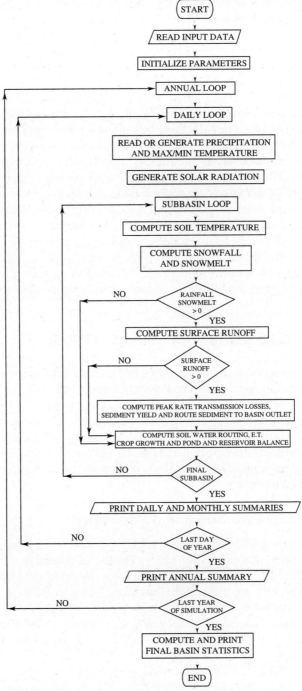

Figure 23.11 Flowchart of the SWRRB hydrologic-process algorithms. (After Arnold.[13])

interviews of one or more experts in some field. The computer can then make "decisions" in much the same way as the experts, applying their judgment and experience and making these available to others through the expert system model.

Streamflow models, especially those that perform continuous simulation, incorporate input parameters that require considerable judgment. Developers and users of the watershed models have accumulated decades of experience in assigning coefficients and parameters. Their experience and judgment can be extracted by an interview process involving hundreds of questions to build an expert system model. The model not only incorporates direct answers but also addresses uncertainties about each. Early applications with this modeling technique show considerable promise.[15]

In addition to streamflow simulation, expert systems have the potential to be useful in the design and management of complex river basin systems of dams, reservoirs, power plants, diversion canals, and flood control structures. Operations for such systems involve independent and collective decisions by dozens of professionals. These experts are normally in radio or telephone contact with numerous other controllers and decision-makers. If ES data could be developed from these teams, the potential for improved management exists. A prime incentive of implementing expert systems in water resources systems involves capturing insights of experienced professionals before they retire or move into other positions.

23.2 CONTINUOUS SIMULATION MODEL STUDIES

This section describes in detail two independent applications of the Kentucky version of the Stanford watershed model to small basins in Kentucky and Nebraska. Results obtained by Clarke[16] in modeling the Cave Creek (CC) watershed in Kentucky and by the authors[17] using KWM for the Big Bordeaux Creek (BBC) watershed in Nebraska are compared. Both are small, homogeneous watersheds having relatively good records of precipitation and runoff. The two case studies are described simultaneously to show how different analysts dealt with the decisions required to develop input data and parameters.

Selection of Watershed Size and Study Period

Several guidelines exist for selecting a watershed subarea size and time period to be modeled in a simulation study. The use of relatively small, homogeneous watersheds or subdivisions of larger watersheds is recommended to minimize any difficulty caused by ignoring spatial variations in precipitation over larger areas. This also minimizes the effects of lumping watershed characteristics such as soil types, soil profiles, impervious areas, and land uses into single parameters representing the entire catchment. Ross suggests an upper limit of 25 mi^2 for study-watershed drainage areas.[18]

One difficulty in restricting watershed size arises from the fact that few streams for small drainage areas are continuously gauged, and reliable streamflow data are difficult to obtain. For example, the BBC watershed was selected because it was one of few small watersheds in Nebraska with sufficient records. The Cave Creek watershed was selected on the basis of the following criteria.[19]

1. A minimum of 10 years of continuous runoff records in order to establish the existing rainfall–runoff relation.
2. A drainage area of less than 5 mi^2 so as to be representative of small drainage basins for which better runoff coefficients are needed.
3. A location in close proximity to a rain guage for which hourly precipitation data are available.
4. The availability of soil surveys for the watershed under study.

Data Sources

Hourly precipitation data were available at sites 1.20 mi from the CC watershed and 8.0 mi from the BBC gauging station. Soil survey records and runoff data were available for both watersheds. Daily pan evaporation data were available approximately 30 mi south of the BBC watershed and 25 mi south of the CC watershed. Drainage areas of 2.53 mi^2 for the CC watershed and 9.22 mi^2 for the BBC watershed were found from U.S. Geological Survey quadrangle maps. Input parameters for the Stanford and Kentucky versions are compared and defined in Table 23.7. Numerical values, using the Stanford version parameter names, are tabulated for both watersheds in Table 23.8. Each parameter is described in detail in the following sections.

Time–Area Histogram Data

The time–area histograms for the BBC and CC watersheds are developed in Figs. 23.12 and 23.13, respectively. Travel times and times of concentration for the Stanford watershed model are found from the Kirpich equation (for watersheds larger than 15 acres),[19] or

$$T_c = 0.0078\left(\frac{L}{S^{0.5}}\right)^{0.77} \tag{23.25}$$

where T_c = the time of concentration (min)
 L = the horizontal projection of the channel length from the most distant point to the basin outlet (ft)
 S = the slope between the two points

In developing the time–area histogram for Big Bordeaux Creek, 30 points and corresponding travel times (min) were plotted as shown in Fig. 23.12. The dashed isochrones (lines of equal travel time to the basin outlet) were constructed by linear interpolation between the plotted points. Areas between paris of isochrones were determined from planimetering. To contrast, the time–area histogram for Cave Creek was constructed, as shown in Fig. 23.13, by assuming that the flow velocity was constant everywhere, equal to the average streamflow velocity obtained by dividing T_c into the channel length L. This procedure simply places all points along each isochrone at equal distances to the basin outlet.

Watershed Parameters

The watershed drainage area (AREA), the impervious fraction of the watershed surface draining directly into the stream (A), the fractional stream and lake surface area (ETL), the average ground slope of overland flow perpendicular to the contours

**TABLE 23.7   PARAMETER NAME COMPARISONS FOR KENTUCKY AND
STANFORD WATERSHED MODELS**

Parameter name		Parameter description
Kentucky version	Stanford version	
NCTRI	Z	Integer number of elements in the time–area histogram
CTRI	C	Time–area histogram ordinate
RMPF	MINH	Discharge value below which no synthesized data are to be printed (cfs)
RGPMB	K1	Multiplication factor for precipitation data for a distant station
AREA	AREA	Watershed drainage area (mi^2)
FIMP	A	Fraction of watershed area that is impervious
FWTR	ETL	Fraction of watershed area in lakes or swamps
VINTMR	EPXM	Maximum rate of vegetative interception for a dry watershed (in./hr)
BUZC	CX	Index for estimating surface storage capacity
SUZC	EDF	Index for estimating soil surface storage capacity during summers
LZC	LZSN	Index of moisture storage in soil profile above water table (in.)
ETLF	K3	Evapotranspiration parameter for lower zone soil moisture
SUBWF	K24L	Subsurface flow from the basin
GWETF	K24EL	Groundwater evapotranspiration by phreatophytes
SIAC	EF	Factor varying infiltration by season
BMIR	CB	Index of infiltration rate
BIVF	CY	Index of rate and quantity of water entering interflow
OFSS	SS	Average basin ground slope (ft/ft)
OFSL	L	Average overland flow distance (ft)
OFMN	NN	Manning roughness coefficient for overland flow
OFMNIS	NNU	Manning roughness coefficient for flow over impervious areas
IFRC	IRC	Daily interflow recession constant
CSRX	KSC	Streamflow routing parameter for low flows
FSRX	KSF	Streamflow routing parameter for flood flows
CHCAP	CHCAP	Index capacity of existing channel, bank-full (cfs)
EXQPV	RFC	Exponent of flow for nonlinear routing
BFNLR	KV24	Daily base flow nonlinear recession adjustment factor
BFRC	KK24	Daily base flow recession constant
GWS	GWS	Index of groundwater storage (in.)
UZS	UZS	Depth of interception and depression storage at beginning of year (in.)
LZS	LZS	Current equivalent depth of moisture in the soil profile (in.)
BENX	—	Current value of BFNLR
IFS	—	Interflow storage
CONOPT	DKN	Control options for input, output, and internal branching
QQQ	QQQ	Title of computer simulation output, alphanumeric input
DIV	DIV	Mean daily diversion into or out of the basin (cfs)

(SS), and the mean length of overland flow (*L*) for the Big Bordeaux Creek watershed
were determined from areas, elevations, and lengths measured from 7.5-min series
USGS topographic maps (see Table 23.8 for BBC values). Other BBC parameter
values are ETL = 0.005 (determined from ETL = half the product of the stream
length and channel width at the outlet), $SS = 0.088$ ft/ft (determined from Fig. 23.14
as the mean of 140 measured values between 20-ft contours at each gridline intersection), and $L = 183.2$ ft (determined from Fig. 23.14 as the average of 140 lengths
measured perpendicular to contour lines from gridline intersection points to the

TABLE 23.8 SUMMARY OF INPUT PARAMETERS FOR BIG BORDEAUX CREEK AND CAVE CREEK WATERSHEDS

Parameter name	Description	BBC value(s)	Units	CC value(s)
TCONC	Time of concentration	105	min	60
TINC	Routing interval	15	min	15
Z	Number of elements in the time–area histogram	7	—	4
C	Time–area histogram ordinates	0.129	—	0.18
		0.158	—	0.29
		0.221	—	0.31
		0.151	—	0.22
		0.126	—	—
		0.145	—	—
		0.070	—	—
AREA	Watershed drainage area	9.22	mi^2	2.53
A	Impervious fraction of the watershed surface	0.0	—	0.0
ETL	Watershed stream and lake surface area as fraction of watershed area	0.005	—	0.0
SS	Average overland flow ground slope	0.088	ft/ft	0.075
L	Average length of overland flow	183.2	ft	300.0
CHCAP	Bank-full flow in channel at gauging station	39	cfs	40
IRC	Daily interflow recession constant	0.485	—	0.75
KK24	Daily base flow recession constant	0.977	—	0.94
KSC	Streamflow routing parameter for low flows	0.989	—	0.90
KSF	Streamflow routing parameter for flood flows	0.989	—	0.90
MINH	Minimum hourly flow rate to be printed	1.0	cfs	0.2
K1	Precipitation adjustment factor for distant gauge	1.0	—	1.0
NN	Manning roughness coefficient for overland flow	0.37	—	0.10
NNU	Manning roughness coefficient for impervious areas	0.013	—	0.015
EMIN	Factor for varying infiltration by seasons	0.5	—	—
EPXM	Maximum interception rate for dry watershed	0.15	in./hr	0.10
CX	Surface storage capacity index	0.80	—	0.90
EDF	Soil surface moisture storage capacity	1.10	—	1.25
LZSN	Soil profile moisture storage index	11.78	in.	4.85
K3	Soil evaporation parameter	0.28	—	0.25
K24L	Index of inflow to deep inactive groundwater	0.0	—	0.0
K24EL	Fraction of watershed area in phreatophytes	0.0	—	0.0
EF	Factor allowing for seasonal infiltration rates	1.0	—	0.15
CB	Factor controlling infiltration rates	0.75	—	0.65
CY	Index controlling water entering interflow	3.0	—	3.50
KV24	Parameter for allowing nonlinear recession	1.0	—	0.99
SGW	Groundwater storage volume parameter	0.1	in.	—
UZS	Equivalent depth of upper zone storage	0.0	in.	—
LZS	Equivalent depth of lower zone storage	7.0	in.	—
GWS	Index of antecedent moisture conditions	0.2	in.	—
VOLUME	Volume of water in swamp storage	0.0	acre-ft	—
EVCR	Monthly evaporation pan coefficients	Jan. 0.911	—	—
		Feb. 0.911	—	—
		Mar. 0.911	—	—
		Apr. 0.911	—	—
		May 0.552	—	—
		June 0.677	—	—
		July 0.654	—	—
		Aug. 0.651	—	—
		Sep. 0.642	—	—
		Oct. 0.911	—	—
		Nov. 0.911	—	—
		Dec. 0.911	—	—

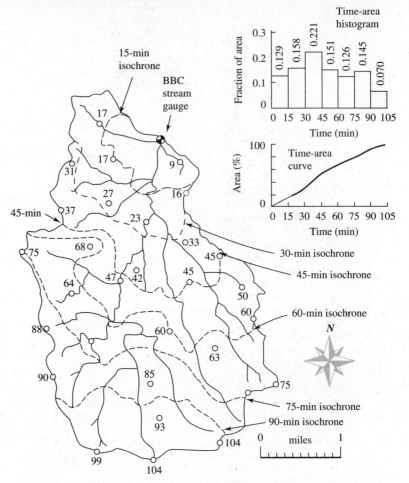

Figure 23.12 Time-area histogram development for Big Bordeaux Creek. The values represent travel times (min.) to the outlet.

nearest channel). The average overland flow distances can also be estimated as the reciprocal of twice the drainage density (see Chapter 10).

The bank-full channel capacity for Big Bordeaux Creek (CHCAP) was assumed to be equal to the discharge corresponding to the annual flood with a return period of 1.5 years. No criteria exist for easily obtaining CHCAP from maps, but Wolman and Leopold[20] have shown that the recurrence interval of a bank-full annual flood for many uncontrolled small and medium streams is nearly 1.5 years. This finding allows a simple determination of the bank-full discharge if a flood frequency curve for the stream is available or can be developed. Regional flood frequency curves were applied resulting in an estimate of the bank-full BBC flood of 39 cfs. Another technique for determining CHCAP involves the use of the stream-gauging-station rating curve if the gauge height for bank-full flow is known or can be estimated from topographic maps. The Cave Creek value of 40 cfs was determined from a hydraulic analysis of the cross section and profile of the main channel. As a rule of thumb, CHCAP is selected to

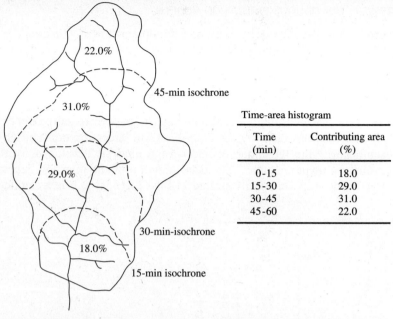

Time-area histogram	
Time (min)	Contributing area (%)
0-15	18.0
15-30	29.0
30-45	31.0
45-60	22.0

Figure 23.13 Derivation of the time–area histogram for the Cave Creek Watershed. Dashed lines represent isochrones. (After Clarke.[16])

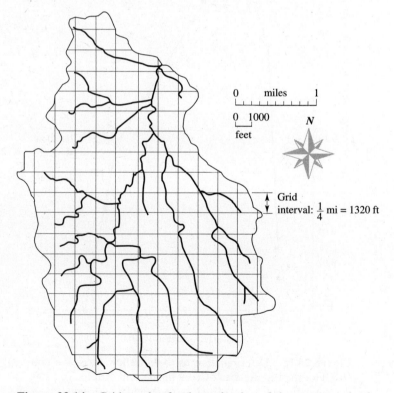

Figure 23.14 Grid overlay for determination of the mean overland slope and distance for Big Bordeaux Creek.

produce well-defined overbank and floodplain flow if Q exceeds twice the value of CHCAP. Similarly, low, shallow flows are assumed whenever Q is less than half of CHCAP.

Streamflow Recession and Routing Parameters

The rate at which the model allows water to pass through the upper soil zones to the channels is controlled by the daily interflow recession constant (IRC). A graphical semilogarithmic technique of hydrograph analysis developed by Barnes is used to estimate this parameter.[21] The determination of the interflow recession constant for Big Bordeaux Creek is illustrated in Fig. 23.15, which is a semilogarithmic plot of

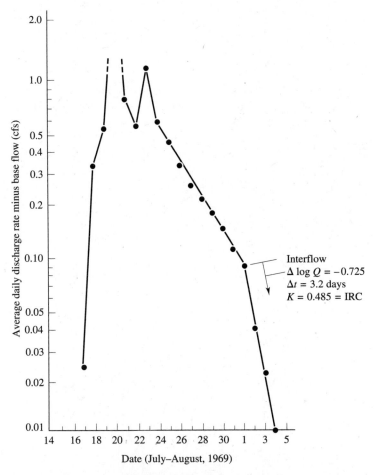

Figure 23.15 Determination of the interflow recession constant (IRC) for the Big Bordeaux Creek flood event of July 20, 1969. The determination of $K = $ IRC, $Q_t = Q_0 K^t$ and $t = 1$ day.

average daily discharge (after deducting base flow) versus time for the flood event of July 20, 1969. The daily interflow recession constant (IRC) was determined as 0.485. Base flow was separated using the technique described in Chapter 11.

The daily base flow recession constant (KK24) controls the rate of discharge to the channels from the groundwater. A graphical technique similar to that used in determining the interflow recession constant is illustrated in Fig. 23.16 using the same flood event. The method reveals that the direct runoff and the interflow ceased on August 4 and that the base flow (by virtue of the linear plot on semilogarithmic paper) has a recession defined adequately by $Q_t = Q_0(0.977)^t$, where Q_t and Q_0 are base flows on successive days after August 4, and t is 1 day for this analysis. The recession constant is simply determined as the slope of the base flow line in Fig. 23.16. An

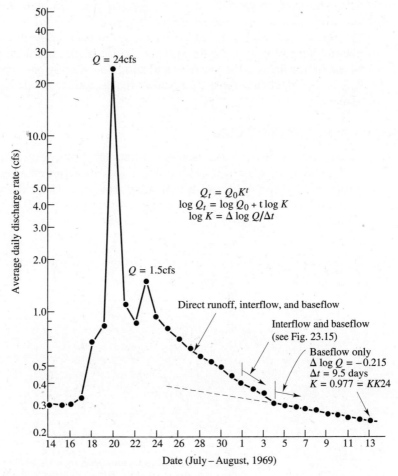

Figure 23.16 Determination of the base flow recession constant (KK24) for the Big Bordeaux Creek flood event of July 20, 1969. The determination of $K = $ KK24.

alternative method for determining KK24 using data collected several days after several flood peaks is illustrated for Cave Creek in Fig. 23.17. The line represents an envelope drawn to the right of the data points.

Two streamflow routing parameters are used by the simulation model to route inflow hydrographs to the point of interest for the basin (in this case to the stream gauging station for comparison with measured flows). The first parameter (KSC) is used in routing only if channel flows are less than half the channel capacity (CHCAP), and the second (KSF) accounts for channel and floodplain storage during flood flows (flows greater than twice the channel capacity). Respective data for Big Bordeaux Creek were determined using the two runoff hydrographs shown in Figs. 23.18 and 23.19, representing low and flood flows, respectively. Because the inflection point on the recession portion of each hydrograph represents the time at which a reversal of flow direction through the channel banks occurs, then the parameters KSC and KSF represent daily recession constants for the water in storage in the channel. The equation used in determining both parameters is derived from a hydrologic routing technique equating $1/K$ times the difference in the average inflow and outflow during the routing period with the time rate of change of the outflow in the channel reach. As shown in the figures, the low flow and flood flow routing parameters for Big Bordeaux Creek are identical. Table 23.8 shows that a similar result was observed for Cave Creek.

Hydrologic Parameters and Data

In addition to the described watershed characteristics, several hydrologic parameters and an impressive amount of hydrologic data are required as input to the simulation program. Due to the excessive bulk of hydrologic data for daily evaporation, hourly precipitation, and daily streamflow for a 4-year period in BBC and a 10-year period

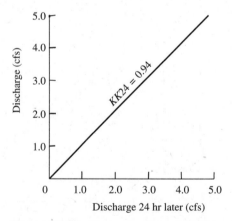

Figure 23.17 Determination of KK24 for Cave Creek. (After Clarke.[16])

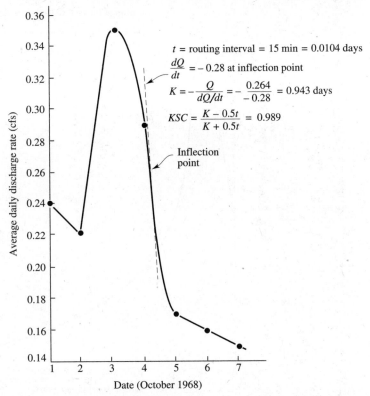

t = routing interval = 15 min = 0.0104 days

$\dfrac{dQ}{dt} = -0.28$ at inflection point

$K = -\dfrac{Q}{dQ/dt} = -\dfrac{0.264}{-0.28} = 0.943$ days

$KSC = \dfrac{K - 0.5t}{K + 0.5t} = 0.989$

Inflection point

Figure 23.18 Determination of low flow streamflow routing parameter (KSC) for the Big Bordeaux Creek low flow event of October 3, 1968.

in CC, the actual hydrologic data values have been compiled but are not included in this text.

One useful hydrologic parameter used as input only in the BBC application is the hourly flow at the gauging station below which no printing of simulated discharges is desired. For the initial simulation of runoff from Big Bordeaux Creek, a MINH of 1.0 cfs was selected to minimize the printed output. This can easily be decreased in successive simulation trials, after the uncertainties in some of the other parameters have been reduced.

A precipitation-weighting adjustment parameter (K1), representing the long-term ratio of average precipitation over the basin to average precipitation at the precipitation gauge, might be greater or less than 1.0 if the gauge were located at any distance from the study basin. The recording gauges for both watersheds were within close proximity, and both K1 parameters were set to 1.0.

Monthly evaporation pan coefficients (EVCR) are used by the simulation program as multipliers in converting input pan-evaporation data to lake evaporation data (and also for determining potential evapotranspiration rates). The nearest evaporation pan for Big Bordeaux Creek is approximately 20 mi south at the Box Butte Reservoir,

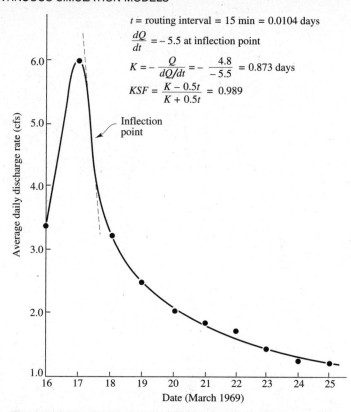

t = routing interval = 15 min = 0.0104 days

$$\frac{dQ}{dt} = -5.5 \text{ at inflection point}$$

$$K = -\frac{Q}{dQ/dt} = -\frac{4.8}{-5.5} = 0.873 \text{ days}$$

$$KSF = \frac{K - 0.5t}{K + 0.5t} = 0.989$$

Inflection point

Figure 23.19   Determination of flood flow streamflow routing parameter (KSF) for the Big Bordeaux Creek flood flow event of March 17, 1969.

providing records of daily and monthly pan-evaporation depths for the months between and including May and September. Estimates of corresponding monthly lake evaporation amounts at Box Butte Reservoir were obtained from maps and charts developed by Shaffer.[22] Values listed in Table 23.8 are assumed to apply to each year of the simulation even though the simulation program allows changes from year to year.

Manning's roughness parameters for flow over soil and impervious surfaces are both required as input to the program. For the Big Bordeaux Creek area, the initial estimates for overland flow (NN) and impervious surface flow (NNU) were 0.37 and 0.013, representing coefficients for dense shrubbery and forest litter for overland flow and smooth concrete for impervious surface flow. The later n is significant only if the fraction of impervious area (A) is nonzero. The Cave Creek analysis incorporated NN = 0.10 for light turf and NNU = 0.015 for concrete pavement.

Model Calibration

Several of the following parameters are determined by trial and adjustment until the comparison between the simulated and recorded streamflows is satisfactory. Guidelines for establishing initial values exist for only a few of the parameters, whereas most are initially determined from suggested ranges. The BBC data in Table 23.8 represent initial, unmodified estimates; those shown for CC are the optimal result of many repetitive runs.

One factor for varying infiltration by seasons (EMIN) ranges between 0.1 and 1.0 and has been shown by Briggs to be significant in matching measured and simulated winter peak flow rates.[23] Because no guidelines for estimating this parameter are presently available, the suggested midvalue of 0.5 was selected for Big Bordeaux Creek.

Several soil moisture and routing parameters require initial input values that are difficult to estimate from available data. Calibration procedures successively improve the initial estimates by trial and adjustment. For Big Bordeaux Creek, the parameters and initial estimates are as follows.

EPXM—the maximum interception rate (in./hr) for a dry watershed. Crawford and Linsley[2] suggest trial values of 0.10, 0.15, and 0.20 for grasslands, moderate forest covers, and heavy forest covers, respectively. The 0.15-in./hr value was selected for the moderate forest cover along Big Bordeaux Creek. Clarke used 0.10 in the Cave Creek study.

CX—an index of the surface capacity to store water as interception and depression storage. This parameter normally ranges from 0.10 to 1.65, and the midvalue of 0.80 was selected for Big Bordeaux Creek although a greater number might be indicative of the forest cover. Clarke independently arrived at a final, similar value (0.90) for Cave Creek.

EDF—an index of soil-surface moisture storage capacity, representing the additional moisture storage capacity available during warmer months due to vegetation. Depending on the soil type, the index ranges from 0.45 to 2.00. Sandy soils similar to those in the BBC area readily give up moisture to vegetation, resulting in increased storage capacity. Initial values of 1.10 and 1.25 were independently selected for the Big Bordeaux and Cave Creek areas.

LZSN—a soil-profile moisture storage index (in.) approximately equal to the volume of water stored above the water table and below the ground surface. This parameter is a major runoff-volume parameter, inversely related to the basin yields, interflow, and groundwater flow. The LZSN index, depending on porosity and the specific yield of the soil, ranges from 2.0 to 20.0, and 4 in. plus half the mean annual rainfall can be used as an initial estimate in areas experiencing seasonal rainfall. By the use of a 1931–1952 average of 15.55 in. of annual precipitation, the Chadron, Nebraska, precipitation station gives an initial LZSN = 11.78 in. for Big Bordeaux Creek. A similar analysis was used for Cave Creek.

K3—a soil evaporation parameter that controls the rate of evapotranspiration losses from the lower soil zone. The parameter ranges from 0.2 to 0.9 depending on the type and extent of the vegetative cover. As an initial estimate for the light forest cover in BBC, 0.28 was selected, which agreed with the estimates suggested by Crawford and Linsley.[2] Also, K3 is approximately equal to the fraction of the basin covered by forest and deep-rooted vegetation. Recommendations[2] for barren ground, grasslands, and heavy forests are, respectively, 0.20, 0.23, and 0.30; 0.25 was optimal for the Cave Creek study.

K24L—a parameter controlling the fraction of moisture lost or diverted from active groundwater storage through transverse flow across the drainage basin boundary. It also represents that portion of the inflow to the groundwater that percolates to the deep or inactive groundwater. The K24L parameter can be estimated from observed changes in deep groundwater levels, or it is often assumed to be zero because these losses are small compared to the magnitudes of rainfall and runoff.

K24EL—the fraction of the total watershed over which evapotranspiration from groundwater storage is assumed to occur at the potential rate. This parameter is assumed zero unless a significant quantity of vegetation draws water directly from the water table. Plants that seek phreatic water, such as cedar or cottonwood or alfalfa, are called phreatophytes.

EF—a factor ranging from 0.1 to 4.0 that relates infiltration rates to evaporation rates. This parameter simply allows a more rapid infiltration rate recovery during warmer seasons. A normal starting value of 1.0 was selected for Big Bordeaux Creek, whereas the Cave Creek value was set much lower.

CB—an index that controls the rate of infiltration, depending on the soil permeability and the volume of moisture that can be stored in the soil. A midvalue of 0.75 (0.3–1.2) was selected for the sandy soils around Big Bordeaux Creek; a smaller value of 0.65 was optimal for the Cave Creek study.

CY—an index controlling the time distribution and quantities of moisture entering interflow. This index ranges from 0.55 to 4.5, and a moderately high value of 3.0 was selected for Big Bordeaux Creek because the watershed contains many pine needle mats. Clarke also selected a moderately high CY for Cave Creek.

KV24—a daily base flow recession adjustment factor used to produce a simulated curvilinear base flow recession. An initial value of 1.0 for Big Bordeaux Creek was selected because the base flow recession for the hydrograph in Fig. 23.16 is linear. Later adjustments might be required in matching simulated and recorded base flow recessions.

SGW—a groundwater storage increment (in.), reflecting the fluctuations in storage volume. Usually, an initial estimate (0.10 was selected for Big Bordeaux Creek) is made and adjusted after several simulation trials; SGW ranges from 0.10 to 3.90 in. This and the following four parameters were used only in the BBC study.

UZS—the current volume (in.) of soil surface moisture as interception and depression storage. Because the simulation begins on October 1 of the first calibration year, the parameter may initially be designated as zero unless precipitation occurs during the last few days of September.

LZS—the current volume (in.) of soil moisture storage between the land surface and the water table. Sixty percent of LZSN, or 7.0 in., was selected to initiate the Big Bordeaux Creek simulation.

GWS—the current groundwater slope index (in.). This index provides an indication of antecedent moisture conditions. Suggested initial values fall between 0.15 and 0.25, or the value assigned to SGW can be used. A midrange 0.20 in. was selected for Big Bordeaux Creek.

VOLUME—the volume (acre-ft) assigned to swamp storage and dry ground recharge, accounting for the runoff required to recharge swamps in late summer. Since no swamps were visible on the USGS 7.5-min topographic maps, an initial value of 0.0 acre-ft was selected for Big Bordeaux Creek.

Hydrologic Data

In addition to the parameter values of Table 23.8, the Stanford model requires a large volume of hydrologic data for each water year of the simulation. The following description of requried input data for each water year illustrates the data that were compiled and reduced to necessary input form for 4 water years beginning on October 1, 1968 and ending on September 30, 1972. Only the input for BBC is described. The input includes the following data for each water year:

1. The new year identification entry contains the water year and the recorded annual streamflow. The values of the annual streamflow for Big Bordeaux Creek are listed in Table 23.9.
2. A description of the stream gauging site.
3. The title to be applied to the ordinate for graphical plots of the simulated and recorded runoff hydrograph; namely, the daily average flow rate (cfs).
4. Pan-evaporation data, read as 365 or 366 single entries containing daily evaporation amounts (in.).
5. The monthly evaporation pan coefficients, comprising the data listed in Table 23.8, beginning with October.
6. Recorded daily streamflows (average flow for the day, cfs) read as 365 or 366 entries for October 1 through September 30.
7. Hourly rainfall data, read for each water year. Two entries per day, each containing an identification of the gauge and date, are used to provide hourly

TABLE 23.9 ANNUAL BIG BORDEAUX CREEK STREAMFLOWS

Water years	Recorded annual streamflow (acre-ft)
1968–1969	434.0
1969–1970	333.2
1970–1971	465.4
1971–1972	296.4

depths in inches before noon on the first and after noon on the second. Values are required only for half-days experiencing precipitation. Because of the variable number of possible rainfall values, a sentinel entry with the year set equal to 2001 is placed at the end of the data, indicating that all the precipitation data for the water year has been read.

Output from the Kentucky Version

Depending on which optimal input, output, and branching parameters are selected, a variety of output data are available from the Kentucky version, including plotted graphs of measured and synthesized daily streamflow rates. Options include the following:

1. A table of synthesized average daily streamflow rates (cfs).
2. A table of monthly and annual totals of synthesized daily flow rates.
3. Synthesized monthly and annual totals of runoff in equivalent inches over the watershed.
4. Synthesized monthly and annual interflow amounts (in.) over the watershed.
5. Synthesized monthly and annual base flow amounts (in.) over the watershed.
6. The volume of synthesized streamflow runoff from the watershed for the entire water year (acre-ft).
7. A summation of all the recorded daily streamflow rates (cfs) for each month and year.
8. The recorded annual total of runoff (in.) over the watershed.
9. The recorded volume of runoff from November through March (in.) over the watershed.
10. The amount of synthesized snow from November through March (in.) over the watershed.
11. The volume of the recorded annual streamflow (acre-ft).
12. The sum of the recorded precipitation for each month and for the year.
13. The synthesized monthly and annual totals of evapotranspiration (in.).
14. The monthly and annual recorded lake evaporation amounts (in.).
15. End-of-the-month levels of UZS, the current surface moisture storage (in.).
16. End-of-the-month levels of LZS, the current soil moisture storage (in.).
17. End-of-the-month values of SGW, the current groundwater storage fluctuation (in.).
18. An annual moisture balance (in.), which represents the moisture not accounted for within the program. This is illustrated in the Cave Creek output.

Cave Creek Model Calibration

Synthetic and actual flow rates at the Cave Creek gauging station are shown for a single day in Fig. 23.20. Other typical output for portions of one water year of the simulation is presented in Tables 23.10 and 23.11. The former provides an hour-by-hour listing of all flow rates in excess of the specified value of MINH, Table 23.8. Note

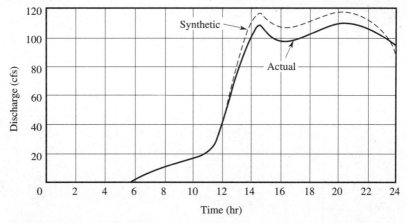

Figure 23.20 Comparison of synthesized and recorded hydrographs for Cave Creek. (After Clarke.[16])

that the streamflow was zero from October through December and exceeded MINH only during two days in January.

Table 23.11 contains most of the information described in the 18 items of output for the Kentucky version. The daily flows are followed by the synthetic and recorded monthly totals, monthly interflow and base flow amounts, monthly precipitation totals, monthly actual and potential ET amounts, and end-of-month storages in the soil profile and groundwater zones in inches. Of particular interest is the annual summary in the lower right. Of the 37.5 in. of precipitation, 23.8 in. went to ET, 11.6 in. ran off or was discharged from storage, and the remaining 2.1 in. recharged the soil profile. Moisture not accounted for during the year was 0.0844 in.

Sensitivity of Model Response to Parameter Changes

One interesting and useful aspect of simulation is the ease with which changes in watershed parameters can be evaluated. Clarke tested the sensitivity of KWM by varying several of the parameters in Table 23.8 over reasonable ranges while holding all other parameters constant. The results of his analysis for Cave Creek on a typical day in March are summarized in Table 23.12. These observations were taken from graphs such as Figs. 23.21 and 23.22, which illustrate the sensitivity of flood magnitude and timing to changes in L and KSC. These results and the summary in Table 23.12 are applicable only to the CC watershed and should not be viewed as generally applicable.

■ Summary

If actual or synthesized precipitation records are available, one of the most effective means of analyzing historical flows or evaluating future possible flows under changing land use patterns is through any of the continuous streamflow simulation models described in this chapter. Hydrologic problems and applications that can be analyzed

TABLE 23.10 TYPICAL STORM HYDROGRAPH OUTPUT FROM STANFORD WATERSHED MODEL FOR CAVE CREEK

CAVE CREEK, PRM G RUN 1, K. CLARKE, OCT. 18, 1966

CAVE CREEK NEAR LEXINGTON, KENTUCKY

WATER YEAR 1963–64   KY. VERSION STANFORD WATERSHED MODEL

October
1	RECORDED FLOW= 0.0

November
1	RECORDED FLOW= 0.0

December
1	RECORDED FLOW= 0.1

JANUARY
1	RECORDED FLOW= 0.0

20	AM	0.2	0.3	1.8	5.5	6.6	6.7	6.7	6.7	6.7	6.7	6.7	
	PM	6.7	6.7	6.7	6.6	6.6	6.5	6.5	6.4	6.4	6.3		5.8
	MAXIMUM= 6.7 C.F.S.	TIME 10.00 A.M.											

21	AM	6.3	6.2	6.1	6.1	6.0	6.0	5.9	5.8	5.8	5.7	5.7	5.6	
	PM	5.5	5.5	5.4	5.3	5.3	5.2	5.1	5.1	5.0	5.0	4.9	4.8	5.6
	MAXIMUM= 6.3 C.F.S.	TIME 0.15 A.M.												

FEBRUARY
1	RECORDED FLOW= 0.8

15	AM	4.3	4.2	4.2	4.2	4.1	4.1	4.0	4.0	4.0	3.9	3.9	3.8	
	PM	3.8	3.7	3.7	4.0	4.1	4.3	4.5	4.6	4.8	4.9	5.0	5.1	4.2
	MAXIMUM= 5.1 C.F.S.	TIME 12.00 P.M.												

16	AM	5.2	5.2	5.3	5.4	5.4	5.4	5.5	5.5	5.5	5.5	5.5	
	PM	5.5	5.5	5.5	5.4	5.4	5.4	5.3	5.3	5.3	5.2	5.2	5.4
	MAXIMUM= 5.5 C.F.S.	TIME 10.30 A.M.											

TABLE 23.10 (Continued)

CAVE CREEK, PRM G RUN 1, K. CLARKE, OCT. 18, 1966

CAVE CREEK NEAR LEXINGTON, KENTUCKY — WATER YEAR 1963-64 KY. VERSION STANFORD WATERSHED MODEL

Day	Period	1	2	3	4	5	6	7	8	9	10	11	12
17	AM	5.1	5.1	5.0	5.0	5.0	4.9	4.9	4.8	4.8	4.7	4.7	4.6
	PM	4.6	4.5	4.5	4.4	4.4	4.3	4.3	4.2	4.2	4.1	4.1	4.0
	MAXIMUM= 5.2 C.F.S.	TIME 0.15 A.M.											4.6
4	AM	3.1	3.1	3.0	3.0	3.0	3.4	6.8	10.2	16.4	23.1	42.3	
	PM	87.2	112.9	112.8	107.4	107.5	112.2	117.0	118.4	112.0	108.5	104.1	
	MAXIMUM= 118.7 C.F.S.	TIME 7.30 P.M.											60.3
5	AM	99.8	95.6	91.7	88.1	84.6	81.4	78.3	75.4	72.6	70.0	67.6	65.2
	PM	63.0	60.9	59.0	57.1	55.3	53.6	52.0	50.5	49.0	47.6	46.3	45.0
	MAXIMUM= 101.4 C.F.S.	TIME 0.15 A.M.											67.1
6	AM	43.8	42.7	41.6	40.5	39.5	38.5	37.6	36.7	35.9	35.1	34.3	33.5
	PM	32.8	32.1	31.4	30.8	30.1	29.5	29.0	28.4	27.8	27.3	26.8	26.3
	MAXIMUM= 44.3 C.F.S.	TIME 0.15 A.M.											33.8
7	AM	25.8	25.3	24.9	24.4	24.0	23.6	23.2	22.8	22.4	22.0	21.6	21.3
	PM	20.9	20.6	20.2	19.9	19.6	19.3	19.0	18.7	18.4	18.1	17.8	17.5
	MAXIMUM= 26.0 C.F.S.	TIME 0.15 A.M.											21.3
8	AM	17.2	17.0	16.7	16.5	16.2	16.0	16.0	16.7	24.6	41.3	47.0	46.0
	PM	44.5	43.2	42.1	41.1	41.5	45.7	47.3	46.4	45.5	44.5	43.6	42.8
	MAXIMUM= 47.4 C.F.S.	TIME 6.30 P.M.											34.2
9	AM	41.9	41.1	40.3	39.6	39.1	40.9	41.4	40.9	40.3	39.8	39.2	38.7
	PM	38.1	37.7	37.2	36.9	36.5	36.3	36.9	45.4	66.5	97.2	118.6	130.9
	MAXIMUM= 134.9 C.F.S.	TIME 12.00 P.M.											50.1

MARCH

1 RECORDED FLOW= 2.0

Source: After Clarke.[16]

587

TABLE 23.11 TYPICAL DAILY AND MONTHLY MOISTURE SUMMARY OUTPUT FROM STANFORD WATERSHED MODEL

CAVE CREEK, PRP G RUN 1, K. CLARKE, OCT. 18, 1966

CAVE CREEK NEAR LEXINGTON, KENTUCKY

KENTUCKY WATERSHED MODEL — WATER YEAR 1963-64

DAY	OCT	NOV	DEC	JAN	FEB	MAR	APR	MAY	JUN	JUL	AUG	SEPT	ANNUAL
1	0.0	0.0	0.0	0.0	0.3	0.6	0.9	1.7	0.1	0.1	0.0	0.0	
2	0.0	0.0	0.0	0.0	0.2	0.6	0.8	1.3	0.1	0.1	0.0	0.0	
3	0.0	0.0	0.0	0.0	0.2	2.6	0.7	1.0	0.1	0.1	0.0	0.0	
4	0.0	0.0	0.0	0.0	0.1	60.3	0.7	0.8	0.1	0.1	0.0	0.0	
5	0.0	0.0	0.0	0.0	0.1	67.1	0.6	0.7	0.1	0.1	0.0	0.0	
6	0.0	0.0	0.0	0.7	1.9	33.8	1.4	0.6	0.1	0.1	0.0	0.0	
7	0.0	0.0	0.0	1.9	3.3	21.3	1.7	0.5	0.1	0.1	0.0	0.0	
8	0.0	0.0	0.0	1.7	2.8	34.2	1.5	0.4	0.1	0.1	0.0	0.0	
9	0.0	0.0	0.0	1.5	2.1	50.1	1.2	0.4	0.1	0.1	0.0	0.0	
10	0.0	0.0	0.0	1.4	1.8	93.3	0.9	0.3	0.1	0.1	0.0	0.0	
11	0.0	0.0	0.0	1.0	2.8	44.9	0.8	0.3	0.1	0.1	0.0	0.0	
12	0.0	0.0	0.0	0.8	2.5	27.7	0.6	0.3	0.1	0.1	0.0	0.0	
13	0.0	0.0	0.0	1.1	3.2	18.9	0.6	0.2	0.1	0.1	0.0	0.0	
14	0.0	0.0	0.0	0.9	4.6	14.0	0.5	0.2	0.2	0.1	0.0	0.0	
15	0.0	0.0	0.0	0.7	4.2	16.4	0.4	0.2	0.4	0.1	0.0	0.0	
16	0.0	0.0	0.0	0.5	5.4	14.5	0.4	0.2	0.4	0.1	0.0	0.0	
17	0.0	0.0	0.0	0.4	4.6	11.0	0.4	0.2	0.3	0.1	0.0	0.0	
18	0.0	0.0	0.0	0.3	3.6	8.1	0.3	0.2	0.4	0.1	0.0	0.0	
19	0.0	0.0	0.0	0.3	3.1	6.0	0.3	0.2	1.1	0.1	0.0	0.0	
20	0.0	0.0	0.0	5.8	2.5	5.7	0.5	0.2	1.0	0.1	0.0	0.0	
21	0.0	0.0	0.0	5.6	1.9	10.5	0.5	0.1	0.8	0.1	0.0	0.0	
22	0.0	0.0	0.0	4.1	1.5	8.9	0.7	0.1	0.6	0.1	0.0	0.0	
23	0.0	0.0	0.0	2.9	1.1	6.8	0.8	0.1	0.5	0.1	0.0	0.0	

TABLE 23.11 *(Continued)*

CAVE CREEK, PRP G RUN 1, K. CLARKE, OCT. 18, 1966

CAVE CREEK NEAR LEXINGTON, KENTUCKY — WATER YEAR 1963-64 — KENTUCKY WATERSHED MODEL

DAY	OCT	NOV	DEC	JAN	FEB	MAR	APR	MAY	JUN	JUL	AUG	SEPT	ANNUAL
24	0.0	0.0	0.0	2.1	0.9	5.0	0.7	0.1	0.4	0.0	0.0	0.0	
25	0.0	0.0	0.0	1.7	0.7	3.8	0.6	0.1	0.3	0.0	0.0	0.0	
26	0.0	0.0	0.0	1.3	0.7	3.1	0.7	0.1	0.2	0.0	0.0	0.0	
27	0.0	0.0	0.0	1.0	0.9	2.5	1.5	0.1	0.2	0.0	0.0	0.0	
28	0.0	0.0	0.0	0.7	0.8	2.0	2.0	0.1	0.1	0.0	0.0	0.5	
29	0.0	0.0	0.0	0.5	0.8	1.6	2.4	0.2	0.1	0.0	0.0	1.4	
30	0.0	0.0	0.0	0.4		1.3	2.1	0.2	0.1	0.0	0.0	1.4	
31	0.0		0.0	0.3		1.1		0.1		0.0	0.0		

	Unit	OCT	NOV	DEC	JAN	FEB	MAR	APR	MAY	JUN	JUL	AUG	SEPT	ANNUAL
SYNTHETIC TOTAL	CFSD	0.	0.	0.	40.	59.	578.	27.	11.	9.	2.	1.	3.	730.
	INCHES/DA	0.004	0.001	0.004	0.589	0.863	8.492	0.401	0.164	0.127	0.029	0.008	0.051	10.73
INTERFLOW	INCHES/DA	0.000	0.000	0.002	0.503	0.747	4.950	0.215	0.054	0.083	0.009	0.000	0.052	6.614
BASE	INCHES/DA	0.004	0.001	0.004	0.047	0.116	0.434	0.195	0.106	0.051	0.029	0.008	0.009	1.003
	ACFT													1448.
RECORDED	CFSD	0.	3.	2.	59.	117.	536.	30.	14.	8.	5.	0.	10.	786.
	INCHES/DA													11.55
	ACFT											(1543.)		1558.
PRECIP	INCHES/DA	0.33	1.81	0.81	2.79	2.55	10.28	2.86	1.68	3.55	3.99	1.96	4.90	37.51
EVP/TRAN-NET	INCHES/DA	1.236	1.265	0.875	0.702	0.576	0.693	1.851	3.149	4.777	4.396	2.925	1.344	23.790
-POTENTIAL	INCHES/DA	4.687	2.137	0.875	0.702	0.576	0.693	1.871	4.346	5.865	5.814	7.841	5.704	41.111
STORAGES-UZS	INCHES/DA	0.050	0.761	0.371	0.440	0.454	0.348	0.854	0.947	0.000	0.000	0.000	2.485	2.485
-LZS	INCHES/DA	0.951	0.784	1.102	2.488	3.524	4.640	4.743	3.150	2.753	2.322	1.355	2.281	2.281
SGW	INCHES/DA	0.001	0.001	0.002	0.024	0.077	0.164	0.120	0.032	0.019	0.008	0.002	0.050	0.050
INDICES-UZSN		1.969	1.065	0.805	0.490	0.361	0.328	0.722	1.579	2.404	2.461	3.113	1.816	1.816
GWS		0.014	0.006	0.005	0.042	0.126	0.329	0.249	0.113	0.067	0.036	0.015	0.058	0.058
INF		1.605	0.899	0.474	0.330	0.330	0.330	0.537	1.191	1.832	1.965	2.421	1.958	1.958
BALANCE		-0.0844 INCHES												

Source: After Clarke.[16]

TABLE 23.12 EFFECTS OF INCREASING VARIOUS PARAMETERS FOR A MARCH STORM OVER CAVE CREEK

KWM parameter name	Representing	Range of increase	Effect on flood peak	Effect on runoff volume	Effect on flood timing
LZSN	Soil moisture	3–24 in.	Decreased	Reduced	Slight
CY	Moisture entering interflow	0.3–3.0	Decreased	Reduced	None
CB	Infiltration	0.2–25	Decreased	Reduced	None
IRC	Moisture leaving interflow	0.2–0.6	Decreased	Slight	None
NN	Overland flow resistance	Not specified	Decreased	Reduced	Delayed
A	Impervious fraction	Not specified	Increased	Increased	Delayed
SS	Overland flow slope	Not specified	Increased	Increased	Hastened
CX	Depression storage	Not specified	None	Reduced	None
EDF	Unfilled depression storage	1–2	Decreased	Slight	None
TCONC	Time of concentration	Not specified	Decreased	None	Delayed
Z	Number of histogram elements	16–62	Decreased	Reduced	Delayed
KSF	Channel routing parameter	0.85–0.99	Decreased	Reduced	Delayed
L	Length of overland flow	100–1000 ft	Decreased	Slight	Delayed

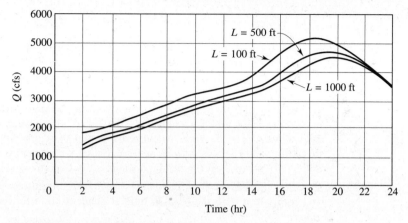

Figure 23.21 Sensitivity of model response to the length-of-overland-flow parameter. (After Clarke.[16])

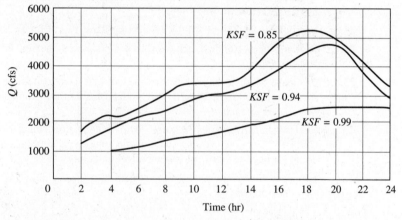

Figure 23.22 Sensitivity of model response to the channel routing parameter. (After Clarke.[16])

using continuous modeling include watershed yield studies, reservoir design and operation studies, sediment yield estimating for reservoir design or analysis of impacts of erosion controls on water quality, water supply studies for municipal, industrial or agricultural demands, litigation over impacts of wellfields or direct diversions, determination of hydropower production potential, evaluations of flows that will pass through critical in-stream or riparian habitat reaches of a stream, identification of flows that will be available for recreational uses of a stream, and, among numerous other applications, analysis of water quantity and quality impacts of removing dams or making other major upstream changes.

The models described in this chapter, and other similar continuous simulation codes, are available from the federal or state agencies that originated the code, or from numerous public outlets or vendors who have been authorized to distribute the software.

Problems

23.1. Assume that a 30-mi^2 rural watershed in your locale receives a 3-in. rain in a 10-day period. Reconstruct the block diagram of Fig. 23.2 and plot approximate percentages to show, for average conditions, how the rain would be distributed (a) initially and (b) after 30 days.

23.2. A sloping, concrete parking lot experiences rain at a rate of 3.0 in./hr for 60 min. The lot is 500 ft deep and has a slope of 0.0001 ft/ft. If the water detention on the lot is zero at the start of the storm, calculate the complete overland flow hydrograph for 1 ft of width using the SWM-IV equations. Use a 5-min routing interval and continue computations until all the detained water is discharged.

23.3. Calculate the SWM-IV overland flow time-to-equilibrium for the lot of Problem 23.2 and compare it with the Kirpich time of concentration for the lot. Should these be equal?

23.4. Compare, by listing traits and capabilites of each, the SWM-IV with its more sophisticated offspring HSP and HSPF.

23.5. Discuss the primary differences among the four versions of the Stanford watershed model described in this chapter.

23.6. Verify Eqs. 23.23 and 23.24 by starting from Eqs. 13.4 and 13.33.

23.7. Discuss the watershed behavior that is depicted in Fig. 23.7. Is this a "typical" watershed?

23.8. Compare the differences between the two U.S. Department of Agriculture continuous simulation models, USDAHL and SWRRB, and discuss the applications that would be best suited to each.

23.9. Review the differences between *water budget* and *simulation* models discussed in Chapter 21 and determine which of the continuous simulation models described here could be used to perform water budget calculations.

23.10. For what applications might the following be best suited?
API model
USDAHL
HSPF
PRMS
SWRRB

23.11. For the continuous simulation model selected by your instructor, describe four different types of problems that could be analyzed if you were given the full, calibrated model.

REFERENCES

1. W. T. Sittner, C. E. Schauss, and J. C. Monro, "Continuous Hydrograph Synthesis with an API-Type Hydrologic Model," *Water Resources Res.* **5**(5), 1007–1022(1969).

2. N. H. Crawford and R. K. Linsley, Jr., "Digital Simulation in Hydrology: Stanford Watershed Model IV," Department of Civil Engineering, Stanford University, Tech. Rep. No. 39, July 1966.

3. L. D. James, "An Evaluation of Relationship Between Streamflow Patterns and Watershed Characteristics Through the Use of OPSET," Research Rep. No. 36, Water Resources Institute, University of Kentucky, Lexington, 1970.

4. E. Y. Liou, "OPSET: Program for Computerized Selection of Watershed Parameter Values for the Stanford Watershed Model," Research Rep. No. 34, Water Resources Institute, University of Kentucky, Lexington, 1970.

5. L. D. James, "Hydrologic Modeling, Parameter Estimation, and Watershed Characteristics," *J. Hydrology* **17,** 283–307(1972).

6. B. J. Claborn and W. Moore, "Numerical Simulation in Watershed Hydrology," Hydraulic Engineering Laboratory, University of Texas Rep. No. HYD 14-7001, 1970.

7. H. N. Holtan and N. C. Lopez, "USDAHL-73 Revised Model of Watershed Hydrology," U. S. Department of Agriculture, Plant Physiology Institute, Rep. No. 1, 1973.

8. U.S. National Weather Service Office of Hydrology, "National Weather Service River Forecast System Forecast Procedures," NOAA Tech. Mem. NWS HYDRO-14, Dec. 1972.

9. U.S. Army Corps of Engineers, "Program Description and User Manual for SSARR Model Streamflow Synthesis and Reservoir Regulation," Program 724-K5-G0010, Dec. 1972.

10. N. H. Crawford, "Studies in the Application of Digital Simulation to Urban Hydrology," Hydrocomp International, Inc., Palo Alto, CA, Sept. 1971.

11. Johanson, R. C. et al. "User Manual for the Hydrologic Simulation Program—Fortran (HSPF)," *Environmental Protection Agency Report EPA-600/9-80-015,* 1980.

12. W. M. Alley, and P. E. Smith, "Distributed Routing Rainfall-Runoff Model, Version II," *U.S. Geological Survey Open File Report 82–344,* 1982.

13. J. G. Arnold, et al. "SWRRB—A Basin Scale Model for Soil and Water Resources Management," Texas A&M University Press, College Station TX, 1990.

14. J. R. Williams, and H. D. Berndt, "Sediment Yield Prediction Based on Watershed Hydrology," *Transactions, American Society of Agricultural Engineers,* v. 20, no. 6, 1977.

15. "Expert Systems in Water Resources," American Society of Civil Engineering, *Civil Eng.* 46(Oct. 1984).

16. D. K. Clarke, "Applications of Stanford Watershed Model Concepts to Predict Flood Peaks for Small Drainage Areas," Division of Research, Kentucky Department of Highways, 1968.

17. Lewis, G. L. and W. Viessman, Jr., "Nebraska Watershed Modeling Program," Nebraska Water Resources Research Institute, Lincoln, NE, 1973.

18. G. A. Ross, "The Stanford Watershed Model: The Correlation of Parameter Values Selected by a Computerized Procedure with Measurable Physical Characteristics of the Watershed," Research Rep. 35, Water Resources Institute, University of Kentucky, Lexington, 1970.

19. Z. P. Kirpich, "Time of Concentration of Small Agricultural Watersheds," *Civil Eng.* **10,** 362(June 1940).

20. M. G. Wolman and L. B. Leopold, "River Flood Plains: Some Observations of Their Formation," U.S. Geological Survey Professional Paper 282-C, 1957.

21. B. S. Barnes, "Discussion of Analysis of Runoff Characteristics," *Trans. ASCE* **105,** (1940).

22. F. B. Shaffer, "Availability and Use of Water in Nebraska, 1970," Nebraska Water Survey Paper No. 31, Conservation and Survey Division, University of Nebraska, Lincoln, Mar. 1972.

23. D. L. Briggs, "Application of the Stanford Streamflow Simulation Model to Small Agricultural Watershed at Coshocton, Ohio," M. S. thesis, Ohio State University, 1969.

Chapter 24

Single-Event Simulation Models

■ Prologue

The purpose of this chapter is to:

- Describe how storm event models are structured and how they are used to simulate direct runoff hydrographs for single storms.
- Describe the five most widely used federal agency single-event models (note that popular single-event urban runoff simulation models are described in Chapter 25).
- Provide a detailed case study using one of the models, HEC-1.
- Introduce the emerging technology of storm surge modeling—the simulation of hydraulic surges resulting from wind energy acting on the ocean surface.

Many severe floods are caused by short-duration, high-intensity rainfall events. A single-event watershed model simulates runoff during and shortly following these discrete rain events. Users of single-event models are normally interested in the peak flow rate, or the entire direct runoff hydrograph if timing or volume of runoff is needed. Single-event models simulate the rainfall-runoff process and make no special effort to account for the rest of the hydrologic cycle. Few, if any, simulate soil moisture, evapotranspiration, interflow, base flow, or other processes occurring between discrete rainfall events.

Models described in this chapter are applicable to studies of watersheds that are primarily rural in makeup. Urbanized subareas are allowed, but for watersheds that are principally urbanized, the single-event and continuous models described in Chapter 25 are more applicable. Coastal flooding that is induced by surges created by wind action on the ocean surface is modeled by a different class of single-event models, described in Section 24.3.

24.1 STORM EVENT SIMULATION

Event simulation model structures closely imitate the rainfall and runoff processes developed in earlier chapters. Lumped parameter approaches, such as unit-hydrograph methods, are generally incorporated even though some use distributed parameter techniques. Preparation for implementing most single-event simulation studies begins with a watershed subdivision into homogeneous subbasins as illustrated in Fig. 24.1. Computations proceed from the most remote upstream subbasin in a downstream direction.

In any single-event model for a typical basin (Fig. 24.1), the runoff hydrographs for each of subbasins A, B, . . . , E are computed independently, and then routed and combined at appropriate points (called nodes) to obtain design hydrographs throughout the basin. The model reads input parameters for the storm; then applies the storm to the first upstream subbasin, B; computes the hydrograph resulting from the storm event; repeats the hydrograph computation for subbasin A; combines the two computed hydrographs into a single hydrograph; routes the hydrograph by conventional techniques through reach C to the upstream end of reservoir R, where it is combined with the computed hydrograph for subbasin C; and so on through the procedure detailed in Fig. 24.1.

Hydrograph computations for subbasins are most often determined using unit-hydrograph procedures as illustrated in Fig. 24.2. The precipitation hyetograph is input uniformly over the subbasin area, and precipitation losses are abstracted, leaving an excess precipitation hyetograph that is convoluted (see Chapter 12) with the

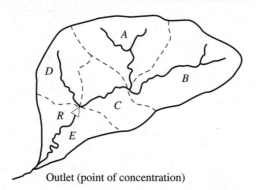

Outlet (point of concentration)

1. Subdivide basin to accommodate reservoir sites, damage centers, diversion points, surface and subsurface divides, gauging stations, precipitation stations, land uses, soil types, geomorphologic features.
2. Computation sequence in event simulation models:
   a. Compute hydrograph for subbasin B.
   b. Compute hydrograph for subbasin A.
   c. Add hydrographs for A and B.
   d. Route combined hydrograph to upstream end of reservoir R.
   e. Compute hydrograph for subbasin C.
   f. Compute hydrograph for subbasin D.
   g. Combine three hydrographs at R.
   h. Route combined hydrograph through reservoir R.
   i. Route reservoir outflow hydrograph to outlet.
   j. Compute hydrograph for subbasin E.
   k. Combine two hydrographs at outlet.

Figure 24.1 Typical watershed subdivision and computation sequence for event-simulation models.

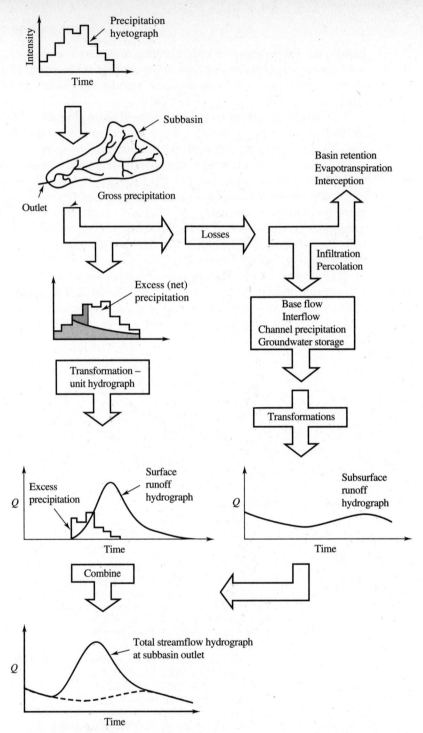

Figure 24.2 Typical lumped parameter event-simulation model of the rainfall–runoff process.

prescribed unit hydrograph to produce a surface runoff hydrograph for the subbasin. The abstracted losses are divided among the loss components on the basis of prescribed parameters. Subsurface flows and waters derived from groundwater storage are transformed into a subsurface runoff hydrograph, which when combined with the surface runoff hydrograph forms the total streamflow hydrograph at the subbasin outlet. This hydrograph can then be routed downstream, combined with another contributing hydrograph, or simply output if this subbasin is the only, or the final, subbasin being considered.

24.2 FEDERAL AGENCY SINGLE-EVENT MODELS

The rainfall–runoff processes depicted in Figs. 24.1 and 24.2 are recognized by most of the event simulation models named in Table 21.1. Specific computation techniques for losses, unit hydrographs, river routing, reservoir routing, and base flow are compared in Table 24.1 for five of the major federal agency rainfall–runoff event simulation models. All the models allow selection among available techniques. Brief descriptions of each of these models are followed by an illustrative example of an application of the HEC-1 model to a single storm occurring over a 250-mi^2 watershed near Lincoln, Nebraska.

U.S. Geological Survey Rainfall–Runoff Model

The USGS model can be used in evaluating short streamflow records and calculating peak flow rates for natural drainage basins.[1] The program monitors the daily moisture content of the subbasin soil and can be used as a continuous streamflow simulation model. The model is classified as an event simulation model because its calibration is based on short-term records of rainfall, evaporation, and discharges during a few documented floods. It has been modified several times and has evolved into the USGS urban continuous simulation model, DR3M, described in detail in Chapter 25.

Input to the model consists of initial estimates of 10 parameters, which are modified by the model through an optimization fitting procedure that matches simulated and recorded flow rates. Other input includes daily rainfall and evaporation, close-interval (5–60 min) rainfall and discharge data, drainage areas, impervious areas, and base flow rates for each flood.

Phillip's[2] infiltration equation is used to determine a rainfall excess hyetograph, which is translated to the subbasin outlet and then routed through a linear reservoir, using the time–area watershed routing technique described in Chapter 13.

The USGS rainfall–runoff model can be used to simulate streamflows for relatively short periods for small basins with approximately linear storage–outflow characteristics in regions where snowmelt or frozen ground is not significant. Output from the model includes a table showing peak discharges, storm runoff volumes, storm rainfall amounts, and an iteration-by-iteration printout of magnitudes of parameters and residuals in fitting volumes and peak flow rates.

TABLE 24.1   HYDROLOGY PROCESSES AND OPTIONS USED BY SEVERAL AGENCY RAINFALL–RUNOFF EVENT SIMULATIONS MODELS

Modeled components	Model code names (see Table 21.1)				
	HEC-1 (Corps)	TR-20 (SCS)	USGS (USGS)	HYMO (ARS)	SWMM (EPA)
Infiltration and losses					
Holtan's equation	X				
Horton's equation					X
Green–Ampt	X				X
Phillip's equation			X		
SCS curve number method	X	X		X	
Exponential loss rate	X				
Standard capacity curves					X
Unit hydrograph					
Input	X	X			
Clark's	X		X		
Snyder's	X				
Two-parameter gamma response				X	
SCS dimensionless unit hydrograph	X	X			
River routing					
Kinematic wave	X				X
Full dynamic wave					X
Muskingum	X				
Muskingum-Cunge	X				
Modified Puls	X				
Normal depth	X				
Variable storage coefficient	X			X	
Att-kin method		X			
Translation only			X		
Reservoir routing					
Storage-indication (Puls)	X	X		X	
Base flow					
Input	X		X		X
Constant value		X	X		X
Recession equation	X				
Snowmelt routine	Yes	No	No	No	Yes

Computer Program for Project Formulation Hydrology (TR-20)

A particularly powerful hydrologic process and water surface profile computer program was developed by CEIR, Inc.[3] and is known by the code name TR-20, which is an acronym for the U.S. Soil Conservation Service Technical Release Number 20. The model is a computer program of methods used by the Soil Conservation Service as presented in the *National Engineering Handbook*.[4]

The program is recognized as an engineer-oriented rather than computer-oriented package, having been developed with ease of use as a purpose. Input data sheets and output data are designed for ease in use and interpretation by field engineers, and the program contains a liberal number of operations that are user-accommodating, even at the expense of machine time.

The TR-20 was designed to use soil and land-use information to determine runoff hydrographs for known storms and to perform reservoir and channel routing of the generated hydrographs. It is a single-event model, with no provision for additional losses or infiltration between discrete storm events. The program has been used in all 50 states by engineers for flood insurance and flood hazard studies, for the design of reservoir and channel projects, and for urban and rural watershed planning.

Surface runoff is computed from an historical or synthetic storm using the SCS curve number approach described in Chapter 4 to abstract losses. The standard dimensionless hydrograph shown in Fig. 12.13 is used to develop unit hydrographs for each subarea in the watershed. The excess rainfall hyetograph is constructed using the effective rain and a given rainfall distribution and is then applied incrementally to the unit hydrograph to obtain the subarea runoff hydrograph for the storm.

As shown in Table 24.1, TR-20 uses the storage-indication method to route hydrographs through reservoirs (see Section 13.2). The base flow is added to the direct runoff hydrographs at any time to produce the total flow rates. The program uses the logic depicted in Fig. 24.1 by computing the total flow hydrographs, routing the flows through stream channels and reservoirs, combining the routed hydrographs with those from other tributaries, and routing the combined hydrographs to the watershed outlet. Prior to 1983, the model routed stream inflow hydrographs by the convex method (Section 13.1), which has since been replaced by the modified att-kin method (Section 13.3). As many as 200 channel reaches and 99 reservoirs or floodwater-retarding structures can be accommodated in any single application of the model. To add to this capability, the program allows the concurrent input of up to 9 different storms over the watershed area.

Subdivision of the watershed is facilitated by determining the locations of control points. Control points are defined as stream locations corresponding to cross-sectional data, reservoir sites, damage centers, diversion points, gauging stations, or tributary confluences where hydrograph data may be desired. Subarea data requirements include the drainage area, the time of concentration, the reach lengths, structure data as described in Section 21.1, and either routing coefficients for each reach or cross-sectional data along the channels. Whenever cross-sectional data are provided, the model calculates the water surface elevations in addition to the peak flow rates and time of occurrence at each section. Subarea sizes are dictated by the locations of control points. To provide routing and flood hazard information, it is necessary to define enough control points so that the hydraulic characteristics of the stream are defined between control sections. Applications with TR-20 normally incorporate control points spaced between a few hundred feet to 2 mi or more apart. The resulting subareas that contribute runoff to a control point are usually less than 5 mi^2. Common subarea sizes for structures are less than 25 mi^2 even though there is no limitation on reach length or subarea size within the program.

Minimal input data requirements to TR-20 include the watershed characteristics, at least one actual or synthetic storm including the depth, duration, and distribution; the discharge, capacity, and elevation data for each structure; and the routing coefficients or cross-sectional data for each reach. Input can be described according to the following outline:

1. Watershed characteristics.
    a. The area (in mi^2) contributing runoff to each reservoir and cross section.
    b. Runoff curve number CN for each subarea. (See Chapter 4.)
    c. The antecedent moisture condition associated with each subarea, coded as dry, normal, or wet.
    d. The time of concentration for each subarea (hr).
    e. The length of each channel routing reach and subarea mainstream.
2. Velocity-routing coefficient table.
    a. A table containing routing coefficients for a range of velocities (ft/sec). This table is contained within the program and need only be entered if the user desires different velocities.
3. Dimensionless hydrograph.
    a. This table is contained within the program and need only be entered if the user desires a different hydrograph.
4. Actual hydrograph.
    a. Actual hydrographs can be introduced at any point in the watershed. Hydrograph ordinates are read as discharge rates (cfs) spaced at equal time increments apart, up to a maximum of 300 entries.
    b. Base flow rates (cfs) can also be specified.
5. Base flow.
    a. In addition to the option of specifying the base flow rates associated with a hydrograph that was input, the base flow can be specified or modified at any other control point.
6. Storm data.
    a. Storms are numbered from 1 to 9 and are input as cumulative depths at equally spaced time increments.
    b. As an alternative to specifying cumulative depths at various time increments, a dimensionless storm can be input, and up to 9 storms can be synthesized by specifying each storm depth and duration.
7. Stream cross-sectional data.
    a. Up to 200 cross sections may be input for a single run. Cross-sectional data consist of up to 20 pairs of values of the discharge versus flow area.
    b. If cross-sectional data are provided, the routing coefficients are determined from them. In the absence of such data, the user must specify a routing coefficient for each reach.
8. Structure data.
    a. The reservoir data consist of up to 20 pairs of outflow discharge rates (cfs) versus storage (acre-ft).
    b. A maximum of 99 structures are allowed in a run.

The desired output from TR-20 must be specified by a set of input file control variables. Hydrographs at each control point for each storm can be printed by specifying the control point identification in the control cards. Any combination of the following items can be produced at each control point:

1. The peak discharge rate, time of peak, and peak water surface elevations.
2. The discharge rates in tabular form for the entire hydrograph.
3. Water surface elevations for the entire duration of runoff.
4. The volume of direct runoff, determined from the area under the hydrograph.
5. Hydrograph ordinates in any specified format.
6. Summary tables containing water balance information.

Basic data needed by the computer program are determined from field surveys. Rainfall frequency data are input from data in the U.S. Weather Bureau TP-40 report.[5] Peak-discharge and area-flooded information for present and future conditions for several return periods are output by TR-20 in a form suitable for direct use in an economic evaluation model.

Problem-Oriented Computer Language for Hydrologic Modeling (HYMO)

A unique computer language designed for use by hydrologists who have no conventional computer programming experience was developed by Williams and Hann.[6] Once the program has been compiled, the user forms a sequence of commands that synthesize, route, store, plot, or add hydrographs for subareas of any watershed. Seventeen commands are available to use in any sequence to transform rainfall data into runoff hydrographs and to route these hydrographs through streams and reservoirs. The HYMO model also computes the sediment yield in tons for individual storms on the watershed.

Watershed runoff hydrographs are computed by HYMO using unit-hydrograph techniques. Unit hydrographs can either be input or synthesized according to the dimensionless unit hydrograph shown in Fig. 24.3. Terms in the equations are

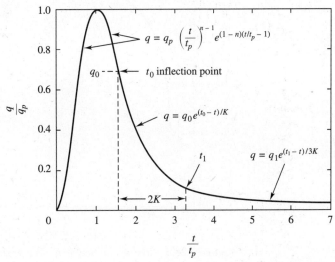

Figure 24.3 Dimensionless unit hydrograph used in HYMO. (After Williams and Hann.[6])

q = flow rate (ft^3/sec) at time t

q_p = peak flow rate (ft^3/sec)

t_p = time to peak (hr)

n = dimensionless shape parameter

q_0 = flow rate at the inflection point (cfs)

t_0 = time at the inflection point (hr)

K = recession constant (hr)

Once K and t_p and q_p are known, the entire hydrograph can be computed from the three segment equations shown in Fig. 24.3. The peak flow rate is computed by the equation

$$q_p = \frac{BAQ}{t_p} \tag{24.1}$$

where B = a watershed parameter, related to n as shown in Fig. 24.4.
  A = watershed area (mi^2)
  Q = volume of runoff (in.), determined by HYMO from the SCS rainfall–runoff equation described in Chapter 4

The duration of the unit hydrograph is equated with the selected time increment. The runoff Q for the unit hydrograph would of course be 1.0 in. The parameter n in Fig. 24.4 is obtained from Fig. 24.5. Parameters K and t_p for ungauged watersheds

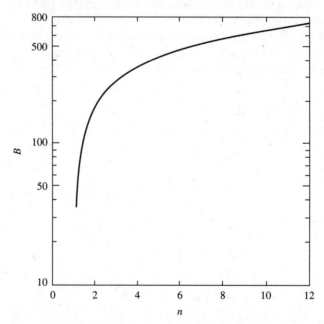

Figure 24.4 Relation between dimensionless shape parameter n and watershed Parameter B. (After Williams and Hann.[6])

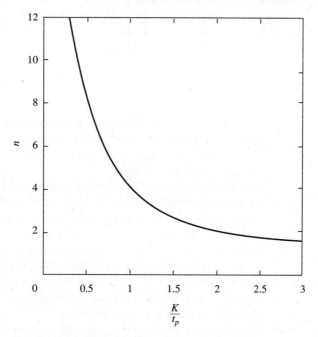

Figure 24.5 Relation between dimensionless shape parameter n and ratio of recession constant to time to peak. (After Williams and Hann.[6])

are determined from regional regression equations based on 34 watersheds located in Texas, Oklahoma, Arkansas, Louisiana, Mississippi, and Tennessee, ranging in size from 0.5 to 25 mi², or

$$K = 27.0A^{0.231}\text{SLP}^{-0.777}\left(\frac{L}{W}\right)^{0.124} \tag{24.2}$$

and

$$t_p = 4.63A^{0.422}\text{SLP}^{-0.46}\left(\frac{L}{W}\right)^{0.133} \tag{24.3}$$

where   SLP = the difference in elevation (ft), divided by floodplain distance (mi), between the basin outlet and the most distant point on the divide

L/W = the basin length/width ratio

River routing is accomplished in HYMO by a revised *variable storage coefficient* (VSC) method.[7] The continuity equation, $I - O = dS/dt$, and the storage equation, $S = KO$, are combined and discretized according to the methods outlined in Chapter 13. The VSC method recognizes the variability in K as the flow leaves the confines of the stream channel and inundates the floodplain and valley area. Relations between K and O are determined by HYMO from the input cross-sectional data, or HYMO will calculate the relation using Manning's equation if the floodplain and channel roughness coefficients are specified. The bed slope and reach length are also part of the required input.

The widely adopted storage-indication method (see Chapter 13) is used to route inflow hydrographs through reservoirs. The storage-outflow curve must be determined externally by the user and is input to the program as a table containing paired storages and outflows, using storage defined as zero whenever the outflow is zero.

The user-oriented commands and the data requirements for each command are as follows:

1. Start: the time rainfall begins on the watershed.
2. Store hydrograph: the time increment to be used, the lowest flow rate to be stored, the watershed area, and the successive flow rates spaced at the specified time increment.
3. Develop hydrograph: the desired time increment, the watershed area, the SCS runoff curve number CN, the watershed channel length and maximum difference in elevation, and the cumulative rainfall beginning with zero and accumulated at the end of each time increment until the end of the storm.
4. Compute the rating curve: cross-sectional identification number, number of points in the cross section, the maximum elevation of the cross section, the main channel and left and right floodplain slopes, Manning's *n* for each segment, and finally pairs of horizontal and vertical coordinates of the points describing the cross section.
5. Reach computations: the number of cross sections in the routing reach, the time increment to be used in routing, the reach length, and the discharge rates for which the variable storage coefficient is to be computed.
6. Print hydrograph: the identification number of the cross section at which hydrographs are to be printed.
7. Plot hydrograph: the identification number of the cross sections at which the hydrographs are to be plotted.
8. Add hydrographs: the identification numbers of the hydrographs to be added.
9. Route reservoir: the identification numbers of the locations of the outflow and inflow hydrographs, and the discharge-storage relation for the reservoir.
10. Compute travel time: the reach identification number, reach length, and reach slope.
11. Sediment yield: several factors describing soil erodibility, cropping management, erosion control practices, slope length, and slope gradient.

Output from HYMO includes the synthesized or user-provided unit hydrographs, the storm runoff hydrographs, the river- and reservoir-routed hydrographs, and the sediment yield for individual storms on each subwatershed. Hydrographs computed by HYMO compared closely with measured hydrographs from the 34 test watersheds.

Storm Water Management Model (SWMM)

The Environmental Protection Agency model, SWMM,[8] is listed in Table 21.1 in two locations corresponding to rainfall–runoff event simulation and urban runoff simulation. The model is primarily an urban runoff simulation model and is described in detail in Chapter 25.

Like most others, the SWMM model has undergone numerous modifications and improvements since its first release in 1972. The initial version[8] was a single-event model, and newer versions[9,10] allow its use in continuous modeling of urban storm water flows and water quality parameters. The latest release includes a new snowmelt routine, a new storm water storage and treatment package, a sediment scour and deposition routine, and a revised infiltration simulation.

SWMM's hydrograph and routing routines are hydraulic rather than hydrologic. A distributed parameter approach is used for subcatchments consisiting of single parking lots, city lots, and so on. Accumulated rainfall on these plots is first routed as overland flow to gutter or storm drain inlets, where it is then routed as open or closed channel flow to the receiving waters or to some type of treatment facility. Of the five event-simulation models compared in Table 24.1, the SWMM gives the greatest detail in simulation, but cannot be used in large rural watershed simulations.

Overland flow depths and flow rates are computed for each time step using Manning's equation along with the continuity equation. The water depth over a subcatchment will increase without inducing an outflow until the depth reaches a specified detention requirement. If and whenever the resulting depth over the sub-catchment, D_r, is larger than the specified detention requirement, D_d, an outflow rate is computed using a modified Manning's equation

$$V = \frac{1.49}{n}(D_r - D_d)^{2/3}S^{1/2} \tag{24.4}$$

and

$$Q_w = VW(D_r - D_d) \tag{24.5}$$

where V = the velocity
   n = Manning's coefficient
   S = the ground slope
   W = the width of the overland flow
   Q_w = the outflow discharge rate

After flow depths and rates from all subcatchments have been computed, they are combined along with the flow from the immediate upstream gutter to form the total flow in each successive gutter.

The gutter and pipe flows are routed by the Manning and continuity equations to any points of interest in the network where they are added to produce hydrograph ordinates for each time step in the routing process. The time step is advanced in increments until the runoff from the storm is no longer being produced. The parameters of the gutter shape, slope, and length must be supplied by the user. Manning's roughness coefficients for the pipes or channels must also be supplied and are available in most hydraulics textbooks.

Other input required for a typical simulation with the SWMM model include the following:

1. Watershed characteristics such as the infiltration parameters, percent impervious area, slope, area, detention storage depth, and Manning's coefficients for overland flow.
2. The rainfall hyetograph for the storm to be simulated.

3. The land-use data, average market values of dwellings in subareas, and populations of subareas.
4. Characteristics of gutters such as the gutter geometry, slope, roughness coefficients, maximum allowable depths, and linkages with other connecting inlets or gutters.
5. Street cleaning frequency.
6. Treatment devices selected and their sizes.
7. Indexes for costs of facilities.
8. Boundary conditions in the receiving waters.
9. Storage facilities, location, and volume.
10. Inlet characteristics such as surface elevations and invert elevations.
11. Characteristics of pipes such as type, geometry, slope, Manning's *n*, and downstream and upstream junction data.

HEC-1 Flood Hydrograph Package (HEC-1)

The U.S. Army Corps of Engineers Hydrologic Engineering Center developed a series of comprehensive computer programs as computational aids for consultants, universities, and federal, state, and local agencies (see Section 21.4). Programs for flood hydrograph computations, water surface profile computations, reservoir system analyses, monthly streamflow synthesis, and reservoir system operation for flood control comprise the series. The single-event flood hydrograph package, HEC-1, is described here.[11]

The HEC-1 model consists of a calling program and six subroutines. Two of these subroutines determine the optimal unit hydrograph, loss rate, or streamflow routing parameters by matching recorded and simulated hydrograph values. The other subroutines perform snowmelt computations, unit-hydrograph computations, hydrograph routing and combining computations, and hydrograph balancing computations. In addition to being capable of simulating the usual rainfall–runoff event processes, HEC-1 will also simulate multiple floods for multiple basin development plans and perform the economic analysis of flood damages by numerically integrating areas under damage–frequency curves for existing and postdevelopment conditions.

HEC-1 underwent revisions in the early 1970s and again in the 1980s. Several features were added (e.g., SCS curve number method, hydraulic routing), and a microcomputer version was developed in 1984. The 1985 release expanded earlier versions to include kinematic hydrograph routing, simulation of urban runoff, hydrograph analysis for flow over a dam or spillway, analysis of downstream impacts of dam failures, multistage pumping plants for interior drainage, and flood control system economics. The 1990 version of HEC-1, available for PCs or Harris minicomputers, incorporates yet other improvements. It adds report-quality graphic and table capability, storage and retrieval of data from other programs, and new hydrologic procedures including the popular Green and Ampt infiltration equation (Chapter 4) and the Muskingum-Cunge flood routing method (Chapter 13).

In addition to the unit-hydrograph techniques of the earlier versions, the modified HEC-1 allows hydrograph syntheses by kinematic-wave overland runoff techniques, similar to those developed for use in SWMM. The runoff can either be concentrated at the outlet of the subarea or uniformly added along the watercourse length through the subarea, distributing the inflow to the channel or gutter in linearly increasing amounts in the downstream direction. The 1990 version allows the use of the Muskingum-Cunge routing method in a land surface runoff calculation mode.

Precipitation can be directly input, or one of three synthetic storms (refer to Chapter 16) can be selected. A standard project storm (SPS) is available for large basins (over 10 mi^2) located east of 105° longitude, using procedures described in Corps of Engineers manuals. A 96-hr duration is synthesized, but the storm has a 6-hr peak during each day.

A second type of storm is the probable maximum precipitation (PMP), using estimates from National Weather Service hydrometeorologic reports available for different locations (Chapter 16 describes these). A minimum duration is 24 hr, and storms up to 96 hr long may be analyzed. The third method allows synthesis of any duration from 5 min to 10 days. The user need only specify the desired duration and depth, and the program balances the depth around the central portion of the duration using the blocked IDF method of Section 16.4.

The later versions of HEC-1 include all the precipitation loss, synthetic unit hydrograph, and routing functions developed for earlier versions. Additional loss methods include both the SCS curve number method and Holtan's loss rate equation (an exponential decay function).

Because of the popularity of SCS techniques, the HEC-1 now includes TR-20 procedures for losses and hydrograph synthesis. The duration of the SCS dimensionless unit hydrograph is interpreted by HEC-1 as approximately 0.2 times the time to peak, but not exceeding 0.25 times the time to peak (this converts to 0.29 times the lag time).

For routing through streams and reservoirs, the newest version of HEC-1 includes all previous methods, and additionally performs kinematic-wave channel routing for several standard geometric cross-section shapes.

In comparison to other event-simulation models, HEC-1 is relatively compact and still able to execute a variety of computational procedures in a single computer run. The model is applicable only to single-storm analysis because there is no provision for precipitation loss rate recovery during periods of little or no precipitation.

After dividing the watershed into subareas and routing reaches as shown in Fig. 24.6, the precipitation for a subarea can be determined by one of four methods: (1) nonrecording and/or recording precipitation station data, (2) basin mean precipitation, (3) standard project or probable maximum hypothetical precipitation distributions, or (4) synthetic balanced storm method using IDF data (Section 16.4). Either actual depths or net rain amounts may be input, depending on the user's choice of techniques for abstracting losses. The HEC-1 loss rate function is easily bypassed if the net rain is available for direct input.

The program logic for HEC-1 is shown in Fig. 24.6. Hydrologic processes such as the subarea runoff computation, routing computation, hydrograph combining, subtracting diverted flow, balancing, comparing, or summarizing are specified in the input using the sequence illustrated in Fig. 24.1.

One loss rate in the HEC-1 model is an exponential decay function that depends on the rainfall intensity and the antecedent losses. The instantaneous loss rate, in in./hr, is

$$L_t = K' P_t^E \tag{24.6}$$

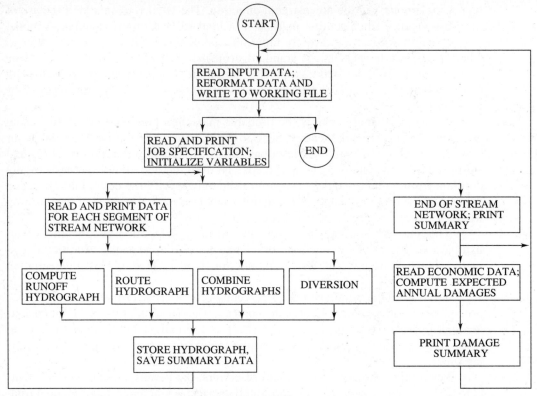

Figure 24.6 HEC-1 Program Operations Overview.[11]

where L_t = the instantaneous loss rate (in./hr)
P_t = intensity of the rain (in./hr)
E = the exponent of recession (range of 0.5–0.9)
K' = a coefficient, decreasing with time as losses accumulate

$$K' = K_0 C^{-\text{CUML}/10} + \Delta K \qquad (24.7)$$

where K_0 = the loss coefficient at the beginning of the storm (when CUML = 0), an average value of 0.6
CUML = the accumulated loss (in.) from the beginning of the storm to time t
C = a coefficient, an average of 3.0

If ΔK is zero, the loss rate coefficient K' becomes a parabolic function of the accumulated loss, CUML, and would thus plot as a straight-line function of CUML on semilogarithmic graph paper if K were plotted on the logarithmic scale. The straight-line relation is depicted in Fig. 24.7, showing the decrease in the loss rate coefficient as the losses accumulate during any storm. Because loss rates typically decrease much more rapidly during the initial minutes of a storm, the loss rate K' is increased above the straight-line amount by an amount equal to ΔK, which in turn is made a function of the amount of losses that will accumulate before the K' value is again equal to the

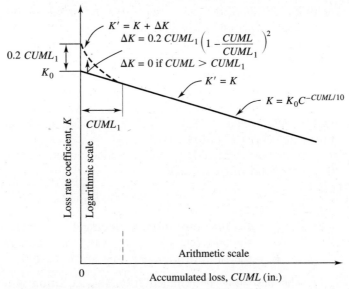

Figure 24.7 Variation of the loss rate coefficient K' with the accumulated loss amount CUML.

straight-line value, K. This initial accumulated loss, $CUML_1$, is user-specified and is related to ΔK in such a fashion that the initial loss rate K' is 20 percent times $CUML_1$ greater than K_0 (see Fig. 24.7). Initial loss coefficients K_0 are difficult to estimate, and standard curves in Chapters 4 and 23 are available to determine initial infiltration rates, L_0. For gauged basins, HEC-1 allows the user to input rainfall and runoff data from which the loss rate parameters are optimized to give a best fit to the information provided. Estimates of parameters for ungauged basins fall in the judgment realm noted in Table 21.1. An alternative to the described loss rate function is available in HEC-1, which is an initial abstraction followed by a constant loss rate, similar to a ϕ index.

The HEC-1 model provides separate computations of snowmelt in up to 10 elevation zones. The precipitation in any zone is considered to be snow if the zone temperature is less than a base temperature, usually 32°F, plus 2°. The snowmelt is computed by the degree-day or energy budget methods whenever the temperature is equal to or greater than the base temperature. The elevation zones are usually considered in increments of 1000 ft although any equal increments can be used.

Unit hydrographs for each subarea can be provided by the user, or Clark's method[13] of synthesizing an instantaneous unit hydrograph (IUH) can be used. Clark's method is more commonly recognized as the time–area curve method of hydrograph synthesis described in Section 12.6. The time–area histogram, determined from an isochronal map of the watershed, is convoluted with a unit design-storm hyetograph using Eq. 12.38, as illustrated in Fig. 12.19. The methods described in Section 13.2 are then used to route the resulting hydrograph through linear reservoir storage using Eq. 12.35 with a watershed storage coefficient K. Input data for Clark's method consists of the time–area curve ordinates, the time of concentration for the Clark unit graph, and the watershed storage coefficient K. If the time–area curve for the watershed under consideration is not available, the model provides a synthetic time–area curve at the user's request.

Because the Corps of Engineers commonly uses Snyder's method (Chapter 12) in unit-hydrograph synthesis for large basins, the Snyder time lag from Eq. 11.5 and Snyder's peaking coefficient C_p from Eq. 12.17 can be input, and Clark's parameters will be determined by HEC-1 from the Snyder coefficients. The actual or synthetic time–area curve is still required.

Base flow is treated by HEC-1 as an exponential recession using an exponent of 0.1 in the following equation:

$$Q_2 = \frac{Q_1}{R^{0.1}} \tag{24.8}$$

where Q_1 = the flow rate at the beginning of the time increment
Q_2 = the flow rate at the end of the time increment
R = the ratio of the base flow to the base flow 10 time increments later

The base flow determined from this equation is added to the direct runoff hydrograph ordinates determined from unit-hydrograph techniques. The starting point for the entire computation is the user-prescribed base flow rate at the beginning of the simulation, which is normally the flow several time increments prior to any direct runoff. If the initial base flow rate is specified as zero, the computer program output contains only direct runoff rates.

The HEC-1 package allows the user a choice of several hydrologic or "storage-routing" techniques for routing floods through river reaches and reservoirs. All use the continuity equation and some form of the storage–outflow relation; some are described in more detail in Chapter 13. The five routing procedures included in the program are the following:

1. **Modified Puls**—this method is also called the storage or storage-indication method and is a level-pool-routing technique normally reserved for use with reservoirs or flat streams. The technique was described in detail in Section 13.2.
2. **Muskingum**—described in detail in Section 13.1.
3. **Muskingum-Cunge**—a blended hydrologic and hydraulic routing method detailed in Section 13.1.
4. **Kinematic wave**—described in Section 13.3.
5. **Straddle-stagger**—also known as the progressive average lag method. The technique simply averages a subset of consecutive inflow rates, and the averaged inflow value is lagged a specified number of time increments to form the outflow rate.

Input to HEC-1 is facilitated by arranging three categories of data in a sequence compatible with the desired computation sequence, summarized in Table 24.2. The individual input records are preceded by a two-character code. The first character

TABLE 24.2 SUBDIVISIONS OF INPUT DATA FOR HEC-1[11]

Job control	Hydrology and hydraulics	Economics and end of job
I _, Job initialization	K_, Job step control	E_, etc., Economics, data
V_, Variable output summary	H_, Hydrograph transformation	ZZ, End of Job
O_, Optimization	Q_, Hydrograph data	
J _, Job type	B_, Basin data	
	P_, Precipitation data	
	L_, Loss (infiltration) data	
	U_, Unit graph data	
	M_, Melt data	
	R_, Routing data	
	S_, Storage data	
	D_, Diversion data	
	W_, Pump withdrawal data	

indicates the type of hydrologic or program operation (P_ for precipitation, U_ for unit hydrograph, etc.), and the second character is reserved for the suboption for the operation (PM for probable maximum storm, PI for IDF balanced storm, etc.).

Output from HEC-1 is both complete and descriptive. Options are available for graphical and tabular report-quality displays of intermediate or summary hydrographs or precipitation hyetographs. The extent of output from HEC-1 is further illustrated in Example 24.1.

EXAMPLE 24.1

In June 1963 the Oak Creek watershed shown in Fig. 24.8 experienced a severe flood-producing storm in a 6-hr period. Average excess rain depths over each of the nine subareas $A, B, \ldots, I$ ranged from 1.0 in. in the southwest to 7.8 in. in the north. Crest-stage flood records show that the storm produced peak flows of 27,500 cfs at Agnew (Point 3) and 21,600 cfs at the watershed outlet (Point 8). The map in Fig. 24.8 gives the subbasin divides, reach lengths, and stream bed elevations (underlined).

A reservoir located at Point 9 will store runoff from virtually any probable storm over Subarea I. For the remaining elongated watershed area A–H within the boldface border, use the June storm to simulate hydrographs at each of Points 1–8 using a single run of HEC-1 and compare peak flows with recorded values at Points 3 and 8.

Solution. The net storm depths shown in the table of Fig. 24.8 are applied uniformly over each subarea. The records provide the time distribution for the 12 successive half-hour periods of the thunderstorm in percent: 3, 4, 5, 6, 9, 37, 10, 8, 6, 4, 5, 3, and 4 percent during the 12 successive half-hour periods of the thunderstorm. The net rain was determined from the measured depths using the SCS runoff equation[4] and a basin-wide composite curve number of 73 (see Chapter 4).

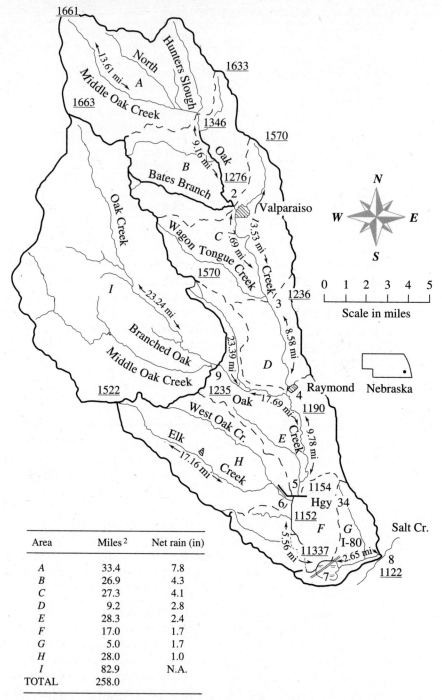

Area	Miles ^{2}	Net rain (in)
A	33.4	7.8
B	26.9	4.3
C	27.3	4.1
D	9.2	2.8
E	28.3	2.4
F	17.0	1.7
G	5.0	1.7
H	28.0	1.0
I	82.9	N.A.
TOTAL	258.0	

Figure 24.8 Oak Creek watershed subarea map and data sheet, June 1963.

The computation logic to simulate runoff for this storm consists of the following 22 steps:

1. Compute the hydrograph for Area *A* at Point 1.
2. Route the *A* hydrograph from Point 1 to Point 2.
3. Compute the hydrograph for Area *B* at Point 2.
4. Combine the two hydrographs at Point 2.
5. Route the combined hydrograph to Point 3.
6. Compute the hydrograph for Area *C* at Point 3.
7. Combine the two hydrographs at Point 3.
8. Route the combined hydrograph to Point 4.
9. Compute the hydrograph for Area *D* at Point 4.
10. Combine the two hydrographs at Point 4.
11. Route the combined hydrograph to Point 5.
12. Compute the hydrograph for Area *E* at Point 5.
13. Combine the two hydrographs at Point 5.
14. Route the combined hydrograph to Point 6.
15. Compute the hydrograph for Area *H* at Point 6.
16. Combine the two hydrographs at Point 6.
17. Route the combined hydrograph to Point 7.
18. Compute the hydrograph for Area *F* at Point 7.
19. Combine the two hydrographs at Point 7.
20. Route the combined hydrograph to Point 8.
21. Compute the hydrograph for Area *G* at Point 8.
22. Combine the two hydrographs at Point 8.

Runoff hydrographs for subareas are simulated by convoluting the net storm hyetograph with unit hydrographs synthesized by Clark's method using Snyder's coefficients (see Chapter 12). A C_p value of 0.8 is applied for Oak Creek because of the moderately high retention capacity of the watershed. Subarea time lag values for each subarea are found from Eq. (11.5) using a C_t value of 2.0.

Hydrograph stream routing is performed using the Muskingum technique (Chapter 13) with $x = 0.15$ and $K =$ the approximate reach travel time, using length divided by the average velocity. A Chézy average velocity determined as 100 times the square root of the average reach slope is used. If K exceeds three routing increments, the reach is further subdivided by HEC-1 into shorter lengths to ensure computational resolution.

A sample of the input and output for this job is shown as Table 24.3. Each of the 22 computational steps are separated in sequence. Only Steps 1–5 are included in the sample. Note that the HEC-1 loss rate function was not used so that the end-of-period excess and rain depths are equal. Note also that hydrograph routing of the *A* hydrograph from Point 1 to Point 2 was facilitated using three equal reach lengths each with a *K*-value (AMSKK) of 1.2 hr.

A summary of HEC-1 peak and time-averaged flow rates for each of Steps 1–22 is given in Table 24.4. Note that the simulated peak at Point 3 is 27,539 cfs, which agrees very well with the recorded value of 27,500. The corresponding simulated and observed peak flows at Point 8 are 22,453 and 21,600 cfs, respectively. ■■

TABLE 24.3 HEC-1 INPUT AND PARTIAL OUTPUT LISTING FOR OAK CREEK PROBLEM IN EXAMPLE 24.1

```
*****************************************          *****************************************
*   FLOOD HYDROGRAPH PACKAGE (HEC-1)    *          *    U.S. ARMY CORPS OF ENGINEERS       *
*          SEPTEMBER 1990               *          *    HYDROLOGIC ENGINEERING CENTER      *
*          VERSION 4.0                  *          *         609 SECOND STREET             *
*                                       *          *    DAVIS, CALIFORNIA 95616            *
*   RUN DATE 02/07/1994   TIME 12:10:01 *          *         (916) 756-1104               *
*****************************************          *****************************************
```

PARTIAL HEC-1 INPUT FOR UPPER THIRD OF WATERSHED

```
 1   ID
 2   ID                   OAK CREEK WATERSHED STUDY
 3   ID                   TEST HEC1 USING STORM OF RECORD ON JUNE 23-24, 1963
                          USE ACTUAL RAIN, SHYDER'S CP = 0.8, VEL = 100 ROOT S, AVG LAG EQ, SNYD, X=.15

 4   IT   30  23 JUN 63  1400   100
 5   IO    0

 6   KK    1                                                        COMPUTE HYDROGRAPH FOR AREA A
 7   BA   33.4
 8   PB    7.8
 9   PI   .234  .390  .468  .702  2.886  .780  .624  .468  .312  .390
10   PI   .234  .312
11   US   2.93

12   KK    2                                                        ROUTE HYDROGRAPH FROM AREA A TO POINT 2
13   RM    3   3.6  0.15

14   KK    2                                                        COMPUTE HYDROGRAPH FOR AREA B
15   BA   26.9
16   PB    4.3
17   PI   .129  .215  .258  .387  1.591  .430  .344  .258  .172  .215
18   PI   .129  .172
19   US   2.51

20   KK    2                                                        ADD HYDROGRAPHS AT 2
21   HC    2

22   KK    3                                                        ROUTE COMBINED HYDROGRAPH TO POINT 3
23   RM    3   3.6  0.15

24   KK    3                                                        COMPUTE HYDROGRAPH FOR AREA C
25   BA   27.3
26   PB    4.1
27   PI   .123  .205  .246  .369  1.517  .410  .328  .246  .164  .205
28   PI   .123  .164
29   US   2.76

30   KK    3                                                        ADD HYDROGRAPHS AT 3
31   HC    2
```

TABLE 24.3 (CONTINUED)

```
***************
6 KK    *  *  *     1     *  *  *          COMPUTE HYDROGRAPH FOR AREA A
        ***************

7 BA      SUBBASIN RUNOFF DATA
          SUBBASIN CHARACTERISTICS
               TAREA    33.40    SUBBASIN AREA
          PRECIPITATION DATA
8 PB           STORM     7.80    BASIN TOTAL PRECIPITATION
9 PI      INCREMENTAL PRECIPITATION PATTERN
               .23   .39   .47   .70   2.89   .78   .62   .47   .31   .39
               .23   .31

11 US     SNYDER UNITGRAPH
               TP    2.93    LAG
               CP     .80    PEAKING COEFFICIENT
          SYNTHETIC ACCUMULATED-AREA VS. TIME CURVE WILL BE USED
                                   ***
```

APPROXIMATE CLARK COEFFICIENTS FROM GIVEN SNYDER CP AND TP ARE TC= 3.74 AND R= 1.46 INTERVALS

```
                         UNIT HYDROGRAPH PARAMETERS
               CLARK    TC= 3.74 HR,        R= 1.46 HR
               SNYDER   TP= 2.92 HR,        CP= .79

                              UNIT HYDROGRAPH
                         20 END-OF-PERIOD ORDINATES

     437.   1546.   2928.   4319.   5397.   5864.   5704.   4815.   3546.   2507.
    1772.   1253.    885.    826.    442.    313.    221.    156.    110.     78.
```

TABLE 24.3 (CONTINUED)

HYDROGRAPH AT STATION   1

DA	MON	HRMN	ORD	RAIN	LOSS	EXCESS	COMP Q		DA	MON	HRMN	ORD	RAIN	LOSS	EXCESS	COMP Q
23	JUN	1400	1	.00	.00	.00	0.		24	JUN	1500	51	.00	.00	.00	0.
23	JUN	1430	2	.23	.00	.23	102.		24	JUN	1530	52	.00	.00	.00	0.
23	JUN	1500	3	.39	.00	.39	532.		24	JUN	1600	53	.00	.00	.00	0.
23	JUN	1530	4	.47	.00	.47	1493.		24	JUN	1630	54	.00	.00	.00	0.
23	JUN	1600	5	.70	.00	.70	3183.		24	JUN	1700	55	.00	.00	.00	0.
23	JUN	1630	6	2.89	.00	2.89	6665.		24	JUN	1730	56	.00	.00	.00	0.
23	JUN	1700	7	.78	.00	.78	12356.		24	JUN	1800	57	.00	.00	.00	0.
23	JUN	1730	8	.62	.00	.62	19107.		24	JUN	1830	58	.00	.00	.00	0.
23	JUN	1800	9	.47	.00	.47	25802.		24	JUN	1900	59	.00	.00	.00	0.
23	JUN	1830	10	.31	.00	.31	31125.		24	JUN	1930	60	.00	.00	.00	0.
23	JUN	1900	11	.39	.00	.39	34078.		24	JUN	2000	61	.00	.00	.00	0.
23	JUN	1930	12	.23	.00	.23	34475.		24	JUN	2030	62	.00	.00	.00	0.
23	JUN	2000	13	.31	.00	.31	32163.		24	JUN	2100	63	.00	.00	.00	0.
23	JUN	2030	14	.00	.00	.00	28114.		24	JUN	2130	64	.00	.00	.00	0.
23	JUN	2100	15	.00	.00	.00	23855.		24	JUN	2200	65	.00	.00	.00	0.
23	JUN	2130	16	.00	.00	.00	19854.		24	JUN	2230	66	.00	.00	.00	0.
23	JUN	2200	17	.00	.00	.00	16164.		24	JUN	2300	67	.00	.00	.00	0.
23	JUN	2230	18	.00	.00	.00	12780.		24	JUN	2330	68	.00	.00	.00	0.
23	JUN	2300	19	.00	.00	.00	9759.		25	JUN	0000	69	.00	.00	.00	0.
23	JUN	2330	20	.00	.00	.00	7176.		25	JUN	0030	70	.00	.00	.00	0.
24	JUN	0000	21	.00	.00	.00	5117.		25	JUN	0100	71	.00	.00	.00	0.
24	JUN	0030	22	.00	.00	.00	3605.		25	JUN	0130	72	.00	.00	.00	0.
24	JUN	0100	23	.00	.00	.00	2527.		25	JUN	0200	73	.00	.00	.00	0.
24	JUN	0130	24	.00	.00	.00	1760.		25	JUN	0230	74	.00	.00	.00	0.
24	JUN	0200	25	.00	.00	.00	1206.		25	JUN	0300	75	.00	.00	.00	0.
24	JUN	0230	26	.00	.00	.00	693.		25	JUN	0330	76	.00	.00	.00	0.
24	JUN	0300	27	.00	.00	.00	447.		25	JUN	0400	77	.00	.00	.00	0.
24	JUN	0330	28	.00	.00	.00	281.		25	JUN	0430	78	.00	.00	.00	0.
24	JUN	0400	29	.00	.00	.00	173.		25	JUN	0500	79	.00	.00	.00	0.
24	JUN	0430	30	.00	.00	.00	105.		25	JUN	0530	80	.00	.00	.00	0.
24	JUN	0500	31	.00	.00	.00	53.		25	JUN	0600	81	.00	.00	.00	0.
24	JUN	0530	32	.00	.00	.00	24.		25	JUN	0630	82	.00	.00	.00	0.
24	JUN	0600	33	.00	.00	.00	0.		25	JUN	0700	83	.00	.00	.00	0.

TOTAL RAINFALL =   7.80,   TOTAL LOSS =   .00,   TOTAL EXCESS =   7.80

PEAK FLOW (CFS)	TIME (HR)
+ 34475.	5.50

MAXIMUM AVERAGE FLOW

	6-HR	24-HR	72-HR	49.50-HR
+ (CFS)	24048.	6974.	3382.	3382.
(INCHES)	6.694	7.766	7.766	7.766
(AC-FT)	11925.	13834.	13834.	13834.

TABLE 24.3 (CONTINUED)

```
12 KK      *      2   *           ROUTE HYDROGRAPH FROM AREA A TO POINT 2

13 RM          HYDROGRAPH ROUTING DATA
               MUSKINGUM ROUTING
                   NSTPS      3    NUMBER OF SUBREACHES
                   AMSKK   3.60    MUSKINGUM K
                       X    .15    MUSKINGUM X
```

HYDROGRAPH AT STATION   2

DA	MON	HRMN	ORD	FLOW
23	JUN	1400	1	0.
23	JUN	1430	2	0.
23	JUN	1500	3	0.
23	JUN	1530	4	5.
23	JUN	1600	5	27.
23	JUN	1630	6	104.
23	JUN	1700	7	302.
23	JUN	1730	8	731.
23	JUN	1800	9	1576.
23	JUN	1830	10	3055.
23	JUN	1900	11	5313.
23	JUN	1930	12	8350.
23	JUN	2000	13	11974.
23	JUN	2030	14	15836.
23	JUN	2100	15	19499.
23	JUN	2130	16	22529.
23	JUN	2200	17	24612.
23	JUN	2230	18	25622.
23	JUN	2300	19	25601.
23	JUN	2330	20	24696.
24	JUN	0000	21	23101.
24	JUN	0030	22	21021.
24	JUN	0100	23	18651.
24	JUN	0130	24	16166.
24	JUN	0200	25	13714.

DA	MON	HRMN	ORD	FLOW
24	JUN	0230	26	11409.
24	JUN	0300	27	9324.
24	JUN	0330	28	7496.
24	JUN	0400	29	5931.
24	JUN	0430	30	4622.
24	JUN	0500	31	3552.
24	JUN	0530	32	2694.
24	JUN	0600	33	2019.
24	JUN	0630	34	1495.
24	JUN	0700	35	1094.
24	JUN	0730	36	791.
24	JUN	0800	37	566.
24	JUN	0830	38	401.
24	JUN	0900	39	281.
24	JUN	0930	40	195.
24	JUN	1000	41	135.
24	JUN	1030	42	92.
24	JUN	1100	43	63.
24	JUN	1130	44	43.
24	JUN	1200	45	29.
24	JUN	1230	46	19.
24	JUN	1300	47	13.
24	JUN	1330	48	8.
24	JUN	1400	49	6.
24	JUN	1430	50	4.

DA	MON	HRMN	ORD	FLOW
24	JUN	1500	51	2.
24	JUN	1530	52	2.
24	JUN	1600	53	1.
24	JUN	1630	54	1.
24	JUN	1700	55	0.
24	JUN	1730	56	0.
24	JUN	1800	57	0.
24	JUN	1830	58	0.
24	JUN	1900	59	0.
24	JUN	1930	60	0.
24	JUN	2000	61	0.
24	JUN	2030	62	0.
24	JUN	2100	63	0.
24	JUN	2130	64	0.
24	JUN	2200	65	0.
24	JUN	2230	66	0.
24	JUN	2300	67	0.
24	JUN	2330	68	0.
25	JUN	0000	69	0.
25	JUN	0030	70	0.
25	JUN	0100	71	0.
25	JUN	0130	72	0.
25	JUN	0200	73	0.
25	JUN	0230	74	0.
25	JUN	0300	75	0.

DA	MON	HRMN	ORD	FLOW
25	JUN	0330	76	0.
25	JUN	0400	77	0.
25	JUN	0430	78	0.
25	JUN	0500	79	0.
25	JUN	0530	80	0.
25	JUN	0600	81	0.
25	JUN	0630	82	0.
25	JUN	0700	83	0.
25	JUN	0730	84	0.
25	JUN	0800	85	0.
25	JUN	0830	86	0.
25	JUN	0900	87	0.
25	JUN	0930	88	0.
25	JUN	1000	89	0.
25	JUN	1030	90	0.
25	JUN	1100	91	0.
25	JUN	1130	92	0.
25	JUN	1200	93	0.
25	JUN	1230	94	0.
25	JUN	1300	95	0.
25	JUN	1330	96	0.
25	JUN	1400	97	0.
25	JUN	1430	98	0.
25	JUN	1500	99	0.
25	JUN	1530	100	0.

PEAK FLOW (CFS)	TIME (HR)
25622.	8.50

MAXIMUM AVERAGE FLOW

	6-HR	24-HR	72-HR	49.50-HR
(CFS)	20848.	6974.	3382.	3382.
(INCHES)	5.803	7.766	7.766	7.766
(AC-FT)	10338.	13833.	13834.	13834.

TABLE 24.3 (CONTINUED)

```
**********
 *      *
 *   2  *
 *      *
**********
```

14 KK   COMPUTE HYDROGRAPH FOR AREA B

SUBBASIN RUNOFF DATA

15 BA   SUBBASIN CHARACTERISTICS

                 TAREA   26.90   SUBBASIN AREA

         PRECIPITATION DATA

16 PB   STORM   4.30   BASIN TOTAL PRECIPITATION

17 PI   INCREMENTAL PRECIPITATION PATTERN

         .13   .21   .26   .39   1.59   .43   .34   .26   .17   .21
         .13   .17

19 US   SNYDER UNITGRAPH

                 TP   2.51   LAG
                 CP   .80   PEAKING COEFFICIENT

         SYNTHETIC ACCUMULATED-AREA VS. TIME CURVE WILL BE USED

***

APPROXIMATE CLARK COEFFICIENTS FROM GIVEN SNYDER CP AND TP ARE TC= 3.40 AND R= 1.06 INTERVALS

                          UNIT HYDROGRAPH PARAMETERS

           CLARK   TC= 3.40 HR,   R= 1.06 HR
           SNYDER   TP= 2.49 HR,   CP= .79

                          UNIT HYDROGRAPH
                     16 END-OF-PERIOD ORDINATES

     528.   1820.   3343.   4722.   5523.   5512.   4681.   3271.   2026.   1254.
     777.   481.   298.   184.   114.   71.
```
```

TABLE 24.3 (CONTINUED)

HYDROGRAPH AT STATION 2

DA	MON	HRMN	ORD	RAIN	LOSS	EXCESS	COMP Q	*
23	JUN	1400	1	.00	.00	.00	0.	*
23	JUN	1430	2	.13	.00	.13	68.	*
23	JUN	1500	3	.22	.00	.22	348.	*
23	JUN	1530	4	.26	.00	.26	959.	*
23	JUN	1600	5	.39	.00	.39	2002.	*
23	JUN	1630	6	1.59	.00	1.59	4135.	*
23	JUN	1700	7	.43	.00	.43	7534.	*
23	JUN	1730	8	.34	.00	.34	11324.	*
23	JUN	1800	9	.26	.00	.26	14700.	*
23	JUN	1830	10	.17	.00	.17	16833.	*
23	JUN	1900	11	.21	.00	.21	17310.	*
23	JUN	1930	12	.13	.00	.13	16128.	*
23	JUN	2000	13	.17	.00	.17	13732.	*
23	JUN	2030	14	.00	.00	.00	11199.	*
23	JUN	2100	15	.00	.00	.00	9031.	*
23	JUN	2130	16	.00	.00	.00	7148.	*
23	JUN	2200	17	.00	.00	.00	5479.	*
23	JUN	2230	18	.00	.00	.00	3991.	*
23	JUN	2300	19	.00	.00	.00	2728.	*
23	JUN	2330	20	.00	.00	.00	1742.	*
24	JUN	0000	21	.00	.00	.00	1062.	*
24	JUN	0030	22	.00	.00	.00	588.	*
24	JUN	0100	23	.00	.00	.00	345.	*
24	JUN	0130	24	.00	.00	.00	199.	*
24	JUN	0200	25	.00	.00	.00	112.	*
24	JUN	0230	26	.00	.00	.00	62.	*
24	JUN	0300	27	.00	.00	.00	29.	*
24	JUN	0330	28	.00	.00	.00	12.	*
24	JUN	0400	29	.00	.00	.00	0.	*
24	JUN	0430	30	.00	.00	.00	0.	*
24	JUN	0500	31	.00	.00	.00	0.	*
24	JUN	0530	32	.00	.00	.00	0.	*
24	JUN	0600	33	.00	.00	.00	0.	*

DA	MON	HRMN	ORD	RAIN	LOSS	EXCESS	COMP Q
24	JUN	1500	51	.00	.00	.00	0.
24	JUN	1530	52	.00	.00	.00	0.
24	JUN	1600	53	.00	.00	.00	0.
24	JUN	1630	54	.00	.00	.00	0.
24	JUN	1700	55	.00	.00	.00	0.
24	JUN	1730	56	.00	.00	.00	0.
24	JUN	1800	57	.00	.00	.00	0.
24	JUN	1830	58	.00	.00	.00	0.
24	JUN	1900	59	.00	.00	.00	0.
24	JUN	1930	60	.00	.00	.00	0.
24	JUN	2000	61	.00	.00	.00	0.
24	JUN	2030	62	.00	.00	.00	0.
24	JUN	2100	63	.00	.00	.00	0.
24	JUN	2130	64	.00	.00	.00	0.
24	JUN	2200	65	.00	.00	.00	0.
24	JUN	2230	66	.00	.00	.00	0.
24	JUN	2300	67	.00	.00	.00	0.
24	JUN	2330	68	.00	.00	.00	0.
25	JUN	0000	69	.00	.00	.00	0.
25	JUN	0030	70	.00	.00	.00	0.
25	JUN	0100	71	.00	.00	.00	0.
25	JUN	0130	72	.00	.00	.00	0.
25	JUN	0200	73	.00	.00	.00	0.
25	JUN	0230	74	.00	.00	.00	0.
25	JUN	0300	75	.00	.00	.00	0.
25	JUN	0330	76	.00	.00	.00	0.
25	JUN	0400	77	.00	.00	.00	0.
25	JUN	0430	78	.00	.00	.00	0.
25	JUN	0500	79	.00	.00	.00	0.
25	JUN	0530	80	.00	.00	.00	0.
25	JUN	0600	81	.00	.00	.00	0.
25	JUN	0630	82	.00	.00	.00	0.
25	JUN	0700	83	.00	.00	.00	0.

TOTAL RAINFALL = 4.30, TOTAL LOSS = .00, TOTAL EXCESS = 4.30

PEAK FLOW	TIME		MAXIMUM AVERAGE FLOW			
(CFS)	(HR)		6-HR	24-HR	72-HR	49.50-HR
17310.	5.00	(CFS)	11207.	3100.	1503.	1503.
		(INCHES)	3.873	4.286	4.286	4.286
		(AC-FT)	5557.	6149.	6149.	6149.

TABLE 24.3 (CONTINUED)

```
************
20  KK  *  2  *
        ************
```

```
21  HC     HYDROGRAPH COMBINATION
    ICOMP   2  NUMBER OF HYDROGRAPHS TO COMBINE
```

ADD HYDROGRAPHS AT 2

HYDROGRAPH AT STATION 2
SUM OF 2 HYDROGRAPHS

DA	MON	HRMN	ORD	FLOW	DA	MON	HRMN	ORD	FLOW	DA	MON	HRMN	ORD	FLOW	DA	MON	HRMN	ORD	FLOW.
23	JUN	1400	1	0.	24	JUN	0230	26	11471.	24	JUN	1500	51	2.	25	JUN	0330	76	0.
23	JUN	1430	2	68.	24	JUN	0300	27	9353.	24	JUN	1530	52	2.	25	JUN	0400	77	0.
23	JUN	1500	3	349.	24	JUN	0330	28	7508.	24	JUN	1600	53	1.	25	JUN	0430	78	0.
23	JUN	1530	4	963.	24	JUN	0400	29	5931.	24	JUN	1630	54	1.	25	JUN	0500	79	0.
23	JUN	1600	5	2029.	24	JUN	0430	30	4622.	24	JUN	1700	55	0.	25	JUN	0530	80	0.
23	JUN	1630	6	4239.	24	JUN	0500	31	3552.	24	JUN	1730	56	0.	25	JUN	0600	81	0.
23	JUN	1700	7	7835.	24	JUN	0530	32	2694.	24	JUN	1800	57	0.	25	JUN	0630	82	0.
23	JUN	1730	8	12055.	24	JUN	0600	33	2019.	24	JUN	1830	58	0.	25	JUN	0700	83	0.
23	JUN	1800	9	16277.	24	JUN	0630	34	1495.	24	JUN	1900	59	0.	25	JUN	0730	84	0.
23	JUN	1830	10	19888.	24	JUN	0700	35	1094.	24	JUN	1930	60	0.	25	JUN	0800	85	0.
23	JUN	1900	11	22624.	24	JUN	0730	36	791.	24	JUN	2000	61	0.	25	JUN	0830	86	0.
23	JUN	1930	12	24478.	24	JUN	0800	37	566.	24	JUN	2030	62	0.	25	JUN	0900	87	0.
23	JUN	2000	13	25706.	24	JUN	0830	38	401.	24	JUN	2100	63	0.	25	JUN	0930	88	0.
23	JUN	2030	14	27035.	24	JUN	0900	39	281.	24	JUN	2130	64	0.	25	JUN	1000	89	0.
23	JUN	2100	15	28530.	24	JUN	0930	40	195.	24	JUN	2200	65	0.	25	JUN	1030	90	0.
23	JUN	2130	16	29677.	24	JUN	1000	41	135.	24	JUN	2230	66	0.	25	JUN	1100	91	0.
23	JUN	2200	17	30092.	24	JUN	1030	42	92.	24	JUN	2300	67	0.	25	JUN	1130	92	0.
23	JUN	2230	18	29613.	24	JUN	1100	43	63.	24	JUN	2330	68	0.	25	JUN	1200	93	0.
23	JUN	2300	19	28328.	24	JUN	1130	44	43.	25	JUN	0000	69	0.	25	JUN	1230	94	0.
23	JUN	2330	20	26438.	24	JUN	1200	45	29.	25	JUN	0030	70	0.	25	JUN	1300	95	0.
24	JUN	0000	21	24163.	24	JUN	1230	46	19.	25	JUN	0100	71	0.	25	JUN	1330	96	0.
24	JUN	0030	22	21609.	24	JUN	1300	47	13.	25	JUN	0130	72	0.	25	JUN	1400	97	0.
24	JUN	0100	23	18996.	24	JUN	1330	48	8.	25	JUN	0200	73	0.	25	JUN	1430	98	0.
24	JUN	0130	24	16364.	24	JUN	1400	49	6.	25	JUN	0230	74	0.	25	JUN	1500	99	0.
24	JUN	0200	25	13826.	24	JUN	1430	50	4.	25	JUN	0300	75	0.	25	JUN	1530	100	0.

	PEAK FLOW	TIME
+	(CFS)	(HR)
+	30092.	8.00

MAXIMUM AVERAGE FLOW

	6-HR	24-HR	72-HR	49.50-HR
(CFS)	26453.	10074.	4885.	4885.
(INCHES)	4.079	6.213	6.213	6.213
(AC-FT)	13117.	19982.	19982.	19982.

TABLE 24.3 (CONTINUED)

```
22 KK   *   3   *    ROUTE COMBINED HYDROGRAPH TO POINT 3
23 RM       HYDROGRAPH ROUTING DATA
            MUSKINGUM ROUTING
            NSTPS    3   NUMBER OF SUBREACHES
            AMSKK  3.60  MUSKINGUM K
            X       .15  MUSKINGUM X
```

HYDROGRAPH AT STATION   3

DA	MON	HRMN	ORD	FLOW	DA	MON	HRMN	ORD	FLOW	DA	MON	HRMN	ORD	FLOW	DA	MON	HRMN	ORD	FLOW
23	JUN	1400	1	0.	24	JUN	0230	26	25320.	24	JUN	1500	51	151.	25	JUN	0330	76	0.
23	JUN	1430	2	0.	24	JUN	0300	27	23988.	24	JUN	1530	52	110.	25	JUN	0400	77	0.
23	JUN	1500	3	0.	24	JUN	0330	28	22346.	24	JUN	1600	53	79.	25	JUN	0430	78	0.
23	JUN	1530	4	3.	24	JUN	0400	29	20482.	24	JUN	1630	54	57.	25	JUN	0500	79	0.
23	JUN	1600	5	18.	24	JUN	0430	30	18482.	24	JUN	1700	55	41.	25	JUN	0530	80	0.
23	JUN	1630	6	68.	24	JUN	0500	31	16428.	24	JUN	1730	56	29.	25	JUN	0600	81	0.
23	JUN	1700	7	195.	24	JUN	0530	32	14394.	24	JUN	1800	57	21.	25	JUN	0630	82	0.
23	JUN	1730	8	469.	24	JUN	0600	33	12438.	24	JUN	1830	58	15.	25	JUN	0700	83	0.
23	JUN	1800	9	1006.	24	JUN	0630	34	10608.	24	JUN	1900	59	10.	25	JUN	0730	84	0.
23	JUN	1830	10	1942.	24	JUN	0700	35	8933.	24	JUN	1930	60	7.	25	JUN	0800	85	0.
23	JUN	1900	11	3369.	24	JUN	0730	36	7432.	24	JUN	2000	61	5.	25	JUN	0830	86	0.
23	JUN	1930	12	5294.	24	JUN	0800	37	6113.	24	JUN	2030	62	3.	25	JUN	0900	87	0.
23	JUN	2000	13	7626.	24	JUN	0830	38	4973.	24	JUN	2100	63	2.	25	JUN	0930	88	0.
23	JUN	2030	14	10211.	24	JUN	0900	39	4003.	24	JUN	2130	64	2.	25	JUN	1000	89	0.
23	JUN	2100	15	12867.	24	JUN	0930	40	3189.	24	JUN	2200	65	1.	25	JUN	1030	90	0.
23	JUN	2130	16	15447.	24	JUN	1000	41	2517.	24	JUN	2230	66	1.	25	JUN	1100	91	0.
23	JUN	2200	17	17869.	24	JUN	1030	42	1968.	24	JUN	2300	67	1.	25	JUN	1130	92	0.
23	JUN	2230	18	20102.	24	JUN	1100	43	1525.	24	JUN	2330	68	0.	25	JUN	1200	93	0.
23	JUN	2300	19	22120.	24	JUN	1130	44	1172.	25	JUN	0000	69	0.	25	JUN	1230	94	0.
23	JUN	2330	20	23868.	24	JUN	1200	45	893.	25	JUN	0030	70	0.	25	JUN	1300	95	0.
24	JUN	0000	21	25273.	24	JUN	1230	46	676.	25	JUN	0100	71	0.	25	JUN	1330	96	0.
24	JUN	0030	22	26255.	24	JUN	1300	47	507.	25	JUN	0130	72	0.	25	JUN	1400	97	0.
24	JUN	0100	23	26759.	24	JUN	1330	48	378.	25	JUN	0200	73	0.	25	JUN	1430	98	0.
24	JUN	0130	24	26759.	24	JUN	1400	49	280.	25	JUN	0230	74	0.	25	JUN	1500	99	0.
24	JUN	0200	25	26265.	24	JUN	1430	50	206.	25	JUN	0300	75	0.	25	JUN	1530	100	0.

```
  PEAK FLOW   TIME
+   (CFS)     (HR)
+  26759.     11.00
```

MAXIMUM AVERAGE FLOW

		6-HR	24-HR	72-HR	49.50-HR
+	(CFS)	24061.	10070.	4885.	4885.
	(INCHES)	3.710	6.211	6.213	6.213
	(AC-FT)	11931.	19973.	19982.	19982.

621

TABLE 24.3 (CONTINUED)

```
********************************************************************************

          ***********
24 KK     **    3    **          COMPUTE HYDROGRAPH FOR AREA C
          ***********

25 BA     SUBBASIN RUNOFF DATA

          SUBBASIN CHARACTERISTICS

                    TAREA    27.30    SUBBASIN AREA

          PRECIPITATION DATA

26 PB     STORM    4.10    BASIN TOTAL PRECIPITATION

27 PI     INCREMENTAL PRECIPITATION PATTERN

          .12    .21    .25    .37    1.52    .41    .33    .25    .16    .20
          .12    .16

29 US     SNYDER UNITGRAPH

          TP    2.76    LAG
          CP     .80    PEAKING COEFFICIENT

SYNTHETIC ACCUMULATED-AREA VS. TIME CURVE WILL BE USED

********************************************************************************

APPROXIMATE CLARK COEFFICIENTS FROM GIVEN SNYDER CP AND TP ARE TC= 3.65 AND R= 1.30 INTERVALS    ***

                              UNIT HYDROGRAPH PARAMETERS

                 CLARK    TC= 3.65 HR,      R= 1.30 HR
                 SNYDER   TP= 2.77 HR,      CP= .80

                              UNIT HYDROGRAPH
                          18 END-OF-PERIOD ORDINATES

          406.   1424.   2670.   3891.   4777.   5075.   4795.   3858.   2685.   1821.
          1234.   837.    567.    385.    261.    177.    120.     81.

********************************************************************************
```

TABLE 24.3 (CONTINUED)

DA	MON	HRMN	ORD	RAIN	LOSS	EXCESS	COMP Q	*	DA	MON	HRMN	ORD	RAIN	LOSS	EXCESS	COMP Q
23	JUN	1400	1	.00	.00	.00	0.	*	24	JUN	1500	51	.00	.00	.00	0.
23	JUN	1430	2	.12	.00	.12	50.	*	24	JUN	1530	52	.00	.00	.00	0.
23	JUN	1500	3	.21	.00	.21	258.	*	24	JUN	1600	53	.00	.00	.00	0.
23	JUN	1530	4	.25	.00	.25	720.	*	24	JUN	1630	54	.00	.00	.00	0.
23	JUN	1600	5	.37	.00	.37	1526.	*	24	JUN	1700	55	.00	.00	.00	0.
23	JUN	1630	6	1.52	.00	1.52	3184.	*	24	JUN	1730	56	.00	.00	.00	0.
23	JUN	1700	7	.41	.00	.41	5873.	*	24	JUN	1800	57	.00	.00	.00	0.
23	JUN	1730	8	.33	.00	.33	9008.	*	24	JUN	1830	58	.00	.00	.00	0.
23	JUN	1800	9	.25	.00	.25	12033.	*	24	JUN	1900	59	.00	.00	.00	0.
23	JUN	1830	10	.16	.00	.16	14308.	*	24	JUN	1930	60	.00	.00	.00	0.
23	JUN	1900	11	.20	.00	.20	15400.	*	24	JUN	2000	61	.00	.00	.00	0.
23	JUN	1930	12	.12	.00	.12	15268.	*	24	JUN	2030	62	.00	.00	.00	0.
23	JUN	2000	13	.16	.00	.16	13880.	*	24	JUN	2100	63	.00	.00	.00	0.
23	JUN	2030	14	.00	.00	.00	11836.	*	24	JUN	2130	64	.00	.00	.00	0.
23	JUN	2100	15	.00	.00	.00	9861.	*	24	JUN	2200	65	.00	.00	.00	0.
23	JUN	2130	16	.00	.00	.00	8061.	*	24	JUN	2230	66	.00	.00	.00	0.
23	JUN	2200	17	.00	.00	.00	6436.	*	24	JUN	2300	67	.00	.00	.00	0.
23	JUN	2230	18	.00	.00	.00	4967.	*	24	JUN	2330	68	.00	.00	.00	0.
23	JUN	2300	19	.00	.00	.00	3679.	*	25	JUN	0000	69	.00	.00	.00	0.
23	JUN	2330	20	.00	.00	.00	2596.	*	25	JUN	0030	70	.00	.00	.00	0.
24	JUN	0000	21	.00	.00	.00	1760.	*	25	JUN	0100	71	.00	.00	.00	0.
24	JUN	0030	22	.00	.00	.00	1180.	*	25	JUN	0130	72	.00	.00	.00	0.
24	JUN	0100	23	.00	.00	.00	780.	*	25	JUN	0200	73	.00	.00	.00	0.
24	JUN	0130	24	.00	.00	.00	445.	*	25	JUN	0230	74	.00	.00	.00	0.
24	JUN	0200	25	.00	.00	.00	279.	*	25	JUN	0300	75	.00	.00	.00	0.
24	JUN	0230	26	.00	.00	.00	171.	*	25	JUN	0330	76	.00	.00	.00	0.
24	JUN	0300	27	.00	.00	.00	102.	*	25	JUN	0400	77	.00	.00	.00	0.
24	JUN	0330	28	.00	.00	.00	60.	*	25	JUN	0430	78	.00	.00	.00	0.
24	JUN	0400	29	.00	.00	.00	30.	*	25	JUN	0500	79	.00	.00	.00	0.
24	JUN	0430	30	.00	.00	.00	13.	*	25	JUN	0530	80	.00	.00	.00	0.
24	JUN	0500	31	.00	.00	.00	0.	*	25	JUN	0600	81	.00	.00	.00	0.
24	JUN	0530	32	.00	.00	.00	0.	*	25	JUN	0630	82	.00	.00	.00	0.
24	JUN	0600	33	.00	.00	.00	0.	*	25	JUN	0700	83	.00	.00	.00	0.

TOTAL RAINFALL = 4.10, TOTAL LOSS = .00, TOTAL EXCESS = 4.10

PEAK FLOW	TIME		MAXIMUM AVERAGE FLOW			
(CFS)	(HR)		6-HR	24-HR	72-HR	49.50-HR
+ 15400.	5.00	(CFS)	10503.	2995.	1452.	1452.
		(INCHES)	3.577	4.080	4.080	4.080
+		(AC-FT)	5208.	5941.	5941.	5941.

623

TABLE 24.3 (CONCLUDE)

```
         ****************
30 KK    *    3    *               ADD HYDROGRAPHS AT 3
         *    3    *
         ****************

31 HC    HYDROGRAPH COMBINATION
         ICOMP    2    NUMBER OF HYDROGRAPHS TO COMBINE
```

SUMMED HYDROGRAPH AT STATION   3

DA	MON	HRMN	ORD	FLOW
23	JUN	1400	1	0.
23	JUN	1430	2	50.
23	JUN	1500	3	259.
23	JUN	1530	4	723.
23	JUN	1600	5	1544.
23	JUN	1630	6	3251.
23	JUN	1700	7	6067.
23	JUN	1730	8	9477.
23	JUN	1800	9	13039.
23	JUN	1830	10	16250.
23	JUN	1900	11	18769.
23	JUN	1930	12	20561.
23	JUN	2000	13	21507.
23	JUN	2030	14	22047.
23	JUN	2100	15	22728.
23	JUN	2130	16	23508.
23	JUN	2200	17	24305.
23	JUN	2230	18	25070.
23	JUN	2300	19	25798.
23	JUN	2330	20	26464.
24	JUN	0000	21	27033.
24	JUN	0030	22	27435.
24	JUN	0100	23	27539.
24	JUN	0130	24	27204.
24	JUN	0200	25	26545.
24	JUN	0230	26	25492.
24	JUN	0300	27	24090.
24	JUN	0330	28	22407.
24	JUN	0400	29	20512.
24	JUN	0430	30	18495.
24	JUN	0500	31	16428.
24	JUN	0530	32	14394.
24	JUN	0600	33	12438.
24	JUN	0630	34	10608.
24	JUN	0700	35	8933.
24	JUN	0730	36	7432.
24	JUN	0800	37	6113.
24	JUN	0830	38	4973.
24	JUN	0900	39	4003.
24	JUN	0930	40	3189.
24	JUN	1000	41	2517.
24	JUN	1030	42	1968.
24	JUN	1100	43	1525.
24	JUN	1130	44	1172.
24	JUN	1200	45	893.
24	JUN	1230	46	676.
24	JUN	1300	47	507.
24	JUN	1330	48	378.
24	JUN	1400	49	280.
24	JUN	1430	50	206.
24	JUN	1500	51	151.
24	JUN	1530	52	110.
24	JUN	1600	53	79.
24	JUN	1630	54	57.
24	JUN	1700	55	41.
24	JUN	1730	56	29.
24	JUN	1800	57	21.
24	JUN	1830	58	15.
24	JUN	1900	59	10.
24	JUN	1930	60	7.
24	JUN	2000	61	5.
24	JUN	2030	62	3.
24	JUN	2100	63	2.
24	JUN	2130	64	2.
24	JUN	2200	65	1.
24	JUN	2230	66	1.
24	JUN	2300	67	1.
24	JUN	2330	68	0.
25	JUN	0000	69	0.
25	JUN	0030	70	0.
25	JUN	0100	71	0.
25	JUN	0130	72	0.
25	JUN	0200	73	0.
25	JUN	0230	74	0.
25	JUN	0300	75	0.
25	JUN	0330	76	0.
25	JUN	0400	77	0.
25	JUN	0430	78	0.
25	JUN	0500	79	0.
25	JUN	0530	80	0.
25	JUN	0600	81	0.
25	JUN	0630	82	0.
25	JUN	0700	83	0.
25	JUN	0730	84	0.
25	JUN	0800	85	0.
25	JUN	0830	86	0.
25	JUN	0900	87	0.
25	JUN	0930	88	0.
25	JUN	1000	89	0.
25	JUN	1030	90	0.
25	JUN	1100	91	0.
25	JUN	1130	92	0.
25	JUN	1200	93	0.
25	JUN	1230	94	0.
25	JUN	1300	95	0.
25	JUN	1330	96	0.
25	JUN	1400	97	0.
25	JUN	1430	98	0.
25	JUN	1500	99	0.
25	JUN	1530	100	0.

PEAK FLOW   TIME
+   (CFS)   (HR)
+ 27539.   11.00

MAXIMUM AVERAGE FLOW

	6-HR	24-HR	72-HR	49.50-HR
(CFS)	25828.	13056.	6337.	6337.
(INCHES)	2.741	5.543	5.549	5.549
(AC-FT)	12807.	25897.	25923.	25923.

TABLE 24.4 RUNOFF SUMMARY OF SIMULATED PEAK AND AVERAGE FLOWS AT POINTS 1 THROUGH 8 FOR THE JUNE, 1963 STORM

		Peak	6-hr	24-hr	72-hr	Area
Hydrograph at	1	34475.	24048.	6974.	3382.	33.40
Routed to	2	25622.	20848.	6974.	3382.	33.40
Hydrograph at	2	17310.	11207.	3100.	1503.	26.90
2 Combined	2	30092.	26453.	10074.	4885.	60.30
Routed to	3	26759.	24061.	10070.	4885.	60.30
Hydrograph at	3	15400.	10503.	2995.	1452.	27.30
2 Combined	3	27539.	25828.	13056.	6337.	87.60
Routed to	4	25912.	24551.	13052.	8409.	87.60
Hydrograph at	4	3362.	2382.	689.	441.	9.20
2 Combined	4	25925.	24606.	13702.	8850.	96.80
Routed to	5	23911.	22858.	13523.	8875.	96.80
Hydrograph at	5	7100.	5604.	1816.	1162.	28.30
2 Combined	5	23911.	22874.	14717.	10038.	125.10
Routed to	6	23911.	22874.	14717.	10039.	125.10
Hydrograph at	6	3349.	2478.	750.	480.	28.00
2 Combined	6	23911.	22874.	15268.	10518.	153.10
Routed to	7	22949.	22024.	14984.	10482.	153.10
Hydrograph at	7	4113.	2764.	773.	495.	17.00
2 Combined	7	22949.	22924.	15112.	10977.	170.10
Routed to	8	22453.	21595.	14994.	10908.	170.10
Hydrograph at	8	1684.	868.	228.	146.	5.00
2 Combined	8	22453.	21595.	14995.	11054.	175.10

24.3 STORM SURGE MODELING

Coastal areas not only experience floods from single-event storms but also from storm surges (short-term changes in sea level) normally caused by hurricanes. Several computer models are available to analyze hurricane-produced storm surges.[12] Most are deterministic, modeling the physical processes of momentum transfer from the atmosphere to the ocean. Parameters in these models can be adjusted to allow analysis of actual or hypothetical storms such as the probable maximum hurricane (PMH) or standard project hurricane (SPH).

Federal agencies, private consultants, and universities have developed surge models. Several nonproprietary, open-coast surge models are listed in Table 24.5. User's manuals are available for all the models listed.

Storm surge models simulate the effects of wind momentum on ocean water masses. This involves principles from meteorology, oceanography, and wave hydrodynamics, which all operate from assumptions about storm, geometry, and water-level conditions. Tide levels are included in most of the models because damage from surges often depends entirely on concurrence with the peak surge level.

The equations solved by the first three models in Table 24.5 are two-dimensional versions of Eq. 13.51 and the equation of continuity, Eq. 13.40. The fourth model disregards the continuity equation and solves Eq. 13.51 through a series of

TABLE 24.5 STORM SURGE MODELS

Program name	User/agency[a]	Data input	Differential equation solution method	Applicable coasts
SPLASH	NWS	Atmospheric pressures, radius to maximum wind, storm speed	Finite-difference	Mildly curved, Gulf and East coasts
SSURGE	COE	Atmospheric pressures, radius to maximum wind, storm speed	Finite-difference	All coasts
FIA model	FIA	Atmospheric pressures, radius to maximum wind, storm speed, maximum wind speed, depth of shelf	Finite-difference	Gulf and Atlantic coasts
BATHYSTROPHIC	CERC	Wind field, pressure differences, radius to maximum wind, forward speed	Finite-difference	All coasts

[a] NWS, National Weather Service; COE, U.S. Army Corps of Engineers; FIA, Federal Insurance Administration; CERC, Coastal Engineering Research Center, Corps of Engineers.

assumptions. The models are not truly dynamic because they treat time as a succession of steady states. Output is a file of water depths at the end of each time step used in the simulation. Storms that can be simulated include the SPH, PMH, or any prescribed wind field.

■ Summary

By far the largest number of hydrologic model applications involves the use of single-event simulation models, whether the general versions described in this chapter or the urban runoff models about to be presented in Chapter 25. Data requirements for single-event models are nominal—far less than for continuous modeling studies. In the majority of cases, the data required for any subarea are easily obtained from readily available topographic and soils maps. Given the basin area, slope, soil types, land use, and location, estimates of peak flow rates and shapes of runoff hydrographs at several locations in the watershed can be obtained for given storms within a few hours' time using these models. They continue to be the primary tool used by practicing engineers in analysis and design of stormwater handling facilities.

PROBLEMS

24.1. Six numbered subareas for a river basin are as shown in the sketch. Prepare a schematic diagram for a model study using boxes as subbasin runoff components, connecting lines as channel routing links, circles as hydrograph combination nodes, and triangles as reservoir routing nodes. Then describe the computation sequence for this basin in the same manner shown in Fig. 24.1. (See sketch on next page).

24.2. Synthesize a unit hydrograph for a watershed in your locale using the HYMO model equations. Compare with corresponding unit hydrographs from Snyder's method and the SCS method in Chapter 12.

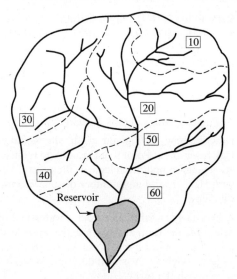

Sketch for Problem 24.1

24.3. Use the HYMO model equations to synthesize a unit hydrograph for the entire 258 sq. mi Oak Creek watershed in Fig. 24.8.

24.4. A watershed experiences a 12-hr rainstorm having a uniform intensity of 0.1 in./hr. Using $E = 0.7$, $K_0 = 0.6$, $C = 3.0$, and $\Delta K = 0.0$, calculate the hourly loss rates L_t as determined by the HEC-1 event-simulation model. Determine the total and percent losses for the storm.

24.5. Repeat Problem 24.4 using $CUML_1 = 0.5$ in.

24.6. Route the inflow hydrograph in Problem 13.7 to the outlet of the 30-mi reach using the HEC-1 straddle-stagger method by lagging averaged pairs of flows two time increments (12 hr). Compare the routed and measured outflow rates.

24.7. Route the inflow hydrograph in Problem 13.7 through the reach by dividing the 30-mi reach into three subreaches and treat the outflow from each as inflow to the next in line. Lag flows one time increment in each subreach and compare the final outflows with the measured values.

24.8. Study Table 24.3 and Fig. 24.8, and then define the following terms from Table 24.3: AMSKK,X, TAREA, NP, STORM, TP, CP, TC, R, RAIN, and EXCESS, COMP Q.

24.9. Search the HEC-1 printout in Table 24.3 to determine values (give units) of the following:
a. The time increment used in the model run.
b. Snyder's C_p, Eq. 12.17 input for Subarea B.
c. The peak flow rate for the synthesized Subarea-A unit hydrograph.
d. The total runoff (in.) from Subarea A.
e. The peak outflow rate from Subarea A.
f. The peak-to-peak time lag in routing the outflow hydrograph from Point 1 to Point 2, Fig. 24.8.
g. The percent attenuation caused by the reach between Points 1 and 2.
h. The Subarea-B peak outflow rate if Subarea A is neglected.
i. The simulated Subarea-B peak outflow rate.

24.10. Refer to the HEC-1 output in Tables 24.3 and 24.4 to answer the following questions:
  a. Are the values labeled "PRECIPITATION PATTERN" actual rainfall depths, or are they fractions of the rainfall?
  b. Name the method used to synthesize the unit hydrograph.
  c. What is the correct value of X for the X-hr unit hydrograph?
  d. How does the storm over Subarea A differ from the storm over Subarea B?
  e. Determine the total Muskingum K and the value of X for the river reach between Points 1 and 2 of Oak Creek.

24.11. Describe how, for a given storm, you would determine the effectiveness of Branched Oak reservoir at Point 9 in Fig. 24.8 to reduce flooding at Point 8. Answer by enumerating your computation logic as illustrated. You are allowed a maximum of two "runs" with HEC-1, and any number of subareas may be used as long as you describe your subareas.

24.12. A hydrologist wishes to model a watershed using the SCS curve number method to determine net rain, and to route the watershed runoff hydrograph through a reservoir using the storage-indication method. Base flow is to be incorporated as a constant value throughout the storm duration. Which event-simulation model or models would accomplish the task?

24.13. The following data were prepared for a Y1 card for routing a hydrograph using the straddle-stagger method of HEC-1. The time increment is 1.0 hr.

Field	Value	Field	Value
0	1	4	0.0
1	1	5	0.0
2	3	6	0.0
3	2	7	−1

Use this information to route the following inflow hydrograph by the straddle-stagger method. The initial outflow is zero. Continue routing until outflow is zero again.

Time (hr)	0	1	2	3	4	5	6	7	8
Inflow (cfs)	0	0	30	60	120	90	30	0	0

REFERENCES

1. P. H. Carrigan, "Calibration of U.S. Geological Survey Rainfall–Runoff Model for Peak-Flow Synthesis—Natural Basins," U.S. Geological Survey Computer Report, U.S. Department of Commerce National Technical Information Service, 1973.
2. J. R. Phillip, "Numerical Solution of Equations of the Diffusion Type with Diffusivity Concentration Dependent," *Aust. J. Phys.* **10,** 29(1957).
3. U.S. Department of Agriculture Soil Conservation Service, "Computer Program for Project Formulation Hydrology," Tech. Release No. 20, June 1973.
4. Hydrology, Section 4, *National Engineering Handbook*. Washington, D.C.: U.S. Department of Agriculture, Soil Conservation Service, 1985.

5. "Rainfall Frequency Atlas of the United States for Durations from 30 Minutes to 24 Hours and Return Periods from 1 to 100 Years," U.S. Weather Bureau Tech. Paper No. 40, May 1961.

6. J. R. Williams and R. W. Hann, "HYMO: Problem-Oriented Computer Language for Hydrologic Modeling," U.S. Department of Agriculture, Agriculture Research Service, May 1973.

7. J. R. Williams, "Flood Routing with Variable Travel Time or Variable Storage Coefficients," *Am. Soc. Agric. Eng. Trans.* **12**(1), 100–103(1969).

8. "SWMM, Volume No. 1, Final Report," Report for U.S. Environmental Protection Agency, Water Resources Engineers, and Metcalf and Eddy Inc., July 1971.

9. W. C. Huber, J. P. Heaney, S. J. Nix, R. E. Dickinson, and D. J. Polmann, "Storm Water Management Model User's Manual, Version III," EPA-600/2-84-109a (NTIS PB84-198423), Environmental Protection Agency, Cincinnati, OH, Nov. 1981.

10. L. A. Roesner, R. P. Shubinski, and J. A. Aldrich, "Storm Water Management Model User's Manual, Version III: Addendum I, Extran," EPA-600/2-84-109b(NTIS PB84-198431), Environmental Protection Agency, Cincinnati, OH, Nov. 1981.

11. U.S. Army Corps of Engineers, "HEC-1 Flood Hydrograph Package," Users Manual, Hydrologic Engineering Center, Sept. 1990.

12. U.S. Water Resources Council, "An Assessment of Storm Surge Modeling," Hydrology Committee, 1980.

13. C. O. Clark, "Storage and the Unit Hydrograph," *Trans. ASCE* **110**, 1419–1488(1945).

Chapter 25

Urban Runoff Simulation Models

■ Prologue

The purpose of this chapter is to:

- Describe nine of the most commonly used computer packages for simulation of urban rainfall-runoff processes.
- Show that the models do more than simulate runoff; they allow the user to analyze and design complete urban stormwater management systems.
- Demonstrate the models' capabilities and precision by comparing results obtained when simulating the same watershed with different models.
- Provide a "shopper's guide" to commercial and public domain urban stormwater software.

Urbanization generally has the effect of increasing the volume of runoff and also tends to result in earlier and greater peak rates of runoff. Early attempts to apply the single-event models of Chapter 24 in urban system analysis and design were successfully accomplished by selecting more intense rainfall-runoff analogs, decreasing total rainfall abstractions by deducting the impervious zones from the total pervious area, increasing the speed of travel over the land surfaces and in improved channels, and by adding components to the codes to allow for kinematic wave approximations of overland flow and hydrologic routing of flow in storm sewers and urban stormwater retention or detention ponds.

Discrepancies between these models' predictions and observed urban watershed response have resulted in the development of a whole class of single-event and continuous streamflow models of the unique processes operating in urbanized systems. These models not only simulate the rainfall-runoff process, but also allow the user to analyze an existing network of interconnected stormwater management facilities or to design new components of the system (underground storm sewers, detention ponds, ditches, street inlet sizes and locations, etc.).

25.1 URBAN STORMWATER SYSTEM MODELS

Nine of the most frequently used public-domain urban stormwater packages are described and compared in this chapter. Some just simulate the urban rainfall-runoff process; others provide specifics on the type, size, and location of drainage and stormwater handling facilities. The model acronyms and dates of original release as software are shown in Table 25.1. These models are periodically updated, and the current version should be requested when acquiring the code. Most are either single-event models or continuous models that are primarily used in a single-event mode.

Chicago Hydrograph Method (CHM)

Tholin's[1] hydrograph method, known as the *Chicago hydrograph method,* is an example of early urban runoff models. The procedure programmed is (1) develop a design storm pattern from local intensity–duration frequency curves and an average chronological storm pattern, (2) compute the overland flow using selected Horton-type infiltration capacity curves, the estimated depth of the rainfall retained in surface depressions, and Izzard's[2] overland flow equations, (3) route overland flow through gutters using the storage equation to obtain the runoff into catchbasins, (4) synthesize hydrographs from roofs and street inlets along a typical sewer lateral to produce a lateral outflow hydrograph, and (5) route the lateral outflow hydrograph by a time-offset method along submains and the main sewer to a point of discharge. The method originally involved a graphical hand computation but was later programmed for digital computer solution by Keifer.[3]

Road Research Laboratory (RRL) Method

An urban runoff model that utilizes the time–area runoff routing method described in Section 12.6 was developed in England and described by Watkins.[4] The technique was developed specifically for the analysis of urban runoff and ignores completely all pervious areas and all impervious areas that are not directly connected to the storm drain system; hence the estimates of peak flow rates and runoff volumes are likely to be low for systems that have these components.

TABLE 25.1   FREQUENTLY USED URBAN STORMWATER SIMULATION MODELS

Code name	Model name	Agency originating	Year
CHM	Chicago Hydrograph Method	City of Chicago	1959
RRL	Road Research Laboratory Method	Road Research Lab	1962
ILLUDAS	Illinois Urban Drainage Area Simulator	Ill. Water Survey	1972
STORM	Storage, Treatment, Overflow Runoff Model	Corps of Engineers	1974
TR-55	SCS Technical Release 55	SCS	1992
DR3M	Distributed Routing Rainfall-Runoff Model	USGS	1978
HYDRA	Hydrologic Component of HYDRAIN Package	FHWA	1990
SWMM	Storm Water Management Model	EPA	1971
UCURM	U. of Cincinnati Urban Runoff Model	U. of Cincinnati	1972

The RRL model could be used for continuous streamflow simulation but tends to be used as an event-simulation model. It has been extensively applied in Great Britain, and moderate success has been reported by Terstriep and Stall[5] for North American applications in the Chicago, Baltimore, and Champaign, Illinois, areas. Other applications are reported in Refs. 6–10. The Illinois urban drainage area simulator (ILLUDAS) (Ref. 10) is an improved version of the RRL model that has a wider range of applications. It incorporates the impervious and pervious areas neglected by the RRL model and is a demonstrated improvement over RRL. ILLUDAS is described in the next section.

The hydrologic processes modeled by RRL are summarized in Column 4 of Table 25.2. Also tabulated are procedures used by the storm water management model (SWMM), the University of Cincinnati urban runoff model (UCURM), and the USGS distributed routing rainfall–runoff model (DR3M). The latter models are described individually in Section 25.2, followed by a comparison applied to several urban areas. Table 25.2 is reproduced in part from a study by Heeps and Mein.[6] The results of their investigation are discussed in a later section along with results of similar investigations by Marsalek et al.[7] and Lager.[8]

The Stall and Terstriep[9] flow diagram of the processes simulated by the RRL model is shown in Fig. 25.1. The major functions of the program involve five principal steps in the development of runoff hydrographs[10] as illustrated in Fig. 25.2. As a first step, the total basin is divided into subbasins similar to the one in Fig. 25.2a, and impervious areas that are directly connected to the storm drain system are identified as shown. The remainder of the basin including surfaces such as lawns, floodways, parks, roofs that are not connected to the storm drainage system, and impervious areas that drain into pervious areas are all ignored by the RRL model, but as shown later, are accounted for in the ILLUDAS model.

After hydraulic characteristics such as lengths, slopes, and roughness coefficients are estimated, the next step is the calculation of flow velocities for all segments. These velocities are then used to construct lines of equal travel time to the outlet of the basin, called isochrones, on the basin map. The areas between isochrones are then determined and plotted against travel time as shown in Fig. 25.2b.

The third step is to apply the specified rainfall pattern to the directly connected impervious area and then determine the translated hydrograph at the subbasin outlet. Excess rainfall hyetograph ordinates are obtained by subtracting the losses from rainfall to give the net supply rate shown in Fig. 25.2c, d, and e. The application is shown in Fig. 25.2f.

Because the routed time–area hydrograph in Fig. 25.2f represents translation effects only, the hydrograph must now be routed through reservoir-type storage to account for the effects of storage within the basin. This is accomplished by routing the hydrograph of Fig. 25.2f through a reservoir using the storage-indication method illustrated in Example 13.4.

The fifth and final step in the RRL method is the routing of the subbasin outflow hydrograph to the next confluence or the next input point by a simple storage routing technique similar to the storage indication method of Chapter 13. The final result is a total basin runoff hydrograph that would result from the specified storm rainfall.

Stall and Terstriep evaluated the merits of the RRL method by applying it to 10 urban watersheds located largely in the east, south, and midwest regions of the

TABLE 25.2 COMPARISON OF URBAN RUNOFF MODEL SIMULATION PROCEDURES

Process (1)	SWMM (2)	UCURM (3)	ILLUDAS (4)	DR3M (5)
Simulation Interception Evaporation	Single-event or continuous Part of depression storage Input by user	Noncontinuous Neglects Neglects	Noncontinuous Neglects Neglects	Continuous or single event Part of depression storage ET from soil zone using pan coefficient
Transpiration Depression storage	Input by user Fills before overland flow begins—part of impervious area assigned zero depression storage Depleted by infiltration	Neglects Exponential filling rate No allowance to be depleted by infiltration on pervious areas	Neglects Variable, defaults to 0.2 in.	One-third of rain on directly connected pervious areas Soil moisture accounted during dry periods
Infiltration	Horton or Green–Ampt equation Satisfied by water on ground surface and rainfall	Horton equation Time offset Satisfied by rainfall only	Holtan's equation Standard curves for SCS soil types A, B, C, D Infiltration reduced for antecedent moisture conditions	Green–Ampt equation on hourly data, percentage of rain on daily data
Overland flow	Uniform depth of detention	Profile with increasing depth	Time–area curve routing Defaults to linear time-area	Unsteady laminar flow by kinematic-wave methods
Reach/reservoir routing	Storage routing using Manning turbulent flow equation and continuity equation Quasisteady state	Solved using an empirical relation, continuity equation, and Manning turbulent flow equation Quasisteady state	Time of entry required as input data Storage routing used in pipes and open channels	Linear reservoir or modified Puls Channels routed by kinematic-wave method Models ponding behind culverts
Gutter flow Inlet pits and junctions	Uniform flow storage routing Outflow = sum of inflows	Outflow = sum of inflows Outflow = sum of inflows	Neglects Outflow = sum of inflows	Treats as open channel Sums, allows external input at nodes
Pipe flow	Storage routing (Manning equation based on the slope of energy line) Kinematic wave or full dynamic wave	No storage routing Lagged using weighted average velocity Quasisteady state	Storage routing (Manning equation for uniform flow) Lagged using full bore velocity Quasisteady state	Kinematic routing in nonpressure pipes Unsteady, nonuniform flow; kinematic wave
Surcharge	Simulates pressure flow conditions	Neglects	Increases pipe diameter in design mode, determines excess volume in analysis mode	Not permitted, uses Manning equation for free-surface flow

Source: Modified from Heeps and Mein.[6]

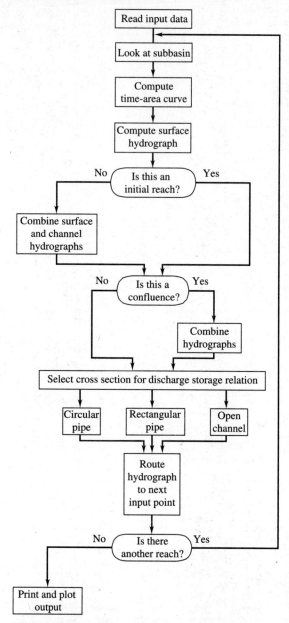

Figure 25.1 Flow diagram of the computer program for the RRL method. (After Stall and Terstriep.[9])

United States.[9] The criteria in selecting basins for the evaluation were (1) basins less than 5 mi² in size, (2) basins that were intensely urbanized, (3) basins that had extensive storm drainage systems, (4) basins with a high amount of paved area, (5) long records of rainfall and runoff, (6) the degree of quality of the data on storm rainfall and runoff, (7) the degree of information available on the storm drainage

(a) Subbasin map (directly
connected paved area shaded)

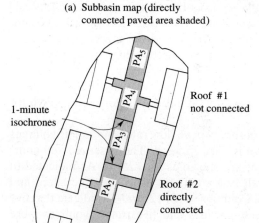

1-minute
isochrones

Roof #1
not connected

Roof #2
directly
connected

Inlets to
storm drain

(c) Rainfall

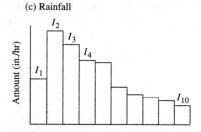

Amount (in./hr)

(d) Losses

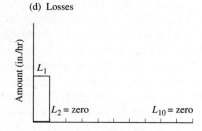

Amount (in./hr)

L_1

L_2 = zero L_{10} = zero

(b) Time versus paved area curve

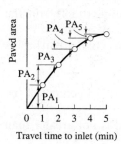

Paved area

PA_4 PA_5

PA_3

PA_2

PA_1

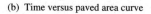

0 1 2 3 4 5

Travel time to inlet (min)

(e) Supply rate

$PASR_2$

$PASR_3$

$PASR_4$

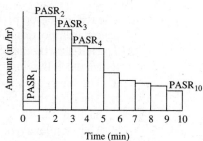

Amount (in./hr)

$PASR_1$

$PASR_{10}$

0 1 2 3 4 5 6 7 8 9 10

Time (min)

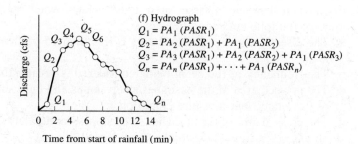

(f) Hydrograph
$Q_1 = PA_1 \,(PASR_1)$
$Q_2 = PA_2 \,(PASR_1) + PA_1 \,(PASR_2)$
$Q_3 = PA_3 \,(PASR_1) + PA_2 \,(PASR_2) + PA_1 \,(PASR_3)$
$Q_n = PA_n \,(PASR_1) + \cdots + PA_1 \,(PASR_n)$

Discharge (cfs)

Q_3 Q_4 Q_5 Q_6

Q_2

Q_1 Q_n

0 2 4 6 8 10 12 14

Time from start of rainfall (min)

Figure 25.2 Elements in the development of paved-area hydrographs. (After Terstriep and
Stall.[10])

system, and (8) data that had not already been published in one form or another. Pertinent data for the selected basins are provided in Table 25.3. The basins represent a variety of hydrologic regimes in the United States, and the storms selected are characteristic of the variable storm rainfall occurring within the United States. As shown, the basins ranged in size from 14.7 to 5326 acres; the percentage of areas directly connected ranged from 14.4 to 61 percent; a variety of hydrologic soil groups and basin slopes were represented; and the number of historical storms simulated for each watershed varied from 2 to 18.

A comparison of the computed and observed peak flow rates and runoff volumes from the Woodoak Drive basin are shown in Fig. 25.3a, b. The observed and computed hydrographs for one of these storms are shown as Fig. 25.3c. As indicated in Table 25.3, this watershed had the overall best comparison between measured and computed peak flows and runoff volumes. Columns 9, 10, 12, and 13 indicate that five computed peaks were higher than observed values and five runoff volumes were greater than the observed volumes. This balance was not experienced in some of the other watersheds. For example, computed peak and runoff volumes are underestimated for practically all storms simulated for the Echo Park Avenue basin. The results for the Echo Park basin are presented in Fig. 25.4. These results are not surprising, since the Echo Park basin had significant grassed area runoff for all 18 of the storms.

Stall and Terstriep arrived at the following conclusions based on their evaluation of the RRL method:[9]

1. The RRL method provides an accurate means of computing runoff from the paved area portion of an urban basin.
2. The RRL method adequately represents the runoff from actual urban basins under the following conditions:
   a. The basin area is less than 5 mi^2.
   b. The directly connected paved area is equal to at least 15 percent of the basin area.
   c. The frequency of the storm event being considered is not greater than 20 yr.
3. The RRL method cannot be used for all urban basins; the methods breaks down when significant grassed area runoff occurs, which happens if one or more of the following conditions exist:
   a. The directly connected paved area is less than 15 percent of the basin area.
   b. The frequency of the event being considered is greater than 20 yr.
   c. The grassed area of the basin has steep slopes and tight soils, regardless of the antecedent moisture condition.
   d. The grassed area of the basin has steep slopes, moderately tight soils, and an antecedent moisture condition of 3 or 4.
   e. The grassed area of the basin has moderate slopes, moderately tight soils, and an antecedent moisture condition of 4.
4. The principal strength of the RRL method is that, by confining runoff calculations to the paved area of a basin, it utilizes hydraulic functions that are largely determinate such as gravity flow from plain sloping concrete

TABLE 25.3 WATERSHED DATA AND RESULTS OF THE STALL AND TERSTRIEP EVALUATION

Basin (1)	Basin area (acres) (2)	Total paved area (acres) (3)	Total paved area (%) (4)	Direct connected paved area (acres) (5)	Direct connected paved area (%) (6)	Hydro-logic soil group (7)	Basin slope (8)	Computed peaks Number high (9)	Computed peaks Number low (10)	Computed peaks Mean absolute error (%) (11)	Computed runoff Number high (12)	Computed runoff Number low (13)	Computed runoff Mean absolute error (%) (14)
Woodoak Drive	14.7	4.9	33.9	2.8	19.4	B	Flat	5	5	28.2	5	5	18.3
Ross-Ade (Upper)	54.0	13.3	24.7	7.8	14.4	B-C	Steep	0	2	50.5	0	2	62.8
Sewer District No. 8	206	43	21.0	37.5	18.2	C-D	Flat	9	1	68.0	10	0	21.4
Echo Park Avenue	252	136	53.8	97.7	38.8	B-C	Steep	0	18	47.2	2	16	30.3
Crane Creek	273	65.5	23.9	39.7	14.5	C-D	Moderate	5	12	41.5	1	16	45.9
Tripps Run tributary	322	100	31.0	56.9	17.7	B-C	Moderate	3	6	37.9	1	8	44.1
Tar Branch	384	227	59.0	195	51.0	B	Moderate	12	5	33.4	13	4	57.3
Third Fork	1075	397	37.0	293	27.0	B-D	Moderate	3	12	31.2	0	15	36.7
Dry Creek	1882	583	31.0	365	19.0	B-C	Flat	2	6	20.3	2	6	27.8
Wingohocking	5326	3246	61.0	3246	61.0	B-D	Moderate	14	2	49.9	13	3	72.5

Source: After Stall and Terstriep.[9]

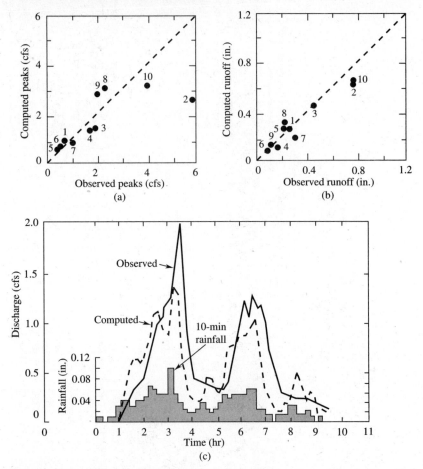

Figure 25.3 RRL results for Woodoak Drive basin, Long Island, New York, storm of October 19, 1966: (a) peaks, (b) volumes, and (c) the hydrographs. (After Stall and Terstriep.[9])

surfaces, gutters, pipes, and open channels. Physical understanding of these flow phenomena is much greater than the present understanding of the many complex phenomena governing runoff from rural areas such as antecedent moisture conditions, infiltration, soil moisture movement, transpiration, evaporation, and so forth.

5. A modification of the RRL method that would provide a function for grassed area contributions to runoff could be developed into a valuable design tool for urban drainage. This is believed to be possible in spite of the many complexities involved. Further flexibility could be offered by the additional provision for routing surface runoff through surface storage.

6. The input data requirements for use of the RRL methods on an urban basin are reasonable for the engineering evaluation of a basin for storm drainage design. The necessary data are no more complex or elaborate than the data usually compiled for a traditional storm drainage design.

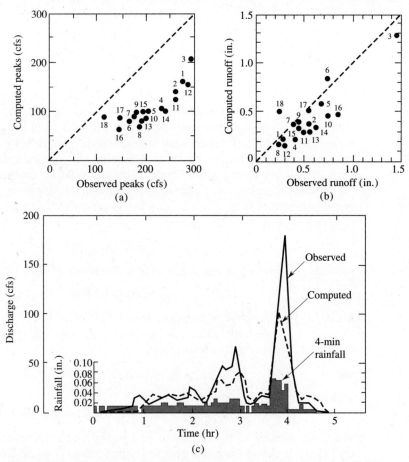

Figure 25.4 RRL results for Echo Park Avenue basin, Los Angeles, California, storm of April 18, 1965: (a) peaks, (b) volumes, (c) the hydrographs. (After Stall and Terstriep.[9])

7. It appears that rainfall occurs in greater amounts in the United States than in Great Britain. This may account for the fact that the RRL method is successful and widely used in Great Britain and yet suffers the above-described breakdowns for some of the basins studied in the United States.
8. Better urban rainfall and runoff data are required for the proper testing of all mathematical models. Research basins that do not have hydraulic problems, such as undersized drains or inadequate inlets, should be selected and instrumented.

Illinois Urban Drainage Area Simulator, ILLUDAS

As mentioned, the RRL method only simulates runoff from paved areas of the basin that are directly connected to the storm drainage system. Grassed areas and nonconnected paved areas are excluded from consideration. The ILLUDAS[10] model incorporates the directly connected paved area technique of the RRL method but also recognizes and incorporates runoff from grassed and nonconnected paved areas.

Computation of grassed area hydrographs for the subbasins is very similar to the approach for paved area hydrographs. Figure 25.5 shows the same subbasin used to illustrate paved area runoff in Fig. 25.2. The shaded area is the contributing grassed area, which is largely the front yards of residences. Rain falling on any not-directly-connected paved area is assumed to run off instantly onto the surrounding grassed area, and grassed area hydrology takes over. Runoff from back and side yards often drains gradually to a common back lot line and then laterally to the nearest street. The travel time required for this virtually eliminates such grassed areas from consideration during relatively short intense storms normally used for drainage design.

After the contributing grassed area has been identified, the curve in Fig. 25.5 can be constructed. Travel times across the grass strips are equivalent to the time of equilibrium from Izzard's equation,[2]

$$t_e = 0.033KLq_e^{-0.67} \qquad (25.1)$$

which is the time when the overland flow discharge reaches 97 percent of q_e, that is,

$$q_e = 0.0000231IL \qquad (25.2)$$

where q_e = discharge of overland flow (cfs/ft of width) at equilibrium
$\quad\quad\quad I$ = rain supply rate (in./hr), assumed to be 1.0 in ILLUDAS
$\quad\quad\quad L$ = length of overland flow (ft)

and $K = (0.0007I + c)S^{-0.33} \qquad (25.3)$

where S = surface slope (ft/ft)
$\quad\quad\quad c$ = coefficient having a value of 0.046 for bluegrass turf.

The time–area curve is assumed to be a straight line. The endpoint, as illustrated in Fig. 25.5 represents the travel time from the farthest point on the contributing grassed area.

In ILLUDAS, depression storage is normally set at 0.20 in. but can be varied. Infiltration is modeled using Holtan's equation,[11]

$$f = a(S - F)^{1.4} + f_c \qquad (25.4)$$

where f = infiltration rate at time t (in./hr)
$\quad\quad\quad\quad a$ = a vegetative factor = 1.0 for bluegrass turf
$\quad\quad\quad\quad S$ = storage available in the soil mantle (in.) (storage at the soil porosity minus storage at the wilting point)
$\quad\quad\quad\quad F$ = water already stored in the soil at time t, in excess of the wilting point (in.) (amount accumulated from infiltration prior to time t)
$\quad\quad S - F$ = storage space remaining in the soil mantle at time t (in.)
$\quad\quad\quad\quad f_c$ = final constant infiltration rate (in./hr), generally equivalent to the saturated hydraulic conductivity (in./hr) of the tightest horizon present in the soil profile

If physical properties of the soil are known, the equation can be used to compute an infiltration curve. Figure 25.6 shows the general interrelation between the various infiltration rates and storage factors involved.

(a) Subbasin map (contributing
     grassed area shaded)

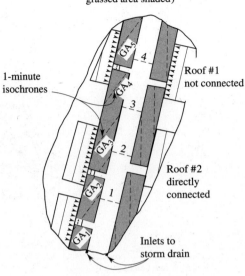

1-minute
isochrones

Roof #1
not connected

Roof #2
directly
connected

Inlets to
storm drain

(b) Time versus grassed area curve

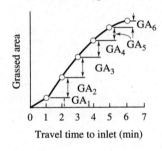

Grassed area

GA_6

GA_5

GA_4

GA_3

GA_2

GA_1

0   1   2   3   4   5   6   7

Travel time to inlet (min)

(c) Rainfall

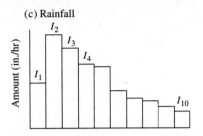

Amount (in./hr)

I_1 I_2 I_3 I_4 I_{10}

(d) Runoff from supplemental paved area

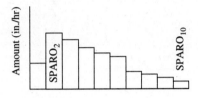

Amount (in./hr)

$SPARO_2$ $SPARO_{10}$

(e) Losses

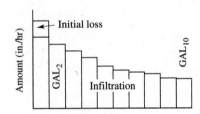

Amount (in./hr)

Initial loss

GAL_2 Infiltration GAL_{10}

(f) Grassed area supply rate

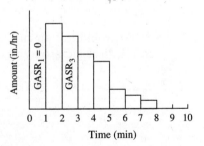

Amount (in./hr)

$GASR_1 = 0$ $GASR_3$

0   1   2   3   4   5   6   7   8   9   10

Time (min)

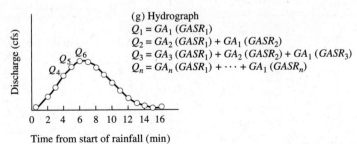

(g) Hydrograph
$Q_1 = GA_1 (GASR_1)$
$Q_2 = GA_2 (GASR_1) + GA_1 (GASR_2)$
$Q_3 = GA_3 (GASR_1) + GA_2 (GASR_2) + GA_1 (GASR_3)$
$Q_n = GA_n (GASR_1) + \cdots + GA_1 (GASR_n)$

Discharge (cfs)

Q_4 Q_5 Q_6

0   2   4   6   8   10 12 14 16

Time from start of rainfall (min)

Figure 25.5   Elements in the development of grassed-area hydrographs. (After Terstriep and
Stall.[10])

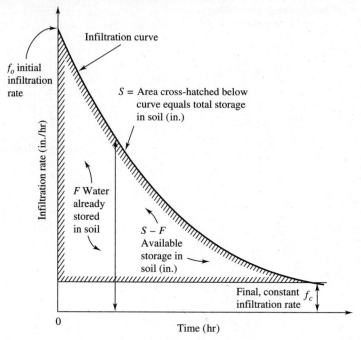

Figure 25.6 Diagram of infiltration relations used in ILLUDAS, Eq. 25.4. (After Terstriep and Stall.[10])

Table 25.4 provides an example computation of an infiltration curve for bluegrass on a silt loam soil in which soil moisture S of 6.95 in. is available. The equation is

$$f = 1(6.95 - F)^{1.4} + 0.50 \qquad (25.5)$$

Standard infiltration curves have been devised for use in ILLUDAS for soils having SCS hydrologic groups A, B, C, and D (Chapter 4). These curves were synthesized

TABLE 25.4   COMPUTATION OF INFILTRATION CURVE FOR SILT LOAM

Available storage, $S - F$ (in.)	ΔF (in.)	Water stored, F (in.)	$(S - F)^{1.4}$	Infiltration rate f (in./hr)	f_{avg} (in./hr)	Δt^a (hr)	t (hr)
6.95		0	15.0	15.5			0
6.00	0.95	0.95	12.3	12.8	14.1	0.07	0.07
5.0	1.0	1.95	9.5	10.0	11.4	0.09	0.16
4.0	1.0	2.95	7.0	7.5	8.7	0.11	0.27
3.0	1.0	3.95	4.65	5.15	6.3	0.16	0.43
2.0	1.0	4.95	2.64	3.14	4.2	0.24	0.67
1.0	1.0	5.95	1.0	1.50	2.3	0.43	1.10
0	1.0	6.95	0	0.50	0.7	1.43	2.53

[a] Incremental time, $\Delta t = \Delta F / f_{avg}$.
Source: Terstriep and Stall.[10]

from the Horton equation

$$f = f_c + (f_0 - f_c)e^{-kt} \qquad (25.6)$$

where f_0 = initial infiltration rate (in./hr)
  f_c = ultimate infiltration rate
  e = base of natural logarithms
  k = a shape factor selected as $k = 2$
  t = time from start of rainfall

This equation is solved in ILLUDAS by the Newton–Raphson technique. The curves are shown in Fig. 25.7.

To account for wet versus dry conditions, ILLUDAS divides the antecedent moisture condition (AMC) into four user-selected ranges, shown in Table 25.5. Each is based on the total 5-day precipitation prior to the storm day. Infiltration from Eq. 25.6 is varied, depending on the AMC value specified.

ILLUDAS allows the user to operate in two modes, analysis and design. For design mode, the model generates hydrographs and provides nominal storm sewer pipe diameters that are adequate, without surcharge, to pass the peak flows. In analysis mode, the model generates hydrographs throughout the basin nodes and links and then alerts the user if any input pipe diameters are too small. It also sums the volume of runoff water backed up at inlets because it could not be accommodated by the undersized storm sewers.

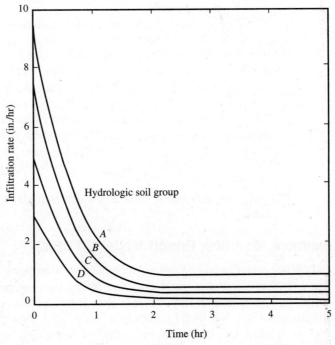

Figure 25.7 Standard infiltration curves for bluegrass turf on four SCS soil types used in ILLUDAS. (After Terstriep and Stall.[10])

TABLE 25.5   ANTECEDENT MOISTURE CONDITIONS
FOR BLUEGRASS LAWNS

AMC number	Description	Total rainfall during 5 days preceding storm (in.)
1	Bone dry	0
2	Rather dry	0–0.5
3	Rather wet	0.5–1
4	Saturated	Over 1

Source: Terstriep and Stall.[10]

The ILLUDAS model requires estimation of several input parameters. Other studies [12,13,14] evaluated the sensitivity of ILLUDAS to variations in parameters. The study in Ref. 13 concludes:

1. The sensitivity of the peak flows to changes in AMC increases as the soil group changes from D to A.
2. The ranges of sensitivity to the soil groups and AMC are approximately the same.
3. A change in the AMC from 2 to 3 and a change in the soil group from B to C are critical for large design return periods, and from C to D for small design return periods.
4. The time to peak for various combinations of soil group and AMC, for different return periods, remains the same.
5. The peak flow and runoff volume increase as the AMC changes from 1 to 4. This increase is particularly important between AMC 2 and 3 in general and between AMC 3 and 4 for soil group A. The peak flow and the runoff volume increase as the soil group changes from A to D for constant AMC. This increase is particularly important between soil groups B and C in general and soil groups C and D for AMC 1.
6. The peak flows decrease markedly for time increments larger than 5 min, and the pipe diameters decrease significantly for time increments larger than 10 min.
7. The time increment should not substantially exceed the paved area inlet time.

Storage, Treatment, Overflow Runoff Model (STORM)

STORM is the Corps of Engineers continuous simulation model of the quantity and quality of urban storm water resulting from single events or continuous daily rainfall.[15] It also simulates dry weather flow from domestic, commercial, or industrial discharges. Wet weather hydrographs, simulated from intermittent or continuous hourly rainfall, can be used for a variety of hydrologic study purposes.

Wet weather pollutographs (hydrographs that also provide water quality characteristics) can be predicted for individual historical or synthetic events and used in assessments of impacts of runoff on receiving streams. The pollutographs consist of hourly runoff rates, amounts of pollutants, and pollutant concentrations.

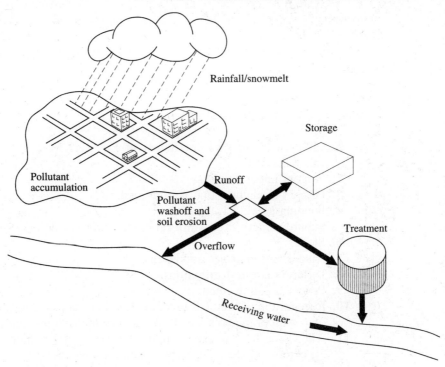

Figure 25.8 Conceptual view of urban system as used in STORM. (After U.S. Army Corps of Engineers.[15])

The model is conceptualized in Fig. 25.8. Snowmelt is simulated by the degree-day method (Chapter 14). Statistical information is output to aid in the selection of storage capacities and treatment rates required to achieve desired control of storm water runoff. Statistics, such as average annual runoff, average annual erosion, average annual overflow volume from storage, and average annual pollutant overflow from storage, are all provided.

The model simulates the interaction of precipitation, air temperature (to signal snowfall), runoff, pollutant accumulation, land surface erosion, dry weather flow, storage, treatment rates, and overflows from the storage or treatment system. The program computes continuous or single-event runoff from rainfall.

Runoff is computed as a fraction of the difference between rainfall and depression storage. The fraction selected depends on land use. Runoff in excess of the specified treatment capacity is diverted into storage for subsequent treatment. Runoff in excess of both the treatment rate and storage capacity becomes overflow and is diverted directly into the receiving waters.

SCS Technical Release No. 55 (TR-55)

The SCS TR-55 procedures for analyzing peak flows and runoff hydrographs from urbanized areas were described in Chapter 15. Even though SCS documentation for these procedures recommends manual rather than computerized applications of the procedures, several vendors have programmed the techniques for PC use and made them available through a number of outlets. A public domain version is available from the U.S. National Technical Information Service.[16]

Users of the vendor-developed renditions of TR-55 should perform initial studies using hand-checks to verify the code. An ideal TR-55 program would be one that uses SCS source code or has SCS endorsement, states all assumptions used, notifies the user of range violations, and incorporates options for making adjustments to account for all or most of the following:

1. Changes in the 484 coefficient of Eq. 12.22 for steep, average, or flat watersheds.
2. Percent imperviousness.
3. Percent of channel that is improved.
4. Ponding area.
5. Subarea length over width ratios that fall outside the assumed range.
6. Slope.
7. Antecedent moisture conditions.
8. Different storm distributions.
9. Proper lag time equation.
10. Recognition of the SCS recommendation that the duration for the derived unit hydrograph be about 13 percent of the subarea time of concentration.
11. Allowance for watersheds that have initial abstractions, I_a, greater than 20 percent of the potential maximum retention, S.

Whether by manual or computer operations, the SCS cautions that TR-55 hydrograph methods should not be used to perform final design if an error of 25 percent in predicted volume cannot be tolerated. Their advice is to use TR-20, after making appropriate parameter adjustments, if the urban wathershed is very complex or if a higher degree of accuracy is required.[17] Other precautions regarding the use of the graphical and tabular methods are identified in Section 15.2.

SCS Urban Time Relationships   The relationships among time parameters in SCS hydrologic methods have not been completely reconciled with observed phenomena or time relationships in other models of urban rainfall-runoff processes. Several formulations for lag time, with miscellaneous adjustments for urban effects, are mentioned in Section 11.7 and elsewhere in SCS literature, and have substantial impact on the shape of the hydrographs. The rational formula (Chapter 15), to illustrate, assumes that the time of concentration, defined as the time for rain falling at the most remote location to reach the outlet, equals the time to peak of the urban hydrograph. Izzard found this to hold approximately true in observing runoff hydrographs from paved areas.[2]

The greatest discrepancy found when comparing SCS and known urban time relationships is the prolonged time base that results when SCS unit hydrograph methods in Chapter 12 are applied. The hydrograph shape is based on observed runoff from undeveloped, rural watersheds. As shown in Fig. 12.13, the time base for the dimensionless unit hydrograph is about 5.0 times the time to peak. It was shown in Chapter 12 that linear superposition of unit hydrographs requires that the release time must equal the time base of the IUH, which in turn is the time of concentration, t_c. Recall

TABLE 25.6 COMPARISON OF TIME RELATIONSHIPS FOR A D-HR UNIT HYDROGRAPH BY SCS AND URBAN RUNOFF METHODS

Rational/Izzard/IUH Models of Urban Runoff	SCS Dimensionless Unit Hydrograph
Given:	
$D = t_c$	$D = 0.2 \times$ time to peak
Time to peak $= t_c$	Time to peak $=$ lag time $+ D/2$
Release time $= t_c$	Lag time $= 0.6\, t_c$ (Mockus Equation)
	Time base $= 5.0 \times$ time to peak
Solving:	
$D = t_c$	$D = 0.133\, t_c$
Time to peak $= t_c$	Time of peak $= 0.666\, t_c$
Release time $= t_c$	Time base $= 3.33\, t_c$
Time base $= D + t_r = 2\, t_c$	Release time $=$ Time base $- D = 3.20\, t_c$

that the *excess-rainfall release time, t_r,* was defined in Chapter 11 as the time from end of excess rain to end of direct runoff. As shown in the Table 25.6, time relationships for the SCS dimensionless unit hydrograph of Fig. 12.13 give prolonged runoff durations compared to other urban runoff models. Only the time to peak is approximately equivalent in this comparison. Urban runoff models based on SCS dimensionless unit hydrograph procedures may result in longer time bases and hydrograph recessions than other methods.

USGS Distributed Routing Rainfall–Runoff Model (DR3M)

The U.S. Geological Survey simulation model for urban rainfall–runoff applications originated in 1978 as a lumped parameter single-event model for small watersheds (described in Chapter 24) and subsequently was expanded to distributed parameter status, intended primarily for urban applicability.[18] Also, a soil moisture routine was added allowing quasicontinuous simulation.

The model can be applied to watersheds from a few acres to several square miles in size (an upper limit of 10 mi² is recommended). It does not simulate subsurface or interflow contributions to streamflow, and these must be externally added if considered important to the simulation.

Routing of rainfall to channels is by unsteady overland flow hydraulics, and routing hydrographs through channel reaches is accomplished by kinematic-wave methods (refer to Chapter 13). The differential routing equations are solved by one of three optional numerical methods. The user may specify an explicit or implicit finite-difference algorithm, or the method of characteristics.

Time may be discretized by the user in as small as 1-min increments. The smallest time increment is used by the program during any days having short-time interval rainfall, called *unit days.* Other days are simulated as 24-hr intervals. Movement of surface water is simulated only during unit days. For the rain-free intervals, daily rainfall is input and used to modify the soil moisture balance leading into the next unit day(s). The format for rain data is compatible with that of the U.S. Geological Survey system, WATSTORE (Water Data Storage and Retrieval System). Input data can also be obtained from any local National Weather Service office.

Practically any basin can be studied by breaking it into several sets of four types of model segments. These include overland flow segments (must be approximately

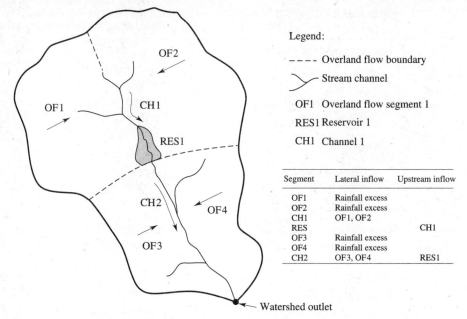

Segment	Lateral inflow	Upstream inflow
OF1	Rainfall excess	
OF2	Rainfall excess	
CH1	OF1, OF2	
RES		CH1
OF3	Rainfall excess	
OF4	Rainfall excess	
CH2	OF3, OF4	RES1

Figure 25.9 Segmentation of watershed for DR3M.

rectangular), detention storage facilities, channels, and nodes. This is illustrated in Fig. 25.9. Each segment in the figure may have inflow from either lateral or upstream sources (or both, as occurs for segment CH2).

Rainfall is uniformly distributed over the overland flow rectangles. Each has a given length, slope, roughness, and percent imperviousness. Laminar flow is assumed to occur over these segments. The values of b and m for Eqs. 13.28 and 13.56 are found from $m = 3$ and

$$b = \frac{8gS_0}{Kv} \tag{25.7}$$

where S_0 is the slope, v is the kinematic viscosity of water, equal to 0.0000141 ft/sec (for 50°F water), and K is a coefficient relating the Reynold's number N_r to Darcy's friction factor f by $K = fN_r$. Flow over rough surfaces is laminar if $K > 24$. The value K is related to rainfall intensity by

$$K = K_0 + 10I \tag{25.8}$$

where K_0 is found from Table 25.7, and I is the rainfall intensity (in./hr).

Channels in Fig. 25.9 represent either natural or artificial gutters or storm sewers (either open channels or nonpressure pipes are allowed). Inflow to channels comes from other channels, overland flow (as lateral inflow), or nodes. Nodes are used when more than three segments contribute to a channel or reservoir, or when the user wishes to specify an input or base flow hydrograph.

Channel routing is by kinematic-wave techniques, described in Section 13.3. Input is the channel length, slope, and routing parameters b and m. These are

TABLE 25.7 ROUGHNESS COEFFICIENTS FOR
OVERLAND AND CHANNEL SEGMENTS

Surface type	Laminar flow K_0	Turbulent flow Manning's n
Concrete asphalt	24–108	0.01–0.013
Bare sand	30–120	0.01–0.016
Graveled surface	90–400	0.012–0.03
Bare clay–loam soil (eroded)	100–500	0.012–0.033
Sparse vegetation	1,000–4,000	0.053–0.13
Short grass prairie	3,000–10,000	0.10–0.20
Bluegrass sod	7,000–40,000	0.17–0.48

Source: Alley and Smith.[18]

developed from the Manning equation and slope of the channel, respectively. The equations are $m = 1.67$ and

$$b = \frac{1.49 S_0^{1/2}}{n} \qquad (25.9)$$

where n is obtained from Table 25.7 or similar information. The model adjusts both b and m for various shapes, including circular and triangular (see Table 13.1 for several m values). If overbank flow is possible, a second set of b and m parameters can be input for all flows in excess of the channel capacity.

Reservoir inflow hydrographs are routed by either of two storage routing methods. If a linear reservoir model is appropriate, the storage coefficient K, relating outflow to storage by $O = KS$, is input. If the modified Puls method is desired, a table of outflows and corresponding storage levels must be input. The model assumes an initial reservoir level equal to that corresponding to an outflow of 0.0 cfs.

Ponding behind culverts can be modeled as a reservoir if an $S–O$ relation is input. This should include data points corresponding to roadway overflow to allow simulation of this common phenomenon.

Excess rainfall (runoff) from pervious areas is developed from the precipitation input, minus several abstractions. Infiltration is simulated by

$$S = K\left(1 + \frac{P(m - m_0)}{\text{SMS}}\right) \qquad (25.10)$$

where K is the hydraulic conductivity, P is the average suction head across the capillary zone, and m_0 and m are the soil moisture contents before and after wetting. The term SMS is the soil moisture storage. The rate of excess rain is found from

$$I = \frac{I^2}{2S} \quad \text{if } I \leq S \qquad (25.11)$$

and

$$I = I - \frac{S}{2} \quad \text{if } I > S \qquad (25.12)$$

where I is the rate of rain supplied to infiltration.

Runoff from impervious areas depends on whether the areas are directly connected to the drainage system. Those not directly connected are assumed to flow immediately onto pervious areas, where they are added as lateral inflow. One third of the rain on directly connected areas is abstracted, and the rest is lateral inflow to the gutter or channel.

Soil moisture is accounted for in a two-layer hypothetical storage zone. The amount in storage affects the infiltration rate and allows continuous soil moisture accounting between rain events. During unit days (days with short-duration rain input) all infiltrated moisture from Eq. 25.10 is added to the upper storage zone. Between unit days, a user-specified proportion of the daily rain is added to SMS. During rainless days, evapotranspiration occurs from SMS, using input pan evaporation rates multiplied by a coefficient. This process continues until the next rain event, at which time infiltration is governed by the amount of soil moisture.

Applications of DR3M have been documented across the continental U.S. and in Alaska and Hawaii.[19] Calibration of computed and measured runoff for almost 400 storms over 37 watersheds reveal a median error in peak flow estimates and volume of 21 and 24 percent, respectively, with the best results obtained for highly impervious watersheds. Indications from verification studies to date are that the model may overestimate the peak flow rates for simulated floods from storms having magnitudes in the design range of flood control facilities, and give better results for smaller storms typically used in runoff quality studies.

FHWA Storm Sewer Design Model (HYDRA)

As part of a package of integrated design computer programs called HYDRAIN,[20] the U.S. Federal Highway Administration developed the HYDRA storm drain design model for use by federal and other engineers. The model is distributed under contract with the FHWA through McTrans Software Center at the Civil Engineering Department of the University of Florida at Gainesville. HYDRA has been linked by commercial vendors to an integrated CAD/GIS system. The program's primary use is analyzing adequacy of existing storm drains or designing new storm drains and inlets by the rational method described in Chapter 15 or by a modified rational method which represents the hydrograph as a trapezoid having a volume equal to the calculated net rain.

Commercial versions of HYDRA allow design by the modified rational method, SCS methods, or revised Santa Barbara hydrograph methods.[21] HYDRA has one advantage over other storm sewer design models in that hydraulic grade lines through the system can be checked by hydraulic backwater computations to determine total system losses and whether inlets, manholes, or junction boxes are surcharged. Another useful feature is that street and gutter flows that exceed the inlet capacity of the storm sewers are routed by HYDRA to the next downstream location and added to the hydrograph at that point.

Storm Water Management Model (SWMM)

A very widely accepted and applied storm runoff simulation model was jointly developed by Metcalf and Eddy, Inc., the University of Florida, and Water Resources Engineers[22] for use by the U.S. Environmental Protection Agency (EPA). This model is designed to simulate the runoff of a drainage basin for any predescribed rainfall pattern. The total watershed is broken into a finite number of smaller units or sub-catchments that can readily be described by their hydraulic or geometric properties. A flowchart for the process is shown in Fig. 25.10.

The SWMM model has the capability of determining, for short-duration storms of given intensity, the locations and magnitudes of local floods as well as the quantity and quality of storm water runoff at several locations both in the system and in the receiving waters. The original SWMM was an event-simulation model, and later versions[23] keep track of long-term water budgets.

The fine detail in the design on the model allows the simulation of both water quantity and quality aspects associated with urban runoff and combined sewer systems. Only the water quantity aspects are described here. Information obtained from SWMM would be used to design storm sewer systems for storm water runoff control. Use of the model is limited to relatively small urban watersheds in regions where seasonal differences in the quality aspects of water are adequately documented.

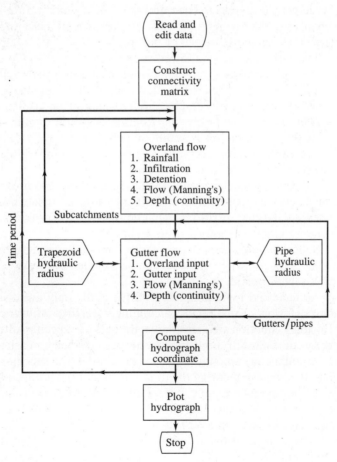

Figure 25.10 Flowchart for SWMM Runoff Block hydrographic computation. (After Metcalf and Eddy, Inc.[22])

The simulation is facilitated by five main subroutine blocks. Each block has a specific function, and the results of each block are entered on working storage devices to be used as part of the input to other blocks.

The main calling program of the model is called the Executive Block. This block is the first and last to be used and performs all the necessary interfacing among the other blocks.

The Runoff Block uses Manning's equation to route the uniform rainfall intensity over the overland flow surfaces, through the small gutters and pipes of the sewer system into the main sewer pipes, and out of the sewer pipes into the receiving streams. This block also provides time-dependent pollutional graphs (pollutographs).

A third package of subroutines, the Transport Block, determines the quality and quantity of dry weather flow, calculates the system infiltration, and calculates the water quality of the flows in the system.

A useful package of subroutines for water quality determination is contained in the Storage Block. The Storage Block allows the user to specify or have the model select sizes of several treatment processes in an optional wastewater treatment facility that receives a user-selected percentage of the peak flow. If used, this block simulates the changes in the hydrographs and pollutographs of the sewage as the sewage passes through the selected sequence of unit processes.

The earlier version[22] allowed simulation of any reservoir for which the outflow could be approximated as either a weir or orifice, or if the water was pumped from the reservoir. The newer version[23] allows input of 11 points of any storage–outflow relation and routes hydrographs through natural or artificial reservoirs, including backwater areas behind culverts. Routing is by the modified Puls method, which assumes that the reservoir is small enough that the water surface is always level.

Evaporation from reservoirs is simulated by a monthly coefficient (supplied by the user) multiplied by the surface area.

The Extran Block[24] completes the hydraulic calculations for overland flows, in channels, and in pipes and culverts. It solves the complete hydrodynamic equations, assesses surcharging, performs dynamic routing, and provides all the depth, velocity, and energy grade line information requested.

Subcatchment areas, slopes, widths, and linkages must be specified by the user. Manning's roughness coefficients can be supplied for pervious and impervious parts of each subcatchment.

As indicated in Table 24.1, SWMM is the only event-simulation model listed that uses Horton's equation for calculating watershed infiltration losses. If parameters for Horton's equation are unavailable, the user can specify ASCE standard infiltration capacity curves. Infiltration amounts thus determined for each time step are compared with instantaneous amounts of water existing on the subcatchment surface plus any rainfall that occurred during the time step, and if the infiltration loss is larger, it is set equal to the amount available. Input for Horton's equation consists of the maximal and minimal infiltration rates and the recession constant k in Eq. 4.1. The Green–Ampt equation is also used in SWMM.

Urban storm drainage components are modeled using Manning's equation and the continuity equation. The hydraulic radius of the trapezoidal gutters and circular pipes is calculated from component dimensions and flow depths. A pipe surcharges if it is full, provided that the inflow is greater than the outflow capacity. In this case, the surcharged amount will be computed and stored in the Runoff and Transport Blocks at the head end of the pipe. The pipe will remain full until the stored water is completely drained. Alternatively, the Extran Block can be used to conduct a dynamic simulation of the system under pressure-flow conditions.

Necessary inputs in the model are the surface area, width of subcatchment, ground slope, Manning's roughness coefficient, infiltration rate, and detention depth. Channel descriptions are the length, Manning's roughness coefficient, invert slope, diameter for pipes, or cross-sectional dimensions. General data requirements are summarized in Table 25.8. A step-by-step process accounts for all inflow, infiltration losses, and flow from upstream subcatchment areas, providing a calculated discharge hydrograph at the drainage basin outlet. The following description of the simulation process incorporated in early versions of SWMM will aid in understanding the logic of the model.[25]

1. Rainfall is added to the subcatchment according to the specified hyetograph:

$$D_1 = D_t + R_t \, \Delta t \qquad (25.13)$$

where D_1 = the water depth after rainfall
   D_t = the water depth of the subcatchment at time t
   R_t = the intensity of rainfall in time interval Δt

2. Infiltration I_t is computed by Horton's exponential function, $I_t = f_c + (f_0 - f_c)e^{-kt}$, and subtracted from the water depth existing on the subcatchment

$$D_2 = D_1 - I_t \, \Delta t \qquad (25.14)$$

where f_c, f_0, k = coefficients in Horton's equation (Eq. 4.1)
   D_2 = the intermediate water depth after accounting for infiltration

TABLE 25.8   GENERAL DATA REQUIREMENTS FOR STORM WATER MANAGEMENT MODEL (SWMM)

Item 1. *Define the Study Area.* Land use, topography, population distribution, census tract data, aerial photos, and area boundaries.

Item 2. *Define the System.* Plans of the collection system to define branching, sizes, and slopes; types and general locations of inlet structures.

Item 3. *Define the System Specialties.* Flow diversions, regulators, and storage basins.

Item 4. *Define the System Maintenance.* Street sweeping (description and frequency), catchbasin cleaning; trouble spots (flooding).

Item 5. *Define the Base Flow (DWF).* Measured directly or through sewerage facility operating data; hourly variation and weekday versus weekend; the DWF characteristics (composited BOD and SS results); industrial flows (locations, average quantities, and quality).

Item 6. *Define the Storm Flow.* Daily rainfall totals over an extended period (6 months or longer) encompassing the study events; continuous rainfall hyetographs, continuous runoff hydrographs, and combined flow quality measurements (BOD and SS) for the study events; discrete or composited samples as available (describe fully when and how taken).

3. If the resulting water depth of subcatchment D_2 is larger than the specified detention depth D_d, an outflow rate is computed using a modified Manning's equation.

$$V = \frac{1.49}{n}(D_2 - D_d)^{2/3}S^{1/2} \qquad (25.15)$$

and
$$Q_w = VW(D_2 - D_d) \qquad (25.16)$$

where V = the velocity
       n = Manning's coefficient
      S = the ground slope
     W = the width
    Q_w = the outflow rate

4. The continuity equation is solved to determine water depth of the subcatchments resulting from rainfall, infiltration, and outflow. Thus

$$D_{t+\Delta t} = D_2 - \frac{Q_w}{A}\Delta t \qquad (25.17)$$

where A is the surface area of the subcatchment.

5. Steps 1–4 are repeated until computations for all subcatchments are completed.
6. Inflow (Q_{in}) to a gutter is computed as a summation of outflow from tributary subcatchments ($Q_{w,i}$) and flow rate of immediate upstream gutters ($Q_{g,i}$)

$$Q_{in} = \sum Q_{w,i} + \sum Q_{g,i} \qquad (25.18)$$

7. The inflow is added to raise the existing water depth of the gutter according to its geometry. Thus

$$Y_1 = Y_t + \frac{Q_{in}}{A_s}\Delta t \qquad (25.19)$$

where Y_1, Y_t = the water depth of the gutter
        A_s = the mean water surface area between Y_1 and Y_t

8. The outflow is calculated for the gutter using Manning's equation:

$$V = \frac{1.49}{n}R^{2/3}S_i^{1/2} \qquad (25.20)$$

and
$$Q_g = VA_c \qquad (25.21)$$

where R = the hydraulic radius
     S_i = the invert slope
    A_c = the cross-sectional area at Y_1

9. The continuity equation is solved to determine the water depth of the gutter resulting from the inflow and outflow. Thus

$$Y_{t+\Delta t} = Y_1 + (Q_{in} - Q_g)\frac{\Delta t}{A_s} \qquad (25.22)$$

10. Steps 6–9 are repeated until all the gutters are finished.
11. The flows reaching the point concerned are added to produce a hydrograph coordinate along the time axis.
12. The processes from 1 to 11 are repeated in succeeding time periods until the complete hydrograph is computed.

Three general types of output are provided by SWMM. If waste treatment processes are simulated or proposed, the capital, land, and operation and maintenance costs are printed. Plots of water quality constituents versus time form the second type of output. These pollutographs are produced for several locations in the system and in the receiving waters. Quality constituents handled by SWMM include suspended solids, settleable solids, BOD, nitrogen, phosphorus, and grease. The third type of output is hydrologic. Hydrographs at any point, for example, the end of a gutter or inlet, are printed for designated time periods. The Statistics Block will provide frequency analysis of storm events from a continuous simulation.

University of Cincinnati Urban Runoff Model (UCURM)

The University of Cincinnati urban runoff model (UCURM) was developed by the Division of Water Resources, the Department of Civil Engineering, of the University of Cincinnati.[26] A flowchart is reproduced in Fig. 25.11. The program consists of three sections: (1) MAIN—infiltration and depression storage, and two subroutines, (2) GUTFL—gutter flow, and (3) PIROU—pipe routing. It is similar to the EPA model and divides the drainage basin into subcatchments whose flows are routed overland into gutters and sewer pipes. The rainfall is read in as a hyetograph. The infiltration and depression storage are summed and subtracted from the rainfall to give overland flow. This is routed through the gutter system to storm water inlets and the pipe network. Starting at the upstream inlet, the flows are calculated in successive segments of the sewer system, including discharges from inlets, to produce the total outflow.

The drainage area is divided into small subcatchments with closely matched characteristics. The rainfall data are introduced and the infiltration is computed for each subcatchment. Principal elements of the modeling process follow:

1. It is assumed that runoff begins whenever the rainfall rate equals the infiltration rate and the mass of precipitation balances infiltration. The equations representing these conditions are

$$t = -\frac{1}{k}\ln\left\{\frac{i(I) + x[i(I+1) - i(I)]/\mathrm{DT} - f_c}{f_0 - f_c}\right\} \qquad (25.23)$$

and

$$\frac{f_c t}{60} + \frac{f_0 - f_c}{60k}(1 - e^{-kt}) = mi(I) + \left\{i(I) + \frac{x}{2\mathrm{DT}}[i(I+1) - i(I)]\right\}\frac{x}{60}$$

$$(25.24)$$

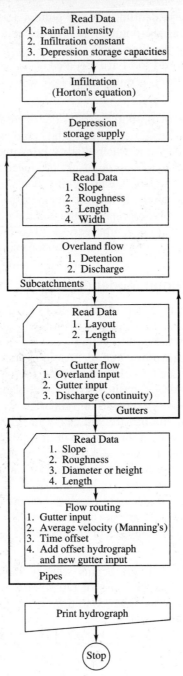

Figure 25.11 UCURM model flowchart. (After Papadakis and Preul.[26])

where $mi(I)$ = the mass precipitated until time t (in.)
   $i(I)$ = the ordinates of rainfall intensity curve
   k = the decay rate of infiltration (units/min)
   f_0 = the initial infiltration capacity (in./hr)
   f_c = the ultimate infiltration capacity (in./hr)
   DT = the time increment of rainfall intensity curve
   t = the time to intersection of rainfall curve and infiltration curve
   x = an increment of DT

The infiltration curve is computed from the equations and t, I, and x are stored.

2. Surface retention is related to depression storage by an equation derived by Linsley et al.,[25]

$$s = (i - f)e^{-(P-F)/S_d} \qquad (25.25)$$

where S_d = the total depression storage (in.)
   P = the accumulated rainfall in storage (in.)
   F = the accumulated infiltration (in.)
   i = the rainfall intensity (in./hr)
   f = the infiltration (in./hr)
   s = the surface retention (in./hr)

The infiltration and surface retention are subtracted from the rainfall intensity to yield the runoff.

3. The hydrograph of the overland flow is derived by solving

$$\frac{r_1 + r_2}{2} - \frac{q_1}{2} + \frac{60D_1}{\Delta t} = \frac{510.35}{nl} s^{1/2} D_2^{5/3} \left[1 + 0.6\left(\frac{D_2}{D_{e_2}}\right)^3\right]^{5/3} + \frac{60D_2}{\Delta t} \qquad (25.26)$$

where $D_e = (0.00979n^{0.6}r^{0.6}l^{0.6})/s^{0.3}$ (25.27)
   $D_{1,2}$ = the detention storage at the beginning and end of time interval t (in./unit area)
   r_1, r_2 = the overland flow supply at the beginning and end of time interval t (cfs/min)
   n = Manning's coefficient
   l = the length of overland flow
   s = the slope (ft/ft)
   q = discharge (in./hr per unit area)

4. For the initial time increment, $q_1 = 0$ and $D_1 = 0$ are substituted, D_2 is calculated, and q_2 is found from

$$q = \frac{1020.7}{nl} s^{1/2} D^{5/3} \left[1 + 0.6\left(\frac{D}{D_e}\right)^3\right]^{5/3} \qquad (25.28)$$

where the symbols are as previously defined. The determined D_2 and q_2 become D_1 and q_1. The overland flow hydrograph is derived by repeating this cycle.

5. The gutter flow is computed using the continuity equation

$$\frac{\partial Q}{\partial x} + \frac{\partial y}{\partial t}T = q_L \tag{25.29}$$

where T is the width of the water surface.

The term $(\partial y/\partial t)T$ is neglected because the change in the depth of the gutter flow is very small with respect to time. After integration, the equation becomes

$$Q = q_L L + Q_0 \tag{25.30}$$

where $\quad Q_0$ = upstream gutter contributions
$\quad\quad L$ = the length of the gutter (ft)
$\quad\quad q_L$ = the overland flow from the hydrograph
$\quad\quad Q$ = the flow from the gutter system

Inlet flows are routed through the pipe network by delaying the hydrographs by the flow time required to reach the next inlet and summing at a terminal point in the network. Manning's equation is used to find the velocity of flow in the sewer and the corresponding time delay. Provision is made for sewers of varying cross section. The model has been found to match gauged flows closely.[26]

TABLE 25.9 COMPARISON OF URBAN RUNOFF MODELS AND METHODS

Model	Surface routing	Pipe flow routing	Quality routing	Degree of sophistication of surface flow routing	Degree of sophistication of pipe flow routing	Accurate modeling of surcharging	Flexibility of modeling of storm drain components
Rational method	Peak flows only	Peak flows only	No	Low	Low	No	Low
Chicago	Yes	No	No	Moderate	NA	NA	NA
Unit hydrograph	Yes	In combination with surface	No	Low	Low	No	Low
STORM	Yes	In combination with surface	Yes	Low	Low	No	Low
RRL	Yes	Yes	No	Moderate	Low-moderate	No	Low
MIT	Yes	No	No	High	NA	NA	NA
EPA-SWMM	Yes	Yes	Yes	High	Moderate	Yes	High
Cincinnati (UCURM)	Yes	Yes	Yes	High	Low	No	Low
HSPF	Yes	Yes	Yes	Moderate	Moderate	No	Low
ILLUDAS	Yes	Yes	No	Moderate	High	No	Low

Source: Modified after Lager and Smith.[8]

25.2 URBAN RUNOFF MODELS COMPARED

Several quantitative comparisons of the RRL, SWMM, UCURM, ILLUDAS, and STORM models have been reported in the literature. A qualitative comparison of several was prepared by Lager and Smith[8] and is shown as Table 25.9. Table 25.10 provides a bullit matrix showing components of most of the same models. Other comparisons involve quantitative analysis of results of the models when applied to actual guaged storm events. One of the first was an application by Heeps and Mein[6] of three models to two urban catchments in Australia for a total of 20 storm events. A similar statistical comparison of the same three models applied to 12 storms over each of three urban watersheds was performed by Marsalek et al.[7] Significant results from these independent evaluations are summarized here. Another quantitative comparison is provided by Huber.[27]

The Heeps and Mein conclusions of model performance are:[6]

1. The degree of subdivision of the catchment has a significant influence on the peak discharge predicted by each of the models. The RRL and SWMM methods give lower peaks and the UCURM gives higher peaks for finer subdivision.

2. The UCURM contains several deficiencies. The major ones, the effects of which can be seen in the predicted hydrographs, are that depression storages

Explicit modeling of in-system storage	Treatment modeling	Receiving model available	Degree of calibration/ verification required	Simulation period	Availability	Documentation	Data requirements
No	NA	No	Usually not verified	Individual storms	Nonproprietary	Good	Low
NA	NA	No	Moderate	Individual storms	Nonproprietary	Fair	Moderate
No	NA	No	High	Individual storms	Nonproprietary	Fair	Moderate
No	Yes	No	Low	Long term	Nonproprietary	Good	Moderate
No	NA	No	Moderate	Individual storms	Nonproprietary	Good	Moderate
NA	NA	No	Moderate	Individual storms	Proprietary	Fair	Moderate
Yes	Yes	Yes	Moderate	Individual or continuous storms	Nonproprietary	Good	Extensive
No	No	No	Moderate	Individual storms	Nonproprietary	Fair	Extensive
No	No	Yes	High	Individual storms or long-term	Nonproprietary	Fair	Extensive
No	NA	No	Moderate	Individual storms	Nonproprietary	Good	Extensive

TABLE 25.10 COMPARISON OF URBAN RUNOFF MODEL COMPONENTS AND CAPABILITIES

	British road research laboratory	Chicago hydrograph method	Corps of engineers STORM	Environmental protection agency SWMM	Hydrocomp HSPF	Massachusetts Institute of Technology MITCAT	University of Cincinnati UCURM	Illinois ILLUDAS
Catchment hydrology								
Multiple catchment inflows	●	●		●	●	●	●	●
Dry weather flow	●	●		●	●	●		●
Input of several hyetographs	●	●		●	●	●		
Snowmelt			●	●	●			●
Runoff from impervious areas	●	●	●	●	●	●	●	●
Runoff from pervious areas		●	●	●	●	●	●	●
Water balance between storms		●	●		●			
Sewer hydraulics								
Flow routing in sewers	●	●		●	●	●	●	●
Upstream and downstream flow control				●	●			●
Surcharging and pressure flow				●				
Diversions			●	●	●			●
Pumping stations				●				●
Storage			●	●	●			
Prints stage				●	●	●		●
Prints velocities				●	●	●		●
Wastewater quality								
Dry weather quality				●	●			
Stormwater quality			●	●	●			
Quality routing				●	●			
Sedimentation and scour				●				
Quality reactions			●	●	●			
Wastewater treatment				●				
Quality balance between storms			●	●	●			
Receiving water flow simulation	●	●		●	●	●	●	
Receiving water quality simulation				●	●			
Miscellaneous								
Continuous simulation			●	●	●			
Can choose time interval	●	●		●	●	●	●	●
Design computations		●		●			●	●
Real-time control					●			
Computer program available	●	●	●	●			●	●

Source: Modified after Brandstetter.[12]

are assigned full when the rainfall intensity falls below the infiltration capacity, and that depression storages are not depleted by infiltration. The use of instantaneous values of the rainfall intensity (difficult to obtain from recorder charts) can cause volume errors.

3. The SWMM was the model with the best overall performance but at the expense of large computer storage and time requirements.

4. The RRL model predicted poorly for storms in which pervious runoff was significant but performed reasonably well for many other types of storms. The results, in general, support those of Stall and Terstriep.[9]

5. A major problem with using noncontinuous models is the prediction of antecedent conditions. This problem is further aggravated by use of the Horton infiltration equation for which prediction of the parameters is virtually impossible.

The Marsalek study results,[7] using the same three models for three watersheds in Illinois, Ontario, and Maryland, indicated that the SWMM model performed slightly better than the RRL model and both these models were more accurate than the UCURM model for the small watersheds studied.

Table 25.11 provides descriptions of the urban watersheds used by Marsalek in the runoff model evaluation. Typical comparisons of observed and calculated times to peak, peak flows, and runoff volumes for the three models are provided in Fig. 25.12 for the Calvin Park watershed in Kingston, Ontario.

Marsalek et al. used the information described to develop the qualitative comparison in Table 25.12 and arrived at the following conclusions:[7]

1. Uncalibrated deterministic models for urban runoff, such as RRL, SWMM, and UCURM, yielded a fairly good agreement between the simulated and measured runoff events on typical urban catchments of small size.

2. On the average, about 60 percent of the simulated peak flows were within 20 percent of the measured values. About the same scatter was found for the simulated times to peak and runoff volumes. The agreement between the measured and simulated values could be further improved by model calibration.

3. When comparing the entire simulated and measured hydrographs using statistical measures, the agreement was found good for SWMM, good to fair in the case of RRL, and fair in the case of UCURM.

TABLE 25.11   DESCRIPTION OF TEST URBAN WATERSHEDS USED FOR THE EVALUATION OF URBAN RUNOFF MODELS

Test urban drainage basin	Location	Area (acres)	Imperviousness	Land use
Oakdale	Chicago, IL	13.0	45.8	Residential
Calvin Park	Kingston, Ontario	89.4	27	Residential and institutional
Gray Haven	Baltimore, MD	23.3	52	Residential

Source: Marsalek et al.[7]

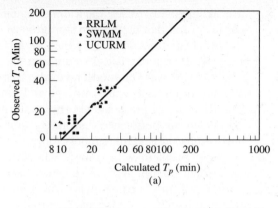

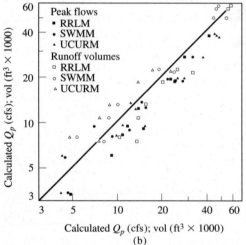

Figure 25.12 Observed and computed values compared for the Calvin Park subcatchment. T_p = time to peak, Q_p = runoff peak flow, and vol = total runoff volume. (After Marsalek et al.[7])

TABLE 25.12   SUBJECTIVE EVALUATION OF THREE URBAN MODELS

	Model		
	RRL	SWMM	UCURM
Effort for input data preparation	Low	Medium	High
Flexibility of schematization	Good	Good	Fair
Accuracy of employed routing scheme	Low	Medium	Low
Model and computer program availability	Good	Excellent	Limited
Availability of a runoff quality submodel	No	Yes	Yes
Computer time required	Low	Moderate	Moderate
Continuous refinement by the corresponding agency	No	Yes	No

Source: Marsalek et al.[7]

4. Out of the three studied models, the SWMM simulations were marginally better than those by RRL, and both these models were more accurate than UCURM, with all models applied in an uncalibrated version. The main advantage of the RRL is its simplicity, since it can be applied without a computer. The SWMM, on the other hand, is much more general and versatile, and is being continuously refined and improved.

5. The SWMM model has the most advanced routing schemes among the considered methods. The accuracy of flow routing becomes particularly important when studying large watersheds.

25.3 VENDOR-DEVELOPED URBAN STORMWATER SOFTWARE

With the profusion of microcomputer and minicomputer software in the urban stormwater field, the urban hydrologist is often frustrated over which to choose for any intended application. Surveys of available software have been conducted to provide help in shopping among the many choices.[28, 29] Federal agency software was inventoried by Jennings,[30] and the American Society of Civil Engineers (ASCE) Task Committee on Microcomputer Software in Urban Hydrology inventoried the makeup, cost, and applicability of packages available from over 40 commercial vendors and public domain sources.[21] Some of the findings of the surveys are summarized in Table 25.13 and are discussed below.

Applications   Computer packages for urban stormwater modeling fall into two categories, those codes intended primarily for the *analysis* of an existing or proposed system, and *design* packages which actually select storm sewer diameters, sizes of channels or ponds, or other features of stormwater management facilities. About 80 percent of the software packages allow the design of detention basin size, and over 70 percent have routines for design of storm sewers or inlets. All but a few (about 10 percent) allow the user to generate full direct runoff hydrographs.

Programming Languages   One third of the codes are written in FORTRAN, another third in BASIC, and the remainder largely "C," PASCAL, assembly language, and spreadsheet format. Some allow integration with computer automated design (CAD) and geographical information software (GIS) systems. A surprising 70 percent of the codes are originally developed by the vendor as opposed to simply adding pre- or post-processors to public domain software. Caution is advised in using any commercial software to verify the assumptions and assure that the procedures coded in these packages replicate the canonized methods.

Users   According to the surveys, the predominant use of urban stormwater software is by consultants (72 percent of reported applications). Another 20 percent of the reported uses are split evenly between local, state, and federal government agencies, with the remainder falling to private industry applications and university instruction.

Hydrologic Processes Modeled   A full range of processes is modeled by the packages.[21] Common among popular renditions are routines available in TR-20, HEC-1, and TR-55. The predominant hydrologic method incorporated in the majority

TABLE 25.13   RESULTS OF ASCE INVENTORY OF 40 SOFTWARE PACKAGES[21]

Software characteristics	No. packages
Number of public domain packages	12
Number of commercial packages with copyright	25
Number of commercial packages without copyright	3
Number offering some technical support	38
Number with full hydrograph option	35
Number with storm sewer option	29
Number with detention basin option	31
Number tested against gauged data	30
Typical costs for complete package	$150-$3500

Hydrologic abstraction methods	No. pkgs. using
SCS Curve Number	17
Rational C coefficient	7
Green and Ampt infiltration equation	6
Horton infiltration equation	6
Constant or uniform loss rate	5
Holtan infiltration equation	3
Other	9

Hydrograph synthesis methods	No. pkgs. using
SCS dimensionless unit hydrograph	14
Rational method	9
Snyder's synthetic unit hydrograph	6
Kinematic wave hydrograph synthesis	5
Clark's IUH/Time-area method	5
TR-55 1986 tabular hydrograph	3
Santa Barbara urban hydrograph method	2
Other	15

Hydrograph routing methods	No. pkgs. using
Translation without attenuation, based on travel time	10
Muskingum/Muskingum-Cunge method	9
Kinematic wave routing	7
Modified-Puls/Storage-Indication routing (channels and reservoirs)	6
SCS Convex/Att-Kin method	2
Full hydrodynamic St. Venant equations	2
Other	10

of commercial codes is SCS techniques—SCS unit hydrographs, SCS rainfall distributions, SCS curve number rainfall abstractions, and, to a lesser extent, SCS stream and channel routing procedures. This change has been progressive as documented by the comparison of three surveys over time shown in Table 25.14. This phenomenon is attributed to the ease of programming SCS methods rather than their particular pertinence to urban rainfall-runoff modeling.

As shown in Tables 25.13 and 25.14, three quarters of the codes now use the CN method for rainfall abstraction; the rest employ infiltration by Horton's equation, the Green and Ampt method, constant losses, exponential losses, and a handful of other

TABLE 25.14   PREDOMINANCE OF SCS METHODS IN VENDOR-DEVELOPED URBAN RUNOFF SOFTWARE

(Tabulated values are percentages of responses falling in the class)

Year of survey: Class of Method	1985 (Ref. 28)	1988 (Ref. 29)	1991 (Ref. 21)
SCS CN Method	40%	70%	75%
SCS NEH-4 Hydrograph Synthesis	30%	60%	70%
SCS TR-55 Methods	10%	20%	40%

methods. Hydrograph synthesis (Chapter 12) is predominantly by the SCS dimensionless unit hydrograph method. Several employ a triangular unit hydrograph having a peak flow rate calculated from Eq. 12.22. This is described in the vendor literature as an SCS-based method; however, this practice has not been endorsed by the SCS. Hydrograph routing in channels (Chapter 13) is about evenly divided among the Muskingum, Muskingum-Cunge, kinematic wave, and storage-indication methods. Reservoir routing is almost universally executed by the storage-indication method. Where needed, detailed reservoir routing has been accomplished by solving the hydrodynamic equations of motion (Section 13.3).

Versatility   In reviewing the software, one finds that some packages have unique features and options that would be highly applicable in certain circumstances. For example, some relatively flat watersheds tend to retain significantly high percentages of the initial rainfall. In comparing observed and calculated runoff, matches are sometimes impossible without revisiting the SCS assumption that the initial abstraction is 20 percent of the potential maximum, or $I_a/S = 0.2$ (see Chapter 4). Increases in the ratio up to values of 0.5 or more are sometimes justified. Among other examples of unique features, commercial packages are available for:

1. Simultaneous tracking and graphic output of hydrographs for the existing *and* proposed conditions.
2. Combined calculation of combined storm drain and sanitary sewer flows in systems that still incorporate this obsolete feature.
3. Tracking and rerouting of flow that is not able to enter underdesigned storm sewers. This overflow generally passes down streets and overland until it reaches a swale or channel, or it may combine with hydrographs from other subareas before being routed to other inlets.
4. Sizing of reservoir capacity by the "bowstring" method, which is essentially the mass-curve analysis method described in Section 22.1. Note the bowstring appearance of Fig. 22.1.

■ Summary

By far the greatest application of single-event rainfall-runoff models is for urbanized or urbanizing watersheds. The developed and developing models described in this chapter are but a few of the many available to the urban stormwater analyst or designer. As in previous chapters, the reader is encouraged to research these descriptions and tabulate the applicability of each, including notes on useful features such as a design mode, full hydrograph synthesis and routing, storm sewer overflow routing, or water quality modeling in streams and ponds.

PROBLEMS

25.1. Determine the maximum and minimum infiltration rates and the recession constant k (give units) for the infiltration capacity curve in Fig. 4.6. Compare these with corresponding values for bluegrass used in ILLUDAS.

25.2. Prepare a table similar to Table 25.13 and identify which of the nine models and methods described in Section 25.1 perform the abstraction, synthesis, and routing methods in Table 25.13. For the three hydrologic categories of Table 25.13, identify any procedures used by the models but not named.

25.3. Using the initial and ultimate infiltration rates from Table 25.4, calculate and plot Horton's (Eq. 25.6) infiltration rates at the intermediate times. Then plot and compare corresponding infiltration curves by Holtan's and Horton's methods. Discuss the differences.

25.4. Repeat Problem 24.1 using the watershed in Fig. 25.9.

25.5. Compare the DR3M method of infiltration in Eqs. 25.10–25.12 with the procedure employed by SWM-IV in Chapter 23.

25.6. Do the Heeps and Mein conclusions regarding RRL, SWMM, and UCURM agree with the conclusions drawn by Marsalek et al.?

25.7. By examining the points in Fig. 25.12b, discuss the most obvious conclusions regarding the three models' ability to reconstitute the peak flow rates for the sample watersheds.

25.8. Solve Problem 25.2, then calculate the percentage of models having SCS attributes as shown in Table 25.14. Discuss why the percentages are different for the models in Section 25.1 and those surveyed in Section 15.3.

REFERENCES

1. A. L. Tholin and C. T. Keifer, "Hydrology of Urban Runoff," *Proc. ASCE J. San. Eng. Div.* **85**(SA2), 47–106(Mar. 1959).
2. C. F. Izzard, "Hydraulics of Runoff from Developed Surfaces," *Proceedings 26th Annual Meeting Highway Research Board,* **26,** 129–146(1946).
3. C. J. Keifer, J. P. Harrison, and T. O. Hixson, "Chicago Hydrograph Method Network Analysis of Runoff Computations," Preliminary Report, City of Chicago, Bureau of Engineering, July 1970.
4. L. H. Watkins, *The Design of Urban Sewer Systems,* Road Research Tech. Paper No. 55, Department of Scientific and Industrial Research London: Her Majesty's Stationery Office, 1962.
5. Michael L. Terstriep and John B. Stall, "Urban Runoff by the Road Research Laboratory Method," *Proc. ASCE J. Hyd. Div.* **95**(HY6), 1809–1834(Nov. 1969).
6. D. P. Heeps and R. G. Mein, "Independent Comparison of Three Urban Runoff Models," *Proc. ASCE J. Hyd. Div.* **100**(HY7), 995–1010(July 1974).

7. J. Marsalek, T. M. Dick, P. E. Wisner, and W. G. Clarke, "Comparative Evaluation of Three Urban Runoff Models," *Water Resources Bull. AWRA* **11**(2), 306–328(Apr. 1975).

8. J. A. Lager and W. G. Smith, "Urban Stormwater Management and Technology—An Assessment," U.S. EPA Rep. EPA-670/2-74-040, Dec. 1974.

9. John B. Stall and Michael L. Terstriep, "Storm Sewer Design—An Evaluation of the RRL Method," EPA Technology Series EPA-R2-72-068, Oct. 1972.

10. M. L. Terstriep and J. B. Stall, "The Illinois Urban Drainage Area Simulator, ILLUDAS," *Illinois State Water Surv. Bull.* **58,** 1–30(1974).

11. H. N. Holtan, "A Concept for Infiltration Estimates in Watershed Engineering," U.S. Department of Agriculture, Agricultural Research Service, 1961.

12. A. Brandstetter, "Comparative Analysis of Urban Stormwater Models," Battelle Memorial Institute, Aug. 1974.

13. H. G. Wenzel and M. L. Terstriep, "Sensitivity of Selected ILLUDAS Parameters," Illinois State Water Survey, Contract Rep. 178, Aug. 1976.

14. J. Han and J. W. Delleur, "Development of an Extension of ILLUDAS for Continuous Simulation of Urban Runoff Quantity and Discrete Simulation of Runoff Quality," Purdue University, July 1979.

15. U.S. Army Corps of Engineers, "Urban Storm Water Runoff, STORM," Computer Program 723-S8-L2520, Hydrologic Engineering Center, Davis, CA, Oct. 1974.

16. National Technical Information Service, "TR-55, Hydrology for Small Urban Watersheds," U.S. Department of Commerce, Springfield VA, 1986.

17. U.S. Soil Conservation Service, "Urban Hydrology for Small Watersheds, Technical Release 55, rev., U.S. Department of Agriculture, Washington D.C., 1986.

18. W. M. Alley and P. E. Smith, "Distributed Routing Rainfall-Runoff Model—Version II, User's Manual," USGS Open-File Rep. 82-344, 1982.

19. W. M. Alley, "Summary of Experience with the Distributed Routing Rainfall-Runoff Model (DR3M)," U.S. Geological Survey, Reston VA, 1986.

20. Federal Highway Administration, *HYDRAIN—Integrated Drainage Design Computer System,* Washington, D.C., 1990.

21. ASCE Task Committee on Urban Stormwater Software, "Microcomputer Software in Urban Hydrology," by D. F. Kibler, M. E. Jennings, G. L. Lewis, B. A. Tschantz, and S. G. Walesh, *HYDATA ,* American Water Resources Association, v. 10, no. 5, Sept. 1991.

22. Metcalf and Eddy, Inc., University of Florida, Gainesville, Water Resources Engineers, Inc., "Storm Water Management Model," Environmental Protection Agency, Vol. 1, 1971.

23. W. C. Huber et al., "Storm Water Management Model User's Manual, Version III," EPA-600/2-84-109a (NTIS PB84-198423), Nov. 1981.

24. L. A. Roesner et al., "Storm Water Management Model User's Manual, Version III: Addendum I, Extran," EPA-600/2-84-109b (NTIS PB84-198431), Nov. 1981.

25. R. K. Linsley, Jr., M. A. Kohler, and J. A. H. Paulhus, *Applied Hydrology.* New York: McGraw-Hill, 1949.

26. C. N. Papadakis and H. C. Preul, "University of Cincinnati Urban Runoff Model," *Proc. ASCE J. Hyd. Div.* **98**(HY 10), 1789–1804(Oct. 1972).

27. W. C. Huber, "Modeling for Storm Water Strategies," *APWA Reporter,* May 1975.

28. G. L. Lewis and D. P. Gilbert, "A Shopper's Guide to Urban Stormwater Micro Software," *Proceedings, ASCE Hydraulics Division Specialty Conference,* Orlando, FL, Aug. 1985.

29. G. L. Lewis, "A Shopper's Guide to Urban Stormwater Software Revisited," *Proceedings, ASCE National Conference on Hydraulic Engineering,* Colorado Springs, CO 1988.

30. M. E. Jennings et al., "Federal Microcomputer Software for Urban Hydrology," *Proceedings, ASCE 1988 National Conference on Hydraulic Engineering,* Colorado Springs, CO, 1988.

PART SIX

STATISTICAL METHODS

Chapter 26

Probability and Statistics

▪ Prologue

The purpose of this chapter is to:

- Introduce the basic tenets of probability theory as applied to random, hydrologic variables, with particular emphasis on the *relative frequency* definition of probability—a concept that is foundational to the frequency analysis procedures presented in Chapter 27 and throughout many other chapters of this text.
- Describe common probability distributions and show how they are applied to hydrologic phenomena.
- Relate the fundamentals of probability theory to hydrologic design criteria described in Chapter 16, Section 16.3.
- Acquaint the reader with the theory behind linear regression and show how this powerful technique is used in hydrology to predict how a study watershed will respond to some change by examining responses of the watershed to past inputs or by statistically scrutinizing responses of other similar watersheds in order to develop a predictive equation for the subject watershed.
- Show how to transform many hydrologic variables that have nonlinear relationships into new variates that can then be analyzed by performing linear regression on the transformed variates.
- Provide the theoretical and practical foundation necessary to fully capitalize on the hydrologic design principles discussed in Chapter 16 and the time-series analysis and modeling procedures described in Chapter 22.

Commonly, the study of hydrology is undertaken by readers who lack the prerequisite background in principles of statistics, probability theory, and frequency analysis. As a consequence, most hydrology courses review these subjects early in the schedule. Practically all hydrology texts include chapters on statistical methods to summarize the basic principles of statistics, probability theory, probability distributions, bivariate and multiple linear correlation and regression, time-series analysis, and frequency analysis. Thus, despite the placement of this material at the end of this text, the authors assume that the reader has this background or will study the material in Part Six prior to beginning a study of Part Three. Readers with an understanding of statistical methods, regression analysis, and the basics of probability distribution functions may wish to turn directly to Chapter 27.

26.1 RANDOM VARIABLES AND STATISTICAL ANALYSIS

A *random variable* is one that demonstrates variability that isn't sufficiently explained by physical processes. Many hydrologic phenomena have this tendency, appearing at times to be fully subject to chance themselves, or driven by some other closely related factor. In practice, hydrologists often analyze problems as systems of connected random and deterministic processes. For example, precipitation is often evaluated statistically as a random variable because of the complexity of understanding and modeling the atmospheric processes that are known to drive the precipitation system. Runoff that results from the precipitation, on the other hand, is viewed deterministically, using the rainfall–runoff analogs that are the nucleus of the majority of this textbook.

Hydrology relies heavily on principles from probability theory, statistics, and information analysis. Whole texts on frequency analysis methods, stochastic generation of data, regression and analysis of variance, and regional analyses are available containing thorough descriptions of the principles.[1,2] Many hydrologic processes are so complex that they can be interpreted and explained only in a *probabilistic* sense. Hydrologic events appear as uncertainties of nature and are the result, it must be assumed, of an underlying process with random or *stochastic* components. The information to investigate these processes is contained in records of hydrologic observations. Methods of *statistical analysis* provide ways to reduce and summarize observed data, to present information in precise and meaningful form, to determine the underlying characteristics of the observed phenomena, and to make predictions concerning future behavior. Statistical analysis deals with methods for drawing inferences about the population based on examination of sample values from the population. These inferences include information about the central tendency, range, distribution within the range, variability around the central tendency, degree of uncertainty, and frequency of occurrence of values.

Statistical analysis involves two basic sets of problems, one *descriptive,* the other *inferential*. The former is a straightforward application of statistical methods, requiring few decisions and representing little risk. The inferential problem, however, entails decisions bearing some risk, and requires an understanding of the methods employed and the dangers involved in predicting and estimating. The most common inferential problem is to describe the whole class of possible occurrences when only a portion of them has been observed. The whole class is the *population* and the portion observed is the *sample*.

The random variables in the process under study are *continuous* if they may take on all values in the range of occurrence, including figures differing only by an infinitesimal amount; they are *discrete* if they are restricted to specific, incremental values. Distribution of the variables over the range of occurrence is defined in terms of the frequency or *probability* with which different values have occurred or might occur.

26.2 CONCEPTS OF PROBABILITY

The laws of probability underlie any study of the statistical nature of repeated observations or trials. The probability of a single event, say E_1, is defined as the relative number of occurrences of the event after a long series of trials. Thus $P(E_1)$, the probability of event E_1, is n_1/N for n_1 occurrences of the same event in N trials if N is sufficiently large. The number of occurrences n_1 is the *frequency,* and n_1/N the *relative frequency.*

Often the probabilities and the rules governing their manipulation are known intuitively or from experience. In the familiar coin-tossing experiment, $P(\text{heads}) = P(\text{tails}) = \frac{1}{2}$. Each outcome of a single toss (a trial) has a finite probability, and the sum of the probabilities of all possible outcomes is 1. Also, the outcomes are *mutually exclusive;* that is, if one occurs, say a head, then a tail cannot occur. In two successive tests, there are four possible outcomes—HH, TT, HT, TH—each with a probability of $\frac{1}{4}$. In this case, because each trial is independent of the other one, probabilities for each outcome are found by $P(\text{first trial}) \times P(\text{second trial}) = \frac{1}{2} \times \frac{1}{2} = \frac{1}{4}$. Again, the sum of the probabilities of the possible outcomes is 1. Note that the probability of getting exactly one head and one tail during the experiment (without any regard to the order) is $P(\text{HT}) + P(\text{TH}) = \frac{1}{2}$.

Summarizing the rules of probability indicated by coin tossing, we find the following:

1. The probability of an event is nonnegative and never exceeds 1.

$$0 \leq P(E_i) \leq 1 \qquad (26.1)$$

2. The sum of the probabilities of all possible outcomes in a single trial is 1.

$$\sum_i P(E_i) = 1 \qquad (26.2)$$

3. The probability of a number of *independent* and *mutually exclusive* events is the sum of the probabilities of the separate events.

$$P(E_i \cup E_2) = P(E_1) + P(E_2) \qquad (26.3)$$

The probability statement, $P(E_1 \cup E_2)$, signifies the probability of the *union* of two events and is read "the probability of E_1 or E_2."

4. The probability of two *independent* events occurring simultaneously or in succession is the product of the individual probabilities.

$$P(E_1 \cap E_2) = P(E_1) \times P(E_2) \qquad (26.4)$$

$P(E_1 \cap E_2)$ is called probability of the *intersection* of two events or *joint probability* and is read "the probability of E_1 and E_2."

Consider the following example of events that are not independent or mutually exclusive: An urban drainage canal reaches flood stage each summer with relative frequency of 0.10; power failures in industries along the canal occur with probability of 0.20; experience shows that when there is a flood the chances of a power failure for whatever reason are raised to 0.40. The probability statements are

$$P(\text{flood}) = P(F) = 0.10 \qquad P(\text{power failure}) = P(P) = 0.20$$

$$P(\text{no flood}) = P(\overline{F}) = 0.90 \qquad P(\text{no power failure}) = P(\overline{P}) = 0.80$$

$$P(\text{power failure given that a flood occurs}) = 0.40$$

The last statement is called a *conditional probability*. It signifies the joint occurrence of events and is usually written $P(P \mid F)$. Rules 3 and 4 no longer are strictly applicable. If Rule 3 applied, $P(F \cup P) = P(F) + P(P) = 0.3$. If the events remained independent, the conditional probability $P(P \mid F)$ would equal the *marginal* probability $P(P)$. Thus the events are independent if the probability of either is not "conditioned by" or changed by knowledge that the other has occurred. For independent events, the joint probabilities would be

$$P(F \cap P) = 0.1 \times 0.2 = 0.02$$

$$P(F \cap \overline{P}) = 0.1 \times 0.8 = 0.08$$

$$P(\overline{F} \cap P) = 0.9 \times 0.2 = 0.18$$

$$P(\overline{F} \cap \overline{P}) = 0.9 \times 0.8 = 0.72$$

The probability of a flood or a power failure during the summer would be the sum of the first three joint probabilities above.

$$P(F \cup P) = P(F \cap P) + P(F \cap \overline{P}) + P(\overline{F} \cap P) = 0.28$$

The events are dependent, however, from the statement of conditional probability: When a flood occurs with $P(F) = 0.1$, a power failure will occur with probability 0.4, and true joint probability is $P(F) \times P(P \mid F) = 0.1 \times 0.4 = 0.04 = P(F \cap P)$. The probability of the union is then $P(F \cup P) = P(F) + P(P) - P(F \cap P) = 0.1 + 0.2 - 0.04 = 0.26$. Note the contrast:

$$P(F \cup P) = 0.30 \quad \text{for mutually exclusive events}$$

$$P(F \cup P) = 0.28 \quad \text{for joint but independent events}$$

$$P(F \cup P) = 0.26 \quad \text{otherwise}$$

The new, more general rule for the union of probabilities is

5. $P(E_1 \cup E_2) = P(E_1) + P(E_2) - P(E_1 \cap E_2)$ \hfill (26.5)

and a sixth rule should be added for conditional probabilities:

6.
$$P(E_1 \mid E_2) = \frac{P(E_1 \cap E_2)}{P(E_2)} \tag{26.6}$$

A very important concept of independence is expressed in a variation of Rule 6, namely, $P(E_1 \mid E_2) = P(E_1)$ if events E_1 and E_2 are independent. This further explains Rule 3 that $P(E_1) \times P(E_2) = P(E_1 \cap E_2)$ for independent events.

The example of flooding can be extended to show some interesting features about probabilities and risks associated with hydrologic phenomena. $P(F) = .10$ implies a 10 percent chance each year for the flood to "occur," meaning that the flood level will be exceeded. Because the probability of any single, exact value of a continuous variable is 0.0, "occur" can also mean the level will be reached or exceeded. In the long run, the level would be reached or exceeded on the average once in 10 years. Thus the average *return period* * T in years is defined as

$$T = \frac{1}{P(F)} = \frac{1}{1 - P(\overline{F})} \tag{26.7}$$

and the following general probability relation hold:

1. The probability that F will be equalled or exceeded in any year

$$P(F) = \frac{1}{T} \tag{26.8}$$

2. The probability that F will not be exceeded in any year

$$P(\overline{F}) = 1 - P(F) = 1 - \frac{1}{T} \tag{26.9}$$

3. The probability that F will not be equalled or exceeded in any of n successive years

$$P_1(\overline{F}) \times P_2(\overline{F}) \times \cdots \times P_n(\overline{F}) = P(\overline{F})^n = \left(1 - \frac{1}{T}\right)^n \tag{26.10}$$

4. The probability R, called *risk*, that F will be equalled or exceeded at least once in n successive years

$$R = 1 - \left(1 - \frac{1}{T}\right)^n = 1 - \{P(\overline{F})\}^n \tag{26.11}$$

Table 26.1 shows return periods associated with various levels of risk.

TABLE 26.1 RETURN PERIODS ASSOCIATED WITH VARIOUS DEGREES OF RISK AND EXPECTED DESIGN LIFE

Risk (%)	Expected design life (years)							
	2	5	10	15	20	25	50	100
75	2.00	4.02	6.69	11.0	14.9	18.0	35.6	72.7
50	3.43	7.74	14.9	22.1	29.4	36.6	72.6	144.8
40	4.44	10.3	20.1	29.9	39.7	49.5	98.4	196.3
30	6.12	14.5	28.5	42.6	56.5	70.6	140.7	281
25	7.46	17.9	35.3	52.6	70.0	87.4	174.3	348
20	9.47	22.9	45.3	67.7	90.1	112.5	224.6	449
15	12.8	31.3	62.0	90.8	123.6	154.3	308	616
10	19.5	48.1	95.4	142.9	190.3	238	475	950
5	39.5	98.0	195.5	292.9	390	488	976	1949
2	99.5	248	496	743	990	1238	2475	4950
1	198.4	498	996	1492	1992	2488	4975	9953

* The terms *return period* and *recurrence interval* are used interchangeably to denote the reciprocal of the annual probability of exceedence.

EXAMPLE 26.1 _____

If T is the recurrence interval for a flood with magnitude Q_a, find the probability (risk) that the peak flow rate will equal or exceed Q_a at least once in two consecutive years. Assume the events are independent.

Solution. The solution is easily obtained by substitution into Eq. 26.11. To assist in understanding the equations, an alternative derivation follows.

The four possible outcomes for the two years are:

a: nonexceedance in both years
b: exceedance in the first year only
c: exceedance in the second year only
d: exceedance in both years

Because these four represent all possible outcomes, the probability of the union of a, b, c, and d is 1.0, or from Eq. 26.2, $P(a \cup b \cup c \cup d) = 1.0$. Exceedance in at least one year is satisfied by b, c, or d, but not a. Thus the risk of at least one exceedance is $P(b \cup c \cup d)$, which is the total less the probability of a. From Eqs. 26.2 and 26.3, we find that

$$2\text{-year risk} = P(b \cup c \cup d) = 1 - P(a)$$

From Eq. 26.3, we find that

$$P(a) = P(Q < Q_a \text{ in Year 1}) \times P(Q < Q_a \text{ in Year 2})$$

$$= \left(1 - \frac{1}{T}\right)\left(1 - \frac{1}{T}\right)$$

and $$\text{Risk} = 1 - P(a) = 1 - \left(1 - \frac{1}{T}\right)^2$$ ∎∎

EXAMPLE 26.2 _____

What return period must a highway engineer use in designing a critical underpass drain to accept only a 10 percent risk that flooding will occur in the next 5 years?

Solution

$$R = 1 - \left(1 - \frac{1}{T}\right)^n$$

$$.10 = 1 - \left(1 - \frac{1}{T}\right)^5$$

$$T = 48.1 \text{ years}$$ ∎∎

26.3 PROBABILITY DISTRIBUTIONS

Random variables, either discrete or continuous, are characterized by the distribution of probabilities attached to the specific values that the variable may assume. A random variable throughout its range of occurrence is generally designated by a capital letter, and a specific value or outcome of the random process is designated by a small letter.

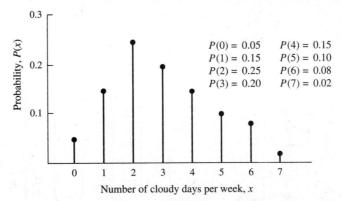

$P(0) = 0.05 \quad P(4) = 0.15$
$P(1) = 0.15 \quad P(5) = 0.10$
$P(2) = 0.25 \quad P(6) = 0.08$
$P(3) = 0.20 \quad P(7) = 0.02$

Figure 26.1 Probability distribution of cloudy days per week.

For example, $P(X = x_1)$ is the probability that random variable X takes on the value x_1. A shorter version is $P(x_1)$. Figure 26.1 shows the probability distribution of the number of cloudy days in a week. It is a discrete distribution because the number of days is exact; in the record from which the relative frequencies were taken, a day had to be described as cloudy or not. Observe that each of the seven events has a finite probability and the sum is 1; that is,

$$\sum_i P(x_i) = 1$$

Another important property of random variables is the *cumulative distribution function,* CDF, defined as the probability that any outcome in X is less than or equal to a stated, limiting value x. The cumulative distribution function is denoted $F(x)$. Thus

$$F(x) = P(X \leq x) \tag{26.12}$$

and the function increases monotonically from a lower limit of zero to an upper bound of unity. Figure 26.2 is the CDF of the number of cloudy days in a week derived from Fig. 26.1 by taking cumulative probabilities. The function shows that the probability is 90% that the number of cloudy days in the week will be 5 or less. Conversely, there is a 10 percent probability that it will be cloudy for 6 or 7 days. This complementary cumulative probability is sometimes called $G(x)$, where[3]

$$G(x) = 1 - F(x) = P(X \geq x) \tag{26.13}$$

Continuous variables present a slightly different picture. Figure 26.3 is the *histogram* of an 85-year record of annual streamflows. The observations were grouped into nine intervals ranging from 0 to 900 cfs and the number falling in each interval was plotted as frequency on the left ordinate. A convenient alternative is to plot the relative frequency as shown by the right ordinate. The CDF for the streamflow record is shown in Fig. 26.4. As the number of observations increase, the continuous distribution will be developed by reducing the size of the intervals. In the limit, the broken curves of Figs. 26.3 and 26.4 will appear as those in Fig. 26.5.

There is a difference between the ordinates of Figs. 26.3 and 26.5a. Since relative frequency is synonymous with probability, it is convenient to reconstitute the

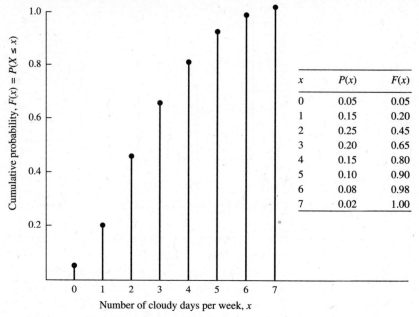

x	P(x)	F(x)
0	0.05	0.05
1	0.15	0.20
2	0.25	0.45
3	0.20	0.65
4	0.15	0.80
5	0.10	0.90
6	0.08	0.98
7	0.02	1.00

Figure 26.2   Cumulative distribution of cloudy days per week.

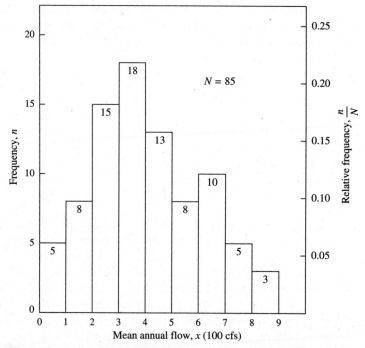

Figure 26.3   Frequency distribution of mean annual flows.

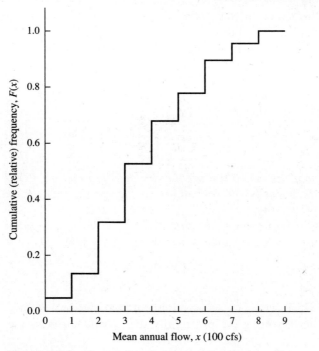

Figure 26.4 Cumulative frequency distribution of mean annual flows.

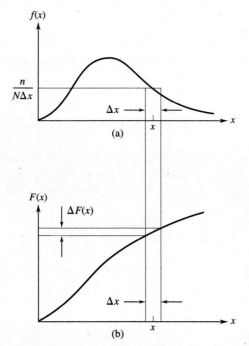

Figure 26.5 Continuous probability distributions: (a) probability density function and (b) cumulative distribution function.

histogram so that the area in each interval represents probability; the total area contained is thus unity. To do this, the ordinate in each interval, say n/N for relative frequency or probability, is divided by the interval width, Δx. The ratio $n/N \, \Delta x$ is literally the probability per unit length in the interval and therefore represents the average density of probability. The probability n/N in the interval is represented on the CDF (before the limiting process) as $\Delta F(x)$, or $F(x + \Delta x/2) - F(x - \Delta x/2)$. We then can define

$$f(x) = \lim_{\Delta x \to 0} \frac{\Delta F(x)}{\Delta x} = \frac{dF(x)}{dx} \tag{26.14}$$

which is called the *probability density function,* PDF.[3] This function is the density (or intensity) of probability at any point; $f(x) \, dx$ is described as the differential probability.

For continuous variables, $f(x) \geq 0$, since negative probabilities have no meaning. Also, the function has the property that

$$\int_{-\infty}^{\infty} f(x) \, dx = 1 \tag{26.15}$$

which again is the requirement that the probabilities of all outcomes sum to 1. Furthermore, the probability that x will fall between the limits a and b is written

$$P(a \leq X \leq b) = \int_{a}^{b} f(x) \, dx \tag{26.16}$$

Note that the probability that x takes on a particular value, say a, is zero; that is,

$$\int_{a}^{a} f(x) \, dx = 0 \tag{26.17}$$

which emphasizes that finite probabilities are defined only as areas under the PDF between finite limits.

The CDF can now be defined in terms of the PDF as

$$P(-\infty \leq X \leq x) = P(X \leq x) = F(x) = \int_{-\infty}^{x} f(u) \, du \tag{26.18}$$

where u is used as a dummy variable to avoid confusion with the limit of integration. The area under the CDF has no meaning, only the ordinates, or the difference in ordinates. For example, $P(x_1 \leq X \leq x_2)$, which is equivalent to Eq. 26.16, can be evaluated as $F(x_2) - F(x_1)$.

For discrete distributions that cannot be summarized in integral form, there are analogous arithmetic statements corresponding to the properties given in Eqs. 26.15, 26.16, and 26.18. In particular, the distribution of sampled date taken from a continuous distribution is a special case of discrete distributions and can be given in the form of arithmetic summations.[3] Thus

$$\sum_{i} f(x_i) = 1 \tag{26.19}$$

$$P(a \leq X \leq b) = \sum_{\substack{x_i \geq a}}^{x_i \leq b} f(x_i) \tag{26.20}$$

$$P(X \leq x_k) = \sum_{i=1}^{k} f(x_i) \qquad (26.21)$$

For a finite number of observations in the sample, $f(x_i)$ is the probability of x_i for each outcome in the sample space and therefore $P(x_i) = P(x_1) = P(x_2) = \cdots = 1/N$. Hence $f(x_i)$ can be replaced with $P(x_i)$ in Eqs. 26.19, 26.20, and 26.21.

EXAMPLE 26.3 _____

Table B.1 contains the area beneath a "standard normal" bell-shaped PDF. Because the distribution is symmetrical, areas are provided only on one side of the center. Use the distribution to determine the values of

1. $P(0 \leq z \leq 2)$.
2. $P(-2 \leq z \leq 2)$.
3. $P(z \geq 2)$.
4. $P(z \leq -1)$

Solution

1. $P(0 \leq z \leq 2) = .4772$.
2. From symmetry, $P(-2 \leq z \leq 0) = P(0 \leq z \leq 2) = .4772$. Since $P(-2 \leq z \leq 2) = P(-2 \leq z \leq 0) + P(0 \leq z \leq 2)$, then $P(-2 \leq z \leq 2) = .4772 + .4772 = .9544$.
3. This is the area under the curve in the right tail beyond $z = 2.0$. Because the area right of center $(z = 0)$ is $.5000$, $P(z \geq 2) = P(z \geq 0) - P(0 \leq z \leq 2)$, or $P(z \geq 2) = .5000 - .4772 = .0228$.
4. From the solution to (3), $P(z \leq -1) = P(z \leq 0) - P(-1 \leq z \leq 0)$. By symmetry, $P(-1 \leq z \leq 0) = P(0 \leq z \leq 1) = .3413$, and $P(z \leq -1) = .5000 - .3413 = .1587$. ■■

26.4 MOMENTS OF DISTRIBUTIONS

The properties of many random variables can be defined in terms of the moments of the distribution. The moments represent parameters that usually have physical or geometrical significance. Readers should recognize the analogy between statistical moments and the moments of areas studied in solid mechanics.

The rth moment of a distribution about the origin is defined as[4]

$$\mu_r' = \int_{-\infty}^{+\infty} x^r f(x) \, dx \qquad (26.22)$$

or

$$\mu_r' = \sum_{i=1}^{n} x_i^r f(x_i) = \frac{1}{n} \sum_{i=1}^{n} x_i^r \qquad (26.23)$$

The first moment about the origin is the *mean,* or as it is commonly known, the *average.* It determines the distance from the origin to the centroid of the distribution frequency function. The prime is normally used to signify moments taken about the origin, but the mean is often written as μ instead of μ'.

Moments can be defined about axes other than the origin; the axis used extensively in defining higher moments is the *mean* or, as given above, the first moment about the origin. Thus

$$\mu_r = \int_{-\infty}^{\infty} (x - \mu)^r f(x)\, dx \tag{26.24}$$

or

$$\mu_r = \frac{1}{n} \sum_{i=1}^{n} (x_i - \mu)^r \tag{26.25}$$

Whenever μ'_r or μ_r are defined for $r = 1, \ldots,$ the distribution $f(x)$ is completely defined. It seldom is necessary to compute more than the first three moments; several important distributions require only two. The moments are used to specify the parameters and descriptive characteristics of distributions that follow in the next section. Because various characteristics of distributions are described by combinations of the moments about the mean and origin, the following relations are occasionally helpful[1,4]:

$$\mu_1 = 0 \tag{26.26}$$

$$\mu_2 = \mu'_2 - \mu^2 \tag{26.27}$$

$$\mu_3 = \mu'_3 - 3\mu'_2\mu + 2\mu^3 \tag{26.28}$$

26.5 DISTRIBUTION CHARACTERISTICS

Characteristics of statistical distributions are described by the parameters of probability functions, which in turn are expressed in terms of the moments. The principal characteristics are *central tendency,* the grouping of observations or probability about a central value; *variability,* the dispersion of the variate or observations; and *skewness,* the degree of asymmetry of the distribution. The theoretical functions shown in Fig. 26.6 exhibit approximately the same grouping about a central value, but f_2 has much greater variability than f_1, and f_2 possesses a pronounced right-skew while f_1 is symmetrical.

Symbol Convention

In introducing the parameters of distributions, the usual sequence of statistical problems will be followed—that is, parameters are derived from the distribution of sample data and used as estimates of the parameters of the population distribution. Summa-

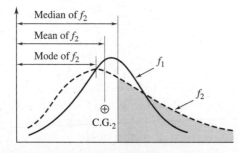

Figure 26.6 Symmetrical and skewed probability distribution for continuous variables.

tion forms of integrals are used to compute moments for samples. For example, the mean of sample data is designated $\bar{x}$ and it is used as the best estimate of the population mean. By convention, Greek letters are used to denote population parameters.

Central Tendency

The familiar arithmetic average, the *mean,* is the most used measure of central tendency. It is the first moment about the origin and is designated

$$\bar{x} = \frac{1}{n} \sum_{i=1}^{n} x_i \tag{26.29}$$

The statistic $\bar{x}$ is the best estimate of the population mean μ.

Means other than the arithmetic mean—for example, the *geometric mean* $\bar{x} = (x_1 x_2 x_3 \cdots x_n)^{1/n}$ or *harmonic mean* $\bar{x} = n/\sum (1/x_i)$—are also used. Two additional measures of central tendency are the *median,* which is the middle value of the observed data and divides the distribution into equal areas, and the *mode,* which in discrete variables is the value occurring most frequently and in continuous variables is the peak value of probability density. All three are illustrated in Fig. 26.6.

Variability

Dispersion can be represented by the total range of values or by the average deviation about the mean; however, the parameter of statistical importance is the mean squared deviation as measured by the second moment about the mean. The parameter is termed the variance and is designated by

$$\sigma^2 = \frac{1}{n} \sum_{i=1}^{n} (x_i - \mu)^2 \tag{26.30}$$

But the population mean μ is not known precisely and therefore it is necessary to compute instead

$$s^2 = \frac{1}{n-1} \sum_{i=1}^{n} (x_1 - \bar{x})^2 \tag{26.31}$$

As the best estimate of σ^2, the quantity s^2 is found using $n - 1$ in place of n in Eq. 26.30. The reasoning for this substitution involves the loss of a degree of freedom by using $\bar{x}$ instead of μ, but a proof is beyond the scope of this text.

The square root of the variance is a statistic known as the *standard deviation* (σ or s), in which form variability is measured in the same units as the variate and the mean, and hence is easier to interpret and manipulate. The *coefficient of variation* C_v, defined as σ/μ or $s/\bar{x}$, is an expression useful in comparing relative variability.

Skewness

A fully symmetrical distribution would exhibit the property that all odd moments equal zero. A skewed distribution, however, would have excessive weight to either side of the center and the odd moments would exist. The third moment α is

$$\alpha = \frac{1}{n} \sum_{i=1}^{n} (x_i - \mu)^3 \tag{26.32}$$

The best estimate of the third moment is computed by

$$a = \frac{n}{(n-1)(n-2)} \sum_{i=1}^{n} (x_i - \bar{x})^3 \qquad (26.33)$$

The *coefficient of skewness* is the ratio α/σ^3 and is estimated by

$$C_s = \frac{a}{s^3} \qquad (26.34)$$

For symmetrical distributions, the third moment is zero and $C_s = 0$; for right skewness (i.e., the long tail to the right side) $C_s > 0$, and for left skewness $C_s < 0$. The PDF for f_2 shown in Fig. 26.6 has a right or positive skew. The property of skewness is of questionable statistical value when it must be estimated from less than 50 sample data points.

EXAMPLE 26.4

Determine the distribution parameters and compare the distributions of annual rainfall for the records shown in Table 26.2.

TABLE 26.2   ANNUAL RAINFALL FOR SELECTED CITIES

	Annual rainfall (in.)		
Year	Anniston, AL	Los Angeles, CA	Richmond, VA
1928	48	9	43
1927	49	19	44
1926	55	19	38
1925	98	9	31
1924	43	8	47
1923	53	6	49
1922	56	15	52
1921	47	20	31
1920	69	11	51
1919	57	9	40
1918	61	18	41
1917	64	8	43
1916	99	23	37
1915	54	17	36
1914	40	23	34
1913	47	17	38
1912	58	10	36
1911	44	18	37
1910	44	5	43
1909	64	24	34
1908	44	19	53
1907	51	15	49
1906	71	21	47

Solution

Parameter	Anniston	Los Angeles	Richmond
Mean, $\bar{x}$	57.2 in.	14.9 in.	41.5 in.
Standard deviation, s	15.5 in.	5.9 in.	6.7 in.
Coefficient of variation, $C_v = s/\bar{x}$	0.27	0.40	0.16
Coefficient of skewness, $C_s = a/s^3$	1.69	−0.16	0.16

Comments. (1) Anniston's record shows a high annual average and a fairly large variability. In particular, Anniston's distribution has a pronounced right skew, caused principally by two very large observed values in this short period of record. (2) Los Angeles has a small annual average but a very large variability and a slightly negative skewness. (3) Richmond has the most uniform distribution: a relatively small variability and only a slight positive skewness. ■■

26.6 TYPES OF PROBABILITY DISTRIBUTION FUNCTIONS

Many standard theoretical probability distributions have been used to describe hydrologic processes. It should be emphasized that any theoretical distribution is not an exact representation of the natural process but only a description that approximates the underlying phenomenon and has proved useful in describing the observed data. Table 26.3 summarizes the common distributions, giving the PDF, mean, and variance of the functions. The distributions presented in the table have experienced wide application and are derived and discussed in many standard textbooks on statistics. In the material to follow, only aspects of the most used distributions are given.

The uses of binomial and Poisson discrete probability distributions in Table 26.3 are restricted generally to those random events in which the outcome can be described either as a success or failure. Furthermore, the successive trials are independent and the probability of success remains constant from trial to trial.[3,4] In a sense, the common discrete distributions are counting or enumerating techniques.

The binomial distribution is frequently used to approximate other distributions, and vice versa. For example, with discrete values, when n is large and p small (such that $np < 5$ preferably), the binomial approaches the Poisson distribution. This is a single-parameter distribution ($\lambda = np$) and is very useful in describing arrivals in queueing theory. When p approaches $\frac{1}{2}$ and n grows large, the binomial becomes indistinguishable from the normal distribution described in the next section.

26.7 CONTINUOUS PROBABILITY DISTRIBUTION FUNCTIONS

Most hydrologic variables are assumed to be continuous random processes, and the common continuous distributions are used to fit historical sequences, as in frequency analysis, for example (Chapter 27). Other applications are also important for continuous distributions. The elementary uniform distribution is the basis for computing

TABLE 26.3   TABLE OF COMMON DISTRIBUTIONS USED IN HYDROLOGY

Distribution of random variable X	Probability density function and CDF	Range	Mean $\bar{x}$ or μ	Variance s^2 or σ^2
Binomial	$P(x) = \dfrac{n!}{x!(n-x)!}p^x(1-p)^{n-x}$	$0 \leq x < n$	np	$np(1-p)$
Poisson	$P(x) = \dfrac{\lambda^x e^{-\lambda}}{x!}$	$0 \leq x \leq \cdots$	λ	λ
Uniform	$f(x) = \dfrac{1}{b-a}$	$a \leq x \leq b$	$\dfrac{b+a}{2}$	$\dfrac{(b-a)^2}{12}$
Exponential	$f(x) = \dfrac{1}{a}e^{-x/a}$	$0 \leq x \leq \infty$	a	a^2
Normal	$f(x) = \dfrac{1}{\sigma\sqrt{2\pi}}e^{-(x-\mu)^2/2\sigma^2}$	$-\infty \leq x \leq \infty$	μ	σ^2
Log–normal $(y = \ln x)$	$f(y) = \dfrac{1}{x\sigma_y\sqrt{2\pi}}\exp\left[\dfrac{-(y-\mu_y)^2}{2\sigma_y^2}\right]$	$-\infty \leq y \leq \infty$ $(0 \leq x \leq \infty)$	μ_y	σ_y^2
Gamma	$f(x) = \dfrac{x^\alpha e^{-x/\beta}}{\beta^{\alpha+1}\Gamma(\alpha+1)}$	$0 \leq x \leq \infty$	$\beta(\alpha+1)$	$\beta^2(\alpha+1)$
Gumbel	$f(x) = \dfrac{1}{\alpha}\exp\left[-\dfrac{x-\xi}{\alpha} - \exp\left(-\dfrac{x-\xi}{\alpha}\right)\right]$ $F(x) = \exp\left[-\exp\left(-\dfrac{x-\xi}{\alpha}\right)\right]$ $x = \xi - \alpha\ln[-\ln F]$	$-\infty \leq x \leq \infty$	$\mu = \xi + 0.5772\alpha$	$\sigma^2 = \dfrac{\pi^2\alpha^2}{6} \approx 1.645\alpha^2$
Weibull	$f(x) = \left(\dfrac{k}{\alpha}\right)\left(\dfrac{x}{\alpha}\right)^{k-1}\exp\left[-\left(\dfrac{x}{\alpha}\right)^k\right]$ $F(x) = 1 - \exp[-(x/\alpha)^k]$	$x \geq 0;\ \alpha, k \geq 0$	$\mu = \alpha\Gamma\left(1+\dfrac{1}{k}\right)$	$\sigma^2 = \alpha^2\left\{\Gamma\left(1+\dfrac{2}{k}\right) - \left[\Gamma\left(1+\dfrac{1}{k}\right)\right]^2\right\}$
Extreme value	$f(x) = \alpha\exp\{-\alpha(x-u) - e^{-\alpha(x-u)}\}$	$-\infty \leq x \leq \infty$	$u + \dfrac{0.5772}{\alpha}$	$\dfrac{\pi^2}{6\alpha^2}$
Log–Pearson III $(y = \ln x)$	$f(x) = \dfrac{(y-\gamma)^\alpha}{\beta^2 x\Gamma(\alpha+1)}\exp\left[\dfrac{-(y-\gamma)}{\beta}\right]$	$-\infty \leq y \leq \infty$ $(0 \leq x \leq \infty)$	$\mu_y = \gamma + \beta(\alpha+1)$	$\sigma_y^2 = \beta^2(\alpha+1)$

random numbers so important in simulation studies. The whole body of material in the area of reliability and estimating depends on derived distributions like Student's t, chi-squared, and the F distribution. The explanations that follow concern the more common distributions applied in fitting hydrologic sequences. The reader is referred to standard texts for more detailed treatment.[3-6]

Normal Distribution

The normal distribution is a symmetrical, bell-shaped frequency function, also known as the Gaussian distribution or the natural law of errors. It describes many processes that are subject to random and independent variations. The whole basis for a large body of statistics involving testing and quality control is the normal distribution. Although it often does not perfectly fit sequences of hydrologic data, it has wide application, for example, in dealing with transformed data that do follow the normal distribution and in estimating sample reliability by virture of the central limit theorem.

The normal distribution has two parameters, the mean μ and the standard deviation σ, for which $\bar{x}$ and s, derived from sample data, are substituted. By a simple transformation, the distribution can be written as a single-parameter function only. Defining $z = (x - \mu)/\sigma$, $dx = \sigma\,dz$, the PDF becomes

$$f(z) = \frac{1}{\sqrt{2\pi}} e^{-z^2/2} \qquad (26.35)$$

and the CDF becomes

$$F(z) = \frac{1}{\sqrt{2\pi}} \int_{-\infty}^{z} e^{-u^2/2}\,du \qquad (26.36)$$

The variable z is called the *standard normal variate;* it is normally distributed with zero mean and unit standard deviation. Tables of areas under the standard normal curve, as given in Appendix B, Table B.1, serve all normal distributions after standardization of the variables. Given a cumulative probability, the deviate z is found in the table of areas and x is found from the inverse transform:

$$x = \mu + z\sigma \quad \text{or} \quad x = \bar{x} + zs \qquad (26.37)$$

EXAMPLE 26.5

Assume that the Richmond, Virginia, annual rainfall in Table 26.2 follows a normal distribution. Use the standard normal transformation to find the rain depth that would have a recurrence interval of 100 years.

Solution.   From example 26.4, the mean is 41.5 in. and the standard deviation is 6.7 in. This gives

$$x = 41.5 + z(6.7)$$

Equation 26.18 shows that the area under the PDF to the right of z is the exceedence probability of the event. For the 100-yr event, Eq. 26.7 gives the exceedence probability $P(z) = 1/T_r = 1/100 = 0.01$. From the figure accompanying Table B.1 in

Appendix B, $F(z) = 0.5 - P(z) = 0.49$, and $z = 2.326$ by interpolating the table. The expected 100-yr rain depth is therefore

$$x = 41.5 + (2.326) \times 6.7 = 57.1 \text{ in.}$$

The 100-yr event for a normal distribution is 2.326 standard deviations above the mean. ■■

Log–Normal Distribution

Many hydrologic variables exhibit a marked right skewness, partly due to the influence of natural phenomena having values greater than zero, or some other lower limit, and being unconstrained, theoretically, in the upper range. In such cases, frequencies will not follow the normal distribution, but their logarithms may follow a normal distribution.[7] The PDF shown in Table 26.3 for the log–normal comes from substituting $y = \ln x$ in the normal. With μ_y and σ_y as the mean and standard deviation, respectively, the following relations have been found to hold between the characteristics of the untransformed variate x and the transformed variate y:[1,7]

$$\mu = \exp(\mu_y + \sigma_y^2/2) \tag{26.38}$$
$$\sigma^2 = \mu^2[\exp(\sigma_y^2) - 1] \tag{26.39}$$
$$\alpha = [\exp(3\sigma_y^2) - 3 \exp(\sigma_y^2) + 2]C_v^3 \tag{26.40}$$
$$C_v = [\exp(\sigma_y^2) - 1]^{1/2} \tag{26.41}$$
$$C_s = 3C_v + C_v^3 \tag{26.42}$$

Also $\mu_y = \ln M$, where M is the median value *and* the geometric mean of the x's.

The log–normal is especially useful because the transformation opens the extensive body of theoretical and applied uses of the normal distribution. Since both the normal and log–normal are two-parameter distributions, it is necessary only to compute the mean and variance of the untransformed variate x and solve Eqs. 26.38 and 26.39 simultaneously. Information on three-parameter or truncated log–normal distributions can be found in the literature.[1,7]

Gamma (and Pearson Type III)

The gamma distribution has wide application in mathematical statistics and has been used increasingly in hydrologic studies now that computing facilities make it easy to evaluate the *gamma function* instead of relying on the painstaking method of using tables of the incomplete gamma function that lead to the CDF, $P(X < x)$. In greater use is a special case of gamma: the *Pearson Type III*. This distribution has been widely adopted as the standard method for flood frequency analysis in a form known as the *log–Pearson III* in which the transform $y = \log x$ is used to reduce skewness.[8-10] Although all three moments are required to fit the distribution, it is extremely flexible in that a zero skew will reduce the log–Pearson III distribution to a log–normal and the Pearson Type III to a normal. Tables of the cumulative function are available and will be explained in a later section.[10,11] A very important property of gamma variates as well as normal variates (including transformed normals) is that the sum of two such variables retains the same distribution. This feature is important in generating synthetic hydrologic sequences.[12,13]

Gumbel's Extremal Distribution

The theory of extreme values considers the distribution of the largest (or smallest) observations occurring in each group of repeated samples. The distribution of the n_1 extreme values taken from n_1 samples, with each sample having n_2 observations, depends on the distribution of the $n_1 n_2$ total observations. Gumbel was the first to employ extreme value theory for analysis of flood frequencies.[14] Chow has demonstrated that the Gumbel distribution is essentially a log–normal with constant skewness.[15] The CDF of the density function given in Table 26.3 is

$$P(X \leq x) = F(x) = \exp\{-\exp[-\alpha(x - u)]\} \qquad (26.43)$$

a convenient form to evaluate the function. Parameters α and u are given as functions of the mean and standard deviation in Table 26.3. Tables of the double exponential are usually in terms of the reduced variate, $y = \alpha(x - u)$.[16] Gumbel also has proposed another extreme value distribution that appears to fit instantaneous (minimum annual) drought flows.[17, 18]

CDFs in Hydrology

Normal and Pearson distributions can often be used to describe hydrologic variables if the variable is the sum or mean of several other random variables. The sum of a number of independent random variables is approximately normally distributed. For example, the annual rainfall is the sum of the daily rain totals, each of which is viewed as a random variable. Other examples include annual lake evaporation, annual pumpage from a well, annual flow in a stream, and mean monthly temperature.

The log–normal CDF has been successfully used in approximating the distribution of variables that are the product of powers of many other random variables. The logarithm of the variable is approximately normally distributed because the logarithm of products is a sum of transformed variables.

Examples of variables that have been known to follow a log–normal distribution include:

1. Annual series of peak flow rates.
2. Daily precipitation depths and stremflow volumes (also monthly, seasonal, and annual).
3. Daily peak discharge rates.
4. Annual precipitation and runoff (primarily in the western United States).
5. Earthquake magnitudes.
6. Intervals between earthquakes.
7. Yield stress in steel.
8. Sediment sizes in streams where fracturing and breakage of larger into smaller sizes is involved.

The Pearson Type III (a form of gamma) has been applied to a number of variables such as precipitation depths in the eastern United States and cumulative watershed runoff at any point in time during a given storm event. The transformed log–Pearson Type III is most used to approximate the CDF for annual flood peaks. If the skew coefficient C_s of the variable is zero, the CDF reverts to a log–normal.

It has also been used with monthly precipitation depth and yield strengths of concrete members.

A useful CDF for values of annual extreme is the Gumbel or extreme value distribution. The mean of the distribution has a theoretical exceedance probability of 0.43 and a recurrence interval T of 2.33 years. Flood peaks in natural streams have exhibited strong conformance to this distribution, including means with 2.33-year recurrence intervals. Graph paper that produces a straight-line fit for Gumbel variables is a available and useful for graphical tests of annual extremes. A sample is shown in Fig. 27.2. The CDF has been applied to peak annual discharge rates, peak wind velocities, drought magnitudes and intervals, maximum rainfall intensities of given durations, and other hydrologic extremes that are independent events.

26.8 BIVARIATE LINEAR REGRESSION AND CORRELATION

Correlation and regression procedures are widely used in hydrology and other sciences.[19] The premise of the methods is that one variable is often conditioned by the value of another, or of several others, or the distribution of one may be conditioned by the value of another. Just as there are probability density functions (PDFs) for evaluating the *marginal* probability of a variable (see Section 26.2), so also are there PDFs for the *conditional* probabilities (also described in Section 26.2) of variables. The concept is illustrated in Fig. 26.7. For two variables, the bivariate density function, $f(y \mid x_1)$, plotted in the vertical on the figure, changes for each value of x. The one shown applies only to variations in y when $x = x_1$. Different distributions might occur for other values of x.

A measure of the degree of linear correlation between two variables x and y is the *linear correlation coefficient*, $\rho_{x,y}$. A value of $\rho_{x,y} = 0.0$ indicates a lack of linear

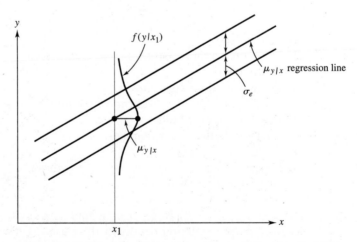

Figure 26.7 Bivariate regression with conditional probability function.

correlation and $\rho_{x,y} = \pm 1.0$ means perfect correlation. The correlation coefficient is found from

$$\rho_{x,y} = \frac{\text{cov}(x, y)}{\sigma_x \sigma_y} = \frac{\sigma_{x,y}}{\sigma_x \sigma_y} \qquad (26.44)$$

where σ_x and σ_y are the variances of each variable, respectively, (see Eq. 26.30), and $\text{cov}(x, y)$ is the *covariance* shared by the two variables, defined as

$$\text{cov}(x, y) = \sigma_{x,y} = \int_{-\infty}^{\infty} \int_{-\infty}^{\infty} (x - \mu_x)(y - \mu_y) f(x, y) \, dy \, dx \qquad (26.45)$$

The sample correlation coefficient, $r = s_{x,y}/s_x s_y$, is used to estimate $\rho_{x,y}$. The sample covariance is found from the square root of

$$s_{x,y}^2 = \frac{\Sigma \, (x_i - \bar{x})(y_i - \bar{y})}{n - 1} \qquad (26.46)$$

The regression line shown on Fig. 26.7 is derived to pass through the mean values of the distributions, so that for any given value of x, the mean value of $y \mid x$ (read "y given x") can be estimated by the regression line. The *standard error* of the estimate of $y \mid x$ is depicted by the line drawn through the conditional distributions at a distance of one standard deviation from the mean. If the conditional distributions at all x-values are normal, it can be shown that the mean value, $\mu_{y\mid x}$, of the conditional distribution is related to the means of x and y, or

$$\mu_{y\mid x} = \mu_y + \rho \frac{\sigma_y}{\sigma_x}(x - \mu_x) \qquad (26.47)$$

and the variance is

$$\sigma_{y\mid x}^2 = \frac{\sigma_e^2}{N}\left[1 + \frac{(x - \mu_x)^2}{\sigma_x^2}\right] \qquad (26.48)$$

where

$$\sigma_e^2 = \sigma_y^2(1 - \rho^2) \qquad (26.49)$$

which is the variance of the residuals of the regression. Just as the mean of the distribution requires substitution of the given value of x into Eq. 26.47, so also does the variance, Eq. 26.48. When the value of x in Fig. 26.7 is set equal to $\bar{x}$, the standard error of the mean is

$$\sigma_{\bar{y}\mid\bar{x}} = \frac{\sigma_e}{\sqrt{N}} \qquad (26.50)$$

Equation 26.47 is linear and expresses the linear dependence between y and x as shown in Fig. 26.7. The mean value of y can be computed for fixed values of x. Also, if the correlation between them is significant, one can predict the values of y with less error than the marginal distribution of y alone. In fact, from Eq. 26.49, the fraction of the original variance explained or accounted by the regression is

$$\rho^2 = 1 - \frac{\sigma_e^2}{\sigma_y^2} \qquad (26.51)$$

It can be seen also from Eq. (26.47) that the slope of the regression line is

$$\rho \frac{\sigma_y}{\sigma_x} = \frac{\mu_{y|x} - \mu_y}{x - \mu_x} \tag{26.52}$$

or, if x and y are standardized, then ρ itself is the slope, where

$$\rho = \frac{(\mu_{y|x} - \mu_y)/\sigma_y}{(x - \mu_x)/\sigma_x} \tag{26.53}$$

The bivariate case can be expanded to cover higher-order, multivariate distributions.

26.9 FITTING REGRESSION EQUATIONS

Regression lines as expressed by Eq. 26.47 and shown in Fig. 26.7 are useful in explaining linear dependence and, where significant correlation exists, in making predictions. For the bivariate case, in general, the procedure is to fit a linear model to a sample of random variables observed in pairs (see Fig. 26.8). The fitting technique is the method of least squares, which minimizes the sum of the residuals squared. Residuals as shown in the figure are the difference, vertically in this instance, from the value of y predicted by the line and the y value observed for the same corresponding value of x. The line to be fitted is

$$y_i = \alpha + \beta x_i \tag{26.54}$$

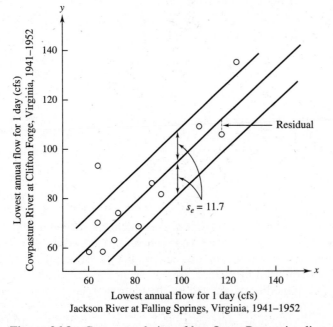

Year	Jackson River	Cowpasture River
41	61	58
42	92	81
43	65	70
44	72	63
45	82	68
46	67	58
47	74	74
48	118	105
49	124	134
50	108	108
51	65	93
52	88	85
Mean =	84.7	83.1
Standard deviation =	21.7	23.2

Figure 26.8   Cross-correlation of low flows. Regression line: $Y = 4.94 + 0.923X$; $r = 0.86$.

The best estimates of α and β are sought. Thus to minimize

$$\sum (y_i - \hat{y}_i)^2 = \sum [y_i - (\alpha + \beta x_i)]^2 \tag{26.55}$$

where y_i are the observed values and $\hat{y}_i$ are the estimated values from Eq. 26.54, take partial derivatives as follows:

$$\frac{\partial}{\partial \alpha} \left\{ \sum [y_i - (\alpha + \beta x_i)]^2 \right\} \tag{26.56}$$

$$\frac{\partial}{\partial \beta} \left\{ \sum [y_i - (\alpha + \beta x_i)]^2 \right\} \tag{26.57}$$

After carrying out the differentiations and summations, two equations result in α and β, called *normal equations*.

$$\sum y_i - n\alpha - \beta \sum x_i = 0 \tag{26.58}$$

$$\sum x_i y_i - \alpha \sum x_1 - \beta \sum x_i^2 = 0 \tag{26.59}$$

Solving Eqs. 26.58 and 26.59 simultaneously yields

$$\alpha = \frac{\sum y_i}{n} - \frac{\beta \sum x_i}{n} = \bar{y} - \beta \bar{x} \tag{26.60}$$

$$\beta = \frac{\sum x_i y_i - \sum x_i \sum y_i / n}{\sum x_i^2 - (\sum x_i)^2 / n} \tag{26.61}$$

Recall the slope is $\rho(\sigma_y / \sigma_x)$, or as estimated from sample data

$$\beta = r \frac{s_y}{s_x} \tag{26.62}$$

Also, the unexplained variance in the regression equation is

$$\sigma_e^2 = \sigma_y^2 (1 - \rho^2) \tag{26.63}$$

the square root of which is the standard deviation of residuals (see Fig. 26.8) and is called the *standard error of estimate*. These can be estimated from

$$s_e^2 = \frac{n - 1}{n - 2} s_y^2 (1 - r^2) \tag{26.64}$$

or

$$s_e^2 = \frac{1}{n - 2} \sum (y_i - \hat{y}_i)^2 \tag{26.65}$$

where y_i and $\hat{y}_i$ are as defined previously (see Eq. 26.55).

Many hydrologic variables are linearly related, and after estimating the regression coefficients, prediction of y can be made for any value of x within the range of observed x values. Extrapolation outside the range is often performed but should be done with caution. Equation 26.48 shows that the variance in the estimate of y for a given x value becomes large when x is several standard deviations above or below the mean.

EXAMPLE 26.6

The lowest annual flows for a 12-yr period on the Jackson and Cowpasture Rivers are tabulated in Fig. 26.8. The stations are upstream of the confluence of the two rivers that form the James River. Find the regression equation and the correlation between low flows.

Solution

1. The basic statistics are $\Sigma\ x = 1016$; $\Sigma\ y = 997$; $\Sigma\ x^2 = 91,216$; $\Sigma\ y^2 = 88,777$; and $\Sigma\ xy = 89,209$.
2. For the two-variable regression α and β are found from Eqs. 26.60 and 26.61.

$$\beta = \frac{[(89,209) - (1016)(997)/(12)]}{(91,216) - (1016)^2/(12)} = 0.923$$

$$\alpha = \frac{997}{12} - \frac{(0.923)(1016)}{12} = 4.91$$

The regression is $y = 4.91 + 0.923x$.
3. The correlation coefficient from Eq. 26.62 is

$$r = \frac{(0.923)(21.7)}{23.2} = 0.86$$

4. From Eq. 26.64 the standard error of estimate, s_e, is 11.7, which is plotted as limits around the regression line in Fig. 26.8. ■■

Coefficient of Determination for the Regression

A regression equation replaces (and extends) the data used in its development. Because it cannot reproduce all the base data, the process results in the loss of some information. This not only includes loss of information about particular pairs of data, but also about the variability of the data. The variance s_y^2 is a statistical measure of the variability of the measured values of y. The greater the value of s_y^2, the wider the spread of points around the mean. The percentage of information about the variance in y that is retained, or explained by, the regression equation is called the *coefficient of determination, C_D*. To determine its value, the residuals or departures (differences between actual and estimated y values) have known variance σ_ϵ^2, which represents the unaccounted variance in the regression equation. The explained variance would be the difference, $\sigma_y^2 - \sigma_\epsilon^2$, and the percentage retained (coefficient of determination) is

$$C_D = \frac{\sigma_y^2 - \sigma_\epsilon^2}{\sigma_y^2} = 1 - \frac{\sigma_\epsilon^2}{\sigma_y^2} \tag{26.66}$$

Comparison with Eq. 26.49 reveals that

$$C_D = \rho^2 \qquad (26.67)$$

Thus the square of the correlation coefficient ρ is the percentage of σ_y^2 explained by the regression. For any sample of data the coefficient of determination r^2 is estimated as $s_{x,y}^2/s_x^2 s_y^2$. A large r^2 indicates a good fit of the regression equation to the data because the equation accounts for or is able to explain a large percentage of the variation in the data.

EXAMPLE 26.7

Determine the coefficient of determination for the regression in Example 26.6.

Solution.   From Eq. 26.67, the coefficient of determination, r^2, is 0.7396. Thus, the regression equation adequately explains or "accounts for" about 74 percent of the original information about y contained in the raw data. Twenty-six percent of the information is lost.   ■ ■

The bivariate example can be extended to multiple linear regressions. For example, the linear model in three variables, with y the dependent variable and x_1 and x_2 the independent variables, has the form

$$y = \alpha + \beta_1 x_1 + \beta_2 x_2 \qquad (26.68)$$

The normal equations are

$$\sum y = \alpha n + \beta_1 \sum x_1 + \beta_2 \sum x_2 \qquad (26.69)$$

$$\sum y x_1 = \alpha \sum x_1 + \beta_1 \sum x_1^2 + \beta_2 \sum x_1 x_2 \qquad (26.70)$$

$$\sum y x_2 = \alpha \sum x_2 + \beta_1 \sum x_1 x_2 + \beta_2 \sum x_2^2 \qquad (26.71)$$

The square of the standard error of estimate is

$$s_e^2 = \frac{1}{n-3} \sum (y_i - \bar{y}_i)^2 \qquad (26.72)$$

where y_i are the observed values and $\bar{y}_i$ are predicted by Eq. (26.68). *The multiple correlation coefficient* is

$$R = \left(1 - \frac{s_e^2}{s_y^2}\right)^{1/2} \qquad (26.73)$$

Linear Transformations in Hydrology

Strong nonlinear bivariate and multivariate correlations are also common in hydrology, and various mathematical models have been used to describe the relations. Parabolic, exponential, hyperbolic, power, and other forms have provided better graphical fits than straight lines. Because of difficulties in the derivation of normal equations using least squares for these models, many can be transformed to linear forms. The most familiar transformation is a linearization of multiplicative nonlinear relations by using logarithms. For example, the equation

$$y = \alpha x_1^{\beta_1} x_2^{\beta_2} \tag{26.74}$$

becomes linear when logarithms are taken, or

$$\log y = \log \alpha + \beta_1 \log x_1 + \beta_2 \log x_2 \tag{26.75}$$

The log transformation procedure results in a linear form when the logarithms of one or both sets of measurements are substituted in Eqs. 26.60 and 26.61. For example, if a bivariate parabolic form $Y = aX^b$ is suggested by the data, logarithms allow use of the linear form $\log Y = \log a + b \log X$. The normal equations can be used by redefining $y = \log Y$, $x = \log X$, $\alpha = \log a$, and $\beta = b$, thereby transforming the equation to $y = \alpha + \beta x$. The regression can now be performed on the logarithms, values of α and β determined, and the estimate of a is found by taking the antilog of α. This transformation is possible for several other nonlinear models, some of which are shown in Table 26.4. The variables x and y must be nonnegative, with values preferably greater than 1.0 to avoid problems with the log transformation.

TABLE 26.4   LINEAR TRANSFORMATIONS OF NONLINEAR FORMS

Equation	Abscissa	Ordinate	Equation in linear form
$Y = A + BX$	X	Y	$[Y] = A + B[X]$
$Y = Be^{AX}$	X	$\log Y$	$[\log Y] = \log B + A(\log e)[X]$
$Y = AX^B$	$\log X$	$\log Y$	$[\log Y] = \log A + B[\log X]$
$Y = AB^X$	X	$\log Y$	$[\log Y] = \log A + (\log B)[X]$

Note: Variables in brackets are the regression variates.

EXAMPLE 26.8

In the following exhibit (Table 26.5) prepared by Beard,[20] the regional correlation is sought of the standard deviation of flow logarithms with the logarithms of the drainage area size and the number of rainy days per year; X_1 is set equal to $(1 + \log s)$ to avoid negative values. Find the regression equation and the multiple correlation coefficient.

Solution

1. From Eqs. 26.69, 26.70, and 26.71, the parameters are

$$\alpha = 1.34; \qquad \beta_1 = -0.013; \qquad \beta_2 = -0.49$$

and the regression equation is

$$X_1 = 1.34 - 0.013X_2 - 0.49X_3$$

or $\qquad \log s = 0.34 - 0.013 \log(\text{DA}) - 0.49 \log(\text{days})$

2. The multiple correlation coefficient from Eqs. 26.72 and 26.73 is $R = 0.56$. ∎∎

TABLE 26.5 LOGARITHMIC DATA FOR 50 GAUGING STATIONS

$X_1 = 1 + \log s$			$X_2 = \log DA$		$X_3 = \log$ number of rainy days per year			
Station number (1)	X_2 (2)	X_3 (3)	X_1 (4)		Station number (5)	X_2 (6)	X_3 (7)	X_1 (8)
1	1.61	2.11	0.29		33	1.94	1.87	0.20
2	2.89	2.12	0.18		34	2.73	1.36	0.58
3	4.38	2.11	0.17		35	3.63	1.81	0.64
4	3.20	2.04	0.44		36	1.91	1.58	0.37
5	3.92	2.07	0.38		37	2.26	1.48	0.27
6	1.61	2.04	0.37		38	2.97	1.89	0.54
7	3.21	2.09	0.30		39	0.70	1.32	0.63
8	3.65	1.99	0.35		40	0.30	1.54	0.78
9	3.23	2.15	0.16		41	3.38	1.62	0.46
10	4.33	2.08	0.11		42	2.87	2.03	0.44
11	1.60	2.09	0.32		43	2.42	2.26	0.24
12	2.82	2.00	0.34		44	4.53	1.93	−0.03
13	2.40	2.00	0.25		45	3.04	1.78	0.30
14	3.69	2.09	0.43		46	4.13	2.00	0.17
15	2.18	2.19	0.27		47	1.49	2.01	0.14
16	2.09	2.17	0.25		48	5.37	1.95	0.10
17	4.48	1.91	0.52		49	1.36	2.11	0.27
18	4.95	1.95	0.18		50	2.31	2.23	0.18
19	2.21	1.97	0.39					
20	3.41	2.08	0.40		ΣX	147.55	96.24	17.89
21	4.82	1.88	0.25		$\overline{X}$	2.951	1.925	0.358
22	1.78	1.93	0.23		ΣXX_2	503.7779	285.5627	51.1527
23	4.39	1.74	0.54		$\Sigma X \Sigma X_2/n$	435.4200	284.0042	52.7934
24	3.23	2.01	0.51		Σxx_2	68.3579	1.5585	−1.6407
25	3.58	2.04	0.45					
26	1.64	1.78	0.63		ΣXX_3		187.5912	33.2598
27	4.58	1.76	0.45		$\Sigma X \Sigma X_3/n$		185.2428	34.4347
28	3.26	1.93	0.59		Σxx_3	1.5585	2.3484	−1.1749
29	4.29	1.81	0.46					
30	1.23	1.89	0.32		ΣXX_1			8.1635
31	3.44	1.48	0.96		$\Sigma X \Sigma X_1/n$			6.4010
32	2.11	1.97	0.12		Σxx_1	−1.6407	−1.1749	1.7625

Note: $x = X - \overline{X}$. (After Beard.[20])

26.10 REGRESSION AND CORRELATION APPLICATIONS

The correlation and regression techniques developed in Section 26.9 provide a powerful means to identify the mathematical dependence between observed values of physically related variables and can account for the additional information contained in correlated sequences of events. Sampling errors are reduced and the reliability of estimates is improved. In addition to predicting the mean or expected value of a hydrologic variable such as rainfall, runoff, or peak flows, the technique can be used to predict the expected value of other statistical parameters, for example, standard deviation, skewness, or autocorrelation. The correlation is determined between the desired statistical parameter as dependent variable, and the appropriate physical and climatic variables within the basin or region as the independent variables. The procedures are significantly better than using relatively short historical sequences and point-frequency analysis. Not only does the method reduce the inherently large sampling errors but it furnishes a means to estimate parameters at ungauged locations.

There are limitations to the techniques of Section 26.9. First, the analyst assumes the form of the model that can express only linear, or logarithmically linear, dependence. Second, the independent variables to be included in the regression analysis are selected. And, third, the theory assumes that the independent variables are indeed independent and are observed or determined without error. Advanced statistical methods that are beyond the scope of this text offer means to overcome some of these limitations but in practice it may be impossible to satisfy them. Therefore, care must be exercised in selecting the model and in interpreting results.

Accidental or casual correlation may exist between variables that are not functionally correlated. For this reason, correlation should be determined between hydrologic variables only when a physical relation can be presumed. Because of the natural dependence between many factors treated as independent variables in hydrologic studies, the correlation between the dependent variable and each of the independent variables is different from the relative effect of the same independent variables when analyzed together in a multivariate model. One way to guard against this effect is by screening the variables initially by graphical methods. Another is to examine the results of the final regression equation to determine physical relevance.

Alternatively, regression techniques themselves aid in screening significant variables. When electronic computation is available, a procedure can be followed in which successive independent variables are added to the multiple regression model, and the relative effect of each is judged by the increase in the multiple correlation coefficient. Although statistical tests can be employed to judge significance, it is useful otherwise to specify that any variable remain in the regression equation if it contributes or explains, say, 1 or 5 percent of the total variance, or of R^2. A frequently used method is to compute the *partial correlation coefficients* for each variable. This statistic represents the relative decrease in the variance remaining $(1 - R^2)$ by the addition of the variable in question. If the variance remaining with the variable included in the regression is $(1 - R^2) = D^2$ and the variance remaining after removal is $(1 - R'^2) = D'^2$, then the partial regression correlation coefficient is $(D'^2 - D^2)/D'^2$.

Most PC spreadsheet software packages have statistical routines for all the analyses described here and many more. Most are extremely flexible, requiring minimal instructions and input data other than raw data. Special manipulations can effect an interchange of dependent and independent variables, bring one variable at a time into the regression equation, rearrange the independent variables in order of significance, and perform various statistical tests.

Extending Hydrologic Records

Regression techniques frequently can be used to extend short records if significant correlation exists between the station of short record and a nearby station with a longer record. In Example 26.6, if the Jackson River records were complete from 1941 to date but the Cowpasture records were incomplete after 1952, the cross-correlation could be used to estimate the missing years by solving the regression equation for Y from 1953 on using the X flows as observed. The reliability of such methods depends on the correlation coefficient and the length of the concurrent records. If the concurrent record is too short or the correlation weak, the standard error of the parameter to be estimated can be increased and nothing is gained. The limiting value of cross-correlation for estimating means is approximately $\rho = 1/\sqrt{n}$, where n is the length of the concurrent record.[21] Thus any correlation above 0.3 would improve the Cowpasture records. Estimates of other parameters with larger standard errors require higher cross-correlation for significant improvement. Extending or filling in deficient records often is necessary for regional studies in which every record used should be adjusted to the same length.

Predicting Regionalized Hydrologic Variables

Cruff and Rantz[22] studied various methods of regional flood analysis and found the multiple regression technique a better predictor than either the index-flood method (Chapter 27) or the fitting of theoretical frequency distributions to individual historical records. They first used regression techniques to extend all records to a common base length. Next they extrapolated by various methods to estimate the 50- and 100-year flood events and with multiple correlation examined several dependent variables including the drainage area A, the basin-shape factor (the ratio of the diameter of a circle of size A to the length of the basin measured parallel to the main channel) S_h, channel slope S, the annual precipitation P, and others. They found only A and S_h to be significant, which resulted in prediction equations of the form $Q_t = cA^aS_h^b$. These equations were superior to those of the other techniques. The multiple correlation coefficient was as high as 0.954. It is interesting that regression techniques were employed in still a third way, that is, to estimate regional values of the mean and standard deviation after adjusting the record length. Example 26.8 illustrated the application of regression analysis to regionalize the standard deviation of annual maximum flow logarithms as a function of the drainage area size and the number of rainy days each year.

■ Summary

Statistics is a diverse subject, and the treatment in this chapter has been nothing more than an introduction. Serious students and practitioners must return again and again to the theory in standard works.[23] They will find that evaluating new developments and techniques must claim a large share of their time. Only certain aspects of statistical hydrology have been presented, principally the common distributions and the methods for analyzing frequency of events observed at a single point. In the next chapter this information is extended to common applications in hydrology.

PROBLEMS

26.1. The probabilities of events E_1 and E_2 are each .3. What is the probability that E_1 or E_2 will occur when (a) the events are independent but not mutually exclusive, and (b) when the probability of E_1, given E_2 is .1?

26.2. Events A and B are independent events having marginal probabilities of .4 and .5, respectively. Determine for a single trial (a) the probability that both A and B will occur simultaneously, and (b) the probability that neither occurs.

26.3. The conditional probability, $P(E_1 \mid E_2)$, of a power failure (given that a flood occurs) is .9, and the conditional probability, $P(E_2 \mid E_1)$, of a flood (given that a power failure occurs) is .2. If the joint probability, $P(E_1 \text{ and } E_2)$, of a power failure and a flood is .1, determine the marginal probabilities, $P(E_1)$ and $P(E_2)$.

26.4. Describe two random events that are (a) mutually exclusive, (b) dependent, (c) both mutually exclusive and dependent, and (d) neither mutually exclusive nor dependent.

26.5. A temporary cofferdam is to be built to protect the 5-year construction activity for a major cross valley dam. If the cofferdam is designed to withstand the 20-year flood, what is the probability that the structure will be overtopped (a) in the first year, (b) in the third year exactly, (c) at least once in the 5-year construction period, and (d) not at all during the 5-year period?

26.6. A 33-year record of peak annual flow rates was subjected to a frequency analysis. The median value is defined as the midvalue in the table of rank-ordered magnitudes. Estimate the following probabilities.
   a. The probability that the annual peak will equal or exceed the median value in any single year.
   b. The average return period of the median value.
   c. The probability that the annual peak in 1993 will equal or exceed the median value.
   d. The probability that the peak flow rate next year will be less than the median value.
   e. The probability that the peak flow rate in all of the next 10 successive years will be less than the median value.
   f. The probability that the peak flow rate will equal or exceed the median value at least once in 10 successive years.
   g. The probability that the peak flow rates in both of two consecutive years will equal or exceed the median value.
   h. The probability that, for a 2-year period, the peak flow rate will equal or exceed the median value in the second year but not in the first.

26.7. What return period must an engineer use in his or her design of a bridge opening if there is to be only a 50 percent risk that flooding will occur at least once in two successive years? Repeat for a risk of 100 percent.

26.8. A temporary flood wall has been constructed to protect several homes in the floodplain. The wall was built to withstand any discharge up to the 20-year flood magnitude. The wall will be removed at the end of the 3-year period after all the homes have been relocated. Determine the probabilities of the following events:
   a. The wall will be overtopped in any year.
   b. The wall will not be overtopped during the relocation operation.
   c. The wall will be overtopped at least once before all the homes are relocated.
   d. The wall will be overtopped exactly once before all the homes are relocated.
   e. The wall will be adequate for the first 2 years and then overtopped in the third year.

26.9. Wave heights and their respective return periods (shown on the next page) are known for a 40-mi long reservoir. Owners of a downstream campsite will accept a 25 percent risk that a protective wall will be overtopped by waves at least once in a 20-year period. Determine the minimum height of the protective wall.

Wave height (ft)	Return period (years)
10.0	100
8.5	50
7.4	30
5.0	10
3.5	5

26.10. Assume that the channel capacity of 12,000 cfs near a private home was equaled or exceeded in 3 of the past 60 years. Find the following:
  a. The frequency of the 12,000-cfs value.
  b. The probability that the home will be flooded next year.
  c. The return period of the 12,000-cfs value.
  d. The probability that the home will not be flooded next year.
  e. The probability of two consecutive, safe years.
  f. The probability of a flood at least once in the next 20 years.
  g. The probability of a flood in the second, but not the first, of two consecutive years.
  h. The 20-year flood risk.

26.11. The distribution of mean annual rainfall at 35 stations in the James River Basin, Virginia, is given in the following summary:

Interval (2-in. groupings)	36 or 37 in.	38 or 39 in.	40 or 41 in.	42 or 43 in.
Number of observations	2	4	7	9

Interval (2-in. groupings)	44 or 45 in.	46 or 47 in.	48 or 49 in.	50 or 51 in.
Number of observations	5	4	2	2

Compute the relative frequencies (see Chapter 27) and plot the frequency distribution and the cumulative distribution. Estimate the probability that the mean annual rainfall (a) will exceed 40 in., (b) will exceed 50 in., and (c) will be between these values.

26.12. Write a simple program to READ in N data points and compute the mean, standard deviation, and skewness coefficient.

26.13. A normally distributed random variable has a mean of 4.0 and a standard deviation of 2.0. Determine the value of

$$\int_{8}^{\infty} f(x)\, dx$$

26.14. For a standard normal density function, use Table B.1 to determine the value of

$$\int_{\mu-2\sigma}^{\mu+\sigma} f(x)\, dx$$

26.15. A normal variable X has a mean of 5.0 and a standard deviation of 1.0. Determine the value of X that has a cumulative probability of 0.330.

26.16. If the mode of a PDF is considerably larger than the median, would the skew most likely be positive or negative?

26.17. Complete the following mathematical statements about the properties of a PDF by inserting in the boxes on the left the correct item number from the right. Assume that X is a series of annual occurrences from a normal distribution.

a. $\displaystyle\int_{-\infty}^{\infty} f(x)\, dx = \Box$ ⠀⠀⠀⠀⠀ 1. Zero

b. $\displaystyle\int_{-\infty}^{m_1} f(x)\, dx = \Box$ ⠀⠀⠀⠀⠀ 2. Unity

c. $\displaystyle\int_{\mu}^{\mu+\Box} f(x)\, dx = .34$ ⠀⠀⠀⠀ 3. Value with 5 percent chance of exceedance each year

d. $\displaystyle\int_{m_1}^{m_2} f(x)\, dx = \Box$ ⠀⠀⠀⠀⠀ 4. 0.68

e. $\displaystyle\int_{-\infty}^{\Box} f(x)\, dx = .5$ ⠀⠀⠀⠀⠀ 5. Value expected every 50 years on the average

f. $\displaystyle\int_{\mu-\sigma}^{\mu+\sigma} f(x)\, dx = \Box$ ⠀⠀⠀⠀ 6. $P(X \le m_1)$

g. $\displaystyle\int_{\Box}^{\infty} f(x)\, dx = .02$ ⠀⠀⠀⠀⠀ 7. $P(m_1 \le X \le m_2)$

h. $\displaystyle\int_{\mu}^{\mu} f(x)\, dx = \Box$ ⠀⠀⠀⠀⠀ 8. $P(m_1 \ge X \ge m_2)$

i. $\displaystyle 1 - \int_{m_1}^{m_2} f(x)\, dx = \Box$ ⠀⠀⠀ 9. Median

j. $\displaystyle\int_{-\infty}^{\Box} f(x)\, dx = 0.95$ ⠀⠀⠀ 10. Standard deviation

26.18. The mean monthly temperature for September at a weather station is found to be normally distributed. The mean is $65.5°$ F, the variance is $39.3°$ F^2, and the record is complete for 63 years. With the aid of Table B.1, find (a) the midrange within which two thirds of all future mean monthly values are expected to fall, (b) the midrange within which 95 percent of all future values are expected, (c) the limit below which 80 percent of all future values are expected, and (d) the values that are expected to be exceeded with a frequency of once in 10 years and once in 100 years. Verify the results by plotting the cumulative distribution on normal probability paper.

26.19. The total annual runoff from a small drainage basin is determined to be approximately normal with a mean of 14.0 in. and a variance of 9.0 in.^{2}. Determine the probability that the annual runoff from the basin will be less than 11.0 in. in all three of the next three consecutive years.

26.20. In the past 60 years, a discharge of 30,000 cfs at a stream gauging station was equaled or exceeded only three times. Determine the average return period (years) of this value.

26.21. Events A and B are independent and have marginal probabilities of .4 and .5, respectively. Determine the following for a single trial:
a. The probability that both A and B occur.
b. The probability that neither occurs.
c. The probability that B, but not A, occurs.

26.22. Existing records reveal the following information about Events A and B, where $A = $ a long March warm spell and $B = $ an April flood:

Year	1	2	3	4	5	6	7	8	9	10
A = warm March?	No	No	Yes	No	Yes	No	Yes	No	Yes	No
B = April flood?	Yes	No	No	Yes	Yes	Yes	No	Yes	Yes	No

On the basis of the 10-year record, answer the following:
a. Are variables A and B independent? Prove.
b. Are variables A and B mutually exclusive? Prove.
c. Determine the marginal probability of an April flood.
d. Determine the probability of having a cold March next year.
e. Determine the probability (one value) of having both a cold March and a flood-free April next year.
f. If a long March warm spell has just ended today, what is the best estimate of the probability of a flood in April?

26.23. Two dependent events are A = a flood will occur in Omaha next year and B = an ice-jam will form near Omaha in the Missouri River next year. Use your judgment to rank from largest to smallest the following probabilities: $P(A)$, $P(A$ and $B)$, $P(A$ or $B)$, $P(A \mid B)$.

26.24. The probability of having a specified return period, T_r, is defined as:

$$P\left(\begin{array}{l}\text{annual value will be equaled or exceeded}\\ \text{exactly once in a period of } t = T_r \text{ years}\end{array}\right) = \left(1 - \frac{1}{T_r}\right)^{t-1}$$

Also,

$$P\left(\begin{array}{l}\text{annual value will be equaled or exceeded}\\ \text{exactly } r \text{ times in a period of } n \text{ years}\end{array}\right) = \frac{n!}{r!(n-r)!} P^{n-r}(1-P)^r$$

a. According to the descriptions in parentheses, the second probability should equal the first when n and r are equal to what values?
b. Show that both equations result in the same probability for an annual value whose frequency is $33\frac{1}{3}$ percent and the return period is $T_r = t = 3$ years. Discusss.

26.25. For the function described below, find (a) the number b that will make the function a probability density function, and (b) the probability that a single measurement of x will be less than $\frac{1}{2}$.

$$f(x) = \begin{cases} 0 & \text{for } x < 0 \\ 3x^2/8 & \text{for } 0 \le x \le b \\ 0 & \text{for } x > b \end{cases}$$

26.26. The random variable x represents depth of precipitation in July. Between values of $x = 0$ and $x = 30$, the probability density function has the equation $f(x) = x/40\mu_x$. In the past, the average July precipitation μ_x was 30 in.
a. Determine the probability that next July's precipitation will not exceed 20 in.
b. Determine the single probability that the July precipitation will equal or exceed 30 in. in all of five consecutive years.

26.27. The random variable x represents depth of precipitation in July. Between values of $x = 0$ and $x = 30$, the probability density function has the equation $f(x) = x/1200$.
a. Determine the probability that next July's precipitation will not exceed 20 in.
b. Determine the probability that next July's precipitation will equal or exceed 30 in.

26.28. Assume that the Anniston, Alabama, data in Table 26.2 follow a normal distribution. Use the standard normal distribution in Table B.1 to determine the rain depth that would have a recurrence interval of 100 years. Use the same table to estimate the recurrence interval of the 1916 annual rain of 99 inches.

26.29. Determine the probability that a measurement of a hydrologic variable with a normal probability density function will fall between the mean and one standard deviation above the mean.

26.30. Determine the probability that a measurement of a hydrologic variable with a normal probability density function will fall in the range of the mean ± 3 standard deviations.

26.31. A hydrologic variable that is the sum of a number of random variables generally may be assumed to follow which theoretical probability density function?

26.32. The mean July precipitation at a station is half as large as the mode. Sketch the probability density function, label the axes, and state: (a) whether the distribution is skewed "left" or "right," and (b) whether the skew is "positive" or "negative."

26.33. A given set of data has a symmetric, zero skew histogram. Determine the frequency and return period of the mode.

26.34. Observations for the past 10 years of withdrawals and estimated recharge of an artesian aquifer are given in the following table. Find the means, variances, standard deviations, covariance, and correlation coefficient.

Measured discharge (1000 acre-ft)	Estimated recharge (1000 acre-ft)
12.2	12.0
10.4	9.8
10.6	11.0
12.6	13.2
14.2	14.6
13.0	14.0
14.0	14.0
12.0	12.4
10.4	10.4
11.4	11.6

26.35. Fit a regression equation to the data in Problem 26.34, treating discharge as the dependent variable. Compute the standard error of estimate. Estimate the expected discharge when recharge is 13 KAF. What would be the estimate of discharge if no information were available on recharge? What is the relative improvement provided by the regression estimate?

26.36. Prepare a computer program for simple, two-variable, linear regression. The program should (a) read in N pairs of observations, Y and X, (b) compute the means, variances, and standard deviations of both Y and X, and (c) find the regression constants, the standard error of estimate, and the correlation coefficient. Verify with the data in Problem 26.34.

26.37. From the following observations of variation of the mean annual rainfall with the altitude of the gauge, determine a linear prediction equation for the catchment. How well correlated are rainfall and altitude?

Gauge number	Mean annual rainfall (in.)	Altitude of gauge (1000 ft)
1	22	4.2
2	28	4.4
3	25	4.5
4	31	5.4
5	32	5.6
6	37	5.6
7	36	5.8
8	35	6.0
9	36	6.6
10	46	6.6
11	41	6.8
12	41	7.0

26.38. Estimate the expected rainfall in Problem 26.37 for a gauge to be installed at an altitude of 5500 ft.

26.39. The least-squares estimates of A and B in the bivariate regression equation $Y = A + BX$ are $A = 2.0$ and $B = 3.0$, where Y is a transformation defined as $\log_{10} y$ and X is defined as $\log_{10} x$. If y and x are related by $y = ax^b$, determine the values of a and b.

26.40. The time of rise of flood hydrographs (T_r), defined as the time for a stream to rise from low water to maximum depth following a storm, is related to the stream length (L) and the average slope (S). From the information given below for 11 watersheds in Texas, New Mexico, and Oklahoma, derive a functional relation of the form $T_r = aL^b S^c$.

Watershed number	T_r (min)	L (1000 ft)	S (ft/1000 ft)
1	150	18.5	7.93
2	90	14.2	19.0
3	60	25.3	12.0
4	60	11.7	13.3
5	100	9.7	11.0
6	75	8.1	15.0
7	90	21.7	16.7
8	30	3.9	146.0
9	30	1.2	20.0
10	45	3.3	64.0
11	50	3.5	33.0

26.41. Repeat the exercise in Problem 26.40 by fitting the relation $T_r = dF^e$, where $F = L/\sqrt{S}$ with L in mi and S in ft/mi. Plot the results on log–log paper.

26.42. The square of the linear correlation coefficient is called the proportion of the variance that is "explained by the regression." Describe the meaning of this phrase by evaluating the equations given in the text. What variance is explained, and what does the term "explained" mean?

26.43. Twenty measured pairs of values of normally distributed variables X and Y are analyzed, yielding values of $\bar{X} = 30$, $\bar{Y} = 20$, $s_x = 20$, and $s_y = 0$. Determine the values of a and b and the standard deviation of residuals for a least-squares fit using the linear equation $Y = a + bX$.

26.44. The least-squares estimates of A and B in the bivariate regression equation $y = A + BX$ are $A = 2.0$ and $B = 1.0$, where y is a transformation defined as $\log_{10} Y$. If Y and X are related by $Y = a(b)^x$, determine the values of a and b.

26.45. Given a table of ten values of mean annual floods and corresponding drainage areas for a number of drainage basins, state how linear regression techniques would be used to determine the coefficient and exponent (p and q) in the equation $Q_{2.33} = pA^q$.

26.46. What choice of transformed variables Y and X would provide a linear transformation for $y = a/(x^3 + b)$? Also, if a regression on these transformed variables yields $Y = 100 + 10X$, determine the corresponding values of a and b. Would the linear transformation be applicable to all possible pairs and values of x and y?

26.47. Which measure of variation in a regression $Y = a + bX$ is generally larger in magnitude, the standard deviation of Y or the standard deviation of residuals? For what condition would the two values be equal?

REFERENCES

1. Ven T. Chow, "Statistical and Probability Analysis of Hydrologic Data," Sec. 8–1, in *Handbook of Applied Hydrology* (V. T. Chow, ed.). New York: McGraw-Hill, 1964.
2. M. B. Fiering, "Information Analysis," in *Water Supply and Waste Water Removal* (G. M. Fair, J. C. Geyer, and D. A. Okun, eds.). New York: Wiley, 1966, Chap. 4.
3. J. R. Benjamin and C. Cornell, *Probability, Statistics and Decision for Civil Engineers.* New York: McGraw-Hill, 1969.
4. A. M. Mood and F. A. Graybill, *Introduction to the Theory of Statistics,* 2nd ed. New York: McGraw-Hill, 1963.
5. A. J. Duncan, *Quality Control and Statistics.* Homewood, IL: Richard D. Irwin, Inc., 1959.
6. P. G. Hoel, *Introduction to Mathematical Statistics,* 3rd ed. New York: Wiley, 1962.
7. J. Aitchison and J. A. C. Brown, *The Log–Normal Distribution.* New York: Cambridge University Press, 1957.
8. H. A. Foster, "Theoretical Frequency Curves," *Trans. ASCE* **87,** 142–203(1924).
9. L. R. Beard, *Statistical Methods in Hydrology,* Civil Works Investigations, U.S. Army Corps of Engineers, Sacramento District, 1962.
10. "A Uniform Technique for Determining Flood Flow Frequencies," Bull. No. 17B, U.S. Geological Survey, 1989.
11. "New Tables of Percentage Points of the Pearson Type III Distribution," Tech. Release No. 38, Central Technical Unit, U.S. Department of Agriculture, 1968.
12. M. B. Fiering, *Streamflow Synthesis.* Cambridge, MA: Harvard University Press, 1967.
13. F. E. Perkins, Simulation Lecture Notes, Summer Institute, "Applied Mathematical Programming in Water Resources," University of Nebraska, 1970.
14. E. J. Gumbel, "The Return Period of Flood Flows," *Ann. Math. Statist.* **12**(2), 163–190(June 1941).
15. Ven T. Chow, "The Log-Probability and Its Engineering Application," *Proc. ASCE* **80,** 1–25(Nov. 1954).
16. "Probability Tables and Other Analysis of Extreme Value Data," Series 22, National Bureau of Standards Applied Mathematics, 1953.
17. E. J. Gumbel, *Statistics of Extremes.* New York: Columbia University Press, 1958.

18. E. J. Gumbel, "Statistical Theory of Extreme Values for Some Practical Application," National Bureau of Standards, Applied Mathematical Series 33, 1954.

19. J. R. Benjamin and C. Cornell, *Probability, Statistics and Decision for Civil Engineers.* New York: McGraw-Hill, 1959.

20. L. R. Beard, *Statistical Methods in Hydrology,* Civil Works Investigations, Sacramento District, U.S. Army Corps of Engineers, 1962.

21. M. B. Fiering, "Information Analysis," in *Water Supply and Waste Water Removal* (G. M. Fair, J. C. Geyer, and D. A. Okun, eds.). New York: Wiley, 1966, Chap. 14.

22. R. W. Cruff and S. E. Rantz, "A Comparison of Methods Used in Flood-Frequency Studies for Coastal Basins in California," in *Flood Hydrology,* U.S. Geological Survey Water-Supply Paper 1580. Washington, DC: U.S. Government Printing Office, 1965.

23. W. Feller, *An Introduction to Probability Theory.* New York: Wiley, Vol. 1, 1957, and Vol. 2, 1966.

Chapter 27

Frequency Analysis

■ Prologue

The purpose of this chapter is to:

- Present methods used in hydrology to evaluate the recurrence of particular magnitudes and durations of random hydrologic variables.
- Elaborate on the definitions of frequency, recurrence interval, return period, and risk analysis introduced in Chapter 10, Section 10.4.
- Illustrate the diverse applications of frequency analysis in hydrology.
- Teach several methods for conducting frequency analyses, including the use of *frequency factors* that allow estimation of recurrence intervals for variables that follow conventional probability distribution functions.
- Introduce methods of point and regional frequency analysis and describe *regional USGS regression equations* that have been adopted throughout the U.S. for estimating flood flows for use in structure design and floodplain analysis.
- Establish how to estimate the reliability of estimates derived from point or regional frequency analyses.
- Explain the widely used Bulletin No. 17B Log-Pearson Type III procedures for performing uniform flood flow frequency analyses.
- Describe how various federal agencies apply frequency methods in design or analysis of water resources systems.

In Chapter 26, probability and statistical characteristics of random variables were introduced, along with common distribution functions and principles of regression and correlation. The present chapter provides applications of these principles to common hydrologic variables.

27.1 FREQUENCY ANALYSIS

The statistical methods presented in Chapter 26 are used most frequently in describing hydrologic data such as rainfall depths and intensities, peak annual discharge, flood flows, low-flow durations, and the like. Frequency analysis was introduced in Section 10.4 and is defined as the investigation of population sample data to estimate recurrence or probabilities of magnitudes of hydrologic variables. Unless stated otherwise, the *frequency* of a hydrologic event is the probability that some value of a discrete variable will occur or some value of a continuous variable will be equalled or exceeded in any given year. The latter is more appropriately called the exceedance probability or *exceedance frequency,* but is often termed the *frequency.* Note that frequency is a probability and has no units of measure. The reciprocal of the exceedance frequency, as shown by Eq. 26.7, is the return period, having units of years.

Two methods of frequency analysis are described: one is a straightforward plotting technique to obtain the cumulative distribution and the other uses frequency factors. The cumulative distribution function provides a rapid means of determining the probability of an event equal to or less than some specified quantity. The inverse is used to obtain recurrence intervals. As a general rule, frequency analysis is cautioned when working with records shorter than 10 years and in estimating frequencies of expected hydrologic events greater than twice the record length.

27.2 GRAPHICAL FREQUENCY ANALYSIS

The frequency of an event can be obtained by use of "plotting position" formulas. When annual maximum values are being analyzed, the recurrence interval is approximated as the mean time in years, with N future trials, for the mth largest value to be exceeded once on the average. The mean number of exceedances for this condition can be shown to be

$$\bar{x} = N\frac{m}{n + 1} \tag{27.1}$$

where $\bar{x}$ = the mean number of exceedances
N = the number of future trials
n = the number of values
m = the rank of descending values, with largest equal to 1

If the mean number of exceedances $\bar{x} = 1$, $N = T$, and

$$T = \frac{n + 1}{m} \tag{27.2}$$

indicating that the recurrence interval is equal to the number of years of record plus 1, divided by the rank of the event.

Several plotting position formulas are available.[1] They give different results as noted in Table 27.1. The range in recurrence intervals obtained for 10 years of record is illustrated in the right-hand column. Most plotting position formulas do not account for the sample size or length of record. One formula that does account for sample size was given by Gringorten[2] and has the general form

$$T = \frac{n + 1 - 2a}{m - a} \tag{27.3}$$

TABLE 27.1   PLOTTING POSITION FORMULAS

Method	Solve for $P(X > x)$	For $m = 1$ and $n = 10$	
		P	T
California	$\dfrac{m}{n}$	.10	10
Hazen	$\dfrac{2m - 1}{2n}$	.05	20
Beard	$1 - (0.5)^{1/n}$	.067	14.9
Weibull	$\dfrac{m}{n + 1}$	.091	11
Chegadayev	$\dfrac{m - 0.3}{n + 0.4}$	.067	14.9
Blom	$\dfrac{m - \frac{3}{8}}{n + \frac{1}{4}}$	.061	16.4
Tukey	$\dfrac{3m - 1}{3n + 1}$	.065	15.5

where n = the number of years of record
m = the rank
a = a parameter depending on n as follows:

n	10	20	30	40	50
a	0.448	0.443	0.442	0.441	0.440

n	60	70	80	90	100
a	0.440	0.440	0.440	0.439	0.439

In general, $a = 0.4$ is recommended in the Gringorten equation. If the distribution is approximately normal, $a = \frac{3}{8}$ is used. A value of $a = 0.44$ is used if the data follows a Gumbel distribution.

The technique in all cases is to arrange the data in increasing or decreasing order of magnitude and to assign order number m to the ranked values. The most efficient formula for computing plotting positions for unspecified distributions,[1] and the one now commonly used for most sample data, is the Weibull equation

$$P = \frac{m}{n + 1} \tag{27.4}$$

When m is ranked from lowest to highest, P is an estimate of the probability of values being equal to or less than the ranked value, that is, $P(X \leq x)$; when the rank is from highest to lowest, P is $P(X \geq x)$. For probabilities expressed in percentages, the value is $100m/(n + 1)$. The probability that $X = x$ is zero for any continuous variable.

EXAMPLE 27.1

Using the 23 years of annual precipitation depths for Los Angeles, California (see data drawn from Table 26.2), estimate the exceedance frequencies and recurrence intervals of the highest ten values using the Weibull equation.

Solution

1. The ten highest flow rates are tabulated below. By ranking them from highest to lowest, the value P in Eq. 27.4 becomes the exceedance probability.
2. Each value can have only one recurrence interval; therefore repeat values have one calculated probability. The greatest value, 24 in., has a calculated recurrence interval of 24 years.

Year	Rain depth, in.	Rank, m	P $m/(n + 1)$	T_r Years
1909	24	1	0.042	24
1916	23	3	—	—
1914	23	3	0.125	8
1906	21	4	0.167	6
1921	20	5	0.208	4.8
1927	19	8	—	—
1926	19	8	—	—
1908	19	8	0.333	3
1918	18	10	—	—
1911	18	10	0.417	2.4

Plotting Paper

Several theoretical cumulative distribution functions plot as straight lines on special graph paper developed for use with Eq. 27.4. This facilitates extrapolation and interpolation of the data. Arithmetic probability paper has an arithmetic ordinate and a probability abscissa scale. It can be used to plot the calculated apparent frequencies of a variable to evaluate whether a normal CDF is approximated by the data. A straight-line plot would identify a normal CDF.

The same paper, but with a logarithmic scale as the ordinate, tests the apparent fit to a log–normal distribution. A third type of paper contains an extreme-value probability scale versus either an arithmetic or a logarithmic scale (both types are available). This allows a test of whether the data approximate a Gumbel or log–Gumbel extreme-value CDF. Extrapolation using any of these graphical aids is not recommended beyond two times the period of record.

27.3 FREQUENCY ANALYSIS USING FREQUENCY FACTORS

Chow[3] has proposed use of

$$x = \bar{x} + Ks \tag{27.5}$$

as the general equation for hydrologic frequency analysis, where K is the *frequency factor* and s is the standard deviation. K is a function of T and varies with the coefficient of skewness in skewed distributions and can be affected greatly by the number of years of record. For the normal distribution, and for the transformed variate in a log–normal distribution, the deviate z given in standard normal tables (Table B.1) is synonymous with the frequency factor.

For a normal distribution, the value of variable Q corresponding to a given recurrence interval T is

$$Q = \overline{Q} + zs_Q \tag{27.6}$$

Values of z for a given T can be obtained from Table B.1 by recognizing that the exceedance probability, $1/T$, is the area under the probability density function to the right of the corresponding value of z.

EXAMPLE 27.2

The mean annual rainfall at Los Angeles is 14.9 in., and the standard deviation is 5.9 in. Find the 10-year rainfall depth, assuming that the annual rain is normally distributed.

Solution. The exceedance probability of the 10-yr annual rain is $1/10$, or 0.1. Because this is the area under the normal curve to the right of the normalized 10-yr rain depth, the value of z_{10} can be found from the standard normal curve shown at the top of Table B.1 in the Appendix. From the figure, the area under the curve between the origin and z_{10} is $F(z) = 0.500 - 0.100 = 0.400$. By interpolating Table B.1, the corresponding value for z_{10} is 1.282. Thus $Q_{10} = 14.9 + 1.282(5.9) = 15.1$ in. This shows that the 10-yr rain depth (or 10-yr value for any other normally distributed variable) is about 1.3 standard deviations above the mean value. ■ ■

Log–Normal

Many variables that plot as curves on normal probability paper plot as straight lines when the logarithms are plotted, or when the values themselves are plotted on logarithmic probability paper. If either occurs, frequency factors for the normal distribution (Table B.1) can be applied using

$$\log Q = \overline{\log Q} + z(s_{\log s}) \tag{27.7}$$

Simply stated, the logarithms of Q follow a normal distribution, and Eq. 27.6 is applied to the logarithms. The mean and standard deviation of the logarithms are both required. Note that the mean and standard deviation of logarithms are not the same as the logarithms of the arithmetic mean and standard deviation. This common error must be avoided.

Log-Pearson III (Bulletin 17B Frequency Method)

Frequency factors for the Pearson Type III (logarithmic or arithmetic) are shown in Appendix B, Table B.2, for various recurrence intervals (or exceedance probabilities) and skew coefficients. In an effort to develop a uniform approach in federal agencies, the U.S. Water Resources Council[4] recommended the log-Pearson III distribution as a base method for flood frequencies. As outlined by the Water Resources Council, the fitting technique involves transforming annual floods to logarithmic values (y_i = log x_i) and finding the mean, standard deviation, and skew coefficients of the logarithms. Flood magnitudes (Q) are estimated from

$$\log Q = \bar{y} + Ks_y \tag{27.8}$$

which is the same form as Eq. 27.5. Note that $K = \phi(T, C_s)$, a function of both recurrence interval and skewness. Because the skewness coefficient has a much greater variability than the mean or standard deviation, Beard[5] has recommended that only average regional coefficients of skew be employed in flood analysis for a single station unless the record exceeds 100 years. In practice this may be impractical to attain, and it is best to compute all parameters and compare results with any other records, experience, or regional studies. The use of logarithms to reduce the skewness of an already skewed distribution helps. Hazen recommended that the computed skewness for Pearson III analysis be multiplied by a factor of $(1 + 8.5/n)$ to obtain an adjusted skewness when dealing wih small samples.[6] Chow has developed K versus T curves for the distribution.[7]

If the skew coefficient falls between -1.0 and 1.0, approximate values of frequency factors for the Pearson Type III can be obtained from

$$K = \frac{2}{C_s}\left\{\left[\left(z - \frac{C_s}{6}\right)\frac{C_s}{6} + 1\right]^3 - 1\right\}$$

where z is the standard normal deviate for the selected recurrence interval T, and C_s is the skew coefficient from Eq. 27.34.

From examining Tables B.1 and B.2, the reader can establish the fact that a Pearson III distribution with a skew C_s of zero is normal. For example, both tables yield a 100-year frequency factor of 2.326. Through logarithmic transformation, this also means that a log–Pearson III CDF with zero skew of logarithms is a log–normal distribution.

EXAMPLE 27.3

For rainfall data developed in Example 26.4, fit distribution functions to the records of Richmond, Virginia, and Los Angeles, California.

Solution. Both distributions exhibit small skewness and are approximately normal. For the purposes of illustration, the Richmond data are fitted with the normal and the Los Angeles data with a Pearson Type III.

1. The data are arrayed for plotting in Table 27.2. The points are plotted in Figs. 27.1 and 27.2 as exceedance probability (left-hand scale) versus inches of rainfall.
2. The theoretical normal of best fit is a straight line through $(\bar{x} - s)$ at 15.9 percent, $\bar{x}$ at 50 percent, and $(\bar{x} + s)$ at 84.1 percent (see Table B.1). Thus for Richmond,

TABLE 27.2

m	Richmond	Los Angeles	100m/(n + 1)
1	53	24	4.2
2	52	23	8.3
3	51	23	12.5
4	49	21	16.7
5	49	20	20.8
6	47	19	25.0
7	47	19	29.2
8	44	19	33.3
9	43	18	37.5
10	43	18	41.7
11	43	17	45.8
12	41	17	50.0
13	40	15	54.7
14	38	15	58.3
15	38	11	62.5
16	37	10	66.7
17	37	9	70.8
18	36	9	75.0
19	36	9	79.7
20	34	8	83.3
21	34	8	87.5
22	31	6	91.7
23	31	5	95.8

x	Plotting position (right-hand scale)
$\bar{x} - s = 41.5 - 6.7 = 34.8$	15.9
$\bar{x} = 41.5$	50.0
$\bar{x} + s = 41.5 + 6.7 = 48.2$	84.1

3. The plotting positions (read as percent chance) come from Table B.2, for the computed skewness. Thus for Los Angeles,

Percent chance	99	95	90	80	50	20	10	4	2	1	0.5
$K(C_s = -0.16)$	−2.44	−1.69	−1.30	−0.83	0.03	0.85	1.26	1.69	1.97	2.21	2.43
$x = 14.9 + K(5.9)$	0.5	4.9	7.2	10.0	15.1	19.9	22.3	24.9	26.5	27.9	29.2

■■

Weighted Skew and Outliers   The Bulletin 17B[4] procedure involves calculation of the mean, standard deviation, and skew of the logarithms of the flows, and then using frequency factors from Table B.2. Because the skew is extremely sensitive to the sample size, a weighted value is recommended from

$$\text{W.S.} = \text{weighted skew} = \alpha S_s + (1 - \alpha)S_m$$

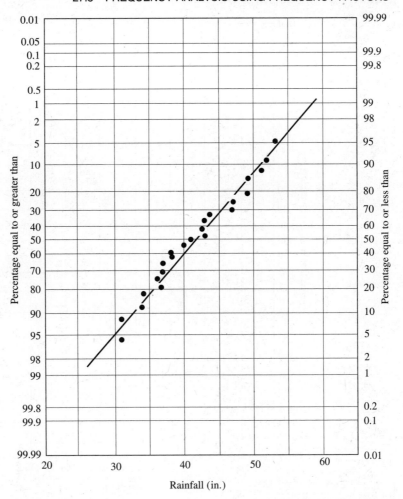

Figure 27.1   Annual rainfall for Richmond, Virginia, 1906–1928, plotted on normal probability paper.

where S_s is the sample skew, S_m is a generalized regional value of the skews in Fig. 16.31, and α is a weighting factor.

The weighting factor is determined from minimizing the variance of the W.S. value,

$$\alpha = \frac{V(S_m)}{V(S_s) + V(S_m)}$$

where V represents the variance of the respective skew coefficients. The variance of map values of skews in Fig. 16.31 is 0.03025, and the variance of sample skews is

$$V(S_s) = 10^{a - b[\log(N/10)]}$$

where N is the number of years in the sample and a and b vary with the calculated

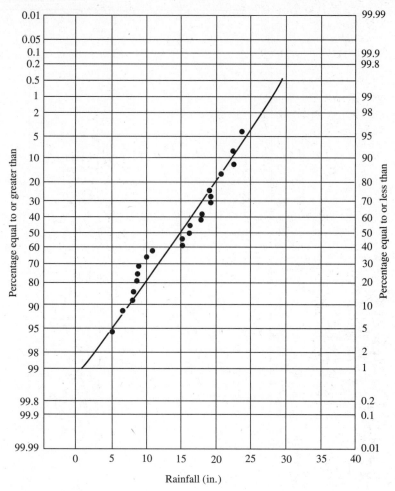

Figure 27.2   Annual rainfall for Los Angeles, California, 1906–1928.

skew, S_s, as follows:

$$\text{For } |S_s| \leq 0.9, \qquad a = -0.33 + 0.08|S_s|$$
$$\text{For } |S_s| > 0.9, \qquad a = -0.52 + 0.30|S_s|$$
$$\text{For } |S_s| \leq 1.5, \qquad b = 0.94 - 0.26|S_s|$$
$$\text{For } |S_s| > 1.5, \qquad b = 0.55$$

Because the largest value in either an annual or partial duration series (see Section 27.6) will have an *apparent* recurrence interval of $N/1$ years (or $N/2$ years for the Weibull plotting formula), it may be that assigning it this frequency underestimates its actual interval. Sufficient time may not have elapsed for an accurate measure of its frequency. This could also hold true for the second, or third, or nth largest flows. Values with incorrect apparent frequencies are called *outliers* because they depart significantly from the rest of the data and are usually removed from the set before completing the frequency analysis.

The U.S. Water Resources Council[4] provides methods for evaluating the existence of high or low outliers for any sample set. Other parts of the procedure involve appropriate methods for handling incomplete records, broken record intervals, zero flow years, transposition of records, and other commonly encountered problems. The method for considering whether a flow is outside the trend of the rest is to evaluate whether it falls outside the maximum permissible range, V, established from

$$V(\text{cfs}) = \text{antilog}(y_0 \pm t s_y)$$

where y_0 = logarithm of the value being tested

s_y = standard deviation of the logarithms of y values

t = 10 percent significance multiplier, depending on the number of years, obtained from:

N	t	N	t	N	t	N	t	N	t
10	2.036	30	2.563	50	2.768	70	2.893	90	2.981
15	2.247	35	2.628	55	2.804	75	2.917	95	3.000
20	2.385	40	2.682	60	2.837	80	2.940	100	3.017
25	2.486	45	2.727	65	2.866	85	2.961	105	3.033

Gumbel's Extreme Value

Equation 26.43 can be solved for the recurrence interval T and for the variate x, as follows:

$$\frac{1}{T} = 1 - F(x) = 1 - \exp\{-\exp[-\alpha(x - u)]\} \tag{27.9}$$

$$x = u - \frac{1}{\alpha}\ln[\ln T - \ln(T - 1)] \tag{27.10}$$

On substituting into Eq. 27.5, the general frequency equation, with u and α as defined in Table 26.3, the frequency factor for the extreme value distribution becomes

$$K = -\frac{\sqrt{6}}{\pi}\left(0.5772 + \ln \ln \frac{T}{T - 1}\right) \tag{27.11}$$

It should be noted that this expression for K is valid only in the limit, that is, as n approaches infinity. For a finite sample, K varies with the sample or length of record as shown in Table 27.3. K versus T curves have also been developed.[1] In Eq. 27.11, when $K = 0$, $T = 2.33$ years; thus in flood frequency analysis the recurrence interval of the mean annual flood is commonly designated the 2.33-year event.

EXAMPLE 27.4

The mean of the annual maximum discharges at a streamflow site with 25 years of record is 1000 cfs. The standard deviation is 400 cfs. Estimate the magnitude of the 50-year flood for a Gumbel extreme-value distribution.

Solution. From Table 27.3, $K = 3.088$; $x = \bar{x} + Ks = 1000 + 3.088(400) = 2235$ cfs. ■■

TABLE 27.3 GUMBEL EXTREME-VALUE FREQUENCY FACTORS

Sample size	Recurrence interval								
	2.33	5	10	20	25	50	75	100	1000
15	0.065	0.967	1.703	2.410	2.632	3.321	3.721	4.005	6.265
20	0.052	0.919	1.625	2.302	2.517	3.179	3.563	3.836	6.006
25	0.044	0.888	1.575	2.235	2.444	3.088	3.463	3.729	5.842
30	0.038	0.866	1.541	2.188	2.393	3.026	3.393	3.653	5.727
40	0.031	0.838	1.495	2.126	2.326	2.943	3.301	3.554	5.476
50	0.026	0.820	1.466	2.086	2.283	2.889	3.241	3.491	5.478
60	0.023	0.807	1.446	2.059	2.253	2.852	3.200	3.446	5.410
70	0.020	0.797	1.430	2.038	2.230	2.824	3.169	3.413	5.359
75	0.019	0.794	1.423	2.029	2.220	2.812	3.155	3.400	5.338
100	0.015	0.779	1.401	1.998	2.187	2.770	3.109	3.349	5.261
∞	-0.067	0.720	1.305	1.866	2.044	2.592	2.911	3.137	4.900

EXAMPLE 27.5

The annual maximum discharge data in Table 27.4 have been obtained from the U.S. Geological Survey Water Resources Division for a small stream in Missouri. Rank the data and plot on extreme-value probability paper.

Solution. The data are plotted in Fig. 27.3. ■■

Effect of Record Length on Flood Prediction

The length of the period of record used in a frequency analysis significantly affects the results. Data from a 68-year record at one gauging station was analyzed in various groupings and subsets using log–Pearson III frequency procedures, with the following results:[8]

1. Use of less than all 68 years caused large increases in the estimated 10-, 50-, and 100-year floods. If only the most recent 10-year record rather than all

TABLE 27.4

Water year	Annual maximum discharge (cfs)	Rank	$(n + 1)/m$
1967	2510	8	1.375
1966	4150	1	11.0
1965	2990	5	2.2
1964	2120	10	1.1
1963	3555	2	5.5
1962	2380	9	1.22
1961	2550	7	1.57
1960	2800	6	1.83
1959	3300	3	3.67
1958	3150	4	2.75

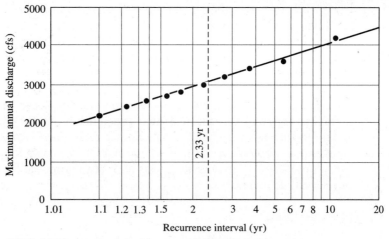

Figure 27.3 Annual floods on a small Missouri stream.

68 years had been used, the estimated 100-year flood was increased by 211 percent. Even the 10-year flood was overestimated by 62 percent.

2. If only the most recent 20-, 30-, 40-, 50-, and 60-year records were available, the overestimate in Q_{100} ranged from 123 percent for 20 years of data to 2 percent for 60 years. The Q_{10} estimates were 51 percent and −5 percent different, respectively, for the 20- and 60-year subsets of data.

3. The choice of wet versus dry sequences during the 68 years also affected the results significantly. A 10-year record in the 1930s (dry period) resulted in Q_{100} that is 61 percent below the 68-year value. Adding 10 more years reduced this error only slightly. On the other hand, use of the 10-year wet cycle in the 1960s produced a Q_{100} that is 181 percent greater than the corresponding 68-year value.

4. Much of the difference is attributed to the sensitivity that the skew coefficient has to the number and type (wet versus dry) of years. For comparison, the 68-year skew of −0.546 changed to −1.827 when 1903–1912 floods were used, and 0.993 when 1943–1952 records were used.

27.4 REGIONAL FREQUENCY ANALYSIS

In dealing with single historical sequences of hydrologic variables, the predictive value of fitted distributions is limited because the records are generally short and the sampling errors correspondingly large.[9-11] Additional and more reliable information often can be obtained within a homogeneous region by correlating dependent hydrologic variables with other causative or physically related factors. In such ways, hydrologic characteristics within the region can be summarized, and estimates of statistical parameters can be derived from general regional relations. Typical examples are (1) the prediction of rainfall depths and frequencies in ungauged or incompletely gauged areas from characteristics at well-gauged sites in the same area; (2) the prediction of peak flood flows from the correlation of observed flows on measurable quantities such as drainage area size, precipitation, and stream slope; and (3) the prediction of the standard deviation of annual streamflows throughout a large basin based on a host of physical and climatic factors. Recourse to multiple regression techniques is apparent in such cases.

Definition of the regional boundaries depends on the parameters or variables to be estimated. In some cases, significant generalizations can be made over large physiographic regions (e.g., mean annual precipitation in the United States); in other situations the region might have to be limited to drainage basins of certain sizes within smaller physiographic zones (e.g., the peak rates of runoff for areas up to 1 mi^2 in the Piedmont Plateau). Homogeneity tests have been proposed for testing the significance of regional delineations.[12,13]

Forms of Regional Data

The rainfall atlases published by the U.S. Weather Bureau represent a convenient method for summarizing mean annual precipitation and storm rainfall depths expected with various frequencies. The controlling features of physiographic regions and climatic patterns are immediately apparent in such summaries, and estimates of mean values are easily interpolated between the isohyetals. Similar summaries can be made for other hydrologic variables—for example, mean annual evapotranspiration or runoff. The longer the records, the more reliable are the data.

Less widely tabulated are measures of variability: the variance or coefficient of variation, for example, of rainfall and runoff. But regional distinctions are still clearly established. Fair et al.[14] report that C_v for the mean annual rainfall varies from 0.1 in well-watered regions to 0.5 in arid regions, whereas C_v for the mean annual runoff ranges from 0.15 to 0.75. Since the runoff is usually considerably less than total rainfall in any region, the greater variability in runoff indicates the effect of other factors and presents difficulty in determining rainfall–runoff relations. Maps showing isolines of the standard deviation of regional flood flows have been prepared by the Corps of Engineers.[9]

As mentioned in Chapter 26, the large sampling errors in computing skewness from short records of hydrologic sequences suggest the use of regional values. Beard[9] proposed the figures in Table 27.5 for the logarithms of flood flows in the absence of

TABLE 27.5   BEARD'S PROPOSED SKEW COEFFICIENTS

Duration	Skew coefficient
Instantaneous	0.00
1 day	−0.04
3 days	−0.12
10 days	−0.23
30 days	−0.32
90 days	−0.37
1 year	−0.40

regional studies. Methods of "smoothing" and averaging regional values of skewness have also been proposed.[10,15]

Many techniques used in the past for generalizing regional characteristics did not rely on statistical considerations. The so-called station-year method of extending rainfall records has proved helpful but has questionable statistical validity, especially if applied to dependent series or to stations in nonhomogeneous areas. The method has been used to combine, say, two 25-year records to obtain a single 50-year sequence. In practice, the analyst may have to use imagination and ingenuity to summarize regional characteristics, while remaining aware of actual and theoretical considerations.

Index Flood Method

The *index-flood* method used in the past by the U.S. Geological Survey is an example of summarizing regional characteristics successfully.[13,16] The method uses statistical data but combines them in graphical summaries. It can be supplemented and generally improved by using statistical methods, employing, for example, the regression techniques explained in Chapter 26. The index method, as illustrated in Fig. 27.4, can be outlined as follows.

1. Prepare single-station flood-frequency curves for each station within the homogeneous region (Fig. 27.4a).
2. Compute the ratio of flood discharges taken from the curves at various frequencies to the mean annual flood for the same station.
3. Compile ratios for all stations and find the median ratio for each frequency (Fig. 27.4b).
4. Plot the median ratios against recurrence interval to produce a regional frequency curve (Fig. 27.4c).

Two statistical considerations involved are (1) a homogeneity test to justify definition of a region, and (2) a method for extending short records to place all stations on the same base period. A somewhat similar technique was developed by Potter for the Bureau of Public Roads.[17] It relies on the graphical correlation of floods with physical and climatic variables and is thus a technique that refers in part to the discussion in Chapter 26 (see also Chapter 16).

U.S.G.S. Regional Regression Equations

Early in the 1950s, the U.S. Geological Survey instituted a process of correlating flood flow magnitudes and frequencies with drainage basin characteristics. Sets of regression equations for the 2-, 5-, 10-, 25-, 50-, and 100-year floods have been developed for practically every hydrologically homogeneous region in every state. The work was largely inaugurated to develop methods for estimating peak flow rates for design of highway structures at ungauged basins. Data from gauged sites was evaluated by regional analysis to provide the best fit of regression models to the data.

Continuous water stage recorders and crest-stage gauge data were consulted to develop frequency curves for all gauged watersheds. Given the frequency curves, a

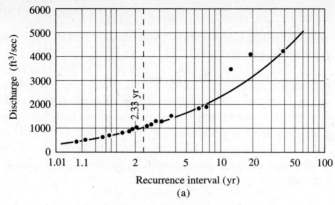

Station	Recurrence intervals (yr)					
	1.1	1.5	5	10	20	50
1	0.49	0.75	1.46	1.93	2.55	3.03
2	0.57	0.78	1.36	1.74	2.18	2.73
3	0.54	0.79	1.32	1.55	1.79	2.09
4	0.57	0.80	1.33	1.62	1.94	2.48
5	0.56	0.79	1.34	1.63	1.97	2.52
6	0.49	0.78	1.37	1.65	1.93	2.28
7	0.50	0.75	1.34	1.70	2.08	2.57
8	0.45	0.76	1.39	1.76	2.21	3.02
9	0.69	0.83	1.37	1.80	2.38	3.00
10	0.62	0.81	1.39	1.77	2.28	2.92
11	0.58	0.78	1.32	1.67	2.10	2.90
12	0.63	0.80	1.36	1.80	2.39	2.91
13	0.47	0.74	1.47	2.13	3.02	4.57
14	0.46	0.76	1.35	1.64	1.91	2.25
15	0.54	0.75	1.35	1.73	2.22	2.80
Median	0.54	0.78	1.36	1.73	2.18	2.80

(b)

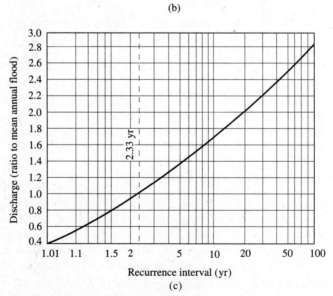

(c)

Figure 27.4 Index-flood method of regional flood frequency analysis: (a) single-station flood frequency curve, (b) ratios Q_t to $Q_{2.33}$ for 15 stations, and (c) regional flood frequency curve.

number of correlation tests were made using multiple linear regression to predict the peak flows from various easily obtained independent parameters such as drainage area, basin slope, watershed aspect, elevation, mean temperature during the snowmelt season, and hundreds of other variables.

Each study was reported by state. The open file or water resource investigation reports are available from the USGS and include discussions of the equations, comments on range of applicability, information on the reliability of the equations, copies of all the gauged basin frequency curves, and sets of equations for estimating floods in ungauged watersheds. Equations for all states have been compiled by the U.S. Geological Survey into a PC software package called NFF (National Flood Frequency), available from the USGS (or FHWA as part of their package, HYDRAIN).

Figure 27.5 shows the six regions for the state of Texas. Regional regression was conducted independently by region using available gauged station data. As in many of the reports, the Texas manual reveals that different independent variables were selected for each region.[18] The equations developed for Region 2 are:

$$Q_2 = 216\, A^{0.574} S^{0.125} \tag{27.12}$$

$$Q_5 = 322\, A^{0.620} S^{0.184} \tag{27.13}$$

$$Q_{10} = 389\, A^{0.646} S^{0.214} \tag{27.14}$$

$$Q_{25} = 485\, A^{0.668} S^{0.236} \tag{27.15}$$

$$Q_{50} = 555\, A^{0.682} S^{0.250} \tag{27.16}$$

$$Q_{100} = 628\, A^{0.694} S^{0.261} \tag{27.17}$$

where Q = peak discharge for given frequency, cfs
A = drainage area, square miles
S = average slope of the streambed between points 10 and 85 percent of the distance along the main stream channel from the mouth to the basin divide, feet per mile

Equations for other regions in Texas include the mean annual precipitation, P, along with the area and slope terms. The equations apply to rural basins with areas from 0.3 to 5000 square miles. Drainage areas in Region 2 ranged in size from 0.33 to 4255 square miles, and slopes from 1.16 to 108.1 feet per mile. Lack of data prevented the development of regression equations for the southern and western parts of the state.

EXAMPLE 27.6 _____

Develop estimates of flood peaks for a 200-square-mile rural watershed near Dallas. The mean slope between the 10 and 85 percent points is 3.4 ft per mile.

Solution.   Dallas is in Region 2. Equations 27.12–27.17 give:

$$Q_2 = 216\, A^{0.574} S^{0.125} = 5{,}270 \text{ cfs}$$

$$Q_5 = 322\, A^{0.620} S^{0.184} = 10{,}770 \text{ cfs}$$

$$Q_{10} = 389\, A^{0.646} S^{0.214} = 15{,}490 \text{ cfs}$$

$$Q_{25} = 485\, A^{0.668} S^{0.236} = 22{,}300 \text{ cfs}$$

$$Q_{50} = 555\, A^{0.682} S^{0.250} = 27{,}800 \text{ cfs}$$

$$Q_{100} = 628\, A^{0.694} S^{0.261} = 34{,}170 \text{ cfs} \quad \blacksquare\blacksquare$$

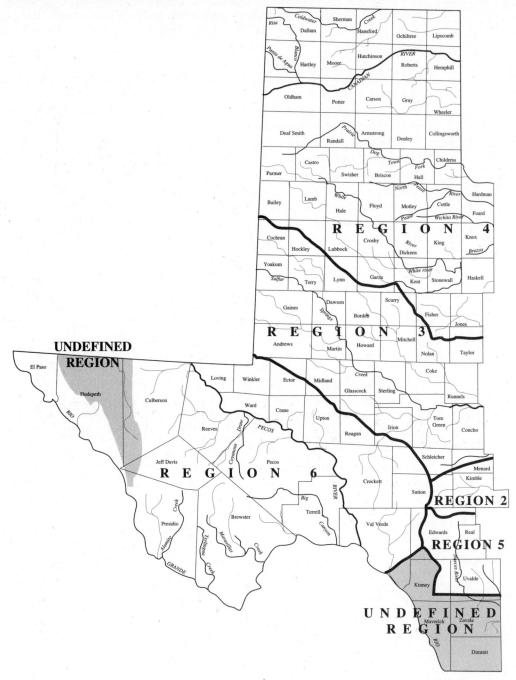

Figure 27.5   Hydrologic regions in Texas for 1976 USGS regional regression equations. (From Ref. 18.).

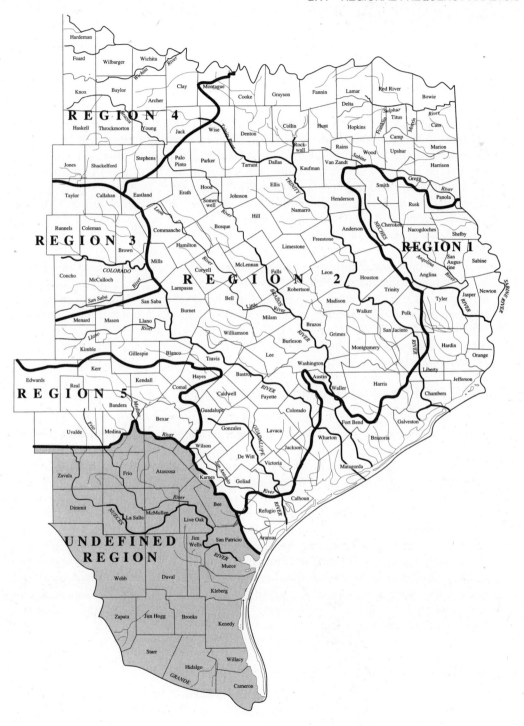

National Flood Frequency (NFF) Program

Since 1973, regression equations like Eqs. 27.12 through 27.17 for estimating flood-peak discharges for rural, unregulated watersheds have been published, at least once, for every state and the Commonwealth of Puerto Rico. In 1993 the USGS, in cooperation with the Federal Highway Administration and the Federal Emergency Management Agency, compiled all of the current statewide and metropolitan area regression equations into a microcomputer program titled the National Flood Frequency (NFF) Program.[19] This program summarizes regression equations for estimating flood-peak discharges for all 52 states. It also addresses techniques for estimating a typical flood hydrograph for a given recurrence interval or exceedence probability peak discharge for unregulated rural and urban watersheds. The program lists statewide regression equations for rural watersheds and provides much of the reference information and input data needed to run the computer program. Regression equations for estimating urban flood-peak discharges for several metropolitan areas in at least 13 states are also available.

Information on computer specifications and the computer program are given.[17] Instructions for installing NFF on a personal computer and a description of the NFF program and the associated data base of regression statistics are also given. The program is available as part of the Federal Highway Administration package, HYDRAIN, or by itself. Though the USGS and FHWA do not distribute or service the software, information about vendors who provide software sales and service can be obtained by contacting the agencies.

Flood Frequency from Channel Geometry

Stream channels in alluvial systems develop their width, depth, slope, and other hydraulic geometry characteristics from the composite hydrographs that flow through their valleys. It has been demonstrated that the shape of some stream channels, if properly evaluated by trained hydrologists, can be correlated with the mean annual flow, peak annual flow, bank-full flow, and the *dominant,* or channel-forming, discharge. Regression equations, similar to Eqs. 27.12–27.17, have been successfully derived for many perennial streams with very reasonable standard errors of estimate.

Measurements for these studies are normally obtained during low flow. The channel of interest is that channel being maintained by the current flow regime. It is characterized by the active channel, limited laterally by the point bars and most recent geologic floodplain deposits. It is felt that these represent the most recent depositions, and are therefore indicative of the width needed by the current flow and sediment transport regime. Figure 27.6 illustrates the principle in determining the active, floodplain-building channel, established for the example as the width A-A'.

Such studies have been conducted in Nevada, California, Kansas, Colorado, and elsewhere. A USGS investigation of 53 gauged streams in mountain regions of Colorado resulted in the following equations.[20]

$$Q_2 = 0.666 \ W^{1.904} D^{-0.201} \tag{27.18}$$

$$Q_5 = 1.53 \ W^{1.682} D^{-0.251} A^{0.077} \tag{27.19}$$

$$Q_{10} = 2.38 \ W^{1.530} D^{-0.259} A^{0.143} \tag{27.20}$$

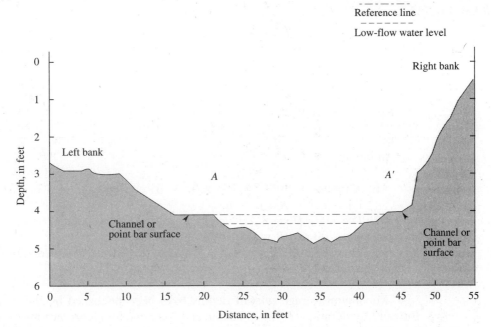

Figure 27.6 Typical stream cross-section, illustrating active channel dimensions. (From Ref. 20.)

$$Q_{25} = 3.70 \ W^{1.372} D^{-0.263} A^{0.215} \qquad (27.21)$$

$$Q_{50} = 4.93 \ W^{1.274} D^{-0.256} A^{0.257} \qquad (27.22)$$

where Q = peak flow for the given frequency, cfs
W = top width of stream at bank-full condition, ft
D = mean depth for bank-full flow, ft
A = cross-section area at bank-full flow, sq ft

The multiple correlation coefficients for these equations ranged from 0.80 for the 50-yr flow to 0.89 for the 2-yr event. Standard errors, respectively, ranged from 42.1 percent to 32.2 percent. These types of investigations offer yet another tool for use in estimating peak flows, and allow the hydrologist to evaluate floods by site-specific conditions versus more uncertain regional parameters.

Regional Rainfall Characteristics

The variation of rainfall frequencies with duration was introduced in Chapter 2. Regression analysis can be used to fit intensity–duration–frequency curves similar to those shown in Chapter 15, and the constants interpreted as regional characteristics. Many formulas have been used in the past to fit these curves, but most of them are in a form with intensity (i) inversely proportional to duration (t). Steel[21] has used a model of the form $i = A/(t + B)$ to fit rainfall data throughout the United States. The constants A and B therefore serve as characteristic features of both the rainfall region and the frequency of occurrence in each area.

EXAMPLE 27.7 _____

Fit the following rainfall data to determine the 10-year intensity–duration–frequency curve.

t = duration (min)	5	10	15	30	60	120
i = intensity (in./hr)	7.1	5.9	5.1	3.8	2.3	1.4
$1/i$	0.14	0.17	0.20	0.26	0.43	0.71

Solution

1. A model of the form $i = A/(t + B)$ can be expressed in linear form as $1/i = t/A + B/A$.
2. The regression of $1/i$ versus t yields $1/i = 0.005t + 0.12$, from which $A = 200$ and $B = 24$.
3. Thus the rainfall formula is $i = 200/(t + 24)$. The correlation coefficient is -0.997. ∎∎

Maximum average rainfall depths have been published by the U.S. Weather Bureau[22] for durations between 30 min and 24 hr and for recurrence intervals between 1 and 100 years. Depth–duration–frequency curves can be constructed for any location by plotting successive values from the various rainfall maps, preferably on logarithmic paper to facilitate fitting flatter curves. Correction factors are given to permit estimates of depths for duration less than 30 min. Special attention must be given to the extrapolation of point rainfall data to account for spatial variations. Duration frequency analyses based on maximum point gauge data exhibit the average area–depth relation shown in Fig. 27.7.

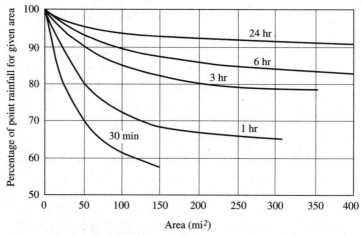

Figure 27.7 Area–depth curves for use with duration frequency values. (U.S. Weather Bureau.)

The accuracy of area rainfall data depends heavily on the density and location of gauges throughout the area considered. The simple averaging of the accumulation in all gauges gives no consideration of the effective area around each gauge or to the storm pattern. Two methods are available in calculating the weighted average of gauge records, the Thiessen polygon method and the isohyetal method. The Thiessen method assumes a linear variation of rainfall between each pair of gauges. Perpendicular bisectors of the connecting lines form polygons around each gauge (or partial polygons within the area boundary). If a sufficient number of gauges are available to construct contours of rainfall depth (isohyets), the weighting process can be carried out by using the average depth between isohyets and the area included between the isohyets and the area boundaries. Figure 27.8 shows both schemes.

An example of the effect of gauge location and density is shown in Fig. 27.9. Figure 27.9a shows the increase in variability between Thiessen-weighted storm rainfalls and rainfall at a single gauge as the distance of single gauges from the watershed center increases. Figure 27.9b shows the effect of gauge density and total area on the standard error of the mean. Complete studies of precipitation patterns over large areas require detailed analysis of depth–area–duration data that depend on the mass curves of accumulation from a network of gauges. The method is described in detail in other references.[23–25]. Figure 27.10 depicts the depth–area relation for the 24-hr storm shown in Fig. 27.8. It also required observations taken at various durations and the successive determination of average depths by the isohyetal method.

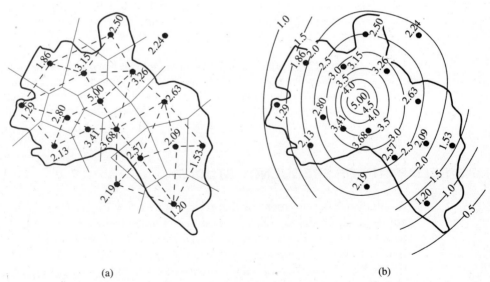

(a)   (b)

Figure 27.8   Methods of determining rainfall averages: (a) Thiessen network (24-hr total; average basin precipitation = 2.54 in.) and (b) isohyetal map (24-hr total; average basin precipitation = 2.50 in.). The arithmetic average over the basin = 39.10/15 = 2.61 in.

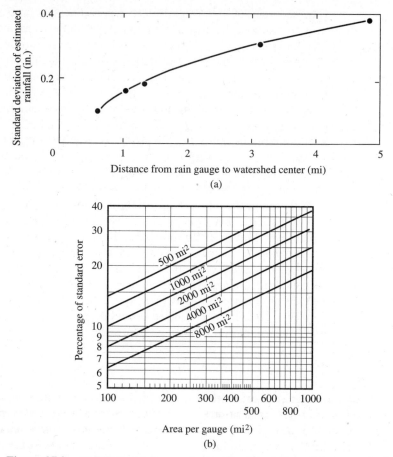

Figure 27.9 (a) Relation between the standard deviation of the watershed and the point rainfall and distance of that point from the center of the watershed. (After Knisel and Baird.[24]) (b) Standard errors of average precipitation as a function of the network density and drainage are for the Muskingum basin. (U.S. Weather Bureau.)

27.5 RELIABILITY OF FREQUENCY STUDIES

A significant development of theoretical statistics is the *central limit theorem*. As a consequence of the law of large numbers, the central limit theorem states that for a population with finite variance σ^2 and a mean μ, the distribution of sample means—that is, a number of equally good means from repeated samples—will be distributed themselves as a normal distribution with mean μ and a variance equal to σ^2/n, where σ is the population standard deviation. This theorem does not limit the type of underlying population distribution but says that the distribution of the sample means will approach a normal distribution as the sample size increases. The statistic $\sigma/\sqrt{n}$ is the standard deviation of the distribution of means and is called the *standard error*

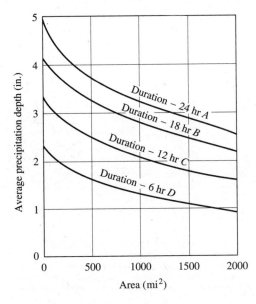

Figure 27.10 Depth–area–duration curves for 24-hr storm of Fig. 27.8. (From Ref. 23.)

of the mean. Listed in Table 27.6 are several parameters of distributions and their standard errors. It is apparent that standard errors, and therefore reliability, are almost completely a function of the sample size.

Confidence Limits

It is possible to place confidence limits on the measurement of a sample mean based on the normal distribution of all means and regardless of the underlying population. As mentioned earlier, approximately two thirds of the observations of a normal variate should fall between the limits of $+1$ and -1 standard deviation. Therefore, two thirds of all sample means should occur between the limits $\pm\sigma/\sqrt{n}$. The 95 percent confidence limits for the mean are approximately $\mu \pm 2\sigma/\sqrt{n}$. Note that this statement requires knowledge of the underlying population variance. However, usually only s^2 instead of σ^2 is known and a slightly different technique is needed to estimate confidence limits for a sample mean with $\bar{x}$ and the standard deviation s. This involves the use of sampling distributions that are beyond the scope of this text. For more information in the field of inferential statistics—in particular, hypothesis testing and statistical decision theory—the reader must turn to other sources.

Approximate error limits or control curves can be placed on frequency curves. A method proposed by Beard[5] involves placing lines above and below the fitted curve

TABLE 27.6

Measure	Standard error
Mean	$\sigma/\sqrt{n}$
Standard deviation	$\sigma/\sqrt{2n}$
Coefficient of variation	$C_v\sqrt{1 + 2C_v^2}/\sqrt{2n}$
Coefficient of skewness	$\sqrt{6n(n-1)/(n+1)(n-2)(n+3)}$

TABLE 27.7 ERROR LIMITS FOR FLOOD FREQUENCY CURVES

Years of record (n)	Exceedance frequency (%, at 5% level)						
	99.9	99	90	50	10	1	0.1
5	1.22	1.00	0.76	0.95	2.12	3.41	4.41
10	0.94	0.76	0.57	0.58	1.07	1.65	2.11
15	0.80	0.65	0.48	0.46	0.79	1.19	1.52
20	0.71	0.58	0.42	0.39	0.64	0.97	1.23
30	0.60	0.49	0.35	0.31	0.50	0.74	0.93
40	0.53	0.43	0.31	0.27	0.42	0.61	0.77
50	0.49	0.39	0.28	0.24	0.36	0.54	0.67
70	0.42	0.34	0.24	0.20	0.30	0.44	0.55
100	0.37	0.29	0.21	0.17	0.25	0.36	0.45
	0.1	1	10	50	90	99	99.9

Exceedance frequency (%, at 95% level)

Note: Tabular values are multiples of the standard deviation of the variate. Five percent error limits are added to the flood value from the fitted curve at the same exceedance frequency and the sum plotted. Ninety-five percent limits are subtracted from the flood value at the same exceedance frequency. Log values are added or subtracted before antilogging and plotting.

to form a reliability band. Table 27.7 shows the factors by which the standard deviations of the variate must be multiplied to mark off a 90 percent reliability band above and below the frequency curve. The 5 percent level, for example, means that only 5 percent of future values should fall higher than the limit, and, similarly, only 5 percent should fall under the 95 percent limit. Nine of ten should fall within the band.

EXAMPLE 27.8 _____

The maximum annual instantaneous flows from the Maury River near Lexington, Virginia, for a 26-year period are listed in Table 27.8.

Plot the log–Pearson III curve of best fit and determine the magnitude of the flood to be equaled or exceeded once in 5, 10, 50, and 100 years. Using Table 27.7, also plot the upper and lower confidence limits.

TABLE 27.8

Water (year)	Discharge (cfs)	Water (year)	Discharge (cfs)	Water (year)	Discharge (cfs)
1926	6,730	1935	13,800	1944	6,680
1927	9,150	1936	40,000	1945	6,540
1928	6,310	1937	10,200	1946	5,560
1929	10,000	1938	13,400	1947	7,700
1930	15,000	1939	8,950	1948	8,630
1931	2,950	1940	11,900	1949	14,500
1932	8,650	1941	5,840	1950	23,700
1933	11,100	1942	20,700	1951	15,100
1934	6,360	1943	12,300		

Solution

1. The statistical calculations are summarized as follows:

	Arithmetic	Log
Mean $\bar{x}$	11,606	4.001
Variance s^2	53.87×10^6	0.0516
Skew coefficient C_s	2.4	0.38

2. After forming an array and computing plotting positions, the data are plotted in Fig. 27.11.

3. Plotting data for log–Pearson III (Table 27.9) are developed from Table B.2. Confidence limits are plotted in Fig. 27.11 using Table 27.7.   ■■

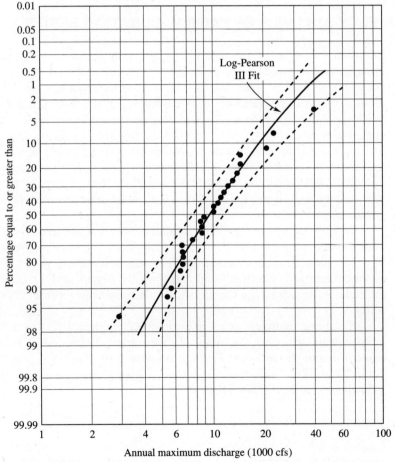

Figure 27.11 Maximum instantaneous annual flows, Maury River, Lexington, Virginia.

TABLE 27.9

Chance (%)	I (yr)	($C_s = 0.38$) K	($\bar{y} = 4.001$) ($s_y = 0.227$) $\bar{y} + Ks_y = \log Q$	Q
99	1.01	−2.044	3.537	3,443
95	1.05	−1.530	3.653	4,498
90	1.11	−1.234	3.721	5,260
80	1.25	−0.855	3.760	5,754
50	2	−0.062	3.987	9,705
20	5	0.818	4.187	15,380
10	10	1.315	4.300	19,950
4	25	1.874	4.426	26,690
2	50	2.251	4.512	32,510
1	100	2.601	4.591	39,030
0.5	200	2.930	4.666	46,360

27.6 FREQUENCY ANALYSIS OF PARTIAL DURATION SERIES

In earlier examples of frequency analysis, only the series of annual maximum or minimum occurrences in the hydrologic record have been described. These extremes constitute an *annual series* that is consistent with frequency analysis and the manipulation of annual probabilities of occurrence. All the observed data—say, all floods or all the daily streamflows—would constitute a *complete series*. Any subset of the complete series is a *partial series*. In selecting the maximum annual events from a record, it often happens that the second greatest event in one year exceeds the annual maximum in some other year. Analysis of the annual series neglects such events. Although they generally contain the same number of events, the extreme values analyzed without regard for the period (i.e., year) of occurrence, is usually termed the *partial duration series*.

In Table 27.10 the maximum rainfall depths that occurred for any 30-min period during excessive rainfalls at Baltimore, Maryland, 1945–1954, are shown in the order of occurrence. The 65 observations represent a complete series. The 11 maximum annual events are underlined and represent the annual series. the greatest 11 events throughout the record are identified by an asterisk and represent the partial duration series.

The larger numbers occur in both series, and hence recurrence intervals for the less-frequent events are the same. The theoretical differences in recurrence intervals based on annual and partial duration series of the same length are shown in Table 27.11. The difference for intervals greater than 10 years is negligible. The following example is illustrative.

EXAMPLE 27.9 _____

Perform a frequency analysis of the 30-min Baltimore rainfall data in Table 27.10 as an annual and a partial duration series and plot the results.

Solution.   See Table 27.12. The data are plotted in Fig. 27.12.   ■■

TABLE 27.10   MAXIMUM 30-MIN RAINFALL DEPTHS, BALTIMORE, MD, 1945–1954

Year	Storm number	RF depth (in.)	Year	Storm number	RF depth (in.)	Year	Storm number	RF depth (in.)
1945	1	0.38	1948	1	1.33*	1953	1	0.40
	2	0.47		2	0.65		2	0.45
	3	0.39		3	0.47		3	0.53
	4	0.76		4	0.84		4	2.50*
	5	0.56		5	0.68		5	1.03
	6	0.35		6	0.63		6	0.75
	7	0.43		7	0.47		7	0.70
	8	0.40					8	1.00*
	9	0.36	1949	1	0.52			
				2	0.49			
1946	1	0.62				1954	1	0.42
	2	0.55					2	0.70
	3	0.88	1950	1	0.55		3	0.85
	4	0.47		2	0.63		4	0.60
	5	0.36		3	0.69			
	6	1.15*		4	1.27*	1955	1	0.70
	7	0.75		5	1.10*		2	0.95
	8	1.53*					3	1.02
	9	0.51	1951	1	0.88		4	0.50
				2	0.97		5	0.65
1947	1	0.88		3	0.59		6	0.55
	2	2.04*		4	0.46		7	0.52
	3	0.76		5	0.50		8	0.45
	4	0.97		6	0.55		9	0.54
	5	0.71					10	0.60
	6	1.07*	1952	1	0.47		11	0.80
	7	0.94		2	1.20*		12	0.95
	8	1.20*		3	0.93			
				4	0.70			
				5	0.57			
				6	0.46			
				7	0.48			
				8	1.30*			

Note: Underlined items are the annual series. Asterisks identify the partial duration series.

TABLE 27.11   RELATION BETWEEN THE PARTIAL DURATION SERIES AND THE ANNUAL SERIES

	Recurrence interval (yr)
Partial duration series	Annual series
0.5	1.2
1.0	1.6
1.5	2.0
2.0	2.5
5.0	5.5
10.0	10.5

TABLE 27.12

Order	Depth (in.)		Recurrence interval $(n + 1)/m$
	Annual series	Partial series	
1	2.50	2.50	12
2	2.04	2.04	6
3	1.53	1.53	4
4	1.33	1.33	3
5	1.30	1.30	2.4
6	1.27	1.27	2
7	1.02	1.20	1.7
8	0.97	1.20	1.5
9	0.85	1.15	1.3
10	0.76	1.10	1.2
11	0.52	1.07	1.1

The preceding example leads to consideration of the frequency analysis of rainfall depth or intensity for various durations of rainfall. Design problems often require the estimation of expected intensities for a critical time period. Frequency analysis of the rainfall record for periods other than the 30-min duration—for example, the maximum 5-, 10-, 20-, and 60-min occurrences—would yield a family of curves similar to those of Fig. 27.10. The usual method of presenting these data is to convert depth in inches to an intensity in in./hr and to summarize the data in intensity–duration–frequency curves as shown in Fig. 27.13. These curves are typical of the point analysis of rainfall data. It should be emphasized that the frequency curves join occurrences that are not necessarily from the same storm; that is, they do not repre-

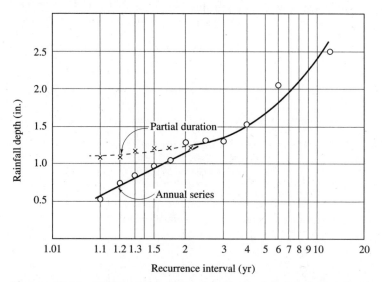

Figure 27.12 Difference in annual and partial duration series 11-year record of maximum 30-min durations, Baltimore, Maryland.

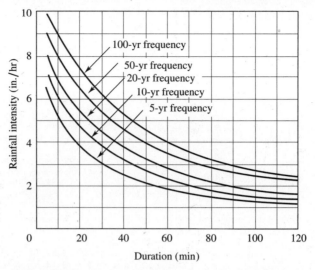

Figure 27.13   Typical intensity–duration–frequency curves.

sent a sequence of intensities during a single storm but only the average intensity expected for a specific duration.

27.7 FLOW DURATION ANALYSIS

Figures 27.14–27.17 illustrate further applications of frequency analysis. Figures 27.14 and 27.15 represent standard point frequency analyses of the annual series of high and low flows for different durations. Figure 27.16 is a low-flow duration–frequency curve based on the same data as Fig. 27.15. Figure 27.17 is based on an

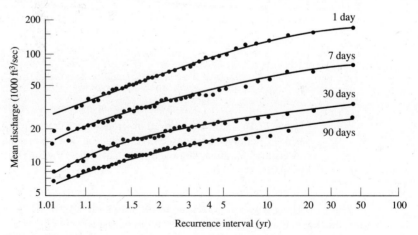

Figure 27.14   High-flow frequency curves in the James River at Cartersville, Virginia. (Virginia Division of Water Resources.)

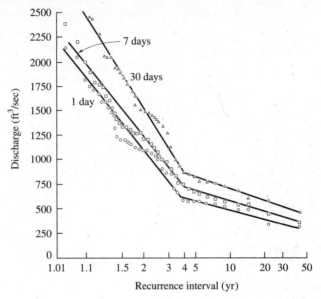

Figure 27.15   Low-flow frequency curves in the James River at Cartersville, Virginia. (Virginia Division of Water Resources.)

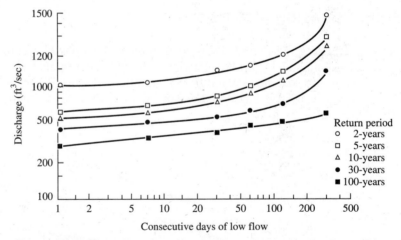

Figure 27.16   Low-flow duration–frequency curves in the James River basin. Discharge to consecutive days of low flow James River at Cartersville, Virginia. Drainage area: 6242 mi². (Virginia Division of Water Resources.)

analysis of the *complete* series of daily flows although all observed values are not plotted. Presented in this form such a curve usually is called a *duration* curve. Note that the probability scale must be labeled "percentage of time," since the annual series was not used. Duration curves are useful in predicting the availability and variability of sustained flows, but, again, they do not represent the actual sequence of flows.

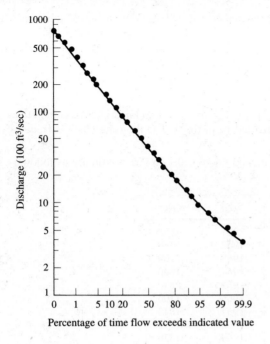

Figure 27.17 Flow duration curve in the James River at Cartersville, Virginia. (Virginia Division of Water Resources.)

■ Summary

Frequency analysis plays an important role in practically every decision regarding allocation of water resources or management of storm water. The procedures described in this chapter have achieved the status of "standards" for the uniform and consistent analysis of the recurrence of magnitudes of all types of hydrologic variables. Modern approaches to evaluating more than single alternatives capitalize on the art and science of frequency and risk analysis to assure that decisions will provide the level of protection and reliability of water supplies desired by society.

PROBLEMS

27.1. The total annual pumpage (in acre-ft) over the last 30 years from a fully developed irrigation well field was observed as follows:

2450	3300	3400	3650	3800
2650	3150	3100	3500	2850
3050	4300	3300	3300	3150
2100	3300	3650	3150	3550
2900	3250	3000	3400	3750
3900	3600	3150	3600	3000

Form an array of the data and plot the apparent frequencies on normal probability paper. Compute and plot the mean of the data and draw the line of best fit through the mean and as close as possible to the other points by eye. Estimate from the line the standard deviation and the highest value expected to be equaled or exceeded once in 50 years.

27.2. Annual floods for a small river are reported to follow a normal probability distribution. The 2-year flood for the basin has been estimated as 40,000 cfs and the 10-year flood as 52,820 cfs. Using normal frequency factors, determine the magnitude of the 25-year flood.

27.3. Discuss how probability paper (graph paper that linearizes the cumulative distribution function) can be constructed from the K–T table using Chow's frequency equation $x = \bar{x} + Ks$.

27.4. The 20-year record of annual flood peaks on Furr's Run are as follows:

Year	Q (cfs)	Year	Q (cfs)
1951	1060	1961	1350
1952	2820	1962	1140
1953	1970	1963	2100
1954	1760	1964	1090
1955	1650	1965	2890
1956	1140	1966	1100
1957	1020	1967	1840
1958	2260	1968	1710
1959	1650	1969	1630
1960	870	1970	1260

Log the values of peak discharge and compute the mean, standard deviation, and skewness coefficient. Plot the data on log–normal probability paper. With the aid of Table B.2, fit both the log–normal and log–Pearson III distributions. Compare estimates of the 50- and 100-year events.

27.5. Test whether the higher flow rates in Problem 27.4 are outliers according to the U.S. Water Resources Council guidelines.

27.6. Compare the results of Problem 27.4 by plotting on extreme-value paper. Compute the mean and variance of the discharges, and, using frequency factors from Table 27.3, find the Gumbel estimates of the 50- and 100-year events.

27.7 The following parameters were computed for a stream near Lincoln, Nebraska:

Period of record = 1940–1959, inclusive

Mean annual flood = 7000 cfs

Standard deviation of annual floods = 1000 cfs

Skew coefficient of annual floods = 1.0

Mean of logarithms (base 10) of annual floods = 3.52

Standard deviation of logarithms = 0.50

Coefficient of skew of logarithms = 2.0

Determine the magnitude of the 25-year flood by assuming that the peaks follow (a) the log–Pearson type III distribution, (b) the Gumbel distribution, and (c) the normal distribution.

27.8. Expand the computer program of Problem 26.12 to include the computation of the mean, standard deviation, and skewness coefficient of the logarithms of the input data. Also, include a routine to sort the data by placing them in descending order and compute the corresponding plotting positions. Verify, using the data in Problem 27.4.

27.9. Perform a complete frequency analysis on one of the three 33-year records given in the table below. Fit a Pearson type III or log–Pearson III and compare with the normal or log–normal of best fit. Plot the data and place control curves around the theoretical curve of best fit using Table 27.7.

Year	Trempeuleau River Dodge, WI (DA = 643 mi^2) Q_{peak} (cfs)	Blow River Banff, Alberta, Canada (DA = 858 mi^2) Q_{peak} (cfs)	James River Scottsville, VA (DA = 4570 mi^2) Q_{peak} (cfs)
1928	3,700	10,200	75,600
1929	1,700	7,590	44,700
1930	3,360	9,280	45,800
1931	1,650	6,610	21,100
1932	3,600	9,850	31,400
1933	11,000	11,000	59,500
1934	2,570	9,490	38,800
1935	4,490	6,940	93,400
1936	7,180	7,720	126,000
1937	1,780	5,210	62,200
1938	3,170	7,770	87,400
1939	6,400	6,270	68,400
1940	3,120	7,220	130,000
1941	2,890	4,450	27,100
1942	5,680	5,850	80,600
1943	5,060	7,380	95,200
1944	2,040	5,590	133,000
1945	8,120	4,450	57,000
1946	4,570	7,210	41,200
1947	5,410	5,880	33,200
1948	4,840	10,320	59,600
1949	1,920	4,290	94,200
1950	3,600	10,080	73,300
1951	4,840	8,570	64,900
1952	6,950	5,460	54,500
1953	4,040	9,180	67,000
1954	5,710	10,120	62,900
1955	10,400	8,680	70,000
1956	17,400	9,060	20,400
1957	713	5,360	64,200
1958	1,140	6,730	44,500
1959	8,000	7,480	29,300
1960	1,480	6,440	64,200

27.10. Compare results of Problem 27.9 with estimates by Gumbel's extreme-value distribution for the 50- and 100-year events.

27.11. The pan-evaporation data (in.) for the month of July at a site in Missouri are

9.7	11.7	11.2	11.3	11.5
11.2	8.8	11.4	11.8	8.9
9.3	9.2	9.3	9.3	10.4
9.8	8.7	11.5	10.9	10.2

Determine the mean, standard deviation, and coefficient of variation. What are the standard errors of these statistics? Establish the approximate 95 percent confidence limits of the mean.

27.12. On which type of plotting paper (probability, log–probability, rectangular coordinate, log–log, semilog, extreme-value, none) would each of the following plot as a straight line?
a. Normal frequency distribution.
b. Gumbel frequency distribution.
c. $Y = 3X + 4$
d. Log–normal frequency distribution.
e. Pearson Type III with a skew of zero.
f. $Q = 43A^{0.7}$.
g. Log–Pearson Type III with a skew of logarithms equal to zero.
h. Pearson Type III with a skew of 3.0.

27.13. Determine the 50-year peak (cfs) for a log–Pearson Type III distribution of annual peaks for a major river if the skew coefficient of logarithms (base 10) is -0.1, the mean of logarithms (base 10) is 3.0, and the standard deviation of base 10 logarithms is 1.0.

27.14. A 40-year record of rainfall indicates that the mean monthly precipitation during April is 3.85 in. with a standard deviation of 0.92. The distribution is normal. With 95 percent confidence, estimate the limits within which (a) next April's precipitation is expected to fall, and (b) the mean April precipitation for the next 40 years is expected to fall.

27.15. Given a table of values of mean annual floods and corresponding drainage areas for a number of basins in a region, describe how regression analysis could be used to determine the coefficients p and q in the relation $Q_{2.33} = pA^q$.

27.16. The 80-year record of annual precipitation at a midwestern gauge location has a range between 14 in. in 1936 and 42 in. in 1965. The record has a mean of 27.6 in. and a standard deviation of 6.06 in. Assuming a normal distribution, (a) plot the frequency curve on probability paper, (b) determine the probability of a drought worse than the 1936 value, and (c) determine the recurrence interval of the 1965 maximum depth and compare it with the apparent recurrence interval.

27.17. A reservoir in the locale of Problem 27.16 will overfill when the annual precipitation exceeds 30 in. Determine the probability that the reservoir will overfill (a) next year, (b) at least once in three successive years, and (c) in each of three successive years.

27.18. Using Eqs. 26.38 and 26.39 and log–probability paper, solve Problems 27.16 and 27.17 assuming that the annual precipitation is log–normal.

27.19. Given the following values of peak flow rates for a small stream, determine the return period (years) for a flood of 100 cfs by first using annual peaks for an annual series and then using all the data for a partial duration series.

Year	Date	Peak (cfs)	Year	Date	Peak (cfs)
1963	June 1	90	1968	May 11	800
	Aug. 3	300		June 8	700
1964	June 7	60		Sept. 4	90
1965	July 2	80	1969	Aug. 8	400
1966	May 18	100	1970	May 9	30
	June 3	90	1971	Sept. 8	700
1967	July 4	40	1972	May 4	80

27.20. Recorded maximum depths (in.) of precipitation for a 30-min duration at a single station are:

Year	Date	Depth (in.)	Year	Date	Depth (in.)
1963	May 3	2.0	1968	Aug. 8	4.0
	June 3	1.0	1969	May 6	6.0
1964	June 7	1.0		June 8	5.0
1965	June 2	1.0		Sept. 4	1.0
1966	June 1	1.0	1970	May 4	1.0
	Aug. 3	3.0	1971	Sept. 8	5.0
1967	July 4	1.0	1977	May 9	1.0

a. Determine the return period (years) for a depth of 2.0 in. using the California method with an annual series.

b. Repeat Part (a) using a partial duration series.

c. Determine from the partial duration series the depth of 30-min rain expected to be equaled or exceeded (on the average) once every 8 years.

27.21. For a 60-year record of precipitation intensities and durations, a 30-min intensity of 2.50 in./hr was equaled or exceeded a total of 85 times. All but 5 of the 60 years experienced one or more 30-min intensities equaling or exceeding the 2.50-in./hr value. Use the Weibull formula to determine the return period of this intensity using (a) a partial series and (b) an annual series.

27.22. From the data given in the accompanying table of low flows, prepare a set of low-flow frequency curves for the daily, weekly, and monthly durations.

LOWEST MEAN DISCHARGE (cfs) FOR THE FOLLOWING NUMBER OF CONSECUTIVE DAYS, MAURY RIVER NEAR BUENA VISTA, VIRGINIA

Year	1-day	7-day	30-day
1939	100.0	103.0	125.0
1940	167.0	171.0	194.0
1941	22.0	59.4	69.1
1942	101.0	127.0	173.0
1943	86.0	93.9	103.0
1944	62.0	65.9	77.4
1945	78.0	80.7	90.3
1946	76.0	78.6	87.1
1947	97.0	102.0	123.0
1948	154.0	176.0	215.0
1949	136.0	138.0	163.0
1950	113.0	125.0	139.0
1951	95.0	95.3	101.0
1952	115.0	116.0	120.0
1953	85.0	86.1	90.8
1954	68.0	70.0	81.7
1955	83.0	96.1	99.9
1956	64.0	66.3	71.7
1957	62.0	64.1	75.8
1958	88.0	92.6	107.0
1959	76.0	80.9	117.0
1960	83.0	91.7	103.0
1961	99.0	103.0	152.0
1962	90.0	95.0	105.0
1963	60.0	60.6	70.8
1964	51.0	54.1	62.0
1965	64.0	68.7	76.2

27.23. For the 7-day low flows at Buena Vista given in Problem 27.22, attempt to fit a straight-line frequency curve on log-normal or extreme-value probability paper, proceeding as follows: From the original plot of the data, estimate the lowest flow (say, q) at the high recurrence intervals; subtract this flow from all observed flows ($Q - q = Q^1$); and replot Q^1 versus the original recurrence intervals. Repeat if necessary. The best fitting curve will be a three-parameter frequency distribution.

27.24. The following table summarizes the number of occurrences of intensities of various durations for a 34-year record of rainfall. Maximum intensities for the given durations were determined for all excessive storms and a count made of the exceedances. Interpolate for the average number of exceedances expected on a 5-year frequency and plot the 5-year intensity–duration–frequency curve.

Duration (min)	Number of times stated intensities were equaled or exceeded						
	Intensity (in./hr)						
	1.0	2.0	3.0	4.0	5.0	6.0	7.0
5			73	48	21	9	2
10		68	51	26	11	3	1
15	72	35	23	11	5	1	
30	29	17	7	3	1		
60	15	6	2	1			
120	8	1					

27.25. The results of a multiple regression analysis of over 200 flood records in Virginia led to the following regional flood frequency equations:

$$Q_{1.2\text{-yr}} = 9.13A^{.909}S^{.293}$$
$$Q_{2.33\text{-yr}} = 20.8A^{.861}S^{.309}$$
$$Q_{5\text{-yr}} = 38.1A^{.830}S^{.300}$$
$$Q_{10\text{-yr}} = 63.0A^{.802}S^{.283}$$
$$Q_{25\text{-yr}} = 104A^{.779}S^{.266}$$
$$Q_{50\text{-yr}} = 118A^{.795}S^{.279}$$

where the flood discharge for the given frequency is in cfs, A is the drainage area in mi^2, and S is the channel slope in ft/mi (measured between the points that are 10 and 85 percent of the total river miles upstream of the gauging station to the drainage divide). Devise a method for graphically portraying these regional flood frequency relations. (Note that there are four factors, Q, T, A, and S.)

27.26. Using the regression equations in Problem 27.25, find the predicted floods for the North Fork, Shenandoah River, at Cootes Store. Drainage area = 215 mi^2 and channel slope = 44.3 ft/mi.

27.27. Compare the predictions from the regression equations in Problem 27.25 with the values estimated by the frequency analysis in Example 27.8. Drainage area = 487 mi^2 and channel slope = 21.1 ft/mi.

27.28. Referring to Fig. 2.6b, compare the average storm rainfall over the city of Baltimore on September 10, 1957, computed by the isohyetal method, with the simple average of total accumulation at the rain gauges within the city. Neglect the area to the south of the 1.0-in. isohyet.

27.29. Fit the formula $i = A/(t + B)$ to the data derived in Problem 27.24 for the 5-year intensity–duration–frequency curve.

27.30. Develop a regional flood index curve for the Rappahannock River basin from the flood frequency data given in the following table:

PEAK FLOOD FREQUENCY DISCHARGES (ft³/sec) FOR STATIONS IN THE RAPPAHANNOCK RIVER BASIN

Station	Drainage area (mi²)	Type of series	2.33 (mean)	Return period in years			
				5	10	25	50
Rappahannock River near Warrenton, VA	192	Annual	4,150	8,350	9,000	14,000	19,250
		Partial	4,600	8,650	9,200	14,000	19,250
Rush River at Washington, VA	15.2	Annual	530	860	1,290	2,100	3,000
		Partial	610	900	1,310	2,100	3,000
Thornton River near Laurel Mills, VA	142	Annual	5,900	11,500	19,900	34,000	48,000
		Partial	7,200	12,500	20,500	34,000	48,000
Hazel River at Rixeyville, VA	286	Annual	7,300	11,800	17,200	25,000	41,000
		Partial	8,300	12,400	18,000	25,500	41,000
Rappahannock River at Remington, VA	616	Annual	11,000	14,500	18,100	24,500	31,000
		Partial	12,000	15,200	18,900	25,000	31,000
Rappahannock River at Kellys Ford, VA	641	Annual	12,300	19,000	26,800	42,000	57,500
		Partial	14,000	20,000	27,500	42,000	57,500
Mountain Run near Culpeper, VA	14.7	Annual	750	1,750	3,350	6,000	10,000
		Partial	950	1,900	3,550	6,000	10,000
Rapidan River near Ruckersville, VA	111	Annual	3,950	7,100	11,600	21,000	34,000
		Partial	4,700	7,700	12,000	21,000	34,000
Robinson River near Locust Dale, VA	180	Annual	4,600	7,000	9,800	15,400	21,500
		Partial	5,150	7,300	10,100	15,800	21,500
Rapidan River near Culpeper, VA	456	Annual	9,100	16,400	26,900	50,000	78,000
		Partial	10,800	17,600	27,600	50,000	78,000
Rappahannock River near Fredericksburg, VA	1,599	Annual	26,000	39,900	55,000	85,000	117,000
		Partial	29,300	42,000	57,500	85,000	117,000

27.31. From the information given in Problem 27.30, find the relation between the mean annual flow and the drainage area. (Note that the functional expression should be of the form $Q_{2.33} = rA^s$.)

27.32. Using the results of Problems 27.30 and 27.31, estimate the 30-year flood for an ungauged watershed with a drainage area of 540 mi².

27.33. Annual flood records for a 10-year period are given by:

Year	1	2	3	4	5	6	7	8	9	10
Flood	300	700	200	400	1000	900	800	500	100	600

Mean = 550 cfs, median = 550 cfs, standard deviation = 300 cfs. Use an annual series and the definition of frequency in a frequency analysis to determine the magnitude of the 4-year flood. Compare this historical value with the analytical 4-year flood obtained assuming floods follow a normal distribution.

27.34. For a 60-year record of precipitation intensities and durations, a 30-min intensity of 2.50 in./hr was equaled or exceeded a total of 85 times. All but 5 of the 60 years experienced one or more 30-min intensities equaling or exceeding the 2.50-in./hr value. Use the Kimball formula to determine the return period of this intensity using (a) a partial duration series and (b) an annual series.

27.35. The total annual runoff from a small drainage basin is determined to be approximately normal with a mean of 14.0 in. and a standard deviation of 3 in. Determine the probability that the annual runoff from the basin will be less than 8.0 in. in the second year only of the next three consecutive years.

27.36. Six years of peak runoff rates are given below. Assume that the floods follow exactly a normal distribution and determine the magnitude of the 50-year peak.

Year	1	2	3	4	5	6
Runoff (cfs)	200	800	500	600	400	500

27.37. Annual floods for a stream are normally distributed with a mean of 30,000 cfs and a variance of 1×10^6 cfs^2. Determine the average return period T_r of a 32,000-cfs flood in the stream.

27.38. Annual floods for a stream have a normal frequency distribution. The 2-year flood is 40,000 cfs and the 10-year flood is 52,820 cfs. Determine the magnitude of the 25-year flood.

27.39. The 80-year record of annual precipitation at Linclon, Nebraska, yields a range of values between 10 and 50 in. with a mean annual value of 25.00 in. and a standard deviation of 5.30 in. Because annual precipitation represents a sum of many random variables (i.e., depth of precipitation for each day of the year), assume that annual precipitation is normally distributed.
   a. In 1936 the precipitation at Lincoln was a mere 14 in. Determine the probability that the annual precipitation will be 14 in. or less next year.
   b. In 1965 Lincoln received 42 in. On the average, this amount would be equaled or exceeded once in how many years?
   c. Compare the theoretical and apparent return periods of the record-high value of 50.00 in.

27.40. Annual floods (cfs) at a particular site on a river follow a zero-skew log–Pearson Type III distributions. If the mean of logarithms (base 10) of annual floods is 2.946 and the standard deviation of base-10 logarithms is 1.000, determine the magnitude of the 50-year flood.

27.41. Annual floods (cfs) at a particular site on a river follow a zero-skew log–Pearson Type III distribution. If the mean of logarithms (base 10) of annual floods is 1.733 and the standard deviation of base-10 logarithms is 1.420, determine the magnitude of the 100-year flood.

27.42 The 100-year record for a drainage basin gives 10- and 50-year flood magnitudes of 12,500 and 22,000 cfs. Determine the magnitude of the mean annual flood if (a) the flood peaks follow the index–flood curve of Fig. 27.4c and (b) the flood peaks follow a Gumbel extreme-value distribution.

27.43. The following parameters were computed for a stream:

Period of record = 1960–1984, inclusive.

Mean annual flood = 7000 cfs

Standard deviation of annual floods = 1000 cfs

Skew coefficient of annual floods = 2.0

Mean of logarithms (base 10) of annual floods = 3.52

Standard deviation of logarithms = 0.50

Coefficient of skew of logarithms = −2.0

Determine the magnitude of the 25-year flood by assuming that the peaks follow a (a) log–Pearson Type III distribution, (b) Gumbel distribution, and (c) log–normal distribution.

27.44. Peak annual discharge rates in the Elkhorn river at Waterloo, Nebraska, yield the following statistics:

Period of record = 1930–1969, inclusive

Mean flood = 16,900 cfs

Standard deviation = 17,600 cfs

Skew of annual floods = 0.8

Mean of logarithms (base 10) = 4.0923

Standard deviation of logarithms = 0.3045

Skew of logarithms = 2.5

a. Determine the 100-year flood magnitude using the uniform technique adopted by the U.S. Water Resources Council for all federal evaluations.
b. Determine the 100-year flood magnitude assuming that the floods follow a two-parameter gamma distribution.

27.45. A Pearson Type III variable X has a mean of 4.0, a standard deviation of 2.0, and a coefficient of skew of 0.0. Determine the value (four significant figures) of $\int_0^8 f(X)\, dX$.

27.46. A timber railroad bridge in Hydrologic Region 2 of Texas at Milepost 738.04 on the railroad system shown in the sketch is to be replaced with a new concrete structure. The 50- and 100-year flood magnitudes are needed to establish the low chord and embankment elevations, respectively. Determine the design flow rates using the USGS Regression Equations. The drainage area is 0.43 sq. mi, and the streambed slope is 62 ft per mi.

REFERENCES

1. M. A. Benson, "Plotting Positions and Economics of Engineering Planning," *Proc. ASCE J. Hyd. Div.* **88,** 57–71(Nov. 1962).
2. I. I. Gringorten, "A Plotting Rule for Extreme Probability Paper," *J. Geophys. Res.* **68**(3), 813–814(Feb. 1963).

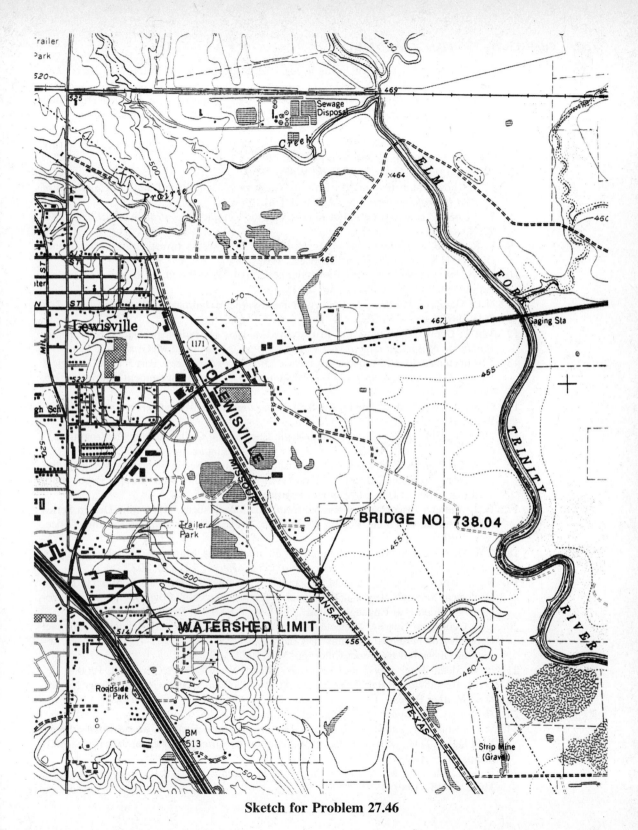

Sketch for Problem 27.46

3. Ven T. Chow, "A General Formula for Hydrologic Frequency Analysis," *Trans. Am. Geophys. Union* **32,** 231–237(1951).

4. Water Resources Council, Hydrology Committee, "Guidelines for Determining Flood Frequency," Bulletin 17B, (Revised) U.S. Water Resources Council, Washington, D.C., Sept., 1981.

5. L. R. Beard, *Statistical Methods in Hydrology,* Civil Works Investigations, U.S. Army Corps of Engineers, Sacramento District, 1962.

6. A. Hazen, *Flood Flows.* New York: Wiley, 1930.

7. V. T. Chow, "Statistical and Probability Analysis of Hydrologic Data," in *Handbook of Applied Hydrology.* New York: McGraw-Hill, 1964.

8. P. Victorov, "Effect of Period of Record on Flood Prediction," *Proc. ASCE J. Hyd. Div.* **97**(Nov. 1971).

9. L. R. Beard, *Statistical Methods in Hydrology,* Civil Works Investigations; Sacramento District, U.S. Army Corps of Engineers, 1962.

10. M. A. Benson and N.C. Matalas, "Synthetic Hydrology Based on Regional Statistical Parameters," *Water Resources Res.* **3**(4)(1967).

11. N. C. Matalas, "Mathematical Assessment of Synthetic Hydrology," *Water Resources Res.* **3**(4)(1967).

12. Ven T. Chow, "Statistical and Probability Analysis of Hydrologic Data," Sec. 8-I, in *Handbook of Applied Hydrology* (V. T. Chow, ed.). New York: McGraw-Hill, 1964.

13. T. Dalrymple, "Flood-Frequency Analysis," *Manual of Hydrology,* Part 3, U.S. Geological Survey Water-Supply Paper 1543-A. Washington, D.C.: U.S. Government Printing Office, 1960.

14. G. M. Fair, J. C. Geyer, and D. A. Okun, *Water and Waste Water Engineering.* New York: Wiley, 1966.

15. "Monthly Stream Simulation," Hydrologic Engineering Center, Computer Program 23-C-L267, Sacramento District, U.S. Army Corps of Engineers, July 1967.

16. R. W. Cruff and S. E. Rantz, "A Comparison of Methods Used in Flood Frequency Studies for Coastal Basins in California," Flood Hydrology, U.S.G.S. Water Supply Paper 1580. Washington, D.C.: U.S. Government Printing Office, 1965.

17. W. D. Potter, "Peak Rates of Runoff from Small Watersheds," Hydraulic Design Series No. 2, Bureau of Public Roads, Washington, D.C.: U.S. Government Printing Office, Apr. 1961.

18. U.S. Geological Survey, "Technique for Estimating the Magnitude and Frequency of Floods in Texas," *Water Resources Investigations Report* 77–110, 1977.

19. M. E. Jennings, W. O. Thomas, Jr., and H. C. Riggs, "Nationwide Summary of U.S. Geological Survey's Regional Regression Equations for Estimating Magnitude and Frequency of Floods at Ungauged Sites," U.S.G.S. WRI 93-1, Reston, VA, 1993.

20. U.S. Geological Survey, "Selected Streamflow Characteristics as Related to Channel Geometry of Perennial Streams in Colorado," *Open-File Report* 72–160, Water Resources Division, Lakewood, Colorado, May 1972.

21. E. W. Steel, *Water Supply and Sewerage,* 4th ed. New York: McGraw-Hill, 1960.

22. D. M. Hershfield, "Rainfall Frequency Atlas of the United States," Tech. Paper No. 40, U.S. Weather Bureau, 1961.

23. *Hydrology Handbook,* ASCE Manual of Practice, No. 28, 1949.

24. W. G. Knisel, Jr. and R. W. Baird, in *ARS Precipitation Facilities and Related Studies.* Washington, D.C.: U.S. Department of Agriculture, Agricultural Research Service, 1971, Chap. 14.

25. R. K. Linsley, Jr., M. A. Kohler, and J. L. H. Paulhus, *Applied Hydrology.* New York: McGraw-Hill, 1949.

Appendixes

TABLE A.1 WATER PROPERTIES, CONSTANTS, AND CONVERSION FACTORS

Gas constants (R)
  $R = 0.0821$ (atm)(liter)/(g-mol)(K)
  $R = 1.987$ g-cal/(g-mol)(K)
  $R = 1.987$ Btu/(lb-mol)(°R)
Acceleration of gravity (standard)
  $g = 32.17$ ft/sec^2 = 980.6 cm/sec^2
Heat of fusion of water
  79.7 cal/g = 144 Btu/Ib

Heat of vaporization of water at 1.0 atm
  540 cal/g = 970 Btu/Ib

Specific heat of air
  $C_p = 0.238$ cal/(g)(°C)
Density of dry air at 0°C and 760 mm Hg 0.001293 g/cm^3

Conversion factors

$$
\begin{aligned}
\text{1 second-foot-day per square mile} &= 0.03719 \text{ inch} \\
\text{1 inch of runoff per square mile} &= 26.9 \text{ second-foot-days} \\
&= 53.3 \text{ acre-feet} \\
&= 2{,}323{,}200 \text{ cubic feet} \\
\text{1 cubic foot per second} &= 0.9917 \text{ acre-inch per hour} \\
&= 1 \text{ sec-ft} = 1 \text{ cusec} \\
\text{1 horsepower} &= 0.746 \text{ kilowatt} \\
&= 550 \text{ foot-pounds per second} \\
e &= 2.71828 \\
\log e &= 0.43429 \\
\ln 10 &= 2.30259
\end{aligned}
$$

Metric equivalents

$$
\begin{aligned}
\text{1 foot} &= 0.3048 \text{ meter} \\
\text{1 mile} &= 1.609 \text{ kilometers} \\
\text{1 acre} &= 0.4047 \text{ hectare} \\
&= 4047 \text{ square meters} \\
\text{1 square mile (mi}^2) &= 259 \text{ hectares} \\
&= 2.59 \text{ square kilometers (km}^2) \\
\text{1 acre foot (acre-ft)} &= 1233 \text{ cubic meters} \\
\text{1 million cubic feet (mcf)} &= 28{,}320 \text{ cubic meters} \\
\text{1 cubic foot per second (cfs)} &= 0.02832 \text{ cubic meters per second} \\
&= 1.699 \text{ cubic meters per minute} \\
\text{1 acre-in. per hour} &= 1.008 \text{ cubic feet per second (cfs)} \\
\text{1 second-foot-day (cfsd)} &= 2447 \text{ cubic meters} \\
\text{1 million gallons (mg)} &= 3785 \text{ cubic meters} \\
&= 3.785 \text{ million liters} \\
\text{1 million gallons per day (mgd)} &= 694.4 \text{ gallons per minute (gpm)} \\
&= 2.629 \text{ cubic meters per minute} \\
&= 3785 \text{ cubic meters per day}
\end{aligned}
$$

TABLE A.2 PROPERTIES OF WATER

Traditional U.S. Units

Temperature (°F)	Specific gravity	Unit weight (lb/ft³)	Heat of vaporization (Btu/lb)	Kinematic viscosity (ft²/sec)	Vapor pressure mb	psi	in.Hg
32	0.99987	62.416	1073	1.93×10^{-5}	6.11	0.09	0.18
40	0.99999	62.423	1066	1.67×10^{-5}	8.36	0.12	0.25
50	0.99975	62.408	1059	1.41×10^{-5}	12.19	0.18	0.36
60	0.99907	62.366	1054	1.21×10^{-5}	17.51	0.26	0.52
70	0.99802	62.300	1049	1.06×10^{-5}	24.79	0.36	0.74
80	0.99669	62.217	1044	0.929×10^{-5}	34.61	0.51	1.03
90	0.99510	62.118	1039	0.828×10^{-5}	47.68	0.70	1.42
100	0.99318	61.998	1033	0.741×10^{-5}	64.88	0.95	1.94

SI Units

Temperature (°C)	Specific gravity	Density (g/cm³)	Heat of vaporization (cal/g)	Kinematic viscosity (cs)	Vapor pressure (mm Hg)	(mb)	(g/cm²)
0	0.99987	0.99984	597.3	1.790	4.58	6.11	6.23
5	0.99999	0.99996	594.5	1.520	6.54	8.72	8.89
10	0.99973	0.99970	591.7	1.310	9.20	12.27	12.51
15	0.99913	0.99910	588.9	1.140	12.78	17.04	17.38
20	0.99824	0.99821	586.0	1.000	17.53	23.37	23.83
25	0.99708	0.99705	583.2	0.893	23.76	31.67	32.30
30	0.99568	0.99565	580.4	0.801	31.83	42.43	43.27
35	0.99407	0.99404	577.6	0.723	42.18	56.24	57.34
40	0.99225	0.99222	574.7	0.658	55.34	73.78	75.23
50	0.98807	0.98804	569.0	0.554	92.56	123.40	125.83
60	0.98323	0.98320	563.2	0.474	149.46	199.26	203.19
70	0.97780	0.97777	557.4	0.413	233.79	311.69	317.84
80	0.97182	0.97179	551.4	0.365	355.28	473.67	483.01
90	0.96534	0.96531	545.3	0.326	525.89	701.13	714.95
100	0.95839	0.95836	539.1	0.294	760.00	1013.25	1033.23

TABLE B.1   AREAS UNDER THE NORMAL CURVE

$$F(z) = \int_0^z \frac{1}{\sqrt{2\pi}} e^{-z^2/2} dz$$

z	.00	.01	.02	.03	.04	.05	.06	.07	.08	.09
0.0	.0000	.0040	.0080	.0120	.0159	.0199	.0239	.0279	.0319	.0359
0.1	.0398	.0438	.0478	.0517	.0557	.0596	.0636	.0675	.0714	.0753
0.2	.0793	.0832	.0871	.0910	.0948	.0987	.1026	.1064	.1103	.1141
0.3	.1179	.1217	.1255	.1293	.1331	.1368	.1406	.1443	.1480	.1517
0.4	.1554	.1591	.1628	.1664	.1700	.1736	.1772	.1808	.1844	.1879
0.5	.1915	.1950	.1985	.2019	.2054	.2088	.2123	.2157	.2190	.2224
0.6	.2257	.2291	.2324	.2357	.2389	.2422	.2454	.2486	.2518	.2549
0.7	.2580	.2611	.2642	.2673	.2704	.2734	.2764	.2794	.2823	.2852
0.8	.2881	.2910	.2939	.2967	.2995	.3023	.3051	.3078	.3106	.3133
0.9	.3159	.3186	.3212	.3238	.3264	.3289	.3315	.3340	.3365	.3389
1.0	.3413	.3438	.3461	.3485	.3508	.3531	.3554	.3577	.3599	.3621
1.1	.3643	.3665	.3686	.3708	.3729	.3749	.3770	.3790	.3810	.3830
1.2	.3849	.3869	.3888	.3907	.3925	.3944	.3962	.3980	.3997	.4015
1.3	.4032	.4049	.4066	.4082	.4099	.4115	.4131	.4147	.4162	.4177
1.4	.4192	.4207	.4222	.4236	.4251	.4265	.4279	.4292	.4306	.4319
1.5	.4332	.4345	.4357	.4370	.4382	.4394	.4406	.4418	.4430	.4441
1.6	.4452	.4463	.4474	.4485	.4495	.4505	.4515	.4525	.4535	.4545
1.7	.4554	.4564	.4573	.4582	.4591	.4599	.4608	.4616	.4625	.4633
1.8	.4641	.4649	.4656	.4664	.4671	.4678	.4686	.4693	.4699	.4606
1.9	.4713	.4719	.4726	.4732	.4738	.4744	.4750	.4756	.4762	.4767
2.0	.4772	.4778	.4783	.4788	.4793	.4798	.4803	.4808	.4812	.4817
2.1	.4821	.4826	.4830	.4834	.4838	.4842	.4846	.4850	.4854	.4857
2.2	.4861	.4865	.4868	.4871	.4875	.4878	.4881	.4884	.4887	.4890
2.3	.4893	.4896	.4898	.4901	.4904	.4906	.4909	.4911	.4913	.4916
2.4	.4918	.4920	.4922	.4925	.4927	.4929	.4931	.4932	.4934	.4936
2.5	.4938	.4940	.4941	.4943	.4945	.4946	.4948	.4949	.4951	.4952
2.6	.4953	.4955	.4956	.4957	.4959	.4960	.4961	.4962	.4963	.4964
2.7	.4965	.4966	.4967	.4968	.4969	.4970	.4971	.4972	.4973	.4974
2.8	.4974	.4975	.4976	.4977	.4977	.4978	.4979	.4980	.4980	.4981
2.9	.4981	.4982	.4983	.4983	.4984	.4984	.4985	.4985	.4986	.4986
3.0	.4987	.4987	.4987	.4988	.4988	.4989	.4989	.4989	.4990	.4990
3.1	.4990	.4991	.4991	.4991	.4992	.4992	.4992	.4992	.4993	.4993
3.2	.4993	.4993	.4994	.4994	.4994	.4994	.4994	.4995	.4995	.4995
3.3	.4995	.4995	.4996	.4996	.4996	.4996	.4996	.4996	.4996	.4997
3.4	.4997	.4997	.4997	.4997	.4997	.4997	.4997	.4997	.4998	.4998
⋮	⋮									
4.0	.499968									

Source: After C. E. Weatherburn, *Mathematical Statistics.* London: Cambridge University Press, 1957 (for $z = 0$ to $z = 3.1$); C. H. Richardson, *An Introduction to Statistical Analysis.* Orlando, FL: Harcourt Brace Jovanovich, 1994 (for $z = 3.2$ to $z = 3.4$); A. H. Bowker and G. J. Lieberman, *Engineering Statistics.* Englewood Cliffs, NJ: Prentice-Hall, 1959 (for $z = 4.0$ and 5.0).

TABLE B.2 K VALUES FOR PEARSON TYPE III DISTRIBUTION

Skew coefficient C_s	Recurrence interval in years										
	1.0101	1.0526	1.1111	1.2500	2	5	10	25	50	100	200
	Percent chance										
	99	95	90	80	50	20	10	4	2	1	0.5
	Positive skew										
3.0	−0.667	−0.665	−0.660	−0.636	−0.396	0.420	1.180	2.278	3.152	4.051	4.970
2.9	−0.690	−0.688	−0.681	−0.651	−0.390	0.440	1.195	2.277	3.134	4.013	4.909
2.8	−0.714	−0.711	−0.702	−0.666	−0.384	0.460	1.210	2.275	3.114	3.973	4.847
2.7	−0.740	−0.736	−0.724	−0.681	−0.376	0.479	1.224	2.272	3.093	3.932	4.783
2.6	−0.769	−0.762	−0.747	−0.696	−0.368	0.499	1.238	2.267	3.071	3.889	4.718
2.5	−0.799	−0.790	−0.771	−0.711	−0.360	0.518	1.250	2.262	3.048	3.845	4.652
2.4	−0.832	−0.819	−0.795	−0.725	−0.351	0.537	1.262	2.256	3.023	3.800	4.584
2.3	−0.867	−0.850	−0.819	−0.739	−0.341	0.555	1.274	2.248	2.997	3.753	4.515
2.2	−0.905	−0.882	−0.844	−0.752	−0.330	0.574	1.284	2.240	2.970	3.705	4.444
2.1	−0.946	−0.914	−0.869	−0.765	−0.319	0.592	1.294	2.230	2.942	3.656	4.372
2.0	−0.990	−0.949	−0.895	−0.777	−0.307	0.609	1.302	2.219	2.912	3.605	4.398
1.9	−1.037	−0.984	−0.920	−0.788	−0.294	0.627	1.310	2.207	2.881	3.553	4.223
1.8	−1.087	−1.020	−0.945	−0.799	−0.282	0.643	1.318	2.193	2.848	3.499	4.147
1.7	−1.140	−1.056	−0.970	−0.808	−0.268	0.660	1.324	2.179	2.815	3.444	4.069
1.6	−1.197	−1.093	−0.994	−0.817	−0.254	0.675	1.329	2.163	2.780	3.388	3.990
1.5	−1.256	−1.131	−1.018	−0.825	−0.240	0.690	1.333	2.146	2.743	3.330	3.910
1.4	−1.318	−1.168	−1.041	−0.832	−0.225	0.705	1.337	2.128	2.706	3.271	3.828
1.3	−1.383	−1.206	−1.064	−0.838	−0.210	0.719	1.339	2.108	2.666	3.211	3.745
1.2	−1.449	−1.243	−1.086	−0.844	−0.195	0.732	1.340	2.087	2.626	3.149	3.661
1.1	−1.518	−1.280	−1.107	−0.848	−0.180	0.745	1.341	2.066	2.585	3.087	3.575
1.0	−1.588	−1.317	−1.128	−0.852	−0.164	0.758	1.340	2.043	2.542	3.022	3.489
0.9	−1.660	−1.353	−1.147	−0.854	−0.148	0.769	1.339	2.018	2.498	2.957	3.401
0.8	−1.733	−1.388	−1.166	−0.856	−0.132	0.780	1.336	1.993	2.453	2.891	3.312
0.7	−1.806	−1.423	−1.183	−0.857	−0.116	0.790	1.333	1.967	2.407	2.824	3.223
0.6	−1.880	−1.458	−1.200	−0.857	−0.099	0.800	1.328	1.939	2.359	2.755	3.132
0.5	−1.955	−1.491	−1.216	−0.856	−0.083	0.808	1.323	1.910	2.311	2.686	3.041
0.4	−2.029	−1.524	−1.231	−0.855	−0.066	0.816	1.317	1.880	2.261	2.615	2.949
0.3	−2.104	−1.555	−1.245	−0.853	−0.050	0.824	1.309	1.849	2.211	2.544	2.856
0.2	−2.178	−1.586	−1.258	−0.850	−0.033	0.830	1.301	1.818	2.159	2.472	2.763
0.1	−2.252	−1.616	−1.270	−0.846	−0.017	0.836	1.292	1.785	2.107	2.400	2.670
0.0	−2.326	−1.645	−1.282	−0.842	0	0.842	1.282	1.751	2.054	2.326	2.576

TABLE B.2 (Continued)

	Recurrence interval in years										
	1.0101	1.0526	1.1111	1.2500	2	5	10	25	50	100	200
Skew coefficient	Percent chance										
C_s	99	95	90	80	50	20	10	4	2	1	0.5
	Negative skew										
−0.1	−2.400	−1.673	−1.292	−0.836	0.017	0.846	1.270	1.716	2.000	2.252	2.482
−0.2	−2.472	−1.700	−1.301	−0.830	0.033	0.850	1.258	1.680	1.945	2.178	2.388
−0.3	−2.544	−1.726	−1.309	−0.824	0.050	0.853	1.245	1.643	1.890	2.104	2.294
−0.4	−2.615	−1.750	−1.317	−0.816	0.066	0.855	1.231	1.606	1.834	2.029	2.201
−0.5	−2.686	−1.774	−1.323	−0.808	0.083	0.856	1.216	1.567	1.777	1.955	2.108
−0.6	−2.755	−1.797	−1.328	−0.800	0.099	0.857	1.200	1.528	1.720	1.880	2.016
−0.7	−2.824	−1.819	−1.333	−0.790	0.116	0.857	1.183	1.488	1.663	1.806	1.926
−0.8	−2.891	−1.839	−1.336	−0.780	0.132	0.856	1.166	1.448	1.606	1.733	1.837
−0.9	−2.957	−1.858	−1.339	−0.769	0.148	0.854	1.147	1.407	1.549	1.660	1.749
−1.0	−3.022	−1.877	−1.340	−0.758	0.164	0.852	1.128	1.366	1.492	1.588	1.664
−1.1	−3.087	−1.894	−1.341	−0.745	0.180	0.848	1.107	1.324	1.435	1.518	1.581
−1.2	−3.149	−1.910	−1.340	−0.732	0.195	0.844	1.086	1.282	1.379	1.449	1.501
−1.3	−3.211	−1.925	−1.339	−0.719	0.210	0.838	1.064	1.240	1.324	1.383	1.424
−1.4	−3.271	−1.938	−1.337	−0.705	0.225	0.832	1.041	1.198	1.270	1.318	1.351
−1.5	−3.330	−1.951	−1.333	−0.690	0.240	0.825	1.018	1.157	1.217	1.256	1.282
−1.6	−3.388	−1.962	−1.329	−0.675	0.254	0.817	0.994	1.116	1.166	1.197	1.216
−1.7	−3.444	−1.972	−1.324	−0.660	0.268	0.808	0.970	1.075	1.116	1.140	1.155
−1.8	−3.499	−1.981	−1.318	−0.643	0.282	0.799	0.945	1.035	1.069	1.087	1.097
−1.9	−3.553	−1.989	−1.310	−0.627	0.294	0.788	0.920	0.996	1.023	1.037	1.044
−2.0	−3.605	−1.996	−1.302	−0.609	0.307	0.777	0.895	0.959	0.980	0.990	0.995
−2.1	−3.656	−2.001	−1.294	−0.592	0.319	0.765	0.869	0.923	0.939	0.946	0.949
−2.2	−3.705	−2.006	−1.284	−0.574	0.330	0.752	0.844	0.888	0.900	0.905	0.907
−2.3	−3.753	−2.009	−1.274	−0.555	0.341	0.739	0.819	0.855	0.864	0.867	0.869
−2.4	−3.800	−2.011	−1.262	−0.537	0.351	0.725	0.795	0.823	0.830	0.832	0.833
−2.5	−3.845	−2.012	−1.250	−0.518	0.360	0.711	0.771	0.793	0.798	0.799	0.800
−2.6	−3.889	−2.013	−1.238	−0.499	0.368	0.696	0.747	0.764	0.768	0.769	0.769
−2.7	−3.932	−2.012	−1.224	−0.479	0.376	0.681	0.724	0.738	0.740	0.740	0.741
−2.8	−3.973	−2.010	−1.210	−0.460	0.384	0.666	0.702	0.712	0.714	0.714	0.714
−2.9	−4.013	−2.007	−1.195	−0.440	0.390	0.651	0.681	0.683	0.689	0.690	0.690
−3.0	−4.051	−2.003	−1.180	−0.420	0.396	0.636	0.660	0.666	0.666	0.667	0.667

Source: After Water Resources Council, Bulletin No.15, December 1967.

TABLE B.3   UNIFORMLY DISTRIBUTED RANDOM NUMBERS

53 74 23 99 67	61 32 28 69 84	94 62 67 86 24	98 33 41 19 95	47 53 53 38 09
63 38 06 86 54	99 00 65 26 94	02 82 90 23 07	79 62 67 80 60	75 91 12 81 19
30 30 58 21 46	06 72 17 10 94	25 21 31 75 96	49 28 24 00 49	55 65 79 78 07
63 43 36 82 69	65 51 18 37 88	61 38 44 12 45	32 92 85 88 65	54 34 81 85 35
98 25 37 55 26	01 91 82 81 46	74 71 12 94 97	24 02 71 37 07	03 92 18 66 75
02 63 21 17 69	71 50 80 89 56	38 15 70 11 48	43 40 45 86 98	00 83 26 91 03
64 55 22 21 82	48 22 28 06 00	61 54 13 43 91	82 78 12 23 29	06 66 24 12 27
85 07 26 13 89	01 10 07 82 04	59 63 69 36 03	69 11 15 83 80	13 29 54 19 28
58 54 16 24 15	51 54 44 82 00	62 61 65 04 69	38 18 65 18 97	85 72 13 49 21
34 85 27 84 87	61 48 64 56 26	90 18 48 13 26	37 70 15 42 57	65 65 80 39 07
03 92 18 27 46	57 99 16 96 56	30 33 72 85 22	84 64 38 56 98	99 01 30 98 64
62 95 30 27 59	37 75 41 66 48	86 97 80 61 45	23 53 04 01 63	45 76 08 64 27
08 45 93 15 22	60 21 54 46 91	98 77 27 85 42	28 88 61 08 94	69 62 03 42 73
07 08 55 18 40	45 44 75 13 90	24 94 96 61 02	57 55 66 83 15	73 42 37 11 61
01 85 89 95 66	51 10 19 34 88	15 84 97 19 75	12 76 39 43 78	64 63 91 08 25
72 84 71 14 35	19 11 58 49 26	50 11 17 17 76	86 31 57 20 18	95 60 78 46 75
88 78 28 16 84	13 52 53 94 53	75 45 69 30 96	73 89 65 70 31	99 17 43 48 76
45 17 75 65 57	28 40 19 72 12	25 12 74 75 67	60 40 60 81 19	24 62 01 61 16
96 76 28 12 54	22 01 11 94 25	71 96 16 16 88	68 64 36 74 45	19 59 50 88 92
43 31 67 72 30	24 02 94 08 63	38 32 36 66 02	69 36 38 25 39	48 03 45 15 22
50 44 66 44 21	66 06 58 05 62	68 15 54 35 02	42 35 48 96 32	14 52 41 52 48
22 66 22 15 86	26 63 75 41 99	58 42 36 72 24	58 37 52 18 51	03 37 18 39 11
96 24 40 14 51	23 22 30 88 57	95 67 47 29 83	94 69 40 06 07	18 16 36 78 86
31 73 91 61 19	60 20 72 93 48	98 57 07 23 69	65 95 39 69 58	56 80 30 19 44
78 60 73 99 84	43 89 94 36 45	56 69 47 07 41	90 22 91 07 12	78 35 34 08 72
84 37 90 61 56	70 10 23 98 05	85 11 34 76 60	76 48 45 34 60	01 64 18 39 96
36 67 10 08 23	98 93 35 08 86	99 29 76 29 81	33 34 91 58 93	63 14 52 32 52
07 28 59 07 48	89 64 58 89 75	83 85 62 27 89	30 14 78 56 27	86 63 59 80 02
10 15 83 87 60	79 24 31 66 56	21 48 24 06 93	91 98 94 05 49	01 47 59 38 00
55 19 68 97 65	03 73 52 16 56	00 53 55 90 27	33 42 29 38 87	22 13 88 83 34
53 81 29 13 39	35 01 20 71 34	62 33 74 82 14	53 73 19 09 03	56 54 29 56 93
51 86 32 68 92	33 98 74 66 99	40 14 71 94 58	45 94 19 38 81	14 44 99 81 07
35 91 70 29 13	80 03 54 07 27	96 94 78 32 66	50 95 52 74 33	13 80 55 62 54
37 71 67 95 13	20 02 44 95 94	64 85 04 05 72	01 32 90 76 14	53 89 74 60 41
93 66 13 83 27	92 76 64 64 72	28 54 96 53 84	48 14 52 98 94	56 07 93 89 30

Source: After L. R. Beard, *Statistical Methods in Hydrology,* U.S. Army Corps of Engineers, 1962.

Index